Digital Communications:
A Discrete-Time Approach

Digital Communications: A Discrete-Time Approach

Michael Rice
Brigham Young University

Upper Saddle River, New Jersey 07458

Library of Congress Cataloging-in-Publication Data

Rice, Michael.
　Digital communications : a discrete-time approach / Michael Rice.
　　p. cm.
　Includes bibliographical references and index.
　ISBN 0-13-030497-2
　1. Digital communications. 2. Signal processing–Digital techniques. I. Title.
　TK5103.7.R53 2009
　621.382—dc22
　　　　　　　　　　　　　　　　　　　　　　　　　　2008004786

Vice President and Editorial Director, ECS: *Marcia J. Horton*
Associate Editor: *Alice Dworkin*
Editorial Assistant: *William Opaluch*
Managing Editor: *Scott Disanno*
Director of Creative Services: *Paul Belfanti*
Art Director: *Jayne Conte*
Art Editor: *Gregory Dulles*
Manufacturing Manager: *Alexis Heydt-Long*
Manufacturing Buyer: *Lisa McDowell*
Senior Marketing Manager: *Tim Galligan*

© 2009 by Pearson Education, Inc.
Pearson Prentice Hall
Pearson Education, Inc.
Upper Saddle River, New Jersey 07458

All rights reserved. No part of this book may be reproduced, in any form, or by any means, without permission in writing from the publisher.

Pearson Prentice Hall® is a trademark of Pearson Education, Inc.

MATLAB and SIMULINK are registered trademarks of The MathWorks, Inc., 3 Apple Hill Drive, Natick, MA.

The authors and publisher of this book have used their best efforts in preparing this book. These efforts include the development, research, and testing of the theories and programs to determine their effectiveness. The authors and publisher make no warranty of any kind, expressed or implied, with regard to these programs or the documentation contained in this book. The authors and publisher shall not be liable in any event for incidental or consequential damages with, or arising out of, the furnishing, performance, or use of these programs.

Printed in the United States of America

10 9 8 7 6 5 4 3 2 1

ISBN 13: 978-0-13-030497-1
ISBN-10:   0-13-030497-2

Pearson Education Ltd., London
Pearson Education Singapore, Pte. Ltd.
Pearson Education Canada, Inc.
Pearson Education–Japan
Pearson Education Australia PTY, Limited
Pearson Education North Asia, Ltd., Hong Kong
Pearson Educación de Mexico, S.A. de C.V.
Pearson Education Malaysia, Pte. Ltd.
Pearson Education, Upper Saddle River, New Jersey

To
Lisa and the A-List

Contents

Foreword   xiii

Preface   xix

Chapter 1   Introduction   1

    1.1   A Brief History of Communications   1
    1.2   Basics of Wireless Communications   10
    1.3   Digital Communications   12
    1.4   Why Discrete-Time Processing Is So Popular   14
    1.5   Organization of the Text   19
    1.6   Notes and References   22

Chapter 2   Signals and Systems 1: A Review of the Basics   23

    2.1   Introduction   23
    2.2   Signals   24
       2.2.1   Continuous-Time Signals   24
       2.2.2   Discrete-Time Signals   26
    2.3   Systems   28
       2.3.1   Continuous-Time Systems   28
       2.3.2   Discrete-Time Systems   29
    2.4   Frequency Domain Characterizations   30
       2.4.1   Laplace Transform   32
       2.4.2   Continuous-Time Fourier Transform   37
       2.4.3   Z Transform   40
       2.4.4   Discrete-Time Fourier Transform   46
    2.5   The Discrete Fourier Transform   50
    2.6   The Relationship Between Discrete-Time and Continuous-Time Systems   55
       2.6.1   The Sampling Theorem   56
       2.6.2   Discrete-Time Processing of Continuous-Time Signals   65
    2.7   Discrete-Time Processing of Band-Pass Signals   67

2.8 Notes and References 70
2.9 Exercises 71

Chapter 3   Signals and Systems 2: Some Useful Discrete-Time Techniques for Digital Communications   114

3.1 Introduction   114
3.2 Multirate Signal Processing   115
    3.2.1 Impulse-Train Sampling   115
    3.2.2 Downsampling   118
    3.2.3 Upsampling   120
    3.2.4 The Noble Identities   122
    3.2.5 Polyphase Filterbanks   122
3.3 Discrete-Time Filter Design Methods   127
    3.3.1 IIR Filter Designs   129
    3.3.2 FIR Filter Designs   134
    3.3.3 Two Important Filters: The Differentiator and the Integrator   149
3.4 Notes and References   159
3.5 Exercises   159

Chapter 4   A Review of Probability Theory   178

4.1 Basic Definitions   178
4.2 Gaussian Random Variables   188
    4.2.1 Density and Distribution Functions   188
    4.2.2 Product Moments   192
    4.2.3 Functions of Random Variables   193
4.3 Multivariate Gaussian Random Variables   195
    4.3.1 Bivariate Gaussian Distribution   196
    4.3.2 Linear Operators and Multivariate Gaussian Random Variables   197
4.4 Random Sequences   198
    4.4.1 Power Spectral Density   199
    4.4.2 Random Sequences and Discrete-Time LTI Systems   200
4.5 Additive White Gaussian Noise   202
    4.5.1 Continuous-Time Random Processes   202
    4.5.2 The White Gaussian Random Process: A Good Model for Noise?   204
    4.5.3 White Gaussian Noise in a Sampled Data System   206
4.6 Notes and References   208
4.7 Exercises   208

Contents   ix

Chapter 5   Linear Modulation 1: Modulation, Demodulation, and Detection   **214**

    5.1   Signal Spaces   215
          5.1.1   Definitions   215
          5.1.2   The Synthesis Equation and Linear Modulation   222
          5.1.3   The Analysis Equation and Detection   223
          5.1.4   The Matched Filter   226
    5.2   M-ary Baseband Pulse Amplitude Modulation   227
          5.2.1   Continuous-Time Realization   229
          5.2.2   Discrete-Time Realization   233
    5.3   M-ary Quadrature Amplitude Modulation   238
          5.3.1   Continuous-Time Realization   246
          5.3.2   Discrete-Time Realization   256
    5.4   Offset QPSK   260
    5.5   Multicarrier Modulation   265
    5.6   Maximum Likelihood Detection   273
          5.6.1   Introduction   273
          5.6.2   Preliminaries   274
          5.6.3   Maximum Likelihood Decision Rule   276
    5.7   Notes and References   279
          5.7.1   Topics Covered   279
          5.7.2   Topics Not Covered   280
    5.8   Exercises   280

Chapter 6   Linear Modulation 2: Performance   **305**

    6.1   Performance of PAM   306
          6.1.1   Bandwidth   306
          6.1.2   Probability of Error   307
    6.2   Performance of QAM   313
          6.2.1   Bandwidth   313
          6.2.2   Probability of Error   314
    6.3   Comparisons   325
    6.4   Link Budgets   331
          6.4.1   Received Power and the Friis Equation   331
          6.4.2   Equivalent Noise Temperature and Noise Figure   334
          6.4.3   The Link Budget Equation   339
    6.5   Projecting White Noise onto an Orthonormal Basis Set   345
    6.6   Notes and References   347
          6.6.1   Topics Covered   347
          6.6.2   Topics Not Covered   347
    6.7   Exercises   348

Chapter 7 Carrier Phase Synchronization — 359

- 7.1 Basic Problem Formulation 360
  - 7.1.1 Approach 1 360
  - 7.1.2 Approach 2 362
- 7.2 Carrier Phase Synchronization for QPSK 365
  - 7.2.1 A Heuristic Phase Error Detector 365
  - 7.2.2 The Maximum Likelihood Phase Error Detector 370
  - 7.2.3 Examples 374
- 7.3 Carrier Phase Synchronization for BPSK 375
- 7.4 Carrier Phase Synchronization for MQAM 381
- 7.5 Carrier Phase Synchronization for Offset QPSK 382
- 7.6 Carrier Phase Synchronization for BPSK and QPSK Using Continuous-Time Techniques 391
- 7.7 Phase Ambiguity Resolution 394
  - 7.7.1 Unique Word 396
  - 7.7.2 Differential Encoding 398
- 7.8 Maximum Likelihood Phase Estimation 409
  - 7.8.1 Preliminaries 409
  - 7.8.2 Carrier Phase Estimation 414
- 7.9 Notes and References 421
  - 7.9.1 Topics Covered 421
  - 7.9.2 Topics Not Covered 423
- 7.10 Exercises 423

Chapter 8 Symbol Timing Synchronization — 434

- 8.1 Basic Problem Formulation 436
- 8.2 Continuous-Time Techniques for M-ary PAM 438
- 8.3 Continuous-Time Techniques for MQAM 443
- 8.4 Discrete-Time Techniques for M-ary PAM 445
  - 8.4.1 Timing Error Detectors 449
  - 8.4.2 Interpolation 462
  - 8.4.3 Interpolation Control 475
  - 8.4.4 Examples 478
- 8.5 Discrete-Time Techniques for MQAM 494
- 8.6 Discrete-Time Techniques for Offset QPSK 497
- 8.7 Dealing with Transition Density: A Practical Consideration 501
- 8.8 Maximum Likelihood Estimation 503
  - 8.8.1 Preliminaries 503
  - 8.8.2 Symbol Timing Estimation 510

Contents   xi

    8.9 Notes and References   514
         8.9.1 Topics Covered   514
         8.9.2 Topics Not Covered   515
    8.10 Exercises   515

Chapter 9 System Components   519

    9.1 The Continuous-Time Discrete-Time Interface   519
         9.1.1 Analog-to-Digital Converter   520
         9.1.2 Digital-to-Analog Converter   529
    9.2 Discrete-Time Oscillators   537
         9.2.1 Discrete Oscillators Based on LTI Systems   538
         9.2.2 Direct Digital Synthesizer   542
    9.3 Resampling Filters   555
         9.3.1 CIC and Hogenauer Filters   557
         9.3.2 Half-Band Filters   562
         9.3.3 Arbitrary Resampling Using Polyphase Filterbanks   565
    9.4 CoRDiC: Coordinate Rotation Digital Computer   578
         9.4.1 Rotations: Moving on a Circle   578
         9.4.2 Moving Along Other Shapes   585
    9.5 Automatic Gain Control   588
    9.6 Notes and References   593
         9.6.1 Topics Covered   593
         9.6.2 Topics Not Covered   597
    9.7 Exercises   597

Chapter 10 System Design   604

    10.1 Advanced Discrete-Time Architectures   604
         10.1.1 Discrete-Time Architectures for
                QAM Modulators   605
         10.1.2 Discrete-Time Architectures for
                QAM Demodulators   611
         10.1.3 Putting It All Together   632
    10.2 Channelization   637
         10.2.1 Continuous-Time Techniques:
                The Superheterodyne Receiver   637
         10.2.2 Discrete-Time Techniques Using Multirate Processing   650
    10.3 Notes and References   658
         10.3.1 Topics Covered   658
         10.3.2 Topics Not Covered   660
    10.4 Exercises   662

Appendix A Pulse Shapes	673
Appendix B The Complex-Valued Representation for QAM	700
Appendix C Phase Locked Loops	718
Bibliography	751
Index	769

Foreword

Another book on Digital Communications! What possessed Michael Rice to write this book? It didn't dawn upon me to ask him because I already knew the answer based on the title: **Digital Communications**, *A Discrete-Time Approach*. The qualifier was the clue! Simply, it was time for this book to be written. It is clear to consumers and practitioners alike that Digital Communications has changed the way we communicate at a distance. The many text books and courses dealing with digital communications as well as the many communication sector companies that populate the Fortune-500 list bare witness to the importance of this technology in our society. Many of us have both witnessed and have guided the evolutionary transformation that has become the fabric of the physical layer in a communication system. This evolution has changed the way we manipulate and process waveforms in communication systems. We have seen Digital Signal Processing quickly replace most of the signal conditioning functions and tasks we perform in modulators and demodulators. Thus the title qualifier: *A Discrete-Time Approach*.

This book is a response to the fact that Digital Signal Processing hardware and DSP algorithms have all but replaced analog hardware as the means to perform the various baseband and intermediate frequency signal processing functions required to perform modulation and demodulation. The change is dramatic. Cable TV modems, Satellite Modems, Local Area Wireless Network Modems, the ubiquitous Cell phone, Global Positioning System (GPS) receivers, and a multitude of consumer entertainment systems such as the audio MP3 player, the High Definition TV player, and the video TiVo player and others have embraced DSP as the core technology that enables the communication process.

An interesting note is that DSP based receivers are replacing analog receivers in non-digital modulation formats such as AM, FM and traditional analog TV. This happens in multi-mode receivers designed to demodulate digital modulation formats such as the OFDM based High Definition Radio signal but for marketing purpose must also demodulate the traditional analog AM and FM signal bands. Similar multimode receivers are required to demodulate digital modulated high definition TV signals as well as the traditional analog modulated National Television Standards Committee (NTSC) TV signals.

Traditionally the resources allocated to the communications task have been signal bandwidth (W) and signal energy (S) to noise (N) ratio as presented in Shannon's Capacity theorem: $C = W \log_2(S/N + 1)$. In the latter half of the 20-th century a third resource was added to the three legged structure we call Modern Communications. This third leg is computational complexity! We have learned that the modem needs a computer! Initially the computer performed transformations on data; *ones and zeros*. These transformations, owing much of their structure to Shannon, include channel coding and decoding, source coding and decoding, and encryption and decryption. In the past two decades the computer was assigned the additional heavy lifting task of applying transformations to sampled wave shapes. The collection of wave shape transformations is generally called Digital Signal Processing. The core of both sets of transformations is the digital computer. This core rides the bow-wave of Moore's law, which paraphrased states that *"for a fixed cost, the complexity, or computational power, of an integrated circuit doubles approximately every two years"*. I regularly remind my students that there are only two justifications for the insertion of DSP in a communication system: these are improved performance and reduced cost. Moore's law offers both options! The second essential core that enables DSP insertion in communication systems is the availability of high performance analog-to-digital and digital-to-analog converters (ADC and DAC respectively). These devices perform the transduction and source coding tasks required to move waveform signals across the boundary separating the continuous analog domain and the discrete sample data domain.

Modern DSP based modulators present sampled data signals at their output ports while DSP demodulators expect sampled data signals at their input ports. Many systems contain mixed signal components in which the converters reside on the same chip performing the DSP functions. Since the industry and the economics of the process favors DSP implementations of the modulation and demodulation process it makes sense to teach communication systems from this implementation perspective.

While discussing perspectives, we have to keep in mind that when we implement a DSP based modulator or demodulator our task is to perform specific functions as opposed to copy a legacy implementation of these functions. Often a conventional receiver already contains an existing analog circuit implementation and we may be tempted to emulate that implementation. We have to keep in mind that our task is not to emulate the analog solution but rather to perform the function currently performed by the analog circuitry. In doing so it is wise to return to first principles with the design tools at our disposal which is likely a richer set of tools than those to which the analog designer had access when he or she derived the legacy analog solution. If we use the DSP based tools to replicate the analog circuit implementation we may also inherit legacy compromises made by the designer which were appropriate for the set of design tools available when the compromises were made. Approaching the processing tasks from first principles may offer solution unique to the DSP approach for which there is not an analog prototype solution. This approach may occasionally lead to the same solution to a problem as that obtained from the analog perspective but often it will lead to a unique DSP based solution. This text differs from earlier texts in that it not only presents the science of what to do in a communication system it also presents some of the modern art of how to do it.

Here we cite a few simple examples in support of the first principle design implementation approach. It is common for an analog prototype receiver to use a pair of recursive Butterworth filters as part of a down conversion signal path. We are not obligated to use a DSP based Butterworth filter to perform the same function in a digital implementation of the same process. We would likely use a linear phase non-recursive filter to avoid the group delay distortion associated with the recursive filter. Here we improve system performance by not emulating the analog design.

In another example, a common analog receiver design translates an intermediate frequency (IF) narrow band signal to baseband using a pair of mixers and low-pass filters. The process can certainly be emulated by DSP techniques but may be implemented more efficiently using a uniquely DSP based option. Such an example is based on multirate signal processing in which the signal can be aliased to baseband by simply reducing the sample rate of the sampled data time series representing the waveform. Aliasing is a viable tool available to the DSP designer that is not available to the analog designer!

One more example is the phase detector in a phase locked loop used for carrier synchronization. In an analog system the phase detector is formed by first hard limiting the input signal (with a clipping circuit) and then translating the, now constant amplitude, input signal to baseband with a pair of mixers. Our interest in the product signal is the phase angle of the resulting ordered pair. In the analog world it is difficult to access the angle directly so we usually use the small angle approximation that the sine of an angle is approximately the angle and use the output of the sine mixer as our approximate angle. In a DSP based solution we would never hard limit the sampled input signal. The resulting harmonics would alias back into the measurement process as undesired artifacts. We recognize that if we want the angle of the ordered pair we can simply compute the angle with an ATAN or by an equivalent but computationally simpler task of applying a CoRDiC rotate to the output of the digital mixers. The ATAN provides a superior phase detection S-curve compared to that offered by the imaginary part of the complex product.

Our final example related to first principles DSP based solution to communication system tasks is the timing recovery process. In timing recovery, samples at the output of the matched filter must be time aligned with the peaks of the periodic correlation function. This corresponds to aligning the output sample points with the maximum eye-opening of the eye-diagram. In legacy analog receivers alignment is accomplished by advancing or retarding the sampling clock by varying the control signal to the voltage controlled oscillator (VCO) supplying the clock signal. Modern receivers do not return to the analog domain via a DAC to affect the same control but rather acquire the time alignment in the DSP domain. We can do so because we trust Nyquist who assured us that values of the signal waveform collected at the Nyquist rate contain all the information residing in the original analog wave shape. In modern receivers the alignment is performed in one of two ways. In one method, an interpolator is used to compute the samples at the desired sample location from the nearby offset sample values. The PLL of the timing recovery system controls the interpolation process. In the second method, the receiver contains a bank of matched filters spanning a range of offsets between the time sample locations and the peak location of the matched filter. Here the PLL of the timing recovery system controls the process of identifying the correct time aligned filter.

Bear in mind that DSP performs other functions beside the replacement of the analog circuitry in a modem. Besides the traditional tasks of AGC, spectral translation, spectral shaping, matched filtering and channelization, timing and carrier synchronization, upsampling and down sampling, and signal synthesis DSP is also used to enhance the performance of the digital and analog hardware. DSP is used to equalize the group delay and amplitude distortion of the receiver analog filters, to balance the gain and phase mismatches between the quadrature signal paths of the modulator and demodulator, and to pre-compensate for the spectral distortion caused by the DAC spectral response. DSP techniques also cancel spectral intrusion components caused by parasitic coupling of sampling clock lines, including the DC generated by self mixing in the analog mixers and DC offsets caused by the ADC. DSP techniques also support a variety of optimal processing options that are difficult to implement in analog realizations. These include non-linear (such as TANH) signal-to-noise ratio dependent gains and signal-to-noise ratio estimators.

A final note of why the discrete-time signal processing has become so entrenched in modulator and demodulator design is the flexibility afforded by the ability to reconfigure the hardware by software upgrades or to modify system parameters by software control. Not the least of the DSP implementation benefits is product manufacturability. DSP parameters do not change with time and temperature and do not have values spanning a tolerance range as do analog components.

Now that I have had a chance to voice my biases on the importance of the *Discrete-Time Approach* to the implementation of *Digital Communication* Systems I would like to share my assessment of Michael Rice's book. I hope you are reading your own copy of this book! Are you?

First I will share with you the fact that I have known Michael Rice for some 10 years. He spent a year at SDSU where we shared common interests related to implementing digital receivers and Multirate Signal Processing. We tried to entice him to join our faculty but he resisted our offer and returned to BYU. Our loss, their gain! He is great instructor and has an interesting sense of humor. When you meet him ask him how Scooby-Doo was named.

I thoroughly enjoyed examining the book. In particular I enjoyed the footnotes distributed throughout the book. I read every one of them! The text starts with Chapter 1, the **Introduction**: a fun section to read. It continues with an initial review of continuous and discrete **Signal and Systems I** in Chapter 2 followed by a review of **Discrete-Time Techniques Useful for Digital Communications** in Chapter 3. Here the emphasis is processes and filters that have application in modem implementation. You can't write a text on communication without a section on probability so Chapter 4 presents a **Review of Probability Theory**. Chapter 5, **Linear Modulation 1: Modulation, Demodulation, and Detection**, presents digital communications using linear modulation techniques. This is followed by Chapter 6 **Linear Modulation 2: Performance** which presents measures of performance of in additive white Gaussian noise.

At this point, the text book swings off the beaten path by presenting the following chapters with an emphasis on discrete-time implementation of the offered material. Chapter 7 does a very nice job presenting the basics and implementation techniques of **Carrier Phase Synchronization**. The nice style continues in Chapter 7 with an introduction and presentation of techniques to obtain **Symbol Timing Synchronization**. Aligned with the interest in Discrete-Time signal processing techniques in communication systems, Chapter 9, titled **System Components**, presents detailed descriptions of important DSP building blocks. These include the ADC and DAC, Discrete-time Oscillators, Resampling Filters, the CoRDiC, and AGC systems. With all the pieces in place, Chapter 10 presents System Design showing modulator and demodulator architectures and channelizers. A number of important appendices follow the chapter text. These include a section on **Pulse Shapes** in Appendix A, an exposition on **Complex-Valued Representation for QAM** signals in Appendix B and a section on **Phase Locked Loops** in Appendix C. The book closes at the end of an extensive **Bibliography**. Of particular interest to the academic community and to the motivated student is the thorough set of problems following each of the nine chapters following the introduction as well as the three appendices. In a style I enjoy, similar to mine, appropriate well annotated MATLAB examples are distributed throughout the text.

I think this is fine book. I also think that the reader will be richer for his or her understanding of the Digital Communications through the modern DSP centric implementation perspectives it presents. Nice piece of work Michael Rice!

fred harris
San Diego State University

Preface

It was the summer of 1998. The Mathworks had introduced the Communications Blockset as part of Simulink the previous summer and I had been trying to use this as a basis for a series of extended design exercises for my senior-level elective course in digital communications at Brigham Young University. I was inspired to develop these exercises by my colleague Brad Hutchings, who had developed a series of laboratory exercises for his computer organization course. In these exercises, the students designed different parts of a simple RISC processor. Each subsystem was tested using a "test vector." The culminating laboratory exercise required the students to connect the subsystems together to form a complete processor and use it to run a simple program. My thinking was, if this could be done with RISC processors, it could be done with a QAM demodulator. Just as a processor has an instruction fetch/decode function, an arithmetic logic unit (ALU), registers, and control, a QAM demodulator has matched filters, a decision device, a carrier phase synchronizer, and a symbol timing synchronizer. In place of test vectors, I envisioned a sequence of known bits used to create a modulated waveform. I struggled to find just the right software environment for the students. Simulink with the Communications Blockset seemed to fit the bill.

I dabbled in the development of QAM demodulators using Simulink throughout the 1997–1998 academic year. My resource was the remarkably large number of text books devoted (in part or in whole) to digital communications. Although these texts explained pulse shapes, matched filtering, and synchronization, most were content to develop the theoretical underpinnings of these concepts in the continuous-time domain. This is not a bad thing. But the material offered little guidance on the proper way to perform these tasks in discrete time: What is the proper sample rate for the matched filter? How is this chosen? What are the trade-offs? How is carrier phase synchronization done in a discrete-time system? How is timing synchronization performed in a discrete-time system (do I really have to form a clock whose edges are aligned with the proper time)? And what about PLLs in discrete time?

Unable to make the proper connections between the continuous-time concepts developed in these texts and my discrete-time environment, I made many false assumptions and errors. Over and over, I ended up with a discrete-time system that was a kluged replica of a continuous-time system described in one of these texts. Something just wasn't right. Like any academic, I searched for conference and journal papers on the subject. Time and time again, I encountered an author who always (when the editors allowed it) spelled his name in lower case letters.

Now back to the summer of 1998. I received a notice that the Institute for Telecommunications Research[1] (ITS) was sponsoring its first International Symposium on Advanced Radio Technologies (ISART) in Boulder Colorado. As I parsed the Advance Program, I was delighted to see that now-somewhat-familiar-name in lower case letters: fred harris. Even more intriguing was the title of his presentation: "A Trap to Avoid: A DSP Radio is Not a Digitized Analog Radio." In an instant, I knew what my problem was and I knew where to find the solution. I attended the ISART in 1998 and listened to fred's talk like I had listened to no other. I really was a magic moment. fred and I became friends and I wound up visiting him on a sabbatical during the 1999–2000 academic year.

I learned a lot that year and I was eager to share it with students here at BYU. I revamped those Simulink exercises and retooled my course. Teaching my new-found knowledge was fun. But having the students retain enough of it to complete Simulink exercises was an entirely different matter. At first, I copied notes from a multitude of sources for use as resource material. (There were a few books devoted to discrete-time processing for digital communications. But these texts presumed a-priori knowledge of digital communications.) Inevitably, these sources used different terminology and different notation that made it more difficult than it needed to be for the students who were seeing for the first time. I produced notes for the class to augment these sources. Over time, we used the notes more and more and the available texts less and less. Well, you see what happened....

There are many challenges associated with writing a text book such as this one. First, the text must teach digital communication theory. Next, the text must present both continuous-time and discrete-time realizations in such a way that the differences between the two can be appreciated. This sounds much easier than it really is. Sufficient background material must also be included to produce a work as self-contained as possible. I also experimented with the development. Continuous-time random processes are an efficient and elegant way to describe most of the concepts fundamental to digital communications. Unfortunately, for seniors and first year graduate students, the newness of continuous-time random processes often gets in the way. In this text, I tried to leverage the experience most modern electrical engineering majors have with random variables. Much of the more theoretical development involves random variables.

This text assumes, as prerequisites, courses in linear systems, transform theory, and probability theory (at least through random variables). Chapters 2–4 present the necessary background material to embark on the journey through this text. Chapters 2 and 3 describe signals and systems while Chapter 4 is devoted to random variables. While most seniors and first year graduate students are familiar with linear systems and transforms in both continuous time and discrete time, the *connection* between the two is usually quite nebulous. For this reason, Chapter 2 devotes substantial attention to this very important topic. In addition, some of the peculiarities of discrete-time processing of band-pass signals are also included. This receives almost no attention in prerequisite courses. Chapter 3 introduces some material that may not be familiar: multi-rate signal processing and discrete-time filter design.

---

[1] ITS is the research and engineering branch of the National Telecommunications and Information Administration (NTIA), a part of the U.S. Department of Commerce.

The principle body of an introductory course in digital communications is embodied in Chapters 5–8. These chapters focus on modulation, detection, performance, and synchronization. I have tried to approach these topics from simultaneous points of view. Modulation, detection, and synchronization are described in both continuous time and discrete time. Practical descriptions as well as theoretical developments from basic principles are also included. Organizing these different points of view in each chapter was a tremendous challenge: I did not have any good examples to follow. The continuous-time systems are those presented in the traditional texts. The discrete-time counterparts start with discrete-time versions of the continuous-time system where appropriate (symbol timing synchronization is the big exception). The full power of discrete-time processing is not exploited until Chapter 10 — and yes, it was hard to wait to describe the good stuff. The practical descriptions use pulse amplitude modulation (PAM) and quaternary phase shift keying (QPSK) as examples. Chapters 5, 7, and 8 end with a derivation of the corresponding maximum likelihood estimator. I realize this is the reverse order from the traditional approach. I have selected the reverse order based on my experience that most students tend to learn specific-to-general rather than the other way around. When the principles are taught using a few special cases, the generalizations tend to be more meaningful.

The description of the digital modulation formats discussed in Chapter 5 depends on pulse shapes, which are described (in both continuous time and discrete time) in Appendix A. The synchronizers described in Chapters 7 and 8 are based on the phase locked loop (PLL) described in Appendix C (in both continuous time and discrete time).

A more extensive treatment of discrete-time processing for digital communications is presented in Chapters 9 and 10. Chapter 9 introduces analog-to-digital and digital-to-analog converters, discrete-time oscillators, resampling filters, the CoRDiC and CoRDiC-like algorithms, and discrete-time automatic gain control. The emphasis is on the performance of these components from a systems level point of view. These components are applied to discrete-time techniques for modulators, demodulators, and channelizers in Chapter 10.

I have tried to write an honest text book. Each chapter ends with a "Notes and References" section that provides some historical notes, references, and a description of related topics that were not discussed in the text. No text presents a completely exhaustive treatment. And this one is no exception. My intent is to describe the basic details that are common to most digital communication systems. Hopefully, this provides a context in which the information provided in the references is more meaningful.

I am indebted to a large number of people who helped make this text possible. I really do feel as though I am standing on the shoulders of giants. First is fred harris, my friend, colleague, and entertainer, who taught me a lot of this stuff. Second is my wife who had to suffer the tribulations of living in paradise (San Diego) during my sabbatical year with fred harris. And my children, who were fatherless on far too many evenings as I toiled away on this project. I am also grateful to my department chair, Prof. Rick Frost, who provided summer support and reduced teaching assignments from time to time as I completed this manuscript. I have many friends in industry and academia took time to explain to me things that I otherwise would not know. You know who you. Of special mention are those who reviewed this text: Xiaohua Li, State University of New York at Binghamton; Mohammad Saquib, University of Texas at Dallas; Jacob Gunther, Utah State University; Tongtong Li,

Michigan State University; John J. Shynk, University of California, Santa Barbara. Their comments and collective wisdom have made this a much better book than I could have produced on my own. Finally, I thank my two editors at Pearson Prentice-Hall: Tom Robbins, who got me into this mess, and Alice Dworkin, who helped me get out.

Michael Rice
Provo, Utah

1

Introduction

1.1 A BRIEF HISTORY OF COMMUNICATIONS   1
1.2 BASICS OF WIRELESS COMMUNICATIONS   10
1.3 DIGITAL COMMUNICATIONS   12
1.4 WHY DISCRETE-TIME PROCESSING IS SO POPULAR   14
1.5 ORGANIZATION OF THE TEXT   19
1.6 NOTES AND REFERENCES   22

1.1 A BRIEF HISTORY OF COMMUNICATIONS

Communications has always been an essential aspect of human society. For much of human history, control of information has been the source of power, and communications is the way information is transmitted. Instantaneous communication over large distances is one of the hall marks of a *modern* human society. Electronic communication is what makes such instantaneous communications possible. Convenient and ubiquitous instantaneous communications has endowed modern human society with its fast pace. This fast pace exhilarates some and causes others to suffer from information overload.

Electronic communications began with the development of the telegraph[1] in the late 18th century. By the mid-19th century, Samuel Morse had demonstrated a working wire-line telegraphy system based on the pioneering work of William Cooke and Sir Charles Wheatstone. Wire-line telegraphy experienced dramatic growth during the second half of the nineteenth century as an ever-increasing number of telegraph lines connected telegraph offices all over North America, Europe, and the Middle East. Transatlantic lines connecting North America and Europe posed a particular challenge: the enormous length of the cable presented a high attenuation, low bandwidth channel to the telegraph system.

After failed attempts in the late 1850s and interruptions caused by the U.S. Civil War, permanent telegraph service between North America and Europe was established in 1866. Transoceanic telegraphy was one of *the* great engineering projects of the 1850s and 1860s.

---

[1] The word *telegraph* comes from the Greek *tele* ($\tau\hat{\eta}\lambda\epsilon$) meaning "afar" and *graph* ($\gamma\rho\alpha\phi$) meaning "that writes" or "writer." It was first used by French inventor Claude Chappe in 1792 to describe a device consisting of an upright post with movable arms whose positions were used for signaling.

Along with the obvious advances in cables, undersea cable laying, and ships to perform the task, electrical engineering began to mature into the scientific- and mathematical-based discipline it is today. One of the trendsetters was William Thompson, who later became Lord Kelvin. Thompson was the first to analyze telegraphy through a long cable as an engineering problem in the modern sense. In an 1854 paper, he applied Fourier analysis to the propagation of voltages and currents[2] in the cable and separated his analysis of the telegraph signal from the analysis of the telegraph cable. This brilliant insight led to better designs of the cable (i.e., the conductor dimensions and its electrical insulation) and to better signaling equipment (i.e., pulse shapes more suitable to the channel and equipment to generate and detect these pulse shapes).

At the same time, *wireless* communications also emerged. James Maxwell's theory of electromagnetic radiation, published in 1864, predicted the existence of propagating electromagnetic waves. Maxwell's theories were confirmed 23 years later when Heinrich Hertz produced electromagnetic propagation using experiments with a spark gap device. During the next few decades, Ambrose Fleming demonstrated an "electronic valve" — a vacuum tube with two electrodes that functioned as a diode — that could be used as a radio wave detector. Lee de Forest added a third electrode, which he called a triode, and formed a vacuum tube that was a key component in radio frequency amplifiers and oscillators. Guglielmo Marconi and Reginald Fressenden were contemporaries (and competitors) who recognized the commercial potential of this work. Fressenden developed a method for producing continuous wave (CW) transmissions at radio frequencies and made significant improvements in receiver and detector technology. He coined the term *heterodyne* to describe the process of frequency translation that is the basis for most modern receivers and *modulate* to describe the variations on a CW carrier due to an external signal. As impressive as these innovations were, they did not produce a practical wireless telegraph. Marconi's talent for the practical made wireless telegraph a commercial reality. His work on wireless telegraph began at the end of the 19th century. By 1900, Marconi had transmitted detectable electromagnetic energy across the Atlantic and, by 1901, had successfully transmitted a wireless telegraph signal (both claims were disputed by Fressenden).

By the early 20th century, wireless telegraphy was a growing business. One of the primary beneficiaries was maritime transportation. While at sea, ships equipped with a wireless telegraph unit were able to stay in communication with land. Three incidents from the maritime world illustrated the value of wireless communications. In early 1909, the *Republic* collided with the steamer *Florida* in dense fog. The *Florida* issued a distress signal using its wireless telegraph. The *Baltic* responded and saved all 1700 from the two ships. The second event involved the passage of King George V and Queen Mary to India in 1911–1912. While in transit, they were able to keep in touch with events at home and even issued a Court Circular in November 1912. The third event was the *Titanic* disaster in April 1912. After striking the iceberg, a distress signal was transmitted using the wireless telegraph. The *Carpathia*, responding to the distress call, arrived after the *Titanic* sank and saved 710 of the survivors who were afloat (1500 went down with the ship). There were several ships closer to the *Titanic*

---

[2]Fourier developed the technique known today as the Fourier transform to analyze heat propagation in a solid.

that could have responded before the ship went down. But these ships were not equipped with wireless telegraph units. Of all the unfortunate ironies that dramatized the disaster, one of the lesser known was that of the *California*. The *California* was very close by when the *Titanic* struck ice. And the *California* was equipped with a wireless telegraph unit. Out of frustration, the crew had shut it off earlier in the day after interference from the *Titanic*'s more powerful telegraph unit rendered the *California*'s useless. (Most wireless telegraph sets operated on the same frequency.)

The wire-line-to-wireless evolution repeated itself with voice communications. Experiments with electronic transmission of voice over wires began in the second half of the 19th century, culminating with the telephone[3] patent issued to Alexander Graham Bell in 1876. Later improvements by Gray, Edison, Hunnings, and others, made the telephone a commercially viable enterprise. By the start of the 20th century, companies began providing regular service to customers in North America and Europe, and by 1915, AT&T had completed a transcontinental telephone line in the United States.

Experiments with wireless transmission of voice began in the late 19th century and were intimately connected with the development of the wireless telegraph. Just after the turn of the century, Marconi and Fressenden demonstrated the reception of transatlantic radio signals. At the end of 1906, Fressenden successfully transmitted the first radio broadcast. Based on de Forest's triode vacuum tube, Howard Armstrong designed oscillator circuits whose performance was good enough to generate a sustained CW radio frequency carrier signal. The stability of Armstrong's design allowed carrier frequency to be defined and maintained, thus enabling different radios to operate on different channels. In addition, Armstrong designed amplifier circuits that improved both the selectivity and sensitivity of radio receivers. With these developments, voice over radio expanded quickly. Radio telephone services were initially developed for maritime communications. Broadcast radio quickly followed, with the first scheduled radio broadcasts by KDKA in Pittsburg in 1920. In 1922, the British Broadcasting Corporation (BBC) began its first radio broadcast as a private company. (The BBC transitioned to a public company in 1926.)

These radio broadcasts used amplitude modulation (AM). With AM, the noise generated by the radio electronics produced static in the audio output. This static limited the quality of the audio signal.

In an effort to improve the noise immunity of a modulated carrier, Howard Armstrong experimented with frequency modulation (FM) in the 1930s even though the Bell Laboratories researcher John Carson, who had developed a mathematical analysis of FM wrote, "I have proved, mathematically, that this type of modulation inherently distorts without any compensating advantages whatsoever. Static, like the poor, will always be with us." Armstrong who is reported to have said, "I could never accept findings based almost exclusively on mathematics. It ain't ignorance that causes all the trouble in this world. It's the things people know that ain't so," continued his work and developed a working FM system whose noise immunity performance was superior to that of AM. FM detectors developed

---

[3]The word *telephone* comes from the Greek *tele* ($\tau\hat{\eta}\lambda\epsilon$) meaning "afar" and *phone* ($\phi\omega\nu\acute{\eta}$) meaning "voice" or "sound" and was first used by the German inventor Philipp Reis in 1861 to describe his device that could "produce tones of all kinds at any desired distance by means of the galvanic current." He called this device "telephon" (in German).

by Foster and Seely (1936) and Peters (1945) made commercial broadcast FM commercially viable. By 1961, stereo FM broadcasts began in the United States.

Television[4] also emerged during the first decades of the 20th century. Experiments with mechanical and electromechanical systems for the transmission of moving pictures had been performed since the late 19th century. By 1928, Philo T. Farnsworth demonstrated the first all-electric television system. The BBC's first experimental broadcast in London occurred in 1932 and television transmissions were broadcast from the Eiffel Tower in Paris in 1932 and from the Empire State Building in New York in 1936. By 1939, the BBC was broadcasting television in the United Kingdom on a commercial basis. In 1941, Federal Communications Commission (FCC), which was formed from the Federal Radio Commission in 1934, authorized television broadcasting in the United States. NTSC color television was introduced in the United States in 1953, but it was not until the improvements offered by Sony's *Trinitron Color Television System* in 1968 that color television became commercially viable. The first decade of the 21st century is witnessing a transition to digital video transmissions in the form of high definition television (HDTV).

Radio and television were not the only developments during the first half of the 20th century. In 1921, Marconi expressed interest in radio waves "coming apparently from outer space." In 1932, Karl Jansky, a Bell Laboratories physicist measuring "static" (radio interference) that might interfere with a planned transatlantic radio telephone service, measured a faint steady hiss of unknown origin. Experiments with a directional antenna showed that these emissions were originating from the center of the Milky Way galaxy (in the constellation Sagittarius). After the Great Depression, Grote Reber and John Kraus followed up on Jansky's work and founded radio astronomy. Kraus started a radio observatory at The Ohio State University and wrote the first textbook on radio astronomy in 1966. Early radio astronomers Arno Penzias and Robert Wilson, using a receiving antenna originally designed by Bell Laboratories for satellite communications, inadvertently discovered[5] background radiation while tracking down a problem with the receiving equipment in 1964.

Important advances in RADAR (Radio Detection and Ranging) took place in the decades leading up to World War II. Sir Robert Watson-Watt demonstrated the first practical RADAR in 1935 and, in 1937, Henry Boot and John Randall developed a device capable of generating high-power pulses at high frequencies. This device, which they called the resonant-cavity magnetron, enabled microwave RADAR.[6] During the war, the first operational RADAR systems were put into service and were used by Germany, France, Great Britain, and the United States to navigate their ships, guide their airplanes, and detect enemy craft before they attacked. James Van-Allen (who later discovered the radiation belts that surround the earth and are named for him) invented a RADAR-based proximity fuse for antiaircraft

---

[4]*Television* was formed from the Greek *tele* (τῆλε) meaning "afar" and the word vision (which English inherited from the Latin *vīsiōn-* or *vīsio*) and was first coined by the Russian scientist Constantine Perskyi in 1900. Its first use in English was in 1904. This is one of those mixed heritage words (*automobile* is another) that linguistic purists intensely dislike. T. S. Elliott, wrote in *Music of Poetry* (1942), "There are words which are ugly because of foreignness or ill-breeding (e.g., television)."

[5]This development is discussed in its communications engineering context in Section 6.4.

[6]Most kitchens in North America, Europe, and Asia are equipped with appliances that contain a magnetron. Its primary function is cooking popcorn and warming leftovers.

shells. The proximity fuse, which uses information from the RADAR to indicate when the shell should explode, was fabricated on a printed circuit board. This was the first use of printed circuit board technology. In addition to its obvious military uses, RADAR is used today to study weather, provide location information for collision avoidance (especially in the shipping and airline industries), study the environment, and as an aide in search and rescue operations.

The commercial prospects for two-way mobile radio, in civilian, government, and military applications, led to significant developments during the 20th century. One of the first developments along this path was in 1920 in Detroit, Michigan, where a police car with radio dispatch was put in use. An early mobile telephone, developed by the Bell Telephone Company, was installed in New York City police cars by 1924. By 1934, 194 municipal police radio systems and 58 state police stations were equipped with radios based on amplitude modulation. Sidney Warner developed a two-way police radio using FM in 1941. The first commercial mobile telephone service was available in St. Louis, Missouri, by 1946, although it would be another 30 years before mobile telephone, in the form of cellular telephony, would become popular. In related developments, W. H. Martin proposed the decibel[7] as a transmission unit in 1929 and E. M. Williams developed the spectrum analyzer in 1946.

During the second half of the 20th century, huge strides were made in satellite communications. By the late 1950s, microwave frequency transmission and reception technology together with rocket technology made space-based communications technologically viable. The Soviet Union launched *Sputnik I* in 1957. Although its only communication function was to transmit telemetry signals (which it did for about five months), *Sputnik I* demonstrated the huge potential of satellite-based systems for communications, surveillance, and other more sinister functions. In response, *Explorer I* was launched by the United States in 1958 followed by *Echo* in 1960. From its 1000-mile-high orbit, *Echo* acted as a passive reflector "bouncing" uplink signals back toward the earth. Because this technique required far too much transmit power to be commercially viable, active satellites equipped with the ability to receive, amplify, and retransmit uplink signals were developed. *Telstar 1* was launched in 1962 and *Telstar 2* and *SYNCOM* were launched in 1963. *Telstar 2*, a product of AT&T, was placed into a medium-altitude orbit and carried several telephone channels and one television channel. *SYNCOM*, designed by the Hughes Aircraft Company under NASA sponsorship, was placed into geostationary orbit (at an altitude of 22,300 miles above the equator) and consisted of a single repeater translating the 7.4 GHz uplink to a 1.8 GHz downlink that was transmitted using a 4-Watt TWTA power amplifier. In 1962, the Communications Satellite Corporation (Comsat), a quasi-public organization representing government and industry, was formed under the Communications Satellite Act. Following this, an international body representing 100 countries was formed in 1964. The purpose of this organization, named INTELSAT, was (and is) to maintain the operation of a global communication satellite system. Under the direction of INTELSAT, *INTELSAT I* was launched into geostationary orbit in 1965. Over the following five years, three more satellites, *INTELSAT II, INTELSAT III,* and *INTELSAT IV* were launched to bring the INTELSAT system up to full operational capacity.

---

[7]The decibel or one-tenth of a bel (after Alexander Graham Bell) was first used in 1923 to express loss in a telephone cable.

By the 1980s, satellite communications expanded to carry television programs and people were able to pick up the satellite signals on their home dish antennas. These antennas were usually large 3-meter dish antennas. In the 2000s, satellites are providing digital television directly to homes equipped with much smaller dish antennas mounted on the sides of homes. In addition to the increased capabilities for telephone and television, satellites have also expanded their roles in many other areas. Satellites were used to discover the ozone hole over Antarctica, helped locate forest fires, and provided photographs of the nuclear power-plant disaster at Chernobyl in 1986. The global positioning system (GPS) consists of 24 satellites arranged in four low-earth orbits at an altitude of 12,000 miles. Astronomical satellites (the *Hubble Space Telescope* is the most well-known example) have provided breathtaking images of the galaxy and universe.

Advances in communication satellite technology were often the result of advances in spacecraft designed to explore the solar system. *Mariner 2* was the first successful interplanetary spacecraft. Launched in 1962, *Mariner 2* was equipped with six scientific instruments and a two-way radio that were used to explore the planet Venus. Over the next eight years, subsequent Mariner missions explored Mars, Venus, and Mercury. Advanced techniques in digital communications were used to transmit the images and other data across the vast interplanetary distances. Numerous missions to explore the solar system, sponsored both by NASA in the United States and by European and Russian entities, have been launched since the 1970s. Perhaps one of the most spectacular were the *Voyager* missions launched in 1977. These missions provided riveting images of the outer planets (Jupiter, Saturn, Neptune, and Uranus) as well as other data that planetary scientists are still analyzing.

Back on earth, the success of two-way mobile radio prompted the development of wireless telephony during the late 1960s and early 1970s. Motorola demonstrated the cellular telephone to the FCC in 1972 and AT&T Bell Laboratories began testing a mobile telephone system based on hexagonal cells in 1978. By the 1980s "cell phones" were relatively common in the United States and Europe. Digital cellular telephony was introduced in Europe in the late 1980s and in the United States in the 1990s. Also in the 1990s, wireless telephony experienced explosive growth in Asia and in less-developed countries where the cost of installing a wireless telephone system was much less than the cost of installing a traditional wired telephone system. Cellular telephony has become so convenient and popular that, for many users, the "cell phone" has become the primary telephone, replacing the wired home telephone.

Three trends emerge from the chaotic history of electronic communications: a trend from wire-line communications to wireless communications, a trend from analog communications to digital communications, and the development of mathematical analysis and communication theory.

The Wireless Trend. Telegraphy started as a wire-line service and quickly transitioned to a wireless service. The appeal of mobile communications, especially for situations where wire-line connections were impossible, was irresistible. Telephony also started as wire-line service, but transitioned to a wireless service much more slowly. Again, the same appeal of mobile connectivity was in play. In the 1990s, the emergence of Internet access for the general public renewed consumers interest in wire-line communications. For much of this time, Internet access has been exclusively a wire-line service. Recent developments hint that the trend to wireless will play out with this service as well. Examples include the development of wireless networking standards, such as the IEEE 802.11 standards, mobile email services via cell phones and personal digital assistants (PDAs), and third and fourth generation mobile telephone standards with increased bandwidth and networking capabilities.

Remarkably, an argument could be made that television is operating in the reverse direction. Historically, the delivery mechanism for television has been broadcast technology: high-powered transmitters are used to provide coverage to a geographical area. The decision to begin this mode of transmission was an economic one: it was less expensive to construct a relatively small number of high-powered transmitters than it was to install high bandwidth wires from the provider to each customer. Since the late 1960s, an increasing number of homes are receiving television content over wires (originally in the form of coaxial cables but more recently in the form of fiber-optic cables). The popularity of this trend, together with the ability to deliver additional services, has had a profound impact on both the dynamics of the marketplace and the regulatory environment in which it operates.

The Digital Trend. It is interesting to note that electronic communications began as a digital communications system. The telegraph used a binary alphabet (the Morse Code with "dots" and "dashes") to transmit our nonbinary, alphanumeric characters. Because the natural input to this system was written correspondence, the telegraph became a quicker (and more expensive) way to send short letters, notes, or other text-based information. The telegraph ushered in the modern era of electronic communication with its ability to provide instantaneous information exchange across long distances. With all of its world-changing successes, it was still impossible to carry on a conversation[8] across these long distances instantaneously.

When real-time voice communication was demonstrated, the disappearance of this limitation created an enormous market for telephone companies all over the world. Both wire-line and wireless voice communications began as analog communications systems, but both have transitioned to digital communications systems. An important development in the migration from analog communications to digital communications occurred in the telephone industry. Until the mid-1960s, all links in the telephone network were analog communication links. Digital communications first appeared in the trunking components (the T-1 carriers) and later in the local and toll switching systems. This migration was motivated in large measure by the improved reliability offered by digital signaling techniques and the economic advantages of digital technology following the development of the transistor.

---

[8]At least as "conversation" was understood at the time. Instant messaging, e-mail, and text-messaging are examples of ways modern readers—at least the younger ones—carry on conversations through a purely written medium. These methods can be thought of as low-delay telegraph systems with improved user interfaces. The 13 May 2005 airing of NBC's *Tonight Show*, however, featured a competition between a pair of amateur radio enthusiasts operating a telegraph and a pair of teenagers, one of whom was the world text-messaging champion. In a test to see which method was faster, the telegraph operators easily defeated the text-messaging youth sending the message "I just saved a bunch of money on my car insurance."

In the 1980s, the desire to increase the capacity and suite of services offered to mobile telephone customers prompted the development of digital communication systems. GSM, the digital mobile telephone standard in Europe, was deployed in the 1980s, whereas digital standards such as IS-54 and IS-136 (TDMA/FDM) and IS-95 (CDMA) were deployed in the United States in the 1990s. In addition to voice, digital radio is becoming more popular and television is increasingly delivered in a digital format (HDTV is an integrated digital video/audio format). The corresponding communications link is a digital communications link.

Data communication systems are also following this trend. One example is aeronautical telemetry.[9] In aeronautical telemetry, the performance of an airborne "test article" is monitored by using a radio link to transmit the measurements output by a set of transducers to a ground-based monitoring station. The first aeronautical telemetry links were analog AM in the 1940s and analog FM in the 1950s. The output of each transducer modulated a separate carrier frequency to form the telemetry downlink. As airborne systems became more complex, more onboard measurements had to be collected and radioed to the ground. The use of separate carriers for each one proved unwieldy and uneconomical. By the 1970s, digital technology had progressed to the point where a new approach was possible. The transducer outputs were sampled to form a bit stream. The bit streams from all transducers were combined to form a composite bit stream that was used to modulate a single carrier. A digital version of FM (known as PCM/FM in the IRIG 106 Standard) became the most popular choice.

Many other communication services are inherently discrete, or digital. Examples include text messages, telemetry, financial data, credit card transactions, and Internet content. Some of the technical reasons for this trend toward digital are discussed in Section 1.3.

The Mathematical Trend. Another broad transition occurred during the 20th century. Essentially all of the work in the second half of the 19th century and the first few decades of the 20th century was devoted to the creation of devices and circuits that performed the functions required for wireless communications. A notable exception is William Thompson's 1854 paper analyzing telegraphy over long undersea cables. His separation of the signal from the channel formed the basis for communications and signal processing. In the 20th century, communications started to migrate away from the study of physics and devices and move toward mathematical analysis. This change was gradual and its approximate beginning was in 1914 with Carl Eglund's development of the mathematical expressions for an amplitude modulated wave. John Carson developed the mathematical

---

[9]Telemetry is the process of telemetering. A telemeter is an instrument for measuring a quantity at a distance from the place where the result is displayed or recorded. The origins of the word are the Greek prefix *tele* (τῆλε) meaning "afar" and the Greek suffix *metron* (μέτρων) meaning "measure" although the grammatical formulation in the 18th century was based on the Latin *metrum* from which other "measurement" words were derived, such as barometer, thermometer, pedometer, and ammeter. It is why the "metric" system uses "meter" as the basic unit of length. The word telemeter was first used in French (*télémètre*) to describe an instrument for measuring the distance to far off objects. Its first use in this sense was in a 1929 AIEE Journal article. A newspaper article appearing in the 10 November 1947 Baltimore *Sun* summed it up best: "Electronic gadgets called telemeters ... are installed in high speed missiles and tuned to send back to the ground by radio whatever information the scientists need."

theory of modulation in 1915 and showed, mathematically, the relationships between AM with a transmitted carrier, AM with a suppressed carrier, and signal sideband AM. His 1922 paper "Notes on the Theory of Modulation" showed that FM requires more bandwidth than AM. In 1923, Ralph Vinton Lyon Hartley showed that the amount of information that can be transmitted at a given time is proportional to the bandwidth of the communication channel, and in 1924, John Carson published "Selective circuit and static interference" which showed that the energy absorbed by a receiver is directly proportional to its bandwidth. R. V. L. Hartley published a mathematical theory of communication in 1927. By 1947 with Stephen O. Rice's publication of a statistical representation for noise, most of the theory for analog modulation had been set.

For digital modulation, Sir Ronald Fisher's development of maximum likelihood estimation in 1925 laid the foundation for the field of estimation theory which became one of the fundamental concepts behind digital communications. Harry Nyquist's classic paper on the theory of signal transmission in telegraphy, published in 1928, was later applied to digital communications. In 1933, J. Neyman and E. S. Pearson published one of the first papers on statistical decision theory. Kotel'nikov's development of a geometric representation of signals, published in 1947, is used extensively in this text. Perhaps the most influential work motivating digital communications was the 1948 paper "A Mathematical Theory of Communication" published by Claude Shannon. This paper laid the foundation for digital communications and established the field of information theory.

Tremendous changes have taken place in the electronic communications area. The invention of the transistor, advances in circuits and battery technology, the explosion of digital systems, together with advances in algorithms and mathematical techniques have all had profound impacts on the way electronic communications shapes society. At the end of World War II, the radio and telephone were the primary electronic communication devices in a typical home in the United States. Today, cellular telephony is the most common mode of voice communication and the Internet is the most common source of information.

These changes have also been reflected in the professional organizations associated with communications engineering. The American Institute of Electrical Engineers (AIEE) was formed in 1884 with one-half of its founding members representing the telegraph and telephone sectors. By the turn of the century, the membership had changed somewhat to reflect the enormous interest in *the* new technology in electrical engineering: power. The Institute of Radio Engineers (IRE) was formed in 1912 by electrical engineers from the wireless sector who did not feel that AIEE, an organization dominated by the electrical engineers from the power and telephone/telegraph industries, was a suitable professional home. For several decades, the AIEE and IRE coexisted with membership among communications engineers roughly following the wire-line–wireless divide. In the years immediately following World War II and its technological advances, electrical engineering as a whole, and communications engineering in particular, experienced tremendous growth in both membership and scope. The expanding scope created the inevitable overlap in interests between AIEE and IRE. In 1952, the Professional Group on Communication Systems (PGCS) was formed within IRE. The focus of the group was "communication activities and related problems in the field of radio and wire telephone, telegraph and facsimile, such as practiced by commercial and governmental agencies in marine, aeronautical, radio relay, coaxial cable and fixed station services." The group sponsored technical meetings and conferences and began publishing *IRE Transactions on Communications Systems*. By 1960, the overlap in interests between AIEE and IRE proved unworkable, and the two organizations merged on 1 January 1963 to form the Institute of Electrical and Electronics Engineers. The IRE PGCS and AIEE Communications Division merged to form the IEEE Communications Society[10] and IRE PGCS flagship publication became the *IEEE Transactions on Communication Technology*. The name was changed to *IEEE Transactions on Communications* in 1970. Most of the references in the bibliography are from this journal.

1.2 BASICS OF WIRELESS COMMUNICATIONS

Early on, it was recognized that transmitting a voice (or other low-bandwidth information-baring signal) directly using electromagnetic propagation was impractical. To do so would require huge antennas and, given the poor propagation characteristics at such low frequencies, a lot of power. To reduce antenna size and exploit the propagation properties of electromagnetic propagation at higher frequencies, carrier modulation was developed. A continuous-wave (CW) carrier is an electromagnetic signal whose time-domain representation is expressed as $A \cos(\omega_0 t + \theta)$. A is the amplitude of the carrier, $\omega_0 = 2\pi f_0$ is the frequency (in radians/second), and θ is the phase. The signal is called a carrier because it "carries" information from transmitter to receiver. This is accomplished by altering the amplitude, and/or phase of the carrier in proportion to the information signal via a process called *modulation*.[11] The resulting modulated carrier is a signal whose spectral energy is concentrated around the carrier frequency f_0 Hz. Efficient antennas have physical dimensions that are proportional to the wavelength of the signal. A signal whose spectrum is centered at f_0 Hz has an average wavelength of about $\lambda = c/f_0$ (c is the speed of light, which is 3×10^8 m/s). Thus, the higher the f_0, the smaller the antenna can be.

The propagation characteristics are also strongly dependent on the wavelength. Propagation in water for submarine communications requires a very low carrier frequency (less than 100 kHz) and hence, very large antenna dimensions. Propagation in the

---

[10]The two groups did not merge at first. Members of the AIEE Communications Division were mostly employed in the telephone industry while members of the IRE PGCS were employed in the wireless and other newer communications industries. Fearing their own interests would be devalued in a combined organization, each side hesitated to fully commit. The cultural differences were eventually ironed out and the new IEEE Group on Communication Technology was formed in mid 1964. The Group was elevated to Society status in 1972 with A. F. Culbertson as President, A. E. Joel, Jr. as Vice-President, A. B. Giordano as Secretary, and D. L. Solomon as Treasurer.

[11]Early on the terms "control," "vary," "mold," and "modify" were used to describe the process by which the information signal changed the amplitude and/or phase of the carrier. Fressenden used the word "modulation" in 1907 to describe fluctuations in radio waves propagating in nonideal conditions. This application was no doubt motivated by the use of the term "modulate" to describe the small pitch variations singers induce on their voices to produce the vibrato effect. By 1910, modulate was used in the wireless communications context in books by Fleming and Pierce. Modulate comes from the Latin verb *modulārī* which means "to measure, adjust to rhythm, or make melody." It comes from the same root word from which *mode* and *modulus* are derived.

Section 1.2 Basics of Wireless Communications

100–3000 kHz range follows the surface of the earth and is used for over-the-horizon communication links. Systems operating with carrier frequencies in this range are capable of providing communication links over distances of a few hundred kilometers. Propagation in the 3–30 MHz band is reflected by the ionosphere and is capable of providing communication links over distances of a few thousand kilometers. At higher frequencies, the propagating electromagnetic wavefront penetrates the ionosphere. Communication systems that use carrier frequencies in this range are usually line-of-sight links. In general, higher frequencies experience more atmospheric attenuation than lower frequencies. Atmospheric attenuation is particularly severe at resonant frequencies of water vapor (22.235 GHz) and the oxygen molecule (53.5–65.2 GHz). A summary of frequency bands is listed in Table 1.2.1.

The *Titanic* disaster illustrated the need for sharing or, to use a more modern term, multiple access. Multiple access was the second motivator for carrier modulation. By assigning each user a *different* carrier frequency, all uses are able to operate simultaneously. Proper selection of the carrier frequencies eliminates the interference a user may experience from the other users. This form of multiple access is called frequency division multiple access.

Table 1.2.1 A Summary of Frequency Bands Used in Wireless Communications

Frequency		Designation	Examples
3–30	kHz	Very low frequency (VLF)	Submarine communications
30–300	kHz	Low frequency (LF)	Marine communications
300–3000	kHz	Medium frequency (MF)	Maritime radio, commercial broadcast AM
3–30	MHz	High frequency (HF)	Amateur radio, military communications
30–300	MHz	Very high frequency (VHF)	VHF television, commercial broadcast FM, two-way radios, aircraft communications
300–3000	MHz	Ultrahigh frequency (UHF)	UHF television, first generation cellular telephony
3–30	GHz	Superhigh frequency (SHF)	Satellite communications, microwave links, radar
30–300	GHz	Extremely high frequency (EHF)	Radar, experimental satellite communications
Letter Designations (RADAR)			
1–2	GHz	L band	GPS, microwave links, low-earth orbit satellite links
2–4	GHz	S band	Microwave links, personal communication systems (PCS)
4–8	GHz	C band	Geostationary satellite communications
8–12	GHz	X band	
12–18	GHz	Ku band	Geostationary satellite communications
18–27	GHz	K band	
27–40	GHz	Ka band	

A modulated radio frequency (RF) carrier is a signal of the form

$$A(t)\cos(\omega_0 t + \theta(t)) = A(t)\cos(\theta(t))\cos(\omega_0 t) - \sin(\theta(t))\sin(\omega_0 t) \quad (1.1)$$
$$= I(t)\cos(\omega_0 t) - Q(t)\sin(\omega_0 t). \quad (1.2)$$

The carrier frequency is $\omega_0 = 2\pi f_0$ rad/s. The amplitude of the modulated carrier is $A(t)$. If $A(t)$ is proportional to the information signal, then the carrier is *amplitude modulated*. The instantaneous phase is $\theta(t)$. (The instantaneous frequency of the modulated carrier is $\omega_0 + \dot{\theta}(t)$ as described in Chapter 9.) In *frequency modulation*, the instantaneous frequency is proportional to the information signal.

The modulation can be either *analog* or *digital*. In *analog modulation*, $A(t)$ and/or $\theta(t)$ are drawn from a continuum of possible waveforms. Common examples are commercial broadcast AM (for amplitude modulation) and FM (for frequency modulation) radio where the source signal is a continuous-time audio signal. (There are an infinite continuum of possibilities.) In these cases, the information *is* the amplitude or phase waveform. If $A(t)$ and/or $\theta(t)$ are drawn from a finite set of waveforms, then *digital modulation* is performed. Cellular telephones, pagers, telemetry links, digital video broadcast (via satellite) are all examples of digital communications. Here the information is conveyed by the waveform selection: *which* waveform is transmitted is the information.

The distinction between analog and digital communications has a profound impact on the structure of processing performed on the receiving end. In general, the term *demodulator* is used to describe the process of removing the carrier from the received signal and usually involves a frequency translation from bandpass to baseband. The term *detector* describes the processing used to extract the information from the noisy, and possibly distorted, received signal. These definitions are not rigorous and are often used interchangeably in the communications literature. This is especially true when the detection is applied directly to the bandpass RF signal. In this case, demodulation and detection occur simultaneously.

Extracting the information from an analog modulated carrier involves a process of producing the best possible replica of $A(t)$ and/or $\theta(t)$ from the received signal. For example, an AM demodulator/detector attempts to recover a replica of $A(t)$ from the noisy and distorted received signal as best it can. Likewise, an FM demodulator attempts to recover $\dot{\theta}(t)$ from the noisy, distorted received signal. In digital communications, the situation is much different. Because there are a finite number of possible waveforms, the receiving-end processing determines *which* one of the possible waveforms was the most likely to have been transmitted. It is often not even necessary to reproduce the $A(t)$ and/or $\theta(t)$ to make this decision.

1.3 DIGITAL COMMUNICATIONS

The trend toward digital modulation has been driven largely by the difference between the performance of analog and digital receivers. Because digital receivers perform a fundamentally different operation, a digital modulation system can usually achieve acceptable levels of performance with much less transmitter power. In addition, many desirable types of services can be offered when the information to be transmitted is in a discrete format. The most influential drivers in this trend are

- *New Services:* Originally, the majority of wireless communications systems were devoted to audio signals (voice, music, etc.). Modern communications services also provide caller ID, e-mail, text messaging, data (e.g., weather, stock prices), electronic financial transactions, etc. These newer services are based on digital data. To be integrated with the traditional audio signal, it is more natural to digitize the audio signal and combine those bits with the information bits from the other services. When the information signal is in the form of bits, it is more natural to use digital modulation than analog modulation.
- *Security:* Protecting the transmitted signal from unauthorized reception is much more natural when the modulating data is discrete (digital) rather than continuous (analog). This is particularly advantageous in cellular telephony, where most users do not want someone else eavesdropping[12] on the conversation. Security will become more important as more financial transactions—particularly those involving credit card numbers or bank account numbers—are performed using portable wireless devices whose radiated RF energy can be detected by anyone. When the information signal is in digital form, powerful algorithms such as encryption and verification can be applied.
- *Transmitter Power:* In a system using analog modulation, the noise accompanying the received signal shows up as distortion in the demodulator/detector output. More noise at the demodulator input produces more noise at the output and there is very little that can be done about it, because the demodulator/detector extracts the amplitude and/or phase of the received modulated carrier. In a digitally modulated system, the detector output is a decision that identifies which of the possible transmitted waveforms is the most likely to have been transmitted. If this decision is correct, the transmitted information is recovered *perfectly.* Error correcting codes can be used to make the bit error rate very small. This fundamental difference in how noise affects the output makes it possible for digitally modulated systems to achieve performance goals with less transmitter power than for a system using analog modulation. This is very important for portable wireless systems that rely on battery power. Lower transmitter power allows a battery to operate longer on a given charge.
- *Flexibility:* Data from digitized voice, audio, or video may be merged with data from other sources to form a composite data stream. The composite data stream forms the input to a digital modulator. The digital demodulator/detector only has to recover the data and pass it on to a processor for parsing, display, and audio output. The ability to reduce different types of signals and data into a common format provides a great deal of flexibility in how multiple access is accomplished and what kind of networks are possible. For example, when the information from multiple sources is in digital format, these sources may be time multiplexed, code multiplexed, or frequency multiplexed

---

[12]One of the most notorious eavesdropping incidents occurred in December 1996 when a Florida couple intercepted and taped a cellular phone call between the then House Speaker Newt Gingrich (who was vacationing in Florida) and some Republican colleagues. The recorded conversation, where Congressman Gingrich's strategy for upcoming hearings was planned, was turned over to his political rivals.

using modulated subcarriers. With analog modulation, the only option is frequency multiplexing.

These advantages are not achieved without a penalty. Relative to analog modulation, digital modulation has the following disadvantages:

- *Synchronization:* The detector for a digitally modulated carrier must know when the data symbols begin and end in order to make a reliable decision. This timing information must be recovered from the received signal using a process called timing synchronization. A system using analog modulation does not have this requirement. The digitally modulated systems with the best performance also require knowledge of the phase of the unmodulated carrier. This information is also extracted from the received signal using a process called carrier phase synchronization. Some systems using analog modulation also require phase synchronization. In this case, carrier phase synchronization is usually accomplished using a transmitted reference.
- *Bandwidth:* A fair bandwidth comparison is often difficult because the behavior of analog and digital modulations are so different. AM and FM may be used to transmit a band-limited continuous-time audio or video signal. To use a digitally modulated carrier, the band-limited continuous-time signal must be sampled, and each sample must be quantized. The direct approach to sampling and quantization usually produces a digitally modulated carrier whose bandwidth is greater than that of the corresponding AM or FM signal. However, powerful source coding and data compression techniques can be applied to reduce the bandwidth of the digitally modulated carrier. In the end, which option has the best bandwidth depends on the details. It should be noted that source coding and data compression are uniquely digital processes.

With analog communications, AM and FM provided a power/bandwidth trade off. FM requires more bandwidth, but less transmitter power because it has superior noise immunity problems. The same power/bandwidth exists with digitally modulated carriers. In most cases, a third dimension is added to the trade-off space: complexity. For example, the use of error-control coding can be used to reduce the error rate without an increase in power or bandwidth. This is achieved at the expense of complexity: the detector must be capable of performing highly complex decoding operations. Many other algorithms (almost all of them discrete-time algorithms) can be applied to improve the performance of a system using digital modulation. Most of these algorithms increase the complexity of the demodulator and detector and do not have a counterpart in analog modulation. At present, it is more expensive to increase power or bandwidth than to increase complexity. For this reason, most of the effort in the past 50 years has been devoted to developing algorithms and signal processing to improve performance.

1.4 WHY DISCRETE-TIME PROCESSING IS SO POPULAR

The use of the terms "analog" and "digital" to describe the two broad classes of modulation are somewhat unfortunate because these words are also used to describe circuits, hardware, and signal processing. The type of processing used in the demodulator/detector for either analog or digital modulation may be either "analog" (continuous-time) or "digital" (discrete-time). Thus, one can speak of an FM demodulator realized with analog hardware or, after sampling, with digital hardware: an "analog FM demodulator" or a "digital FM demodulator."

When speaking of the modulation type in general, an awkward situation occurs as illustrated in Figure 1.4.1. Here the words "analog" and "digital" refer to both the modulation type and to the hardware realization. This dual use of the words produces confusing phrases such as "a digital digital demodulator" or "digital analog demodulator." In this text, the terms "analog" and "digital" are reserved for the modulation type. "Continuous-time" and "discrete-time" are used to describe the processing performed by the modulator and/or demodulator/detector. This situation is illustrated in Figure 1.4.2. This usage produces the somewhat clearer expressions "a discrete-time digital demodulator" or "a discrete-time analog demodulator." This nontraditional terminology recognizes that the application of discrete-time processing is independent of the modulation type. As long as there are samplers (analog-to-digital converters) that can sample the signal fast enough to satisfy the sampling theorem, discrete-time processing can be applied.

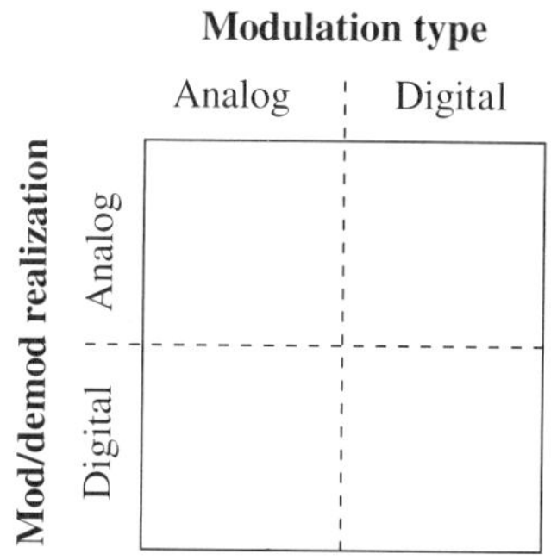

Figure 1.4.1 The possible combinations of labels using "analog" and "digital" to describe both the modulation type and the hardware and signal processing.

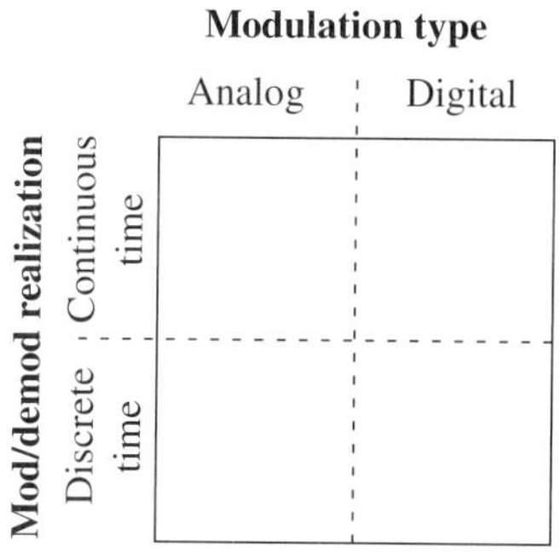

Figure 1.4.2 The possible combination of labels using "analog" and "digital" to describe the modulation type and "continuous-time" and "discrete-time" to describe the hardware and signal processing.

The earliest demodulators/detectors for analog modulation were continuous-time circuits. The early systems using digitally modulated carriers were also implemented in mostly continuous-time processing. It was not until the development of high-speed discrete-time hardware in the form of VLSI circuits and programmable processors that discrete-time processing became practical in this application. Of course, analog-to-digital and digital-to-analog converter technology also had to be in place.

Most systems today use a combination of continuous-time and discrete-time processing. The trend has been to move as much processing as possible to the discrete-time side. The primary drivers for this trend are

- *Improved Design Cycle:* Systems based mostly on continuous-time processing (analog circuits) took a long time to develop, debug, and prepare for manufacture. RF circuits are hard to design, and it takes years of experience for engineers to master the art of analog circuit design. In systems where the received RF signal is down-converted and sampled, the system designer has two options to realize the processing: "digital hardware" in the form of a custom application-specific integrated circuit (ASIC), off-the-shelf parameterizable chip, field programmable gate array (FPGA), etc., or a programmable processor such as a DSP chip. Although custom ASIC design and real-time code development have their own problems, it appears that the issues are resolvable in shorter design cycles. Designing circuits to "move bits around" is quicker than designing circuits to "move voltages and currents around" without noise and interference from other system components.

    Another important reason design cycles are shorter with discrete-time processing is the fact that the basic system architecture can be reused for many different product lines. A good RF front end can be used with many different discrete-time processing algorithms. This significantly shortens design cycles because different radios may be designed by making small changes to the continuous-time processor and reconfiguring the discrete-time processing.
- *Improved Manufacturing:* The realities of manufacturing are that no two circuits have exactly the same behavior. This is a consequence of the fact that circuit parameter values vary with time, temperature, age, and vagaries of the manufacturing process that produced them. This is the case for both "analog" circuits and "digital" circuits. For systems based on continuous-time processing, the components in the system must be carefully calibrated to meet the system performance requirements. This can be a tedious and time-consuming process for very complex systems. For discrete-time systems, the variances due to limited manufacturing tolerances are less of an issue. As long as a voltage is above a threshold, it is a one; otherwise it is a zero. This makes the system more robust to component variations and improves the efficiency of the manufacturing process. Discrete-time processing also makes computer-controlled calibration easy to perform. Hence, much of the residual calibration can be automated.

    A more subtle issue is what manufacturers call "parts obsolescence." This is a particularly troublesome issue for product lines with long operational lives (such as public radio systems used by law enforcement, public safety, and municipalities). If a key "analog component" goes out of production before the product line reaches the end of its life, then the manufacturer must either purchase enough of these parts in advance (and store them) or redesign the system every time one of the analog components has to be replaced. Systems based on discrete-time processing avoid this inventory conundrum. If a processor becomes obsolete and unavailable, it is usually replaced by a processor that has more capability. And it often has the same footprint on printed circuit board. If the footprint is different, the redesign is not as complicated as it is to replace an "analog part."

- *Advanced Signal Processing Techniques:* Many of the signal processing techniques that make digital communications more attractive than analog communications can only be performed using discrete-time processing.
- *Flexibility:* The realities of today's wireless marketplace are that most units must be capable of operating on more than one standard. Multimode operation is important in cellular telephony because there are multiple standards in place across the United States and the world. "Global mobility" requires multimode operation. In military communications, proper coordination of different operational units requires communication with a multitude of incompatible radio standards. Response to natural disasters presents a tremendous logistical challenge as search and rescue, emergency medical treatment, fire, and other safety efforts involve multiple agencies with different incompatible communication networks. Systems based solely on continuous-time processing have only one way to realize multimode operation: the unit must be equipped with multiple "analog circuits," each dedicated to one of the desired modes. Systems based on discrete-time processing require one system and the ability to reconfigure or reprogram the discrete-time operations.

The advantages of shortened design cycles and multifucntionality are enhanced when the discrete-time processing is programmable and controlled by software. In this case, the operational mode of the radio is defined by its software. This notion is embodied in the *software radio* concept. The ideal software radio is a system consisting of an antenna, a sampler, and a programmable software as illustrated in Figure 1.4.3 (a) for the case of a receiver. It appears that limitations in sampling and processor technology will keep the true software radio from becoming a commercial reality in the foreseeable future. The current generation of programmable radios, called *software defined radios,* is a mixture of flexible continuous-time processing, dedicated discrete-time hardware, and programmable processors as illustrated in Figure 1.4.3 (b) for the case of a receiver. The programmable portion of the system is what endows the software defined radio with the flexibility needed for multifunctionality, global mobility, short design cycles.

Most modern communication systems are based, more or less, on a structure like that of Figure 1.4.3 (b). Consequently, this text emphasizes discrete-time processing to perform the functions required for demodulation and detection. Whether these functions are realized in the dedicated discrete-time hardware or in the programmable processor is unimportant—the algorithm is the same. As technological advancements alter the economic landscape of discrete-time hardware and programmable processors, the boundary separating algorithms in the two will shift. The biggest challenge facing engineers will be the broad knowledge-base required to make intelligent decisions about this partition. The complexity

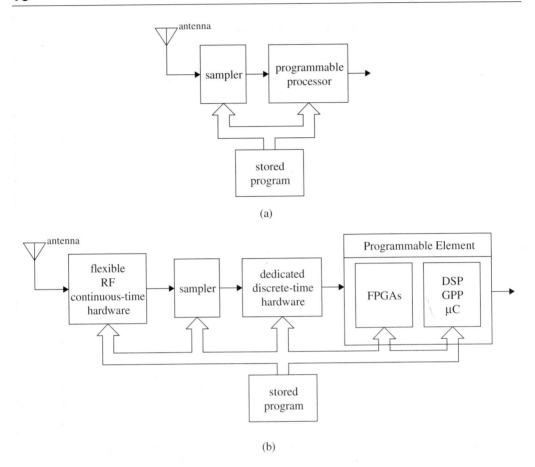

Figure 1.4.3 A block diagram of software radio receiver (a) and a software defined radio receiver (b).

and performance of RF circuits, discrete-time processing, digital algorithms, and software development methodologies all influence what the best partition is.

In the end, many different things have made modern wireless communications what it is. As John Anderson put it

> Modern telecommunications is the confluence of three great trends over the last two centuries: First, the invention of electromagnetic signaling technology in the form of the telegraph, the telephone, and the radio; second, the development of mathematical theories that made these inventions practical and efficient; and finally microcircuitry, the "chip," which made these inventions small, fast, reliable, and very cheap.

Communications has been and will continue to be an exciting subdiscipline of electrical engineering and, with the increased reliance on discrete-time processing and software, an important subdiscipline of computer engineering. Even as the study of communications has transitioned from analog communications to digital communications, from continuous-time processing to discrete-time processing, from RF circuits to algorithms, from ad hoc techniques to theoretical and mathematical development, it has proved to be an exciting topic for students to study, a fruitful area for advanced research, and a commercial/industrial market with vast income potential.

1.5 ORGANIZATION OF THE TEXT

This text assumes, as prerequisites, courses in linear systems, transform theory, and probability theory (at least through random variables). Chapters 2 and 4 were written to review these concepts and are really not appropriate for teaching these subjects to students for the first time. Chapter 3 introduces some material that may not be familiar to undergraduates or first-year graduate students. These topics include multi-rate signal processing and discrete-time filter design.

Chapter 5 introduces linear digital modulation. Pulse amplitude modulation (PAM) is introduced as a baseband method for digital modulation, and quadrature amplitude modulation (QAM) is introduced as a band-pass carrier modulation. The development of both PAM and QAM relies heavily on geometrical interpretations of signals using the signal space concept. Continuous-time modulators and demodulators/detectors are covered first followed discrete-time realizations. Where appropriate, the differences between the continuous-time and discrete-time realizations are pointed out. However, the discrete-time systems presented in this chapter are merely discrete-time versions of their continuous-time prototypes. In general, this is an inefficient design method. Efficient variants are described in Chapter 10 where some of the real advantages of discrete-time processing come to light.

Chapter 6 outlines the performance analysis of linear digital modulation. Two figures of merit are presented: bandwidth and probability of bit error. These two figures of merit are combined to produce the spectral efficiency analysis. The chapter ends with an introduction to link budgets. This topic is included to connect the probability of error calculations with component characteristics and link topology.

Chapters 7 and 8 develop carrier phase synchronization and symbol timing synchronization. Continuous-time techniques are briefly reviewed and the emphasis is on discrete-time techniques based on a discrete-time phase locked loop. (Continuous-time and discrete-time phase locked loops are outlined in Appendix C.) Quaternary phase shift keying (QPSK) is the primary example for carrier phase synchronization and binary PAM is the primary example for symbol timing synchronization. Extensions to the other modulations introduced in Chapter 5 are explained.

The last section of Chapters 5–8 is a general treatment of the chapter's main subject from a more abstract, theoretical point of view. The maximum likelihood formulation is developed and used to derive the principles used in the chapter. This treatment is the usual starting point, especially in graduate courses, because the tradition is to treat the general case first and explore special cases. It is the author's experience that most students learn the other way around. When the principles are taught using a few special cases, the generalizations tend to be more meaningful.

A more extensive treatment of discrete-time processing for digital communications is presented in Chapters 9 and 10. Chapter 9 covers analog-to-digital and digital-to-analog converters, discrete-time oscillators, resampling filters, the CoRDiC and CoRDiC-like algorithms, and discrete-time automatic gain control. The emphasis is on the performance of these components from a systems level point of view. These components are applied to discrete-time techniques for modulators, demodulators, and channelizers in Chapter 10. Chapter 10 also includes a section on link budgets.

Appendices on pulse shapes, the complex-valued representation for QAM, and phase-locked loops are provided to compliment the material in Chapters 5–10.

Each chapter ends with a "Notes and References" section that provides some historical notes and a large number of references to other text books and papers published in the research journals dealing with signal processing, information theory, communication theory, and wireless system design. Not only are references provided for the subjects covered in the text, but also for closely related topics that are not covered. This is the case because no text with a reasonable page count can be completely comprehensive in what it covers.

The relationship between the material in the chapters and appendices is illustrated in Figure 1.5.1. The relationship is somewhat complex given the nature of the subject matter. Students of digital communications need to be familiar with linear systems, frequency domain representations, signal spaces, probability and random variables, detection and estimation theory, discrete-time filter design, multi-rate processing, analog-to-digital converter (ADC) and digital-to-analog converter (DAC), and continuous-time signal processing all at the same time. This is often a challenging task, but is what makes digital communications such an interesting and rewarding topic.

This text was written to be used as the basis for a senior level elective course in digital communications or as a first-year graduate course in digital communications (with an emphasis on the discrete-time processing algorithms). In undergraduate programs without a strong signals and systems preparation, the following organization is suggested:

1. Chapter 2
2. Chapter 5 (Sections 5.1–5.5) with Appendix A (with Chapter 3 as a resource)
3. Chapter 7 (Sections 7.1–7.7) with Appendix C (with Chapter 3 as a resource)
4. Chapter 8 (Sections 8.1–8.6)

In undergraduate programs with a strong signals and systems preparation, the following organization is appropriate:

1. A quick review of Chapter 2
2. A more thorough review of Chapter 4
3. Chapter 5 with Appendix A (with Chapter 3 as a resource)
4. Chapter 6
5. Chapter 7 with Appendix C (with Chapter 3 as a resource)
6. Chapter 8

Section 1.5 Organization of the Text

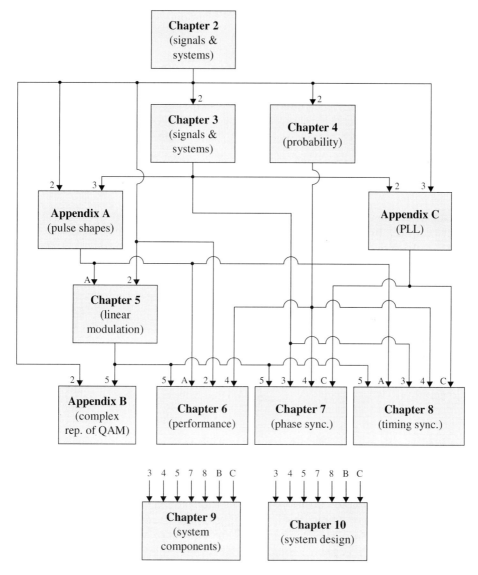

Figure 1.5.1 An illustration of the relationship between the chapters and appendices.

A first-year graduate course may be organized in the following way:

1. A quick review of Chapter 4
2. Chapter 5 (Section 5.6 with Sections 5.1–5.5 optional, Appendix A as needed)
3. Chapter 6 (Section 6.5 with Sections 6.1–6.4 optional, Appendix C as needed)
4. Chapter 7 (Section 7.8 with Sections 7.1–7.7 optional)

5. Chapter 8 (Section 8.7 with Sections 8.1–8.6 optioal)
6. Chapter 9 with Appendix B as needed
7. Chapter 10

This text has been used at BYU to teach a one-semester senior-level elective course on digital communication theory. This course covers Chapters 5–8 and Appendices A–C. This course includes about 15 Simulink Exercises that reinforce the material covered in each section. A follow-on senior project course focusing on the design of a DSP-based demodulator/detector relies heavily on the material in Chapters 9 and 10.

1.6 NOTES AND REFERENCES

The historical overview discussed here was taken largely from the book by Sarker et al. [1] and a booklet produced by the IEEE Communications Society celebrating its 50th anniversary in 2002 [2]. Other texts consulted include Anderson [3], Couch [4], Frerking [5], Harada and Prasad [6], Haykin [7], Haykin and Moher [8], Lewis [9], Proakis [10], Proakis and Salehi [11], Rappaport [12], Reed [13], Roden [14], Sklar [15], Stüber [16], Tsui [17], Tuttlebee [18], Wilson [19], and Ziemer and Tranter [20]. The data on propagation was obtained from the NASA report by Ippolito, Kual, and Wallace [21].

The definitions for software radio and software defined radio embodied in Figure 1.4.3 were inspired by the discussions of Reed [13] and Blust, who wrote Chapter 1 of Ref. [18].

This text focuses on linear modulation and discrete-time techniques for demodulation, synchronization, and detection. These topics do not form a complete picture of digital communications. Important topics not covered in the text include source coding and quantization, characterization of multipath interference, equalization, diversity techniques, error control coding, multiple-input/multiple-output (MIMO) systems, space–time coding, and multiple access. Important modulation types not covered in this text are frequency shift keying (FSK), continuous phase modulation (CPM), orthogonal frequency division multiplexing (OFDM), and direct sequence spread spectrum. Source coding and quantization and covered in most texts on information theory such as those by Blahut [22], Cover and Thomas [23], and Gallager [24]. Multipath propagation is described in books by Jakes [25] and Parsons [26] and in Chapter 14 of Proakis [10]. Error control coding is thoroughly developed in many textbooks. Examples include Lin and Costello [27], Moon [28], Schlegel [29], and Wicker [30]. FSK and CPM are described in the text by Anderson, Aulin, and Sundberg [31] and in most advanced text books on digital communications such as Anderson [3], Proakis [10], Proakis and Salehi [11], and Stüber [16]. OFDM, spread spectrum, and multiple access techniques are also described in these texts. MIMO and space–time coding are relatively new. These topics have not completely worked themselves into the standard texts in digital communication theory. A representative example of books dealing with these topics are Goldsmith [32], Larsson and Stoica [33], Jankiraman [34], Paulraj, Nabar, and Gore [35], Giannakis, Liu, Ma, and Zhou [36], Giannakis, Hua, Stoica, and Tong [37], and Vucetic and Yuan [38].

2

Signals and Systems 1: A Review of the Basics

2.1 INTRODUCTION   23
2.2 SIGNALS   24
2.3 SYSTEMS   28
2.4 FREQUENCY DOMAIN CHARACTERIZATIONS   30
2.5 THE DISCRETE FOURIER TRANSFORM   50
2.6 THE RELATIONSHIP BETWEEN DISCRETE-TIME
     AND CONTINUOUS-TIME SYSTEMS   55
2.7 DISCRETE-TIME PROCESSING OF BAND-PASS SIGNALS   67
2.8 NOTES AND REFERENCES   70
2.9 EXERCISES   71

2.1 INTRODUCTION

Wireless communications is fundamentally an information transmission problem. The transmission of information through physical media involves the transmission of signals through systems. The signals transmitted and received by antennas are *waveforms* that are examples of continuous-time signals. They are continuous-time signals because the transmission media, the antenna/free-space combination, is a continuous-time system. As such an understanding of continuous-time signals, both in the time and frequency domains, is required to design and analyze a communication system.

Most modern detection techniques sample the received waveform and use discrete-time processing to recover the data. The sampling process converts the bandlimited continuous-time signal to a discrete-time signal, and the algorithm that processes the samples of the discrete-time signal is a discrete-time system. Thus, an understanding of discrete-time signals and systems in both the time domain and frequency domain is required.

There is a temptation to be familiar with either the continuous-time world or the discrete-time world, but not both. This division is, in part, a result of natural divisions in the professional world where RF circuit designers (continuous-time systems) and DSP algorithm developers (discrete-time systems) rarely have to forge a close working relationship. After the decision of "where to put the A/D converter" has been made, the two groups often work independently from one another. A good system designer, however, will have equal expertise in the time-domain and frequency-domain characteristics for both continuous-time and discrete-time systems. Not only does the system designer have to *know* both worlds, he must also *understand the relationship* between the two. The importance of the relationship cannot be overlooked. It must be remembered that the samples being processed in discrete-time were once continuous-time waveforms and subject to all the noise and distortion the continuous-time world has to offer.

This chapter assumes the student has already had a junior level course in signals and systems and understands frequency domain concepts for both continuous-time signals and systems and discrete-time signals and systems. As such the basics of signals, systems, and frequency domain concepts are reviewed only briefly. The focus of the chapter is on the relationship between continuous-time signals and discrete-time signals.

2.2 SIGNALS

2.2.1 Continuous-Time Signals

A continuous-time signal is a function of the continuous-time variable t and is denoted as $x(t)$. In electrical engineering, a signal can be thought of as a waveform whose amplitude is measured in volts or amperes.

Signals may be classified in a variety of ways. One way is in terms of energy and power. The energy in a signal $x(t)$ is

$$E = \lim_{T \to \infty} \int_{-T}^{T} |x(t)|^2 dt. \qquad (2.1)$$

If $x(t)$ is exactly zero outside the range $T_1 \leq t \leq T_2$, then the energy is

$$E = \int_{T_1}^{T_2} |x(t)|^2 dt. \qquad (2.2)$$

The power of the signal is

$$P = \lim_{T \to \infty} \frac{1}{2T} \int_{-T}^{T} |x(t)|^2 dt. \qquad (2.3)$$

Signals with finite nonzero energy are sometimes called *energy signals* and signals with finite nonzero power are sometimes called *power signals*. Note that energy signals have $P = 0$ and power signals have infinite energy. Signals that are exactly zero outside the range $T_1 \leq t \leq T_2$ have zero power and are thus energy signals as long as $E > 0$. For an energy signal $x(t)$, the

Section 2.2 Signals

autocorrelation function is

$$r_x(\tau) = \lim_{T \to \infty} \int_{-T}^{T} x(t)x^*(t-\tau)dt \qquad (2.4)$$

where $x^*(t)$ is the complex conjugate of $x(t)$. Observe that $E = r_x(0)$. For a power signal $x(t)$, the autocorrelation function is

$$\phi_x(\tau) = \lim_{T \to \infty} \frac{1}{2T} \int_{-T}^{T} x(t)x^*(t-\tau)dt. \qquad (2.5)$$

Note that $P = \phi_x(0)$.

Signals may also be classified as either periodic or aperiodic. The signal $x(t)$ is *periodic* if and only if $x(t) = x(t + T_0)$ for all t for some constant T_0. The smallest constant T_0 for which this is true is called the period. If a signal is not periodic, then it is aperiodic. Periodic signals are power signals (i.e., they have infinite energy). The power of a periodic signal is given by (2.3), which reduces to

$$P = \frac{1}{T_0} \int_{t_0}^{t_0+T_0} |x(t)|^2 dt \qquad (2.6)$$

for any t_0. If a periodic signal has a finite number of maxima and minima and a finite number of discontinuities in a period, then the periodic signal has an alternate representation as a Fourier series

$$x(t) = \sum_{n=-\infty}^{\infty} a_n e^{jn\omega_0 t} \qquad (2.7)$$

where

$$\omega_0 = \frac{2\pi}{T_0} \qquad (2.8)$$

$$a_n = \frac{1}{T_0} \int_{t_0}^{t_0+T_0} x(t) e^{-jn\omega_0 t} dt. \qquad (2.9)$$

The coefficients a_n are often called the *Fourier series coefficients*.

Formally, only energy signals have a Laplace or Fourier transform. However, if the existence of singularity functions is allowed, then the class of power signals that are periodic have these transforms. The two singularity functions of special interest in communication theory are the impulse function and the unit-step function.

The impulse function is defined by the integral expression

$$\int_{-\infty}^{\infty} x(t)\delta(t)dt = x(0) \tag{2.10}$$

for any signal $x(t)$ that is continuous at $t = 0$. This definition of the impulse function is an operational definition. The impulse function may also be described using a constructive definition and mathematical limits. The impulse function is interpreted as a function that is zero everywhere except $t = 0$. At $t = 0$ it has zero width, infinite height, and unit area. This function results by taking the limit $\Delta \to 0$ of an appropriate functional shape whose width is Δ and whose height is proportional to $1/\Delta$. (A rectangle, triangle, and Gaussian shape are common choices.)

The impulse function has the following properties:

1. $\delta(at) = \dfrac{1}{|a|}\delta(t)$
2. $\delta(-t) = \delta(t)$
3. The *sifting property*: $\int_{t_1}^{t_2} x(t)\delta(t - t_0)dt = \begin{cases} x(t_0) & t_1 \leq t_0 \leq t_2 \\ 0 & \text{otherwise} \end{cases}$
4. $x(t)\delta(t - t_0) = x(t_0)\delta(t - t_0)$

The unit-step function, $u(t)$, is defined as the time integral of the impulse function:

$$u(t) = \int_{-\infty}^{t} \delta(\tau)d\tau. \tag{2.11}$$

The unit-step function may also be described using the following definition:

$$u(t) = \begin{cases} 0 & t < 0 \\ \text{undefined} & t = 0 \\ 1 & t > 0 \end{cases}. \tag{2.12}$$

The unit-step function is most often used to provide a mathematical representation of switching a signal on or off.

2.2.2 Discrete-Time Signals

A discrete-time signal is a function of an index n and is denoted as $x(n)$. Where continuous-time signals are a function of the continuous variable t, discrete-time signals are defined only at discrete instances in time and are undefined in between. In electrical engineering and signal processing, $x(n)$ is often the result of sampling a bandlimited continuous-time waveform at T-spaced intervals. In this case, the notation $x(nT)$ is used. There are applications where there is no underlying continuous-time waveform (such as the bits conveying the data for a banking transaction). In these cases $x(n)$ will simply be used. The relationship between $x(t)$ and $x(nT)$ is described in detail in Section 2.6.

Discrete-time signals may also be classified as either energy signals or power signals in a discrete-time analog of the continuous-time case. The energy and power of a discrete-time signal are

$$E = \lim_{N \to \infty} \sum_{n=-N}^{N} |x(n)|^2 \qquad (2.13)$$

$$P = \lim_{N \to \infty} \frac{1}{2N+1} \sum_{n=-N}^{N} |x(n)|^2. \qquad (2.14)$$

An energy signal is a signal with finite nonzero energy whereas a power signal is a signal with finite nonzero energy. Just as was the case with continuous-time signals, an energy signal has zero power and a power signal has infinite energy.

A discrete-time signal is periodic if $x(n) = x(n + N_0)$ for all n for some fixed integer N_0. N_0 is called the *period* if it is the smallest integer for which this is true. Otherwise, the signal is aperiodic. A periodic discrete-time signal with period N_0 has an alternate representation as a discrete-time Fourier series:

$$x(n) = \sum_{k=-\infty}^{\infty} c_k e^{jk\Omega_0 n} \qquad (2.15)$$

where

$$\Omega_0 = \frac{2\pi}{N_0} \qquad (2.16)$$

$$c_k = \frac{1}{N_0} \sum_{n=n_0}^{n_0+N_0-1} x(n) e^{-jk\Omega_0 n} \qquad (2.17)$$

for any integer n_0.

The discrete-time counterparts of the singularity functions are convenient functions to define because they can be used to describe a wide variety of useful discrete-time functions. The discrete-time impulse and step functions are defined by

$$\delta(n) = \begin{cases} 1 & n = 0 \\ 0 & n \neq 0 \end{cases} \qquad u(n) = \begin{cases} 1 & n \geq 0 \\ 0 & n < 0 \end{cases}. \qquad (2.18)$$

Observe that, unlike its continuous-time counterpart, there is no need to formulate a limiting argument to define the unit-impulse function because it can be defined directly. Due to the discrete-time nature of the signal, there is no fussing about continuity. The unit-impulse and unit-step functions are related to each other by

$$u(n) = \sum_{k=-\infty}^{n} \delta(n) \qquad \delta(n) = u(n) - u(n-1). \qquad (2.19)$$

One of the more useful properties of the discrete-time impulse function is the discrete-time sifting property

$$\sum_{n=N_1}^{N_2} x(n)\delta(n - N_0) = \begin{cases} x(N_0) & N_1 \leq N_0 \leq N_2 \\ 0 & \text{otherwise} \end{cases}. \qquad (2.20)$$

2.3 SYSTEMS

A system is a process that operates on an input signal to produce an output signal that is, in some sense, a function of the input signal. Systems may be characterized in a variety of ways, such as in terms of invertibility, stability, causality, linearity, and time invariance.

2.3.1 Continuous-Time Systems

Continuous-time systems are usually a collection of continuous-time components such as resistors, capacitors, inductors, and transistors. This collection of components operates on a continuous-time input signal $x(t)$ and produces a continuous-time output signal $y(t)$.

The most useful class of continuous-time systems is the class of linear, time-invariant (LTI) systems. A system with input $x(t)$ and output $y(t)$ is *time invariant* if a delayed input produces an output delayed by the same amount, that is, $x(t - t_0)$ applied at the input produces $y(t - t_0)$ at the output. A *linear system* obeys the principle of superposition. If the input $x_1(t)$ produces the output $y_1(t)$ and the input $x_2(t)$ produces the output $y_2(t)$, then the input $\alpha x_1(t) + \beta x_2(t)$ produces the output $\alpha y_1(t) + \beta y_2(t)$ for any choice of the constants α and β.

The input/output relationship for LTI systems is completely characterized by the system impulse response $h(t)$. In concept, the impulse response may be obtained experimentally by applying an impulse function at the input terminals to the system and recording the output. Because the impulse function is only an ideal abstraction, a real experimental process of this nature is only approximate.

Given the impulse response of a continuous-time LTI system, the output for any other input can be computed mathematically using convolution. If $x(t)$ is the signal at the input to a continuous-time LTI system with impulse response $h(t)$, then the output is given by

$$y(t) = h(t) * x(t) = \int_{-\infty}^{\infty} h(\lambda) x(t - \lambda) d\lambda = \int_{-\infty}^{\infty} h(t - \lambda) x(\lambda) d\lambda. \qquad (2.21)$$

A very important class of LTI systems is described by the linear, constant-coefficient differential equation

$$a_0 y(t) + a_1 \frac{dy(t)}{dt} + \cdots + a_N \frac{d^N y(t)}{dt^N} = b_0 x(t) + b_1 \frac{dx(t)}{dt} + \cdots + b_M \frac{d^M x(t)}{dt^M}. \qquad (2.22)$$

Section 2.3 Systems

An equation of this kind results from a voltage/current analysis of a continuous-time system composed of linear elements such as inductors, capacitors, differentiators, and integrators.

2.3.2 Discrete-Time Systems

Discrete-time systems take one of two basic forms:

1. Memory elements coupled to a processor that executes a computer program
2. A digital logic circuit that performs the equivalent operations

The most useful class of discrete-time systems is the class of LTI systems. Sometimes the terminology linear *shift*-invariant (LSI) or linear *delay*-invariant (LDI) systems is used because the discrete-time index n is usually thought of as a shift (or delay) and may or may not reference time.

A discrete-time system with input $x(n)$ and output $y(n)$ is *time invariant* if a delayed input produces an output delayed by the same amount, that is, $x(n - n_0)$ applied at the input produces $y(n - n_0)$ at the output. A *linear* discrete-time system produces the superposition of the inputs at the output. If $x_1(n)$ at the input produces $y_1(n)$ at the output and $x_2(n)$ at the input produces $y_2(n)$ at the output, then $\alpha x_1(n) + \beta x_2(n)$ at the input produces $\alpha y_1(n) + \beta y_2(n)$ at the output.

Just as with continuous-time systems, the input/output relationship for discrete-time LTI systems is characterized by the impulse response of the system. The impulse response may be obtained experimentally by applying the unit impulse at the input to the system (either the algorithm or the hardware) and recording the output. This is not an abstraction because the discrete-time unit impulse function is easily produced.

Given the impulse response of a discrete-time LTI system, the output for any other input sequence may be computed using the convolution sum. If $x(n)$ is the input to a discrete-time LTI system with impulse response $h(n)$, then the output is

$$y(n) = h(n) * x(n) = \sum_{k=-\infty}^{\infty} h(k)x(n-k) = \sum_{k=-\infty}^{\infty} h(n-k)x(k). \qquad (2.23)$$

For discrete-time LTI systems, an important classification is based on the character of the impulse response. Finite impulse response (FIR) systems have an impulse response that is nonzero over a finite time interval. That is, $h(n) = 0$ for $n < N_1$ and $n > N_2$ where N_1 and N_2 are finite. If either N_1 or N_2 is infinite, the impulse response is nonzero over an infinite time interval, and the system is called an infinite impulse response (IIR) system.

An important class of LTI systems are those whose input/output relationship is described by a linear, constant-coefficient difference equation of the form

$$a_0 y(n) + a_1 y(n-1) + \cdots + a_N y(n-N) = b_0 x(n) + b_1 x(n-1) + \cdots + b_M x(n-M). \qquad (2.24)$$

This equation applies to systems based on multipliers, adders, and memory elements that compute an output sample for each input sample using a finite number of operations.

2.4 FREQUENCY DOMAIN CHARACTERIZATIONS

The frequency domain characterizations for continuous-time systems and discrete-time systems are described in this section. As a preface, it is helpful to pause and consider the context: Why are frequency domain characterizations helpful? Which frequency domain characterization should be used? What, if any, is the relationship between the transforms that produce the various frequency domain characterizations?

Both continuous-time and discrete-time LTI systems may be represented in a number of equivalent forms. Continuous-time systems may be described in the time domain, the complex frequency domain (or s-domain) using the Laplace transform, or in the frequency domain using the Fourier transform. The value of multiple representations is that a characteristic difficult to describe or analyze in one domain may be easier to describe in one of the others.

- The time-domain description for continuous-time systems is the impulse response $h(t)$. The impulse response describes the input/output relationship using the convolution integral and defines how the output is computed from the input. For LTI systems that can be described by a linear, constant-coefficient differential equation, an equation of the form (2.22) is an alternate characterization. In continuous-time processing, this equivalent characterization is seldom used.
- The Laplace transform of $h(t)$, $H(s)$, is a description of the LTI system in the complex frequency domain. The benefit of this description is that differentiation and integration in the time-domain transform to simple polynomial operations in s. As a consequence, the convolution in t transforms to polynomial multiplication in s. The essential characteristics of the impulse response are defined by the poles of $H(s)$. As a consequence, the locations of the poles and zeros is an important characterization of $H(s)$. The locations of the poles and zeros relative to the $s = j\omega$ axis determines the frequency response of the system. This relationship, however, is often hard to see in the s-domain. The Fourier transform is much better suited to examining the frequency response of the LTI system.
- The Fourier transform of $h(t)$, $H(j\omega)$, is used to characterize the frequency selectivity of the LTI system. Frequency selectivity is usually characterized by pass bands and stop bands. Although the frequency selectivity may be obtained from $H(s)$ by evaluation at $s = j\omega$, the resulting functions often have unique analytical qualities that make it easier to move between $h(t)$ and $H(j\omega)$ directly using the Fourier and inverse Fourier transforms.

A graphical summary of the three characterizations is illustrated in the top portion of Figure 2.4.1.

A similar situation exists for a discrete-time LTI system. The system may be described in the time domain, the complex frequency domain (or z-domain), using the z-transform, or the frequency domain using the discrete-time Fourier transform.

- The time domain representation for an LTI system is the impulse response $h(n)$. The impulse response and convolution define the input/output relationship for the LTI system. For systems that can be characterized by a linear, constant-coefficient

Section 2.4　Frequency Domain Characterizations　　　　　　　　　　　　　　　　　　　**31**

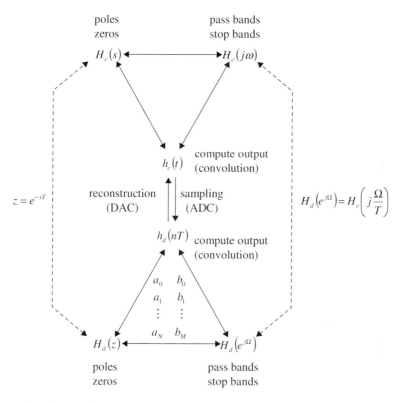

Figure 2.4.1 The relationships between the three domains for describing continuous-time LTI systems and discrete-time LTI systems. The connections between continuous-time systems and discrete-time systems apply only to strictly bandlimited continuous-time systems sampled at or above the minimum rate defined by the sampling theorem.

difference equation, an equation of the form (2.24) is an alternate characterization of the input/output relationship.

- The z-transform of $h(n)$, $H(z)$, is a description of the LTI system in the discrete-time complex frequency domain. The benefit of this description is that discrete-time convolution transforms to polynomial multiplication. Because the poles of $H(z)$ define the essential characteristics of the impulse response $h(n)$, the locations of the poles and zeros of $H(z)$ is one of the important characterizations in the z-domain. The locations of the poles and zeros relative to the unit circle determines the frequency response of the LTI system. Because this relationship is often hard to see in the z-domain, the discrete-time Fourier transform is used.

- The discrete-time Fourier transform of $h(n)$, $H(e^{j\Omega})$, is used to characterize the frequency selectivity of the LTI system. Although the discrete-time Fourier transform may be obtained from the z-transform by evaluating $H(z)$ on the unit circle (i.e., at $z = e^{j\Omega}$), the resulting functions often have unique analytical qualities that make direct evaluation of $H(e^{j\Omega})$ more convenient. In addition, there are efficient algorithms for computing sampled versions of $H(e^{j\Omega})$ directly from $h(n)$.

A graphical summary of the three characterizations is illustrated in the bottom portion of Figure 2.4.1. For systems characterized by a linear, constant-coefficient difference equation of the form (2.24), the coefficients $a_0, \ldots, a_N$ and $b_0, \ldots, b_M$ play a key role in all three representations.

In digital communications, much of the discrete-time processing is dedicated to sampled waveforms. For strictly bandlimited LTI systems, the sampling theorem provides the link between the continuous-time version $h(t)$ and the discrete-time version $h(nT)$. Because $h(t)$ and $h(n)$ are related, their transforms are also related. These relationships are illustrated by the dashed line in Figure 2.4.1. In the interval $-\pi/T \le \omega \le \pi/T$, the continuous-time Fourier transform and the discrete-time Fourier transform are related by

$$H_d\left(e^{j\Omega}\right) = H_c\left(j\frac{\Omega}{T}\right). \tag{2.25}$$

The Laplace and z-transforms are also related. Each pole s_0 in the s-plane maps to a pole in the z-plane at $z_0 = e^{s_0 T}$. Care must be taken in interpreting the information in Figure 2.4.1. The connections between the continuous-time descriptions and the discrete-time descriptions apply only to continuous-time systems that are strictly bandlimited and sampled at or above the minimum sampling rate defined by the sampling theorem. If the continuous-time system is not strictly bandlimited, then approximations are used. These approximations are described in Section 3.3.1.

2.4.1 Laplace Transform

The Laplace transform decomposes a time-domain signal into a continuum of complex exponentials in the complex frequency variable s. Convolution in t is transformed to multiplication in s. Consequently, the s-domain is ideal for the analysis of cascaded LTI subsystems. The Laplace transform and inverse Laplace transform are given by

$$X(s) = \int_{-\infty}^{\infty} x(t)e^{-st}dt \tag{2.26}$$

$$x(t) = \frac{1}{j2\pi} \oint X(s)e^{st}ds. \tag{2.27}$$

Some of the more important properties of the Laplace transform are summarized in Table 2.4.1. Some common Laplace transform pairs are listed in Table 2.4.2. The *region of convergence* or ROC is the set of all s for which the integral (2.26) converges.

The inverse transform (2.27) is a contour integral because integration is with respect to a complex variable. The contour is a straight line in the complex plane parallel to the imaginary (or $j\omega$) axis that is contained in the ROC. Formal evaluation of (2.27) can be performed using the Cauchy residue theorem.

Section 2.4 Frequency Domain Characterizations

Table 2.4.1 Some properties of the Laplace transform

Property	Signal	Laplace Transform	ROC
	$x(t)$	$X(s)$	R_x
	$y(t)$	$Y(s)$	R_y
Linearity	$ax(t) + by(t)$	$aX(s) + bY(s)$	at least $R_x \cap R_y$
Time Shifting	$x(t - t_0)$	$e^{-st_0} X(s)$	R_x
Time Scaling	$x(at)$	$\frac{1}{\|a\|} X\left(\frac{s}{a}\right)$	all s for which s/a is in R_x
Conjugation	$x^*(t)$	$X^*(s^*)$	R_x
Convolution	$x(t) * y(t)$	$X(s)Y(s)$	at least $R_x \cap R_y$
Differentiation	$\frac{d}{dt} x(t)$	$sX(s)$	at least R_x
Integration	$\int_{-\infty}^{t} x(\tau) d\tau$	$\frac{1}{s} X(s)$	at least $R_x \cap \{\text{Real}\{s\} > 0\}$
Initial Value Theorem[a]		$\lim_{t \to 0^+} x(t) = \lim_{s \to \infty} sX(s)$	
Final Value Theorem[b]		$\lim_{t \to \infty} x(t) = \lim_{s \to 0} sX(s)$	

[a] The initial value theorem is valid for signals $x(t)$ that satisfy the following conditions: $x(t) = 0$ for $t < 0$ and $x(t)$ contains no impulses or higher-order singularities at $t = 0$.
[b] The final value theorem is valid for signals $x(t)$ that satisfy the following conditions: $x(t) = 0$ for $t < 0$ and $x(t)$ has a finite limit as $t \to \infty$.

For the important class of LTI systems defined by a linear, constant-coefficient differential equation of the form (2.22), the Laplace transform is *rational*, that is, it is the ratio of two polynomials in s. Applying the linearity and differentiation properties of Table 2.4.1 to (2.22) produces

$$\left[a_0 + a_1 s + \cdots + a_N s^N \right] Y(s) = \left[b_0 + b_1 s + \cdots + b_M s^M \right] X(s) \tag{2.28}$$

from which the transfer function is easily obtained:

$$H(s) = \frac{Y(s)}{X(s)} = \frac{b_M s^M + \cdots + b_1 s + b_0}{a_N s^N + \cdots + a_1 s + a_0} = \frac{B(s)}{A(s)}. \tag{2.29}$$

The Laplace transform is characterized by M zeros—the roots of $B(s)$—and by N poles—the roots of $A(s)$. Note that if the coefficients $a_p, a_{p-1}, \ldots, a_0$ are real, then a complex-valued pole of $X(s)$ must also be accompanied by another pole that is its complex-conjugate. As an example, the Laplace transform

$$H(s) = \frac{s}{s^2 + 2s + 5} = \frac{s}{(s + 1 - j2)(2 + 1 + j2)} \tag{2.30}$$

Table 2.4.2 Some Laplace transform pairs

$x(t)$	$X(s)$	ROC
$\delta(t)$	1	all s
$u(t)$	$\dfrac{1}{s}$	Real $\{s\} > 0$
$\dfrac{t^{n-1}}{(n-1)!}u(t)$	$\dfrac{1}{s^n}$	Real $\{s\} > 0$
$e^{-at}u(t)$	$\dfrac{1}{s+a}$	Real $\{s\} > -a$
$\dfrac{t^{n-1}}{(n-1)!}e^{-at}u(t)$	$\dfrac{1}{(s+a)^n}$	Real $\{s\} > 0$
$\delta(t-T)$	e^{-sT}	all s
$\cos(\omega_0 t)u(t)$	$\dfrac{s}{s^2+\omega_0^2}$	Real $\{s\} > 0$
$\sin(\omega_0 t)u(t)$	$\dfrac{\omega_0}{s^2+\omega_0^2}$	Real $\{s\} > 0$
$e^{-at}\cos(\omega_0 t)u(t)$	$\dfrac{s+a}{(s+a)^2+\omega_0^2}$	Real $\{s\} > -a$
$e^{-at}\sin(\omega_0 t)u(t)$	$\dfrac{\omega_0}{(s+a)^2+\omega_0^2}$	Real $\{s\} > -a$
$\dfrac{d^n \delta(t)}{dt}$	s^n	all s
$\underbrace{u(t) * \cdots * u(t)}_{n \text{ times}}$	$\dfrac{1}{s^n}$	Real $\{s\} > 0$

has one zero at $s = 0$ and two poles at $s = -1 \pm j2$. The locations of the poles and zeros in the s-plane are indicated using ×'s for the poles and ○'s for the zeros. The resulting figure is called a *pole-zero plot*. The pole-zero plot for $H(s)$ is illustrated in Figure 2.4.2 (a). An alternate graphical representation of $H(s)$ is a plot of $|H(s)|$ as a function of s. The plot is a 3-dimensional plot because s is a complex variable and requires two axis to represent the real and imaginary parts of s. This plot for $H(s)$ is illustrated in Figure 2.4.2 (b). Note that at the poles $|H(s)| = \infty$ and at the zeros $|H(s)| = 0$.

The inverse Laplace transform of rational transforms is obtained using the partial fraction expansion. The partial fraction expansion is an algebraic decomposition that produces a sum of terms of the form

$$\frac{1}{(s+a)^n}.$$

As a consequence, the causal inverse Laplace transform is a weighted sum of terms of the form

$$\frac{t^{n-1}}{(n-1)!}e^{-at}u(t).$$

Section 2.4 Frequency Domain Characterizations

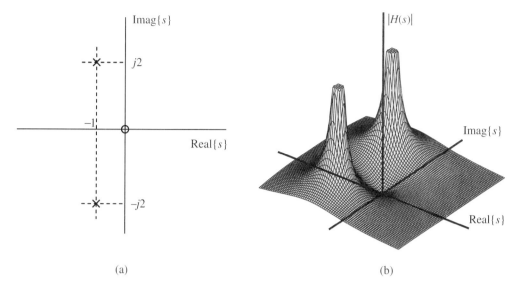

Figure 2.4.2 An example of S-domain representations of $H(s)$ given by (2.30): (a) The pole-zero plot; (b) a plot of $|H(s)|$ versus s.

In partial fraction expansions, it is customary to consider two cases: first, the case where their are p distinct poles (i.e., none of the roots are repeated) and the case where one or more of the poles are repeated. If the poles of $X(s)$ are distinct, then $X(s)$ is of the form

$$X(s) = \frac{B(s)}{(s+r_0)(s+r_2)\cdots(s+r_{p-1})} \quad (2.31)$$

and the corresponding partial fraction expansion is

$$X(s) = \frac{A_0}{s+r_0} + \frac{A_1}{s+r_1} + \cdots + \frac{A_{p-1}}{s+r_{p-1}} \quad (2.32)$$

where

$$A_i = (s+r_i)X(s)\Big|_{s=-r_i} \quad (2.33)$$

for $i = 0, 1, \ldots, p-1$. If one of the poles is repeated, say, r_0 with multiplicity n_0, then $X(s)$ is of the form

$$X(s) = \frac{B(s)}{(s+r_0)^{n_0}(s+r_1)\cdots(s+r_{p-n_0})} \quad (2.34)$$

and the corresponding partial fraction expansion is

$$X(s) = \frac{A_{0,0}}{(s-r_0)^{n_0}} + \frac{A_{0,1}}{(s-r_0)^{n_0-1}} + \cdots + \frac{A_{0,n_0-1}}{(s-r_0)} + \frac{A_1}{s+r_1} + \cdots + \frac{A_{p-n_0}}{s+r_{p-n_0}}. \quad (2.35)$$

The coefficients $A_{0,0}, A_{0,1}, \ldots, A_{0,n_0-1}$ are given by

$$A_{0,0} = (s+r_0)^{n_0} X(s)\Big|_{s=-r_0}$$

$$A_{0,1} = \frac{d}{ds}\left[(s+r_0)^{n_0} X(s)\right]\Big|_{s=-r_0}$$

$$A_{0,2} = \frac{1}{2!}\frac{d^2}{ds^2}\left[(s+r_0)^{n_0} X(s)\right]\Big|_{s=-r_0}$$

$$\vdots$$

$$A_{0,n_0-1} = \frac{1}{(n_0-1)!}\frac{d^{n_0-1}}{ds^{n_0-1}}\left[(s+r_0)^{n_0} X(s)\right]\Big|_{s=-r_0}.$$

The other coefficients, $A_1, A_2, \ldots, A_{p-n_0}$, are given by (2.33). Generalizations to the case of multiple repeated poles are straightforward.

Returning to the example, the partial fraction expansion is

$$H(s) = \frac{s}{(s+1-j2)(2+1+j2)} = \frac{\frac{2+j}{4}}{s+1-j2} + \frac{\frac{2-j}{4}}{s+1+j2} \quad (2.36)$$

from which the causal inverse transform is easily obtained:

$$h(t) = \left[\frac{2+j}{4}e^{(1-j2)t} + \frac{2-j}{4}e^{-(1+j2)t}\right]u(t) \quad (2.37)$$

$$= e^{-t}\left[\cos(2t) - \frac{1}{2}\sin(2t)\right]u(t) = \frac{\sqrt{5}}{2}e^{-t}\cos\left(2t + 26.56°\right)u(t). \quad (2.38)$$

For rational Laplace transforms, the following properties of the ROC are important:

1. The ROC is a strip, parallel to the $j\omega$ axis, in the s-plane.
2. The ROC may not contain any poles of $H(s)$.
3. The Fourier transform (reviewed in Section 2.4.2) converges if and only if the ROC includes the $j\omega$ axis.
4. If the time-domain function has finite support ($h(t) = 0$ except in the finite interval $-\infty < T_1 \leq t \leq T_2 < \infty$), then the ROC is the entire s-plane.
5. If the time-domain function $h(t)$ is right-sided ($h(t) = 0$ for $t < T_1 < \infty$), then the ROC is to the right of the right-most pole of $H(s)$.
6. If the time-domain function $h(t)$ is left-sided ($h(t) = 0$ for $t > T_2 > -\infty$), then the ROC is to the left of the left-most pole of $H(s)$.
7. If the ROC of $H(s)$ contains $s = 0$, then the system is stable.
8. $H(s)$ is stable if and only if the $j\omega$ axis is contained in the ROC.
9. $H(s)$ is causal and stable if and only if the $H(s)$ has no poles in the "right-half plane" (the region of the s-plane for which $\text{Re}\{s\} > 0$).

2.4.2 Continuous-Time Fourier Transform

The continuous-time Fourier transform decomposes a signal into a continuum of sinusoids (represented by the complex exponential $e^{j\omega t}$). As a consequence, it is a powerful analytic tool in analyzing the properties of a signal in the frequency domain. Examples include the following: Most continuous-time filters are described almost exclusively in terms of their amplitude and phase and as a function of ω. The sampling theorem is best understood in the frequency domain. Frequency planning and channel assignments in frequency division multiplexed or frequency division multiple access systems (such as commercial broadcast radio (both AM and FM), broadcast television, cable television, and many cellular telephone standards) are described in the frequency domain. As a purely mathematical analysis tool, the Fourier transform is most useful characterizing the steady-state response of an LTI system to a sinusoid.

The Fourier transform and inverse Fourier transform are

$$X(j\omega) = \int_{-\infty}^{\infty} x(t) e^{-j\omega t} dt \qquad (2.39)$$

$$x(t) = \frac{1}{2\pi} \int_{-\infty}^{\infty} X(j\omega) e^{j\omega t} d\omega. \qquad (2.40)$$

The notation $X(j\omega)$ is used because the continuous-time Fourier transform is often thought of as the Laplace transform evaluated along the imaginary (or $j\omega$) axis in the complex plane. To illustrate this concept, reconsider the example of the causal LTI system from the previous section whose Laplace transform is

$$H(s) = \frac{s}{s^2 + 2s + 5}. \qquad (2.41)$$

This function is illustrated in Figure 2.4.2 (b). The Fourier transform is obtained by substituting $s = j\omega$:

$$H(j\omega) = \frac{j\omega}{5 - \omega^2 + j2\omega} = \frac{j\omega}{[1 + j(\omega - 2)][1 + j(\omega - 2)]}. \qquad (2.42)$$

The factored expression provides a clue to the shape of $|H(j\omega)|$ as a function of ω. $|H(j\omega)|$ is largest at $\omega \approx \pm j2$, is zero at $\omega = 0$, and decreases as $|\omega|$ grows large. A plot of $|H(j\omega)|$ versus ω may be obtained by evaluating $|H(s)|$, shown in Figure 2.4.2 (b), along the imaginary axis. This is illustrated in Figure 2.4.3 (a). The magnitude of $H(s)$ along the slice $s = j\omega$ is clearly visible and is plotted in the traditional way in Figure 2.4.3 (b). This plot of $|H(j\omega)|$ shows that the system is a bandpass filter[1] with pass band approximately 2 rad/s.

---
[1] This simple example illustrates how a frequency selective filter—whose requirements are defined as a function of ω—may be designed by proper placement of poles in the s-plane. In classical filter design, an n-th order Butterworth filter places n poles on a semicircle in the left-half s-plane and an n-order Chebyshev filter places n poles on a semi-ellipse in the left-half s-plane.

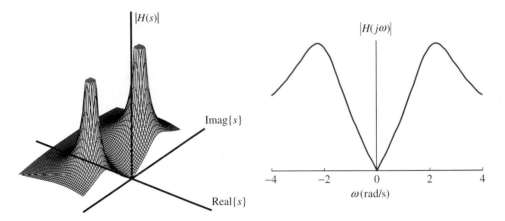

Figure 2.4.3 A plot of |H(s)| given by (2.30). (a) A break-away of the plot of Figure 2.4.2 (b) revealing the slice along $s = j\omega$; (b) a plot of the slice from part (a) in the more traditional form showing |H(jω)| versus ω.

One of the consequences of the notion that the Fourier transform is a special case of the Laplace transform is that the Fourier transform inherits the properties of the Laplace transform. These properties are summarized in Table 2.4.3.

For historical reasons, an alternate form of the Fourier transform is used in most wireless communication engineering applications. This form uses $f = \omega/2\pi$ with units cycles/second, or Hertz (Hz), as the frequency variable instead of ω which has units radians/second. Using this form, the Fourier transform and inverse Fourier transform are

$$X(f) = \int_{-\infty}^{\infty} x(t) e^{-j2\pi f t} dt \quad (2.43)$$

$$x(t) = \int_{-\infty}^{\infty} X(f) e^{j2\pi f t} df. \quad (2.44)$$

In this form, it is customary to not include j in the notation for the transformed signal. Note the absence of the factor 2π in the definition of the inverse transform. Some commonly used Fourier transform pairs are listed in Table 2.4.4.

The Fourier transform provides insight into the frequency content of $x(t)$. Parseval's theorem shows that the magnitude squared of the Fourier transform quantifies the energy distribution of $x(t)$ as a function of frequency. The magnitude squared of the Fourier transform is used to classify signals as baseband or band-pass signals. If $|X(f)|^2$ is nonzero only over an interval $-B \leq f \leq +B$, then $x(t)$ is a *baseband* signal. An example is illustrated in Figure 2.4.4 (a). In this simple example, the *bandwidth* of the signal is B Hz. *Band-pass* signals are characterized by a Fourier transform that is nonzero only over an interval that does not include $f = 0$. An example of $|X(f)|^2$ for a real valued band-pass signal $x(t)$ is illustrated in Figure 2.4.4

Section 2.4 Frequency Domain Characterizations

Table 2.4.3 Some properties of the Fourier transform in both ω and f

Property	Signal	Fourier Transform in ω	Fourier Transform in f
	$x(t)$	$X(j\omega)$	$X(f)$
	$y(t)$	$Y(j\omega)$	$Y(f)$
Linearity	$ax(t) + by(t)$	$aX(j\omega) + bY(j\omega)$	$aX(f) + bY(f)$
Time Shifting	$x(t - t_0)$	$e^{-j\omega t_0} X(j\omega)$	$e^{-j2\pi f t_0} X(f)$
Time Scaling	$x(at)$	$\dfrac{1}{\|a\|} X\left(\dfrac{j\omega}{a}\right)$	$\dfrac{1}{\|a\|} X\left(\dfrac{f}{a}\right)$
Conjugation	$x^*(t)$	$X^*(-j\omega)$	$X^*(-f)$
Convolution	$x(t) * y(t)$	$X(j\omega)Y(j\omega)$	$X(f)Y(f)$
Differentiation	$\dfrac{d}{dt} x(t)$	$j\omega X(j\omega)$	$j2\pi f X(f)$
Integration	$\int_{-\infty}^{t} x(\tau) d\tau$	$\dfrac{1}{j\omega} X(j\omega) + \pi X(0)\delta(\omega)$	$\dfrac{1}{j2\pi f} X(f) + \dfrac{1}{2} X(0)\delta(f)$
Frequency Shifting ($\omega_0 = 2\pi f_0$)	$e^{j\omega_0 t} x(t)$	$X(j(\omega - \omega_0))$	$X(f - f_0)$
Multiplication	$x(t)y(t)$	$\dfrac{1}{2\pi} X(j\omega) * Y(j\omega)$	$X(f) * Y(f)$
Parseval's Theorem	$\int_{-\infty}^{\infty} \|x(t)\|^2 dt = \dfrac{1}{2\pi} \int_{-\infty}^{\infty} \|X(j\omega)\|^2 d\omega = \int_{-\infty}^{\infty} \|X(f)\|^2 df$		

(b). In this case, $|X(f)|^2$ is centered about a frequency f_c (the center frequency) and has a bandwidth $2B$ Hz. Note that although the shape $|X(f)|^2$ is the same in baseband and band-pass cases of Figure 2.4.4, the bandwidth of band-pass signal is twice that of the baseband signal.

Strictly speaking, only energy signals have a Fourier transform. However, if the impulse function is allowed to be used, the class of power signals that is periodic have a Fourier transform. The Fourier transform of a periodic signal may be obtained by representing the signal as a Fourier series and using the Fourier transform pair on the sixth row of Table 2.4.4. In this case, the Fourier transforms consist of weighted impulses whose spacing is the inverse of the period. For this reason periodic signals are often said to have a "discrete spectrum."

Unlike the inverse Laplace transform, partial fraction expansion is not the usual method for computing the inverse Fourier transform. This is a consequence of the fact that the inverse Fourier transform (2.40)—and the f-form (2.44)—are integrals with respect to the real variable ω—or f—and are not encumbered by the usual complications associated with computing a contour integral with respect to a complex variable. As a consequence, direct evaluation of the inverse Fourier transform integral is a commonly used technique. Another very popular technique is to use a table of Fourier transform pairs. The use of such a table, coupled with judicious use of the Fourier transform properties, has proven to be a powerful technique capable of producing the inverse Fourier transform for most of the signals commonly encountered in communication theory.

Table 2.4.4 Some Fourier transform pairs in both ω and f

$x(t)$	$X(j\omega)$	$X(f)$
$\delta(t)$	1	1
$\delta(t - t_0)$	$e^{-j\omega t_0}$	$e^{-j2\pi f t_0}$
$u(t)$	$\dfrac{1}{j\omega} + \pi\delta(\omega)$	$\dfrac{1}{j2\pi f} + \dfrac{1}{2}\delta(f)$
$\dfrac{t^{n-1}}{(n-1)!}e^{-at}u(t)$	$\dfrac{1}{(a+j\omega)^n}$	$\dfrac{1}{(a+j2\pi f)^n}$
$\displaystyle\sum_{k=-\infty}^{\infty} a_k e^{jk\omega_0 t}$ $(\omega_0 = 2\pi f_0)$	$2\pi \displaystyle\sum_{n=-\infty}^{\infty} a_k \delta(\omega - k\omega_0)$	$\displaystyle\sum_{n=-\infty}^{\infty} a_k \delta(f - k f_0)$
$\displaystyle\sum_{k=-\infty}^{\infty} \delta(t - kT)$	$\dfrac{2\pi}{T} \displaystyle\sum_{n=-\infty}^{\infty} \delta\left(\omega - \dfrac{2\pi n}{T}\right)$	$\dfrac{1}{T} \displaystyle\sum_{n=-\infty}^{\infty} \delta\left(f - \dfrac{n}{T}\right)$
1	$2\pi\delta(\omega)$	$\delta(f)$
$\begin{cases} 1 & \|t\| < T \\ 0 & \|t\| > T \end{cases}$	$2T\dfrac{\sin(\omega T)}{\omega T}$	$2T\dfrac{\sin(2\pi f T)}{2\pi f T}$
$e^{j\omega_0 t}$ $(\omega_0 = 2\pi f_0)$	$2\pi\delta(\omega - \omega_0)$	$\delta(f - f_0)$
$\cos(\omega_0 t)$ $(\omega_0 = 2\pi f_0)$	$\pi\delta(\omega - \omega_0) + \pi\delta(\omega + \omega_0)$	$\dfrac{1}{2}\delta(f - f_0) + \dfrac{1}{2}\delta(f + f_0)$
$\sin(\omega_0 t)$ $(\omega_0 = 2\pi f_0)$	$\dfrac{\pi}{j}\delta(\omega - \omega_0) - \dfrac{\pi}{j}\delta(\omega + \omega_0)$	$\dfrac{1}{j2}\delta(f - f_0) - \dfrac{1}{j2}\delta(f + f_0)$
$\exp\{-a\|t\|\}$	$\dfrac{2a}{a^2 + \omega^2}$	$\dfrac{2a}{a^2 + (2\pi f)^2}$
$\exp\{-\pi t^2\}$	$\exp\left\{-\pi\left(\dfrac{\omega}{2\pi}\right)^2\right\}$	$\exp\{-\pi f^2\}$

2.4.3 z-Transform

The z-transform is used to characterize a discrete-time system in the complex frequency domain. The complex frequency variable is z and z^{-1} is the z-domain operator for a unit sample delay in the time domain. The z-transform converts discrete-time convolution to polynomial multiplication in z. As a consequence, the z-transform is ideal for analyzing a cascade of discrete-time LTI subsystems.

The z-transform and inverse z-transform are given by

$$X(z) = \sum_{n=-\infty}^{\infty} x(n) z^{-n} \tag{2.45}$$

$$x(n) = \frac{1}{j2\pi} \oint X(z) z^{n-1} dz. \tag{2.46}$$

Section 2.4 Frequency Domain Characterizations

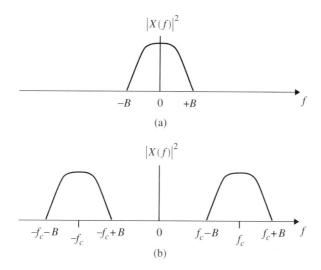

Figure 2.4.4 An illustration of the Fourier transforms of a baseband signal (a) and band-pass signal (b).

Table 2.4.5 Some properties of the z-transform

Property	Signal	z-transform	ROC
	$x(n)$	$X(z)$	R_x
	$y(n)$	$Y(z)$	R_y
Linearity	$ax(n) + by(n)$	$aX(z) + bY(z)$	at least $R_x \cap R_y$
Time Shifting	$x(n - n_0)$	$z^{-n_0} X(z)$	R_x
Time Scaling (Upsampling)	$\begin{cases} x(n/K) & n \text{ is a multiple of } K \\ 0 & \text{otherwise} \end{cases}$	$X(z^K)$	$R_x^{1/K}$
Conjugation	$x^*(n)$	$X^*(z^*)$	R_x
Convolution	$x(n) * y(n)$	$X(z) Y(z)$	at least $R_x \cap R_y$
First Difference	$x(n) - x(n-1)$	$(1 - z^{-1}) X(z)$	at least $R_x \cap \{|z| > 0\}$
Accumulation	$\sum_{k=-\infty}^{n} x(k)$	$\dfrac{1}{1 - z^{-1}} X(z)$	at least $R_x \cap \{|z| > 1\}$
Initial Value Theorem		$x(0) = \lim_{z \to \infty} X(z)$	

Some important properties of the z-transform are summarized in Table 2.4.5. The inverse transform (2.46) is a contour integral in the complex plain where the contour is a closed circular contour, centered at the origin and evaluated in the counterclockwise direction. The contour is usually taken to be a circle whose radius is such that the contour is contained in the region of convergence (ROC). Formal evaluation of (2.46) can be performed using the Cauchy residue theorem. Some common z-transform pairs are listed in Table 2.4.6.

Table 2.4.6 Some z-transform pairs

Signal	Transform	ROC
$\delta(n)$	1	all z
$u(n)$	$\dfrac{1}{1-z^{-1}}$	$\|z\| > 1$
$a^n u(n)$	$\dfrac{1}{1-az^{-1}}$	$\|z\| > \|a\|$
$na^n u(n)$	$\dfrac{az^{-1}}{(1-az^{-1})^2}$	$\|z\| > \|a\|$
$\delta(n-m)$	z^{-m}	all z†
$\cos(\Omega_0 n)\, u(n)$	$\dfrac{1-\cos(\Omega_0)z^{-1}}{1-[2\cos(\Omega_0)]z^{-1}+z^{-2}}$	$\|z\| > 1$
$\sin(\Omega_0 n)\, u(n)$	$\dfrac{\sin(\Omega_0)z^{-1}}{1-[2\cos(\Omega_0)]z^{-1}+z^{-2}}$	$\|z\| > 1$
$r^n \cos(\Omega_0 n)\, u(n)$	$\dfrac{1-r\cos(\Omega_0)z^{-1}}{1-[2r\cos(\Omega_0)]z^{-1}+z^{-2}}$	$\|z\| > r$
$r^n \sin(\Omega_0 n)\, u(n)$	$\dfrac{r\sin(\Omega_0)z^{-1}}{1-[2r\cos(\Omega_0)]z^{-1}+z^{-2}}$	$\|z\| > r$

†except 0 if $m > 0$ or ∞ if $m > 0$

For the important class of systems characterized by a linear, constant-coefficient difference equation of the form (2.24), the z-transform is rational (i.e., the ratio of two polynomials in z). Applying the linearity and time-shifting properties of Table 2.4.5 to (2.24) produces

$$\left[a_0 + a_1 z^{-1} + \cdots + a_N z^{-N}\right] Y(z) = \left[b_0 + b_1 z^{-1} + \cdots + b_M z^{-M}\right] X(z) \tag{2.47}$$

from which the transfer function is easily obtained:

$$H(z) = \frac{Y(z)}{X(z)} = \frac{b_0 + b_1 z^{-1} + \cdots + b_M z^{-M}}{a_0 + a_1 z^{-1} + \cdots + b_N z^{-N}} = \frac{B(z)}{A(z)}. \tag{2.48}$$

The system is characterized by M zeros—the roots of $B(z)$—and by N poles—the roots of $A(z)$. As an example, the z-transform

$$H(z) = \frac{z^{-1} - z^{-2}}{1 - \tfrac{2}{3}z^{-1} + \tfrac{1}{3}z^{-2}} = \frac{z - 1}{(z - \tfrac{1}{3} - j\tfrac{1}{2})(z - \tfrac{1}{3} + j\tfrac{1}{2})} \tag{2.49}$$

has one zero at $z = 1$ and two poles at $z = \tfrac{1}{3} \pm j\tfrac{1}{2}$. The locations of the poles and zeros can be plotted in the z-plane by marking the locations of the zeros with ○'s and the locations of the poles with ×'s. The resulting plot is called a *pole-zero plot*. The pole-zero plot for $H(z)$ given by (2.49) is illustrated in Figure 2.4.5 (a). The pole-zero plot helps visualize $|H(z)|$ as

Section 2.4 Frequency Domain Characterizations   43

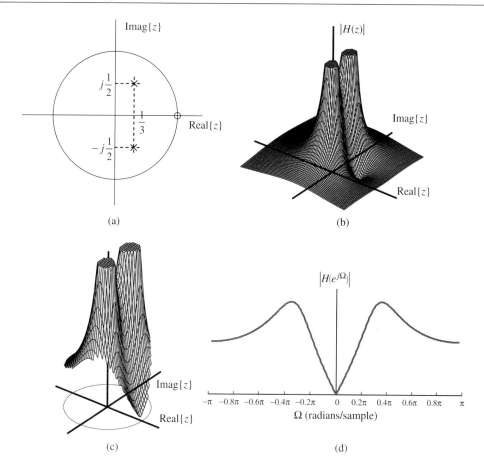

Figure 2.4.5 An example of discrete-time frequency domain representations for the example $H(z)$ given by (2.49). (a) The pole-zero plot of $H(z)$ in the z-plane; (b) $|H(z)|$ versus z in the complex z-plane; (c) a cut-way view of $|H(z)|$ along the unit circle revealing $|H(e^{j\Omega})|$; (d) a more traditional plot of $|H(e^{j\Omega})|$ versus Ω.

a function of z: $|H(z)| = 0$ at each of the zeros and $|H(z)| = \infty$ at each of the poles. This is illustrated for the example in Figure 2.4.5 (b).

The inverse z-transform of rational functions is usually obtained using a partial fraction expansion of $H(z)$. The procedure for performing the partial fraction expansion was outlined in Section 2.4.1. The partial fraction expansion produces a sum of terms of the form

$$\frac{1}{z-a}.$$

The linearity property of Table 2.4.5 shows that the resulting (right-sided) inverse z-transform consists of a sum of exponentials of the form

$$a^n u(n).$$

Returning to the example, the partial fraction expansion is

$$H(z) = \frac{\frac{1}{2} - j\frac{1}{3}}{z - \frac{1}{3} - j\frac{1}{2}} + \frac{\frac{1}{2} + j\frac{1}{3}}{z - \frac{1}{3} + j\frac{1}{2}} \qquad (2.50)$$

$$= \frac{\left(\frac{1}{2} - j\frac{1}{3}\right)z^{-1}}{1 - \left(\frac{1}{3} + j\frac{1}{2}\right)z^{-1}} + \frac{\left(\frac{1}{2} + j\frac{1}{3}\right)z^{-1}}{1 - \left(\frac{1}{3} - j\frac{1}{2}\right)z^{-1}}. \qquad (2.51)$$

The inverse z-transform is obtained by applying transform pair on the third row of Table 2.4.6 and the Time-Shifting Property from the fourth row of Table 2.4.5

$$h(n) = \left(\frac{1}{2} - j\frac{1}{3}\right)\left(\frac{1}{3} + j\frac{1}{2}\right)^{n-1} u(n-1) + \left(\frac{1}{2} + j\frac{1}{3}\right)\left(\frac{1}{3} - j\frac{1}{2}\right)^{n-1} u(n-1) \qquad (2.52)$$

$$= \left(\frac{1}{3}\right)^{\frac{n-1}{2}} \left[\cos\left(\theta_0(n-1)\right) + \frac{2}{3}\sin\left(\theta_0(n-1)\right)\right] u(n-1) \qquad (2.53)$$

where $\theta_0 = \tan^{-1}\left(\frac{3}{2}\right) = 0.9828$ rad/sample.

Several comments on the region of convergence (ROC) are in order. These properties are taken from Ref. [41]:

1. The ROC is a ring or disk in the z-plane centered at the origin: $0 \leq r_R < |z| < r_L \leq \infty$.
2. The DTFT (covered in Section 2.4.4) converges absolutely if and only if the ROC includes the unit circle.
3. The ROC cannot contain any poles.
4. If $x(n)$ is a finite-duration sequence (a sequence that is zero except in a finite interval $-\infty < N_1 \leq n \leq N_2 < \infty$) then the ROC is the entire z-plane except possibly $z = 0$ or $z = \infty$.
5. If $x(n)$ is a right-sided sequence (a sequence that is zero for $n < N_1 < \infty$), then the ROC extends outward from the outermost finite pole in $X(z)$ to (and possibly including) $z = \infty$.
6. If $x(n)$ is a left-sided sequence (a sequence that is zero for $n > N_2 > -\infty$) then the ROC extends inward from the innermost nonzero pole in $X(z)$ to (and possibly including) $z = 0$.
7. A two-sided sequence is an infinite-duration sequence that is neither right-sided nor left-sided. If $x(n)$ is a two-sided sequence, then the ROC is a ring bounded on the interior and exterior by poles and does not contain any poles.
8. The ROC must be a connected region.

As a consequence of these properties, two important observations can be made: first, the z-transform of the impulse response of a stable system must include the unit circle. Second, if the z-transform of a causal, stable system has poles, then those poles must lie *inside* the unit circle.

Another way of visualizing the region of convergence is to consider a causal signal of the form

$$x(n) = a^n u(n).$$

Section 2.4 Frequency Domain Characterizations

A series of numbers of the form a^n is called a *geometric series*. The z-transform is

$$X(z) = \sum_{n=0}^{\infty} a^n z^{-n} = \sum_{n=0}^{\infty} (az^{-1})^n$$

which is of the form

$$\sum_{n=0}^{\infty} r^n$$

(where $r = az^{-1}$). The infinite sum converges to

$$\sum_{n=0}^{\infty} r^n = \frac{1}{1-r}$$

if $|r| < 1$. This implies that the infinite sum converges only for $|a| < |z|$. The values of z for which $|a| < |z|$ is the ROC. The ROC is the entire z-plane *outside* the circle of radius $|a|$. If the system represented by $x(n)$ is to be stable, the ROC must also contain the unit circle. The ROC contains the unit circle only when $|a| < 1$. Observe that the z-transform has a pole at $1 - r = 0$ or $z = a$. This shows that a geometric series $a^n u(n)$ produces a pole at $z = a$. Because stability requires $|a| < 1$, the pole must reside *inside* the unit circle.

The partial fraction expansion of a z-transform exploits the connection between a single pole and the geometric series. For the case of no repeated poles, the partial fraction expansion produces a sum of terms, each with a single pole. The inverse z-transform of each of these terms produces a geometric series. The geometric series plays an important role in z-domain analysis of discrete-time LTI systems. Some important summation formulae for the geometric series are (for $0 < |r| < 1$)

$$\sum_{n=0}^{\infty} r^n = \frac{1}{1-r}$$

$$\sum_{n=0}^{N-1} r^n = \frac{1-r^N}{1-r}$$

$$\sum_{n=1}^{\infty} r^n = \frac{r}{1-r}$$

$$\sum_{n=0}^{\infty} r^{n-1} = \frac{1}{r(1-r)}$$

$$\sum_{n=0}^{\infty} nr^{n-1} = \frac{1}{(1-r)^2}$$

$$\sum_{n=0}^{\infty} nr^n = \frac{r}{(1-r)^2}.$$

2.4.4 Discrete-Time Fourier Transform

The discrete-time Fourier transform (DTFT) is the z-transform evaluated on the unit circle. As such, the DTFT is a special case of the z-transform where $z = e^{j\Omega}$ and is denoted $X\left(e^{j\Omega}\right)$ in recognition of this fact. Note the use of the uppercase Greek letter to represent the frequency variable for a discrete-time sequence. This is done to distinguish Ω with units radians/sample from the frequency variable for continuous-time signals ω which has units radians/second. The DTFT and inverse DTFT are defined by

$$X\left(e^{j\Omega}\right) = \sum_{n=-\infty}^{\infty} x(n) e^{-j\Omega n} \qquad (2.54)$$

$$x(n) = \frac{1}{2\pi} \int_{-\pi}^{\pi} X\left(e^{j\Omega}\right) e^{j\Omega n} d\Omega. \qquad (2.55)$$

As a special case of the z-transform, the DTFT inherits the properties of the z-transform as summarized in Table 2.4.7. Some commonly used DTFT pairs are listed in Table 2.4.8. As a consequence of evaluating the z-transform on the unit circle, the DTFT is periodic in Ω with period 2π.

Table 2.4.7 Some properties of the DTFT

Property	Signal	DTFT
	$x(n)$	$X\left(e^{j\Omega}\right)$
	$y(n)$	$Y\left(e^{j\Omega}\right)$
Linearity	$ax(n) + by(n)$	$aX\left(e^{j\Omega}\right) + bY\left(e^{j\Omega}\right)$
Time Shifting	$x(n - n_0)$	$e^{-j\Omega n_0} X\left(e^{j\Omega}\right)$
Time Scaling (Upsampling)	$\begin{cases} x(n/K) & n \text{ is a multiple of } K \\ 0 & \text{otherwise} \end{cases}$	$X\left(e^{jK\Omega}\right)$
Conjugation	$x^*(n)$	$X^*\left(e^{-j\Omega}\right)$
Convolution	$x(n) * y(n)$	$X\left(e^{j\Omega}\right) Y\left(e^{j\Omega}\right)$
Multiplication	$x(n) y(n)$	$\frac{1}{2\pi} X\left(e^{j\Omega}\right) * Y\left(e^{j\Omega}\right)$
First Difference	$x(n) - x(n-1)$	$\left(1 - e^{-j\Omega}\right) X\left(e^{j\Omega}\right)$
Accumulation	$\sum_{k=-\infty}^{n} x(k)$	$\frac{1}{1 - e^{-j\Omega}} X\left(e^{j\Omega}\right)$
Parseval's Relation	$\sum_{n=-\infty}^{\infty} |x(n)|^2 = \frac{1}{2\pi} \int_{2\pi} \left|X\left(e^{j\Omega}\right)\right|^2 d\Omega$	

Section 2.4 Frequency Domain Characterizations **47**

Table 2.4.8 Some DTFT pairs

Signal	DTFT
$x(n)$	$X\left(e^{j\Omega}\right)$
$\delta(n)$	1
$\delta(n - n_0)$	$e^{-j\Omega n_0}$
$u(n)$	$\dfrac{1}{1 - e^{-j\Omega}} + \pi \sum_{l=-\infty}^{\infty} \delta(\Omega - 2\pi l)$
$a^n u(n), |a| < 1$	$\dfrac{1}{1 - ae^{-j\Omega}}$
$\sum_{k=k_0}^{k_0+N-1} c_k e^{jk(2\pi/N)n}$	$2\pi \sum_{k=-\infty}^{\infty} c_k \delta\left(\Omega - \dfrac{2\pi k}{N}\right)$
$\sum_{k=-\infty}^{\infty} \delta(n - kN)$	$\dfrac{2\pi}{N} \sum_{l=-\infty}^{\infty} \delta\left(\Omega - \dfrac{2\pi l}{N}\right)$
1	$2\pi \sum_{l=-\infty}^{\infty} \delta(\Omega - 2\pi l)$
$\begin{cases} 1 & |n| \le N \\ 0 & |n| > N \end{cases}$	$\dfrac{\sin\left(\Omega\left[N + \dfrac{1}{2}\right]\right)}{\sin\left(\dfrac{\Omega}{2}\right)}$
$e^{j\Omega_0 n}$	$2\pi \sum_{l=-\infty}^{\infty} \delta(\Omega - \Omega_0 - 2\pi l)$
$\cos(\Omega_0 n)$	$\pi \sum_{l=-\infty}^{\infty} \left\{\delta(\Omega - \Omega_0 - 2\pi l) + \delta(\Omega + \Omega_0 - 2\pi l)\right\}$
$\sin(\Omega_0 n)$	$\dfrac{\pi}{j} \sum_{l=-\infty}^{\infty} \left\{\delta(\Omega - \Omega_0 - 2\pi l) - \delta(\Omega + \Omega_0 - 2\pi l)\right\}$

Some of the more important attributes are the following:

- $X(e^{j\Omega})$ is periodic in Ω with period 2π. This is a consequence of periodicity of the path around the unit circle in the z-plane. The point $z = e^{j\Omega}$ on this path is the point on the unit circle whose angle with respect to the positive real z-axis is Ω. It is customary to plot only the first period of the DTFT for $-\pi \le \Omega \le \pi$ such as that illustrated in Figure 2.4.6 (a). The values of $X(e^{j\Omega})$ at other values of Ω are defined by periodicity. However, the alternate forms shown in Figures 2.4.6 (b) and (c) are equivalent.
- The Ω-axis is the frequency axis and has units radians/sample. The values $\Omega = \pm\pi$ correspond to the half sample rate frequency or Nyquist[2] rate. Another common scaling

---

[2]Harry Nyquist was born in Sweden in 1889 and was raised in the United States. He received his B.S. and M.S. in electrical engineering from the University of North Dakota and his Ph.D. from Yale University

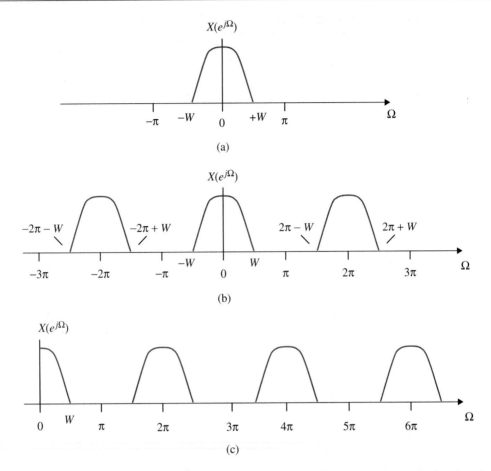

Figure 2.4.6 An illustration of the periodicity of the DTFT. (a) The conventional plot showing only the one period between $-\pi \leq \Omega \leq \pi$; (b) a plot similar to (a), showing three periods between $-3\pi \leq \Omega \leq 3\pi$; (c) a plot showing four periods over $0 \leq \Omega \leq 7\pi$.

for the discrete-time frequency axis uses units cycles/sample and is obtained from the Ω-axis by dividing by 2π. In this case, $\Omega = \pm\pi$ rad/sample becomes $\pm\frac{1}{2}$ cycles/sample and makes the connection between the frequency axis and the sampling rate somewhat

---

in 1917. He spent his entire 37-year career in the Bell System, first at AT&T and later at Bell Telephone Laboratories. He focused on data transmission over telegraph and telephone systems. In recognition of his seminal work, the sampling theorem described in Section 2.6.1 is often called the Nyquist sampling theorem and the minimum sampling rate for a bandlimited signal is called the Nyquist rate. He developed two criteria for ISI free data transmission through bandlimited channels which are called the first and second Nyquist conditions. The first Nyquist condition is the basis for the development of bandlimited pulse shapes outlined in Appendix A. Nyquist also made significant contributions to feedback control theory resulting in the Nyquist stability criterion and "Nyquist plots." He received 138 patents and published 12 technical articles during his remarkable career.

Section 2.4 Frequency Domain Characterizations   49

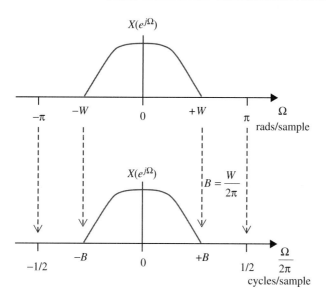

Figure 2.4.7 The relationship between the DTFT using the Ω with units radians/sample as the frequency axis and using $\Omega/2\pi$ with units cycles/sample as the frequency axis.

more explicit. The relationship between the Ω-axis and the scaled axis is illustrated in Figure 2.4.7 and very similar to the relationship between the ω and f axes of the continuous-time Fourier transform.

- The DTFT of a signal can be used to characterize the signal as a baseband or band-pass signal in much the same way this was done using the continuous-time Fourier transform with continuous-time signals. The formal definitions have to be adjusted somewhat to account for the periodicity of the DTFT. A discrete-time *baseband* signal $x(n)$ is characterized by a DTFT $|X(e^{j\Omega})|^2$ that is nonzero only over width-$2W$ intervals centered on multiples of $\Omega = 2\pi$ as illustrated in Figure 2.4.8 (a). In contrast a discrete-time *band-pass* signal $x(n)$ is characterized by a DTFT $|X(e^{j\Omega})|^2$ that is nonzero only over a width-$2W$ intervals that do not include multiples of $\Omega = 2\pi$ (i.e., $|X(e^{j\Omega})|^2$ is centered at $\Omega_c + k2\pi$ for $k = 0, \pm 1, \pm 2, \ldots$). This situation is illustrated in Figure 2.4.8 (b).

A related characterization is the frequency response of an LTI system (or filter). Frequency selective filters are described as low-pass, band-pass, high-pass, band-stop, etc., as a function of $|H(e^{j\Omega})|^2$, the magnitude squared of the DTFT of the impulse response $h(n)$. The relationship between the frequency response and the locations of the poles and zeros of the z-transform $H(z)$ is illustrated in Figure 2.4.5. A cutaway of $|H(z)|$, given by (2.49), along the unit circle is shown in Figure 2.4.5 (c). The value of $|H(z)|$ along this contour is $|H(e^{j\Omega})|$, which is plotted in the more traditional fashion in Figure 2.4.5 (d). This example demonstrates how the poles of $H(z)$, especially those near the unit circle, cause $|H(e^{j\Omega})|$ to be large and how zeros near the unit circle force $|H(e^{j\Omega})|$ to be small.

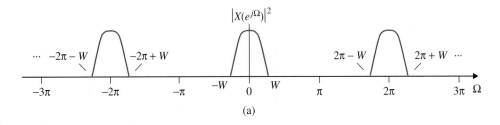

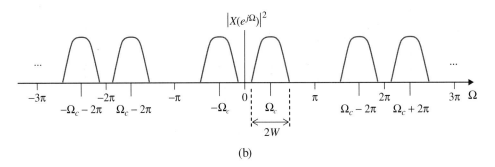

Figure 2.4.8 The definitions of baseband and band-pass signals based on the DTFT. (a) A baseband signal; (b) a band-pass signal.

2.5 THE DISCRETE FOURIER TRANSFORM

Let $x(n)$ be a finite-length sequence which is zero outside the interval $0 \leq n \leq N-1$. Let the DTFT of this sequence be

$$X\left(e^{j\Omega}\right) = \sum_{n=0}^{N-1} x(n)e^{-j\Omega n}. \tag{2.56}$$

The DTFT is a function of the *continuous* variable Ω. Computation in discrete-time processing (either in a computer or using "digital" hardware) can only be performed at discrete values of Ω. In fact, very efficient algorithms for computing $X\left(e^{j\Omega}\right)$ at $2\pi/N$-spaced samples of Ω have been developed. This leads to the question: What happens to $x(n)$ when $X(e^{j\Omega})$ is sampled? The answer is a pair of relationships known as the *Discrete Fourier Transform* or DFT.

The DFT and inverse DFT are defined by

$$X[k] = \sum_{n=0}^{N-1} x(n)e^{-j\frac{2\pi}{N}nk} \quad n = 0, 1, \ldots, N-1 \tag{2.57}$$

$$x(n) = \frac{1}{N}\sum_{k=0}^{N-1} X[k]e^{j\frac{2\pi}{N}kn} \quad k = 0, 1, \ldots, N-1. \tag{2.58}$$

Section 2.5 The Discrete Fourier Transform

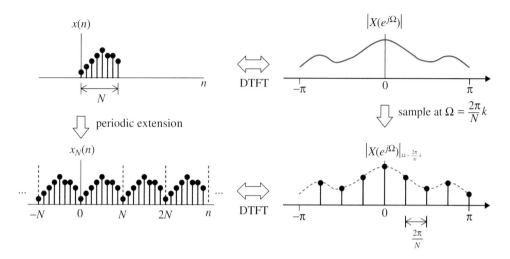

Figure 2.5.1 The relationship between a length-N sequence, its periodic extension, the DTFT, and the DFT.

The development of these relationships proceeds as follows: Let $x_N(n)$ be the length-N *periodic extension* of $x(n)$. The relationship between $x(n)$ and $x_N(n)$ is illustrated in Figure 2.5.1. Because $x_N(n)$ is periodic, it may also be represented as a discrete-time Fourier series:

$$x_N(n) = \sum_{k=0}^{N-1} a_k e^{j\frac{2\pi}{N}kn} \tag{2.59}$$

where the Fourier series coefficients are given by

$$a_k = \frac{1}{N}\sum_{n=0}^{N-1} x_N(n)e^{-j\frac{2\pi}{N}nk} = \frac{1}{N}\sum_{n=0}^{N-1} x(n)e^{-j\frac{2\pi}{N}nk} \tag{2.60}$$

for $k = 0, 1, \ldots, N-1$. Comparing (2.56) and (2.60) shows that the k-th scaled Fourier series coefficient Na_k can be obtained from the DTFT of $x(n)$ by evaluating $X(e^{j\Omega})$ at $\Omega = \frac{2\pi}{N}k$ for $k = 0, 1, \ldots, N-1$. Let $X[k]$ be the DTFT of $x(n)$ evaluated at $\Omega = \frac{2\pi}{N}k$:

$$X[k] = X\left(e^{j\Omega}\right)\bigg|_{\Omega = \frac{2\pi}{N}k} = \sum_{n=0}^{N-1} x(n)e^{-j\frac{2\pi}{N}nk} \tag{2.61}$$

for $k = 0, 1, \ldots, N-1$. This establishes the forward DFT (2.57). Equation (2.59) shows that the periodic extension of $x(n)$ may be computed from the Fourier series coefficients. Using the relationship $X[k] = Na_k$, (2.59) may be reexpressed as

$$x_N(n) = \sum_{k=0}^{N-1} a_k e^{j\frac{2\pi}{N}kn} = \sum_{k=0}^{N-1} \frac{X[k]}{N} e^{j\frac{2\pi}{N}kn}. \tag{2.62}$$

If n is restricted to the interval $0 \leq n \leq N-1$, then $x_N(n) = x(n)$ and the inverse DFT relationship (2.58) is established.

The relationship between a finite length sequence, its periodic extension, the DTFT, and the DFT are illustrated in Figure 2.5.1. Sampling the DTFT at N-points creates the length-N periodic extension in the time domain. The DFT inherits some of the important properties of the DTFT. Application of these properties needs to be done with care. One example is convolution. With the DTFT, convolution in the n-domain corresponds to multiplication in the Ω-domain. However, multiplication of two DFTs almost produces convolution in the n-domain. Because of the periodic extensions, the convolution is a *circular* convolution.

A length-L DFT may be computed from a length-N sequence. If $L < N$, then aliasing occurs in the n-domain as the periodic extensions overlap. If $L > N$, $L-N$ zeros are appended to the end of $x(n)$ and the periodic extension is applied to the zero-padded sequence. It should be pointed out that the appending zeros do not change the DTFT; they simply reduce the spacing between the samples of the DTFT. This concept is explored in Exercise 2.89.

The fast Fourier transform (FFT) is an efficient algorithm for computing the DFT. It is especially efficient when the length is a power of 2. An FFT to compute a length-N DFT when N is a power of 2 is called a *radix*-2 FFT. Consider the length-2 DFT whose two values are given by

$$X[0] = x(0) + x(1)$$
$$X[1] = x(0) - x(1).$$

Observe that this computation requires no multiplications and may be accomplished using the simple system illustrated in Figure 2.5.2 (a). This block diagram represents the 2-point FFT.

Now consider the length-4 DFT. The computation for $X[k]$ may be decomposed as follows:

$$\begin{aligned} X[k] &= x(0) + x(1)e^{-j\frac{\pi}{2}k} + x(2)e^{-j\pi k} + x(3)e^{-j\frac{3\pi}{2}k} \\ &= x(0) + x(2)e^{-j\pi k} + x(1)e^{-j\frac{\pi}{2}k} + x(3)e^{-j\frac{3\pi}{2}k} \\ &= \underbrace{\left[x(0) + x(2)e^{-j\pi k} \right]}_{\text{2-point FFT}} + e^{-j\frac{\pi}{2}k} \underbrace{\left[x(1) + x(3)e^{-j\pi k} \right]}_{\text{2-point FFT}}. \end{aligned} \tag{2.63}$$

This shows that the 4-point DFT may be computed using two 2-point FFTs: one operating on the even indexed-points and the other operating on the odd-indexed points. The four outputs of the two 2-point FFTs are combined as defined by (2.63). A block

Section 2.5 The Discrete Fourier Transform

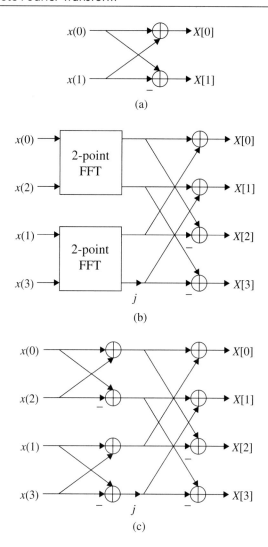

Figure 2.5.2 Structures of some short FFTs: (a) Structure of the length-2 FFT; (b) structure of the length-4 FFT in terms of length-2 FFTs; (c) structure of the length-4 FFT.

diagram of this approach is illustrated in Figure 2.5.2 (b). Replacing the 2-point FFT blocks by the structure illustrated in Figure 2.5.2 (a) produces the final result shown in Figure 2.5.2 (c). This is the 4-point FFT. Note that the only multiplication required is the one by j. Because this is not really a multiplication, the 4-point FFT does not require any multiplication.

Equipped with the 2-point and 4-point FFTs, the length-8 DFT may be examined in a way that illustrates the basic idea of the general radix-2 FFT. The computation for $X[k]$ may be decomposed as follows:

$$X[k] = x(0) + x(1)e^{-j\frac{\pi}{4}k} + x(2)e^{-j\frac{\pi}{2}k} + x(3)e^{-j\frac{3\pi}{4}k}$$
$$+ x(4)e^{-j\pi k} + x(5)e^{-j\frac{5\pi}{4}k} + x(6)e^{-j\frac{3\pi}{2}k} + x(7)e^{-j\frac{7\pi}{4}k}$$
$$= x(0) + x(2)e^{-j\frac{\pi}{2}k} + x(4)e^{-j\pi k} + x(6)e^{-j\frac{3\pi}{2}k} + x(1)e^{-j\frac{\pi}{4}k}$$
$$+ x(3)e^{-j\frac{3\pi}{4}k} + x(5)e^{-j\frac{5\pi}{4}k} + x(7)e^{-j\frac{7\pi}{4}k}$$
$$= \underbrace{\left[x(0) + x(2)e^{-j\frac{\pi}{2}k} + x(4)e^{-j\pi k} + x(6)e^{-j\frac{3\pi}{2}k}\right]}_{\text{4-point FFT}}$$
$$+ e^{-j\frac{\pi}{4}k}\underbrace{\left[x(1) + x(3)e^{-j\frac{\pi}{2}k} + x(5)e^{-j\pi k} + x(7)e^{-j\frac{3\pi}{2}k}\right]}_{\text{4-point FFT}}. \qquad (2.64)$$

The decomposition (2.64) shows that the length-8 DFT may be computed using two length-4 FFTs: one operating on the even-indexed samples and the other operating on the odd-indexed samples. Each of the 4-point FFTs may be decomposed into a pair to 2-point FFTs as follows:

$$X[k] = \underbrace{\left[x(0) + x(4)e^{-j\pi k}\right]}_{\text{2-point FFT}} + e^{-j\frac{\pi}{2}k}\underbrace{\left[x(2) + x(6)e^{-j\pi k}\right]}_{\text{2-point FFT}}$$
$$+ e^{-j\frac{\pi}{4}k}\underbrace{\left[x(1) + x(5)e^{-j\pi k}\right]}_{\text{2-point FFT}} + e^{-j\frac{3\pi}{4}k}\underbrace{\left[x(3) + x(7)e^{-j\pi k}\right]}_{\text{2-point FFT}}. \qquad (2.65)$$

Note the ordering of the input indexes applied to each 2-point FFT. Using the basic 2-point FFT illustrated in Figure 2.5.2 (a), the structure of the 8-point FFT, illustrated in Figure 2.5.3 results. Ignoring multiplications by ± 1 and $\pm j$, the 8-point FFT requires two complex-valued multiplications! Observe that the input samples must be reordered to produce the desired result. There are efficient ways to perform this reordering such as the one examined in Exercise 2.91. For the general case where N is a power of 2, the procedure introduced above is repeated $\log_2 N$ times to produce a structure similar to that of Figure 2.5.3, except with $\log_2 N$ stages. At each stage, the outputs of the half-length FFTs must be combined to form the N values needed for the next computational stage. At each stage, no more than N complex-valued multiplications are needed so that the number of complex-valued multiplications required to compute the length-N FFT is upper bounded by $N \log_2 N$. By comparison, direct computation of the DFT given by (2.57) requires N^2 complex-valued multiplications. This represents a substantial savings in resources, especially for large N.

Section 2.6 The Relationship between Discrete-Time and Continuous-Time Systems 55

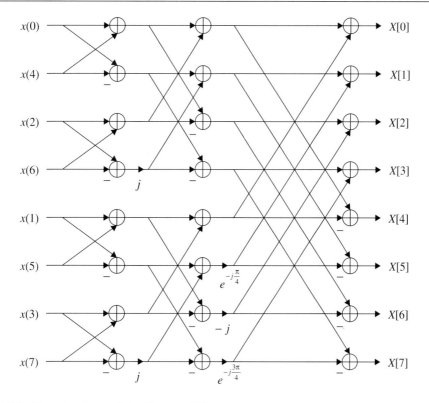

Figure 2.5.3 The structure of the 8-point FFT.

2.6 THE RELATIONSHIP BETWEEN DISCRETE-TIME AND CONTINUOUS-TIME SYSTEMS

The relationship between a continuous-time signal and its samples is of tremendous importance in wireless communications. Because the antenna and electromagnetic propagation environment are inherently continuous time, continuous-time signals are used to transmit the information from transmitter to receiver. Most modern receivers sample the received signal and use the samples to recover the data. The fact that the samples correspond to a continuous-time signal cannot be forgotten. Noise, time delays, phase shifts, and other distortions occur in the continuous-time domain. If these impairments are to be overcome with discrete-time processing, how the signals and the impairments pass through to the discrete-time world must be clearly understood.

For bandlimited continuous-time signals and systems, there exists an exact discrete-time representation. This relationship is expressed mathematically by the sampling theorem described in Section 2.6.1. When the need arises to perform continuous-time functions in discrete-time, the technique of matching the continuous-time and the corresponding discrete-time Fourier transforms of bandlimited systems outlined in Section 2.6.2 may be used.

2.6.1 The Sampling Theorem

The sampling theorem states a bandlimited continuous-time signal $x_c(t)$ whose Fourier transform is zero for $\omega > W$ and is uniquely determined by its T-spaced samples $x_c(nT)$ as long as the sample spacing satisfies

$$\frac{2\pi}{T} > 2W. \qquad (2.66)$$

The relationship between the samples $x_c(nT)$ and the waveform $x_c(t)$ is

$$x_c(t) = \frac{WT}{\pi} \sum_{n=-\infty}^{\infty} x_c(nT) \frac{\sin(W(t-nT))}{W(t-nT)}. \qquad (2.67)$$

The key step in the development of this result is developing a relationship between $x_c(t)$ and its samples $x_c(nT)$. This relationship is based on the notion of *impulse sampling*. Impulse sampling, illustrated in Figure 2.6.1, represents the samples of $x(t)$ as the product of $x(t)$ with an impulse train $p(t)$. The impulse train and its Fourier transform, given by

$$p(t) = \sum_{n=-\infty}^{\infty} \delta(t - nT) \qquad (2.68)$$

$$P(j\omega) = \frac{2\pi}{T} \sum_{k=-\infty}^{\infty} \delta\left(\omega - k\frac{2\pi}{T}\right), \qquad (2.69)$$

are illustrated in Figure 2.6.1. The product $x_p(t) = x_c(t)p(t)$ may be expressed as

$$x_p(t) = \sum_{n=-\infty}^{\infty} x_c(nT)\delta(t - nT) \qquad (2.70)$$

because the impulse function $\delta(t)$ is zero everywhere except $t = 0$. The weight on each impulse in the sum is the sample value and the delay of each impulse is used to denote the sampling instant.

The sampling theorem follows from a direct analysis of the Fourier transform of $x_p(t)$. Using the convolution property of Fourier transforms, the Fourier transform of $x_p(t)$ may be expressed as

$$X_p(j\omega) = \frac{1}{2\pi} \int_{-\infty}^{\infty} X_c(ju) P(j(\omega - u)) du \qquad (2.71)$$

$$= T \sum_{k=-\infty}^{\infty} X_c\left(\omega - k\frac{2\pi}{T}\right). \qquad (2.72)$$

Section 2.6 The Relationship between Discrete-Time and Continuous-Time Systems

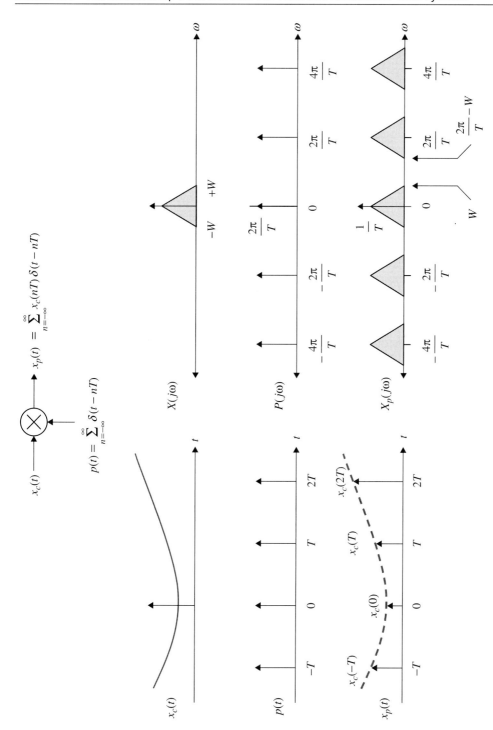

Figure 2.6.1 A system for generating T-spaced samples of a bandlimited continuous-time signal.

The product signal $x_p(t)$ and its Fourier transform $X_p(j\omega)$ are illustrated at the bottom of Figure 2.6.1. It is clear that $X_c(j\omega)$ can be obtained from $X_p(j\omega)$ as long as

$$W < \frac{2\pi}{T} - W. \qquad (2.73)$$

Condition (2.66) follows from this observation. If the sampling theorem is satisfied, then $x_c(t)$ can be recovered from $x_p(t)$ by using a low-pass filter with pass band $|\omega| \leq W$, transition band $W < |\omega| < \frac{2\pi}{T} - W$, and stop band $|\omega| > \frac{2\pi}{T} - W$. An ideal low-pass filter with bandwidth W rad/s and amplitude T could be used as illustrated in Figure 2.6.2. The impulse response of this filter is

$$h(t) = \frac{WT}{\pi} \frac{\sin(Wt)}{Wt}. \qquad (2.74)$$

Convolving $x_p(t)$ with $h(t)$ produces the result (2.67)

The T-spaced samples of $x_c(t)$ produce a discrete-time sequence $x_d(n) = x_c(nT)$. The relationship between the DTFT of $x_d(n)$ and the Fourier transform of $x_c(t)$ is a tremendously important concept that is essential to analyzing and designing discrete-time processors for digital communications. First the relationship between $X_d(e^{j\Omega})$ and $X_p(j\omega)$ is established. From this relationship, the connection between $X_d(e^{j\Omega})$ and $X_c(j\omega)$ is obtained using the relationship (2.72).

Direct computation of the Fourier transform of (2.70) produces an alternate expression for $X_p(j\omega)$:

$$X_p(j\omega) = \int_{-\infty}^{\infty} \sum_{n=-\infty}^{\infty} x_c(nT)\delta(t - nT)e^{-j\omega t} dt$$

$$= \sum_{n=-\infty}^{\infty} x_c(nT)e^{-j\omega Tn}. \qquad (2.75)$$

The DTFT of the discrete-time sequence $x_d(n) = x_c(nT)$ is

$$X_d\left(e^{j\Omega}\right) = \sum_{n=-\infty}^{\infty} x_c(nT)e^{-j\Omega n}. \qquad (2.76)$$

The transforms (2.75) and (2.76) are of the same form and define the relationship between the frequency variable for the continuous-time Fourier transform and the frequency variable for the discrete-time Fourier transform:

$$\Omega = \omega T. \qquad (2.77)$$

Equating (2.75) and (2.76) using this relationship produces the intermediate result

$$X_d\left(e^{j\Omega}\right) = X_p\left(j\frac{\Omega}{T}\right). \qquad (2.78)$$

Section 2.6 The Relationship between Discrete-Time and Continuous-Time Systems   **59**

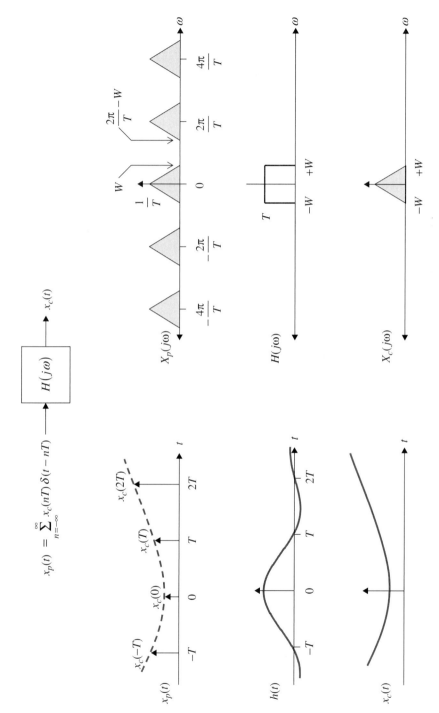

Figure 2.6.2 Using a low-pass filter to produce a continuous-time bandlimited signal from its T-spaced samples using an ideal low-pass filter.

Substituting (2.72) for $X_p(j\omega)$ produces the desired result:

$$X_d\left(e^{j\Omega}\right) = \frac{1}{T} \sum_{k=-\infty}^{\infty} X_c\left(j\left(\frac{\Omega - k2\pi}{T}\right)\right). \tag{2.79}$$

Two observations are important:

- $X_d\left(e^{j\Omega}\right)$ is periodic in Ω with period 2π. This result was observed earlier when the DTFT was interpreted as the z-transform evaluated on the unit circle in the z-plane. In this case, the same property is inherited through the periodicity of $X_p(j\omega)$. $X_p(j\omega)$ is periodic because it is the Fourier transform of discrete-time sequence $x_c(nT)$.
- $X_d\left(e^{j\Omega}\right)$ is obtained by using a periodic and scaled repetition of $X_c(j\omega)$. Frequency scaling follows from the normalization that occurs in converting the samples of $x_c(t)$ to a discrete-time sequence. If $x_c(t)$ is sufficiently bandlimited so that the sampling theorem is satisfied, $X_c(j\omega)$ is identical to $X_d\left(e^{j\Omega}\right)$ in the interval $-\pi \leq \Omega \leq \pi$ with appropriate frequency-axis scaling. In equation form, this observation is

$$X_d\left(e^{j\Omega}\right) = \frac{1}{T} X_c\left(j\frac{\Omega}{T}\right) \quad -\pi \leq \Omega \leq \pi. \tag{2.80}$$

Some important observations are illustrated in the following two examples.

Example 2.6.1
Sampling a Baseband Signal

As an example of the sampling theorem and the properties of the DTFT of the corresponding discrete-time sequence, consider the bandlimited baseband signal $x_c(t)$ whose Fourier transform $X_c(f)$ is the top plot in Figure 2.6.3. The bandwidth of the signal is 3 Hz. The sampling theorem states that the minimum sample rate for sampling without loss of information is 6 samples/s. Suppose the signal is sampled at 8 samples/s. The Fourier transform of the intermediate signal $x_p(t)$ is shown in the second plot of Figure 2.6.3. Observe that $X_p(j\omega)$ consists of periodic copies of $X_c(j\Omega)$ spaced by the sampling rate. Applying the frequency-axis scaling (2.77) produces the DTFT $X_d(e^{j\Omega})$ shown in the third plot of Figure 2.6.3. Note that it is periodic in Ω with period 2π as expected and that the bandwidth is $3\pi/4$ rad/sample. A magnified plot of $X_d(e^{j\Omega})$ over $-\pi \leq \Omega \leq \pi$ is shown in the bottom plot of Figure 2.6.3. Observe that the bandwidth of the sampled signal is the ratio of the bandwidth of $x_c(t)$ to the sample rate:

$$BW = 2\pi \times \frac{3}{8} = \frac{3\pi}{4} \quad \text{rad/sample.} \tag{2.81}$$

Example 2.6.2
Sampling a Band-pass Signal

Although the sampling theorem and DTFT-Fourier transform relationship were developed for baseband signals, the same principles apply to band-pass signals.

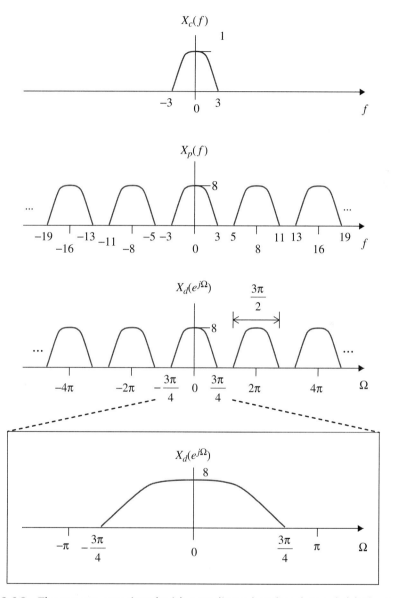

Figure 2.6.3 The spectra associated with sampling a baseband signal. (a) The Fourier transform of the continuous time signal $x_c(t)$; (b) the Fourier transform of the intermediate signal $x_p(t)$; (c) the DTFT of the sequence $x_d(n) = x_c(nT)$; (d) a close-up of (c) over the interval $-\pi \leq \Omega \leq \pi$.

The key to understanding is proper construction of the Fourier transform of the intermediate signal $x_p(t)$. As an example, consider the band-pass signal whose Fourier transform $X_c(f)$ is illustrated in the top plot of Figure 2.6.4. $X_c(f)$ is centered at 455 Hz and has a bandwidth of 10 Hz. This signal could be sampled by a naive application of the sampling theorem that identifies 460 Hz as the highest frequency in the signal and uses a sample rate that is greater than $2 \times 460 = 920$ samples/s. Suppose this signal is sampled at 1000 samples/s. The Fourier transform of the intermediate signal, $X_p(f)$, is obtained from (2.75) by producing periodic copies of $X_c(f)$ spaced 1000 Hz apart. The result is shown in the second plot of Figure 2.6.4. Note that it is periodic, but unlike the baseband example, there is no spectrum at multiples of the sample rate. Applying the frequency-axis scaling (2.77) produces the DTFT $X_d(e^{j\Omega})$ shown in the third plot of Figure 2.6.4. Note that it is periodic in Ω with period 2π as expected. A magnified plot of $X_d(e^{j\Omega})$ over $-\pi \leq \Omega \leq \pi$ is shown in the bottom plot of Figure 2.6.4. Note that the sampled signal is also a band-pass signal centered at $\Omega_c = 455\pi/500$ rad/sample and a bandwidth of $\frac{460\pi}{500} - \frac{450\pi}{500} = \frac{\pi}{50}$ rad/sample. Two characteristics are important. First, the center frequency Ω_c is determined by the ratio of the continuous-time center frequency and the sample rate:

$$\Omega_c = 2\pi \times \frac{455}{1000} = \frac{455}{500}\pi. \tag{2.82}$$

(In general, this is not the case, as illustrated below.) The second important observation is the bandwidth that is obtained from the ratio of the bandwidth of $x_c(t)$ and the sample rate:

$$\text{BW} = 2\pi \times \frac{10}{1000} = \frac{\pi}{50}. \tag{2.83}$$

(In general, this will *always* be the case.)

The rather small bandwidth shows that the signal is highly oversampled. The conclusion is that a lower sample rate could have been used. Any sample rate that does not alias the spectral copies of $X_p(j\omega)$ can be used. Clearly the sample rate must be greater than 10 samples/s. (Note the relationship to the bandwidth of the band-pass signal.) Even if the sample rate is greater than 10 samples/s, aliasing may still occur. The sample rate for band-pass signals must take into account not only the bandwidth of the continuous-time signal, but also the center frequency.

A second example using a much lower sample rate is illustrated in Figure 2.6.5. Starting with the same continuous-time signal whose Fourier transform is shown in the top plot, consider the Fourier transform of the intermediate signal corresponding to a sample rate of 140 samples/s. The Fourier transform, $X_p(j\omega)$ is constructed exactly as before by placing copies of $X_c(j\omega)$ every 140 Hz apart. A portion of $X_p(j\omega)$ is illustrated in the second plot of Figure 2.6.5. Applying the frequency-axis scaling (2.77) produces the DTFT $X_d(e^{j\Omega})$ shown in the third plot of Figure 2.6.5.

Section 2.6 The Relationship between Discrete-Time and Continuous-Time Systems   63

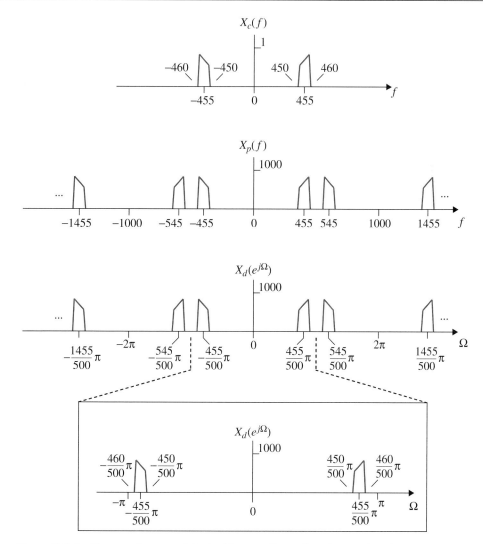

Figure 2.6.4 The spectra associated with sampling a band-pass signal.
(a) The Fourier transform of the continuous time signal $x_c(t)$; (b) the Fourier transform of the intermediate signal $x_p(t)$; (c) the DTFT of the sequence $x_d(n) = x_c(nT)$; (d) a close-up of (c) over the interval $-\pi \leq \Omega \leq \pi$.

Note that it is periodic in Ω with period 2π as expected. A magnified plot of $X_d(e^{j\Omega})$ over $-\pi \leq \Omega \leq \pi$ is shown at the bottom of Figure 2.6.5. Note that $X_d(e^{j\Omega})$ is a band-pass signal centered at $\Omega_c = \frac{\pi}{2}$ rad/sample and has a bandwidth $\frac{4\pi}{7} - \frac{3\pi}{7} = \frac{\pi}{7}$ rad/sample. Observe that in this case the center frequency is not determined by the simple ratio of the center frequency to the sample rate, but rather by its value

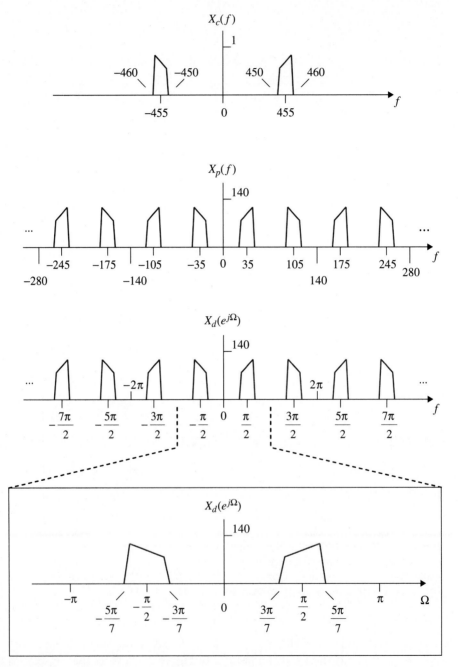

Figure 2.6.5 The spectra associated with sampling a band-pass signal. (a) The Fourier transform of the continuous time signal $x_c(t)$; (b) the Fourier transform of the intermediate signal $x_p(t)$; (c) the DTFT of the sequence $x_d(n) = x_c(nT)$; (d) a close-up of (c) over the interval $-\pi \leq \Omega \leq \pi$.

Section 2.6   The Relationship between Discrete-Time and Continuous-Time Systems   **65**

modulo 2π:

$$\Omega_c = \left(2\pi \times \frac{455}{140}\right) \;(\text{mod } 2\pi) = \left(\frac{13\pi}{2}\right) \;(\text{mod } 2\pi) = \frac{\pi}{2}. \tag{2.84}$$

As before, the bandwidth is determined from the ratio of the bandwidth of $x_c(t)$ to the sample rate:

$$\text{BW} = 2\pi \times \frac{10}{140} = \frac{\pi}{7}. \tag{2.85}$$

Other sample rates that may be used for this band-pass signal are explored in Exerices 2.92–2.94.

2.6.2 Discrete-Time Processing of Continuous-Time Signals

With the sampling theorem defined and the relationship between the continuous-time Fourier transform of a continuous-time signal and the DTFT of its samples established, discrete-time processing of continuous-time signals may now be explored. The most common application of this technique is the design of an equivalent discrete-time system to perform a function that is expressed in the continuous-time frequency domain. The transfer function of the continuous-time system is translated to an equivalent discrete-time system transfer function by equating the DTFT of the discrete-time system to the Fourier transform of the continuous-time system.

Let $x_c(t)$ be a bandlimited continuous-time signal that is sampled and processed in discrete-time as illustrated in Figure 2.6.6. Proper design of the discrete-time processing requires knowledge of the relationship between $x_c(t)$ and its samples $x_c(nT)$ as well as the relationship between the Fourier transform $X_c(j\omega)$ and the DTFT $X_d(e^{j\Omega})$. The relationship between $x_c(t)$ and $x_c(nT)$ is defined by the sampling theorem while the relationship between $X_c(j\omega)$ and $X_d(e^{j\Omega})$ is given by (2.80).

In many situations, the desired processing is described in the continuous-time frequency domain in the form of the transfer function $H_c(j\Omega)$. To perform equivalent processing in using a system such as that illustrated in Figure 2.6.7, the relationship between $H_c(j\Omega)$ and the discrete-time LTI system defined by $H_d(e^{j\Omega})$ must also be known. The relationship (2.80) suggests the system relationship

$$H_d\left(e^{j\Omega}\right) = H_c\left(j\frac{\Omega}{T}\right) \quad -\pi \leq \Omega \leq \pi. \tag{2.86}$$

The overall processing represented by the top system in Figure 2.6.7 may also be expressed in terms of the transfer function of the discrete-time system as follows:

$$H_c(j\omega) = \begin{cases} H_d(e^{j\omega T}) & |\omega| \leq \frac{\pi}{T} \\ 0 & \text{otherwise} \end{cases}. \tag{2.87}$$

The application of these ideas is best illustrated by an example.

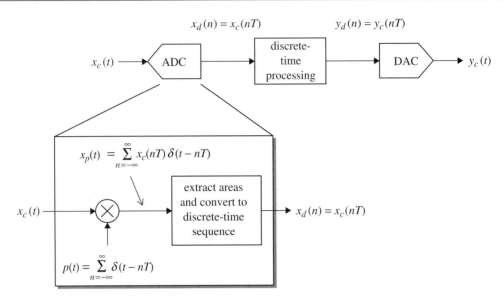

Figure 2.6.6 A system for performing discrete-time processing of a continuous-time signal.

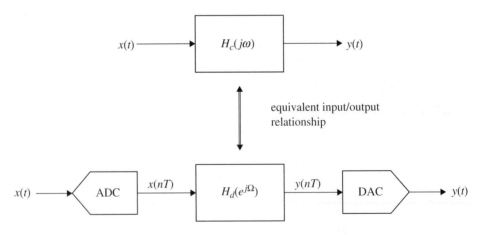

Figure 2.6.7 Two linear systems with equivalent functionality. A continuous-time linear time-invariant system operating on the continuous-time input signal (top). A discrete-time linear shift-invariant system operating on the samples of the continuous-time input signal.

Example 2.6.3
Discrete-Time Low-Pass Filter
This technique can be applied to the design of a low-pass filter. Suppose $x(t)$ is be filtered by an ideal low-pass filter with bandwidth W rad/s as illustrated

Section 2.7 Discrete-Time Processing of Band-Pass Signals 67

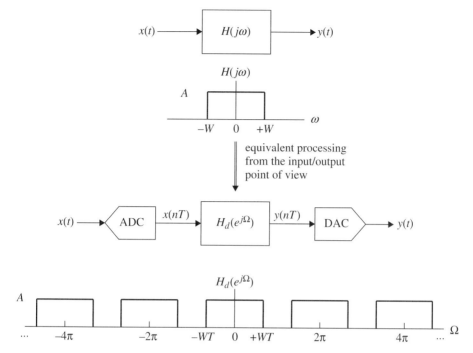

Figure 2.6.8 Continuous-time low-pass filter (top) and equivalent discrete-time low-pass filter (bottom).

in Figure 2.6.8. The requirement is to perform the filtering with discrete-time processing. To do this, $x(t)$ is sampled at a rate $1/T$ samples/s by the ADC to produce the sequence $x(nT)$. The sequence $x(nT)$ is processed by a discrete-time low-pass filter with DTFT $H_d\left(e^{j\Omega}\right)$ to produce a sequence of samples $y(nT)$. The filtered waveform $y(t)$ is constructed from the samples $y(nT)$ by the DAC. Applying (2.86) to $H(j\omega)$ produces the discrete-time filter requirements shown at the bottom of Figure 2.6.8.

2.7 DISCRETE-TIME PROCESSING OF BAND-PASS SIGNALS

In digital communication systems, it is often desirable to filter samples of a band-pass signal $x(n)$ (centered at Ω_0 rad/sample). In continuous-time processing, the most commonly used method is to frequency translate the band-pass signal to a lower frequency and filter. A discrete-time system that frequency translates the band-pass signal $x(n)$ to baseband and filters the frequency-translated signal with the low-pass filters $h(n)$ is illustrated in Figure 2.7.1 (a).

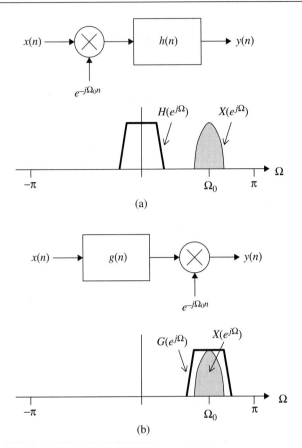

Figure 2.7.1 Equivalent processing for a band-pass signal: (a) Moving the signal to the filter; (b) moving the filter to the signal.

The baseband signal, $y(n)$ is given by

$$\begin{aligned}
y(n) &= \left[x(n)e^{-j\Omega_0 n}\right] * h(n) \\
&= \sum_k x(k)e^{-j\Omega_0 k} h(n-k) \\
&= \sum_k x(k) \underbrace{h(n-k)e^{j\Omega_0(n-k)}}_{g(n-k)} e^{-j\Omega_0 n} \\
&= \left[x(n) * g(n)\right] e^{-j\Omega_0 n}.
\end{aligned} \qquad (2.88)$$

This shows that an equivalent sequence of operations is to filter the band-pass signal $x(n)$ with a band-pass filter $g(n)$ followed by a frequency translation to baseband as illustrated in

Section 2.7 Discrete-Time Processing of Band-Pass Signals

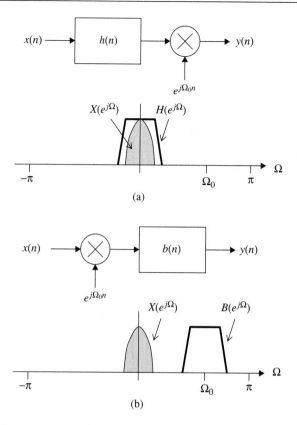

Figure 2.7.2 Equivalent processing for a low-pass signal: (a) Low-pass filter followed by frequency translation; (b) frequency translation followed by band-pass filter.

Figure 2.7.1 (b). The system in Figure 2.7.1 (a) moves the signal to the filter while the system in Figure 2.7.1 (b) moves the filter to the signal. Both operations are equivalent.

Another sequence of operations common to digital communications is to low-pass filter a baseband signal $x(n)$ and then to frequency translate the filter output to produce a band-pass signal $y(n)$. Such a system is illustrated in Figure 2.7.2 (a). The band-pass signal may be expressed as

$$\begin{aligned}
y(n) &= \left[x(n) * h(n)\right] e^{j\Omega_0 n} \\
&= \sum_k x(k) h(n-k) e^{j\Omega_0 n} \\
&= \sum_k x(k) e^{j\Omega_0 k} \underbrace{h(n-k) e^{j\Omega_0(n-k)}}_{b(n-k)} \\
&= \left[x(n) e^{j\Omega_0 n}\right] * b(n).
\end{aligned} \qquad (2.89)$$

This shows that an equivalent sequence of operations is to frequency translate the baseband signal $x(n)$ first, followed by a filtering operation with a band-pass filter $b(n)$ as shown in Figure 2.7.2 (b).

2.8 NOTES AND REFERENCES

There are many excellent textbooks devoted to frequency-domain analysis of continuous-time and discrete-time signals and systems, and they also offer a seemingly inexhaustible supply of exercises. Among these are Bracewell [39], Oppenheim and Willsky [40], Oppenheim and Schafer [41], and McClellan, Shafer, and Yoder [42]. The inverse Laplace and z-transforms may be computed using the Cauchy residue theorem or using the more common partial fraction expansion. The Cauchy residue theorem is described in most textbooks devoted to complex analysis. Churchill and Brown [43] is a popular example. Partial fraction expansions are described in great detail in most textbooks on transforms. Appendix A of Oppenheim and Willsky [40] presents a rather thorough treatment.

The FFT has been studied extensively and is a common topic in most textbooks devoted to discrete-time signal processing, such as Oppenheim and Schafer [41]. The interested reader may also find Cochran's 1967 article [44] of interest. In many of these texts, the assumed application of the FFT is spectral analysis of finite length sequences. They are used to compute a periodogram that forms the basis of nonparametric power spectral density estimation techniques. The FFT has other applications such as filtering and transmultiplexing as described in Section 10.2.

The FFT algorithm presented in this chapter is a special case of the Cooley–Tukey algorithm published by Cooley and Tukey in 1965 [45]. The basic operation of the Cooley–Tukey algorithm is a divide-and-conquer approach that recursively divides a DFT of length $N = N_1 N_2$ into smaller DFTs of length N_1 and N_2 along with the approximately N multiplications by complex-valued constants required to assemble the pieces into the desired result. The N complex-valued constants are the N roots of unity, called "twiddle factors" by Gentleman and Sande in 1966 [46]. The basic step of the algorithm presented in this chapter is to divide the transform into two pieces of size $N/2$ at each step, and is therefore limited to cases where N is a power of 2 (the radix-2 FFTs). Any factorization can be used in general to produce mixed-radix FFT. Although the basic idea is recursive, most traditional implementations rearrange the algorithm to avoid explicit recursion. Because the Cooley–Tukey algorithm breaks the DFT into smaller DFTs, it can be combined with other algorithms for computing the DFT. The other FFT algorithms include the prime-factor algorithm, due to Good [47] and Thomas [48], that uses the Chinese remainder theorem to factor the length $N = N_1 N_2$ (for relatively prime N_1 and N_2) DFT into smaller DFTs that do not require multiplication by twiddle factors to combine the results. The Rader–Brenner algorithm, by Brenner and Rader [49], also factors the DFT length N, but does so in such a way to produce purely imaginary twiddle factors. This algorithm requires more complex-valued additions than the Cooley–Tukey algorithm. The Winograd algorithm (by Winograd [50] who developed much of the theory for efficient FFT algorithms) factors $z^N - 1$ into cyclotomic polynomials, which form the basis of a factorization that requires the minimum number of multiplications at the expense of increased additions. Rader's algorithm [51] is based on advanced concepts from finite field theory and expresses a length-N DFT (for N prime) as a cyclic convolution of DFTs of composite size $N-1$. A nice history of the development of the FFT is by Heideman, Johnson, and Burrus [52].

2.9 EXERCISES

2.1 Determine the energy and power of each of the following signals:

(a) $x(t) = \begin{cases} 1 & -T \leq t \leq T \\ 0 & \text{otherwise} \end{cases}$

(b) $x(t) = \begin{cases} 0 & t < -T_1 \\ 1 & -T_1 \leq t < -T_2 \\ 2 & -T_2 \leq t \leq T_2 \\ 1 & T_1 < t \leq T_2 \\ 0 & t > T_1 \end{cases}$

(c) $x(t) = \cos(\omega_0 t)$

(d) $x(t) = A\cos(\omega_0 t + \theta)$

(e) $x(t) = \cos^2(\omega_0 t)$

(f) $x(t) = \begin{cases} \cos(\omega_0 t) & -T \leq t \leq T \\ 0 & \text{otherwise} \end{cases}$

(g) $x(t) = e^{-at}\cos(\omega_0 t)u(t)$

(h) $x(t) = Ate^{-t/T}u(t)$

(i) $x(t) = t^{-1/3}u(t-3)$

2.2 Evaluate the following integrals:

(a) $\displaystyle\int_{-\infty}^{\infty} (x^3 - 1)\delta(3 - x)dx$

(b) $\displaystyle\int_{-\infty}^{\infty} \frac{\sin(\pi t)}{\pi t}\delta(1 - t)dt$

(c) $\displaystyle\int_{-\infty}^{\infty} \frac{\sin^2(\pi t)}{(\pi t)^2}\delta(t - 1/2)dt$

(d) $\displaystyle\int_{-\infty}^{\infty} \frac{\sin^3(\pi t)}{(\pi t)^3}\delta(t)dt$

2.3 Compute the Laplace transforms of the following signals:
 (a) $x(t) = 5e^{-t}u(t)$
 (b) $x(t) = 5^{-t}u(t)$
 (c) $x(t) = 5e^{t}u(t)$
 (d) $x(t) = 5e^{t}u(-t)$

2.4 Compute the Laplace transforms of the following signals:
 (a) $x(t) = (t-2)^2 u(t)$
 (b) $x(t) = (t-2)^2 e^{-3t} u(t)$
 (c) $x(t) = \cos(\omega_0 t - \pi/4)\, u(t)$
 (d) $x(t) = e^{-3t} \cos(\omega_0 t + \pi/3)\, u(t)$

2.5 Compute the inverse Laplace transforms of the following:
 (a) $X(s) = \dfrac{s}{s^2 + 4s + 5}$
 (b) $X(s) = \dfrac{4}{s^3 + 4s^2 + 5s + 2}$
 (c) $X(s) = \dfrac{4}{s^3 + 5s^2 + 8s + 4}$

2.6 Compute the inverse Laplace transforms of the following:
 (a) $X(s) = \dfrac{6s^2 + 22s + 18}{s^3 + 6s^2 + 11s + 6}$
 (b) $X(s) = \dfrac{108}{s^6 + 14s^5 + 80s^4 + 238s^3 + 387s^2 + 324s + 108}$
 (c) $X(s) = \dfrac{54s^5 + 668s^4 + 3225s^3 + 7504s^2 + 8311s + 3426}{s^6 + 14s^5 + 80s^4 + 238s^3 + 387s^2 + 324s + 108}$

2.7 For stability reasons, many high-order systems are constructed as cascades of first- and second-order systems. The general form for a second-order system is

$$H(s) = \dfrac{b_1 s + b_0}{s^2 + a_1 s + a_0}$$

where $b_0, b_1, a_0,$ and a_1 are real-valued constants. This exercise explores some of the properties of the general second-order system.

(a) Write the time-domain differential equation corresponding to the second-order system $H(s)$. Use $x(t)$ for the input and $y(t)$ for the output.

(b) $H(s)$ has two poles. There are three cases to consider: real and distinct poles, real and repeated poles, and complex conjugate poles.

 i. What conditions must a_1 and a_0 satisfy to produce real and distinct poles? What are the poles?
 ii. What conditions must a_1 and a_0 satisfy to produce real and repeated poles? What are the poles?

iii. What conditions must a_1 and a_0 satisfy to produce complex conjugate poles? What are the poles?
(c) The conditions for the three cases can be simplified by expressing the denominator in the form $s^2 + 2\zeta\omega_n s + \omega_n^2$.
   i. Express ζ and ω_n in terms of a_1 and a_0.
   ii. What conditions must ζ and ω_n satisfy to produce real and distinct poles? What are the poles?
   iii. What conditions must ζ and ω_n satisfy to produce real and repeated poles? What are the poles?
   iv. What conditions must ζ and ω_n satisfy to produce complex conjugate poles? What are the poles?
   v. Why is this form of the denominator polynomial preferred?
(d) Second-order systems are categorized by the properties of its poles. There are three categories corresponding to the three cases considered in parts (b) and (c).
   i. The *overdamped system* corresponds to the case of real and distinct poles. Determine the impulse response $h(t)$ for the overdamped system. (Assume $b_1 = 0$.)
   ii. The *critically damped system* corresponds to the case of real and repeated poles. Determine the impulse response $h(t)$ for the overdamped system. (Assume $b_1 = 0$.)
   iii. The *underdamped system* corresponds to the case of complex conjugate poles. Determine the impulse response $h(t)$ for the overdamped system. (Assume $b_1 = 0$.)
(e) Based on the impulse responses obtained in part (d), can you explain the names given to each type of second-order system?

2.8 Consider the feedback system below with input $x(t)$.

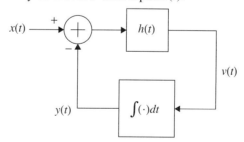

Let $h(t) = 6\delta(t)$ and $x(t) = 5u(t)$.
(a) Compute $y(t)$.
(b) Sketch a plot of $x(t)$ and $y(t)$ on the same set of axes.
(c) Compute $v(t)$.
(d) Sketch a plot of $v(t)$.

2.9 Repeat Exercise 2.8, except use $h(t) = 6\delta(t)$ and $x(t) = 5tu(t)$.

2.10 Repeat Exercise 2.8, except use $h(t) = 6e^{-5t}u(t)$ and $x(t) = 5u(t)$.

2.11 Repeat Exercise 2.8, except use $h(t) = 6e^{-5t}u(t)$ and $x(t) = 5tu(t)$.

2.12 Repeat Exercise 2.8, except use $h(t) = 125e^{-10t}u(t)$ and $x(t) = 5u(t)$.

2.13 Repeat Exercise 2.8, except use $h(t) = 125e^{-10t}u(t)$ and $x(t) = 5tu(t)$.

2.14 Consider the feedback system of Exercise 2.8 where $x(t)$ is the input and $y(t)$ is the output. If $h(t) = k\delta(t)$, what values of k produce a causal, stable system?

2.15 Consider the feedback system of Exercise 2.8 where $x(t)$ is the input and $y(t)$ is the output. Let $h(t) = e^{-at}u(t)$.
(a) What values of a produce a causal, stable system?
(b) What values of a produce oscillations in the system impulse response?

2.16 Consider the feedback system of Exercise 2.8 where $x(t)$ is the input and $y(t)$ is the output. Let $h(t) = ae^{-at}u(t)$.
(a) What values of a produce a causal, stable system?
(b) What values of a produce oscillations in the system impulse response?

2.17 Consider the signal $x(t) = A\cos(\omega_0 t + \theta)$.
(a) Compute the Fourier transform $X(j\omega)$.
(b) Compute the Fourier transform $X(f)$.

2.18 Compute the Fourier transform (in both ω and f) of $x(t) = e^{-a|t|}$ for $a > 0$.

2.19 Compute the Fourier transform (in both ω and f) of

$$x(t) = \begin{cases} 0 & t < -T \\ \frac{t}{T} + 1 & -T \leq t \leq 0 \\ -\frac{t}{T} + 1 & 0 \leq t \leq T \\ 0 & T < t \end{cases}.$$

2.20 Compute the Fourier transform (in both ω and f) of the signal $x(t)$ plotted below.

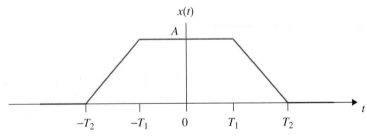

2.21 Consider the signal $x(t) = \delta(t) + \Gamma\delta(t - \tau)$ where Γ is a complex-valued constant.
(a) Determine $X(f)$.
(b) Determine $|X(f)|^2$. Express your answer in terms of the magnitude of Γ, $|\Gamma|$, and the phase of Γ, $\angle\Gamma$.

(c) Sketch a plot of $|X(f)|^2$ versus f.

2.22 Consider a signal $x(t)$ with Fourier transform

$$X(j\omega) = \frac{j\omega}{1 + j\omega}.$$

Find the Fourier transform $Y(j\omega)$ for each of the following signals.
(a) $y(t) = 3x(t - 5)$
(b) $y(t) = e^{-j5t}x(t - 3)$
(c) $y(t) = x(-t)$
(d) $y(t) = \dfrac{d}{dt}x(t)$

2.23 Consider the Fourier transform $X(f)$ shown below.

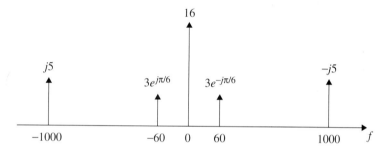

(a) Compute $x(t)$, the inverse Fourier transform of $X(f)$.
(b) Is $x(t)$ periodic? If so, what is its period?

2.24 Compute the inverse Fourier transform of the signal

$$X(f) = \frac{1}{2}\delta(f + 2) + \frac{1}{2}\delta(f - 2) + \frac{e^{j4\pi f}}{2 + 2\pi f}$$

2.25 The 3 dB bandwidth B_{3dB} of a baseband signal $x(t)$ is defined using the Fourier transform $X(f)$. B_{3dB} is the value of f for which $|X(f)|^2 = |X(0)|^2/2$. Determine B_{3dB} for each of the following signals.
(a) $x(t) = e^{-at}u(t)$
(b) $x(t) = e^{-a|t|}$
(c) $x(t) = e^{-\pi t^2}$

2.26 Use Parseval's theorem to evaluate the following integral

$$\int_{-\infty}^{\infty} \frac{\sin^2(\pi x)}{(\pi x)^2} dx$$

2.27 The 90% bandwidth $B_{90\%}$ of a baseband signal $x(t)$ is defined using the Fourier transform $X(f)$. $B_{90\%}$ is defined as the bandwidth that captures 90% of the total power in $x(t)$. The mathematical definition is

$$\int_{-B_{90\%}}^{B_{90\%}} |X(f)|^2 df = 0.90 \times \int_{-\infty}^{\infty} |X(f)|^2 df$$

where the integral on the right is the power contained in the interval $-B_{90\%} \leq f \leq B_{90\%}$ and the integral on the right is the total power given by Parseval's theorem. Determine $B_{90\%}$ for each of the following signals.
(a) $x(t) = e^{-at}u(t)$
(b) $x(t) = e^{-a|t|}$
(c) $x(t) = e^{-\pi t^2}$ (Express your answer in terms of the function $Q(x) = \frac{1}{\sqrt{2\pi}} \int_x^\infty e^{-t^2/2} dt$.)

2.28 Consider a nonlinear system with input $x(t)$ and output $y(t)$ defined by the following relationship

$$y(t) = x(t) + \frac{1}{2}x^2(t) + \frac{1}{3}x^3(t).$$

One of the methods used to quantify nonlinear distortion in an audio system is the *total harmonic distortion* or THD. THD is computed by applying a pure 1 kHz sinusoid to the input of the nonlinear system and comparing the output power at 1 kHz to the power at other frequencies in the audio range 20 Hz to 20 kHz. In equation form, the THD is

$$\text{THD} = \frac{\text{power at all other frequencies}}{\text{power at 1 kHz}} \times 100\%$$

When $x(t)$ is a pure 1 kHz sinusoid, the nonlinear system produces components only at multiples (or harmonics) of 1 kHz. Hence the name total *harmonic* distortion.
(a) Let $x(t) = A\cos(2\pi f_0 t)$ where $f_0 = 1000$. Compute $y(t)$.
(b) Compute $Y(f)$. Sketch $|Y(f)|^2$.
(c) From the plot obtained in part (b), compute the THD.

2.29 Consider a nonlinear system with input $x(t)$ and output $y(t)$ defined by the following relationship

$$y(t) = x(t) + \frac{1}{2}x^2(t) + \frac{1}{3}x^3(t).$$

One of the methods used to quantify nonlinear distortion in an audio system is the *intermodulation distortion* or ID test. The ID test signal is the sum of two pure sinusoids each oscillating at a different frequency. In addition to harmonics of the two input frequencies, a nonlinear system also produces frequency components at linear combinations of the two frequencies. These frequency components are called *intermodulation products* because the effect of the nonlinearity is to cause one of the input sinusoids (and its harmonics) to amplitude modulate the sinusoid (and its harmonics). A commonly used test signal is the SMPTE IM test signal[3] defined by

$$x(t) = \frac{A}{4}\cos(2\pi f_0 t) + A\cos(2\pi f_1 t)$$

where $f_0 = 60$ Hz and $f_1 = 7000$ Hz. The ID is calculated by comparing the power at the input frequencies of 60 Hz and 7 kHz to the power at all other frequencies in the audio range 20 Hz to 20 kHz. When this signal is used as the input, the intermodulation distortion is defined as

$$\text{ID} = \frac{\text{power at all other frequencies}}{\text{power at 60 Hz and 7 kHz}}$$

(a) Compute $y(t)$.
(b) Compute $Y(f)$. Sketch $|Y(f)|^2$.
(c) From the plot obtained in part (b), compute the ID.

2.30 Determine the Fourier series representation of the periodic square wave signal $x(t)$ shown below. Make a sketch of $|X(f)|$.

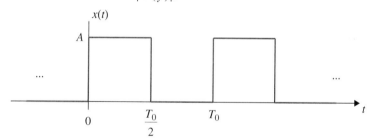

2.31 Determine the Fourier series representation of the periodic square wave signal $x(t)$ shown below. Make a sketch of $|X(f)|$.

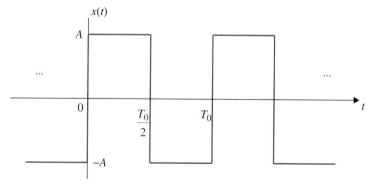

---

[3]SMPTE is the acronym for the Society of Motion Picture and Television Engineers. The EIA 560 Standard specifies the exact procedure for performing the intermodulation (IM) test using the SMPTE IM signal.

2.32 Determine the Fourier series representation of the periodic square wave signal $x(t)$ shown below. Make a sketch of $|X(f)|$.

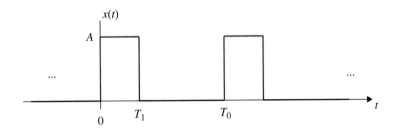

2.33 Determine the Fourier series representation of the periodic triangle wave signal $x(t)$ shown below. Make a sketch of $|X(f)|$.

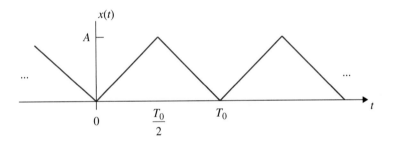

2.34 Determine the Fourier series representation of the periodic triangle wave signal $x(t)$ shown below. Make a sketch of $|X(f)|$.

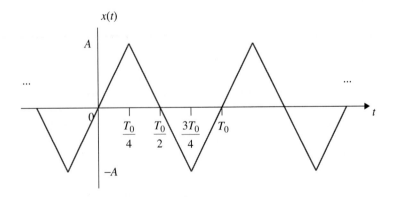

Section 2.9 Exercises

2.35 Determine the Fourier series representation of the periodic sawtooth signal $x(t)$ shown below. Make a sketch of $|X(f)|$.

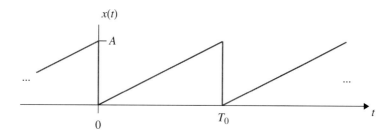

2.36 Determine the Fourier series representation of the half-rectified sine wave $x(t)$ shown below. Make a sketch of $|X(f)|$.

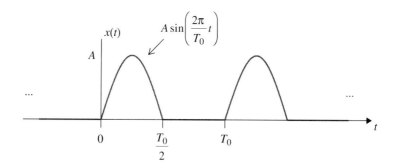

2.37 Determine the Fourier series representation of the full-rectified sine wave $x(t)$ shown below. Make a sketch of $|X(f)|$.

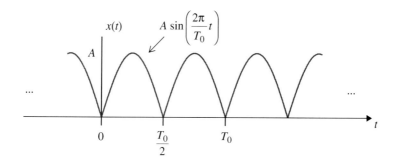

2.38 Derive the Fourier transform for the impulse train. The Fourier transform pair is given by (2.68) and (2.69).

2.39 Consider the system shown below with input $x(t)$ and output $y(t)$.

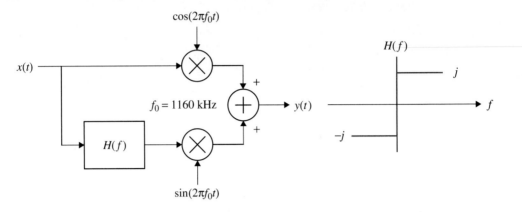

Sketch $Y(f)$, the Fourier transform of $y(t)$ if the input is

$$x(t) = \cos(2\pi 1000 t) + \cos(2\pi 2000 t).$$

2.40 Repeat Exercise 2.39 except replace $x(t)$ with the signal whose Fourier transform is

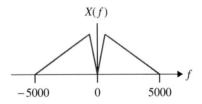

2.41 Consider the system shown below with input $x(t)$ and output $y(t)$.

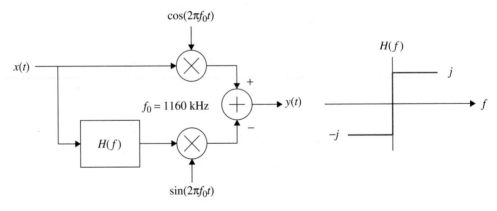

Sketch $Y(f)$, the Fourier transform of $y(t)$ if the input is

$$x(t) = \cos(2\pi 1000 t) + \cos(2\pi 2000 t).$$

If you worked Exercise 2.39, compare your answer to the one you obtained in Exercise 2.39.

2.42 Repeat Exercise 2.41, except replace $x(t)$ with the signal whose Fourier transform is

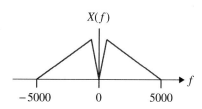

If you worked Exercise 2.40, compare your answer to the one you obtained in Exercise 2.40.

2.43 Consider the LTI system shown below with input $x(t)$, output $y(t)$, and transfer function $H(f)$.

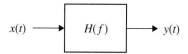

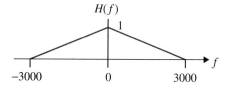

(a) Determine $y(t)$ when the input is

$$x(t) = 3\cos(2\pi 1000 t) + 6\cos(2\pi 2000 t).$$

(b) Sketch $Y(f)$, the Fourier transform of $y(t)$, when the Fourier transform of the input is

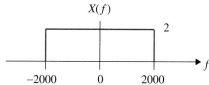

2.44 Consider the LTI system shown below with input $x(t)$, output $y(t)$, and transfer function $H(f)$.

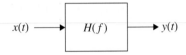

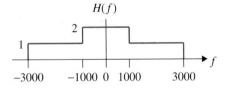

Determine $y(t)$ when the input is
(a) $x(t) = \cos(2\pi 440t) + 2\cos(2\pi 2000t)$
(b) $x(t) = \cos(2\pi 440t) + \frac{1}{2}\cos(2\pi 880t)$
(c) $x(t) = \cos(2\pi 600t) + 2\cos(2\pi 2600t)$
(d) The SMTP-IM test signal defined in Exercise 2.29
(e) $x(t) = 3\cos(2\pi 900t) \times \cos(2\pi 2700t)$

2.45 Consider the band-pass signal $x(t)$ whose Fourier transform, $X(f)$, is shown below.

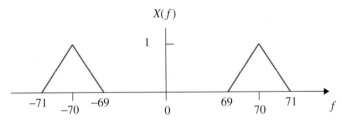

The signal $x(t)$ is processed by the system shown below to produce the output $y(t)$.

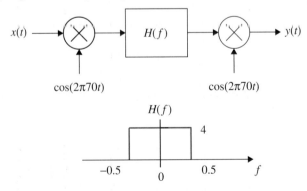

Sketch $Y(f)$, the Fourier transform of the output $y(t)$.

2.46 Consider the band-pass signal $x(t)$ whose Fourier transform, $X(f)$, is shown below.

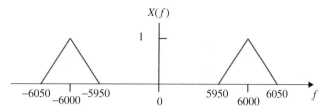

The signal $x(t)$ is processed by the system shown below to produce the output $y(t)$.

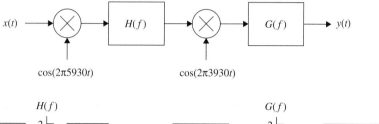

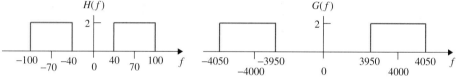

Sketch $Y(f)$, the Fourier transform of the output $y(t)$.

2.47 Consider the band-pass signal $x(t)$ whose Fourier transform, $X(f)$, is shown below.

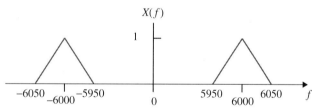

The signal $x(t)$ is processed by the system shown below to produce the output $y(t)$.

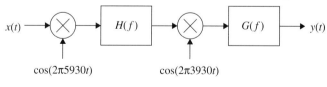

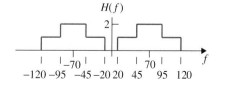 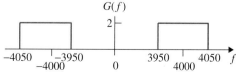

Sketch $Y(f)$, the Fourier transform of the output $y(t)$.

2.48 Consider the band-pass signal $r(t)$ whose Fourier transform, $R(f)$, is shown below.

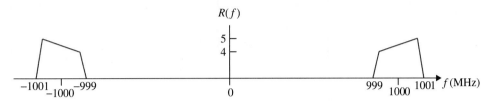

The signal $r(t)$ is processed by the system shown below to produce the output $x(t)$.

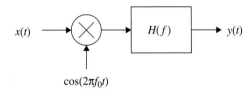

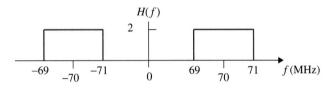

(a) Sketch $X(f)$, the Fourier transform of $x(t)$, when $f_0 = 930$ MHz.
(b) Sketch $X(f)$, the Fourier transform of $x(t)$, when $f_0 = 1070$ MHz.
(c) Comment on the differences between the answers for parts (a) and (b). What is the difference in the time domain? Hint: write

$$r(t) = I(t)\cos(2\pi f_1 t) - Q(t)\sin(2\pi f_1 t)$$

($f_1 = 1000$ MHz) and compute $x(t)$ for both cases in terms of $I(t)$ and $Q(t)$.

2.49 Consider the band-pass signal $r(t)$ whose Fourier transform, $R(f)$, is shown below.

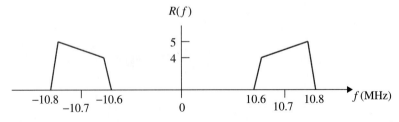

The signal $r(t)$ is processed by the system shown below to produce the outputs $x(t)$ and $y(t)$.

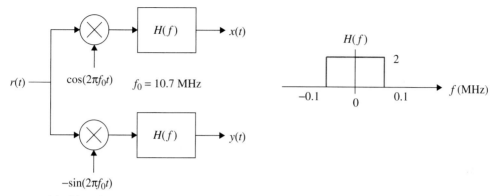

(a) Sketch $|X(f)|$, the magnitude of the Fourier transform of $x(t)$.
(b) Sketch $|Y(f)|$, the magnitude of the Fourier transform of $y(t)$.

2.50 Consider the band-pass signal $r(t)$ whose Fourier transform, $R(f)$, is shown below.

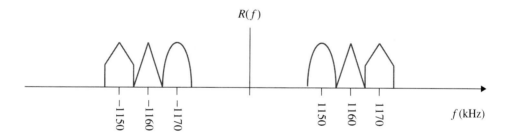

The signal $r(t)$ is processed by the system shown below to produce the output $x(t)$.

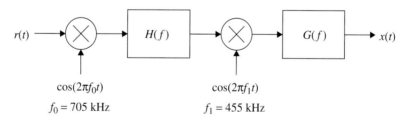

(a) Determine $H(f)$ and $G(f)$ to produce an output whose Fourier transform is $X(f)$

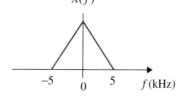

(b) Now suppose the Fourier transform of $r(t)$ is given by the system shown below.

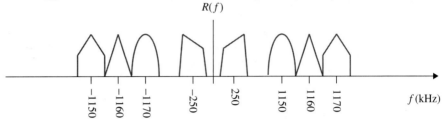

Is it possible to design an $H(f)$ and $G(f)$ to produce the desired output? If not, what changes to the system would you recommend to make it possible to produce the desired output?

2.51 Consider the band-pass signal $r(t)$ whose Fourier transform, $R(f)$, is shown below.

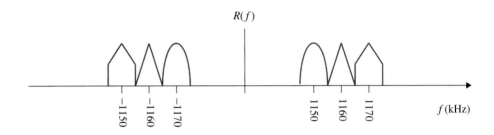

The signal $r(t)$ is processed by the system shown below to produce the output $x(t)$.

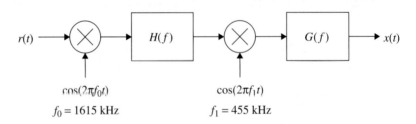

(a) Determine $H(f)$ and $G(f)$ to produce an output whose Fourier transform is

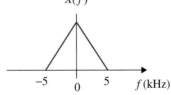

(b) Now suppose the Fourier transform of $r(t)$ is given by the system shown below.

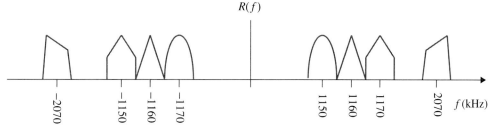

Is it possible to design an $H(f)$ and $G(f)$ to produce the desired output? If not, what changes to the system would you recommend to make it possible to produce the desired output?

2.52 This exercise introduces the most common method used to transmit stereophonic (or stereo) signals. In its simplest form, a stereophonic[4] signal consists of two audio signals usually designated "left" and "right" and denoted $l(t)$ and $r(t)$ for this exercise. One of the challenges faced by systems engineers developing a method for transmitting stereophonic signals in the 1950s was *backward compatibility*: the stereophonic signal had to be compatible with existing receivers that were not capable of handling stereophonic signals. The system that emerged from this effort is illustrated below.

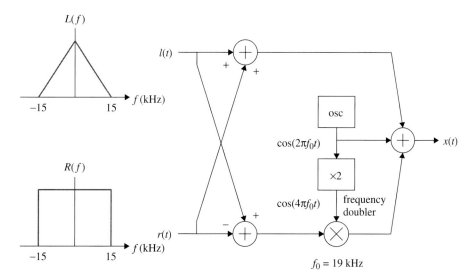

---

[4]The word *stereo* comes from the Greek *stereos* (στερεός) which means solid. It was first used in English in the 19th century as part of *stereotype* which carried the modern connotation of a description that applies the entire group as if they were a "solid." In the late 1800s, the word was used to form *stereoscope* which was a device that produced a "solid" (or 3-dimensional) image and was based on two 2-dimensional images of the same object from different angles. In the 1950s *stereophonic* was applied in the same sense, except to audio signals.

Shown are the Fourier transforms of the left and right audio components, $L(f)$ and $R(f)$, respectively.
(a) Sketch $X(f)$.
(b) Sketch a block diagram of a system that recovers $l(t)$ and $r(t)$ from $x(t)$.
(c) For the purposes of backward compatibility, the signal $l(t) + r(t)$ must be available. Does your answer to part (b) make this signal available?

2.53 In amplitude modulation (AM), an information bearing signal $m(t)$ is used to change, or *modulate*, the amplitude of an RF carrier $A_c \cos(2\pi f_c t)$. The carrier frequency is f_c. The block diagram of a system that performs amplitude modulation is illustrated below.

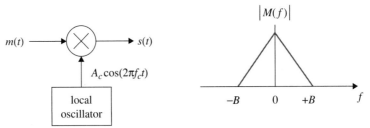

The modulation theorem of Fourier transform theory (given as the Frequency Shifting property in Table 2.4.3) can be used to analyze an AM signal in the frequency domain.

(a) In many communications textbooks, the modulation theorem is stated in the following form: Let $x(t) \leftrightarrow X(f)$ be a Fourier transform pair. Then

$$x(t) \cos(2\pi f_0 t) \leftrightarrow \frac{1}{2} X(f + f_0) + \frac{1}{2} X(f - f_0)$$

are a Fourier transform pair. Use the Frequency Shifting property of the Fourier transform defined in Table 2.4.3 to prove the modulation theorem. (Hint: use Euler's identity for the cosine and apply the Linearity and Time Scaling properties.)

(b) Given the magnitude of the Fourier transform of $m(t)$ illustrated above, sketch the magnitude of the Fourier transform of the AM signal $s(t)$.

(c) Based on your answer to part (a), determine conditions on B and f_c so that $s(t)$ does not possess any spectral aliasing.

(d) If you were designing an AM broadcast system where each station is assigned to a different carrier frequency, what frequency spacing (difference between adjacent channels) would you suggest? Why?

2.54 A common method used to test AM systems is to use a special test signal as the modulating signal and observe the output. A popular choice is an audio frequency sinusoid:

$$m(t) = A_m \cos(2\pi f_m t)$$

where $f_m \ll f_c$.

(a) Determine the Fourier transform $M(f)$ of the test signal.
(b) Use the Frequency Translation property of Table 2.4.3 (or the modulation theorem described in Exercise 2.53) to determine the Fourier transform $S(f)$ of the AM modulated carrier.
(c) Sketch the Fourier Transform $S(f)$.

2.55 An *AM demodulator* is a system that recovers $m(t)$ from an AM modulated carrier. An obvious method for recovering $m(t)$ is to "remodulate" an AM modulated carrier, as illustrated in the block diagram below.

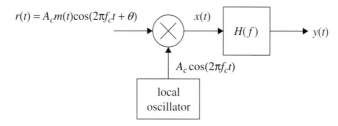

This type of demodulator is called a *coherent demodulator* because the local oscillator must be *phase coherent* with the phase of RF carrier. The reason the RF carrier phase must be known is explored in this problem. The demodulator may be analyzed in both the time domain and in the frequency domain.
(a) Frequency-domain analysis: Let $\theta = 0$ and use $|M(f)|$ illustrated in Exercise 2.53.
  i. Sketch $|R(f)|$ and $|X(f)|$.
  ii. State the conditions $|H(f)|$ must satisfy to produce $y(t) = m(t)$.
(b) Time-domain analysis: Let $\theta = 0$.
  i. Obtain an expression for $x(t)$ in terms of $m(t)$.
  ii. Use the properties of $|H(f)|$, derived in part (a), to obtain an expression for $y(t)$.
(c) Effect of uncompensated phase shift: Now let $\theta \neq 0$.
  i. Derive an expression for $x(t)$ and $y(t)$.
  ii. What effect does θ have on the output? What would you suggest to overcome this effect?

2.56 It was shown in Exercise 2.55 that an unknown RF carrier phase limits the performance of the coherent AM demodulator. One solution is to use a noncoherent demodulation technique (i.e., one that does require knowledge of the RF carrier phase). One such technique is the *envelope detector*. As illustrated below, the envelope detector produces $|A(t)|$ from $A(t)\cos(2\pi f_c t + \theta)$.

Noncoherent detection using the envelope detector imposes certain constraints on the amplitude to achieve acceptable performance.

(a) Suppose $m(t)$ is the audio test signal of Exercise 2.54. If the input to the envelope detector is
$$A_c m(t) \cos(2\pi f_c t + \theta),$$
what is the envelope detector output?

(b) Sketch an example of the envelope detector output.

(c) What is the problem with using the envelope detector as a demodulator for this signal? Are there changes to the modulating signal together with some additional post processing that can fix the problem?

2.57 In Exercise 2.56, it was shown that the envelope detector is unable to distinguish between positive and negative values of the amplitude and that this seriously limits its usefulness in recovering $m(t)$ from an AM modulated carrier of the form
$$A_c m(t) \cos(2\pi f_c t).$$

This shortcoming can be overcome be adding a DC bias to $m(t)$ and using this biased signal to amplitude modulate the RF carrier as illustrated by the block diagram below.

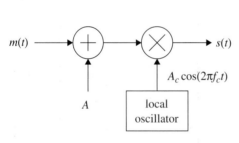
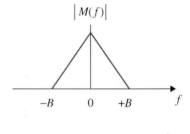

(a) Given the magnitude of the Fourier transform of $m(t)$ illustrated above, sketch the magnitude of the Fourier transform of the AM signal $s(t)$.

(b) This type of AM modulated signal is often called AM transmitted carrier or AM-TC for short. Why is this name used?

(c) If the envelope detector described in Exercise 2.56 is used to demodulate the AM-TC signal, what additional post-processing is required to recover $m(t)$ from the envelope detector output? What condition must A satisfy for $m(t)$ to be recovered from the envelope detector output?

2.58 The audio frequency sinusoid of Exercise 2.54 can also be used as a test signal for the AM-TC system described in Exercise 2.57. Let the modulated carrier be
$$s(t) = A_c[A + A_m \cos(2\pi f_m t)] \cos(2\pi f_c t) = A(t) \cos(2\pi f_c t).$$

(a) What condition must A satisfy to guarantee $A(t) \geq 0$ for all t?

(b) Determine the Fourier transform $A(f)$ of $A(t)$.

(c) Use the Frequency Translation property of Table 2.4.3 (or the modulation theorem described in Exercise 2.53) to determine the Fourier transform $S(f)$ of the AM-TC signal.

(d) Sketch the Fourier transform $S(f)$.

2.59 An AM suppressed carrier or AM-SC signal is of the form

$$s_{\text{AM-SC}}(t) = A_c m(t) \cos(2\pi f_c t)$$

and an AM-TC signal is of the form

$$s_{\text{AM-TC}}(t) = A_c [A + m(t)] \cos(2\pi f_c t).$$

The AM-SC signal cannot be demodulated using an envelope detector (see Exercise 2.56) and thus requires a coherent demodulator. On the other hand, the AM-TC signal is suitable for use with an envelope detector (provided an appropriate value for A has been used—see Exercise 2.57). It is tempting to conclude that the AM-TC signal is superior to the AM-SC signal. This problem derives the power ratio for the two signals and shows that the AM-SC signal is more power efficient than the AM-TC signal. For this problem let $m(t)$ be the test signal

$$m(t) = A_m \cos(2\pi f_m t)$$

and assume the period $T_m = 1/f_m$ is an integer multiple of $1/f_c$. (This assumption is not required in general, but simplifies that time-domain mathematics.) The power ratio is defined as the ratio of power in $s(t)$ due to $m(t)$ to the total power in $s(t)$.

(a) Using (2.6) in the time domain or Parseval's theorem in the frequency domain, compute the power of $s_{\text{AM-SC}}(t)$. Because all of the power is due to $m(t)$, the power ratio is 1.

(b) Using (2.6) in the time domain or Parseval's theorem in the frequency domain, compute the power of $s_{\text{AM-TC}}(t)$. Show that the power ratio is

$$\text{power ratio} = \frac{1}{2\left(\dfrac{A}{A_m}\right)^2 + 1}.$$

What is the largest value the power ratio can be?

2.60 In frequency modulation (FM), an information-bearing signal $m(t)$ is used to change, or *modulate*, the instantaneous frequency of an RF carrier $A_c \cos(2\pi f_c t)$. Because the instantaneous frequency of a sinusoid is the time derivative of the argument—see Section 9.2.2 for the formal definitions—the FM modulated carrier is

$$s(t) = A_c \cos\left(2\pi f_c t + 2\pi \Delta f \int_{-\infty}^{t} m(\tau) d\tau\right).$$

(a) Sketch $s(t)$ versus t for the $m(t)$ shown below. Use $f_c = 2$ Hz, and $\Delta f = 1/10$.

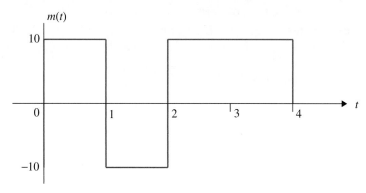

(b) Sketch $s(t)$ versus t for the $m(t)$ shown below. Use $f_c = 2$ Hz, and $\Delta f = 1/10$.

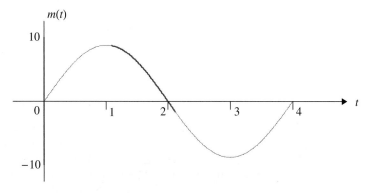

2.61 This exercise explores the properties of an FM modulated carrier when $m(t)$ is the test signal

$$m(t) = A_m \cos(2\pi f_m t).$$

The relationship between $m(t)$ and $s(t)$, the FM modulated carrier, is nonlinear. As a consequence, traditional frequency domain techniques cannot be used to characterize $s(t)$. However, in the case where $m(t)$ is the sinusoid shown above, a frequency domain analysis is possible.

(a) Show that when $m(t)$ is the test signal defined above, the FM modulated carrier may be expressed as

$$s(t) = A_c \cos\left(2\pi f_c t + \beta \sin(2\pi f_m t)\right).$$

What is β? What is the peak frequency deviation?

(b) Show that $s(t)$ given above may be expressed as

$$s(t) = A_c \text{Real}\left\{e^{j\beta \sin(2\pi f_m t)} e^{j(2\pi f_c t)}\right\}.$$

(c) For convenience, define the *complex envelope* of $s(t)$ as

$$\tilde{s}(t) = e^{j\beta \sin(2\pi f_m t)}.$$

Because $\tilde{s}(t)$ is periodic with period $T_0 = 1/f_m$, it may also be expressed as a Fourier series of the form (2.7):

$$\tilde{s}(t) = \sum_{k=-\infty}^{\infty} c_k e^{j2\pi k f_m t}.$$

Using the Fourier series representation for $\tilde{s}(t)$, the FM modulated carrier may be expressed as

$$s(t) = A_c \text{Real}\left\{\sum_{k=-\infty}^{\infty} c_k e^{j2\pi k f_m t} e^{j2\pi f_c t}\right\} = A_c \sum_{k=-\infty}^{\infty} c_k \cos\left(2\pi (f_c + k f_m) t\right).$$

Derive an expression for the Fourier series coefficients. Hint: use the relationship

$$J_\nu(z) = \frac{1}{2\pi} \int_0^{2\pi} e^{j[z \sin(x) - \nu x]} dx$$

where $J_\nu(z)$ is the ν-th order Bessel function of the first kind.[5]

(d) Derive an expression for the Fourier transform $S(f)$ in terms of β, f_m, and f_c using the answer to part (c) Sketch the $S(f)$. What is the bandwidth of the FM modulated carrier?

(e) Are there values of β that produce the curious and interesting situation of no power at the carrier frequency?

---

[5]Bessel functions are solutions to Bessel's equation

$$x^2 \frac{d^2 y}{dx^2} + x \frac{dy}{dx} + (x^2 - \nu^2) y = 0$$

for a real-valued constant ν. The differential equation arises in the analysis of such phenomena as electromagnetic propagation in a cylindrical waveguide and in heat propagation in a cylindrical object. It was first posited by Daniel Bernoulli and later refined by Friedrich Bessel, a German mathematician and astronomer who worked during the first half of the 19th century. A contemporary of Carl Gauss, Bessel made improvements in the calculations describing the orbit of Halley's Comet and, in 1838, was the first to show that parallax could produce accurate measurements of interstellar distances. (He used the method to calculate the distance to 61 Cygni.) Bessel solved the equation by investigating the application of the function defined by the integral

$$\frac{1}{\pi} \int_0^\pi \cos\left(n\theta - x \sin(\theta)\right) d\theta$$

as a solution. This integral first arose in calculations involving, not surprisingly, celestial mechanics. The integral form provided in the hint is a generalized version of this integral.

2.62 This exercise explores the bandwidth of an FM modulated carrier using the test signal

$$m(t) = A_m \cos(2\pi f_m t)$$

as the modulating signal so that the modulated carrier is

$$s(t) = A_c \cos\left(2\pi f_c t + \beta \sin(2\pi f_m t)\right) = A_c \sum_{k=-\infty}^{\infty} c_k \cos\left(2\pi (f_c + k f_m) t\right)$$

where the c_k are functions of β as shown in Exercise 2.60. Most meaningful definitions of bandwidth are based on the power. As a consequence, this exercise starts with finding an expression for power as a function of frequency. Note that the total power in the FM modulated carrier above is $P = A_c^2/2$.

(a) Derive an expression for $|S(f)|^2$ and use Parseval's theorem to show that the power in $s(t)$ may also be expressed as

$$P = \frac{A_c^2}{2} \sum_{k=-\infty}^{\infty} |c_k|^2.$$

(b) When the sum in part (a) is truncated to $\pm K$, the power in the frequency band $f_c - K f_m \le |f| \le f_c + K f_m$ (corresponding to a bandwidth of $2K f_m$) is obtained. The ratio

$$R = \frac{\dfrac{A_c^2}{2} \sum_{k=-K}^{K} |c_k|^2}{\dfrac{A_c^2}{2}} = \sum_{k=-K}^{K} |c_k|^2$$

is used to quantify the fraction of the total carrier power contained in the band. The 98% bandwidth $B_{98\%}$ is defined as

$$B_{98\%} = 2 K_{98} f_m$$

where K_{98} is obtained by setting $R = 0.98$:

$$0.98 = \sum_{k=-K_{98}}^{K_{98}} |c_k|^2.$$

Create a table of c_k for $\beta = 1, 2,$ and 5 and $k = 0, 1, \ldots, 8$. From this table, determine the value of K_{98} for each case. Hint: use the fact that $|c_k|^2 = |c_{-k}|^2$ to write

$$R = |c_0|^2 + 2 \sum_{k=1}^{K} |c_k|^2.$$

(c) Based on the results of part (b), write down the relationship between K_{98} and β. Use this relationship to express the $B_{98\%}$ in terms of β and f_m. The result is a simple form of *Carson's rule* which gives the 98% bandwidth of an FM modulated carrier. It is named for John Carson, who was introduced in Chapter 1.

2.63 An *FM demodulator* is a system that recovers $m(t)$ from an FM modulated carrier. A common method for accomplishing this is shown by the block diagram below.

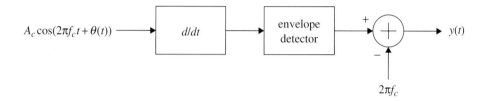

The block diagram uses the shorthand notation

$$\theta(t) = 2\pi \Delta f \int_{-\infty}^{t} m(\tau)d\tau.$$

Hence, the goal is to recover the time derivative of $\theta(t)$ from the input signal. This type of demodulator uses a differentiator to perform "FM-to-AM conversion" and an envelope detector (described in Exercise 2.56) to recover $m(t)$ from the converted signal.

(a) Derive the expression for the output of the differentiator.
(b) Derive the expression for the output of the envelope detector. State any conditions necessary to obtain a scaled version of $m(t)$ from $y(t)$.
(c) Let the modulating signal be the test signal $m(t) = A_m \cos(2\pi f_m t)$. Sketch the envelope detector output. Sketch the output $y(t)$.

2.64 Another method for recovering $m(t)$ from an FM modulated carrier is based on the phase locked loop (or PLL). This method has better noise performance and dynamic range than the FM-to-AM conversion method described in the previous exercise. A general block diagram of a PLL is shown below and consists of phase detector, a loop filter with s-domain transfer function $F(s)$, and a voltage controlled oscillator (VCO) arranged in a feedback system.

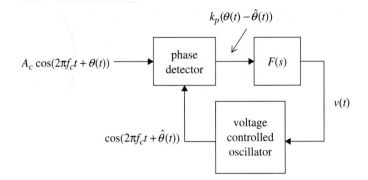

The input signal is a sinusoid with phase $\theta(t)$ that is shorthand for

$$\theta(t) = 2\pi \Delta f \int_{-\infty}^{t} m(\tau)d\tau.$$

The VCO is an FM modulator that uses the loop filter output $v(t)$ as the modulating signal. The VCO output is a sinusoid with phase $\hat{\theta}(t)$ where the relationship between $\hat{\theta}(t)$ and $v(t)$ is

$$\hat{\theta}(t) = k_0 \int_{-\infty}^{t} v(\tau)d\tau.$$

A detailed description and analysis of the PLLs is presented in Appendix C. An abstraction used to analyze PLLs is the phase equivalent PLL shown below.

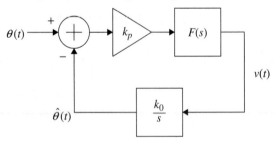

The goal of the PLL is to force the phase error $\theta(t) - \hat{\theta}(t) = 0$. Because this occurs when the VCO output matches the input, the VCO input $v(t)$ is the desired output. The goal is to produce a $v(t)$ that is proportional to the time derivative of $\theta(t)$.

(a) Derive the transfer function $H_{FM}(s) = \dfrac{V(s)}{\Theta(s)}$.

(b) It is claimed that $H_{FM}(s) = s$ is the desired transfer function. Why?

(c) Find the loop filter transfer function $F(s)$ that forces $H_{FM}(s) = s$. What is the corresponding impulse response $f(t)$? Are there any values of the constants k_p and k_0 that should be avoided?

2.65 Compute the inverse z-transform of the following stable, causal LTI systems.

(a) $H(z) = \dfrac{1 - z^{-1}}{1 + \dfrac{1}{2}z^{-1}}$

(b) $H(z) = \dfrac{1 + \dfrac{1}{3}z^{-1}}{1 + \dfrac{1}{2}z^{-1}}$

(c) $H(z) = \dfrac{z^{-2}}{1 + \dfrac{1}{2}z^{-1}}$

(d) $H(z) = -z^{-1} + 2z^{-2} - 3z^{-3} + 4z^{-4}$

2.66 Compute the inverse z-transform of the following stable, causal LTI systems.

(a) $H(z) = \dfrac{2z^{-1}}{1 - \dfrac{1}{2}z^{-1}}$

(b) $H(z) = \dfrac{6z^{-1}}{1 - \dfrac{2}{3}z^{-1} + \dfrac{25}{144}z^{-2}}$

(c) $H(z) = \dfrac{2z^{-1}}{1 - \dfrac{1}{2}z^{-1} + \dfrac{25}{144}z^{-2}}$

(d) $H(z) = \dfrac{2z^{-1}}{1 - \dfrac{5}{4}z^{-1} + \dfrac{1}{2}z^{-2} - \dfrac{1}{16}z^{-3}}$

2.67 The celebrated Fibonacci sequence[6]

$$0, 1, 2, 3, 5, 8, 13, 21, 34, 55, 89, 144, 233, 377, 610, 987, \ldots$$

may be "solved" using the z-transform. Let $y(n)$ be the n-th Fibonacci number. Each number in the sequence is the sum of the previous two. Thus, the Fibonacci sequence

---

[6]This sequence is named after Leonardo da Pisa who was also known as Fibonacci. He lived and worked in Pisa during the early 13th century and traveled extensively through north Africa and the Middle East. His publication *Liber Abaci* ("Book of Calculations") accelerated the adoption of Hindu-Arabic decimal numbers in Europe. In *Liber Abaci*, Fibonacci posed the following problem:

Beginning with a single pair of rabbits, if every month each productive pair bears a new pair, which becomes productive when they are 1 month old, how many rabbits will there be after n months?

The recursion in the problem statement is Fibonacci's solution for the number of pairs of rabbits. Incidentally, the z-transform produces a second order polynomial in z with two real roots. The positive root, usually denoted ϕ, is also known as the *golden section* and has been known since antiquity. The Greeks used the ratio $1 : \phi$ in the design of the Parthenon and ϕ is the ratio of the side of a regular pentagon to its diagonal. Author Dan Brown endows ϕ with magic and mystery in *The Da Vinci Code*.

is defined by the recursion

$$y(n) = y(n-1) + y(n-2)$$

with initial conditions $y(0) = 1$ and $y(1) = 1$. The initial conditions may be incorporated into the recursion as

$$y(n) = y(n-1) + y(n-2) + \delta(n-1).$$

Solving the Fibonacci sequence means finding a formula for $y(n)$.
(a) Compute the z-transform of the recursion.
(b) Solve the answer in part (a) for $Y(z)$.
(c) Compute the causal inverse z-transform of $Y(z)$.

2.68 Define the discrete-time signal $x(n)$ as

$$x(n) = \sum_{k=0}^{5} \delta(n-k).$$

(a) Sketch $x(n)$.
(b) Find $X(z)$.
(c) Let $g(n) = x(n) - x(n-1)$. Sketch $g(n)$.
(d) Find $G(z)$ directly from $g(n)$.
(e) Using the Linearity and Time Shifting properties in Table 2.4.5, express $G(z)$ in terms of $X(z)$. Substitute your answer from part (b) for $X(z)$ to obtain an expression for $G(z)$. Compare this answer for $G(z)$ with your answer for part (d).
(f) Note that $x(n) = \sum_{k=-\infty}^{n} g(n)$. Using the Accumulation property from Table 2.4.5, express $X(z)$ in terms of $G(z)$. Substitute your answer from part (d) for $G(z)$ to obtain an expression for $X(z)$. Compare with your answer to part (b).

2.69 Consider and LTI system with input $x(n)$ and output $y(n)$. Suppose experiments show that the input/output relationship is

$$y(n) = \frac{1}{3}y(n-1) + x(n).$$

(a) Determine the transfer function $H(z)$.
(b) Sketch the pole-zero plot.
(c) Determine the causal output $y(n)$ when the input is

$$x(n) = \begin{cases} 1 & n = 0, 1, 2 \\ 0 & \text{otherwise} \end{cases}.$$

Section 2.9 Exercises

2.70 Consider and LTI system with input $x(n)$ and output $y(n)$. Suppose experiments show that the input/output relationship is

$$y(n) = -\frac{1}{4}y(n-1) + x(n).$$

(a) Determine the transfer function $H(z)$.
(b) Sketch the pole-zero plot.
(c) Determine the causal output $y(n)$ when the input is

$$x(n) = \begin{cases} 1 & 0 \le n \le 2 \\ 0 & \text{otherwise} \end{cases}.$$

2.71 Consider the cascade of two LTI systems $H_1(z)$ and $H_2(z)$ illustrated below with input $x(n)$ and output $y_2(n)$.

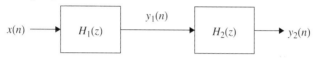

$$H_1(z) = \frac{z^{-1}}{1 - \frac{11}{6}z^{-1} + z^{-1} - \frac{1}{6}z^{-1}} \qquad H_2(z) = 6 - 8z^{-1} + \frac{7}{3}z^{-2}$$

(a) Is $H_1(z)$ FIR, IIR, neither, both?
(b) Is $H_2(z)$ FIR, IIR, neither, both?
(c) Determine the transfer function $H(z)$ for the overall cascade.
(d) Is $H(z)$ FIR, IIR, neither, both?
(e) Write the time-domain difference equation defining the input/output expression.
(f) Sketch the pole-zero plot for $H(z)$.
(g) Determine $h(n)$ corresponding to the causal (stable?) system.
(h) Determine the output $y_2(n)$ when the input $x(n)$ is

$$x(n) = \begin{cases} 1 & 0 \le n \le 2 \\ 0 & \text{otherwise} \end{cases}.$$

2.72 Consider the LTI system shown below with input $x(n)$ and output $y(n)$.

Suppose experiments have shown that the relationship between $x(n)$ and $y(n)$ is

$$y(n) = x(n) - \alpha x(n-8)$$

where $0 < \alpha < 1$.

(a) Determine $H(z)$. Sketch the pole-zero plot for $H(z)$ and specify the possible ROCs.
(b) Now suppose you want to recover $x(n)$ from $y(n)$ using an LTI system $G(z)$ arranged as shown below.

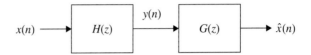

Determine $G(z)$ and identify the ROC corresponding to a causal stable system.
(c) Compute causal stable impulse response $g(n)$.

2.73 For stability reasons, many high-order systems are constructed as cascades of first- and second-order systems. The general form of a second-order system is

$$H(z) = \frac{b_0 + b_1 z^{-1} + b_2 z^{-2}}{1 - a_1 z^{-1} - a_2 z^{-2}}$$

where $b_0, b_1, b_2, a_1,$ and a_2 are real-valued constants. This exercise explores some of the properties of second-order systems.
(a) Write the time-domain difference equation describing the input/output relationship corresponding to $H(z)$. Use $x(n)$ for the input and $y(n)$ for the output.
(b) $H(z)$ has two poles. There are three cases to consider: real and distinct poles, real and repeated poles, and complex conjugate poles.

  i. What conditions must a_1 and a_2 satisfy to produce real distinct poles? What are the poles? Determine the impulse response. (Assume $b_1 = b_2 = 0$.)
  ii. What conditions must a_1 and a_2 satisfy to produce real repeated poles? What are the poles? Determine the impulse response. (Assume $b_1 = b_2 = 0$.)
  iii. What conditions must a_1 and a_2 satisfy to produce complex conjugate poles? What are the poles? Determine the impulse response. (Assume $b_1 = b_2 = 0$.)

(c) The expression for the poles in part (b)-iii was messy. The expression can be cleaned up by expressing each pole in polar form: pole 1 = $re^{j\theta}$ and pole 2 = $re^{-j\theta}$.

  i. Express r and θ in terms of a_1 and a_2.
  ii. Express $B(z) = 1 - a_1 z^{-1} - a_2 z^{-2}$ in terms for r and θ.
  iii. Plot the poles in the z-plane. Describe the easy relationship between the pole locations and the coefficients of the polynomial obtained in the previous part.
  iv. Express the impulse response $h(n)$ in terms of r and θ. (Assume $b_1 = b_2 = 0$.)

(d) The case of real distinct roots looks like the cascade of two first-order systems. In this case, most of the system properties can be deduced from the properties of the first-order system. Thus, the second-order systems of interest are those whose z-transform possesses complex conjugate poles or real and repeated poles. In part (c), an expression for $H(z)$ in terms of r and θ was obtained. This form was much more instructive for the case of complex conjugate poles. Does this form also cover the case of real repeated poles? If so, state the conditions on r and θ that correspond to this case.

2.74 Consider the feedback system shown below with input $x(n)$.

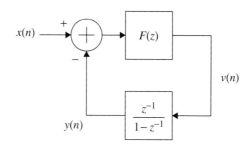

(a) Let $F(z) = 1/2$ and $x(n) = u(n)$. Plot $y(n)$ and $x(n)$ on the same set of axes.
(b) Let $F(z) = 1$ and $x(n) = u(n)$. Plot $y(n)$ and $x(n)$ on the same set of axes.
(c) Let $F(z) = 3/2$ and $x(n) = u(n)$. Plot $y(n)$ and $x(n)$ on the same set of axes.
(d) Compare the answers in parts (a)–(c). What conclusions can you draw?

2.75 Consider the feedback system shown below with input $x(n)$.

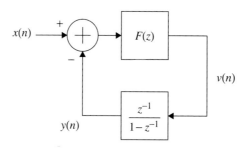

(a) Let $F(z) = 1/2$ and $x(n) = u(n)$. Plot $v(n)$.
(b) Let $F(z) = 1$ and $x(n) = u(n)$. Plot $v(n)$.
(c) Let $F(z) = 3/2$ and $x(n) = u(n)$. Plot $v(n)$.
(d) Compare the answers in parts (a)–(c).
What conclusions can you draw?

2.76 Consider the feedback system shown below.

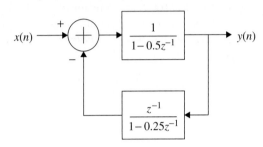

(a) Determine the transfer function $H(z) = \dfrac{Y(z)}{X(z)}$.

(b) Sketch a pole-zero plot of $H(z)$.

(c) Derive the output $y(n)$ when the input is $x(n) = u(n)$.

2.77 Consider the feedback system shown below.

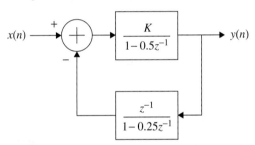

This system is identical to the one in Exercise 2.76, except the constant 0.5 in the feedforward part has been replaced by the constant K. What values for K produce a stable system?

2.78 Consider the feedback system shown below.

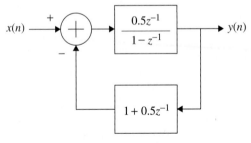

(a) Determine the transfer function $H(z) = \dfrac{Y(z)}{X(z)}$.

(b) Sketch a pole-zero plot of $H(z)$.

(c) Derive the output $y(n)$ when the input is $x(n) = u(n)$.

Section 2.9 Exercises

2.79 Consider the feedback system shown below.

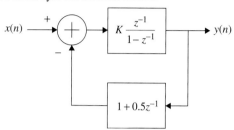

This system is identical to the one in the Exercise 2.78, except the constant 0.5 in the feedforward part has been replaced by the constant K. What values for K produce a stable system?

2.80 An LTI system is determined to have the input/output relationship

$$y(n) = \frac{2}{3}y(n-1) - \frac{2}{3}x(n) + x(n-1).$$

(a) Determine the transfer function $H(z)$.
(b) Sketch the pole-zero plot for $H(z)$.
(c) From $H(z)$ determine $H(e^{j\Omega})$.
(d) Based on the properties of $|H(e^{j\Omega})|^2$, determine whether the LTI system is a low-pass, band-pass or high-pass filter. (It could be that it is none of these.)

2.81 Determine the DTFT of each of the following sequences:
(a) $x(n) = u(n-1) - u(n-2)$
(b) $x(n) = u(n-1) - u(n-4)$
(c) $x(n) = \left(\frac{1}{2}\right)^n \cos\left(\frac{\pi}{8}n - \frac{\pi}{4}\right)$
(d) $x(n) = \begin{cases} 1 & -4 \leq n \leq 4 \\ 0 & \text{otherwise} \end{cases}$
(e) $x(n) = \cos(n)$
(f) $x(n) = \cos\left(\frac{5\pi}{3}\right) + \sin\left(\frac{7\pi}{3}\right)$

2.82 Determine the inverse DTFT for each of the following:

(a) $X(e^{j\Omega}) = \begin{cases} 0 & -\pi \leq \Omega < -\frac{\pi}{8} \\ 1 & -\frac{\pi}{8} \leq \Omega \leq \frac{3\pi}{8} \\ 0 & \frac{3\pi}{8} < \Omega \leq \pi \end{cases}$.

(b) $X(e^{j\Omega}) = e^{j\Omega/2}$.
(c) $X(e^{j\Omega}) = 10 - e^{-j\Omega} + 2e^{-j2\Omega} - 3e^{-j3\Omega}$.

2.83 Consider the signal
$$x(n) = \cos\left(\frac{\pi}{3}n\right) + 2\sin\left(\frac{\pi}{4}n\right).$$
Sketch the DTFT $Y(e^{j\Omega})$ for each of the following:
(a) $y(n) = x(n)\cos\left(\frac{\pi}{8}n\right)$
(b) $y(n) = x(n)e^{j\frac{\pi}{8}n}$
(c) $y(n) = x(n)\cos\left(\frac{\pi}{2}n\right)$
(d) $y(n) = x(n)e^{j\frac{\pi}{2}n}$
(e) $y(n) = x(n)\cos\left(\frac{7\pi}{8}n\right)$
(f) $y(n) = x(n)e^{j\frac{7\pi}{8}n}$.

2.84 Repeat Exercise 2.83, except use for $x(n)$ the sequence with the DTFT shown below.

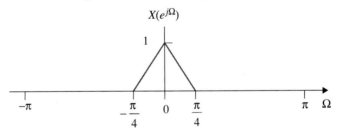

2.85 Consider a system composed of two LTI systems $H_1(e^{j\Omega})$ and $H_2(e^{j\Omega})$ arranged in parallel as shown below.

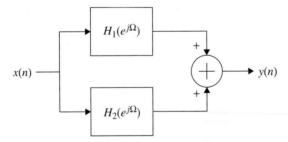

It is known that the impulse response of the upper system is
$$h_1(n) = \left(\frac{1}{3}\right)^n u(n)$$
and that the input/output relationship is
$$12y(n) = 7y(n-1) - y(n-2) - 12x(n) + 5x(n-1).$$
Determine $h_2(n)$.

2.86 Design an LTI system that produces the output

$$y(n) = \left(\frac{1}{3}\right)^n u(n)$$

when the input is

$$x(n) = \left(\frac{1}{2}\right)^n u(n) - \frac{1}{4}\left(\frac{1}{2}\right)^{n-1} u(n-1).$$

Determine the frequency response and impulse response of the system.

2.87 An all-pass system is an important tool in discrete-time signal processing. An all-pass system is defined by $|H(e^{j\Omega})| = 1$ for all Ω. This exercise explores the relationship between $H(z)$ and $H(e^{j\Omega})$ for an all-pass system. An examination of the relationship between the poles and zeros of $H(z)$ and the magnitude of $H(e^{j\Omega})$ shows that each pole of $H(z)$ must be accompanied by a conjugate reciprocal zero. A simple first-order all-pass system has a pole at $z = a$ and a zero at $z = 1/a^*$. There are several possibilities for the form of the first-order all-pass system. For each system below, sketch the pole-zero plot, determine $|H(e^{j\Omega})|^2$, and rewrite $H(z)$ as a polynomial in z^{-1}. Which one is the desired all-pass system?

(a) $H(z) = \dfrac{z - \dfrac{1}{a^*}}{z - a}$

(b) $H(z) = \dfrac{z - a^*}{z - \dfrac{1}{a}}$

(c) $H(z) = a\dfrac{z - \dfrac{1}{a^*}}{z - a}$

(d) $H(z) = \dfrac{a^* z - 1}{z - a}$

2.88 For each of the following systems, determine whether it is a low-pass, band-pass, or high-pass filter.

(a) $H(z) = \dfrac{z^{-1}}{1 + \frac{8}{9}z^{-1}}$

(b) $H(z) = \dfrac{1 + \frac{8}{9}z^{-1}}{1 - \frac{16}{9}z^{-1} + \frac{64}{81}z^{-2}}$

(c) $H(z) = \dfrac{1}{1 + \frac{64}{81}z^{-2}}$

2.89 This exercise explores the relationship between the length of a sequence and the length of its DFT. Consider the length-4 sequence

$$x(n) = \begin{cases} 0 & n < 0 \\ 2 & n = 0 \\ 1 & n = 1 \\ 0 & n = 2 \\ 1 & n = 3 \\ 0 & n > 3 \end{cases}.$$

(a) Sketch $x(n)$.
(b) Compute the DTFT $X(e^{j\Omega})$ and determine $|X(e^{j\Omega})|^2$. Sketch $|X(e^{j\Omega})|^2$ for $-\pi \leq \Omega \leq \pi$.
(c) Sketch a plot of the length-4 periodic extension of $x(n)$.
(d) Compute the length-4 DFT of $x(n)$ directly from the $x(n)$.
(e) Sample $X(e^{j\Omega})$ at $\Omega = 0, \pi/2, \pi, 3\pi/4$. Compare these numbers with those obtained in part (d).
(f) Sketch a plot of the length-8 periodic extension of $x(n)$. You will need to zero-pad $x(n)$ by appending 4 zeros to the end of $x(n)$.
(g) Compute the length-8 DFT of $x(n)$ directly from the $x(n)$.
(h) Sample $X(e^{j\Omega})$ at $\Omega = 0, \pi/4, \pi/2, 3\pi/4, \pi, 5\pi/4, 3\pi/2, 7\pi/4$. Compare these numbers with those obtained in part (g).
(i) Compare your answers from parts (d) and (g). What do you conclude about the relationship between the length-4 and length-8 DFTs?

2.90 Consider the two sequences $x_1(n)$ and $x_2(n)$ shown below.

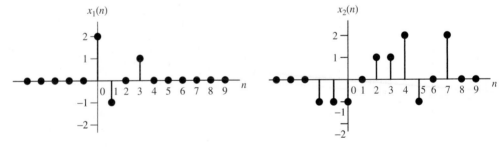

Show that both series have the same length-4 DFT. (Hint: do the analysis in the time domain.)

2.91 Consider the length-8 FFT illustrated in Figure 2.5.3. Observe that the input samples must be reordered to take advantage of this structure. This exercise examines an efficient method for reordering the input samples.

(a) Make a table with three columns. In the first column, write the decimal numbers 0, 1, ..., 7. In the second column, write the 3-bit binary equivalents (treat each index as an unsigned integer). In the third column, write the 3-bit binary equivalents for the indexes in the permuted order required by the FFT.

(b) Based on second and third columns in your table from part (a), suggest an operation for generating indexes in the permuted order. (Hint: most DSP processors have a built-in "bit reversal" operation that reverses the order of the bits in a binary word. Why?)

2.92 Consider the continuous-time, bandlimited band-pass signal $x_c(t)$ with Fourier transform shown in the top plot of Figure 2.6.4. Assume that this signal is sampled at 100 samples/s.

(a) Sketch the DTFT $X_d(e^{j\Omega})$ for $-\pi \leq \Omega \leq \pi$ corresponding to the samples $x_d(n) = x_c(nT)$.

(b) Compare this DTFT with the one shown in the bottom plot of Figure 2.6.4. Note any similarities and differences.

2.93 Consider the continuous-time, bandlimited band-pass signal $x_c(t)$ with Fourier transform shown in the top plot of Figure 2.6.4. Assume that this signal is sampled at 102 samples/s.

(a) Sketch the DTFT $X_d(e^{j\Omega})$ for $-\pi \leq \Omega \leq \pi$ corresponding to the samples $x_d(n) = x_c(nT)$.

(b) Compare this DTFT with the one shown in the bottom plot of Figure 2.6.4. Note any similarities and differences.

2.94 Consider the continuous-time, bandlimited band-pass signal $x_c(t)$ with Fourier transform shown in the top plot of Figure 2.6.5. Assume that this signal is sampled at 52 samples/s.

(a) Sketch the DTFT $X_d(e^{j\Omega})$ for $-\pi \leq \Omega \leq \pi$ corresponding to the samples $x_d(n) = x_c(nT)$.

(b) Compare this DTFT with the one shown in the bottom plot of Figure 2.6.5. Note any similarities and differences.

2.95 Let $x(n)$ be the length-8 sequence defined by

$$x(n) = \begin{cases} 0 & n < 0 \\ \cos\left(\frac{\pi}{4}n\right) & 0 \leq n \leq 7 \\ 0 & n \geq 8 \end{cases}.$$

(a) Derive an expression for $X(e^{j\Omega})$, the DTFT of $x(n)$. Plot $|X(e^{j\Omega})|$.

(b) Compute the length-8 DFT of $x(n)$. Compare this answer with $X(e^{j\Omega})$ from part (a).

(c) Compute the length-16 DFT of $x(n)$ by zero-padding $x(n)$. Compare this answer with $X(e^{j\Omega})$ from part (a).

2.96 Let $x(n)$ be the length-8 sequence defined by

$$x(n) = \begin{cases} 0 & n < 0 \\ \cos\left(\frac{\pi}{4}n\right) & 0 \leq n \leq 7 \\ 0 & n \geq 8 \end{cases}.$$

and let $y(n)$ be the length-8 sequence defined by

$$y(n) = \begin{cases} 0 & n < 0 \\ \cos\left(\frac{\pi}{3}n\right) & 0 \leq n \leq 7 \\ 0 & n \geq 8 \end{cases}.$$

(a) Compute the length-8 DFT of $x(n)$ and sketch a plot of the magnitude of this DFT.
(b) Compute the length-8 DFT of $y(n)$ and sketch a plot of the magnitude of this DFT.
(c) Compare your answers from parts (a) and (b). Explain the differences.

2.97 Consider the real-valued signal $s(t)$ whose Fourier transform is

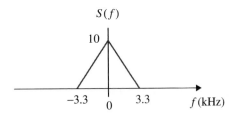

Sketch the DTFT of the sequence resulting from sampling $s(t)$ at 8 ksamples/s.

2.98 Consider the real-valued signal $s(t)$ whose Fourier transform is

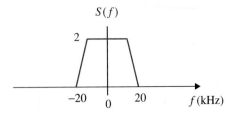

(a) Sketch the DTFT of the sequence resulting from sampling $s(t)$ at 48 ksamples/s.
(b) Sketch the DTFT of the sequence resulting from sampling $s(t)$ at 44.1 ksamples/s.
(c) Compare the DTFTs obtained in parts (a) and (b).

Section 2.9 Exercises

2.99 Consider the real-valued signal $s(t)$ whose Fourier transform is

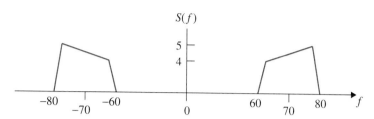

Sketch the DTFT of the sequence resulting from sampling $s(t)$ at 100 samples/s.

2.100 Consider the signal $s(t)$ described in Exercise 2.99.
(a) Sketch the DTFT of the sequence resulting from sampling $s(t)$ at 56 samples/s.
(b) Sketch the DTFT of the sequence resulting from sampling $s(t)$ at 40 samples/s.
(c) Compare the DTFTs obtained in parts (a) and (b).

2.101 This exercise explores the relationship between band-pass sampling and frequency translation.
(a) Consider the complex-valued signal $\tilde{x}(t)$ whose Fourier transform is

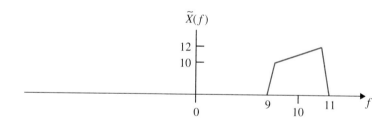

Suppose the signal is to be translated to baseband and sampled. The block diagram of a system that accomplishes this is illusrated below.

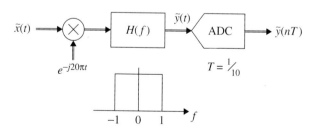

Sketch $\tilde{Y}(f)$, the Fourier transform of $\tilde{y}(t)$, and $\tilde{Y}(e^{j\Omega})$, the DTFT of the sequence $\tilde{y}(nT)$.

(b) Show that the system illustrated below accomplishes the same task as the system in part (a) by showing that the DTFT of the sequence $\tilde{z}(nT)$ is the same as the DTFT of the sequence $\tilde{y}(nT)$ from part (a).

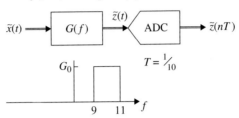

(c) Now let $x(t)$ be the real-valued version of $\tilde{x}(t)$. Fourier transform of $x(t)$ is illustrated below.

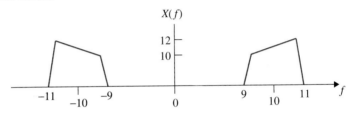

Motivated by the system illustrated in part (b), the system illustrated below is proposed as one capable of producing the desired samples $z(nT)$.

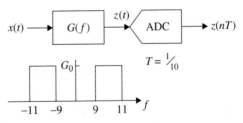

Sketch the DTFT of the sequence $z(nT)$. Show that this procedure does not produce the desired samples. What is the fundamental problem with this approach? What general conclusions can be drawn?

2.102 A continuous-time band-pass signal $s(t)$ may be represented as the sum of two amplitude modulated carriers that are 90° out of phase:

$$s(t) = I(t)\cos(\omega_c t) - Q(t)\sin(\omega_c t).$$

A common requirement in many communication systems is to produce samples of $I(t)$ and $Q(t)$ from samples of $s(t)$. This exercise explores the use of discrete-time processing to perform the required task. Suppose $s(t)$ is the bandpass signal centered at $f_c = 70$ Hz as shown below.

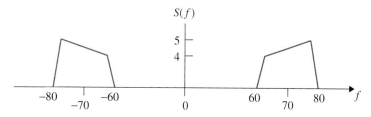

The system used to obtain samples of $I(t)$ and $Q(t)$ is shown below.

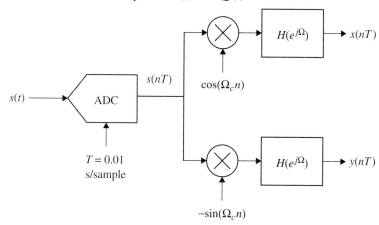

The bandpass signal $s(t)$ is sampled at 100 samples/s, multiplied by discrete-time sinusoids in parallel, and filtered to produce the outputs $x(nT)$ and $y(nT)$. Determine the frequency Ω_c and the frequency response $H(e^{j\Omega})$ required to make $x(nT) = I(nT)$ and $y(nT) = Q(nT)$. (Note: this exercise is a trick problem. A small adjustment must be made to the system to obtain the desired result. If you are not careful, you will overlook this.)

2.103 CD-quality music is a bandlimited signal whose Fourier transform is zero for frequencies greater than 20 kHz. Commercial broadcast AM standards in the United States require the modulating signal to be bandlimited to 5 kHz. Thus, a commercial broadcast AM radio station must low-pass filter a signal with CD quality before modulating. This low-pass filtering could be performed using either continuous-time processing or discrete-time processing.

(a) Consider the case of using continuous-time processing to perform the low-pass filtering shown below.

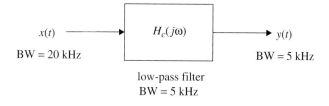

The input signal $x(t)$ is a bandlimited continuous-time signal with a 20 kHz bandwidth. Design the requirements for $H_c(j\omega)$, an ideal low-pass filter.

(b) Consider the case of using discrete-time processing to perform the low-pass filtering shown below.

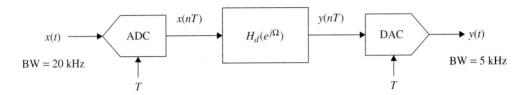

The input signal $x(t)$ is a bandlimited continuous-time signal with a 20 kHz bandwidth that is sampled at a rate $1/T$. The discrete-time low-pass filter $H_d\left(e^{j\Omega}\right)$ is the discrete-time equivalent of the ideal low-pass filter derived in part (a). Design the requirements for $H_d\left(e^{j\Omega}\right)$.

(c) What is the lowest sample rate that can be used to produce the desired output?

2.104 Consider the design of a discrete-time system that produces a time delay that is a fraction of the sample time as illustrated below.

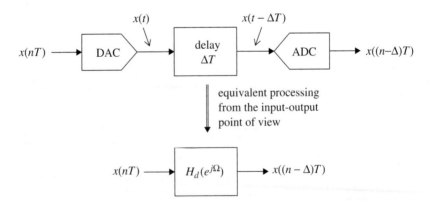

A bandlimited fractional sample delay could be realized using continuous-time processing with a DAC–ADC pair. However, if the system is to operate over a finite bandwidth $-W \leq \omega \leq +W$, then the entire operation can by performed in discrete-time processing using the technique outlined in Section 2.6.2.

(a) Derive the transfer function $H_c(j\omega)$ that produces a time delay of ΔT.

(b) Assuming that this transfer function operates over a finite bandwidth $-W \leq \omega \leq +W$, derive the transfer function of the equivalent discrete-time system $H_d(e^{j\Omega})$.

(c) Compute the inverse DTFT of $H_d\left(e^{j\Omega}\right)$ to produce the desired impulse response $h_d(n)$.

(d) An important special case is a fractional delay system that operates over the entire bandwidth. What is the impulse response $h_d(n)$ for this case?

(e) Another important special case is a 1/2-sample delay that operates over the entire bandwidth. What is the impulse response $h_d(n)$ for this case?

3

Signals and Systems 2: Some Useful Discrete-Time Techniques for Digital Communications

3.1 INTRODUCTION 114
3.2 MULTIRATE SIGNAL PROCESSING 115
3.3 DISCRETE-TIME FILTER DESIGN METHODS 127
3.4 NOTES AND REFERENCES 159
3.5 EXERCISES 159

3.1 INTRODUCTION

The review of continuous-time and discrete-time signals and systems outlined in the previous chapter summarized topics routinely covered in junior-level courses in signals and systems. There are many discrete-time techniques that are widely used in the processing of samples of a digitally modulated carrier. This chapter reviews some of these techniques to provide a foundation for the topics outlined in Chapters 9 and 10 and Appendices A and C.

This chapter begins with the important concept of *multirate signal processing* by reviewing the basic operations of downsampling and upsampling. These operations are characterized in both the time and frequency domains. The polyphase filterbank implementations of downsampling and upsampling are also described. Section 3.3 summarizes discrete-time *filter design techniques*. It is not the intent to provide a complete description of all the details of discrete-time filter design, but rather to familiarize the reader with the terminology, design criteria, and basic steps involved in some of the more common techniques. This discussion is included because some of the more novel discrete-time techniques used in digital communication follow directly from discrete-time filter design. The chapter ends with the design of two very important filters in digital communications: the discrete-time differentiator and the discrete-time integrator.

3.2 MULTIRATE SIGNAL PROCESSING

Multirate processing is unique to discrete-time processing. The two basic operations of multirate processing are *downsampling* and *upsampling*. In general, these operations are special cases of *resampling*. Resampling essentially rescales the time axis which, in turn, scales the frequency axis. The consequences of this rescaling are derived in the following sections.

3.2.1 Impulse-Train Sampling

Consider the discrete-time signal $x(n)$ with DTFT $X(e^{j\Omega})$ as illustrated in Figure 3.2.1. Samples of the discrete-time sequence $x(n)$ may be obtained by multiplying $x(n)$ by the discrete-time impulse train

$$p(n) = \sum_{n=-\infty}^{\infty} \delta(n - kN) \tag{3.1}$$

with DTFT

$$P(e^{j\Omega}) = \frac{2\pi}{N} \sum_{k=-\infty}^{\infty} \delta\left(\Omega - k\frac{2\pi}{N}\right). \tag{3.2}$$

Examples of $p(n)$ and $P(e^{j\Omega})$ for $N = 4$ and $N = 10$ are illustrated in Figures 3.2.1 and 3.2.2, respectively. The desired sampled sequence may be expressed as the product of the original sequence $x(n)$ and the impulse train $p(n)$:

$$x_p(n) = x(n)p(n) \tag{3.3}$$

which is related to $x(n)$ by

$$x_p(n) = \begin{cases} x(n) & n = \text{integer} \times N \\ 0 & \text{otherwise} \end{cases}. \tag{3.4}$$

The DTFT of $x_p(n)$ may be obtained using frequency domain convolution along with the sifting property for impulses:

$$X_p(e^{j\Omega}) = \frac{1}{2\pi} \int_{-\pi}^{\pi} P(e^{j\theta}) X(e^{j(\Omega-\theta)}) d\theta$$

$$= \frac{1}{N} \sum_{k=0}^{N-1} X\left(e^{j\left(\Omega - k\frac{2\pi}{N}\right)}\right). \tag{3.5}$$

This shows that sampling $x(n)$ by taking every N-th sample produces N copies of $X(e^{j\Omega})$ evenly spaced in the interval $-\pi \leq \Omega \leq \pi$. The amplitude of each copy is scaled by $1/N$. This is

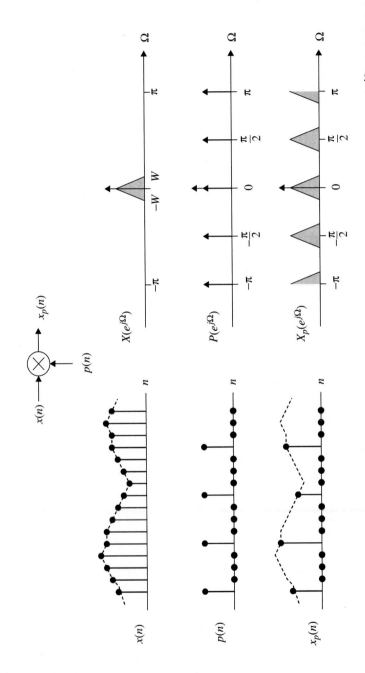

Figure 3.2.1 Discrete-time impulse sampling for $N = 4$: (top) A discrete-time sequence $x(n)$ and its DTFT $X(e^{j\Omega})$; (middle) the discrete-time impulse train $p(n)$ and its DTFT $P(e^{j\Omega})$; (bottom) the sampled discrete-time signal $x_p(n)$ and its DTFT $X_p(e^{j\Omega})$. Note that N is sufficiently small that no aliasing occurs. Compare with Figure 3.2.2.

Section 3.2 Multirate Signal Processing

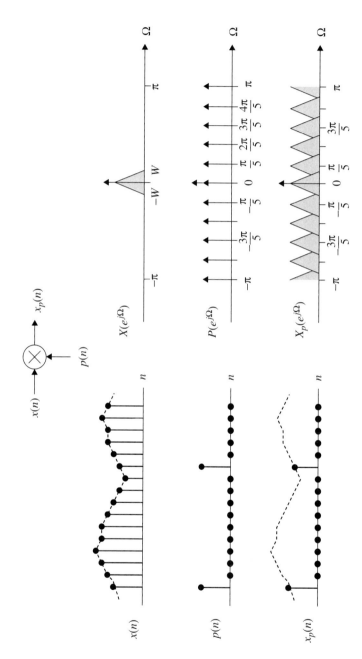

Figure 3.2.2 Discrete-time impulse sampling for $N = 10$: (top) A discrete-time sequence $x(n)$ and its DTFT $X(e^{j\Omega})$; (middle) the discrete-time impulse train $p(n)$ and its DTFT $P(e^{j\Omega})$; (bottom) the sampled discrete-time signal $x_p(n)$ and its DTFT $X_p(e^{j\Omega})$. Note that N is too large to avoid aliasing. Compare with Figure 3.2.1.

illustrated in Figure 3.2.1 for $N = 4$ and in Figure 3.2.2 for $N = 10$. Observe that in the example using $N = 4$, the spectral copies of $X(e^{j\Omega})$ do not overlap and no aliasing results from the sampling operation. This is not the case for the example using $N = 10$. The spectral copies of $X(e^{j\Omega})$ overlap, thereby producing aliasing. Even in discrete-time processing, undersampling can produce aliasing similar to the way undersampling a continuous-time signal produces aliasing. Close investigation of the spectra in Figures 3.2.1 and 3.2.2 shows that in order to avoid aliasing, the parameter N must satisfy

$$N < \frac{\pi}{W}. \tag{3.6}$$

3.2.2 Downsampling

Downsampling the sequence $x(n)$ by N is the process of decreasing the sampling rate by N by retaining every N-th sample of $x(n)$ to produce a new sequence $x_D(m)$. This process is illustrated in Figure 3.2.3 Because each sample increment in $x_D(m)$ steps through $x(n)$ N samples at a time, the two are related by the mathematical expression

$$x_D(m) = x(mN). \tag{3.7}$$

Note that because only every N-th sample of $x(n)$ is of interest, $x_p(n)$, given by (3.3), could also be downsampled (at the proper starting point) to obtain the same sequence $x_D(m)$. The DTFT of $x_D(m)$ is obtained as follows:

$$X_D(e^{j\Omega}) = \sum_{m=-\infty}^{\infty} x_d(m) e^{-j\Omega m} = \sum_{m=-\infty}^{\infty} x_p(mN) e^{-j\Omega m}. \tag{3.8}$$

Substituting $m = n/N$, the summation may be reexpressed as

$$X_D(e^{j\Omega}) = \sum_{n=\text{integer} \times N} x_p(n) e^{-j\Omega n/N} = \sum_{n=-\infty}^{\infty} x_p(n) e^{-j\Omega n/N} = X_p(e^{j\Omega/N})$$

$$= \frac{1}{N} \sum_{k=0}^{N-1} X\left(e^{j\left(\frac{\Omega - k 2\pi}{N}\right)}\right). \tag{3.9}$$

The relationship (3.9) defines the following procedure to produce $X_D(e^{j\Omega})$ from $X(e^{j\Omega})$:

1. Draw N shifted copies of $X(e^{j\Omega})$, each shifted by $2\pi k/N$ for $k = 0, 1, \ldots, N - 1$.
2. Add the shifted copies together and scale the amplitude by $1/N$.
3. Stretch the frequency axis by N by multiplying each of the "tick marks" on the frequency axis by N.

Technically, the Ω in $X(e^{j\Omega})$ is different from the Ω in $X_D(e^{j\Omega})$, because the sample rate for $x(n)$ and $x_D(m)$ is different. The notation only hints that this might be the case. This difference manifests itself by "stretching" the spectrum of $x(n)$ by a factor of N as illustrated in Figure 3.2.3.

Section 3.2 Multirate Signal Processing

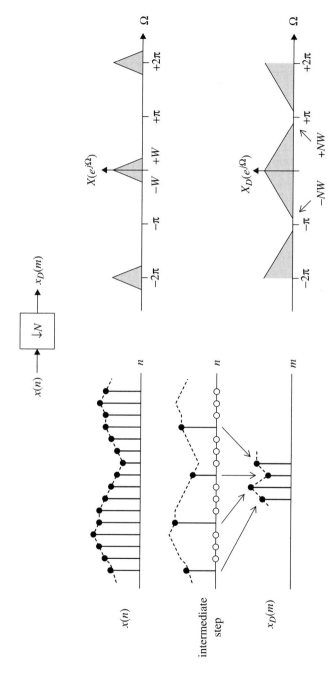

Figure 3.2.3 The downsampling operation in the time domain (left) and frequency domain (right).

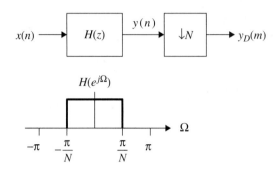

Figure 3.2.4 A block diagram of the filter-downsample operation.

Because the bandwidth of the signal, relative to the sample rate, increases with downsampling, filtering is often used to limit the spectrum of $x(n)$ to avoid spectral aliasing in $x_D(m)$. The basic system is shown in Figure 3.2.4. The filter must eliminate the frequency components that overlap as a result of the downsample operation. Close inspection of the spectra in Figure 3.2.3 shows that an ideal low-pass filter with bandwidth π/N performs the task. The frequency response of the ideal downsampling filter is illustrated in Figure 3.2.4.

3.2.3 Upsampling

Upsampling the sequence $x(n)$ by N is the process of increasing the sample rate by N by inserting $N - 1$ zeros between each sample of $x(n)$ to produce a new sequence $x_U(m)$. This procedure is illustrated in Figure 3.2.5 The mathematical relationship between $x(n)$ and $x_U(m)$ is

$$x_U(m) = \begin{cases} x\left(\dfrac{m}{N}\right) & m = \text{integer} \times N \\ 0 & \text{otherwise} \end{cases}. \tag{3.10}$$

Note that $x_U(m)$ looks like $x_p(m)$ given by (3.3). In other words, if $x(n)$ is a downsampled version of a fictitious, high-rate version $xx(m)$, then $x_U(m)$ can be thought of as the product $xx_p(m) = xx(m)p(m)$. Applying the relationship (3.5) to $xx_p(n)$ produces

$$X_U(e^{j\Omega}) = XX_p(e^{j\Omega}) = \frac{1}{N}\sum_{k=0}^{N-1} XX\left(e^{j\left(\Omega - k\frac{2\pi}{N}\right)}\right). \tag{3.11}$$

Applying the downsampling result (3.9) establishes the relationship between the DTFTs of $xx(m)$ and $x(n)$:

$$X(e^{j\Omega}) = \frac{1}{N}\sum_{k=0}^{N-1} XX\left(e^{j\left(\frac{\Omega - k2\pi}{N}\right)}\right) \Rightarrow X(e^{j\Omega N}) = \frac{1}{N}\sum_{k=0}^{N-1} XX\left(e^{j\left(\Omega - k\frac{2\pi}{N}\right)}\right). \tag{3.12}$$

Section 3.2 Multirate Signal Processing

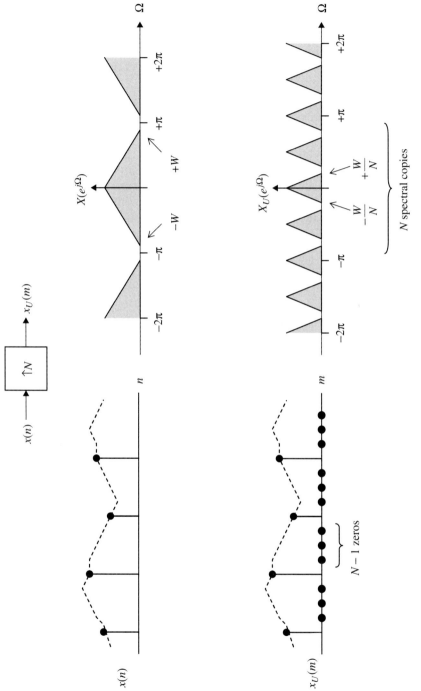

Figure 3.2.5 The upsampling operation in the time domain (left) and frequency domain (right).

Putting this all together produces the desired result:
$$X_U(e^{j\Omega}) = X(e^{j\Omega N}). \tag{3.13}$$
This relationship defines the following procedure for obtaining $X_U(e^{j\Omega})$ from $X(e^{j\Omega})$:
1. Starting with $X(e^{j\Omega})$, compress the frequency axis by N by dividing all the "tick marks" on the frequency axis by N.
2. Draw $N-1$ shifted copies of the compressed spectrum. Each copy is shifted by $2\pi k/N$ for $k = 1, 2, \ldots, N-1$.

As was the case for the downsample operation, the Ω in $X(e^{j\Omega})$ is different from the Ω in $X_U(e^{j\Omega})$ due to the different sample rates of $x(n)$ and $x_U(m)$. This difference manifests itself as a compression of the Ω axis by a factor N. As a consequence, N copies of the compressed spectrum of $x(n)$ alias into the first Nyquist zone as illustrated in Figure 3.2.5.

Often, filtering is used to "fill in" the zeros between each nonzero sample of the original signal. This filtering process is informally known as *interpolation*. As illustrated in Figure 3.2.6, a low-pass filter with cutoff frequency greater than W/N but less than $2/N\pi - W/N$ isolates the baseband replica of the compressed spectral copies to produce a sample sequence that looks like a high-rate sampling of the underlying continuous-time waveform.

3.2.4 The Noble Identities

The Noble identities capture an important relationship for upsampling and downsampling. These identities are illustrated in Figure 3.2.7 for downsampling and upsampling. These relationships show that when the z-transform of the filter is a polynomial in z^N, it may be exchanged as shown. The power of this result is that the filtering operations of this form may be placed at the low-clock-rate of each operation (on the output side of the downsample operation and the input side of the upsample operation) without changing the signal processing that occurs. This allows the filters to operate at the lowest sample rate. The Noble identities are exploited in Section 3.2.5 to derive the polyphase filterbank.

3.2.5 Polyphase Filterbanks

An efficient method for performing the filter-downsample operation can be derived with the aid of the Noble identities. The resulting structure is known as polyphase filterbank. The starting point is the filter-downsample system illustrated in Figure 3.2.4. This filter is described by the z-transform
$$H(z) = h(0) + h(1)z^{-1} + h(2)z^{-2} + \cdots$$
The output of the filter is
$$y(n) = \sum_k h(k)x(n-k)$$
and the downsampled signal is
$$y_D(m) = y(mD) = \sum_k h(k)x(mD-k).$$

Section 3.2 Multirate Signal Processing

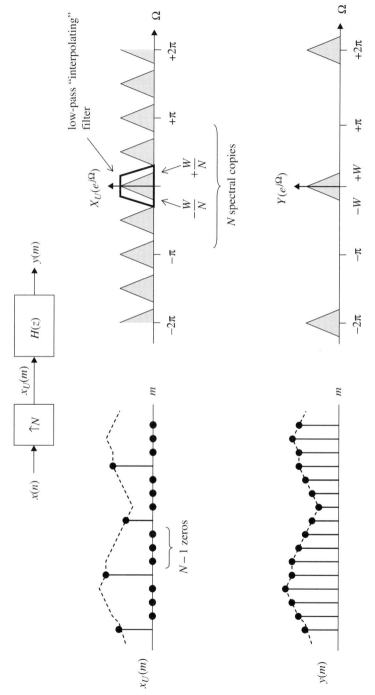

Figure 3.2.6 The upsample-filter operation in the time domain (left) and frequency domain (right).

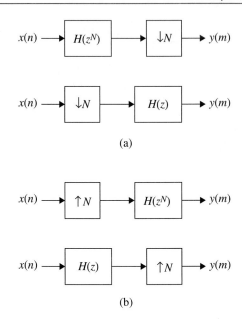

Figure 3.2.7 The Noble identities: (a) For downsampling; (b) for upsampling.

In this formulation, one filter output sample is computed for each filter input sample. Every $N-1$ of the filter output samples, however, will be discarded by the downsampling operation. There is no need to even compute these samples. This characteristic can easily be accommodated by writing $H(z)$ as

$$
\begin{aligned}
H(z) = \quad & h(0) & + \ & h(N)z^{-N} & + \ & h(2N)z^{-2N} & + \cdots \\
+ \ & h(1)z^{-1} & + \ & h(N+1)z^{-N-1} & + \ & h(2N+1)z^{-2N-1} & + \cdots \\
+ \ & h(2)z^{-2} & + \ & h(N+2)z^{-N-2} & + \ & h(2N+2)z^{-2N-2} & + \cdots \\
& \vdots \\
+ \ & h(N-1)z^{-N+1} & + \ & h(2N-1)z^{-2N+1} & + \ & h(3N-1)z^{-3N+1} & + \cdots
\end{aligned}
$$
(3.14)

Each row of (3.14) consists of every N-th tap of the filter, but with a different offset. Each row defines a subfilter:

$$H_0(z^N) = h(0) + h(N)z^{-N} + h(2N)z^{-2N} + \cdots$$
$$H_1(z^N) = h(1) + h(N+1)z^{-N} + h(2N+1)z^{-2N} + \cdots$$

$$H_2(z^N) = h(2) + h(N+2)z^{-N} + h(2N+2)z^{-2N} + \cdots \qquad (3.15)$$

$$\vdots$$

$$H_{N-1}(z^N) = h(N-1) + h(2N-1)z^{-N} + h(3N-1)z^{-2N} + \cdots$$

Using the definitions in (3.15), $H(z)$ can be expressed in terms of every N-th filter tap as

$$H(z) = H_0(z^N) + z^{-1}H_1(z^N) + z^{-2}H_2(z^N) + \cdots + z^{-N+1}H_{N-1}(z^N). \qquad (3.16)$$

A system based on the partition (3.16) is illustrated in Figure 3.2.8 (a). The downsample operation commutes with the addition operation and may be moved to the input side of the subfilters by replacing z^N with z as illustrated in Figure 3.2.8 (b). Note that the delay operator $z-1$ on the input side of the downsample operations is referenced to the high sample rate, whereas the delay operator $z-1$ in the subfilters is referenced to the low sample rate. The delay-downsample operations can be realized with a commutator as shown in Figure 3.2.8 (c).

The filter structure shown in Figure 3.2.8 (c) is known as a *polyphase filterbank* and provides a computationally efficient method for performing the filter-downsample combination. The data arrive at the input to the filterbank at the high sample rate. The commutator presents every N-th sample, with appropriate offset, to each of the subfilters in the filterbank. Thus, each subfilter operates at the low sample rate. In this way, clock rate and area can be exchanged.

The polyphase filter partition can also be applied to upsample-filter (or upsample-interpolate) function so that the filtering can operate at the lower clock rate. The output $y(m)$ is the convolution of $x_U(m)$ with the filter impulse response $h(m)$:

$$y(m) = \sum_k h(k) x_U(m-k). \qquad (3.17)$$

The polyphase decomposition follows from the fact that not all of the multiplications suggested by (3.17) are required. This is due to the fact that only every N-th sample of $x_U(m)$ is nonzero. When the index m is a multiple of N, the nonzero values of $x_U(m)$ coincide with the filter taps

$$h(0) \quad h(N) \quad h(2N) \quad \cdots$$

so that the filter output may be expressed as

$$y(m) = \sum_k h(kN) x\left(\frac{m}{N} - kN\right).$$

At the next sample index, $m+1$, the nonzero values of $x_U(m)$ coincide with the filter taps

$$h(1) \quad h(N+1) \quad h(2N+1) \quad \cdots$$

so that the filter output may be expressed as

$$y(m+1) = \sum_k h(kN+1) x\left(\frac{m-1}{N} - kN\right).$$

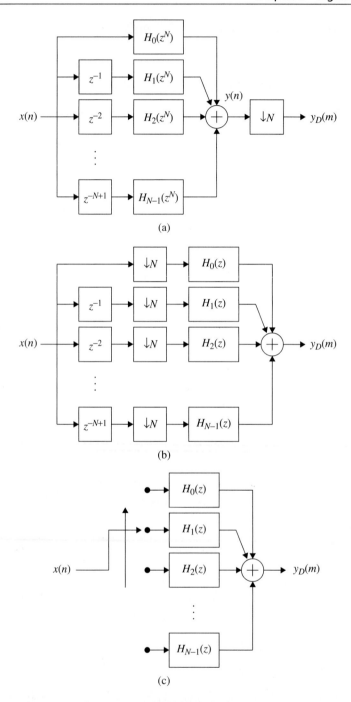

Figure 3.2.8 Polyphase filterbank development for the filter-downsample operation. (a) Basic operations defined by convolution; (b) result of moving the downsample operation from the output to the input of the filterbank; (c) result of replacing the delay-downsample operation with a commutator.

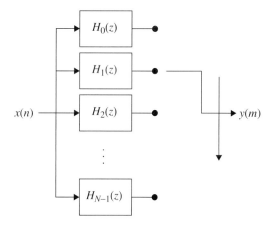

Figure 3.2.9 The upsample-filter (or upsample-interpolate) operation performed by a polyphase filterbank.

Continuing on to the index $m + r$, the nonzero values of $x_U(m)$ coincide with the filter taps

$$h(r) \quad h(N + r) \quad h(2N + r) \quad \cdots$$

so that the filter output may be expressed as

$$y(m + r) = \sum_k h(kN + r) x\left(\frac{m - r}{N} - kN\right).$$

This shows that successive filter outputs at the high sample rate use every N-th filter tap of $H(z)$ but with a different offset. Using the polyphase partition (3.14) and the subfilter definitions (3.15), the upsample-filter (or upsample-interpolate) function may be realized by the system shown in Figure 3.2.9. The data are clocked into the filterbank at the low sample rate. N samples are produced in parallel by the subfilters operating at the low sample rate. A commutator, operating at the high sample rate, clocks the N available subfilter outputs into the output for each sample at the input.

3.3 DISCRETE-TIME FILTER DESIGN METHODS

The development of good filter design techniques is an important component in the history of discrete-time signal processing. Initially, the development focused on infinite impulse response (IIR) filters derived, via an appropriate transformation, from a continuous-time filter. This approach leveraged the well-established continuous-time filter design techniques[1] that produced stable filters using relatively simple design formulas.

---

[1] Because these continuous-time approaches were essentially pole-placement procedures in the s-plane, the resulting filters are IIR systems. In fact, it is quite challenging to construct a continuous-time finite impulse response (FIR) filter.

Later, the design of finite impulse response (FIR) discrete-time filters became popular. Because discrete-time FIR filters have no continuous-time counterpart, the transformation techniques used to produce discrete-time IIR filters could not be used and new techniques had to be developed.

Discrete-time filters are defined by the parameters illustrated in Figure 3.3.1 for the case of a low-pass filter. The frequency response consists of pass-band, a transition band, and a stop-band. The ideal, or desired, filter has a pass-band amplitude of one and a stop-band amplitude of zero. The transition band is the bandwidth required for the frequency response to transition from unity gain to zero gain and is required of all realizable filters. These filter design parameters are

W_p = the edge of the pass band. The pass band starts at $\Omega = 0$ and extends to $\Omega = W_p$. The bandwidth of the low-pass filter is W_p.

W_s = the edge of the stop band. The stop band starts at $\Omega = W_s$ and extends to $\Omega = \pi$. Note that the transition band is defined by W_p and W_s. The width of the transition band is $W_s - W_p$.

δ_p = the pass-band ripple. The pass-band ripple measures the error between the frequency response of the desired filter and the frequency response of the actual filter in the pass band. This parameter is usually specified in decibels.

δ_s = the stop-band ripple. The stop-band ripple measures the error between the frequency response of the desired filter and the actual filter in the stop band. Because the frequency response of the desired filter is zero in the stop band, the stop-band ripple measures the stop-band attenuation as shown.

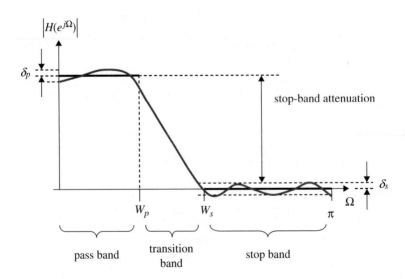

Figure 3.3.1 Frequency domain parameters specifying the performance of a discrete-time low-pass filter.

Section 3.3 Discrete-Time Filter Design Methods

These filter design parameters are used by filter design techniques to produce discrete-time filters that approximate the desired frequency response.

Inspired by the historical order of development, discrete-time IIR filter design is summarized in Section 3.3.1 followed by a summary of discrete-time FIR filter design in Section 3.3.2.

3.3.1 IIR Filter Designs

Basic Structure. The most important class of discrete-time IIR filters are those whose transfer function $H(z)$ is of the form

$$H(z) = \frac{B(z)}{A(z)} = \frac{b_0 + b_1 z^{-1} + \cdots + b_q z^{-q}}{1 + a_1 z^{-1} + \cdots + a_p z^{-p}}. \tag{3.18}$$

Note the convention that $a_0 = 1$. The roots of the polynomial $B(z)$ are the zeros of the filter, and the roots of the polynomial $A(z)$ are the poles of the filter. If the input to the filter, $x(n)$, has z-transform $X(z)$, then the filter output, $y(n)$, has z-transform $Y(z) = H(z)X(z)$. Expressing $H(z)$ in terms of the polynomials $A(z)$ and $B(z)$ produces the relationship

$$Y(z)A(z) = X(z)B(z)$$
$$Y(z)[1 + a_1 z^{-1} + \cdots + a_p z^{-p}] = X(z)[b_0 + b_1 z^{-1} + \cdots + b_q z^{-q}] \tag{3.19}$$
$$Y(z) + a_1 z^{-1} Y(z) + \cdots + a_p z^{-p} Y(z) = b_0 X(z) + b_1 z^{-1} X(z) + \cdots + b_q z^{-q} X(z).$$

From this, the time-domain relationship follows:

$$y(n) + a_1 y(n-1) + \cdots + a_p y(n-p) = b_0 x(n) + b_1 x(n-1) + \cdots + b_q x(n-q). \tag{3.20}$$

Writing the time-domain relationship as

$$y(n) = -a_1 y(n-1) - \cdots - a_p y(n-p) + b_0 x(n) + b_1 x(n-1) + \cdots + b_q x(n-q) \tag{3.21}$$

shows that filters of this form are *recursive*: the current output depends on the previous outputs. This recursive property is what makes the impulse response infinite in length. Note that the poles in $H(z)$ are the source of the recursive nature of the filter.

The most common basic realizations of IIR filters are illustrated by the structures in Figures 3.3.2 and 3.3.3. The realization shown in Figure 3.3.2 (a) is referred to as the direct form I because it is the result of a direct realization of (3.20) or (3.21). The realization shown in Figure 3.3.2 (b) is referred to as the direct form II and is a canonical form for an IIR filter (or any discrete-time LTI systems with poles and zeros) because it requires the fewest number of memory elements. These realizations can also be thought of as weighted signal graphs. The rich theory of signal flow graph transformations can be applied to produce some appealing alternatives to the realizations of Figure 3.3.2. The most popular of these transformations are the "transposed" forms of the direct form I and direct form II realizations. These transposed direct form I and transposed direct form II realizations are illustrated in Figures 3.3.3 (a) and (b), respectively. The direct form I and transposed direct form I realize the zeros of the filter

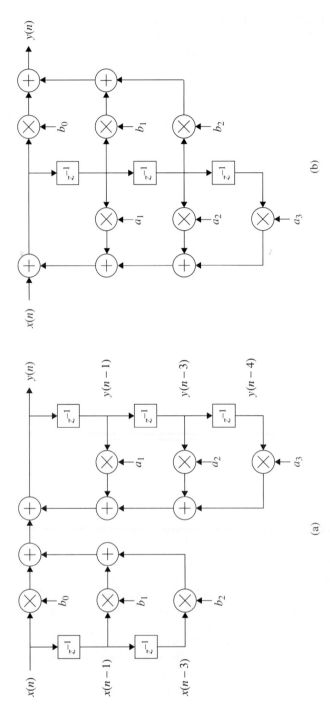

Figure 3.3.2 Direct form I (a) and direct form II (b) realizations of an IIR filter with $q = 3$ zeros and $p = 4$ poles.

Section 3.3 Discrete-Time Filter Design Methods

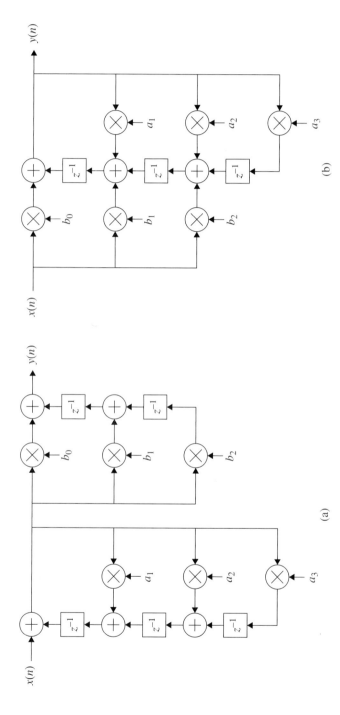

Figure 3.3.3 Transposed direct form I (a) and transposed direct form II (b) realizations of an IIR filter with $q = 3$ zeros and $p = 4$ poles.

first. The processing corresponding to the poles follows. The direct form II and transposed direct form II first realize the poles of the filter, followed by the realization of the zeros. Although mathematically equivalent, these differences can have a profound impact on the performance and stability of the filter in the presence of quantization noise due to finite precision arithmetic.

Impulse Invariance. The impulse invariance filter design technique starts with the design of a continuous-time filter, with impulse response $h_c(t)$, that meets the desired specifications. The discrete-time impulse response, $h_d(n)$, is obtained from $h_c(t)$ by sampling and scaling $h_c(t)$:

$$h_d(n) = Th_c(t)|_{t=nT}. \qquad (3.22)$$

Applying the principles of Section 2.6, the relationship between $H_c(j\omega)$ and $H_d(e^{j\Omega})$ is

$$H_d(e^{j\Omega}) = \frac{1}{T} \sum_{k=-\infty}^{\infty} H_c\left(j\left(\frac{\Omega - k2\pi}{T}\right)\right). \qquad (3.23)$$

This relationship is shown in Figure 3.3.4 for the case of a low-pass filter. Figure 3.3.4 illustrates the difficulty with this approach. All practical continuous-time filters are not strictly bandlimited. As a consequence, some aliasing occurs, as shown. Usually the sampling rate is selected to be high enough to render the distortion due to this aliasing negligibly small.

Note that if $H_c(j\omega)$ is truly bandlimited, that is, $H_c(j\omega) = 0$ for ω outside the interval $-W_c \leq \omega \leq W_c$, then it is possible to entirely eliminate the aliasing by selecting $T < \pi/W_c$.

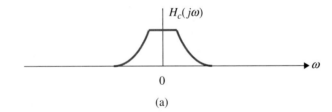

(a)

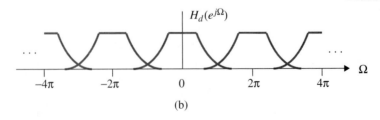

(b)

Figure 3.3.4 A graphical representation of the relationship (3.23), the relationship between the continuous-time prototype filter $H_c(j\omega)$ and the discrete-time filter $H_d(e^{j\Omega})$ using the impulse invariance technique.

Section 3.3 Discrete-Time Filter Design Methods

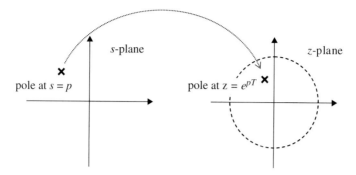

Figure 3.3.5 An illustration of the pole mapping resulting from the impulse invariance technique. A pole of the continuous-time prototype filter at $s = -p$ in the s-plane maps to a pole of the discrete-time filter at $z = e^{-pT}$ in the z-plane.

In this case,

$$H_d(e^{j\Omega}) = \frac{1}{T} H_c\left(j\left(\frac{\Omega}{T}\right)\right) \quad \text{for} \quad -\pi \leq \Omega \leq \pi. \tag{3.24}$$

The continuous-time prototype filter and the resulting discrete-time filter are related by a linear scaling of the frequency axis.

A causal, stable continuous-time prototype filter produces a causal, stable discrete-time filter under impulse invariance. This is a consequence of the way the impulse invariance technique maps a pole in the s-plane to a pole in the z-plane. A pole of $H_c(s)$ at $s = p$ in the s-plane maps to a pole of $H_d(z)$ at $z = e^{pT}$ in the z-plane. This mapping is illustrated in Figure 3.3.5 and derived in Exercise 3.31. A causal, stable continuous-time filter only has poles in the left-half of the s-plane. Thus, each pole is of the form $p = -\sigma + j\omega$ with $\sigma > 0$. The corresponding pole in the z-plane is

$$e^{pT} = e^{-\sigma T} e^{j\omega T}.$$

The magnitude of this pole is

$$|e^{pT}| = |e^{-\sigma T} e^{j\omega T}| = e^{-\sigma T} < 1$$

which shows that this pole is inside the unit circle. Because $h_d(n)$ is causal, the ROC is exterior to this pole and includes the unit circle. Hence the discrete-time filter is stable.

The mapping also maps the imaginary axis in the s-plane to the unit circle in the z-plane. Each length-2π segment of the $j\omega$-axis, however, maps to the entire unit circle. Thus, an infinite number of periodically spaced segments of the continuous-time frequency axis are mapped to the discrete-time frequency axis. This characteristic is manifested in Figure 3.3.4 and is the source of aliasing when $H_c(j\omega)$ is not strictly bandlimited.

The impulse invariance technique is used to produce the discrete-time square-root raised-cosine pulse-shaping filter described in Appendix A.

Bilinear Transform. The bilinear transform uses a different mapping from the s-plane to the z-plane that avoids the aliasing problem associated with the impulse invariance technique, but introduces a nonlinear "warping" of the frequency axis. This filter design technique also starts with a continuous-time prototype filter design $H_c(s)$. The bilinear transform uses the substitution

$$s = \frac{2}{T}\frac{1-z^{-1}}{1+z^{-1}} \qquad (3.25)$$

to produce the discrete-time filter

$$H_d(z) = H_c\left(\frac{2}{T}\frac{1-z^{-1}}{1+z^{-1}}\right). \qquad (3.26)$$

After some algebra, $H_d(z)$ may be expressed in the standard form (3.18). Any of the structures illustrated in Figures 3.3.2 and 3.3.3 may be used to realize the filter.

The bilinear transform has been the transform of choice in discrete-time filter design because the mapping has two desirable properties. The first is that the transform maps the imaginary axis in the s-plane to the unit circle in the z-plane as illustrated in Figure 3.3.6 (a). In other words, the entire continuous-time frequency axis is mapped to the discrete-time frequency axis. This mapping is not a linear one. Substituting $z = e^{j\Omega}$ in (3.25) produces the relationship

$$\omega = \frac{2}{T}\tan\left(\frac{\Omega}{2}\right). \qquad (3.27)$$

When the pass-band and stop-band frequencies are specified in the discrete-time frequency domain, these frequencies should be "prewarped" according to (3.27) to give the continuous-time frequencies that should be used in the continuous-time filter design procedure. When the resulting continuous-time filter is transformed using (3.25), a discrete-time filter with the correct pass-band and stop-band frequencies is obtained.

The second property is that all points in the left-half s-plane are mapped to points inside the unit circle in the z-plane. As a consequence, a pole in the left-half s-plane is mapped to a z-plane pole inside the unit circle as illustrated in Figure 3.3.6 (b). Thus, a causal, stable continuous-time filter $H_c(s)$ is transformed to a causal, stable discrete-time filter $H_d(z)$.

The bilinear transform is used to transform continuous-time phase-locked loops to discrete-time phase locked loops as described in Appendix C. The technique can also be used to design discrete-time integrators as described in Section 3.3.3.

3.3.2 FIR Filter Designs

Basic Structure. Conceptually, the FIR filter is a special case of the IIR filter with no poles. As such, the transfer function $H(z)$ could be written in the form (3.18) with $a_1 = a_2 = \cdots = a_p = 0$. It is more customary, however, to express the transfer function as

$$H(z) = h(0) + h(1)z^{-1} + h(2)z^{-2} + \cdots + h(M)z^{-M}. \qquad (3.28)$$

Section 3.3 Discrete-Time Filter Design Methods

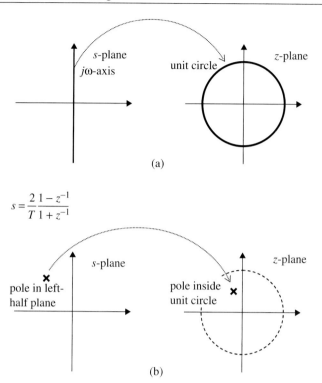

Figure 3.3.6 The mapping of points from the s-plane to the z-plane performed by the bilinear transform (3.25): (a) The $s = j\omega$ axis is mapped to $z = e^{j\Omega}$; (b) the left-half of the s-plane is mapped to the region inside the unit circle in the z-plane.

If the sequence $x(n)$ is the input to the filter, then the output sample $y(n)$ is

$$y(n) = \sum_{k=0}^{M} h(k)x(n-k) \qquad (3.29)$$

which is the convolution of the input sequence with the filter impulse response. The length of the impulse response is $M + 1$.

The two expressions above assume the FIR filter is causal. Sometimes, FIR filter designs produce the coefficients for a noncausal FIR filter. In this case, it is more convenient to express the filter as a noncausal LTI system. For an odd-length filter, the expression is

$$H(z) = h(-L)z^L + h(-L+1)z^{L-1} + \cdots + h(0) + \cdots + h(L-1)z^{-L+1} + h(L)z^{-L}. \quad (3.30)$$

The length of the filter is $2L + 1$. The output $y(n)$ corresponding to an input sequence $x(n)$ is

$$y(n) = \sum_{k=-L}^{L} h(k)x(n-k) \qquad (3.31)$$

Table 3.3.1 FIR filter types

	Symmetric [Eq. (3.32)]	Antisymmetric [Eq. (3.33)]
Odd-length (M is even)	Type I	Type III
Even-length (M is odd)	Type II	Type IV

The coefficient $h(0)$ is often called the "center tap" of the filter and multiplies the input sample with zero delay relative to the current output sample.

One of the advantages of FIR filters is that a linear phase frequency response can be guaranteed when the filter coefficients satisfy either of the symmetries (for the causal formulation)

$$h(M - n) = h(n) \qquad \text{for } n = 0, 1, \ldots, M \qquad (3.32)$$

or

$$h(M - n) = -h(n) \qquad \text{for } n = 0, 1, \ldots, M. \qquad (3.33)$$

Linear phase FIR filters are traditionally categorized in terms of their length and type of symmetry. These categories are summarized in Table 3.3.1.

There are many FIR filter realizations. Two of the more popular are illustrated in Figure 3.3.7. The realization illustrated in Figure 3.3.2 (a) is a direct application of (3.28). It corresponds to either of the direct forms for IIR filters illustrated in Figure 3.3.2 when $a_1 = a_2 = \cdots = a_p = 0$. Two important features should be noted. First is the series of delay elements across the top of the diagram. Because of this structure, this type of filter is often called a "tapped delay-line filter" or a "transversal filter." The second feature is that each two-input adder is coupled with a constant-coefficient multiplier to form a processing block commonly referred to as a multiply-accumulate or MAC. This structure is the picture that comes to the mind of a DSP engineer or embedded systems programmer when asked to realize (3.28).

The transposed form for the FIR filter is illustrated in Figure 3.3.7 (b). It corresponds to either of the transposed forms for IIR filters illustrated in Figure 3.3.3 when $a_1 = a_2 = \cdots = a_p = 0$. This represents an alternate form for the FIR filter. Observe that the adders are now inserted in-line with the memory registers and that all of the multipliers operate on the same input sample. Each register stores what is called a "partial sum." As such, the registers operate as partial sum accumulators. This structure is the picture that comes to the mind of a VLSI engineer when asked to realize (3.28).

When the FIR filter is a linear phase filter, the filter coefficients possess one of the two symmetry conditions (3.32) or (3.33). These symmetries may be exploited to reduce the number of multipliers needed to compute each output sample. An example of a realization of a length-7 filter that incorporates the symmetry (3.32) is illustrated in Figure 3.3.8. This

Section 3.3 Discrete-Time Filter Design Methods

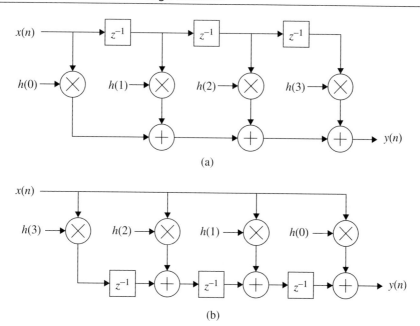

Figure 3.3.7 Alternate realizations of an FIR filter for $M = 3$ (i.e., a length-4 filter): (a) The "tapped delay-line" or "transversal" filter based on the multiply-accumulate operation; (b) the realization based on the partial sum accumulator.

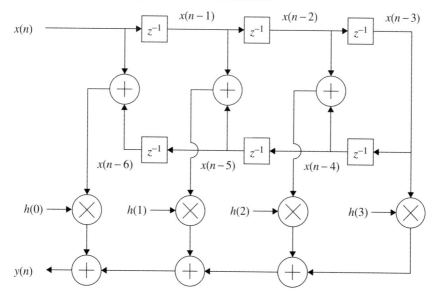

Figure 3.3.8 An FIR filter realization that exploits the linear phase symmetry (3.32) for the special case of a length-7 ($M = 6$) filter.

realization is based on the multiply-accumulate realization of Figure 3.3.7 (a). Observe that the filter of Figure 3.3.8 requires three multipliers, whereas the same filter realized along the lines of Figure 3.3.7 (a) requires seven multipliers.

FIR Filter Design Based on Windowing. The simplest method for designing a discrete-time FIR filter is windowing. This method is used exclusively in the discrete-time domain. The starting point is the ideal (or desired) frequency response $H_{\text{ideal}}(e^{j\Omega})$. The corresponding discrete-time impulse response, $h_{\text{ideal}}(n)$ is obtained using the inverse DTFT (2.55). Most ideal frequency responses are piecewise linear with discontinuities at the band edges. In these cases, $h_{\text{ideal}}(n)$ is noncausal and infinitely long. The windowing method is a smart way of "truncating" the infinitely long impulse response of the ideal filter to produce an FIR filter.

This method is best described by the example of designing a low-pass filter. The frequency response of the ideal low-pass filter is given by

$$H_{\text{ideal}}(e^{j\Omega}) = \begin{cases} 1 & -W \leq \Omega \leq W \\ 0 & \text{otherwise} \end{cases} \qquad (3.34)$$

and the corresponding impulse response is

$$h_{\text{ideal}}(n) = \frac{1}{2\pi} \int_{-W}^{W} e^{j\Omega n} d\Omega = \frac{W}{\pi} \frac{\sin(Wn)}{Wn}. \qquad (3.35)$$

Observe that $h_{\text{ideal}}(n)$ is infinitely long and noncausal. As such, it defines an IIR filter.

Now suppose an FIR version of this filter was constructed by simply truncating the filter by keeping only the $2L + 1$ samples corresponding to $-L \leq n \leq L$ as illustrated at the top of Figure 3.3.9. To see the effect of this truncation on the frequency response, the truncation is represented as a multiplication:

$$h_{\text{FIR}}(n) = h_{\text{ideal}}(n) \times w(n) \qquad (3.36)$$

where

$$w(n) = \begin{cases} 1 & -L \leq n \leq L \\ 0 & \text{otherwise} \end{cases} \qquad (3.37)$$

is a "window." An illustration of this windowing operation is illustrated in the left-half column of Figure 3.3.9. In the frequency domain, the multiplication property listed in the eighth row of Table 2.4.7 is applied. The frequency response of the FIR filter produced by windowing is given by

$$H_{\text{FIR}}(e^{j\Omega}) = \frac{1}{2\pi} H_{\text{ideal}}(e^{j\Omega}) * W(e^{j\Omega}) \qquad (3.38)$$

Section 3.3 Discrete-Time Filter Design Methods

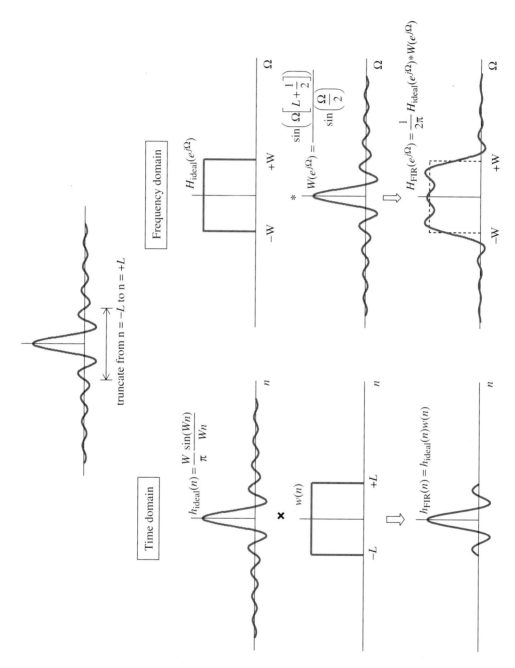

Figure 3.3.9 Windowing in the time domain and frequency domains.

where

$$W(e^{j\Omega}) = \sum_{n=-L}^{L} e^{-j\Omega n} = \frac{\sin(\Omega(L+1/2))}{\sin(\Omega/2)} \qquad (3.39)$$

is the frequency response of the window $w(n)$. Examples of $H_{\text{ideal}}(e^{j\Omega})$ and $W(e^{j\Omega})$ are illustrated in the right-hand column of Figure 3.3.9.

The convolution operation "slides" a frequency-reversed version of $W(e^{j\Omega})$ past $H_{\text{ideal}}(e^{j\Omega})$ and measures the area under the product. The pass band of $H_{\text{FIR}}(e^{j\Omega})$ is obtained when the main lobe of $W(e^{j\Omega})$ coincides with the pass band of $H_{\text{ideal}}(e^{j\Omega})$. The transition band of $H_{\text{FIR}}(e^{j\Omega})$ is produced as the main lobe of $W(e^{j\Omega})$ slides past the band edge of $H_{\text{ideal}}(e^{j\Omega})$. The stop band of $H_{\text{FIR}}(e^{j\Omega})$ is produced as the side lobes of $W(e^{j\Omega})$ slide through the pass band of $H_{\text{ideal}}(e^{j\Omega})$. The final result is illustrated at the bottom of the right-hand column in Figure 3.3.9.

The properties of $W(e^{j\Omega})$ determine many of the important filter characteristics. The side lobes of $W(e^{j\Omega})$ cause the pass-band and stop-band ripples in $H_{\text{FIR}}(e^{j\Omega})$. The relatively high side lobes of $W(e^{j\Omega})$ result in only modest stop-band attenuation. The transition band of $H_{\text{FIR}}(e^{j\Omega})$ is determined largely by the width of the main lobe of $W(e^{j\Omega})$. As L increases, the width of the main lobe decreases. As a result, the width of the transition band decreases. This is another way of saying that as the length of $h_{\text{FIR}}(n)$ increases, $H_{\text{FIR}}(e^{j\Omega})$ more closely approximates $H_{\text{ideal}}(e^{j\Omega})$.

These features are illustrated by the simple examples shown in Figures 3.3.10 and 3.3.11. The truncated impulse responses for three different lengths are shown in Figure 3.3.10. The corresponding frequency response is plotted in Figure 3.3.11. Observe that as the length increases, the width of the transition band decreases, but that the highest lobe in the stop band remains fixed.

The stop-band attenuation can be reduced by using a window with a more smooth transition from unity amplitude to zero amplitude. The smoother time-domain signal has a DTFT with lower side lobe levels. The lower side lobe levels produce better stop-band attenuation. Unfortunately, lower side lobe levels are achieved at the expense of a wider main lobe. A wider main lobe produces a filter with a larger transition bandwidth.

Most of the smoother windows commonly used in FIR filter design are of the (noncausal, symmetric) form

$$w(n) = a_0 + a_1 \cos\left(\frac{2\pi}{M}n\right) + a_2 \cos\left(\frac{4\pi}{M}n\right) + a_3 \cos\left(\frac{6\pi}{M}n\right) \quad \text{for } -L \leq n \leq L \qquad (3.40)$$

where the length of the window is $M + 1 = 2L + 1$. A summary of the windows of this form, along with the pertinent spectral features, is listed in Table 3.3.2. Time-domain plots of the Hann, Hamming, Blackman, Blackman–Harris, and Nuttall windows of length 31 are shown in Figure 3.3.12. The corresponding frequency-domain plots are shown in Figure 3.3.13.

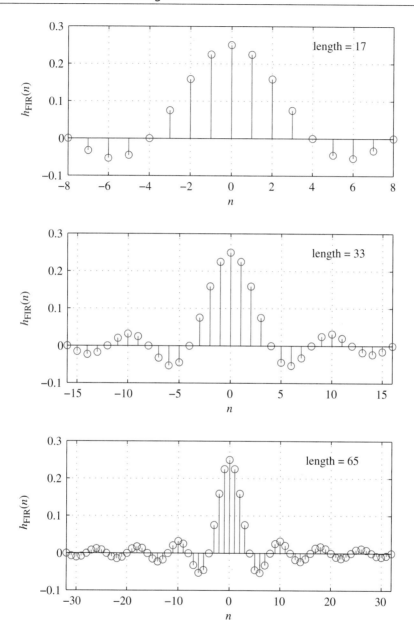

Figure 3.3.10 Impulse responses for an FIR low-pass filter design using the rectangular window for $W = \pi/4$ and truncation lengths of 17 (top), 33 (middle), and 65 (bottom). The corresponding frequency responses are plotted in Figure 3.3.11.

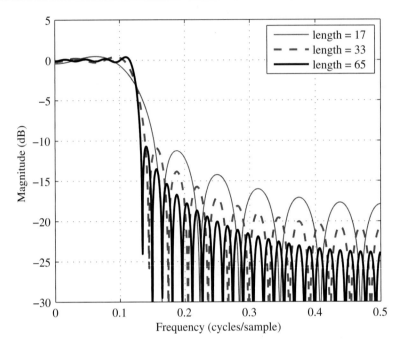

Figure 3.3.11 Frequency responses for an FIR low-pass filter design using the rectangular window for $W = \pi/4$ and truncation lengths of 17, 33, and 65. The corresponding impulse responses are illustrated in Figure 3.3.10.

Another popular window is the Kaiser–Bessel window. The Kaiser–Bessel window is of the form

$$w(n;\beta) = \begin{cases} \dfrac{I_0\left(\pi\beta\sqrt{1-\left(\dfrac{n}{M/2}\right)^2}\right)}{I_0(\pi\beta)} & -L \leq n \leq +L \\ 0 & \text{otherwise} \end{cases} \qquad (3.41)$$

where $I_0(x)$ is the zero-order modified Bessel function of the first kind. The window is parameterized by β, which is the time/bandwidth product of the window. The Kaiser–Bessel window is plotted in Figure 3.3.12 for $\beta = 2, 5$, and 10. The corresponding frequency domain plots are shown in Figure 3.3.13.

The performance of an FIR filter designed by windowing is a function of its length, the width of its transition band, and the stop-band attenuation. For a fixed length, a narrow transition band and good stop-band attenuation present competing demands on the filter (i.e., one improves at the expense of the other). This trade-off is controlled by the choice of window used. For a fixed filter length, the rectangular window produces the narrowest transition band, but the worst stop-band attenuation. The Blackman window is able to produce a filter with much better stop-band attenuation, but requires a much larger transition band. An empirical

Table 3.3.2 Summary of properties for windows of the form (3.40) commonly used in FIR filter design. The length of the window is $M + 1$. The main lobe width is measured from the peak of $W(e^{j\Omega})$ at $\Omega = 0$ to the first null. The peak side lobe level is measured relative to the amplitude of $W(e^{j\Omega})$ at $\Omega = 0$

Window	Coefficients	Main Lobe Width	Peak Side Lobe (dB)
Rectangular	$a_0 = 1$ $a_1 = 0$ $a_2 = 0$ $a_3 = 0$	$2\pi/(M+1)$	-13.5
Hann	$a_0 = 0.5$ $a_1 = 0.5$ $a_2 = 0$ $a_3 = 0$	$4\pi/M$	-32
Hamming	$a_0 = 0.54$ $a_1 = 0.46$ $a_2 = 0$ $a_3 = 0$	$4\pi/M$	-43
Blackman	$a_0 = 0.42$ $a_1 = 0.5$ $a_2 = 0.08$ $a_3 = 0$	$6\pi/M$	-58
Blackman–Harris	$a_0 = 0.35875$ $a_1 = 0.48829$ $a_2 = 0.14128$ $a_3 = 0.01168$	$8\pi/M$	-92
Nuttall	$a_0 = 0.3635819$ $a_1 = 0.4891775$ $a_2 = 0.1365995$ $a_3 = 0.0106411$	$8\pi/M$	-92

formula that captures this trade-off was derived by Kaiser and is

$$M \approx \frac{1}{\Delta F} \frac{\text{Attenuation (dB)} - 8}{14} \quad (3.42)$$

where ΔF is the width of the transition band (in cycles/sample) and Attenuation (dB) is the stop-band attenuation in dB.

The windowing method of FIR filter design is used in Section 3.3.3 to create an FIR differentiator.

FIR Filter Design Using Approximations. FIR filter designs based on windowing produce filters with equal pass-band and stop-band ripple. The ripple represents the approximation error of the filter (i.e., the departure of the frequency response of the actual filter

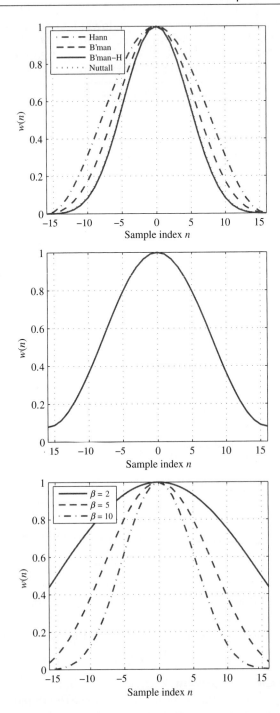

Figure 3.3.12 Time series plots of windows commonly used for FIR filter design: (top) The Hann, Blackman, Blackman–Harris, and Nuttall windows of length 31; (middle) the Hamming window of length 31; (bottom) the Kaiser window of length 31 for different values of β.

Section 3.3 Discrete-Time Filter Design Methods

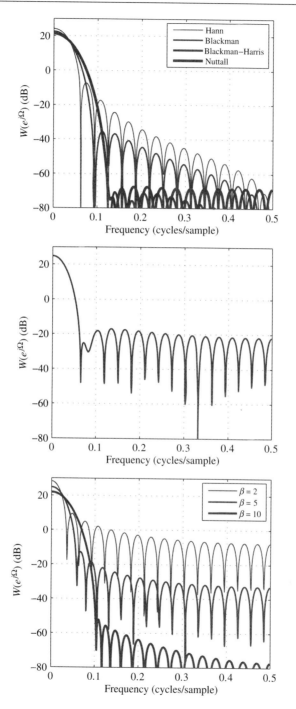

Figure 3.3.13 DTFTs of the length-31 windows plotted in Figure 3.3.12.

from that of the ideal, desired filter). In many applications, a larger pass-band ripple can be tolerated than stop-band ripple. The windowing method does not allow the pass-band and stop-band ripples to be independently adjusted. When the pass-band and stop-band ripples can be independently adjusted, the filter requirements can often be met with a shorter filter.

This characteristic has motivated the search for other FIR filter approximations. In general, these FIR filter approximations produce FIR filters whose frequency response $H(e^{j\Omega})$ minimizes the function

$$\int_{\Theta} (W(\Omega)[|H(e^{j\Omega})| - H_{\text{ideal}}(e^{j\Omega})]^p) d\Omega \qquad (3.43)$$

where Θ is the set of frequencies of interest, $W(\Omega)$ is a frequency-domain error-weighting function, $H_{\text{ideal}}(e^{j\Omega})$ is the ideal (or desired) frequency response, and p is the norm of the approximation error. The complexity of the minimization is reduced by a factor of two when linear phase, either of the conditions (3.32) or (3.33) for the filter coefficients, is imposed.

When $p = 2$, the FIR filter offering the best least squares approximation to $H_{\text{ideal}}(e^{j\Omega})$ is produced. When $p = \infty$, the FIR filter offering the best minimax approximation error is produced (i.e., the filter that minimizes the maximum approximation error). A very efficient minimax algorithm, known as the Parks–McClellan algorithm, has been developed for this case for linear phase filters. As a consequence, it has emerged as the most popular design method for linear phase FIR filters.

The Parks–McClellan algorithm starts by expressing the frequency response for a length-$(M + 1) = 2L + 1$ linear phase symmetric FIR filter as

$$H(e^{j\Omega}) = \sum_{n=-L}^{L} h(n) e^{j\Omega n} = h(0) + \sum_{n=1}^{L} h(n) \cos(\Omega_0 n). \qquad (3.44)$$

This shows that the frequency response is a weighted sum of harmonically related cosines. The weighed sum of harmonically related cosines may also be expressed as a Chebyshev polynomial in Ω. The Parks–McClellan algorithm applies the results of the "alternation theorem" from the theory of polynomial optimization to produce an extraordinarily efficient method for minimizing (3.43). The weighting function $W(\Omega)$ controls the degree to which pass-band and stop-band ripples contribute to the error.

The resulting filter possesses a property known as the "equiripple" property: the pass-band ripples all have the same peak amplitude and the stop-band ripples all have the same peak amplitude (but the two peak amplitudes do not have to be equal). This property is illustrated by the low-pass filter frequency response illustrated in Figure 3.3.14. The filter was designed to meet the following specifications:

$$F_p = 0.125 \text{ cycles/sample}$$
$$F_s = 0.225 \text{ cycles/sample}$$
$$\delta_p \leq 0.01$$
$$\delta_s \leq 0.001$$

Section 3.3 Discrete-Time Filter Design Methods

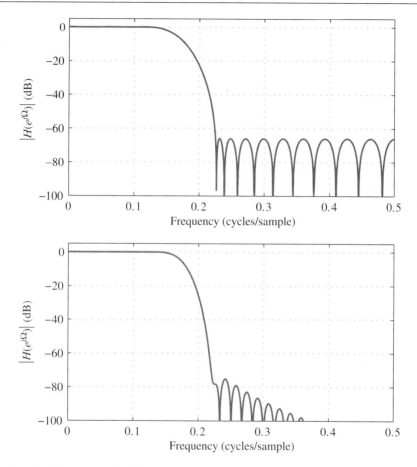

Figure 3.3.14 FIR low-pass filter frequency responses with $F_p = 0.125$ cycles/sample, $F_s = 0.225$ cycles/sample, and $\delta_s < 0.001$: (top) The length-31 FIR filter produced by the Parks–McClellan algorithm; (bottom): a length-61 FIR filter produced using the Blackman window.

The length-31 linear phase FIR filter produced by the Parks–McClellan algorithm exhibits a stop-band attenuation of 66 dB or more as shown in the top part of Figure 3.3.14. Observe that the minimum stop-band attenuation (or peak stop-band ripple) is constant over the entire stop-band. This is in contrast to the stop-band attenuation of FIR filters produced using the windowing method. For comparison, the frequency response of the FIR filter produced using the windowing method[2] is shown in the lower portion of Figure 3.3.14. This filter is a length-61

---

[2]Using the window method with (3.34) produces a filter whose frequency response is −3 dB at the pass-band frequency F_p. This violates the requirement that the pass-band ripple should not exceed $\delta_p = 0.01$. To address this issue, and to make the comparison fair with the result obtained using the Parks–McClellan algorithm, the parameter W of the ideal prototype filter $H_{\text{ideal}}(e^{j\Omega})$ was increased by the width of the main

filter and was designed using the Blackman window. In this case, the stop-band attenuation decreases with frequency.

The example also illustrates another important feature of the Parks–McClellan algorithm: the filter length required to meet a given set of specifications is often much smaller using the result of the Parks–McClellan algorithm than with the windowing method. The filter length is related to the pass-band ripple δ_p, stop-band ripple δ_s, and the width of the transition band ΔF (cycles/sample). The form of this relationship is

$$M = \frac{K(\delta_p, \delta_s, \Delta F)}{\Delta F} \quad (3.45)$$

where $K(\delta_p, \delta_s, \Delta F)$ is a function of the filter specifications. Various empirically based formulas for $K(\delta_p, \delta_s, \Delta F)$ have been proposed. The original, by Herrmann, is

$$K(\delta_p, \delta_s, \Delta F) = f_1(\delta_p) \log_{10}(\delta_s) - f_2(\delta_p) - f_3(\delta_p, \delta_s)(\Delta F)^2 \quad (3.46)$$

$$f_1(\delta_p) = [0.0729 \log_{10}(\delta_p)]^2 + 0.07114 \log_{10}(\delta_p) - 0.4761 \quad (3.47)$$

$$f_2(\delta_p) = [0.0518 \log_{10}(\delta_p)]^2 + 0.59410 \log_{10}(\delta_p) + 0.4278 \quad (3.48)$$

$$f_3(\delta_p, \delta_s) = 11.01217 + 0.51244[\log_{10}(\delta_p) - \log_{10}(\delta_s)]. \quad (3.49)$$

A simpler approximation put forth by Kaiser is

$$K(\delta_p, \delta_s) = \frac{-10 \log_{10}(\delta_p \delta_s) - 13}{2\pi \times 2.324}. \quad (3.50)$$

Harris offers an even simpler approximation

$$K(\delta_s) = \frac{-20 \log_{10}(\delta_s)}{22}. \quad (3.51)$$

The relationship between these approximations is illustrated by the plots in Figure 3.3.15. In general, these equations tend to underestimate the length of the filter required to meet the specifications. Nonetheless, these equations provide an excellent starting point for trial-and-error iterations.

The Parks–McClellan algorithm is very flexible and easily generalizable. The desired frequency response can be arbitrary. For example, in Chapter 9, the need for predistortion filters to compensate for sample-and-hold distortion in digital-to-analog converters is introduced. The example filters were produced using the Parks–McClellan algorithm where $H_{\text{ideal}}(e^{j\Omega})$ is set to

$$H_{\text{ideal}}(e^{j\Omega}) = \frac{\frac{\Omega}{2}}{\sin\left(\frac{\Omega}{2}\right)} \quad \text{for } -\pi \leq \Omega \leq \pi. \quad (3.52)$$

In Appendix A, the Parks–McClellan algorithm is used to generate pulse shapes that have improved stop-band attenuation over that available using the window method.

---

lobe of the window. The length was selected to produce a transition band approximately 0.10 cycles/sample wide. Some trial-and-error iteration was required.

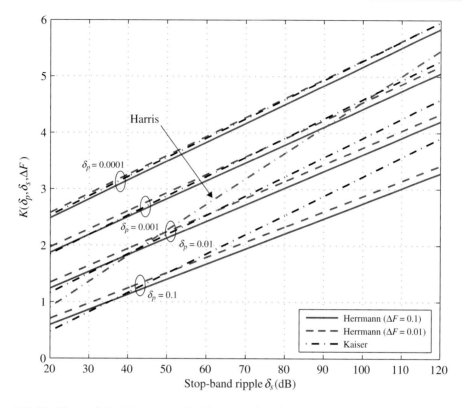

Figure 3.3.15 Plots of the Herrmann (3.49), Kaiser (3.50), and Harris (3.51) approximations for the multiplier $K(\delta_p, \delta_s, \Delta F)$ of (3.45). (Reproduced from Harris [53].)

3.3.3 Two Important Filters: The Differentiator and the Integrator

Two important discrete-time filters are used extensively in this text: the discrete-time differentiator and the discrete-time integrator.

Discrete-Time Differentiator. Suppose $x(nT)$ represents T-spaced samples of a band-limited signal $x(t)$ and that samples of the time-derivative $\dot{x}(t)$ are desired. This could be accomplished as shown in the top part of Figure 3.3.16. The samples $x(nT)$ are converted to the corresponding waveform $x(t)$. The waveform is differentiated to form $\dot{x}(t)$, which is then sampled to produce the sequence $\dot{x}(nT)$. If the sample rate used to produce the samples $x(nT)$ satisfies the sampling theorem, then all the information about the waveform $x(t)$ is contained in the samples $x(nT)$. Clearly, the process of converting back to continuous-time to use continuous-time processing followed by a sampling is not necessary. The goal then is to design a discrete-time system $D(e^{j\Omega})$ to produce the sequence $\dot{x}(nT) = \dot{x}(t)|_{t=nT}$.

In this example, the relationship (2.86) is applied to formulate the infinite impulse response of a discrete-time differentiator and the windowing method is applied to produce

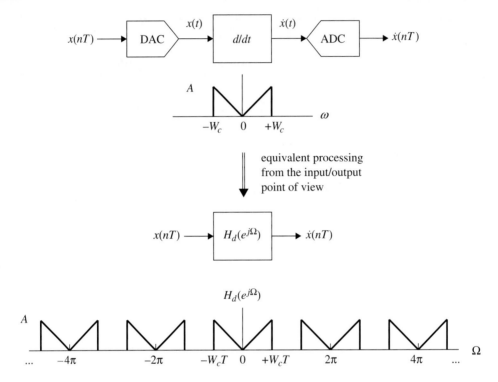

Figure 3.3.16 Computing samples of the time-derivative of a discrete-time signal: (top) An approach using continuous-time processing; (bottom) an approach using discrete-time processing.

an FIR version. To apply (2.86), the transfer function of the continuous-time system is required. A continuous-time differentiator has transfer function $H(j\omega) = j\omega$. A bandlimited continuous-time differentiator with bandwidth W rad/s has transfer function

$$H(j\omega) = \begin{cases} j\omega & -W_c \leq \omega \leq +W_c \\ 0 & \text{otherwise} \end{cases}. \qquad (3.53)$$

Applying (2.86) produces the desired transfer function for the equivalent discrete-time system:

$$H_d(e^{j\Omega}) = \begin{cases} j\dfrac{\Omega}{T} & |\Omega| \leq W_c T \\ 0 & W_c T < |\Omega| \leq \pi \end{cases}. \qquad (3.54)$$

Section 3.3 Discrete-Time Filter Design Methods

The impulse response $h_d(n)$ is given by

$$h_d(n) = \frac{1}{2\pi} \int_{-W_cT}^{W_cT} j\frac{\Omega}{T} e^{j\Omega n} d\Omega \tag{3.55}$$

$$= \frac{W_cT}{\pi T} \frac{\cos(W_c Tn)}{n} - \frac{1}{\pi T} \frac{\sin(W_c Tn)}{n^2}. \tag{3.56}$$

The impulse response has infinite support in n. As a consequence, the discrete-time system is an IIR system.

Observe that this same impulse response is obtained using the impulse invariance method because the continuous-time impulse response is assumed to be strictly bandlimited. The impulse response of the bandlimited system with frequency response (3.53) is

$$h_c(t) = \frac{1}{2\pi} \int_{-W_c}^{W_c} j\omega e^{j\omega t} d\omega = \frac{W_c}{\pi t} \cos(W_c t) - \frac{1}{\pi t^2} \sin(W_c t). \tag{3.57}$$

Applying the impulse invariance principle, $h_d(n)$ is obtained from $h_c(t)$ by replacing t with nT and scaling the amplitude by T:

$$h_d(n) = Th_c(nT) = \frac{W_cT}{\pi nT} \cos(W_c Tn) - \frac{T}{\pi(nT)^2} \sin(W_c Tn) \tag{3.58}$$

which is identical to (3.56).

An important special case is the one where the differentiator is to operate over the entire bandwidth of the first Nyquist zone. Setting $W_cT = \pi$ in (3.56) produces

$$h_d(n) = \begin{cases} \dfrac{1}{T}\dfrac{(-1)^n}{n} & n \neq 0 \\ 0 & n = 0 \end{cases}. \tag{3.59}$$

The first few samples of $h_d(n)$ are plotted in Figure 3.3.17.

If this impulse response is truncated to $n = -1, 0, +1$, then the differentiator consists of the three center coefficients. These coefficients should be scaled by 1/2 to preserve the gain of the longer versions of the filter. If the input sequence is $x(nT)$, then the output sequence is

$$\dot{x}(nT) \approx \sum_{k=-1}^{+1} h_d(k) x((n-k)T) \tag{3.60}$$

$$= \frac{x((n+1)T) - x((n-1)T)}{2T} \tag{3.61}$$

which is the familiar first central difference.[3]

---
[3]The first central difference approximates the derivative at $t = nT$ by computing the slope of $x(t)$ at $t = nT$ using the values of the waveform at adjacent time instants $(n-1)T$ and $(n+1)T$.

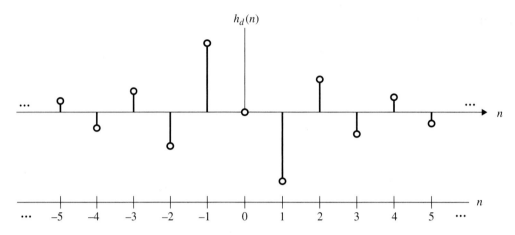

Figure 3.3.17 The first few samples of the impulse response for the full-bandwidth differentiator given by (3.59).

An FIR version of the differentiator is obtained by applying a window, $w(n)$, to truncate the infinite impulse response in an appropriately smooth manner. Following the development in the previous section, the finite impulse response is

$$h_{d,\text{FIR}}(n) = h_d(n)w(n) \qquad (3.62)$$

where $w(n)$ is the window and $h_d(n)$ is given by (3.56) or the special case (3.59). Figure 3.3.18 is a plot of $H_{d,\text{FIR}}(e^{j\Omega})$, the DTFT of (3.62) where $h_d(n)$ is given by (3.59) and $w(n)$ is the Blackman window for filter lengths of $3, 7, 11, \ldots, 31$. (Note: the filter length is always odd to provide a "center-tap" for the FIR filter.) The length-3 FIR approximation is the first central difference and is a poor approximation to the differentiator for $\Omega > 0.2\pi$. In other words, the first central difference is a good approximation to differentiation as long as the signal is at least five times oversampled. The usable bandwidth of the differentiator increases as the filter length increases.

Discrete-Time Integrator. Let $x(t)$ and $y(t)$ be continuous-time signals defined by the relationship

$$y(t) = \int_{-\infty}^{t} x(u)du. \qquad (3.63)$$

Now suppose $x(t)$ is bandlimited and is sampled to produce a sequence of T-spaced samples $\ldots, (x(n-1)T), x(nT), x((n+1)T), \ldots$. The goal is to compute samples of $y(t)$ from samples of $x(t)$. One way this could be accomplished is illustrated in Figure 3.3.19. The waveform $x(t)$ could be reconstructed from the samples $x(nT)$ by the DAC. The waveform $x(t)$ would then be integrated by a continuous-time integrator to produce the waveform $y(t)$ which, in turn, would be sampled by the ADC to produce the desired sequence $y(nT)$. Because

Section 3.3 Discrete-Time Filter Design Methods   **153**

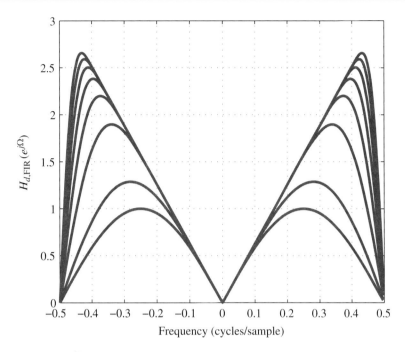

Figure 3.3.18 $H_{d,\text{FIR}}(e^{j\Omega})$ versus Ω for the full-bandwidth FIR differentiator for filter lengths 3, 7, 11, ..., 31. The Blackman window was used to perform the windowing.

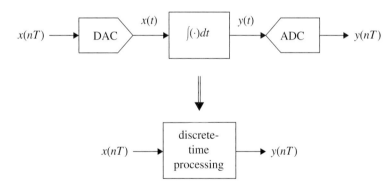

Figure 3.3.19 Computing samples of the integral of a continuous-time signal: (top) An approach using continuous-time processing; (bottom) an approach using discrete-time processing.

$x(t)$ was assumed bandlimited, all the information in $x(t)$ is contained in the sample sequence $x(nT)$. As a consequence, the conversion to continuous-time, continuous-time processing, and the conversion back to discrete-time is unnecessary.

This problem is a natural application of the bilinear transform filter design technique outlined in Section 3.3.1. The continuous time system is characterized by the nonbandlimited transfer function

$$H_c(s) = \frac{1}{s}. \tag{3.64}$$

Applying the bilinear transform relationship (3.25), the discrete-time integrator is

$$H_{d,1}(z) = H_c(s)\big|_{s=\frac{2}{T}\frac{1-z^{-1}}{1+z^{-1}}} = \frac{T}{2}\frac{1+z^{-1}}{1-z^{-1}}. \tag{3.65}$$

A system that performs this function is illustrated in Figure 3.3.20. Note that this system is the direct form II realization (see Section 3.3.1) of (3.65). The corresponding time-domain relationship between the input $x(n)$ and the integrator output $y(n)$ is a special case of (3.21):

$$y(n) = y(n-1) + \frac{T}{2}x(nT) + \frac{T}{2}x((n-1)T). \tag{3.66}$$

This relationship defines an accumulator: the current accumulator value $y(nT)$ is the previous value plus the average of the current and previous inputs.

In practice, this form of the integrator is rarely used. It, however, is very closely related to the form of the integrator that is used. The most commonly used discrete-time integrator and its relationship to (3.65) and (3.66) is developed as follows. The relationship (3.63) is sampled and the resulting discrete-time relationship is written as a recursion. An approximation for the definite integral that is part of the recursion is substituted to produce the desired result.

Sampling the relationship (3.63) is performed by replacing t by nT:

$$y(nT) = \int_{-\infty}^{nT} x(u)du$$

$$= \int_{-\infty}^{(n-1)T} x(u)du + \int_{(n-1)T}^{nT} x(u)du$$

$$= y((n-1)T) + \int_{(n-1)T}^{nT} x(u)du. \tag{3.67}$$

Note that (3.67) defines a recursion on $y(nT)$. The integral on the right-hand side of (3.67) may be replaced by an operation on the samples $x(nT)$. Interpreting the integral in (3.67) as the area under $x(t)$ in the interval $(n-1)T \leq t \leq nT$, three common approximations could be used. These three approximations, illustrated in Figure 3.3.21, are termed the backward difference, the forward difference, and the trapezoid rule. The properties of the three approximations are summarized in Table 3.3.3.

The recursion resulting from the backward difference is

$$y(nT) = y((n-1)T) + Tx((n-1)T). \tag{3.68}$$

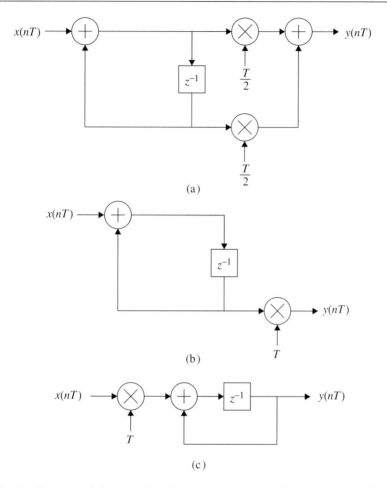

Figure 3.3.20 Realization of discrete-time integrators: (a) A realization of the discrete-time integrator based on the bilinear transform. This system realizes the transfer function (3.65) and time-domain relationship (3.66). (b) A realization of the discrete-time integrator based on the backward difference. This system realizes the transfer function (3.70) and time-domain relationship (3.68). (c) A rearrangement of (b) to produce the more traditional system block diagram.

The z-transform of the recursion is

$$Y(z) = z^{-1}Y(z) + Tz^{-1}X(z). \tag{3.69}$$

Solving for the transfer function $H_{d,2}(z)$ produces the desired result:

$$H_{d,2}(z) = \frac{Y(z)}{X(z)} = T\frac{z^{-1}}{1 - z^{-1}}. \tag{3.70}$$

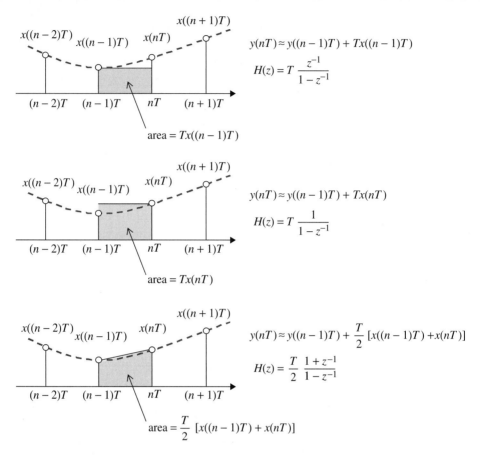

Figure 3.3.21 An illustration of the backward difference, forward difference, and trapezoid rules for approximating the integral of a continuous-time signal using discrete-time processing.

A realization of $H_{d,2}(z)$ is illustrated in Figure 3.3.20 (b). Note that this system is a direct application of direct form II realization of IIR systems described in Section 3.3.1. A minor rearrangement, illustrated in Figure 3.3.20 (c) produces the most common form of the discrete-time integrator.

The recursions and transfer functions for the forward difference and trapezoid rules are listed in Table 3.3.3. The substitution resulting from the trapezoid rule is

$$\frac{1}{s} = \frac{T}{2}\frac{1+z^{-1}}{1-z^{-1}}. \tag{3.71}$$

This substitution, known as *Tustin's equation*, is identical to the bilinear transform. The integrator based on the bilinear transform and the integrator based on the backward difference are related in the sense that both use a simple approximation for integration.

Section 3.3 Discrete-Time Filter Design Methods

Table 3.3.3 Summary of discrete-time approximations to integration

Rule	Approximation	Recursion	$H_I(z)$	$H_I(e^{j\Omega})$
Backward Difference	$\int_{(n-1)T}^{nT} x(t)dt \approx Tx((n-1)T)$	$y(nT) = y((n-1)T) + Tx((n-1)T)$	$T\dfrac{z^{-1}}{1-z^{-1}}$	$T\dfrac{e^{-j\Omega/2}}{j2\sin(\Omega/2)}$
Forward Difference	$\int_{(n-1)T}^{nT} x(t)dt \approx Tx(nT)$	$y(nT) = y((n-1)T) + Tx(nT)$	$T\dfrac{1}{1-z^{-1}}$	$T\dfrac{e^{j\Omega/2}}{j2\sin(\Omega/2)}$
Trapezoid	$\int_{(n-1)T}^{nT} x(t)dt \approx \dfrac{T}{2}[x((n-1)T) + x(nT)]$	$y(nT) = y((n-1)T) + \dfrac{T}{2}[x((n-1)T) + x(nT)]$	$\dfrac{T}{2}\dfrac{1+z^{-1}}{1-z^{-1}}$	$T\dfrac{\cos(\Omega/2)}{j2\sin(\Omega/2)}$

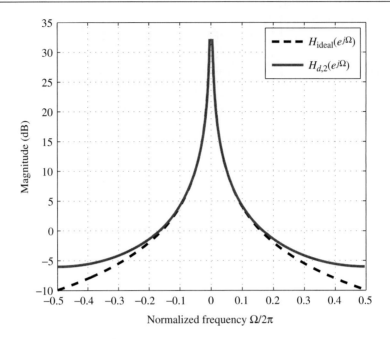

Figure 3.3.22 A comparison of the DTFT of the integrator based on the backward difference (solid line) and the DTFT of the ideal bandlimited integrator (dashed line).

In fact, the trapezoid rule can be thought of as the average of the backward difference and forward difference approximations. In the frequency domain, the two approaches differ in only a phase shift.

An indication of how well these approximations hold is revealed in the frequency domain. The DTFT of $H_{d,2}(z)$ for the backward difference is

$$H_{d,2}(e^{j\Omega}) = T\frac{e^{j\Omega/2}}{j2\sin(\Omega/2)}. \tag{3.72}$$

Figure 3.3.22 shows a plot of (3.72). For comparison, a plot of the ideal integrator

$$H_{\text{ideal}}(e^{j\Omega}) = \frac{T}{j\Omega} \tag{3.73}$$

is included. Comparing the two plots shows that the backward difference rule is a good approximation for integration in the interval $-0.1 \leq \Omega/2\pi \leq 0.1$. Thus, the backward difference performs ideal integration on a bandlimited signal as long as the signal bandwidth does not exceed 1/10 of the sample rate.[4] The frequency domain properties of forward difference and trapezoid rule are identical.

---

[4]Compare this observation with conditions under which the first central difference performs ideal differentiation.

The discrete-time integrator, illustrated in Figure 3.3.20 (c) is the foundation on which the direct digital synthesizer (DDS) is built. The DDS is the discrete-time version of the voltage controlled oscillator and is described in Section 9.2.2. The DDS is one of the core components of the discrete-time phase-locked loop as described in Appendix C. This is important in both carrier-phase synchronization and symbol-timing synchronization, which are discussed in Chapters 7 and 8, respectively.

3.4 NOTES AND REFERENCES

Multirate processing is described in much more detail in the classic text by Vaidyanathan [54] and other textbooks such as Oppenheim and Schafer [41], Proakis [55], and Harris [53]. Discrete-time filter design has produced a considerable number of published results and continues to be an active area of research. Most of the commonly used filter design techniques are described in textbooks for discrete-time signal processing. The impulse invariance and bilinear transform techniques for IIR filter design are described in the text by Oppenheim and Schafer [41], among others. These techniques are based on continuous-time filter design. Continuous-time filter design is described in many textbooks. The classic text by Humpherys [56] is the primary example. At present, there are many software packages available that automate the continuous-time filter design process. The windowing technique for FIR filter design is also described in Oppenheim and Schafer [41] and in Harris [53]. A detailed description of windows is the classic paper by Harris [57]. The Parks–McClellan algorithm was first published by Parks and McClellan in 1972 [58] and a more general version followed in 1973 [59]. A FORTRAN code listing of the algorithms was published in Refs. [60,61]. The original formula (3.49) relating FIR filter length to pass-band and stop-band ripples and transition bandwidth was published by Herrmann, Rabiner, and Chan [62]. The simplified version (3.50) by Kaiser appeared in Ref. [63], and the simplified version by Harris (3.51) appeared in Ref. [53]. The connection between the bilinear transform and the trapezoidal rule for approximating a definite integral was observed by Kaiser in Ref. [64].

3.5 EXERCISES

3.1 Consider the real-valued sequence $x(n)$ whose DTFT $X(e^{j\Omega})$ is shown in the plot below.

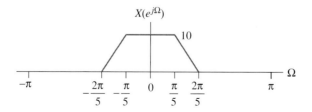

(a) Sketch the DTFT of the sequence obtained by downsampling $x(n)$ by 2.
(b) Sketch the DTFT of the sequence obtained by downsampling $x(n)$ by 4.

3.2 Consider the real-valued sequence $x(n)$ whose DTFT $X(e^{j\Omega})$ is shown in the plot below.

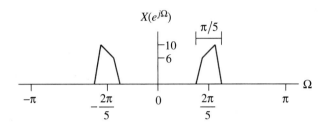

(a) Sketch the DTFT of the sequence obtained by downsampling $x(n)$ by 2.
(b) Sketch the DTFT of the sequence obtained by downsampling $x(n)$ by 4.

3.3 Consider the real-valued sequence $x(n)$ whose DTFT $X(e^{j\Omega})$ is shown in the plot below.

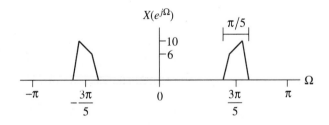

(a) Sketch the DTFT of the sequence obtained by downsampling $x(n)$ by 2.
(b) Sketch the DTFT of the sequence obtained by downsampling $x(n)$ by 4.

3.4 (a) Consider the real-valued sequence $x(n)$ whose DTFT is shown in the plot below.

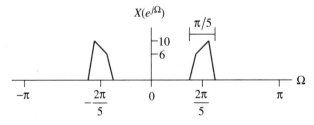

Sketch the DTFT of the sequence obtained by downsampling $x(n)$ by 2.

(b) Consider the real-valued sequence $x(n)$ whose DTFT is shown in the plot below.

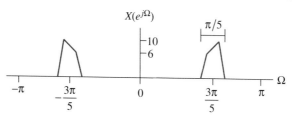

Sketch the DTFT of the sequence obtained by downsampling $x(n)$ by 2.

(c) Compare your answers to parts (a) and (b).

3.5 (a) Consider the real-valued sequence $x(n)$ whose DTFT is shown in the plot below.

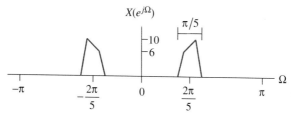

Sketch the DTFT of the sequence obtained by downsampling $x(n)$ by 4.

(b) Consider the real-valued sequence $x(n)$ whose DTFT is shown in the plot below.

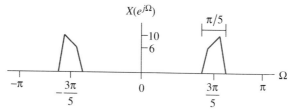

Sketch the DTFT of the sequence obtained by downsampling $x(n)$ by 4.

(c) Compare your answers to parts (a) and (b).

3.6 Consider the complex-valued sequence $x(n)$ whose DTFT $X(e^{j\Omega})$ is shown in the plot below.

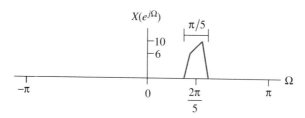

(a) Sketch the DTFT of the sequence obtained by downsampling $x(n)$ by 2.
(b) Sketch the DTFT of the sequence obtained by downsampling $x(n)$ by 4.

3.7 Consider the complex-valued sequence $x(n)$ whose DTFT $X(e^{j\Omega})$ is shown in the plot below.

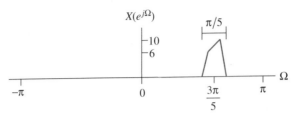

(a) Sketch the DTFT of the sequence obtained by downsampling $x(n)$ by 2.
(b) Sketch the DTFT of the sequence obtained by downsampling $x(n)$ by 4.

3.8 (a) Consider the complex-valued sequence $x(n)$ whose DTFT is shown in the plot below.

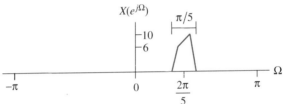

Sketch the DTFT of the sequence obtained by downsampling $x(n)$ by 2.

(b) Consider the complex-valued sequence $x(n)$ whose DTFT is shown in the plot below.

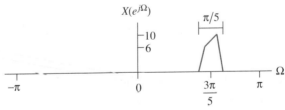

Sketch the DTFT of the sequence obtained by downsampling $x(n)$ by 2.

(c) Compare your answers to parts (a) and (b).

3.9 (a) Consider the complex-valued sequence $x(n)$ whose DTFT is shown in the plot below.

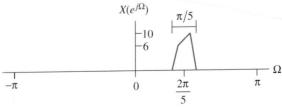

Sketch the DTFT of the sequence obtained by downsampling $x(n)$ by 4.

Section 3.5 Exercises   163

(b) Consider the complex-valued sequence $x(n)$ whose DTFT is shown in the plot below.

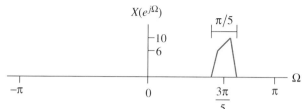

Sketch the DTFT of the sequence obtained by downsampling $x(n)$ by 4.

(c) Compare your answers to parts (a) and (b).

3.10 Consider the real-valued sequence $x(n)$ whose DTFT is shown below.

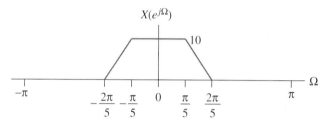

This sequence is to be downsampled by N using the system shown below.

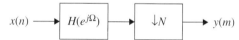

(a) Determine the requirements for the filter $H(e^{j\Omega})$ if $N = 2$ and sketch the DTFT of the output sequence $y(m)$.
(b) Determine the requirements for the filter $H(e^{j\Omega})$ if $N = 3$ and sketch the DTFT of the output sequence $y(m)$.
(c) Determine the requirements for the filter $H(e^{j\Omega})$ if $N = 4$ and sketch the DTFT of the output sequence $y(m)$.
(d) Determine the requirements for the filter $H(e^{j\Omega})$ if $N = 5$ and sketch the DTFT of the output sequence $y(m)$.

3.11 Consider the real-valued sequence $x(n)$ whose DTFT $X(e^{j\Omega})$ is shown in the plot below.

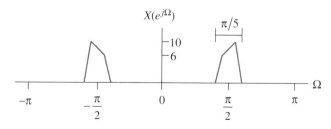

(a) Sketch the DTFT of the sequence obtained by downsampling $x(n)$ by 2.
(b) Sketch the DTFT of the sequence obtained by downsampling $x(n)$ by 4.

3.12 Consider the complex-valued sequence $x(n)$ whose DTFT $X(e^{j\Omega})$ is shown in the plot below.

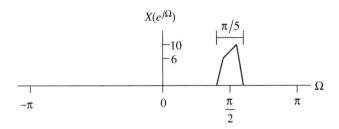

(a) Sketch the DTFT of the sequence obtained by downsampling $x(n)$ by 2.
(b) Sketch the DTFT of the sequence obtained by downsampling $x(n)$ by 4.

3.13 (a) Consider the real-valued sequence $x(n)$ whose DTFT is shown in the plot below.

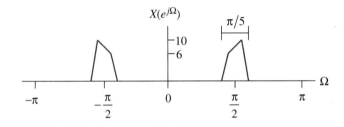

Sketch the DTFT of the sequence obtained by downsampling $x(n)$ by 2.

(b) Consider the complex-valued sequence $x(n)$ whose DTFT is shown in the plot below.

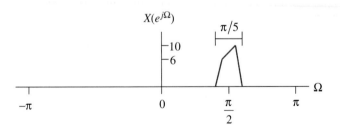

Sketch the DTFT of the sequence obtained by downsampling $x(n)$ by 2.

(c) Compare your answers to parts (a) and (b).

3.14 (a) Consider the real-valued sequence $x(n)$ whose DTFT is shown in the plot below.

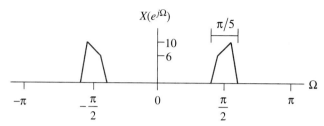

Sketch the DTFT of the sequence obtained by downsampling $x(n)$ by 4.

(b) Consider the complex-valued sequence $x(n)$ whose DTFT is shown in the plot below.

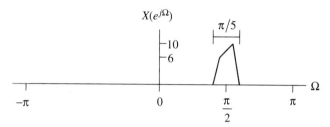

Sketch the DTFT of the sequence obtained by downsampling $x(n)$ by 4.

(c) Compare your answers to parts (a) and (b).

3.15 Consider the complex-valued sequence $x(n)$ whose DTFT is shown below.

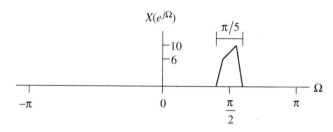

(a) Sketch the DTFT of the sequence obtained by downsampling $x(n)$ by 4.

(b) Compute the DTFT of the sequence $y(m)$ obtained using the system shown below.

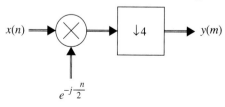

(c) Compare the answers in parts (a) and (b). What are your conclusions?

3.16 Consider the complex-valued sequence $x(n)$ whose DTFT is shown below.

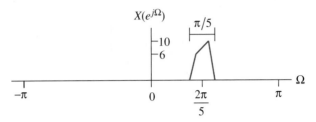

This sequence is processed by the system shown below to produce a new sequence $y(m)$.

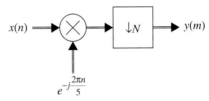

Is it possible to select a value for N so that a simple "downsample-by-N" operation produces the same output $y(m)$? If so, what is the value for N?

3.17 Let $s(n)$ be the complex-valued sequence whose DTFT is shown below.

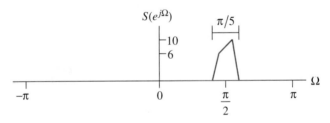

Because $s(n)$ is a complex-valued band-pass sequence whose DTFT is centered at $\pi/2$ rad/sample, it may be written in the form

$$s(n) = z(n)e^{j\pi n/2}.$$

The goal is to extract $z(n)$ from $s(n)$ at the lowest possible sample rate.

(a) Consider the system shown below.

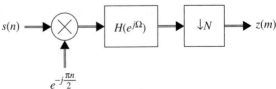

Specify the requirements for the LTI system $H(e^{j\Omega})$ and the value of N.

Section 3.5 Exercises

(b) Consider the system shown below.

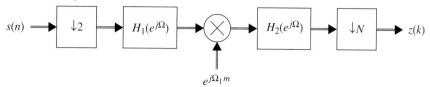

Can a system like this be used to extract $z(n)$ from $s(n)$? If so, specify the requirements for the LTI systems $H_1(e^{j\Omega})$ and $H_2(e^{j\Omega})$ and determine the value of N.

(c) Consider the system shown below.

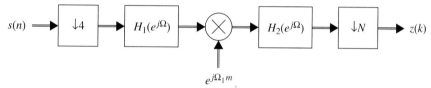

Can this system be used to extract $z(n)$ from $s(n)$? If so, specify the requirements for the LTI systems $H_1(e^{j\Omega})$ and $H_2(e^{j\Omega})$ and determine the value of N.

3.18 Consider the real-valued sequence $x(n)$ whose DTFT $X(e^{j\Omega})$ is shown in the plot below.

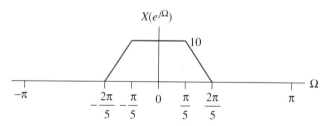

(a) Sketch the DTFT of the sequence obtained by upsampling $x(n)$ by 2.
(b) Sketch the DTFT of the sequence obtained by upsampling $x(n)$ by 4.

3.19 Consider the real-valued sequence $x(n)$ whose DTFT $X(e^{j\Omega})$ is shown in the plot below.

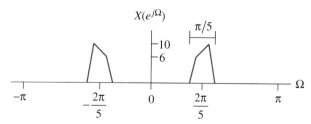

(a) Sketch the DTFT of the sequence obtained by upsampling $x(n)$ by 2.
(b) Sketch the DTFT of the sequence obtained by upsampling $x(n)$ by 4.

3.20 Consider the real-valued sequence $x(n)$ whose DTFT $X(e^{j\Omega})$ is shown in the plot below.

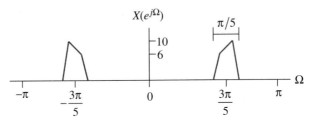

(a) Sketch the DTFT of the sequence obtained by upsampling $x(n)$ by 2.
(b) Sketch the DTFT of the sequence obtained by upsampling $x(n)$ by 4.

3.21 (a) Consider the real-valued sequence $x(n)$ whose DTFT is shown in the plot below.

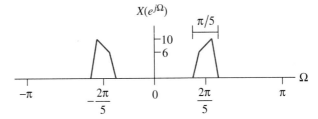

Sketch the DTFT of the sequence obtained by upsampling $x(n)$ by 2.

(b) Consider the real-valued sequence $x(n)$ whose DTFT is shown in the plot below.

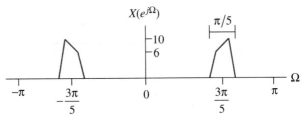

Sketch the DTFT of the sequence obtained by upsampling $x(n)$ by 2.

(c) Compare your answers to parts (a) and (b).

3.22 (a) Consider the real-valued sequence $x(n)$ whose DTFT is shown in the plot below.

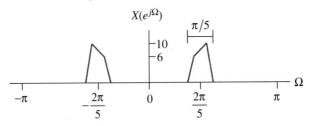

Sketch the DTFT of the sequence obtained by upsampling $x(n)$ by 4.

(b) Consider the real-valued sequence $x(n)$ whose DTFT is shown in the plot below.

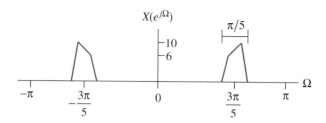

Sketch the DTFT of the sequence obtained by upsampling $x(n)$ by 4.

(c) Compare your answers to parts (a) and (b).

3.23 Consider the complex-valued sequence $x(n)$ whose DTFT $X(e^{j\Omega})$ is shown in the plot below.

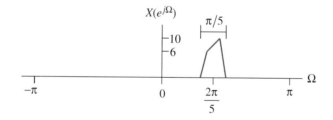

(a) Sketch the DTFT of the sequence obtained by upsampling $x(n)$ by 2.
(b) Sketch the DTFT of the sequence obtained by upsampling $x(n)$ by 4.

3.24 Consider the complex-valued sequence $x(n)$ whose DTFT $X(e^{j\Omega})$ is shown in the plot below.

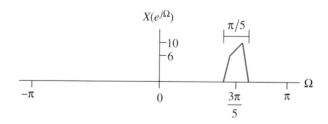

(a) Sketch the DTFT of the sequence obtained by upsampling $x(n)$ by 2.
(b) Sketch the DTFT of the sequence obtained by upsampling $x(n)$ by 4.

3.25 (a) Consider the complex-valued sequence $x(n)$ whose DTFT is shown in the plot below.

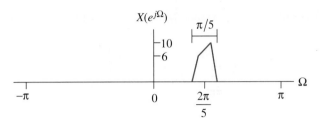

Sketch the DTFT of the sequence obtained by upsampling $x(n)$ by 2.

(b) Consider the complex-valued sequence $x(n)$ whose DTFT is shown in the plot below.

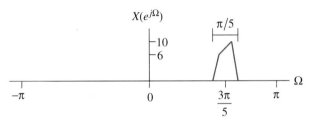

Sketch the DTFT of the sequence obtained by upsampling $x(n)$ by 2.

(c) Compare your answers to parts (a) and (b).

3.26 (a) Consider the complex-valued sequence $x(n)$ whose DTFT is shown in the plot below.

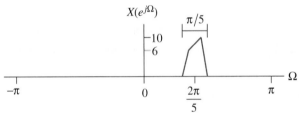

Sketch the DTFT of the sequence obtained by upsampling $x(n)$ by 4.

(b) Consider the complex-valued sequence $x(n)$ whose DTFT is shown in the plot below.

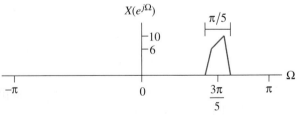

Sketch the DTFT of the sequence obtained by upsampling $x(n)$ by 4.

(c) Compare your answers to parts (a) and (b).

3.27 Complex-valued frequency translations are a common operation performed in most communication systems. The general structure of a frequency translation performed in a discrete-time modulator is illustrated by the system below. The input sequence is upsampled (to create space on the discrete-time frequency axis) and filtered by a low-pass (interpolating) filter. The upsampled and filtered sequence is frequency translated using multiplication by the complex exponential.

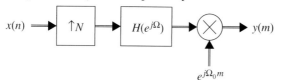

(a) Suppose the input sequence $x(n)$ has the DTFT illustrated below and that $N = 10$ and $\Omega_0 = 3\pi/5$ rad/sample.

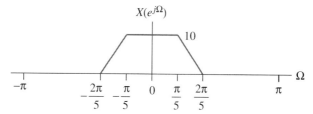

Determine the requirements for the filter $H(e^{j\Omega})$.

(b) Can a system such as the one shown below be used to accomplish the same thing? (Again, $N = 10$ and $\Omega_0 = 3\pi/5$ rad/sample.)

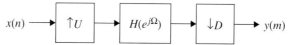

If so, specify the requirements for the LTI system $G(e^{j\Omega})$.

3.28 The real-valued sequence $x(n)$ whose DTFT is

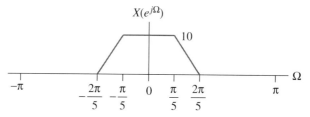

is to be resampled using the system shown below.

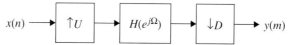

If the sample rate of $y(m)$ is 2/5 the sample rate of $x(n)$, determine the filter specifications and sketch the DTFT of the output sequence $y(m)$.

3.29 The real-valued sequence $x(n)$ whose DTFT is

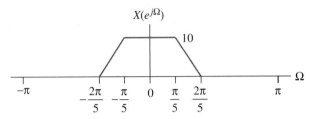

is to be resampled using the system shown below.

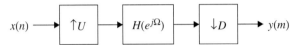

If the sample rate of $y(m)$ is 3/5 the sample rate of $x(n)$, determine the filter specifications and sketch the DTFT of the output sequence $y(m)$.

3.30 The sample rate used for the popular audio compact-disk (CD) is 44.1 ksamples/s. A competing technology is the digital audio tape (DAT) whose sample rate is 48 ksamples/s. One of the classic problems in multirate discrete-time processing is the design of an efficient resampler to convert a DAT signal to a CD signal. The (continuous-time) Fourier transform of the audio signal $x(t)$ is illustrated below.

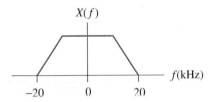

(a) Sketch the DTFT of the sequence $x(nT)$ if $1/T = 48 \times 10^3$.
(b) The desired resampling could be accomplished using the simple system shown below.

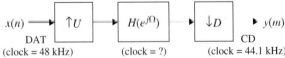

Determine U, D, and the requirements of the filter $H(e^{j\Omega})$. Assuming the filter is implemented as an FIR filter, estimate the length of the filter if the stop-band attenuation is to be 96 dB with a pass-band ripple not exceeding 0.1 dB. At what clock rate must the filter operate?

(c) The desired resampling could be accomplished using the system shown below.

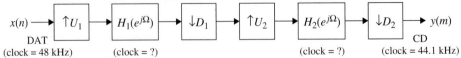

Determine suitable values for U_1, U_2, D_1, and D_2. For these values, determine the requirements for $H_1(e^{j\Omega})$ and $H_2(e^{j\Omega})$. Assuming the filters are implemented as FIR filters, estimate the length of the filters if the stop-band attenuation is to be 96 dB with a pass-band ripple not exceeding 0.1 dB. At what clock rates must the filters operate?

(d) The desired resampling could be accomplished using the system shown below.

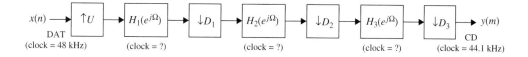

Determine suitable values for U, D_1, D_2, and D_3. For these values, determine the requirements for $H_1(e^{j\Omega})$, $H_2(e^{j\Omega})$, and $H_3(e^{j\Omega})$. Assuming the filters are implemented as FIR filters, estimate the length of the filters if the stop-band attenuation is to be 96 dB with a pass-band ripple not exceeding 0.1 dB. At what clock rates must the filters operate?

(e) To compare the systems in parts (b)–(d), compute the number of multiplications per output sample for each option. Which option is the most efficient?

(f) Describe how a polyphase filterbank could be used to accomplish the desired resampling task.

3.31 This exercise derives the relationship between the poles of a continuous-time prototype filter $H_c(s)$ and the poles of the corresponding discrete-time filter $H_d(z)$ designed using the impulse invariance technique.

(a) Suppose the continuous-time filter consists of N distinct, non-repeated poles $s_1, s_2, \ldots, s_N$. Using the partial fraction expansion technique, $H_c(s)$ may be written as

$$H_c(s) = \frac{A_1}{s - s_1} + \frac{A_2}{s - s_2} + \cdots + \frac{A_N}{s - s_N}.$$

Express the impulse response $h_c(t)$ in terms of $A_1, A_2, \ldots, A_N$ and $s_1, s_2, \ldots, s_N$.

(b) Derive the expression for the discrete-time impulse response $h_d(n)$. Your answer should be in terms of $T, A_1, A_2, \ldots, A_N$ and $s_1, s_2, \ldots, s_N$.

(c) Derive the transfer function $H_d(z)$. Your answer should be in terms of T, $A_1, A_2, \ldots, A_N$ and $s_1, s_2, \ldots, s_N$. (Hint: Write $e^{an}u(n)$ as $(e^a)^n u(n)$ and use the z-transform relationship on the third line of Table 2.4.6.)

(d) What are the poles of $H_d(z)$? What is the relationship between these poles and the poles $s_1, s_2, \ldots, s_N$.

3.32 Consider the design of a discrete-time system that produces a time delay that is a fraction of the sample time as illustrated below.

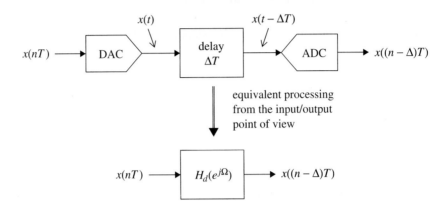

A bandlimited fractional sample delay could be realized using continuous-time processing with a DAC–ADC pair. However, if the system is to operate over a finite bandwidth $-W \leq \omega \leq +W$, then the entire operation can be performed in discrete-time processing using the technique outlined in Section 2.6.2.

(a) Derive the transfer function $H_c(j\omega)$ that produces a time delay of ΔT.
(b) Assuming that this transfer function operates over a finite bandwidth $-W \leq \omega \leq +W$, derive the transfer function of the equivalent discrete-time system $H_d(e^{j\Omega})$.
(c) Compute the inverse DTFT of $H_d(e^{j\Omega})$ to produce the desired impulse response $h_d(n)$.
(d) An important special case is a fractional delay system that operates over the entire bandwidth. What is the impulse response $h_d(n)$ for this case?
(e) Another important special case is a 1/2-sample delay that operates over the entire bandwidth. What is the impulse response $h_d(n)$ for this case?

3.33 Consider the continuous-time system with transfer function

$$H_c(s) = \frac{a}{s+a}.$$

Because this system is not bandlimited, the technique outlined in Section 2.6.2 for conversion to a discrete-time system cannot be used. This exercise explores the impulse invariance technique.

(a) Compute the impulse response $h_c(t)$.
(b) Sample the impulse response at instants $t = nT$ to produce a discrete-time impulse response $h_d(n) = h_c(nT)$.

Section 3.5 Exercises

(c) Compute the DTFT of $h_d(n)$ and compare $H_d(e^{j\Omega})$ to the Fourier transform $H_c(j\omega)$. Is this the same result obtained by using the technique outlined in Section 2.6.2?

3.34 Consider the continuous-time system with transfer function

$$H_c(s) = \frac{a}{s+a}.$$

Because this system is not bandlimited, the technique outlined in Section 2.6.2 for conversion to a discrete-time system cannot be used. This exercise explores the step invariance technique.

(a) Compute the step response $s_c(t)$.

(b) Sample the step response at instants $t = nT$ to produce a discrete-time step response $s_d(n) = s_c(nT)$.

(c) Compute the transfer function $H_d(e^{j\Omega})$. (Warning: $H_d(e^{j\Omega})$ is not the DTFT of $s_d(n)$!).

(d) Compare $H_d(e^{j\Omega})$ to the Fourier transform $H_c(j\omega)$. Is this the same result obtained by using the technique outlined in Section 2.6.2?

3.35 Consider the continuous-time system with transfer function

$$H_c(s) = \frac{a}{s+a}.$$

Because this system is not bandlimited, the technique outlined in Section 2.6.2 for conversion to a discrete-time system cannot be used. This exercise explores the use of Tustin's equation.

(a) Use Tustin's equation to convert $H_c(s)$ into an equivalent $H_d(z)$.

(b) Derive the impulse response $h_d(n)$ by computing the inverse z-transform of $H_d(z)$. How does this compare to the continuous-time impulse response?

(c) Compute corresponding DTFT $H_d(e^{j\Omega})$. Compare $H_d(e^{j\Omega})$ to the Fourier transform $H_c(j\omega)$. Is this the same result obtained by using the technique outlined in Section 2.6.2?

3.36 Evaluate the integral (3.55) to produce the expression (3.56).

3.37 Compute the DTFT of the 3-tap FIR approximation to the differentiator using a window $w(-1) = 0.5$, $w(0) = 1$, $w(1) = 0.5$. Show that for small Ω, the DTFT is approximately $j\Omega$.

3.38 Compute the DTFT of the integrator using the forward difference. Show that for small Ω, the DTFT is approximately $T/j\Omega$.

3.39 Compute the DTFT of the integrator using the Tustin's equation. Show that for small Ω, the DTFT is approximately $T/j\Omega$.

3.40 In frequency modulation (FM), the instantaneous phase of a carrier $\cos(2\pi f_c t)$ is proportional to the modulating signal $m(t)$. The relationship between the modulated carrier $s(t)$ and the modulating signal $m(t)$ is

$$s(t) = \cos(2\pi f_c t + \phi(t)) \qquad \phi(t) = 2\pi f_d \int m(x)dx.$$

The goal of an FM demodulator is to recover $\phi'(t) = d\phi(t)/dt$ from $s(t)$. (Methods for doing this with continuous-time processing are summarized in Exercises 2.63 and 2.64.) In this exercise, a discrete-time method for recovering samples of $\phi'(t)$ from samples of $s(t)$ is explored. Suppose the modulated carrier is sampled at T-spaced intervals to produce the sequence

$$s(nT) = \cos(\Omega_c n + \phi(nT))$$

where $\Omega_c = (2\pi f_c T) \mod (2\pi)$.

(a) Show that $s(nT)$ may be expressed as

$$s(nT) = \cos(\phi(nT))\cos(\Omega_c n) - \sin(\phi(nT))\sin(\Omega_c n)$$

(b) Show that the system below can produce $x(nT) = \cos(\phi(nT))$ and $y(nT) = \sin(\phi(nT))$.

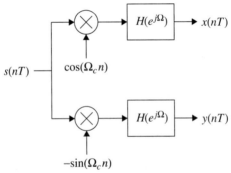

Specify, in general terms, what kind of filter the LTI system $H(e^{j\Omega})$ must be.

(c) Show that

$$\phi(nT) = \tan^{-1}\left(\frac{y(nT)}{x(nT)}\right).$$

(d) Derive a formula for $\phi'(nT)$ in terms of $x(nT)$ and $y(nT)$. Hint: use

$$\frac{d}{dx}\tan^{-1}(u) = \frac{du/dx}{1+u^2}.$$

(e) Sketch a block diagram of a system that produces $\phi'(nT)$ from $x(nT)$ and $y(nT)$.

(f) Your answer to the previous part required the use of a "derivative filter" (i.e., one that produces samples of the derivative of the discrete-time input such as the one described in Section 3.3.3). Suppose the Fourier transform of $s(t)$ is given by the plot below.

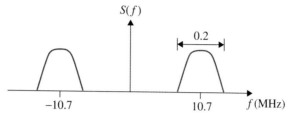

If $s(t)$ is sampled at 50 Msamples/s, design the derivative filter using the method outlined in Section 3.3.3.

4

A Review of Probability Theory

4.1 BASIC DEFINITIONS 178
4.2 GAUSSIAN RANDOM VARIABLES 188
4.3 MULTIVARIATE GAUSSIAN RANDOM VARIABLES 195
4.4 RANDOM SEQUENCES 198
4.5 ADDITIVE WHITE GAUSSIAN NOISE 202
4.6 NOTES AND REFERENCES 208
4.7 EXERCISES 208

The principles of probability and statistics are used to describe the random nature of information and noise. Because an understanding of the randomness of both signals and noise is essential for understanding the performance of a communications system, these principles are reviewed here. This review is by no means exhaustive and focuses almost exclusively on Gaussian random variables.

4.1 BASIC DEFINITIONS

The notions of probability are based on sets and an appropriately defined measure. The most fundamental set is the list of all possible outcomes of an experiment where there is uncertainty regarding *which* outcome will occur. Examples of such an experiment are tossing a coin or rolling a die. The set of all possible experimental outcomes is the *sample space* and is denoted by Ω. A given experimental outcome ω is an element of the sample space Ω. This is expressed mathematically as $\omega \in \Omega$. Let A be a subset of Ω. This is expressed mathematically as $A \subset \Omega$. The subset A is called an *event*. The set of all possible events is called the *event space* $\mathcal{E}$, that is, $\mathcal{E}$ is the set of all possible subsets of Ω.

Now define a measure P that maps members of $\mathcal{E}$ to the real interval [0, 1]. The measure P is a probability measure if the following axioms are satisfied:

1. $P(A) \geq 0$ for event $A \in \mathcal{E}$.
2. $P(\Omega) = 1$.

3. If A and B are mutually exclusive elements in $\mathcal{E}$ (i.e., are disjoint subsets of Ω), then
$$P(A \cup B) = P(A) + P(B).$$
The triple $(\Omega, \mathcal{E}, P)$ defines a *probability space*. Two examples illustrate these concepts.

Example 4.1.1
Let the experiment be the tossing of a fair coin. The possible outcomes are "heads," denoted H, and "tails," denoted T. The sample space is
$$\Omega = \{H, T\}$$
where the notation $\{\cdot\}$ means a set of objects (outcomes in this case). The event space is the set of all subsets of Ω and is
$$\mathcal{E} = \left\{\emptyset, \{H\}, \{T\}, \{H, T\}\right\}.$$
Because the coin is fair, the probability measures on the elements of $\mathcal{E}$ are
$$P(\emptyset) = 0$$
$$P(\{H\}) = \frac{1}{2}$$
$$P(\{T\}) = \frac{1}{2}$$
$$P(\{H, T\}) = 1.$$
The triple $(\Omega, \mathcal{E}, P)$ defines the probability space for this experiment.

Example 4.1.2
Let the experiment be the rolling of a fair die. The possible outcomes are one spot, two spots, etc. The sample space is
$$\Omega = \{\boxed{\cdot}, \boxed{\cdot\cdot}, \boxed{\cdot\cdot\cdot}, \boxed{::}, \boxed{:\cdot:}, \boxed{:::}\}$$
The event space $\mathcal{E}$ consists of the 64 possible subsets of Ω, which are too numerous to list here. Because the die is fair, the probability of any elementary event is $1/6$. From this, the probabilities of the events in the event space can be derived. Some examples are
$$P(\{\boxed{\cdot\cdot}\}) = \frac{1}{6}$$
$$P(\{\boxed{:\cdot:}\}) = \frac{1}{6}$$
$$P(\{\boxed{\cdot\cdot}, \boxed{:\cdot:}\}) = \frac{2}{6} = \frac{1}{3}$$
$$P(\{\boxed{\cdot}, \boxed{::}, \boxed{:::}\}) = \frac{3}{6} = \frac{1}{2}$$
Again the triple $(\Omega, \mathcal{E}, P)$ defines the probability space for this experiment.

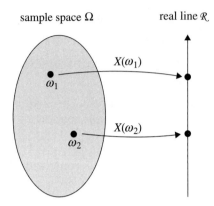

Figure 4.1.1 A graphical representation of the concept of a random variable as a mapping from the sample space Ω to the real line $\Re$.

In these examples, the number of outcomes is finite. As a consequence, the size of the sample space Ω is finite and the size of the event space $\mathcal{E}$ is finite. The elements of $\mathcal{E}$ operating under complements, unions, and intersections form what is called a Boolean algebra. When the size of the sample space is *infinite*, the event space generalizes to what is called a σ-field.

These examples illustrate that the formal definitions are based on operations involving sets of outcomes. In electrical engineering, these formal definitions are often not very useful because the familiar tools from calculus cannot be applied. For this reason, the notion of a *random variable* was developed. The random variable translates the set-based concepts to intervals on the real line. From there, more familiar mathematical operations can be applied.

A *random variable* is a mapping from the sample space Ω to the real line $\Re$. For each $\omega \in \Omega$, the random variable $X(\omega)$ is a point $x \in \Re$. This concept is illustrated in Figure 4.1.1 for two elements in the sample space. The probability space generated by this random variable is defined by the triple $(\Re, \mathcal{B}, P_X)$ where $\Re$ is the set of real numbers; $\mathcal{B}$ is the set of all countable unions, complements, and intersections of open intervals in $\Re$ (this set is called the Borel[1] set); and P_X is a probability measure for the random variable $X(\omega)$. P_X maps elements of $\mathcal{B}$ to the interval $[0, 1]$.

The probability measure P_X is derived from the probability measure P associated with the original probability space. The starting point in understanding this relationship is to consider an interval B on the real line $\Re$. (Note that $B \in \mathcal{B}$.) Because the random variable $X(\omega)$ maps Ω to $\Re$, the interval $B \in \Re$ corresponds to a set of $\omega \in \Omega$. This set is called the *inverse image* of B and is denoted $X^{-1}(B)$. The formal definition is

$$X^{-1}(B) = \left\{\omega : X(\omega) \in B\right\}. \qquad (4.1)$$

---

[1] The Borel set is named after Émile Borel, a French mathematician who worked during the first half of the 20th century. He made important contributions to measure theory and probability theory. Later in life, he was a politician and was a member of the French Resistance during World War II. A crater on the moon is named after him.

Section 4.1 Basic Definitions	**181**

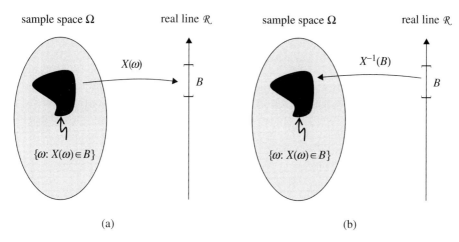

Figure 4.1.2 The relationship between a random variable and an interval on the real line: (a) the mapping from a subset of Ω to the interval $B \in \Re$; (b) the inverse image of the interval $B \in \Re$.

A graphical representation of this concept is illustrated in Figure 4.1.2. The probability measure P_X inherits its properties through the probability measure P through the inverse image. In the probability space generated by the random variable $X(\omega)$, the probability measure of B, denoted $P_X(B)$, is

$$P_X(B) = P(X^{-1}(B)) = P\left(\{\omega : X(\omega) \in B\}\right). \tag{4.2}$$

In words, this probability measure quantifies the probability that the random variable $X(\omega)$ takes on a value in the interval B.

One of the most convenient intervals to consider is the open interval $(-\infty, x]$ where x is a real number. This interval is desirable because it can be defined using only the end point x and all other intervals in $\Re$ can be expressed as complements, intersections, or unions of intervals of this form. The probability measure of this interval is

$$P_X\left((-\infty, x]\right) = P\left(X^{-1}\left((-\infty, x]\right)\right) = P\left(\{\omega : X(\omega) \leq x\}\right). \tag{4.3}$$

This probability measure is a function of the end point x. It is called the *cumulative distribution function* and is usually denoted $F_X(x)$. The formal definition of the cumulative distribution function is

$$F_X(x) = P_X\left((-\infty, x]\right) = P\left(\{\omega : X(\omega) \leq x\}\right). \tag{4.4}$$

Most textbooks on probability theory for engineers use an abbreviated notation. The argument ω is dropped from the random variable and the definition of the cumulative distribution function is shortened to

$$F_X(x) = P(X \leq x). \tag{4.5}$$

If the cumulative distribution function is differentiable, its derivative exists and is called *probability density function*:

$$f_X(x) = \frac{d}{dx} F_X(x). \tag{4.6}$$

For discrete random variables, the cumulative distribution function has a "stair-step" shape. As a consequence of these discontinuities, such a cumulative distribution function is not differentiable. If impulse functions are permitted, then the derivative exists and a probability density function may be defined. The probability density function possesses impulses. The following examples illustrate these concepts.

Example 4.1.3

Returning the fair coin toss experiment of Example 4.1.1, let the random variable X be defined as the mapping

$$X(H) = +1 \quad X(T) = -1. \tag{4.7}$$

X is a random variable that assumes one of two discrete values on the real line and inherits the probabilities associated with the elementary events H and T. Some examples of the probability measure P_X are

$$P_X(+1) = P\left(X^{-1}(+1)\right) = P(\{H\}) = \frac{1}{2}$$

$$P_X(-1) = P\left(X^{-1}(-1)\right) = P(\{T\}) = \frac{1}{2}$$

$$P_X(3.14) = P\left(X^{-1}(3.14)\right) = P(\emptyset) = 0.$$

The last line follows from the fact that 3.14 is not one of the values X can assume. Hence, its inverse image is the empty set $\emptyset$. The probability measure for the empty set is always 0. The cumulative distribution function $F_X(x)$ and the corresponding probability density function $f_X(x)$ are shown below.

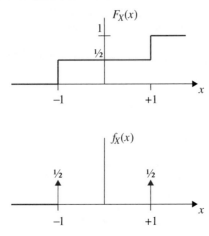

Section 4.1 Basic Definitions

The cumulative distribution function has a stair-step shape. This shape is the source of the impulse functions in the probability density function.

Example 4.1.4
Consider the fair die toss of Example 4.1.2. Let Y be the random variable defined by the mapping

$Y(⚀) = 1$ $Y(⚁) = 2$ $Y(⚂) = 3$ $Y(⚃) = 4$ $Y(⚄) = 5$ $Y(⚅) = 6$.

The random variable is a discrete random variable because it can assume a value from a finite set of possible values. The probability measure P_Y is obtained from the probability measure P as described above. Some examples are

$$P_Y(1) = P\left(Y^{-1}(1)\right) = P(\{⚀\}) = \frac{1}{6}$$

$$P_Y(2) = P\left(Y^{-1}(2)\right) = P(\{⚁\}) = \frac{1}{6}$$

$$P_Y(3) = P\left(Y^{-1}(3)\right) = P(\{⚂\}) = \frac{1}{6}$$

$$P_Y(4) = P\left(Y^{-1}(4)\right) = P(\{⚃\}) = \frac{1}{6}$$

$$P_Y(5) = P\left(Y^{-1}(5)\right) = P(\{⚄\}) = \frac{1}{6}$$

$$P_Y(6) = P\left(Y^{-1}(6)\right) = P(\{⚅\}) = \frac{1}{6}$$

$$P_Y(7) = P\left(Y^{-1}(7)\right) = P(\emptyset) = 0.$$

The probability measure for $Y = 7$ is 0 because the random variable Y cannot assume the value 7. For this reason, its inverse image is the empty set $\emptyset$ which has a probability measure 0. The cumulative distribution function $F_X(x)$ and the corresponding probability density function $f_X(x)$ are shown below.

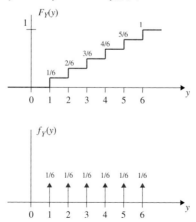

The cumulative distribution function has a stair-step shape. As before, note that the presence of step discontinuities in the cumulative distribution function produces impulse functions in the probability density function.

Example 4.1.5
The mappings that define the random variable do not have to be one-to-one as suggested by the previous two examples. For example, consider another random variable Z that maps the sample space of the fair die toss of Example 4.1.2 to the numbers -1 and $+1$ as follows:

$$Z(\boxdot) = -1 \quad Z(\boxdot) = +1 \quad Z(\boxdot) = -1$$
$$Z(\boxdot) = +1 \quad Z(\boxdot) = -1 \quad Z(\boxdot) = +1.$$

Even though there are six elements in the sample space Ω, the random variable only has two possible values. The probabilities that the random variable Z assumes its possible values is determined from the probability measure P_Z.

$$P_Z(-1) = P\left(Z^{-1}(-1)\right) = P\left(\{\boxdot,\boxdot,\boxdot\}\right) = \frac{1}{2}$$

$$P_Z(+1) = P\left(Z^{-1}(+1)\right) = P\left(\{\boxdot,\boxdot,\boxdot\}\right) = \frac{1}{2}.$$

The cumulative distribution function and the probability density function for the random variable Z are shown below.

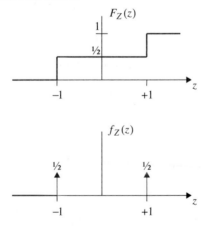

Observe that the cumulative distribution function and the probability density function for Z are identical to those of X in Example 4.1.3.

These examples illustrate that once a probability assignment is defined on the outcomes in the sample space, the probability that a random variable assumes a certain *numerical* value can be defined. This definition takes the form of the cumulative distribution function or the probability density function. The probability that a random variable X assumes a value in the

Section 4.1 Basic Definitions

interval $[a, b) \in \Re$ can be expressed in terms of either the probability density function or the cumulative distribution function:

$$P(a \leq X < b) = \int_a^b f_X(x)\,dx = F_X(a) - F_X(b). \tag{4.8}$$

Note the convention of using the uppercase letter to denote the random variable (in the subscript) and the lowercase letter to denote the value of the random variable (in the parentheses).

The probability density functions from Examples 4.1.3–4.1.5 are listed in the examples. They may also be expressed mathematically as

$$f_X(x) = \frac{1}{2}\delta(x+1) + \frac{1}{2}\delta(x-1) \tag{4.9}$$

$$f_Y(y) = \frac{1}{6}\delta(y-1) + \frac{1}{6}\delta(y-2) + \frac{1}{6}\delta(y-3)$$
$$+ \frac{1}{6}\delta(y-4) + \frac{1}{6}\delta(y-5) + \frac{1}{6}\delta(y-6) \tag{4.10}$$

$$f_Z(z) = \frac{1}{2}\delta(z+1) + \frac{1}{2}\delta(z-1). \tag{4.11}$$

Note that the whole mass of the probability density functions is concentrated at the discrete values the discrete random variables are allowed to assume. This is an important characteristic of discrete random variables. One of the more significant consequences is that the probability that a discrete random variable can assume a given numerical value may be nonzero. For example, the probability that the random variable X defined in Example 4.1.3 takes on the value $+1$ is $1/2$. That is,

$$P(X = +1) = \frac{1}{2}.$$

The probability that the random variable Y in Example 4.1.4 assumes the value 2 is $1/6$:

$$P(Y = 2) = \frac{1}{6}.$$

This is not the case with continuous random variables. The probability that a continuous random variable X assumes a particular value $x \in \Re$ is 0. (The fact that this might be true is evident when considering that X can assume an infinite number of possibilities and that the probability that it assumes any particular value from among the infinite possibilities must be arbitrarily small.)

Signal processing and communications engineers find it much easier to work in the probability space generated by the random variable rather than in the probability space generated by the underlying sample space. This is because an operation on a random variable can be expressed as a function of the random variable. As a consequence, the familiar tools of calculus can be applied to the distribution and density functions to quantify the result.

Mathematical functions can be defined and used to describe the randomness. A commonly used class of functions are generally called *moments* but have the special name *expectation* when applied to random variables. The *mean* of a random variable X is the first moment of X and is defined by

$$E\{X\} = \int_{-\infty}^{\infty} x f_X(x) dx. \tag{4.12}$$

For discrete random variables, $f_X(x)$ is a sum of impulse functions so that

$$E\{X\} = \sum_{x \in R_X} x_n f_X(x) \tag{4.13}$$

where R_X is the set of discrete values in $\Re$ that X can assume. The notation $E\{X\}$ means "the expected value of X." The mean of a discrete random variable can be interpreted as a weighted average of all possible values of X. The weighting assigned to each value is given by the probability density function.

The *variance* of a random variable X is the second central moment of X:

$$E\left\{(X - \mu_X)^2\right\} = \int_{-\infty}^{\infty} (x - \mu_X)^2 f_X(x) dx \tag{4.14}$$

where $\mu_X = E\{X\}$. Again, for discrete random variables, the probability density function consists of a set of weighted impulses so that

$$E\left\{(X - \mu_X)^2\right\} = \sum_{x \in R_X} (x - \mu_X)^2 f_X(x) \tag{4.15}$$

where R_X is the set of discrete values in $\Re$ that X can assume. The notation $E\left\{(X - \mu_X)^2\right\}$ means "the expected value of $(X - \mu_X)^2$."

Higher order moments can also be defined. The n-th central moment of the random variable X is the expected value of $(X - \mu_X)^n$:

$$E\left\{(X - \mu_X)^n\right\} = \int_{-\infty}^{\infty} (x - \mu_X)^n f_X(x) dx \tag{4.16}$$

which, for the discrete case reduces to

$$E\left\{(X - \mu_X)^n\right\} = \sum_{x \in R_X} (x - \mu_X)^n f_X(x) \tag{4.17}$$

where $\mu_X = E\{X\}$ and R_X is the set of discrete values in $\Re$ that X can assume.

Three of the most common continuous random variables are the uniform, exponential, and Gaussian random variables. The probability density functions for these random variables are

$$\text{Uniform: } f_X(x) = \begin{cases} 0 & x < a \\ \dfrac{1}{b-a} & a \leq x \leq b \\ 0 & x > b \end{cases} \qquad (4.18)$$

$$\text{Exponential: } f_X(x) = \begin{cases} ae^{-ax} & x \geq 0 \\ 0 & \text{otherwise} \end{cases} \qquad (4.19)$$

$$\text{Gaussian: } f_X(x) = \frac{1}{\sqrt{2\pi\sigma^2}} \exp\left\{-\frac{(x-\mu)^2}{2\sigma^2}\right\} \qquad (4.20)$$

Examples of these probability density functions are illustrated in Figure 4.1.3. Gaussian random variables play a tremendously important role in digital communications. For this reason, a thorough treatment of Gaussian random variables is presented in the next section.

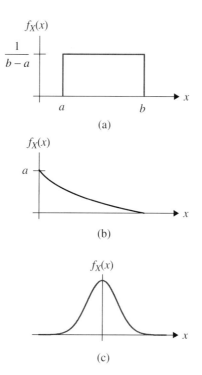

Figure 4.1.3 Examples of probability density functions for continuous random variables: (a) Uniform random variable; (b) Exponential random variable; (c) Gaussian random variable.

4.2 GAUSSIAN RANDOM VARIABLES

4.2.1 Density and Distribution Functions

Let X be a Gaussian (or normal) random variable. Then X has the probability density function

$$f_X(x) = \frac{1}{\sqrt{2\pi\sigma^2}} \exp\left\{-\frac{(x-\mu)^2}{2\sigma^2}\right\}. \tag{4.21}$$

The mean is

$$E\{X\} = \int_{-\infty}^{\infty} x f_X(x) dx = \mu \tag{4.22}$$

and the variance is

$$E\left\{(X-\mu)^2\right\} = \int_{-\infty}^{\infty} (x-\mu)^2 f_X(x) dx = \sigma^2. \tag{4.23}$$

The Gaussian probability density function is completely specified by the two parameters μ and σ^2. As a consequence, knowledge of the mean and variance of a Gaussian random variable allows one to write down the probability density function. For this reason, the notation

$$X \sim N(\mu, \sigma^2) \tag{4.24}$$

is often used to mean X is a Gaussian (or normal — hence the "N") random variable with mean μ and variance σ^2. When $X \sim N(0, 1)$, X is said to have the *standard normal distribution*.

The cumulative distribution function of the Gaussian random variable X is given by

$$F_X(x) = \int_{-\infty}^{x} f_X(t) dt \tag{4.25}$$

and is used to compute the probability that the random variable X is less than a value x. Unfortunately, the cumulative distribution function given by the integral (4.25) has no closed form and therefore must be evaluated numerically. Historically, the most common method was to numerically compute (4.25), or some closely related integral, and list the values in a table. One of the most commonly used related integrals was the so-called error function $\text{erf}(x)$ defined by

$$\text{erf}(x) = \frac{2}{\sqrt{\pi}} \int_{0}^{x} e^{-t^2} dt. \tag{4.26}$$

Section 4.2 Gaussian Random Variables

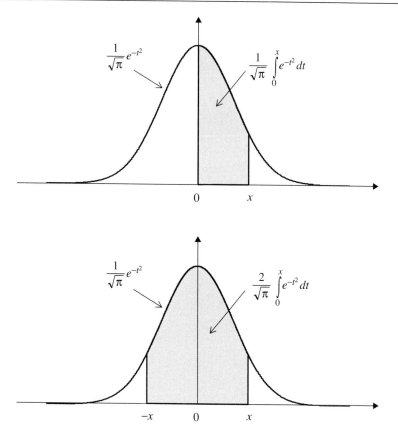

Figure 4.2.1 A graphical interpretation of the error function.

A graphical interpretation of erf(x) is illustrated in Figure 4.2.1. Note that

$$\frac{1}{\sqrt{\pi}}e^{-t^2}$$

is the probability density function of a Gaussian random variable with zero mean and variance 1/2. The integral of this probability density function over the interval $0 \leq t \leq x$ is shown. This integral represents one-half the area under the main "hump" of the probability density function. This means that the integral (4.26) is the area under the main hump of the probability density function.

In many applications, the probability that a Gaussian random variable X *exceeds* a certain value x is of interest. The probability corresponds to the area in the "upper tail" of the Gaussian probability density function as illustrated in Figure 4.2.2. Because this area is the complement

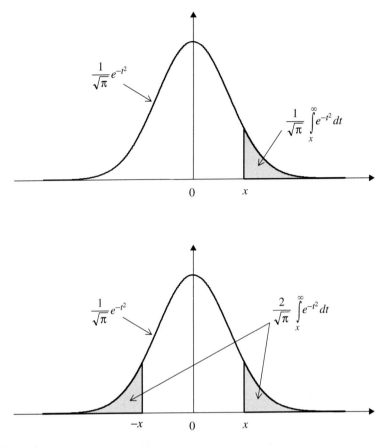

Figure 4.2.2 A graphical interpretation of the complementary error function.

of the area represented by the error function erf(x), a function called the *complementary error function* erfc(x) is defined:

$$\text{erfc}(x) = 1 - \text{erf}(x) = \frac{2}{\sqrt{\pi}} \int_x^\infty e^{-t^2} dt. \qquad (4.27)$$

The graphical interpretation of this integral is that it is the area under the two "tails" of the probability density function of a Gaussian random variable with zero mean and variance 1/2.

Another common form is the Q-function defined by

$$Q(x) = \frac{1}{\sqrt{2\pi}} \int_x^\infty \exp\left\{-\frac{t^2}{2}\right\} dt \qquad (4.28)$$

Section 4.2 Gaussian Random Variables

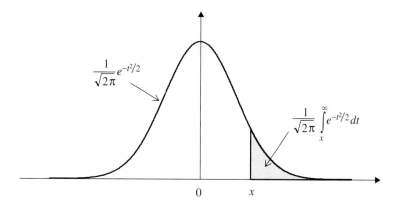

Figure 4.2.3 A graphical interpretation of the Q-function.

which is the area under the upper tail of the standard normal probability density function as illustrated in Figure 4.2.3. It is easy to show, using the appropriate substitutions in the integral definitions, that the Q-function and the complementary error function are related by

$$Q(x) = \frac{1}{2}\text{erfc}\left(\frac{x}{\sqrt{2}}\right). \tag{4.29}$$

The probability that $X \sim N(\mu, \sigma^2)$ is greater than x can be expressed in terms of the upper tail probability of a standard Gaussian random variable using the Q-function. Starting from basic principles, the desired probability is expressed in integral form and is reexpressed using the substitution

$$v = \frac{t - \mu}{\sigma}$$

as follows:

$$P(X > x) = \frac{1}{\sqrt{2\pi\sigma^2}} \int_x^\infty \exp\left\{-\frac{(t-\mu)^2}{2\sigma^2}\right\} = \frac{1}{\sqrt{2\pi}} \int_{\frac{x-\mu}{\sigma}}^\infty \exp\left\{\frac{v^2}{2}\right\} dv = Q\left(\frac{x-\mu}{\sigma}\right). \tag{4.30}$$

Many computing tools used by communications engineers contain built-in functions to compute integrals associated with the Gaussian probability density function. MATLAB® has built in functions for both the error function (4.26) and the complementary error function (4.27). The Q-function may be implemented in MATLAB by writing a simple script based on the relationship (4.29). An example of such a script is

```
function y = Q(x)
y = 0.5*erfc(x/sqrt(2));
```

The HP calculators (models HP 28 and HP 48) have a built-in variation of the Q-function called UTPN for Upper Tail Probability Normal. $Q(x)$ may be computed as follows in HP-speak:

>   0   :the mean of the standard normal distribution
>   1   :the variance of the standard normal distribution
>   x   :the x of $Q(x)$
>   UTPN :executes the function

UTPN, however, is even more general. To compute the probability that $X \sim N(\mu, \sigma^2)$ is greater than x, enter the following:

> μ :the mean
> σ^2 :the variance
>   x   :the x that X is greater than
>   UTPN :executes the function

The Texas Instruments scientific calculators also have a built-in routine to find the area under the Gaussian probability density function with mean μ and variance σ^2. For the TI-83, the function is called `normalcdf`. To find the area under the probability density function for $X \sim N(\mu, \sigma^2)$ from $x1$ to $x2$, use

$$\text{normalcdf}(x1, x2, \mu, \sigma^2).$$

The function `normalcdf` can be found by pressing 2nd VARS (DIST) and choosing 2:normalcdf(. To use the command to calculate $Q(x)$ execute the following command:

$$\text{normalcdf}(x, 1\text{e}99, 0, 1).$$

For the TI-89, the function is `tistat.normcdf`, which can be found by pressing CATALOG F3 and choosing normCdf(. To calculate $P(x1 \leq X \leq x2)$ for $X \sim N(\mu, \sigma^2)$, use

$$\text{tistat.normcdf}(x1, x2, \mu, \sigma).$$

To calculate $Q(x)$, use

$$\text{tistat.normcdf}(x, 1\text{e}99, 0, 1).$$

4.2.2 Product Moments

Let X and Y be two random variables with respective means μ_X and μ_Y and with joint probability density function $f_{X,Y}(x, y)$. The *correlation function* corr(X, Y) of X and Y is defined as

$$\text{corr}(X, Y) = E\{XY\} \tag{4.31}$$

$$= \int_{-\infty}^{\infty} \int_{-\infty}^{\infty} xy f_{X,Y}(x, y) dx dy \tag{4.32}$$

Note that when X and Y are independent

$$\text{corr}(X, Y) = \mu_X \mu_Y. \tag{4.33}$$

Section 4.2 Gaussian Random Variables

The *covariance function* cov(X, Y) of the random variables X and Y is defined as

$$\text{cov}(X, Y) = E\{(X - \mu_X)(Y - \mu_Y)\} \quad (4.34)$$

$$= \int_{-\infty}^{\infty} \int_{-\infty}^{\infty} (x - \mu_X)(y - \mu_Y) f_{X,Y}(x, y) dx dy \quad (4.35)$$

Note that when X and Y are independent, $\text{cov}(X, Y) = 0$. The covariance function is indicative of the relationship between the values of X and Y. When the covariance is positive, there is a high probability that large values of X occur with large values of Y, and small values of X occur with small values of Y. When the covariance is negative, there is a high probability that large values of X occur with small values of Y, and small values of X occur with large values of Y. When the covariance is zero, the two random variables are said to be *uncorrelated*.[2]

4.2.3 Functions of Random Variables

In the analysis of communications systems, functions of random variables are often encountered. Let X and Y be two Gaussian random variables and let Z be a random variable that is a function of both X and Y:

$$Z = g(X, Y).$$

There are several methods for determining the probability density function of Z as outlined in most introductory texts on probability and random variables: the distribution function technique, the transformation technique, and the moment-generating function technique. Here, the results for a few important cases of $g(\cdot, \cdot)$ are stated without proof.

Z = aX + bY

When $X \sim N(\mu_X, \sigma_X^2)$ and $Y \sim N(\mu_Y, \sigma_Y^2)$ are independent, $Z \sim N(\mu_Z, \sigma_Z^2)$ where

$$\mu_Z = a\mu_X + b\mu_Y$$
$$\sigma_Z^2 = a^2 \sigma_X^2 + b^2 \sigma_Y^2$$

Z = X² + Y²

When $X \sim N(0, \sigma^2)$ and $Y \sim N(0, \sigma^2)$ are independent (with common variance σ^2), Z is a *central chi-square* random variable with probability density function given by

$$f_Z(z) = \frac{1}{2\sigma^2} \exp\left\{-\frac{z}{2\sigma^2}\right\} \quad z \geq 0. \quad (4.36)$$

---

[2] Note that, in general, independence implies uncorrelated but not vice versa. Gaussian random variables are an exception. Two Gaussian random variables are independent if and only if they are uncorrelated.

When $X \sim N(\mu_X, \sigma^2)$ and $Y \sim N(\mu_Y, \sigma^2)$ are independent, Z is a *noncentral chi-square* random variable with probability density function given by

$$f_Z(z) = \frac{1}{2\sigma^2} \exp\left\{-\frac{z+s^2}{2\sigma^2}\right\} I_0\left(\sqrt{z}\frac{s}{\sigma^2}\right) \quad z \geq 0 \tag{4.37}$$

where s^2 is the noncentrality parameter defined by

$$s^2 = \mu_X^2 + \mu_Y^2 \tag{4.38}$$

and $I_0(\cdot)$ is the zeroth order modified Bessel function of the first kind.

$Z = \sqrt{X^2 + Y^2}$

When $X \sim N(0, \sigma^2)$ and $Y \sim N(0, \sigma^2)$ are independent, Z is a *Rayleigh* random variable with probability density function

$$f_Z(z) = \frac{z}{\sigma^2} \exp\left\{-\frac{z^2}{2\sigma^2}\right\} \quad z \geq 0. \tag{4.39}$$

Note that Z is really the square root of a central chi-square random variable.

When $X \sim N(\mu_X, \sigma^2)$ and $Y \sim N(\mu_Y, \sigma^2)$ are independent, Z is a *Rice* random variable with probability density function

$$f_Z(z) = \frac{z}{\sigma^2} \exp\left\{-\frac{z^2 + s^2}{2\sigma^2}\right\} I_0\left(\frac{zs}{\sigma^2}\right) \quad z \geq 0 \tag{4.40}$$

where s^2 is the noncentrality parameter given by (4.38) and $I_0(\cdot)$ is the zeroth order modified Bessel function of the first kind. Note that Z is really the square root of a noncentral chi-square random variable.

Rectangular to Polar Conversion. Let $X \sim N(\mu_x, \sigma^2)$ and $Y \sim N(\mu_y, \sigma^2)$ be independent Gaussian random variables and interpret them as Cartesian coordinates. The rectangular to polar conversion defines two new random variables R and Θ using the transform

$$R = \sqrt{X^2 + Y^2} \tag{4.41}$$

$$\Theta = \tan^{-1}\frac{Y}{X}. \tag{4.42}$$

The joint probability density function of R and Θ is given by

$$f_{R,\Theta}(r, \theta) = \frac{r}{2\pi\sigma^2} \exp\left\{-\frac{r^2 + \mu_x^2 + \mu_y^2 - 2r(\mu_x \cos\theta + \mu_y \sin\theta)}{2\sigma^2}\right\}. \tag{4.43}$$

4.3 MULTIVARIATE GAUSSIAN RANDOM VARIABLES

Another frequently encountered scenario in signal processing and communications involves a sequence of N random variables that are not only marginally Gaussian but also jointly Gaussian. A workable form for the joint probability density function of these random variables is required to characterize the response of a system to these random variables. This workable form is called the *multivariate* Gaussian probability density function.

Let $X_1, X_2, \ldots, X_N$ be N random variables with means $\mu_1, \mu_2, \ldots, \mu_N$; variances $\sigma_1^2, \sigma_2^2, \ldots, \sigma_N^2$; and covariances

$$\begin{aligned} m_{i,j} &= \text{cov}(X_i, X_j) \\ &= E\{(X_i - \mu_i)(X_j - \mu_j)\} \\ & i = 1, 2, \ldots, N \quad j = 1, 2, \ldots, N \end{aligned}$$

arranged into the $N \times N$ covariance matrix $\mathbf{M}$

$$\mathbf{M} = \begin{bmatrix} m_{1,1} & m_{1,2} & \cdots & m_{1,N} \\ m_{2,1} & m_{2,2} & \cdots & m_{2,N} \\ \vdots & \vdots & & \vdots \\ m_{N,1} & m_{N,2} & \cdots & m_{N,N} \end{bmatrix}$$

For notational convenience, the random variables, the values they may assume, and the means are organized into the following length-N vectors:

$$\mathbf{X} = \begin{bmatrix} X_1 \\ X_2 \\ \vdots \\ X_N \end{bmatrix} \quad \mathbf{x} = \begin{bmatrix} x_1 \\ x_2 \\ \vdots \\ x_N \end{bmatrix} \quad \boldsymbol{\mu} = \begin{bmatrix} \mu_1 \\ \mu_2 \\ \vdots \\ \mu_N \end{bmatrix}$$

The N random variables are referred to as *jointly Gaussin (normal)* means that the joint probability density function is of the form

$$f_\mathbf{X}(\mathbf{x}) = \frac{1}{(2\pi)^{N/2} |\mathbf{M}|^{1/2}} \exp\left\{-\frac{1}{2}(\mathbf{x} - \boldsymbol{\mu})^T \mathbf{M}^{-1} (\mathbf{x} - \boldsymbol{\mu})\right\} \quad (4.44)$$

where $|\mathbf{M}|$ denotes the determinant of $\mathbf{M}$, $\mathbf{M}^{-1}$ denotes the inverse of $\mathbf{M}$, and $\mathbf{x}^T$ denotes the transpose of $\mathbf{x}$.

Note that the covariance matrix can also be expressed in terms of $\mathbf{X}$ as

$$\mathbf{M} = E\{\mathbf{X}\mathbf{X}^T\} \quad (4.45)$$

where the expectation of a matrix should be interpreted as the expectation applied to each element in the matrix. The vector product $\mathbf{X}\mathbf{X}^T$ is sometimes called the *outer product* to distinguish it from the *inner product* $\mathbf{X}^T\mathbf{X}$. The outer product is an $N \times N$ matrix whereas the inner product is a scalar.

4.3.1 Bivariate Gaussian Distribution

The special case corresponding to $N = 2$ produces what is commonly referred to as the *bivarate Gaussian probability density function*. This case is a very important one and merits additional comments. The covariance matrix for the bivariate Gaussian probability density function is expressed in the special form

$$\mathbf{M} = \begin{bmatrix} \sigma_1^2 & \sigma_1 \sigma_2 \rho \\ \sigma_1 \sigma_2 \rho & \sigma_2^2 \end{bmatrix} \quad (4.46)$$

where ρ is the normalized covariance given by

$$\rho = \frac{\text{cov}\{X_1 X_2\}}{\sigma_1 \sigma_2}. \quad (4.47)$$

Using these definitions the vectors and matrices of (4.44) are

$$|\mathbf{M}| = \sigma_1^2 \sigma_2^2 (1 - \rho^2)$$

$$(\mathbf{x} - \mu)^T \mathbf{M}^{-1} (\mathbf{x} - \mu) = [(x_1 - \mu_1) \ (x_2 - \mu_2)] \begin{bmatrix} \sigma_1^2 & \sigma_1 \sigma_2 \rho \\ \sigma_1 \sigma_2 \rho & \sigma_2^2 \end{bmatrix}^{-1} \begin{bmatrix} (x_1 - \mu_1) \\ (x_2 - \mu_2) \end{bmatrix}$$

$$= \frac{\sigma_2^2 (x_1 - \mu_1)^2 - 2\sigma_2 \sigma_2 (x_1 - \mu_1)(x_2 - \mu_2) + \sigma_1^2 (x_2 - \mu_2)}{\sigma_1^2 \sigma_2^2 (1 - \rho^2)}.$$

Putting it all together produces the bivariate Gaussian probability density function

$$f_{X_1, X_2}(x_1, x_2) = \frac{1}{2\pi \sigma_1 \sigma_2 \sqrt{1 - \rho^2}} \exp \left\{ -\frac{1}{2(1 - \rho^2)} \right.$$

$$\left. \times \left[\left(\frac{x_1 - \mu_1}{\sigma_1} \right)^2 - 2\rho \left(\frac{x_1 - \mu_1}{\sigma_1} \right) \left(\frac{x_2 - \mu_2}{\sigma_2} \right) + \left(\frac{x_2 - \mu_2}{\sigma_2} \right)^2 \right] \right\}. \quad (4.48)$$

In this representation, the correlation between the random variables is made explicit through the use of the correlation coefficient ρ. The correlation coefficient measures how the two random variables X_1 and X_2 vary together (i.e., it is a measure of the statistical dependency between the two). The bivariate Gaussian random variables are independent if and only if $\rho = 0$.

It is easy to show that the two marginal distributions are also Gaussian. The converse, however, is not true: two random variables with Gaussian distributions are not necessarily jointly Gaussian.

An illustration of (4.48) is shown in Figure 4.3.1 where the z-axis is the function $f_{X_1, X_2}(x_1, x_2)$ evaluated at the point (x_1, x_2). The bivariate surface has the following properties:

- $f_{X_1, X_2}(x_1, x_2)$ attains a maximum at the point (μ_1, μ_2).
- The intersection of any plane parallel to the z-axis and the bivariate normal surface has the Gaussian shape.

Section 4.3 Multivariate Gaussian Random Variables **197**

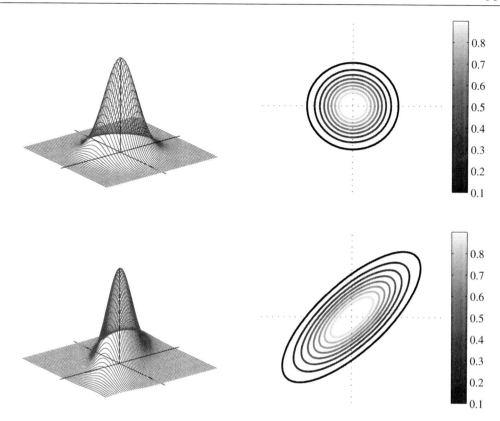

Figure 4.3.1 Examples of the bivariate Gaussian probability density function: (upper left) Surface plot for the case $\rho = 0$; (upper right) contour plot for the case $\rho = 0$; (lower left) surface plot for the case $\rho = 0.75$; (lower right) contour plot for the case $\rho = 0.75$.

- The intersection of a plane parallel to the xy-plane and the bivariate normal surface is an ellipse that is called a *contour of constant probability density*. When $\rho = 0$ and $\sigma_1^2 = \sigma_2^2$, these contours are circles and the bivariate density is sometimes called the *circular normal distribution*.

4.3.2 Linear Operators and Multivariate Gaussian Random Variables

The multivariate Gaussian probability density function is completely specified by the covariance matrix **M** and the mean vector $\boldsymbol{\mu}$. This means that knowledge of **M** and $\boldsymbol{\mu}$ are all that is required to write down the joint probability density function. As a consequence, when analyzing linear combinations of N jointly Gaussian random variables, all the analysis focuses on computing the covariance matrix and the mean vector for the result.

To illustrate how this works, consider the linear operation

$$Y = aX_1 + bX_2 \tag{4.49}$$

applied to a jointly Gaussian random variables X_1 and X_2 for real-valued constants a and b. Let μ_1 and μ_2 be the means of X_1 and X_2, respectively, and let $\mathbf{M}$ be the covariance matrix. The linear operation can be expressed as a matrix equation. Y is the product of a 1×2 matrix $\mathbf{A}$ and the vector $\mathbf{X}$ as follows:

$$Y = \mathbf{AX} = [a \quad b] \begin{bmatrix} X_1 \\ X_2 \end{bmatrix}. \tag{4.50}$$

Because Y is a linear combination of Gaussian random variables, Y is also a Gaussian random variable. All that remains to be done is to compute the mean vector $\boldsymbol{\mu}'$ and the covariance matrix $\mathbf{M}'$. A straightforward application of the definitions produces

$$\boldsymbol{\mu}' = \mathbf{A}\boldsymbol{\mu} = [a \quad b] \begin{bmatrix} \mu_1 \\ \mu_2 \end{bmatrix} = a\mu_1 + b\mu_2 \tag{4.51}$$

$$\mathbf{M}' = E\{(Y - \boldsymbol{\mu}')(Y - \boldsymbol{\mu}')^T\} = E\{(\mathbf{AX} - \mathbf{A}\boldsymbol{\mu})(\mathbf{AX} - \mathbf{A}\boldsymbol{\mu})^T\}$$
$$= \mathbf{A}E\{(\mathbf{X} - \boldsymbol{\mu})(\mathbf{X} - \boldsymbol{\mu})^T\}\mathbf{A}^T = \mathbf{A}\mathbf{M}\mathbf{A}^T \tag{4.52}$$

Because $\mathbf{A}$ is a 1×2 matrix, Y is a 1×1 matrix or scalar. Thus, $\boldsymbol{\mu}'$ and $\mathbf{M}'$ are scalars and Y is a Gaussian random variable with mean and variance given by (4.51) and (4.52), respectively. Note that this is a generalization of the special case of a linear combination of two independent Gaussian random variables described in Section 4.2.3.

If the joint probability density function of X_1 and X_2 is expressed in the standard bivariate form (4.48), then the covariance matrix $\mathbf{M}$ is of the form (4.46) and the variance of Y may be expressed as

$$\mathbf{M}' = \mathbf{A}\mathbf{M}\mathbf{A}^T = [a \quad b] \begin{bmatrix} \sigma_1^2 & \sigma_1\sigma_2\rho \\ \sigma_1\sigma_2\rho & \sigma_2^2 \end{bmatrix} \begin{bmatrix} a \\ b \end{bmatrix}$$
$$= a^2\sigma_1^2 + 2ab\sigma_1\sigma_2\rho + b^2\sigma_2^2. \tag{4.53}$$

Note that when X_1 and X_2 are independent, $\rho = 0$ and (4.53) reduces to $a^2\sigma_1^2 + b^2\sigma_2^2$ as described in Section 4.2.3.

In general, a linear transformation of N Gaussian random variables into K random variables can be represented by a $K \times N$ matrix $\mathbf{A}$. The resulting K random variables are jointly Gaussian with mean vector $\boldsymbol{\mu}'$ and covariance matrix $\mathbf{M}'$ are given y

$$\boldsymbol{\mu}' = \mathbf{A}\boldsymbol{\mu} \tag{4.54}$$

$$\mathbf{M}' = \mathbf{A}\mathbf{M}\mathbf{A}^T. \tag{4.55}$$

4.4 RANDOM SEQUENCES

Let $X(1), X(2), \ldots$ be a sequence of indexed random variables. The previous definitions can be applied except the result is a function of the index n:

Mean function. $\mu_X(n) = E\{X(n)\}$

Section 4.4 Random Sequences

Variance function. $\sigma_X^2(n) = E\{(X(n) - \mu(n))^2\}$

Correlation function. $R_{XX}(n,k) = E\{X(n)X(k)\}$

Covariance function. $C_{XX}(n,k) = E\{(X(n) - \mu(n))(X(k) - \mu(k))\}$

The correlation function defined above is usually called the *autocorrelation function* because it computes the expected value of the product of two random variables from the same sequence.[3] Similarly, the covariance function defined above is often referred to as the *autocovariance function*. The autocovariance function measures the relationship between $X(n)$ and $X(k)$. When the autocovariance is positive, there is a high probability that large values of $X(n)$ occur with large values of $X(k)$ and that small values of $X(n)$ occur with small values of $X(k)$. When the autocovariance is negative, there is a high probability that large values of $X(n)$ occur with small values of $X(k)$ and that small values of $X(n)$ occur with large values of $X(k)$. In other words, the sequence has *memory* and the random sequence is said to be correlated over the range of indices n and k where the autocovariance is either positive or negative. The instantaneous power contained in the random sequence X is given by $R_{XX}(n,n)$.

A *Gaussian random sequence* is a sequence of random variables for which any finite number of members of the sequence are jointly Gaussian. This means that the joint probability function of any N members of the sequence $X(1), X(2), \ldots$ is of the form (4.44).

4.4.1 Power Spectral Density

Sometimes the autocorrelation function is a function only of the difference between n and k. In this case

$$R_{XX}(n,k) = R_{XX}(n-k). \tag{4.56}$$

Setting $m = n - k$, the autocorrelation function can be thought of as a function of time via the time delay index m as follows:

$$R_{XX}(n-k) = R_{XX}(m) = E\{X(n)X(n-m)\}. \tag{4.57}$$

The frequency domain properties of this discrete-time sequence can be quantified using the DTFT. The DTFT of the autocorrelation function is denoted

$$S_X(e^{j\Omega}) = \sum_{m=-\infty}^{\infty} R_{XX}(m)e^{-j\Omega m}. \tag{4.58}$$

$S_X(e^{j\Omega})$ is called the *power spectral density* of the random sequence $X(n)$ and is a measure of the power contained in the sequence as a function of frequency.

An extremely important special case is the *white* random sequence. Let $X(n)$ for $n = 1, 2, \ldots$ be a random sequence with autocorrelation function

$$R_{XX}(k) = E\{X(n)X(n-k)\} = \sigma^2 \delta(k). \tag{4.59}$$

---

[3]The expected value of the product of two random variables can be computed using members of two different random sequences. In this case the function is called the *cross correlation function*.

Because $R_{XX}(k)$ is zero for $k \neq 0$, this sequence is completely uncorrelated. That is, each element in the sequence is uncorrelated from all the other elements. For a Gaussian random sequence, this means that each element in the sequence is independent from all other elements in the sequence. The power spectral density is the DTFT of (4.59)

$$S_X(e^{j\Omega}) = \sigma^2 \qquad (4.60)$$

which shows that all the average power in the sequence is equally shared by all frequencies in the sequence.[4]

For zero-mean Gaussian random sequences, the power spectral density uniquely specifies the autocovariance matrix **M** through the inverse DTFT. In other words, the power spectral density is just another way of specifying the autocovariance matrix of the multivariate Gaussian probability density function that describes the sequence.

4.4.2 Random Sequences and Discrete-Time LTI Systems

Let $X(n)$ be a sequence of Gaussian random variables with autocorrelation function $R_{XX}(k)$ and power spectral density $S_X(e^{j\Omega})$. Now, suppose $X(n)$ is the input to a discrete-time LTI system with impulse response $h(n)$ as shown in Figure 4.4.1. What can be said about the autocorrelation $R_Y(k)$ and power spectral density of the output $Y(n)$? First, it is known that $Y(n)$ is related to $X(n)$ through convolution:

$$Y(n) = \sum_{k=-\infty}^{\infty} X(k)h(n-k).$$

Second, it is known that $Y(n)$ will be a sequence of random variables because the input is a sequence of random variables. Further, for fixed n, $Y(n)$ is a linear combination of the X's. Because each $X(k)$ is Gaussian, $Y(n)$ is also Gaussian.

Using these observations, the autocorrelation of $Y(n)$ may be expressed as

$$R_Y(k) = E\{Y(n)Y(n-k)\}$$

$$= E\left\{\sum_{i=-\infty}^{\infty} h(i)X(n-i) \sum_{m=-\infty}^{\infty} h(m)X(n+k-m)\right\}$$

$$= \sum_{i=-\infty}^{\infty} \sum_{m=-\infty}^{\infty} h(i)h(m) E\{X(n-i)X(n+k-i)\}$$

$$= \sum_{i=-\infty}^{\infty} \sum_{m=-\infty}^{\infty} h(i)h(m) R_{XX}(k-m+i)$$

$$= \sum_{i=-\infty}^{\infty} h(i) \sum_{m=-\infty}^{\infty} h(m) R_{XX}(k-m+i)$$

---

[4] In physics *white light* is light where all colors (or frequencies) have the same power. This terminology has been borrowed here and applied to random sequences where all frequencies have the same power.

Section 4.4 Random Sequences

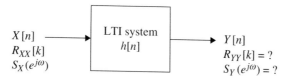

Figure 4.4.1 LTI system with a random sequence input. What is the output?

$$= \sum_{i=-\infty}^{\infty} h(i) \, (h(k) * R_{XX}(k+i))$$

$$= h(-k) * h(k) * R_{XX}(k)$$

The power spectral density of $Y(n)$ is thus

$$S_Y(e^{j\Omega}) = \sum_{k=-\infty}^{\infty} R_Y(k) e^{-j\Omega k}$$

$$= \sum_{k=-\infty}^{\infty} \sum_{i=-\infty}^{\infty} \sum_{m=-\infty}^{\infty} h(i) h(m) R_{XX}(k-m+i) e^{-j\Omega k}$$

$$= \sum_{i=-\infty}^{\infty} \sum_{m=-\infty}^{\infty} h(i) h(m) \sum_{k=-\infty}^{\infty} R_{XX}(k-m+i) e^{-j\Omega k}$$

$$= \sum_{i=-\infty}^{\infty} \sum_{m=-\infty}^{\infty} h(i) h(m) \sum_{l=-\infty}^{\infty} R_{XX}(l) e^{-j\Omega(l+m-i)}$$

$$= \sum_{i=-\infty}^{\infty} \sum_{m=-\infty}^{\infty} h(i) h(m) e^{-j\Omega(m-i)} \sum_{l=-\infty}^{\infty} R_{XX}(l) e^{-j\Omega l}$$

$$= \underbrace{\sum_{i=-\infty}^{\infty} h(i) e^{j\Omega i}}_{H^*(e^{j\Omega})} \underbrace{\sum_{m=-\infty}^{\infty} h(m) e^{-j\Omega m}}_{H(e^{j\Omega})} \underbrace{\sum_{l=-\infty}^{\infty} R_{XX}(l) e^{-j\Omega l}}_{S_X(e^{j\Omega})}$$

$$= |H(e^{j\Omega})|^2 S_X(e^{j\Omega}).$$

The picture of Figure 4.4.1 can now be completed as shown in Figure 4.4.2.

Example 4.4.1
Let $X(n)$ be a white Gaussian random sequence with

$$\mu_X(n) = 0 \quad \sigma_X^2(n) = \sigma^2 \quad R_X(k) = \sigma^2 \delta(k) \quad S_X(e^{j\omega}) = \sigma^2$$

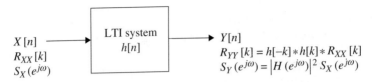

Figure 4.4.2 LTI system with random input showing the properties of the resulting random output.

$X(n)$ is the input to a discrete-time LTI system with impulse response

$$h(n) = a^n u(n).$$

If $Y(n)$ is the output of the LTI system, then the power spectral density of $Y(n)$ is

$$S_Y(e^{j\Omega}) = |H(e^{j\Omega})|^2 S_X(e^{j\Omega}) = \left|\frac{1}{1 - ae^{-j\Omega}}\right|^2 \sigma^2 = \frac{\sigma^2}{1 + a^2 - 2a\cos\Omega}. \qquad (4.61)$$

The autocorrelation function $R_{YY}(k)$ is the inverse DTFT of $S_Y(e^{j\Omega})$:

$$R_{YY}(k) = \frac{\sigma^2}{1 - a^2} a^{|k|}. \qquad (4.62)$$

$R_{XX}(k), S_X(e^{j\Omega}), S_Y(e^{j\Omega})$ and $R_{YY}(k)$ are illustrated in Figure 4.4.3.

4.5 ADDITIVE WHITE GAUSSIAN NOISE

The performance of digital communication schemes is almost always quantified by the probability of error in the additive white Gaussian noise (AWGN) environment. In a wireless communication system, most of the additive noise is the result of the addition of random waveforms by the receiver electronics. As a consequence, additive noise is best understood and analyzed in the continuous-time world. Thus, a very brief overview of continuous-time random processes is included to help understand the essential properties of the additive noise and how it manifests itself in a sampled data system.

4.5.1 Continuous-Time Random Processes

A continuous-time random process is a generalization of a continuous random variable. Just as a continuous random variable is a mapping from a continuous sample set Ω to a real number, a continuous time *random process* is a mapping from a continuous sample set Ω to a real function of time. This concept is illustrated in Figure 4.5.1. Formally, this mapping should be denoted $X(\omega, t)$ for $\omega \in \Omega$ to convey the notion that a particular realization of the random process is both a function of the sample space and time. The convention is to drop the dependence on the sample space and to use the notation $X(t)$.

Section 4.5 Additive White Gaussian Noise

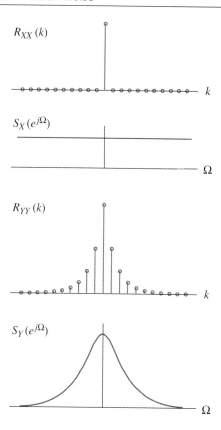

Figure 4.4.3 Autocorrelation functions and power spectral densities for the input and output random sequences of Example 4.4.1.

The basic property of a random process is that at any fixed time, say t_0, $X(t_0)$ is a *random variable*. This concept is also illustrated in Figure 4.5.1. When $X(t_0)$ is a Gaussian random variable, $X(t)$ is called a *Gaussian random process*. Because $X(t_0)$ is a random variable, the mean and variance of $X(t_0)$ are defined in a straightforward way:

$$\mu_X(t_0) = E\{X(t_0)\} \tag{4.63}$$

$$\sigma_X^2(t_0) = E\left\{\left(X(t_0) - \mu_X(t_0)\right)^2\right\}. \tag{4.64}$$

In the most general case, the mean and variance are a function of choice of t_0.

For two instances of time, say t_0 and t_1, $X(t_0)$ and $X(t_1)$ are two random variables. The correlation and autocovariance functions may be defined in a way very closely related to how these functions were defined for random sequences:

$$R_{XX}(t_0, t_1) = E\{X(t_0)X(t_1)\} \tag{4.65}$$

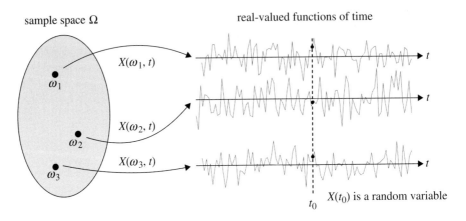

Figure 4.5.1 A graphical representation of notion that a random process is a mapping from the sample space Ω to a real-valued waveform. Compare this picture with the one for a random variable in Figure 4.1.1.

$$C_{XX}(t_0, t_1) = E\left\{\left(X(t_0) - \mu_X(t_0)\right)\left(X(t_1) - \mu_X(t_1)\right)\right\}. \qquad (4.66)$$

Sometimes the autocorrelation function is a function only of the difference $\tau = t_0 - t_1$. In this case, the autocorrelation function may be expressed as

$$R_{XX}(\tau) = E\{X(t_0)X(t_0 - \tau)\} \qquad (4.67)$$

so that the temporal dependencies of the random process are a function of a signal variable of time.

The spectral properties of the random process $X(t)$ may be quantified by computing the Fourier transform of the autocorrelation function $R_{XX}(\tau)$. The Fourier transform of the autocorrelation is called the *power spectral density* of the random process and is denoted $S_X(f)$:

$$S_X(f) = \int_{-\infty}^{\infty} R_{XX}(\tau) e^{j2\pi f \tau} d\tau. \qquad (4.68)$$

When the power spectral density $S_X(f)$ is a constant, the power in the random process is spread uniformly in frequency and the random process is called a *white random process*.

4.5.2 The White Gaussian Random Process: A Good Model for Noise?

Thermal noise is the result of the random motion of charge carriers in a conducting or semiconducting device. Atomic level motion occurs in all matter at temperatures above absolute zero. Because the motion of the charge carriers is random, the tiny voltage produced by each moving charge carrier is also random. At the device terminals, the measured noise voltage is the sum of the random voltages due to the individual charge carriers. Given the large number of individual charge carriers in any device, the law of large numbers predicts that the measured voltage signal $V(t)$ will tend toward a Gaussian random process.

Nyquist was one of the first to study the properties of thermal noise. Using the principles of thermodynamics and quantum mechanics, Nyquist argued that the power spectral density of thermal noise is

$$S_V(f) = \frac{hf}{\exp\left\{\frac{hf}{kT}\right\} - 1} \quad \text{W/Hz} \qquad (4.69)$$

where

$$h \text{ (Planck's constant)} = 6.6261 \times 10^{-34} \quad \text{Js}$$
$$k \text{ (Boltzmann's constant)} = 1.3807 \times 10^{-23} \quad \text{J/K}$$

and T is the temperature in Kelvin. The power spectral density is plotted in Figure 4.5.2 for various temperatures. Observe that for the frequencies of interest in wireless communications (approximately 100 kHz to 100 GHz), $S_V(f)$ is constant for practical temperatures.[5] It can be shown that this constant value is kT W/Hz (see Exercise 4.30). The magnitude of the power spectral density is proportional to the equivalent temperature, hence the name *thermal noise*.

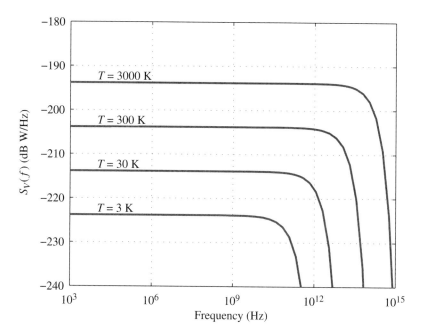

Figure 4.5.2 The power spectral density of thermal noise.

---

[5]Absolute zero is $-273.15°$C. The Kelvin scale uses 0 for absolute zero and increments on the centigrade (or Celsius) scale. Room temperature is $72°$F $= 22.2°$C $= 295.4$ K.

Note that the power spectral density is not constant for all frequencies. Any wireless communication system, however, involves filtering that removes most of the spectral energy outside the spectral band occupied by the signal. Thus, if the random process $V(t)$ were replaced by a fictitious random process $\tilde{W}(t)$ that is a zero-mean Gaussian random process with power spectral density

$$S_{\tilde{W}}(f) = kT \quad \text{W/Hz}, \tag{4.70}$$

an observation taken *after* the filtering would not be able to tell the difference between the output corresponding to $V(t)$ and the output corresponding to $\tilde{W}(t)$. For this reason, the thermal noise is usually modeled using the fictitious random process $\tilde{W}(t)$. Because the power spectral density of $\tilde{W}(t)$ is a constant, $\tilde{W}(t)$ is a white Gaussian random process. For historical reasons, this constant value is designated $N_0/2$:

$$S_{\tilde{W}}(f) = N_0/2 \quad \text{W/Hz}. \tag{4.71}$$

The autocorrelation function is the inverse Fourier transform of (4.71):

$$R_{\tilde{W}\tilde{W}}(\tau) = \frac{N_0}{2}\delta(\tau). \tag{4.72}$$

Observe that $\tilde{W}(t)$ has infinite power ($P = R_{\tilde{W}\tilde{W}}(0) = \infty$) and as such is not a physically realizable signal. As discussed above, this signal is a mathematical abstraction used to simplify the analysis of filtered continuous-time waveforms in noise.

As a final note, the zero-mean white Gaussian random process described above is not a good model for many types of noise and interference. It is intended only to model the thermal noise that affects the waveforms being processed by the receiver electronics. Other types of noise, such as shot noise, phase noise, and flicker noise, use other random models.

4.5.3 White Gaussian Noise in a Sampled Data System

To assess the impact of additive white Gaussian noise on the performance of a sampled-data receiver, it must be understood how the noise waveform shows up in the samples after conversion to discrete time. To accomplish this, sampling of the random process $\tilde{W}(t)$, described in the previous section, is analyzed.

Because $\tilde{W}(t)$ has infinite bandwidth, the first step in the sampling process is to band limit $\tilde{W}(t)$ by using an anti-aliasing low-pass filter with impulse response $h(t)$ and transfer function $H(f)$. The purpose of the low-pass filter is to limit the spectrum of $\tilde{W}(t)$ to $f \leq \pm\frac{1}{2T}$ to prevent aliasing in the T-spaced samples. Let $W(t)$ be the bandlimited output of this filter when $\tilde{W}(t)$ is the input. Using the basic definitions of convolution, autocorrelation, and power spectral density, it can be shown that the power spectral density of $W(t)$ may be expressed as[6]

$$S_W(f) = |H(f)|^2 S_{\tilde{W}}(f). \tag{4.73}$$

---

[6]This relationship is the continuous-time analog of the discrete-time results presented in Section 4.4.2.

Section 4.5 Additive White Gaussian Noise

If the low-pass filter is an ideal low-pass filter with

$$H(f) = \begin{cases} 1 & |f| \leq \dfrac{1}{2T} \\ 0 & |f| > \dfrac{1}{2T} \end{cases} \qquad (4.74)$$

then the power spectral density and corresponding autocorrelation function of the filter output are

$$S_W(f) = \begin{cases} \dfrac{N_0}{2} & |f| \leq \dfrac{1}{2T} \\ 0 & |f| > \dfrac{1}{2T} \end{cases} \qquad R_{WW}(\tau) = \dfrac{N_0}{2T} \dfrac{\sin\left(\dfrac{\pi \tau}{T}\right)}{\dfrac{\pi \tau}{T}}. \qquad (4.75)$$

This situation is illustrated in Figure 4.5.3. Conversion to discrete-time produces T-spaced samples of $W(t)$ denoted $W(nT)$ for integers n. Because $\tilde{W}(t)$ is a Gaussian random process, $W(t)$ is also a Gaussian random process. This is because filtering by an LTI system is a linear operation. By the definition of a random process, the samples $W(nT)$ are Gaussian random variables. The sequence of random variables forms a random sequence of Gaussian random variables. This sequence is completely described by the joint probability density function of all the elements in the sequence. The multivariate Gaussian density function is completely specified using the mean vector and autocorrelation matrix. It is easy to show that each $W(nT)$ has zero mean (this follows from the fact that $\tilde{W}(t)$ is a zero-mean random process). The entries in the autocorrelation matrix are the values of the autocorrelation function for the random sequence. The autocorrelation function for this sequence of random variables is

$$R_{WW}(k) = E\left\{W(nT)W((n-k)T)\right\} = R_{WW}(kT) = \dfrac{N_0}{2T}\delta(k). \qquad (4.76)$$

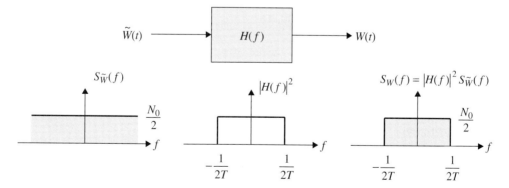

Figure 4.5.3 The use of an ideal low-pass filter to produce an ideally bandlimited Gaussian random process suitable for sampling.

Thus, the samples of a zero-mean white Gaussian random process are uncorrelated zero-mean Gaussian random variables with common variance

$$\sigma^2 = R_{WW}(0) = \frac{N_0}{2T}. \qquad (4.77)$$

4.6 NOTES AND REFERENCES

There are many textbooks devoted to probability theory. These books range from introductory treatments through advanced expositions to impenetrable abstractions. Examples of introductory texts include Hogg and Tanis [65] and Ross [66]. Many texts are written for engineers, which include the texts by Beckmann [67], Davenport and Root [68], Gray and Davisson [69], Helstrom [70], Leon-Garcia [71], Papoulis [72], Stark and Woods [73], and Ziemer [74]. More advanced treatments of the topic are found in Tucker [75], Karr [76], Wong [77], and Gray and Davisson [78].

4.7 EXERCISES

4.1 Let $X \sim N(0, 1)$. Express in terms of the Q-function each of the following:
  (a) $\Pr\{X < -A\}$
  (b) $\Pr\{X < +A\}$
  (c) $\Pr\{X > -A\}$
  (d) $\Pr\{-A < X < +A\}$

4.2 Prove that the picture shown in Figure 4.2.1 is correct by proving

$$\frac{2}{\sqrt{\pi}} \int_0^x e^{-t^2} dt = \frac{1}{\sqrt{\pi}} \int_{-x}^x e^{-t^2} dt.$$

4.3 Prove that the picture shown in Figure 4.2.2 is correct by proving

$$\frac{2}{\sqrt{\pi}} \int_x^\infty e^{-t^2} dt = \frac{1}{\sqrt{\pi}} \int_{-\infty}^{-x} e^{-t^2} dt + \frac{1}{\sqrt{\pi}} \int_x^\infty e^{-t^2} dt.$$

4.4 Using the basic integral definitions and a wise choice for variable substitution, prove (4.29).

4.5 Let X be a zero-mean Gaussian random variable with variance σ^2.
  (a) Express the probability that $-\sigma \leq X \leq \sigma$ using the Q-function. Evaluate your answer.
  (b) Express the probability that $-2\sigma \leq X \leq 2\sigma$ using the Q-function. Evaluate your answer.

Section 4.7 Exercises

(c) Express the probability that $-3\sigma \leq X \leq 3\sigma$ using the Q-function. Evaluate your answer.

(d) Express the probability that $-4\sigma \leq X \leq 4\sigma$ using the Q-function. Evaluate your answer.

4.6 Let X be a Gaussian random variable with mean μ and variance σ^2.
  (a) Express the probability that $\mu - \sigma \leq X \leq \mu + \sigma$ using the Q-function. Evaluate your answer.
  (b) Express the probability that $\mu - 2\sigma \leq X \leq \mu + 2\sigma$ using the Q-function. Evaluate your answer.
  (c) Express the probability that $\mu - 3\sigma \leq X \leq \mu + 3\sigma$ using the Q-function. Evaluate your answer.
  (d) Express the probability that $\mu - 4\sigma \leq X \leq \mu + 4\sigma$ using the Q-function. Evaluate your answer.

4.7 Let X be a Gaussian random variable with zero mean and variance 9. What is the probability that X is *not* in the interval $-6 < X < 6$?

4.8 Derive the central chi-square probability density function (4.36).

4.9 Derive the noncentral chi-square probability density function (4.37).

4.10 Derive the Rayleigh probability density function (4.39).

4.11 Derive the Rice probability density function (4.40).

4.12 Derive the joint probability density function (4.43) resulting from the rectangular to polar conversion of two independent Gaussian random variables.

4.13 Consider two jointly Gaussian random variables X_1 and X_2 with means 1 and 2, respectively, and covariance matrix

$$\mathbf{M} = \begin{bmatrix} 5 & 0 \\ 0 & 5 \end{bmatrix}.$$

  (a) Compute the probability that $X_1 > 3$ and $X_2 > 3$.
  (b) Compute the probability that $X_1 > 1$ and $X_2 < 2$.

4.14 Consider two jointly Gaussian random variables X_1 and X_2 with means 2 and 5, respectively, and covariance matrix

$$\mathbf{M} = \begin{bmatrix} 10 & 0 \\ 0 & 10 \end{bmatrix}.$$

Compute the probability that $X_1 > X_2$.

4.15 Consider two jointly Gaussian random variables X_1 and X_2 with means 2 and 5, respectively, and covariance matrix

$$\mathbf{M} = \begin{bmatrix} 10 & 2 \\ 2 & 10 \end{bmatrix}.$$

Compute the probability that $X_1 > X_2$.

4.16 Consider three jointly Gaussian random variables $X_1, X_2,$ and X_3 with joint probability density function given by (4.48) where the means are 0, 4, and 2, respectively, and the autocovariance matrix is given by

$$\mathbf{M} = \begin{bmatrix} 1 & 0.5 & 0.1 \\ 0.5 & 2 & 0.5 \\ 0.1 & 0.5 & 1 \end{bmatrix}$$

(a) Evaluate the joint probability density function at the point $\mathbf{x} = [0.25, 5, 2]^T$.
(b) Compute the probability that $X_1 + X_2 + X_3 > 5$.

4.17 Consider three jointly Gaussian random variables $X_1, X_2,$ and X_3 with joint probability density function given by (4.48) where the means are 1, 1, and 1, respectively, and the autocovariance matrix is given by

$$\mathbf{M} = \begin{bmatrix} 1 & 0 & 0 \\ 0 & 1 & 0 \\ 0 & 0 & 1 \end{bmatrix}$$

(a) Evaluate the joint probability density function at the point $\mathbf{x} = [0.25, 5, 2]^T$.
(b) Compute the probability that $X_1 > 2$ and $X_2 > 2$ and $X_3 > 2$.

4.18 Consider five jointly Gaussian random variables $X_1, X_2, \ldots X_5$ with joint probability density function given by (4.48) where the means are 2, 1, 1, 2, and 1, respectively, and the autocovariance matrix is given by

$$\mathbf{M} = \begin{bmatrix} 1 & 0.3 & 0.4 & 0.2 & 0.1 \\ 0.3 & 1 & 0.3 & 0.4 & 0.2 \\ 0.4 & 0.3 & 1 & 0.3 & 0.4 \\ 0.2 & 0.4 & 0.3 & 1 & 0.3 \\ 0.1 & 0.2 & 0.4 & 0.3 & 1 \end{bmatrix}$$

(a) Determine the probability density function for $\mathbf{Y} = \mathbf{A}\mathbf{X}$ where

$$\mathbf{A} = \begin{bmatrix} 5 & 1 & 2 & 4 & 3 \\ 2 & 4 & 6 & 1 & 2 \end{bmatrix}$$

(b) Evaluate the probability density function of $\mathbf{Y}$ at $\mathbf{y} = [0.5, 0.5]^T$.
(c) Compute the probability that $Y_1 > Y_2$.

4.19 A rotation matrix $\mathbf{R}$ is the linear operator that rotates two points in a Cartesian space by θ. The rotation matrix may be expressed as

$$\mathbf{R} = \begin{bmatrix} \cos\theta & -\sin\theta \\ \sin\theta & \cos\theta \end{bmatrix}. \tag{4.78}$$

In other words, if the point $\mathbf{x} = [x_1, x_2]^T$ is rotated by θ to the new point $\mathbf{y} = [y_1, y_2]^T$, then the points are related by $\mathbf{y} = \mathbf{Rx}$. Let

$$\mathbf{X} = \begin{bmatrix} X_1 \\ X_2 \end{bmatrix} \quad (4.79)$$

be two jointly Gaussian random variables with probability density function (4.48) where

$$\mu = \begin{bmatrix} 0 \\ 0 \end{bmatrix} \quad \mathbf{M} = \begin{bmatrix} \sigma^2 & 0 \\ 0 & \sigma^2 \end{bmatrix}.$$

Show that when $\mathbf{X}$ is rotated by θ, the resulting point $\mathbf{Y} = \mathbf{RX}$ is composed of two jointly Gaussian random variables that have zero mean and are still uncorrelated.

4.20 A unitary operator is a linear operator, defined by the square matrix $\mathbf{U}$ that satisfies the property

$$\mathbf{UU}^T = \mathbf{U}^T\mathbf{U} = \mathbf{I}.$$

One of the interesting consequences of this property is that $\mathbf{U}^{-1} = \mathbf{U}^T$. Let

$$\mathbf{X} = \begin{bmatrix} X_1 & X_2 & \cdots & X_N \end{bmatrix}^T$$

be N jointly Gaussian random variables with density function (4.48) where

$$\mu = \begin{bmatrix} 0 \\ 0 \\ \vdots \\ 0 \end{bmatrix} \quad \mathbf{M} = \begin{bmatrix} \sigma^2 & 0 & \cdots & 0 \\ 0 & \sigma^2 & \cdots & 0 \\ & & \ddots & \\ 0 & \cdots & 0 & \sigma^2 \end{bmatrix}.$$

In other words, X consists of N zero-mean, uncorrelated Gaussian random variables with a common variance. Show that the random vector

$$\mathbf{Y} = \mathbf{UX}$$

also consists of N zero-mean, uncorrelated Gaussian random variables with a common variance.

4.21 Express the covariance function $C_{XX}(n, k)$ of a random sequence in terms of the correlation function $R_{XX}(n, k)$.

4.22 Consider a zero-mean Gaussian random sequence consisting of four random variables

$$X_1, X_2, X_3, X_4.$$

The four random variables are jointly Gaussian with probability density function given by (4.44). Assuming the covariance matrix is given by

$$\mathbf{M} = \begin{bmatrix} 1 & 0.35 & 0.13 & 0.04 \\ 0.35 & 1 & 0.35 & 0.13 \\ 0.13 & 0.35 & 1 & 0.35 \\ 0.04 & 0.13 & 0.35 & 1 \end{bmatrix}$$

Determine the power spectral density $S_X\left(e^{j\Omega}\right)$.

4.23 Consider a zero-mean Gaussian random sequence consisting of four random variables

$$X_1, X_2, X_3, X_4.$$

Let the power spectral density be

$$S_X\left(e^{j\Omega}\right) = 1 + \cos(\Omega) + 0.5\cos(2\Omega) + 0.25\cos(3\Omega).$$

(a) Determine the joint probability density function of $\mathbf{X} = [X_1, X_2, X_3, X_4]^T$.
(b) Compute the probability that $X_1 + X_2 + X_3 + X_4 > 0$.

4.24 Consider a zero-mean Gaussian random sequence consisting of four random variables

$$X_1, X_2, X_3, X_4$$

with autocorrelation function

$$R_{XX}(m) = (-2)^m.$$

(a) Determine the covariance matrix $\mathbf{M}$ of the joint probability density function of $\mathbf{X} = [X_1, X_2, X_3, X_4]^T$.
(b) Evaluate the joint probability density function at the point $\mathbf{x} = [0.5, 0, 0.25, 1]^T$.
(c) Compute the probability that $X_1 + X_2 + X_3 + X_4 > 0$.
(d) Compute the power spectral density for this random sequence.

4.25 Consider a system that computes the three-point moving average of the input. This system is modeled as an LTI system with impulse response given by

$$h(n) = \frac{1}{3}\left(\delta(n+1) + \delta(n) + \delta(n-1)\right).$$

Suppose the input to this system is a Gaussian random sequence with zero mean and power spectral density σ^2.
(a) Compute the power spectral density of the output.
(b) Compute the autocorrelation function of the output.
(c) What does this tell you about the moving average of a white random sequence?

4.26 Consider an LTI system described by the impulse response

$$h(n) = \frac{\sin Wn}{\pi n}.$$

Suppose the input to this system is a Gaussian random sequence with zero mean and power spectral density σ^2.
(a) Compute the power spectral density of the output.
(b) Compute the autocorrelation function of the output.

4.27 Consider an LTI system described by the impulse response

$$h(n) = (n+1)a^n u(n).$$

Suppose the input to this system is a Gaussian random sequence with zero mean and power spectral density σ^2.
(a) Compute the power spectral density of the output.
(b) Compute the autocorrelation function of the output.

4.28 Consider an LTI system described by the impulse response

$$h(n) = \begin{cases} 1 & |n| < N_1 \\ 0 & |n| > N_1 \end{cases}.$$

Suppose the input to this system is a Gaussian random sequence with zero mean and power spectral density σ^2.
(a) Compute the power spectral density of the output.
(b) Compute the autocorrelation function of the output.

4.29 Consider an LTI system described by the impulse response

$$h(n) = \delta(n - n_0).$$

Suppose the input to this system is a Gaussian random sequence with zero mean and power spectral density σ^2.
(a) Compute the power spectral density of the output.
(b) Compute the autocorrelation function of the output.

4.30 Prove that for $hf \ll kT$, the power spectral density given by (4.69) is approximated by

$$S_n(f) \approx kT. \tag{4.80}$$

(Hint: use the series expansion

$$e^x = 1 + x + \frac{1}{2!}x^2 + \frac{1}{3!}x^3 + \frac{1}{4!}x^4 + \cdots$$

and truncate appropriately.)

5

Linear Modulation 1: Modulation, Demodulation, and Detection

5.1 SIGNAL SPACES   215
5.2 M-ARY BASEBAND PULSE AMPLITUDE MODULATION   227
5.3 M-ARY QUADRATURE AMPLITUDE MODULATION   238
5.4 OFFSET QPSK   260
5.5 MULTICARRIER MODULATION   265
5.6 MAXIMUM LIKELIHOOD DETECTION   273
5.7 NOTES AND REFERENCES   279
5.8 EXERCISES   280

A memoryless M-ary digital communication system uses M waveforms to transmit $\log_2(M)$ bits over a waveform channel during each symbol interval T_s. Let

$$\mathcal{S} = \{s_0(t), s_1(t), \ldots s_{M-1}(t)\} \qquad (5.1)$$

represent the set of M waveforms. The M possible $\log_2(M)$–bit patterns are mapped to *symbols* that alter the amplitudes of basis waveforms that are combined to construct each of the waveforms in the signal set $\mathcal{S}$. When the combination is linear (i.e., a weighted sum of the basis waveforms), the resulting signal set belongs to the general class of waveforms known as *linear modulation*.

Linear modulation formats are very popular in wireless communications as a result of their good bit-error-rate performance and bandwidth performance. These signal sets are conveniently described using the concept of a *signal space*. A signal space is a generalized version of a vector space. Just as a vector space is a set of vectors, a signal space is a set of signals (or functions of time). As each vector in a vector space can be described as a linear combination of basis vectors, so too can each signal in a signal space be described as a linear combination of basis functions. It will be shown that a signal set can be thought of as a subset in an appropriately chosen signal space.

The properties of the signals in a signal space are fundamentally tied to the properties of the basis functions. Because the signal set is a subset in the signal space, the properties of the signal set are also linked to the properties of the basis functions. If the basis functions are low-pass waveforms, the resulting signal set is also low pass. Signal sets based on a single baseband basis function are called *baseband pulse amplitude modulation* or baseband PAM. If two sinusoids, 90° out of phase, are used as basis functions, then the resulting signal set is a band-pass signal set called *quadrature amplitude modulation* or QAM.

This chapter focuses on PAM and QAM signal sets. Many other signal sets are possible by selecting other basis functions. However, these two signal sets represent the most commonly used types of signal sets in wireless communications. The chapter begins with an overview of signal spaces and then applies these principles to the generation and detection of PAM and QAM. In these sections, the application of both continuous-time processing and discrete-time processing to realize modulators, demodulators, and detectors is summarized. Short descriptions of offset QPSK and multicarrier modulation follow. The chapter concludes with a derivation of the maximum likelihood decision rule for additive linear modulation (PAM or QAM) in additive white Gaussian noise.

5.1 SIGNAL SPACES

5.1.1 Definitions

Let $\mathcal{B} = \{\phi_0(t), \phi_1(t), \ldots, \phi_{K-1}(t)\}$ be a set of K real-valued waveforms defined on the interval $T_1 \leq t \leq T_2$. The following definitions are important:

Linear Independence. The waveforms in the set $\mathcal{B}$ are *linearly independent* if

$$\sum_{i=0}^{K-1} c_i \phi_i(t) = 0$$

only if $c_0 = c_1 = \cdots = c_{K-1} = 0$. Otherwise, the set of waveforms $\mathcal{B}$ is *linearly dependent*.

Span. The *span of* $\mathcal{B}$, denoted Span$\{\mathcal{B}\}$ is the set of all waveforms that are linear combinations of the waveforms in the set $\mathcal{B}$. In other words

$$y(t) \in \text{Span}\{\mathcal{B}\} \leftrightarrow y(t) = \sum_{k=0}^{K-1} a_k \phi_k(t)$$

where the $a_k, k = 0, 1, \ldots, K-1$ are constants.

Energy. The energy of $\phi_k(t)$ is

$$E_k = \int_{T_1}^{T_2} \phi_k^2(t) dt.$$

Orthogonality. The waveforms $\phi_i(t)$ and $\phi_j(t)$ are *orthogonal* if

$$\int_{T_1}^{T_2} \phi_i(t)\phi_j(t)dt = 0.$$

Note that a real-valued signal cannot be orthogonal to itself. The above equation reduces to the energy in $\phi_j(t)$ when $i = j$. As an extension, let $x(t)$ and $y(t)$ be two signals in Span$\{\mathcal{B}\}$. $x(t)$ and $y(t)$ are orthogonal if

$$\int_{T_1}^{T_2} x(t)y(t)dt = 0.$$

Orthonormal Set. $\mathcal{B}$ is an orthonormal set of waveforms if

$$\int_{T_1}^{T_2} \phi_i(t)\phi_j(t) = \delta(i-j).$$

Because

$$\delta(i-j) = \begin{cases} 1 & i = j \\ 0 & i \neq j \end{cases}$$

orthonormality means the set $\mathcal{B}$ is orthogonal and that each waveform in $\mathcal{B}$ has unit energy.

Let $s(t)$ be a waveform in the span of a set of K orthonormal functions defined on the interval $T_1 \leq t \leq T_2$. By definition, $s(t)$ may be expressed as

$$s(t) = \sum_{k=0}^{K-1} a_k \phi_k(t) \tag{5.2}$$

where the a_k are constants. These constants can be thought of as the components of a K-dimensional vector

$$\mathbf{s} = \begin{bmatrix} a_0 & a_1 & \ldots & a_{K-1} \end{bmatrix}. \tag{5.3}$$

In this way, the waveform $s(t) \in$ Span$\{\mathcal{B}\}$ has an equivalent representation as a K-dimensional point in the signal space Span$\{\mathcal{B}\}$.

Section 5.1 Signal Spaces

The energy in $s(t)$ may be expressed in terms of the vector **s**:

$$E = \int_{T_1}^{T_2} s^2(t)\,dt$$

$$= \int_{T_1}^{T_2} \sum_{i=0}^{K-1} a_i \phi_i(t) \sum_{j=0}^{K-1} a_j \phi_j(t)\,dt$$

$$= \sum_{i=0}^{K-1}\sum_{j=0}^{K-1} a_i a_j \int_{T_1}^{T_2} \phi_i(t)\phi_j(t)\,dt$$

$$= \sum_{i=0}^{K-1}\sum_{j=0}^{K-1} a_i a_j \delta(i-j)$$

$$= \sum_{i=0}^{K-1} a_i^2. \tag{5.4}$$

The geometric interpretation of this result is that the energy in the waveform $s(t)$ is equal to the square of the Euclidean distance between the corresponding point in the signal space and the origin of the signal space.

Now suppose $r(t) = s(t) + w(t)$ and that $w(t)$ is such that even though $s(t)$ is in Span$\{\mathcal{B}\}$, $r(t)$ is not. What is the best approximation of $r(t)$ in Span$\{\mathcal{B}\}$? Let $\hat{r}(t)$ be the best approximation. Then, by definition

$$\hat{r}(t) = \sum_{k=0}^{K-1} x_k \phi_k(t) \tag{5.5}$$

for some vector of constants

$$\mathbf{x} = \begin{bmatrix} x_0 & x_1 & \cdots & x_{K-1} \end{bmatrix}. \tag{5.6}$$

Because $r(t)$ is not in Span$\{\mathcal{B}\}$, there will be an approximation error

$$e(t) = r(t) - \hat{r}(t). \tag{5.7}$$

The best approximation depends on the criterion chosen to measure "best." A commonly used criterion is the mean-square criterion. Using the mean-square criterion, the "best" approximation is the one that minimizes the mean-square error

$$\mathcal{E} = \int_{T_1}^{T_2} |r(t) - \hat{r}(t)|^2\,dt. \tag{5.8}$$

Choosing the best approximation $\hat{r}(t)$ reduces to selecting the coefficients $\mathbf{x} = \begin{bmatrix} x_0 & x_1 & \ldots & x_{K-1} \end{bmatrix}$ that minimize the energy in the approximation error. Substituting (5.5) for $\hat{r}(t)$ in (5.8), computing K derivatives with respect to x_k for $k = 0, 1, \ldots, K-1$, and setting the K derivatives equal to zero produces

$$x_k = \int_{T_1}^{T_2} r(t)\phi_k(t)dt \quad \text{for} \quad k = 0, 1, \ldots, K-1 \tag{5.9}$$

The integral is often called a *projection*. The geometric interpretation is as follows: the best approximation of $r(t)$ in Span{$\mathcal{B}$} in the mean-square sense results from projecting $r(t)$ onto each of the orthonormal basis functions in $\mathcal{B}$. The geometric interpretation is illustrated in Figure 5.1.1. In a signal space, vectors and points represent waveforms. (Points imply a vector that begins at the origin.)

The two key equations from this section are the *synthesis equation* and the *analysis equation*:

$$s(t) = \sum_{k=0}^{K-1} a_k \phi_k(t) \quad \text{synthesis equation} \tag{5.10}$$

$$\hat{x}_k = \int_{T_1}^{T_2} r(t)\phi_k(t)dt \quad \text{analysis equation} \tag{5.11}$$

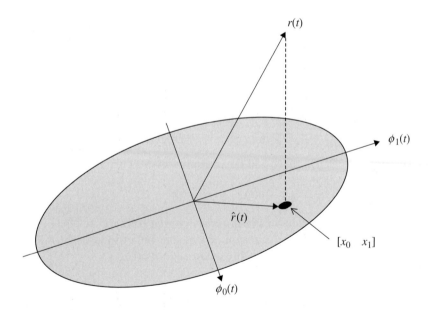

Figure 5.1.1 The geometric interpretation of signal space projections.

Section 5.1 Signal Spaces

The synthesis equation defines how waveforms in Span{$\mathcal{B}$} are constructed as a weighted sum of the orthonormal basis functions. The analysis equation defines the projection of a signal $r(t)$ that is not in Span{$\mathcal{B}$} onto the signal space spanned by $\mathcal{B}$.

Example 5.1.1
Consider the two orthonormal waveforms $\phi_0(t)$ and $\phi_1(t)$ illustrated in Figure 5.1.2 (a). The span of $\phi_0(t)$ and $\phi_1(t)$ forms a 2-dimensional signal space. One interpretation of the synthesis equation is that waveforms in this signal space have two equivalent representations: waveforms (functions of time) and points in a 2-dimensional coordinate system where the coordinate axes represent the basis functions. An example of this concept is illustrated in Figure 5.1.2 (b) where the points

$$\mathbf{s}_0 = [1, 1]$$
$$\mathbf{s}_1 = [3, 1]$$

are shown. The signal $s_0(t)$ corresponding to the point $\mathbf{s}_0$ is

$$s_0(t) = 1 \times \phi_0(t) + 1 \times \phi_1(t)$$

and may be constructed as illustrated in Figure 5.1.2 (b). Likewise, the point $\mathbf{s}_1$ corresponds to the waveform

$$s_1(t) = 3 \times \phi_0(t) + 1 \times \phi_1(t)$$

and may be constructed as shown.

Close inspection of the functions $\phi_0(t)$ and $\phi_1(t)$ shows that all waveforms in Span{$\phi_0(t), \phi_1(t)$} have a stair-step shape. The waveform

$$s(t) = a_0 \phi_0(t) + a_1 \phi_1(t)$$

is of the form

$$s(t) = \begin{cases} a_0 + a_1 & 0 \le t < \frac{1}{2} \\ a_0 - a_1 & \frac{1}{2} \le t < 1 \end{cases}.$$

This structure plays an important role in the application of the analysis equation.

Consider the signal $r(t)$ plotted in Figure 5.1.3 (a). Clearly, $r(t)$ is not in Span{$\phi_0(t), \phi_1(t)$} because it is not of the proper form. The best approximation of $r(t)$ that is in Span{$\phi_0(t), \phi_1(t)$} is defined by the analysis equation. The first step is to project $r(t)$ onto $\phi_0(t)$ and $\phi_1(t)$ as follows:

$$x_0 = \int_0^1 r(t)\phi_0(t)dt = 2$$

$$x_1 = \int_0^1 r(t)\phi_1(t)dt = \frac{2}{3}.$$

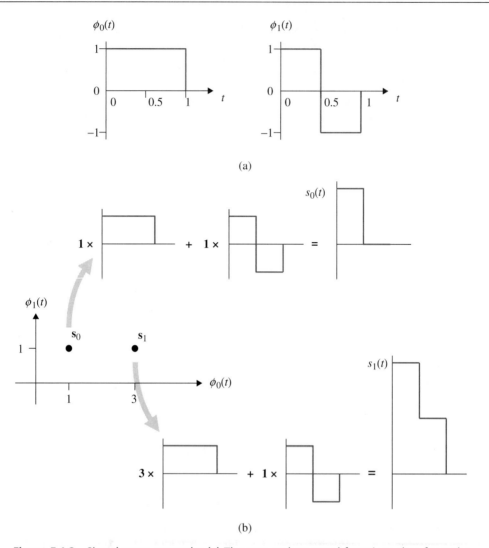

Figure 5.1.2 Signal space example: (a) The two orthonormal functions that form the basis of a 2-dimensional signal space; (b) two points in the signal space and the corresponding waveforms.

A graphical representation of these projections is illustrated in Figure 5.1.3 (b). The corresponding waveform $\hat{r}(t)$ is

$$\hat{r}(t) = x_0\phi_0(t) + x_1\phi_1(t) = 2\phi_0(t) + \frac{2}{3}\phi_1(t)$$

Section 5.1 Signal Spaces

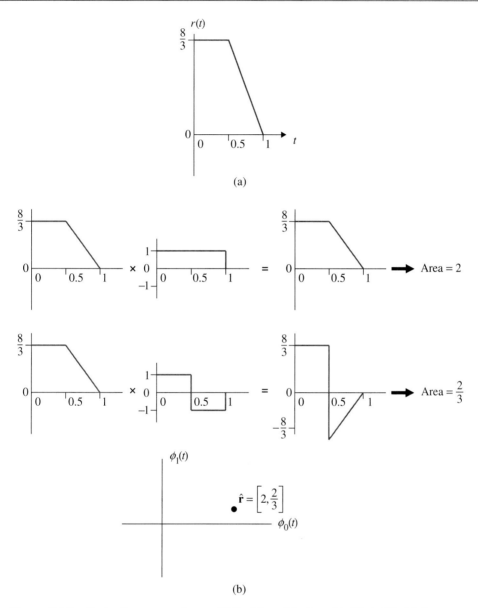

Figure 5.1.3 Projecting a waveform $r(t)$ that is not in $\text{Span}\{\phi_0(t), \phi_1(t)\}$: (a) An example of a waveform that is not in $\text{Span}\{\phi_0(t), \phi_1(t)\}$; (b) a graphical representation of the projections onto $\phi_0(t)$ and $\phi_1(t)$.

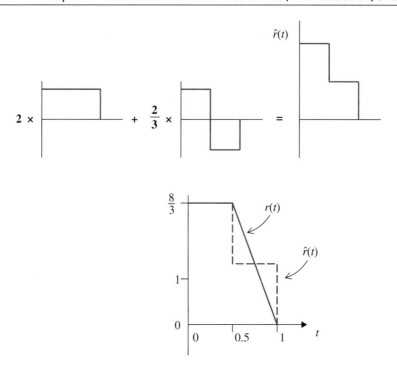

Figure 5.1.4 Constructing $\hat{r}(t)$, the projection of $r(t)$ onto $\phi_0(t)$ and $\phi_1(t)$.

and is illustrated in Figure 5.1.4. Of all the stair-step waveforms in Span$\{\phi_0(t), \phi_1(t)\}$, this waveform is the one with the minimum approximation error energy.

5.1.2 The Synthesis Equation and Linear Modulation

Let
$$\mathcal{S} = \{s_0(t), s_1(t), \ldots, s_{M-1}(t)\} \tag{5.12}$$
represent the set of M waveforms used to transmit $\log_2(M)$ bits over a waveform channel during each symbol interval T_s. What is an appropriate set of waveforms $\mathcal{B}$ that can be used to represent the signal set $\mathcal{S}$? Clearly, $\mathcal{B}$ must be chosen so that $\mathcal{S}$ is a subset of Span$\{\mathcal{B}\}$. In addition, if $\mathcal{B}$ is orthonormal, then the modulator and detector structures are greatly simplified. With these constraints in place, the question can be asked in a more refined way: What is an appropriate orthonormal set $\mathcal{B}$ such that $\mathcal{S}$ is a subset of Span$\{\mathcal{B}\}$? Sometimes the set of M waveforms is given without any clue as to what a good orthonormal set $\mathcal{B}$ might be. In this case, a systematic procedure for constructing an appropriate orthonormal set of basis functions must be used. The Gram–Schmidt procedure is such a method. Fortunately, it is often the case that the orthonormal basis set $\mathcal{B}$ is defined and M waveforms are an appropriately selected subset of Span$\{\mathcal{B}\}$.

Section 5.1 Signal Spaces

Let S be a subset of Span$\{B\}$ where B consists of K orthonormal waveforms defined over the interval $T_1 \leq t \leq T_2$. Using the synthesis equation, each of the M waveforms in S may be expressed as a linear combination of the orthonormal basis functions:

$$\begin{aligned}
s_0(t) &= a_{0,0}\phi_0(t) + a_{0,1}\phi_1(t) + \cdots + a_{0,K-1}\phi_{K-1}(t) \\
s_1(t) &= a_{1,0}\phi_0(t) + a_{1,1}\phi_1(t) + \cdots + a_{1,K-1}\phi_{K-1}(t) \\
&\vdots \\
s_m(t) &= a_{m,0}\phi_0(t) + a_{m,1}\phi_1(t) + \cdots + a_{m,K-1}\phi_{K-1}(t) \\
&\vdots \\
s_{M-1}(t) &= a_{M-1,0}\phi_0(t) + a_{M-1,1}\phi_1(t) + \cdots + a_{M-1,K-1}\phi_{K-1}(t).
\end{aligned} \quad (5.13)$$

A second subscript has been added to the basis function weights. The first subscript indexes the waveform membership in S and the second subscript indexes the dimension in B. This representation demonstrates that each of the waveforms in S can be represented by a K-tuple composed of the K weights defined in (5.13). That is,

$$\begin{aligned}
s_0(t) &\leftrightarrow \mathbf{s}_0 = \begin{bmatrix} a_{0,0} & a_{0,1} & \cdots & a_{0,K-1} \end{bmatrix} \\
s_1(t) &\leftrightarrow \mathbf{s}_1 = \begin{bmatrix} a_{1,0} & a_{1,1} & \cdots & a_{1,K-1} \end{bmatrix} \\
&\vdots \\
s_m(t) &\leftrightarrow \mathbf{s}_m = \begin{bmatrix} a_{m,0} & a_{m,1} & \cdots & a_{m,K-1} \end{bmatrix} \\
&\vdots \\
s_{M-1}(t) &\leftrightarrow \mathbf{s}_{M-1} = \begin{bmatrix} a_{M-1,0} & a_{M-1,1} & \cdots & a_{M-1,K-1} \end{bmatrix}.
\end{aligned} \quad (5.14)$$

The K-tuple $\mathbf{s}_m$ representing the waveform $s_m(t)$ is a point in a K-dimensional space that is called the *signal space*. The set of M K-dimensional points is called the *constellation*. Each point in the constellation is a K-dimensional *symbol*. Because $\log_2 M$ bits are required to select one of the M symbols, each symbol transmits $\log_2 M$ bits of information.

A modulator based on this development is illustrated in Figure 5.1.5. A serial bit stream is divided into segments consisting of $\log_2 M$ bits by the serial-to-parallel converter. These bits form a common address to K look-up tables; one for each basis function in B. The contents of the k-th look-up table are the k-th components from each vector in (5.14). Figure 5.1.5 illustrates the case where the bits select $s_m(t)$ from the set S. The outputs of the K look-up tables are the components of the vector $\mathbf{s}_m$ in (5.14). The outputs of the K look-up tables weight each of the basis functions in B and are summed to form $s_m(t)$ following the synthesis equation. This architecture is particularly attractive for those cases where K is small and M is large.

5.1.3 The Analysis Equation and Detection

Let S be the set of M possible waveforms used to transmit $\log_2 M$ bits of information across a waveform channel. Let $B = \{\phi_0(t), \phi_1(t), \ldots, \phi_{K-1}(t)\}$ be a set of K real-valued waveforms defined on the interval $T_1 \leq t \leq T_2$ whose span contains the waveforms in S. Suppose $s_m(t) \in S$ was transmitted and the received waveform is $r(t) = s_m(t) + w(t)$ where $w(t)$ represents

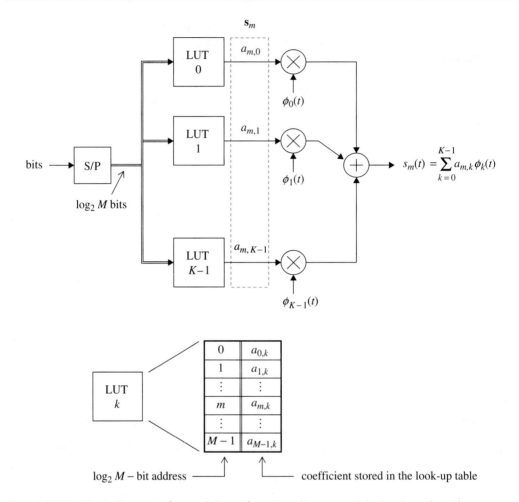

Figure 5.1.5 Block diagram of a modulator for *M*-ary linear modulation based on the synthesis equation.

additive noise introduced by the environment. The goal of the detector is to determine which of the M possible waveforms in S was sent on the basis of processing $r(t)$.

When the additive noise is white and Gaussian, the optimum detector[1] (in the maximum likelihood sense) selects the waveform in S that minimizes the error energy

$$\mathcal{E}_m = \int_{T_1}^{T_2} |r(t) - s_m(t)|^2 \, dt. \tag{5.15}$$

---

[1]The optimum detection rule in the maximum likelihood sense is derived in Section 5.6.

Expressed mathematically, this decision rule is

$$\hat{\mathbf{s}} = \arg\min_{s_m(t) \in \mathcal{S}} \left\{ \int_{T_1}^{T_2} |r(t) - s_m(t)|^2 \, dt \right\}. \quad (5.16)$$

Using $s_m(t) = \sum_{k=0}^{K-1} a_{m,k} \phi_k(t)$ and expanding produces

$$\hat{\mathbf{s}} = \arg\min_{s_m(t) \in \mathcal{S}} \left\{ \int_{T_1}^{T_2} |r(t)|^2 \, dt - 2 \sum_{k=0}^{K-1} a_{m,k} \int_{T_1}^{T_2} r(t)\phi_k(t) dt + \sum_{k'=0}^{K-1}\sum_{k'=0}^{K-1} a_{m,k} a_{m,k'} \int_{T_1}^{T_2} \phi_k(t)\phi_{k'}(t) dt \right\}.$$
(5.17)

The first term in the argument of (5.17) is not a function of m and can be omitted from the argument. The second integral is the projection of $r(t)$ onto the basis function $\phi_k(t)$, which is denoted x_k. Using the orthonormal property of the basis functions, the third term in (5.17) may be expressed as $\sum_{k=0}^{K-1} a_{m,k}^2$. Putting all this together gives

$$\hat{\mathbf{s}} = \arg\min_{\mathbf{s}_m \in \mathcal{S}} \left\{ -2 \sum_{k=0}^{K-1} a_{m,k} x_k + \sum_{k=0}^{K-1} a_{m,k}^2 \right\}. \quad (5.18)$$

Adding a positive constant (constant with respect to m) to the argument of (5.18) does not change the answer. A convenient positive constant is $\sum_{k=0}^{K-1} x_k^2$. Now the decision rule is

$$\hat{\mathbf{s}} = \arg\min_{\mathbf{s}_m \in \mathcal{S}} \left\{ \sum_{k=0}^{K-1} x_k^2 - 2 \sum_{k=0}^{K-1} a_{m,k} x_k + \sum_{k=0}^{K-1} a_{m,k}^2 \right\}$$

$$= \arg\min_{\mathbf{s}_m \in \mathcal{S}} \left\{ \sum_{k=0}^{K-1} \left(x_k - a_{m,k} \right)^2 \right\}. \quad (5.19)$$

The decision rule (5.19) shows that the maximum likelihood decision is based on a comparison of the K-dimensional vector $\mathbf{x} = \begin{bmatrix} x_0 & x_1 & \ldots & x_{K-1} \end{bmatrix}$ with the M possible K-dimensional points $\mathbf{s}_0, \mathbf{s}_1, \ldots, \mathbf{s}_{M-1}$ given by (5.14). The K-dimensional point closest to $\mathbf{x}$ in Euclidean distance is selected as the decision $\hat{\mathbf{s}}$. A block diagram of the detector based on this rule is illustrated in Figure 5.1.6. The received waveform during the interval $T_1 \leq t \leq T_2$ is projected onto the signal space $\text{Span}\{\mathcal{B}\}$ to produce a K-dimensional vector $\mathbf{x} = \begin{bmatrix} x_0 & x_1 & \ldots & x_{K-1} \end{bmatrix}$. Note that the result of this projection is the best approximation of $r(t)$ in $\text{Span}\{\mathcal{B}\}$ in the mean-square sense. To implement the decision rule (5.19), a list of Euclidean distances between $\mathbf{x}$ and the M constellation points is produced. The constellation point corresponding to the smallest entry in this list is output as the decision.

The format of the decision depends on the required interface. The possibilities include the index m, the K coefficients corresponding to $\hat{\mathbf{s}}$: $\begin{bmatrix} \hat{a}_0 & \hat{a}_1 & \ldots & \hat{a}_{K-1} \end{bmatrix}$, or the $\log_2 M$-bit pattern associated with $\hat{\mathbf{s}}$. All are related by a one-to-one relationship. Thus, knowing one format means any other format is known. In this text, the decision output format is the K-dimensional vector format.

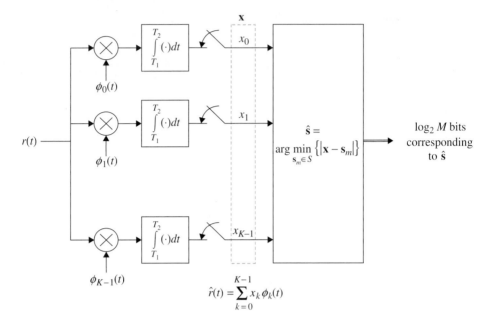

Figure 5.1.6 Block diagram of a detector for *M*-ary linear modulation based on the analysis equation.

5.1.4 The Matched Filter

The detector shown in Figure 5.1.6 computes K signal space projections x_k given by the analysis equation

$$x_k = \int_{T_1}^{T_2} r(t)\phi_k(t)dt \quad k = 0, 1, \ldots, K-1 \tag{5.20}$$

where $\phi_k(t)$ is defined over the interval $T_1 \leq t \leq T_2$ for $k = 0, 1, \ldots, K-1$. As it turns out, the projection x_k could also be computed as the output of a filter with impulse response

$$h(t) = \phi_k(-t). \tag{5.21}$$

The convolution integral for the filter output is

$$x(t) = \int_{t+T_1}^{t+T_2} r(\lambda)\phi_k(-t+\lambda)d\lambda. \tag{5.22}$$

Section 5.2 M-ary Baseband Pulse Amplitude Modulation

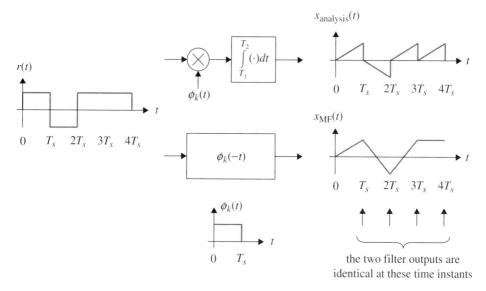

Figure 5.1.7 Comparison of two approaches for generating the signal space projection. A direct application of the analysis equation (top). Using the matched filter (bottom).

Evaluating this integral at $t = 0$ produces

$$x(0) = \int_{T_1}^{T_2} r(\lambda)\phi_k(\lambda)d\lambda = x_k. \tag{5.23}$$

This result does not mean that (5.20) and (5.22) are the same for all t as illustrated in Figure 5.1.7. What it does say is that the two are the same at key instants in time, that is, (5.20) at $t = nT_s + T_2$ is equal to (5.22) at $t = nT_s$ for $n = 0, 1, 2, \ldots$. The matched filter equivalent of the linear detector in Figure 5.1.6 is illustrated in Figure 5.1.8.

5.2 M-ARY BASEBAND PULSE AMPLITUDE MODULATION

M-ary baseband pulse amplitude modulation (PAM) is a 1-dimensional signal set with basis function

$$\phi_0(t) = p(t) \tag{5.24}$$

where $p(t)$ is any unit energy pulse such as those described in Appendix A. The notation developed in the preceding sections is extended here to denote a symbol sequence with a new symbol occurring every T_s seconds. The resulting PAM signal is

$$s(t) = \sum_n a(n)p(t - nT_s) \tag{5.25}$$

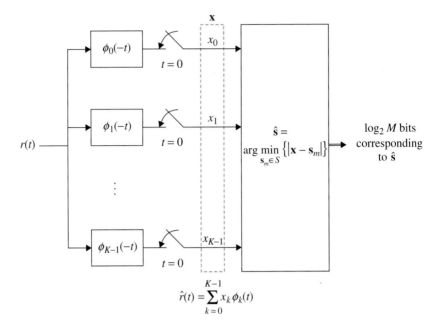

Figure 5.1.8 Block diagram of a matched filter detector for *M*-ary linear modulation based on the analysis equation.

where the subscript "0" has been dropped on the symbol value a and an index n has been added to denote the n-th symbol in a sequence of symbols. The subscript n on the summation symbol means that the sum is over n and may be finite (for a finite data sequence) or infinite if an infinite data sequence is assumed. When it is important, limits on the summation will be explicit.

Some example constellations for baseband PAM are shown in Figure 5.2.1. Note that except for $M = 2$, the constellations points do not all have the same energy. For $M = 4$ the average symbol energy is

$$E_{\text{avg}} = \frac{(-3)^2 + (-1)^2 + (+1)^2 + (+3)^2}{4} A^2 = 5A^2 \tag{5.26}$$

and for $M = 8$ the average symbol energy is

$$E_{\text{avg}} = \frac{(-7)^2 + (-5)^2 + (-3)^2 + (-1)^2 + (+1)^2 + (+3)^2 + (+5)^2 + (+7)^2}{8} A^2 = 21A^2. \tag{5.27}$$

For the general case of M-ary PAM, the constellation points are evenly spaced along the constellation axis at locations

$$-(M-1)A, -(M-3)A, \ldots, -A, +A, \ldots, +(M-3)A, +(M-1)A \tag{5.28}$$

Section 5.2 M-ary Baseband Pulse Amplitude Modulation

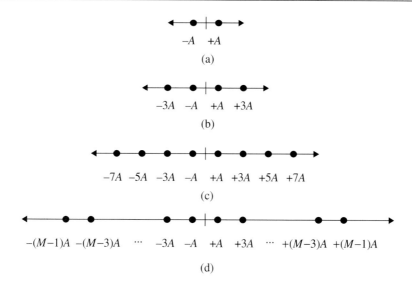

Figure 5.2.1 Baseband PAM constellations for (a) $M = 2$, (b) $M = 4$, (c) $M = 8$, and (d) general M.

The average energy of the general M-ary PAM constellation is

$$E_{\text{avg}} = \frac{M^2 - 1}{3} A^2. \tag{5.29}$$

For each constellation, the minimum Euclidean distance is 2.

5.2.1 Continuous-Time Realization

The PAM modulator using continuous-time processing is illustrated in the upper portion of Figure 5.2.2. The input is a serial bit stream where new bits arrive every T_b seconds. Bits are converted to $\log_2 M$ bit symbols by the serial to parallel converter that outputs a new symbol every $T_s = \log_2(M) \times T_b$ seconds. Thus, the symbol rate is $1/(\log_2 M)$ times the bit rate. The $\log_2 M$ bits form the address to a single look-up table that stores the symbol values specified by the constellation. Conceptually, the series of look-up table outputs forms a weighted impulse train

$$i(t) = \sum_n a(n)\delta(t - nT_s) \tag{5.30}$$

that drives the pulse shaping filter with impulse response $p(t)$. The output of the pulse shaping filter is the pulse train given by (5.25).

The matched filter detector is illustrated in the lower portion of Figure 5.2.2. The received waveform is the pulse train (5.25) plus noise. The matched filter is a filter whose impulse response is a time reversed version of the pulse shape as described in Section 5.1.4. The matched filter output, $x(t)$ is sampled at T_s-spaced intervals to produce a sequence of signal space projections that are used for detection. Detection uses the decision rule (5.16).

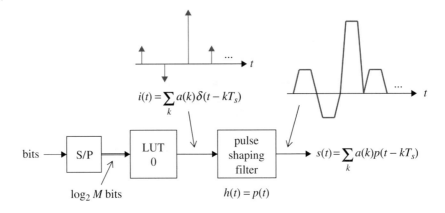

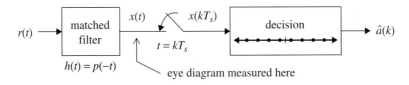

Figure 5.2.2 A PAM modulator (top) and detector (bottom) using continuous-time processing.

The mathematical expressions defining the operation of the detector are as follows. Let

$$r(t) = s(t) + w(t) \tag{5.31}$$

be the received signal where $s(t)$ is the PAM pulse train given by (5.25) and $w(t)$ is the additive noise.[2] The matched filter output is

$$x(t) = r(t) * p(-t) = \int_{T_1+t}^{T_2+t} r(\lambda)p(\lambda - t)d\lambda. \tag{5.32}$$

Using (5.31) and (5.25), the matched filter output is

$$x(t) = \sum_l a(l) \int_{T_1+t}^{T_2+t} p(\lambda - lT_s)p(\lambda - t)d\lambda + \int_{T_1+t}^{T_2+t} w(\lambda)p(\lambda - t)d\lambda \tag{5.33}$$

which may be expressed as

$$x(t) = \sum_l a(l) r_p(lT_s - t) + v(t) \tag{5.34}$$

---

[2]The noise term $w(t)$ will be modeled as a zero-mean white Gaussian random process with power spectral density $N_0/2$ W/Hz. See Section 4.5.

Section 5.2 M-ary Baseband Pulse Amplitude Modulation

where $v(t)$ is the output of the matched filter due to the additive noise at the input (the second integral in (5.33)) and $r_p(\tau)$ is the pulse shape autocorrelation function defined by

$$r_p(\tau) = \int_{T_1}^{T_2} p(t)p(t-\tau)dt \qquad (5.35)$$

as described in Appendix A.

The matched filter output is sampled at $t = kT_s$ to produce the signal space projection

$$x(kT_s) = \sum_l a(l)r_p((l-k)T_s) + v(kT_s). \qquad (5.36)$$

As discussed in Appendix A, the autocorrelation function for all full response unit-energy pulse shapes satisfies

$$r_p(mT_s) = \begin{cases} 1 & m = 0 \\ 0 & m \neq 0 \end{cases}. \qquad (5.37)$$

Condition (5.37) also holds for partial response pulse shapes that satisfy the Nyquist no-ISI condition. As a consequence, (5.36) may be expressed as

$$x(kT_s) = a(k) + v(kT_s). \qquad (5.38)$$

This shows that the signal space projection consists of the true point plus a noise term. The effect of the noise term is to perturb the position of $x(kT)$ in the signal space. The symbol decision follows the rule given by (5.16) or the equivalent form given by (5.17) as described in Section 5.1.3. The detector produces an erroneous output (i.e., a *symbol error*) when the noise term forces $x(kT_s)$ to be closer to a point in the signal space other than the point corresponding to the waveform that was sent. The probability of error for M-ary PAM is derived in Section 6.1.

Example 5.2.2

As an example of binary PAM, consider the modulator shown in Figure 5.2.3 (a). The modulator is defined by the constellation included in the illustration. The constellation shows that the relationship between the input bit $b(k)$ and the symbol $a(k)$ is

$$a(k) = \begin{cases} -1 & \text{if } b(k) = 0 \\ +1 & \text{if } b(k) = 1 \end{cases}. \qquad (5.39)$$

Using this information, the bit sequence 1011 produces the sequence of binary symbols

k	0	1	2	3
$b(k)$	1	0	1	1
$a(k)$	+1	−1	+1	+1

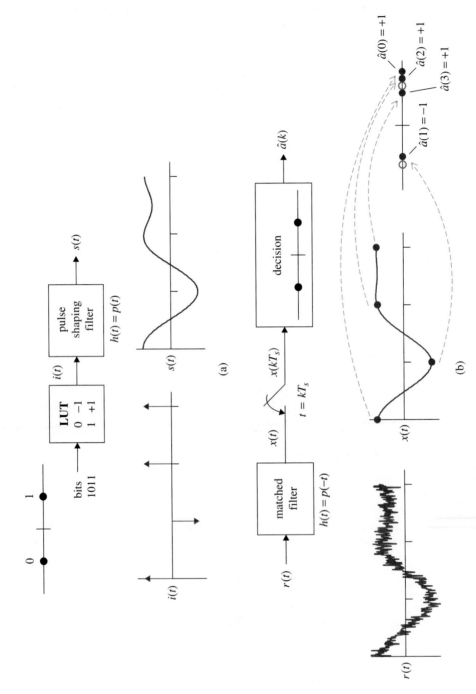

Figure 5.2.3 An example of binary PAM corresponding to the bit sequence 1011: (a) The modulator; (b) the detector.

These symbols are used to form the weighted impulse train $i(t)$ that forms the input to the pulse shaping filter. The output of the pulse shaping filter, $s(t)$, corresponding to this bit sequence is also shown.

In the additive white Gaussian noise (AWGN) environment, the received signal $r(t)$ is a noisy version of $s(t)$ as shown in Figure 5.2.3 (b). The noisy waveform $r(t)$ is filtered by a filter matched to the pulse shape to produce the output $x(t)$ as shown. The matched filter output $x(t)$ is sampled at the required sampling instants to produce the sequence of signal space projections. The relationship between the matched filter output $x(t)$, its samples $x(kT_s)$, and the signal space projections is illustrated in Figure 5.2.3 (b). Signal space projection $x(kT_s)$ is used to produce an estimate of the k-th symbol $\hat{a}(k)$. The symbols estimates are listed along with the signal space projections below.

k	0	1	2	3
$x(k)$	+1.02	−0.98	+1.01	+0.97
$\hat{a}(k)$	+1	−1	+1	+1

The detectors presented in Figure 5.2.2 and in the example (see Figure 5.2.3) are idealized systems. It is assumed that the instant the matched filter output is to be sampled is known precisely. In a real system, this knowledge is provided by a *symbol timing synchronization* subsystem. Continuous-time techniques for symbol timing synchronization are summarized in Chapter 8.

An important conceptual tool is a modulo-T_s plot of the matched filter output $x(t)$. Examples of these plots for binary PAM using the NRZ, MAN, HS, and SRRC pulse shapes, described in Appendix A, are shown in Figure 5.2.4. Note that the time axis on the plots are shifted so that the optimum sampling instant occurs in the middle of the plot. The overlay is used to present all the possible trajectories of $r_p(\tau)$ as determined by the symbol sequence. These plots for the HS and SRRC pulse shapes resemble a human eye. For this reason, these plots are called *eye diagrams*.

Eye diagrams can also serve an important diagnostic function. An example is illustrated in Figure 5.2.5 for binary PAM. In this case, the eye diagram spans two symbol times. The important characteristics are the eye height, the eye width, the peak overshoot, and the noise margin. The noise margin is defined relative to the decision threshold as shown. The diagram also illustrates the effect of a timing error. In the presence of a timing error, the eye is not sampled at the maximum average eye opening, but at another point. For some trajectories through the eye, the different sampling point *improves* the noise margin. For many other trajectories, however, the different sampling point *reduces* the noise margin. The decrease in the noise margin is a function of the shape of the "inner eye opening" as shown. Channel distortions are often quantified using the eye diagram. In general, channel distortion decreases the eye opening at the optimum sampling instant and decreases the width of the eye opening.

5.2.2 Discrete-Time Realization

The PAM modulator using discrete-time processing is illustrated in the upper portion of Figure 5.2.6. The input bit stream is segmented into nonoverlapping $\log_2 M$-bit symbols that

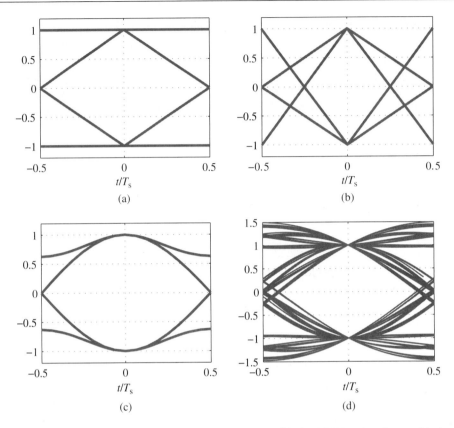

Figure 5.2.4 Eye diagrams for (a) The NRZ pulse shape, (b) the MAN pulse shape, (c) the HS pulse shape, and (d) the SRRC pulse shape with 50% excess bandwidth.

access M-ary symbol values stored in the look-up tables. Here the pulse shaping is performed in discrete-time by using an FIR filter whose impulse response consists of T-spaced samples of the pulse shape $p(t)$. The symbol sequence $\{\ldots, a(n-2), a(n-1), a(n), \ldots\}$ is converted to a discrete-time impulse train by upsampling by N where N is the ratio of sample rate to symbol rate:

$$N = \frac{T_s}{T}. \tag{5.40}$$

The upsampling operation inserts $N-1$ zeros in between each symbol as shown to produce the sequence $s_0(nT)$. Interpolation is performed by the pulse shaping filter which processes the samples $s_0(nT)$ to produce the discrete-time pulse train $s(nT)$. The discrete-time pulse train is converted to a continuous-time pulse train by the ADC for transmission over the waveform channel.

Three clock rates are used in the modulator. The first clock rate is the bit-rate clock used to clock the bits into the serial-to-parallel converter. The output of the serial-to-parallel

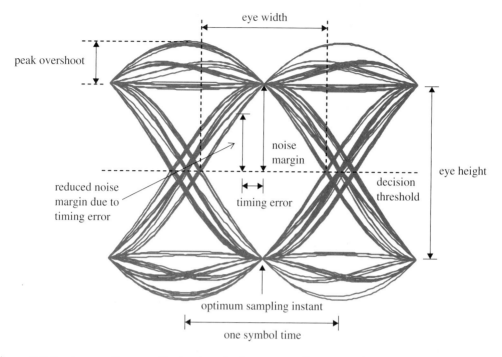

Figure 5.2.5 An eye diagram illustrating the important diagnostic characteristics. The diagram was generated using 200 random binary PAM symbols and the SRRC pulse shape with 50% excess bandwidth.

converter is clocked at $1/\log_2(M)$ times the bit rate. This is the symbol-rate clock. The pulse shaping occurs at a rate that is N times the symbol clock rate. At a minimum, the high sample rate should satisfy the sampling theorem. In practice, this rate is as high as the system constraints will allow it to be.

The modulator illustrated in Figure 5.2.6 is intended to show *what* operations need to be performed. There are more efficient methods for realizing a modulator. This is discussed in much more detail in Section 10.1.1. Even though the material in Chapter 10 is devoted to QAM, many of the same principles apply to PAM modulators.

The PAM detector using discrete-time processing is illustrated in the lower portion of Figure 5.2.6. The received signal is the transmitted pulse train (5.25) plus noise. T-spaced samples of $r(t)$ are produced by the ADC. These samples are filtered by a matched filter whose impulse response consists of T-spaced samples of the time reversed pulse shape. The resultant matched filter output is given by

$$x(nT) = \sum_{m=\frac{T_1}{T}+n}^{\frac{T_2}{T}+n} r(mT)p((m-n)T) \tag{5.41}$$

236 Chapter 5 Linear Modulation 1: Modulation, Demodulation, and Detection

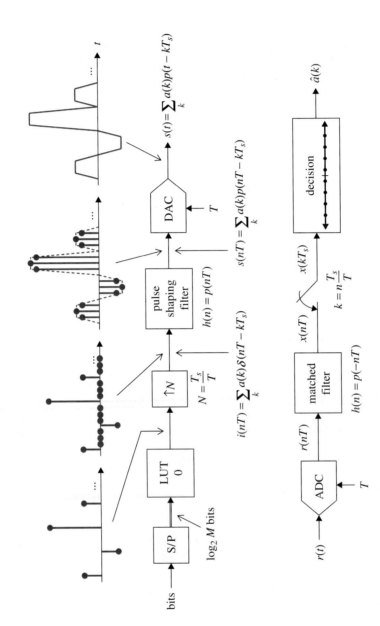

Figure 5.2.6 A PAM modulator using discrete-time processing and a DAC (top) and a PAM matched filter detector using discrete-time processing with an ADC (bottom).

Section 5.2 M-ary Baseband Pulse Amplitude Modulation

where T_1 and T_2 define the time support of the pulse shape $p(t)$. The matched filter output is downsampled at $n = k\frac{T_s}{T}$ to produce the sequence of signal space projections $x(kT_s)$. The k-th symbol estimate is made using the decision rule (5.16).

This detector is an idealized system in two ways. First, it is assumed that the ADC produces an integer number of samples per symbol, in other words, T_s/T is an integer. Because this assumption is rarely true in practice, a resampling filter is often included in the system block diagram. (See Section 9.3 for a description of resampling filters and Section 10.1.2 for their application in a QAM demodulator.) The second assumption is that the exact symbol timing instant is known and coincides with a sample. In reality, this will not be the case and a discrete-time symbol synchronizer must be used to estimate the phase and frequency of the downsampling operation. (See Chapter 8 for a description of discrete-time techniques for symbol timing synchronization.)

There are at least two clock rates in this system. The first clock rate is the sampling rate used by the ADC to sample the received waveform. The sample rate must satisfy the sampling theorem. Unlike the modulator, however, there is little need to sample the signal any higher than the minimum rate given by the sampling theorem.[3] The matched filtering occurs at the high clock rate and the output of the matched filter is downsampled by N to 1 sample/symbol. The clock rate here is the symbol rate. Symbol decisions are made at the clock rate. There is a third clock rate, the bit clock, if the symbol decisions are mapped to $\log_2 M$ bits.

The mathematical expressions for the receiver operations mimic those for the continuous-time case. The T-spaced samples of the received signal are

$$r(nT) = s(nT) + w(nT) = \sum_l a(l)p(nT - lT_s) + w(nT) \qquad (5.42)$$

where $w(nT)$ are samples of the appropriately bandlimited additive noise. The matched filter output is

$$x(nT) = \sum_{m=n+\frac{T_1}{T}}^{n+\frac{T_2}{T}} \left(\sum_l a(l)p(mT - lT_s) \right) p(mT - nT) + \sum_{m=n+\frac{T_1}{T}}^{n+\frac{T_2}{T}} w(mT)p(mT - nT)$$

$$= \sum_l a(l) \sum_{m=n+\frac{T_1}{T}}^{n+\frac{T_2}{T}} p(mT - lT_s)p(mT - nT) + \sum_{m=n+\frac{T_1}{T}}^{n+\frac{T_2}{T}} w(mT)p(mT - nT). \qquad (5.43)$$

Using the substitution $i = m - n$, the first summation in (5.43) can be expressed as

$$\sum_{m=n+\frac{T_1}{T}}^{n+\frac{T_2}{T}} p(nT - lT_s)p(mT - nT) = \sum_{i=\frac{T_1}{T}}^{\frac{T_2}{T}} p(iT)p(iT - (lT_s - nT)) \qquad (5.44)$$

---

[3] The complexity of symbol timing synchronization can be reduced somewhat in certain cases by sampling at two or four times the minimum sampling rate given by the sampling theorem.

Using the relationship

$$r_p(\tau) = \int_{T_1}^{T_2} p(t)p(t-\tau)dt \approx T \sum_{n=T_1/T}^{T_2/T} p(nT)p(nT-\tau), \quad (5.45)$$

the right-hand side of (5.44) may be expressed as

$$\sum_{i=\frac{T_1}{T}}^{\frac{T_2}{T}} p(iT)p(iT - (lT_s - nT)) = \frac{1}{T} r_p(lT_s - nT). \quad (5.46)$$

Using $v(nT)$ to represent the second term in (5.43), the matched filter output may be expressed as

$$x(nT) = \frac{1}{T} \sum_l a(l) r_p(lT_s - nT) + v(nT). \quad (5.47)$$

The signal space projection is produced by taking every $k\frac{T_s}{T}$-th sample of the matched filter output. Evaluating (5.47) at $n = k\frac{T_s}{T}$ gives

$$x(kT_s) = \frac{1}{T} \sum_l a(l) r_p((l-k)T_s) + v(kT_s). \quad (5.48)$$

When the pulse shape satisfies the Nyquist no-ISI condition, the signal space projection is

$$x(kT_s) = \frac{a(k)}{T} + v(kT_s). \quad (5.49)$$

The additive noise term moves the signal space projection $x(kT_s)$ away from the true value $a(k)/T$ as was the case with continuous-time processing. A decision error occurs when the noise moves $x(kT_s)$ closer to a signal space point other than the one that was sent. The probability that this occurs is derived in Section 6.1.

5.3 M-ARY QUADRATURE AMPLITUDE MODULATION

M-ary Quadrature Amplitude Modulation (MQAM) is a 2-dimensional signal set using orthonormal basis functions

$$\begin{aligned}\phi_0(t) &= \sqrt{2}p(t)\cos(\omega_0 t) \\ \phi_1(t) &= -\sqrt{2}p(t)\sin(\omega_0 t)\end{aligned} \quad (5.50)$$

where $p(t)$ is any unit energy pulse (such as those described in Appendix A) with finite time support on the interval $T_1 \leq t \leq T_2$. The orthonormal basis functions are based on two

Section 5.3 M-ary Quadrature Amplitude Modulation

sinusoids that are 90° out of phase with each other. The term quadrature is borrowed from astronomy where it is used to describe the situation where the positions of two celestial objects are 90° apart.

That $\phi_0(t)$ and $\phi_1(t)$ as defined in (5.50) are orthogonal is sometimes challenging for students to believe at first. Orthogonality can be demonstrated by first considering the case where $p(t)$ is a constant. The correlation between $\phi_0(t)$ and $\phi_1(t)$ is thus

$$\rho = -\int_{T_1}^{T_2} 2\cos(\omega_0 t)\sin(\omega_0 t)\,dt \tag{5.51}$$

$$= \frac{\cos(2\omega_0 T_2) - \cos(2\omega_0 T_1)}{2\omega_0} \tag{5.52}$$

$$= \frac{\sin(\omega_0(T_2+T_1))\sin(\omega_0(T_2-T_1))}{4\omega_0}. \tag{5.53}$$

When $T_2 - T_1$ spans an integer number of periods (i.e., $\omega_0(T_2 - T_1) = 2\pi \times$ integer), $\rho = 0$ and the sinusoids are orthogonal. When the $T_2 - T_1$ does not span an integer number of periods, the sinusoids are only approximately orthogonal. Using $-1 \le \sin(X) \le +1$, the correlation ρ is bounded by

$$-\frac{1}{4\omega_0} \le \rho \le \frac{1}{4\omega_0}. \tag{5.54}$$

This shows that the sinusoids are orthogonal in the limit as $\omega_0 \to \infty$. In most applications, the carrier frequency ω_0 is much larger than the symbol rate (which is proportional to $1/(T_2+T_1)$). This makes orthogonality a good approximation.

For the case where $p(t)$ is not a constant, the correlation is

$$\rho = -2\int_{T_1}^{T_2} p^2(t)\cos(\omega_0 t)\sin(\omega_0 t)\,dt \tag{5.55}$$

$$= -2\int_{T_1}^{T_2} p^2(t)\sin(2\omega_0 t)\,dt. \tag{5.56}$$

Let $T_0 = \pi/\omega_0$ be the period of the sinusoid in the integrand. The integration interval may be partitioned into $L+1$ nonoverlapping intervals of width T_0 as illustrated in Figure 5.3.1. Because the intervals are nonoverlapping, the integral (5.56) may be expressed as the sum of integrals over each interval. Referring to Figure 5.3.1, the l-th partition may be expressed as $[T_1 + lT_0, T_1 + (l+1)T_0]$ for $l = 0, 1, \ldots, L-1$, where[4]

$$L = \left\lfloor \frac{T_2 + T_1}{T_0} \right\rfloor. \tag{5.57}$$

---

[4]The notation $\lfloor x \rfloor$ means the largest integer not greater than x.

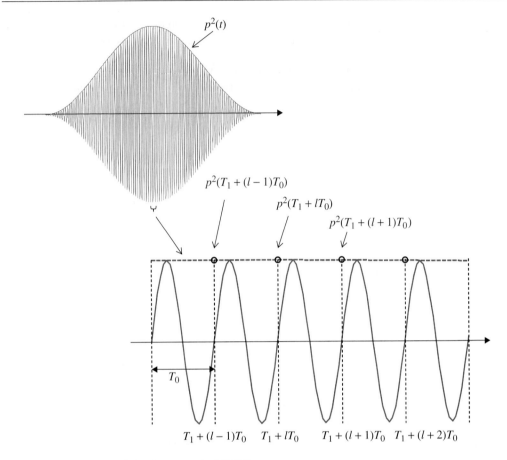

Figure 5.3.1 A plot of the integrand of (5.58).

The last partition is $[T_1 + LT_0, T_2]$. Using these partitions, the correlation is

$$\rho = -\sum_{l=0}^{L-1} \int_{T_1+lT_0}^{T_1+(l+1)T_0} p^2(t)\sin(2\omega_0 t)dt - \int_{T_1+LT_0}^{T_2} p^2(t)\sin(2\omega_0 t)dt. \quad (5.58)$$

Because $p(t)$ is a low-pass waveform, $p^2(t)$ is approximately constant over the period T_0, i.e.,

$$p^2(t) \approx p^2(T_1 + lT_0) \quad \text{for } T_1 + lT_0 \leq t \leq T_1 + (l+1)T_0. \quad (5.59)$$

Section 5.3 M-ary Quadrature Amplitude Modulation

Using this approximation, the correlation is

$$\rho \approx -\sum_{l=0}^{L-1} p^2(T_1 + lT_0) \int_{T_1+lT_0}^{T_1+(l+1)T_0} \sin(2\omega_0 t)\,dt - p^2(T_1 + LT_0) \int_{T_1+LT_0}^{T_2} \sin(2\omega_0 t)\,dt \quad (5.60)$$

$$= -p^2(T_1 + LT_0) \int_{T_1+LT_0}^{T_2} \sin(2\omega_0 t)\,dt \quad (5.61)$$

where (5.61) follows from (5.60) because each integral in the summation of the first term in (5.60) is the integral of a sinusoid over its period. If $(T_2 + T_1)/T_0$ is an integer, (5.61) is zero because the interval of integration is over one period of $\sin(2\omega_0 t)$. When $(T_2 + T_1)/T_0$ is not an integer, then, proceeding as before, it is straightforward to show that

$$-\frac{p^2(T_1 + LT_0)}{4\omega_0} \leq \rho \leq \frac{p^2(T_1 + LT_0)}{4\omega_0}. \quad (5.62)$$

As $\omega_0 \to \infty$, the approximation (5.59) improves and $\rho \to 0$.

The general MQAM signal is a pulse train of the form

$$s(t) = \sqrt{2} \sum_k a_0(k) p(t - kT_s) \cos(\omega_0 t) - a_1(k) p(t - kT_s) \sin(\omega_0 t). \quad (5.63)$$

For notational convenience, $s(t)$ is often expressed as

$$s(t) = I(t)\sqrt{2}\cos(\omega_0 t) - Q(t)\sqrt{2}\sin(\omega_0 t) \quad (5.64)$$

where

$$\begin{aligned} I(t) &= \sum_k a_0(k) p(t - kT_s) \\ Q(t) &= \sum_k a_1(k) p(t - kT_s) \end{aligned} \quad (5.65)$$

are PAM pulse trains. $I(t)$ is the "inphase" component of $s(t)$ and $Q(t)$ is the "quadrature" component of $s(t)$. It can be shown (see the Exercises) that the symbol energy for the k-th symbol is

$$\int_{T_1+kT_s}^{T_2+kT_s} s^2(t)\,dt = a_0^2(k) + a_1^2(k). \quad (5.66)$$

The k-th symbol, of the form

$$a_0(k) p(t - kT_s)\sqrt{2}\cos(\omega_0 t) - a_1(k) p(t - kT_s)\sqrt{2}\sin(\omega_0 t) \quad (5.67)$$

may also be expressed as

$$\sqrt{a_0^2(k) + a_1^2(k)}\, p(t - kT_s)\sqrt{2}\cos\left(\omega_0 t + \tan^{-1}\frac{a_1(k)}{a_0(k)}\right). \tag{5.68}$$

The expression (5.67) presents the transmitted waveform as a combination of amplitude modulated sinusoids. Because the amplitudes on the inphase and quadrature components are controlled by the data bits through the symbols $a_0(k)$ and $a_1(k)$, this form is often called the *rectangular* form. The expression (5.68) presents the transmitted waveform as a sinusoid whose magnitude and phase are controlled by the data through the symbols $a_0(k)$ and $a_1(k)$. This form is often called the *polar* form.

There are a wide variety of MQAM signal sets as the following examples illustrate. Some other constellations are explored in the Exercises.

Example 5.3.3
M-ary phase shift keying (MPSK) is characterized by constellations consisting of M points equally spaced around a circle of radius $\sqrt{E_{\text{avg}}}$ as illustrated in Figure 5.3.2. Each of the signals in the signal set differ from each other only in phase. All signals have the same energy.

Example 5.3.4
Square MQAM is characterized by constellations whose points are placed on an evenly spaced grid and whose boundary forms a square, as illustrated in Figure 5.3.3. These constellations only exist for the case where M is an even power of 2. These signal sets are often thought of as composed of $\sqrt{M}$-ary PAM signal sets on the inphase and quadrature components. The corresponding constellations are formed as the outer product of two $\sqrt{M}$-ary PAM constellations.

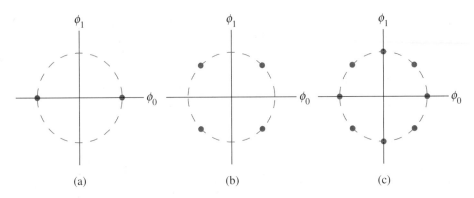

Figure 5.3.2 MPSK constellations for (a) $M = 2$, (b) $M = 4$, and (c) $M = 8$.

Section 5.3 M-ary Quadrature Amplitude Modulation

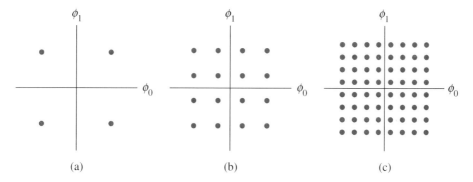

Figure 5.3.3 Square MQAM constellations for (a) $M = 4$, (b) $M = 16$, and (c) $M = 64$.

Not all signals in this signal set have the same energy. Let the points along the ϕ_0 axis be

$$-A(\sqrt{M}-1), -A(\sqrt{M}-3), \ldots -A, +A, \ldots, +A(\sqrt{M}-3), +A(\sqrt{M}-1) \quad (5.69)$$

and let the points along the ϕ_1 axis be the same. The average signal energies are

$$\begin{aligned} E_{\text{avg}} &= 2A^2 & M &= 4 \\ E_{\text{avg}} &= 10A^2 & M &= 16 \\ E_{\text{avg}} &= 42A^2 & M &= 64 \\ E_{\text{avg}} &= \tfrac{2}{3}(M-1)A^2 & &\text{general } M. \end{aligned} \quad (5.70)$$

Cable modems, defined by DOCSIS[5] 1.0, use 64 and 256 square QAM for the "downstream data channel" (to the home) and QPSK and square 16 QAM for the "upstream data channel" (from the home to the network).

Example 5.3.5
When M is an odd power of 2, the constellation points on a square grid do not form a square, but form that is commonly called a "cross constellation" as illustrated in Figure 5.3.4. The cross constellations consist of points on a square grid with vacancies in the corners as shown. If the minimum separation between points is $2A$, then the average signal energies are

$$\begin{aligned} E_{\text{avg}} &= 6A^2 & M &= 8 \\ E_{\text{avg}} &= 20A^2 & M &= 32 \\ E_{\text{avg}} &= 82A^2 & M &= 128 \end{aligned} \quad (5.71)$$

---

[5]The Data Over Cable Service Interface Specification (DOCSIS) standard was ratified by the International Telecommunication Union (ITU-TS) in March 1998. Although the DOCSIS continues to be used, the specification is now known as CableLabs Certified Cable Modems to emphasize the use of the specification as a certification definition for the modems produced by the makers of cable modems.

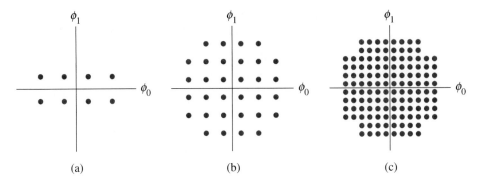

Figure 5.3.4 Cross MQAM constellations for (a) $M = 8$, (b) $m = 32$, and (c) $M = 128$.

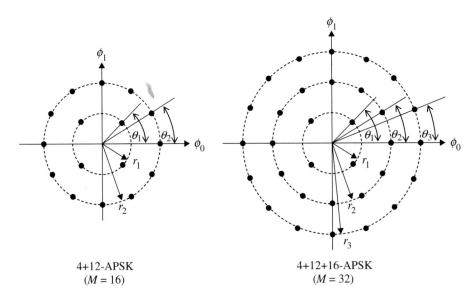

4+12-APSK
($M = 16$)

4+12+16-APSK
($M = 32$)

Figure 5.3.5 Examples of APSK Constellations for $M = 16$ and 32.

Example 5.3.6
Amplitude phase shift keying (APSK) constellations are characterized by points equally spaced on concentric rings in the signal space. Examples for 16-APSK and 32-APSK are shown in Figure 5.3.5. There are four parameters that define the 4+12-APSK constellation: the radii r_1 and r_2 and the phase angles θ_1 and θ_2 as shown. (Often, the 4+12-APSK is defined by only two parameters: the ratio of the radii r_2/r_1

Section 5.3   M-ary Quadrature Amplitude Modulation

and the phase angle offset $\Delta\theta = \theta_2 - \theta_1$.) There are six parameters that define the 4+12+16-APSK constellation shown: the radii r_1, r_2, and r_3 and the phase angles θ_1, θ_2, and θ_3 as shown. (The 4+12+16-APSK constellation may also be defined using a smaller number of ratios and differences. In this case, the four parameters are ratios r_2/r_1 and r_3/r_1 and the phase angle offsets $\Delta\theta_1 = \theta_2 - \theta_1$ and $\Delta\theta_2 = \theta_3 - \theta_1$). These parameters may be chosen to maximize the minimum Euclidean distance between any two points or to minimize the peak to average power ratio for the constellation. These constellations are currently being considered for the digital video broadcast (DVB) standard by the European Telecommunications Standards Institute (ETSI) [79].

Example 5.3.7
Dial-up Modems use some interesting constellations. The 9600 bits/s dial-up Modem standard is defined by the CCITT[6] V.29 standard and is illustrated in Figure 5.3.6. The constellation contains $M = 16$ points arranged as shown. This particular constellation is more tolerant of phase jitter than the square constellation with 16 points illustrated in Figure 5.3.3.

Sometimes, the signal-to-noise ratio is not high enough to allow reliable communications using 16 points. In this case, a smaller constellation can be used as a fallback constellation. The CCITT V.29 standard defines an 8-ary QAM constellation as the fallback constellation as shown in Figure 5.3.6. Note that the fallback constellation is a subset of the full rate constellation.

Both of these constellations can also be thought of as APSK constellations. This equivalence is explored in Exercise 5.57.

Example 5.3.8
It will be shown in Section 6.2 that the probability of error for MQAM can be expressed in terms of the Euclidean distances, normalized to the average energy, between the constellation points—see (6.76) and (6.77). To minimize the probability of error, the constellation points should be positioned to maximize the normalized Euclidean distances between the points. Constellations which do this for $M = 4, 8$, and 16 are illustrated in Figure 5.3.7.

When the constellation points are restricted to lie on a regular grid or in concentric circles, the constrained optimization described above can also be applied. The minimum probability of error constellations in this case are illustrated in Figure 5.3.8 for $M = 4, 8, 16, 32$, and 64 and in Figure 5.3.9 for $M = 128$. The probability of error for the square or cross QAM constellations is very close to the

---

[6]The Comite Consultatif Internationale de Telegraphie et Telephonie is an international committee based in Geneva, Switzerland, that recommends telecommunications standards, including the audio compression/decompression standards (codecs) and the famous V. standards for modem speed and compression (V.34 and so on). Although this organization changed its name to ITU-T (International Telecommunications Union-Telecommunication) in 1990, the old French name lives on.

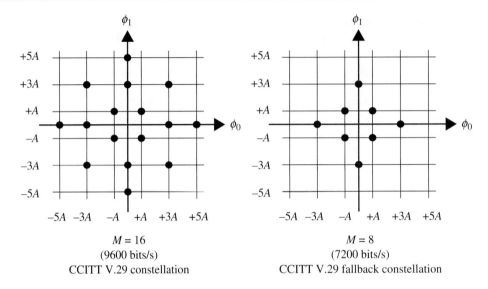

Figure 5.3.6 The CCITT V.29 constellations.

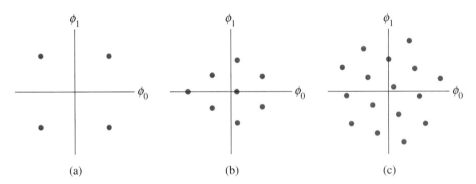

Figure 5.3.7 Minimum probability of error constellations for (a) $M = 4$, (b) $M = 8$, and (c) $M = 16$.

probability of error for each of these constellations except for the case $M = 8$. The performance penalty for using the square (or cross) constellation instead of the minimum probability of error constellation is quite small. Furthermore, the decision regions for square (or cross) constellations makes the square (or cross) much simpler than those of the minimum probability of error constellations. For this reason, the square (or cross) constellations are often used in practice.

Other considerations may also be taken into account when choosing a constellation. In some applications, a constant envelope is desirable. In these cases, MPSK with an NRZ pulse shape is preferred. On other applications, some envelope

Section 5.3 M-ary Quadrature Amplitude Modulation

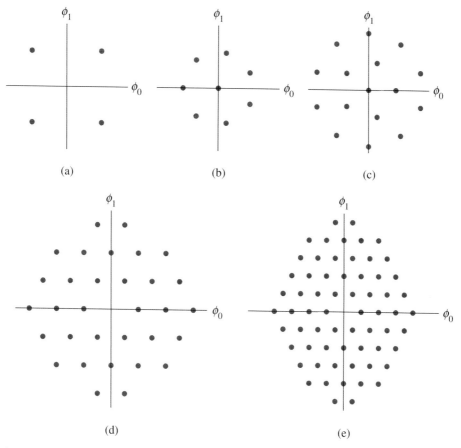

Figure 5.3.8 Minimum probability of error constellations when the points are constrained to lie on a grid or concentric circles for (a) $M = 4$, (b) $M = 8$, (c) $M = 16$, (d) $M = 32$, (e) $M = 64$.

variation can be tolerated, and constellations with a peak-to-average energy ratio as close to unity as possible are used.

5.3.1 Continuous-Time Realization

The MQAM modulator using continuous-time processing is illustrated in Figure 5.3.10 and is a direct application of the general linear modulator illustrated in Figure 5.1.5. The basis functions (5.50) are formed in two steps. First, the weighted pulse trains for the two basis functions are formed using a pulse shaping filter with an impulse train at the input. Second, the weighted pulse trains are multiplied by the sinusoids using mixers.

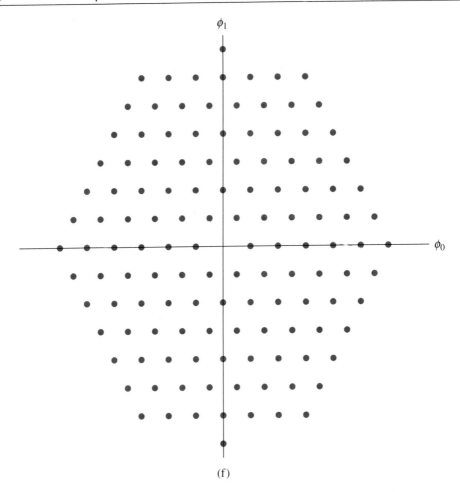

(f)

Figure 5.3.9 Minimum probability of error constellations when the points are constrained to lie on a grid or in concentric circles (continued from Figure 5.3.8) for (f) $M = 128$.

The input is a serial bit stream where a new bit arrives every T_b seconds. The input bits are partitioned into nonoverlapping segments of $\log_2 M$ bits by the serial-to-parallel converter. The output of the serial-to-parallel converter is a $\log_2 M$-bit address at the input of two look-up tables that store the symbol values specified by the constellation. The symbol values stored in the look-up tables are output in parallel every $T_s = T_b/\log_2 M$ seconds. The inphase and quadrature pulse trains given by (5.65) are formed in the same way the PAM pulse train was formed. The two pulse trains amplitude modulate two carriers in quadrature to form the two components of the 2-dimensional signals. The two components are summed to form the QAM signal.

Section 5.3 M-ary Quadrature Amplitude Modulation

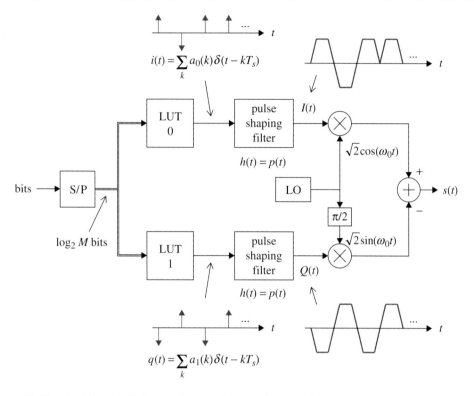

Figure 5.3.10 An M-ary QAM modulator using continuous-time processing.

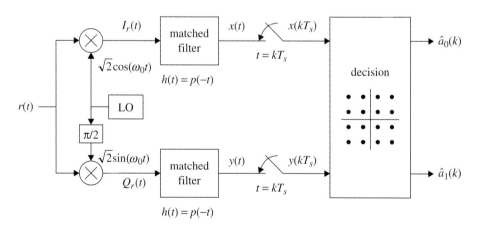

Figure 5.3.11 The matched filter detector for M-ary QAM using continuous-time processing.

Note that the sign on the quadrature sinusoid at the input of the lower mixer is positive but that the sign of $\phi_1(t)$ in (5.50) is negative. This is because $+\sin(\omega_0 t)$ is easily produced from $+\cos(\omega_0 t)$ in continuous-time processing as described in Exercises 5.37–5.39. The negative sign required by the basis function definition (5.50) is moved to the summing junction.

The matched filter detector is illustrated in Figure 5.3.11. The signal space projections onto $\phi_0(t)$ and $\phi_1(t)$ given by (5.50) are performed in two steps using a mixer and a filter whose impulse response is *matched to the pulse shape*. The two parallel matched filter outputs, denoted $x(t)$ and $y(t)$, are sampled at T_s-spaced intervals to produce a sequence of signal space projections that are used for detection. Detection uses the decision rule (5.16) or (5.19). Again note that the sign on the quadrature sinusoid $\sin(\omega_0 t)$ is positive instead of negative as required by (5.50). The sign change can be accounted for either by an inverter or in the decision rule.

The mathematical expressions for QAM are as follows. For notational convenience, let $I_r(t)$ and $Q_r(t)$ represent the inphase and quadrature components of the signal portion of the received waveform so that the received signal $r(t)$ may be expressed as

$$r(t) = I_r(t)\sqrt{2}\cos(\omega_0 t) - Q_r(t)\sqrt{2}\sin(\omega_0 t) + w(t) \tag{5.72}$$

where $w(t)$ represents the additive noise. Using the product formulae for sines and cosines, the inputs to the two parallel matched filters are

$$r(t)\sqrt{2}\cos(\omega_0 t) = I_r(t) + \underbrace{I_r(t)\cos(2\omega_0 t) - Q_r(t)\sin(2\omega_0 t)}_{\text{double frequency terms}} + \text{noise} \tag{5.73}$$

$$r(t)\sqrt{2}\sin(\omega_0 t) = -Q_r(t) + \underbrace{I_r(t)\sin(2\omega_0 t) + Q_r(t)\cos(2\omega_0 t)}_{\text{double frequency terms}} + \text{noise} \tag{5.74}$$

Because the matched filter is a low-pass filter, the double frequency terms need not be considered. The matched filter outputs are

$$
\begin{aligned}
x(t) &= \int_{T_1+t}^{T_2+t} I_r(\lambda)p(\lambda - t)d\lambda + v_0(t) \\
y(t) &= -\int_{T_1+t}^{T_2+t} Q_r(\lambda)p(\lambda - t)d\lambda + v_1(t)
\end{aligned}
\tag{5.75}
$$

where $v_0(t)$ and $v_1(t)$ are the matched filter outputs due to the noise.

When the receiver local oscillator (LO) is in perfect frequency and phase synchronism with the received carrier, $I_r(t)$ and $Q_r(t)$ are given by (5.65) and the matched filter outputs may be expressed as

$$
\begin{aligned}
x(t) &= \sum_m a_0(m) r_p(mT_s - t) + v_0(t) \\
y(t) &= -\sum_m a_1(m) r_p(mT_s - t) + v_1(t).
\end{aligned}
\tag{5.76}
$$

Section 5.3 M-ary Quadrature Amplitude Modulation

Sampling the matched filter outputs at $t = kT_s$ produces the signal space projections

$$x(kT_s) = \sum_m a_0(m) r_p((m-k)T_s) + v_0(kT_s) = a_0(k) + v_0(kT_s)$$
$$y(kT_s) = -\sum_m a_1(m) r_p((m-k)T_s) + v_1(kT_s) = -a_1(k) + v_1(kT_s)$$
(5.77)

where the second equality in each line is true if the pulse shape satisfies the Nyquist no-ISI condition. This shows that the signal space projections consist of the true point plus a noise term in each dimension. (Note that the negative sign on $a_1(k)$ follows from using $\sin(\omega_0 t)$ instead of $-\sin(\omega_0 t)$ as described previously.) The symbol decision follows the rule (5.16) or the equivalent form given by (5.17) as described in Section 5.1.3.

Example 5.3.9
As an example, consider the QPSK modulator illustrated in Figure 5.3.12. The constellation consists of four points as shown. Because QPSK is defined by 4 symbols, 2 bits are required to select one of the 4 symbols for transmission. The bit-to-symbol mapping is also shown on the constellation and defines the symbol $(a_0(k), a_1(k))$ corresponding to each pair of bits. This information is encoded in the two look-up tables shown in the modulator. In this example, the 8 input bits 10110100 are partitioned into the four nonoverlapping segments 10 11 01 00 by the serial-to-parallel converter. Each two-bit segment forms an address for use with the two look-up tables. The upper look-up table stores the values for $a_0(k)$ while the lower look-up table stores the values for $a_1(k)$. The bits and the look-up table outputs are listed below.

n	0	1	2	3	4	5	6	7
$b(n)$	1	0	1	1	0	1	0	0
k	0		1		2		3	
$a_0(k)$	+1		+1		−1		−1	
$a_1(k)$	−1		+1		+1		−1	

Note the relationship between the bit index n and the symbol index k. This relationship follows from the relationship $T_s = 2T_b$. The outputs of the look-up tables are the weighted impulse trains $i(t)$ and $q(t)$ plotted in Figure 5.3.12. These impulse trains are the inputs to pulse shaping filters that produce the pulse trains $I(t)$ and $Q(t)$ as shown. The pulse trains $I(t)$ and $Q(t)$ are multiplied by a pair of quadrature sinusoids and added to form the transmitted signal $s(t)$ which is also plotted.

The demodulator and detector are illustrated in Figure 5.3.13. The received signal $r(t)$ is a noisy version of the transmitted signal $s(t)$ as shown in the figure.

252 Chapter 5 Linear Modulation 1: Modulation, Demodulation, and Detection

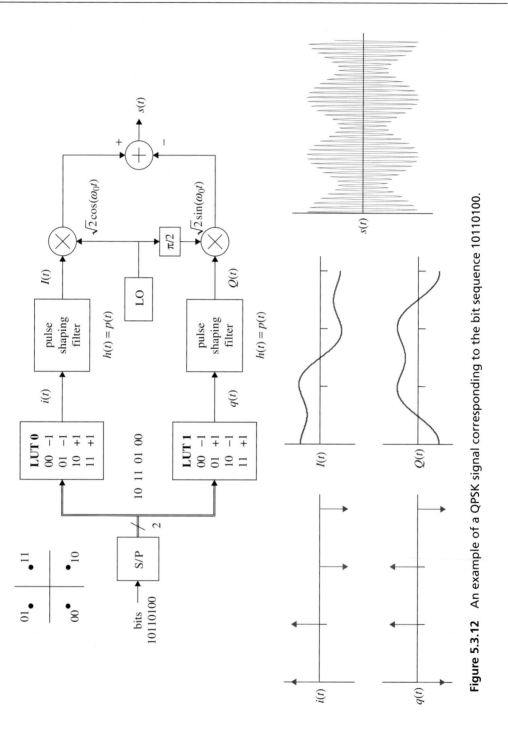

Figure 5.3.12 An example of a QPSK signal corresponding to the bit sequence 10110100.

Section 5.3 M-ary Quadrature Amplitude Modulation

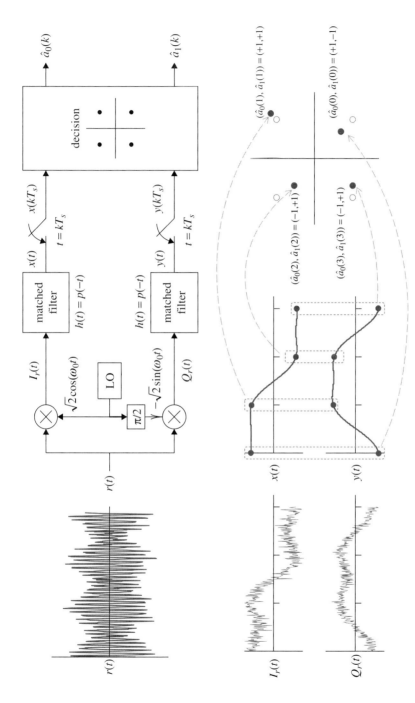

Figure 5.3.13 An example of a QPSK demodulator/detector corresponding to the QPSK modulator in Figure 5.3.12.

The inphase and quadrature components $I_r(t)$ and $Q_r(t)$ are also plotted in the figure.[7] $I_r(t)$ and $Q_r(t)$ are the inputs to filters matched to the pulse shape. Plots of the matched filter outputs, $x(t)$ and $y(t)$, are included in the figure. The matched filter outputs are sampled at the appropriate times to produce four signal space projections as shown. The signal space projection $(x(kT_s), y(kT_s))$ is used to make the symbol estimate $(\hat{a}_0(k), \hat{a}_1(k))$. In this example, the four matched filter outputs and their relationships to the symbol estimates are listed below.

k	0	1	2	3
$x(kT_s)$	+1.00	+1.02	−0.97	−0.98
$y(kT_s)$	−0.95	+1.02	+0.98	−1.00
$a_0(k)$	+1	+1	−1	−1
$a_1(k)$	−1	+1	+1	+1

As with PAM, the eye diagram is an important diagnostic tool for assessing the quality of the communications channel. What is different in this case is that *two* eye diagrams are needed. The first is the modulo-T_s plot of $x(t)$, the matched filter output corresponding to the inphase component of $r(t)$, and the second is the modulo-T_s plot of $y(t)$, the matched filter output corresponding to the quadrature component[8] of $r(t)$. Another useful diagnostic is a plot of $Q(t)$ versus $I(t)$. Such a plot is called a *phase trajectory* because the angle of the plot with respect to the $I(t)$ axis is the instantaneous phase of the modulated carrier. (The instantaneous magnitude of the modulated carrier is the Euclidean distance from the origin to the point on the plot.) Example trajectories for QPSK and 16 QAM are illustrated in Figure 5.3.14. Observe that the constellation points are easily identifiable in the phase trajectory plot. Note that signals with a small amount of excess bandwidth exhibit substantial overshoot in transitioning from one constellation point to another.

The demodulator/detector of Figure 5.3.11 and that of the example (Figure 5.3.13) are idealized in two ways. First, it is assumed that local oscillators are in perfect frequency and phase alignment with the received signal carrier. In real systems, this information is provided by a *carrier phase synchronization* subsystem. Continuous-time methods for performing carrier phase synchronization are described in Chapter 7. The effect of carrier phase offset is derived below. Second, it is assumed that the exact time instant for sampling the matched filter outputs is known. In real systems, this information is provided by the *symbol timing synchronization* subsystem. Continuous-time methods for performing symbol timing synchronization are described in Chapter 8.

---

[7]Note that, for this example, the double frequency terms that result from multiplication by the sinusoids have been omitted to make the plots of $I_r(t)$ and $Q_r(t)$ in Figure 5.3.13 clear. Also note that it has been assumed that $+\cos(\omega_0 t)$ and $-\sin(\omega_0 t)$ are capable of being produced. For the case where $+\cos(\omega_0 t)$ and $+\sin(\omega_0 t)$ can only be produced, $Q_r(t)$ and $y(t)$ should be replaced by their negatives. As a consequence, quadrature component of each signal space projection has the incorrect sign. This can be accounted for by either changing the sign of $y(kT_s)$ or by altering the symbol-to-bits mapping associated with the decision rule.

[8]For BPSK, there is no quadrature component so that only one eye diagram is needed.

Section 5.3 M-ary Quadrature Amplitude Modulation

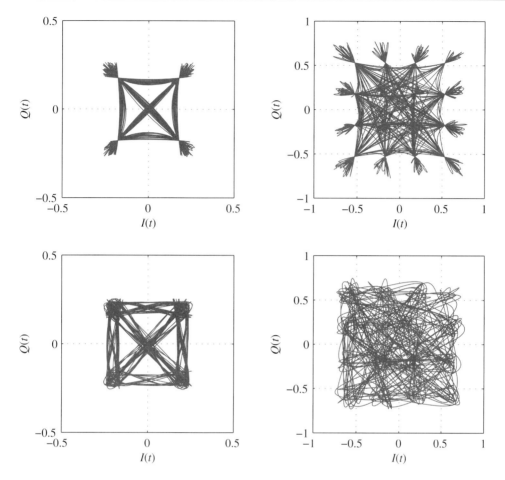

Figure 5.3.14 Phase trajectory plots for QPSK and 16-QAM using the SRRC pulse: (upper left) QPSK with 100% excess bandwidth; (upper right) 16-QAM with 100% excess bandwidth; (lower left) QPSK with 50% excess bandwidth; (lower right) 16-QAM with 50% excess bandwidth.

When the receiver LO is not in perfect phase synchronism with the received carrier, the received signal is, neglecting noise,

$$r(t) = I(t)\sqrt{2}\cos(\omega_0 t + \theta) - Q(t)\sqrt{2}\sin(\omega_0 t + \theta) \tag{5.78}$$

where $I(t)$ and $Q(t)$ are given by (5.65) and θ is the phase difference between the received carrier and the LO. Using the identities for $\cos(A + B)$ and $\sin(A + B)$, the received signal may be expressed as

$$r(t) = \underbrace{\left[I(t)\cos\theta - Q(t)\sin\theta\right]\sqrt{2}\cos(\omega_0 t)}_{I_r(t)} - \underbrace{\left[I(t)\sin\theta + Q(t)\cos\theta\right]\sqrt{2}\sin(\omega_0 t)}_{Q_r(t)} \quad (5.79)$$

from which $I_r(t)$ and $Q_r(t)$ are defined as shown. Substituting these expressions into (5.75) and applying (5.65) produces

$$x(kT_s) = \sum_m \left[a_0(m)\cos\theta - a_1(m)\sin\theta\right] r_p(mT_s - t)$$

$$y(kT_s) = -\sum_m \left[a_0(m)\sin\theta + a_1(m)\cos\theta\right] r_p(mT_s - t). \quad (5.80)$$

If the pulse shape satisfies the Nyquist no-ISI condition and sign compensation on the quadrature component is assumed, the signal space projection is

$$x(kT_s) = a_0(k)\cos\theta - a_1(k)\sin\theta$$

$$y(kT_s) = a_0(k)\sin\theta + a_1(k)\cos\theta. \quad (5.81)$$

These equations may be expressed in matrix form

$$\begin{bmatrix} x(kT_s) \\ y(kT_s) \end{bmatrix} = \begin{bmatrix} \cos\theta & -\sin\theta \\ \sin\theta & \cos\theta \end{bmatrix} \begin{bmatrix} a_0(k) \\ a_1(k) \end{bmatrix}. \quad (5.82)$$

The square matrix is the rotation matrix that rotates a point in two-space by an angle θ in the counterclockwise (CCW) direction. This shows that the signal space projection is a rotated version of the transmitted constellation point. This is illustrated in Figure 5.3.15 for 16 QAM.

5.3.2 Discrete-Time Realization

A natural starting point for the discrete-time realization of the modulator is the discrete-time equivalent of the continuous-time modulator outlined in Figure 5.3.10. The discrete-time version is illustrated in Figure 5.3.16 (a). The two sampled pulse trains, $I(nT)$ and $Q(nT)$ are produced using the upsampling-filter operation described in Section 3.2.3. The symbol streams $a_0(k)$ and $a_1(k)$ are upsampled by N (by inserting $N-1$ zeros between each symbol) to produce the inphase and quadrature signals

$$i(n) = \sum_m a_0\left(\frac{m}{N}\right) \delta(n - mN) \quad (5.83)$$

$$q(n) = \sum_m a_1\left(\frac{m}{N}\right) \delta(n - mN) \quad (5.84)$$

where $a_0\left(\frac{m}{N}\right)$ is understood to mean

$$a_0\left(\frac{m}{N}\right) = \begin{cases} a_0\left(\frac{m}{N}\right) & m = \text{integer} \times N \\ 0 & \text{otherwise} \end{cases}. \quad (5.85)$$

Section 5.3 M-ary Quadrature Amplitude Modulation   **257**

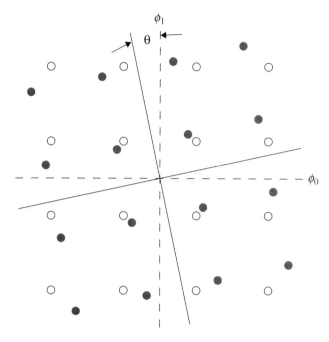

Figure 5.3.15 An illustration of the effect of uncompensated carrier phase offset on the signal space projections for 16 QAM. The empty circles are the points in the unrotated constellation. The solid circles are the points in the rotated constellation.

The corresponding understanding applies to $a_1\left(\frac{m}{N}\right)$. Interpolation is performed by the discrete-time pulse shaping filters whose impulse responses are samples of the pulse shape $p(t)$. The resulting discrete-time pulse trains, $I(nT)$ and $Q(nT)$, are multiplied by discrete-time quadrature sinusoids operating at

$$\Omega_0 = \omega_0 T \quad \text{rad/sample.} \tag{5.86}$$

The products are combined to form the sampled signal $s(nT)$, which is then converted to a continuous-time signal by the DAC operating at $1/T$ samples/s.

It is important to note that the description above is not the most efficient method for generating the samples $s(nT)$. This method is outlined only to document *what* processing has to be performed. *How* it should be performed is discussed in Section 10.1.1.

Three clock rates are used in the modulator. The first clock rate is the bit-rate clock used to clock the bits into the serial-to-parallel converter. The output of the serial-to-parallel converter is clocked at $1/\log_2(M)$ times the bit rate. This is the symbol-rate clock. The pulse shaping occurs at a rate that is N times the symbol clock rate. Because the sequences $I(nT)$ and $Q(nT)$ are translated from baseband to Ω_0 rad/sample, N must be chosen to be large enough to create enough space in the first Nyquist zone to accommodate the band-pass signals. As a consequence, the high sample rate (N times the symbol rate) is much higher than the

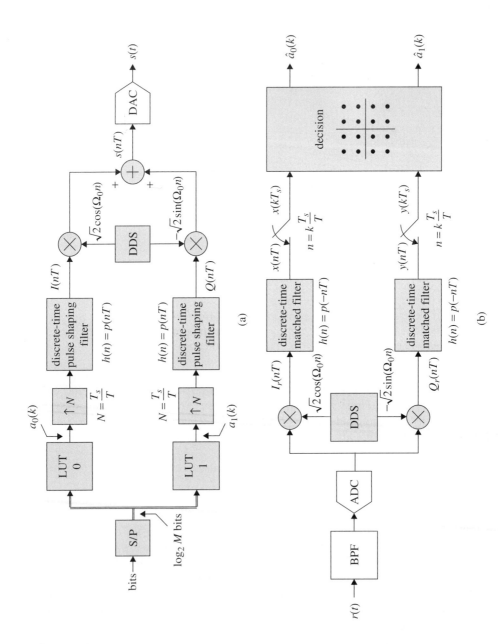

Figure 5.3.16 Discrete-time processing for QAM: (a) A discrete-time QAM modulator; (b) a discrete-time QAM demodulator.

Section 5.3 M-ary Quadrature Amplitude Modulation

minimum sample rate that satisfies the sampling theorem for the pulse shapes. How high it needs to be is determined by system constraints on clock rates and the relationship between Ω_0 and the desired continuous-time center frequency ω_0 rad/s.

There are several receiver architectures that could be applied for detecting QAM. The most straightforward (and least efficient) is a discrete-time version of the continuous-time detector such as the one illustrated in Figure 5.3.16 (b). After appropriate band-limiting by a continuous-time band-pass filter (not shown), the band-pass signal is sampled by the ADC operating at $1/T$ samples/s to produce the sampled signal

$$r(nT) = I_r(nT)\sqrt{2}\cos(\Omega_0 n) - Q_r(nT)\sqrt{2}\sin(\Omega_0 n). \tag{5.87}$$

These samples are multiplied by discrete-time quadrature sinusoids and filtered by an FIR filter matched to the pulse shape. The products on the inphase and quadrature portions of the receiver are

$$r(nT)\sqrt{2}\cos(\Omega_0 n) = I_r(nT) + \underbrace{I_r(nT)\cos(2\Omega_0 n) - Q_r(nT)\sin(2\Omega_0 n)}_{\text{double frequency terms}} + \text{noise}$$

$$-r(nT)\sqrt{2}\sin(\Omega_0 n) = Q_r(nT) + \underbrace{I_r(nT)\sin(2\Omega_0 n) + Q_r(nT)\cos(2\Omega_0 nT)}_{\text{double frequency terms}} + \text{noise} \tag{5.88}$$

The matched filter outputs are

$$x(nT) = \sum_{l=\frac{T_1}{T}+n}^{\frac{T_2}{T}+n} I_r(lT)p(lT - nT) + v_0(nT)$$

$$y(nT) = \sum_{l=\frac{T_1}{T}+n}^{\frac{T_2}{T}+n} Q_r(lT)p(lT - nT) + v_1(nT) \tag{5.89}$$

where $v_0(nT)$ and $v_1(nT)$ are samples of the matched filter outputs due to the noise.

When the DDS used to generate the quadrature sinusoids is in perfect frequency and phase synchronism with the received carrier, $I_r(t)$ and $Q_r(t)$ are given by (5.65) and the matched filter outputs may be expressed as

$$x(nT) = \frac{1}{T}\sum_m a_0(m)r_p(mT_s - nT) + v_0(nT)$$

$$y(nT) = \frac{1}{T}\sum_m a_1(m)r_p(mT_s - nT) + v_1(nT). \tag{5.90}$$

Sampling the matched filter outputs at $n = kT_s/T$ produces the signal space projections

$$x(kT_s) = \sum_m a_0(m)r_p((m-k)T_s) + v_0(kT_s) = \frac{a_0(k)}{T} + v_0(kT_s)$$

$$y(kT_s) = \sum_m a_1(m)r_p((m-k)T_s) + v_1(kT_s) = \frac{a_1(k)}{T} + v_1(kT_s)$$

(5.91)

where the second equality in each line is true if the pulse shape satisfies the Nyquist no-ISI condition. This shows that the signal space projections consist of the true point plus a noise term in each dimension. The decisions are made by finding the constellation point closest to the point $(x(kT_s), y(kT_s))$ as outlined in Section 5.1.3.

This demodulator/detector is idealized in three ways. First, it is assumed that local oscillators are in perfect frequency and phase alignment with the received signal carrier. In real systems, this information is provided by a *carrier phase synchronization* subsystem. Discrete-time methods for performing carrier phase synchronization are described in Chapter 7. Second, it is assumed that the ADC sample rate is an integer multiple of the symbol rate. This is very rarely the case in practice (see Chapter 10). Most demodulators include a resampling system that changes the sample rate from the ADC rate to a rate that is an integer multiple of the symbol rate. See Chapter 9 for a brief overview of resampling filters. Third, it is assumed that the desired signal space projection corresponds to one of the samples available at the matched filter output. In real systems, this information is provided by the *symbol timing synchronization* subsystem. Discrete-time methods for performing symbol timing synchronization are described in Chapter 8. As was the case with the modulator, this is not the most efficient method for generating the signal space projections. More efficient alternatives are outlined in Section 10.1.2.

5.4 OFFSET QPSK

Offset (or staggered) QPSK is a variation on QPSK where the quadrature component $Q(t)$ is delayed by half a symbol time relative to the inphase component $I(t)$. This is done to offset the transitions of $I(t)$ and $Q(t)$ so that these transitions do not occur simultaneously. The reason this is done lies in the behavior of nonlinear power amplifiers that are used in most wireless systems.

Consider the input/output characteristic for two classes of typical power amplifiers shown in Figure 5.4.1. These curves are called "AM/AM" curves because they characterize the amplitude of the modulated carrier at the power amplifier output as a function of the amplitude of the modulated carrier at the power amplifier input. Note that for small input amplitudes, the power amplifier behaves as a linear amplifier. As the input amplitude increases, however, the output amplitude departs from the linear behavior and "flattens out" a characteristic called *saturation*.

Figure 5.4.2 shows the power spectral density of QPSK modulated carrier when distorted by the AM/AM characteristics of Figure 5.4.1. The distortion manifests itself as an increase in

Section 5.4 Offset QPSK

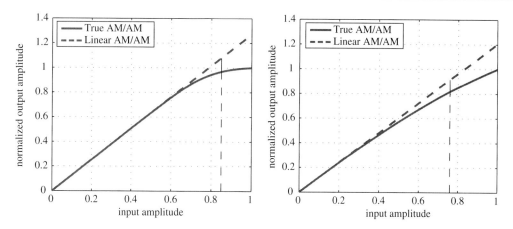

Figure 5.4.1 Examples of AM/AM curves for nonlinear power amplifiers: (left) A solid-state power amplifier (SSPA) commonly used in mobile terminals (cellular telephony, handheld radios, etc.); (right) a traveling wave tube amplifier (TWTA) used in satellite communications.

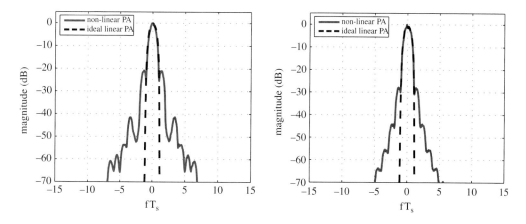

Figure 5.4.2 Power spectral densities of a QPSK modulated carrier at the output of the two non-linear power amplifiers of Figure 5.4.1: (left) SSPA; (right) TWTA. The pulse shape is the SRRC with 100% excess bandwidth.

the bandwidth of the modulated carrier. This increase is called, for historical reasons, *spectral regrowth*. In these examples, the -60 dB bandwidth of the modulated carrier increases from $2/T_s$ to $12/T_s$ for the SSPA and from $2/T_s$ to $7.6/T_s$ for the TWTA.

The mechanisms causing the spectral regrowth are best illustrated by the phase trajectory, such as those illustrated in Figure 5.3.14. The important characteristics of the magnitude (or amplitude) of the modulated carrier are

1. Amplitude variations in the modulated carrier: these variations manifest themselves as trajectory paths that are not on a circle.

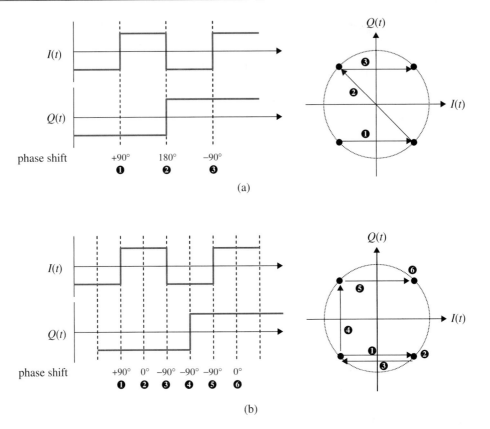

Figure 5.4.3 Illustration of the phase trajectories for the NRZ pulse shape: (a) QPSK; (b) offset QPSK.

2. Instantaneous zero input amplitudes: this characteristic manifests itself in the phase trajectory plot as transitions through the origin.

Offset QPSK addresses the second mechanism by disallowing the inphase and quadrature components to change sign at the same time. This feature is illustrated in Figure 5.4.3. In Figure 5.4.3 (a), there are three phase transitions corresponding to the four symbols for QPSK. The phase trajectory is also shown. Observe that a phase trajectory through the origin corresponds to simultaneous sign transitions on the inphase and quadrature components. When the quadrature component is delayed (or offset) by one-half symbol time, as illustrated in Figure 5.4.3 (b), the phase trajectory does not contain any transitions through the origin. The phase trajectory for offset QPSK using the SRRC pulse shape is illustrated in Figure 5.4.4.

The benefits of offset QPSK are readily apparent in the frequency domain. Figure 5.4.5 depicts QPSK and offset QPSK side-by-side for the purposes of comparison for the SSPA. Observe that the spectral regrowth is reduced dramatically and the -60 dB bandwidth is reduced from $12/T_s$ to $6.3/T_s$.

Section 5.4 Offset QPSK

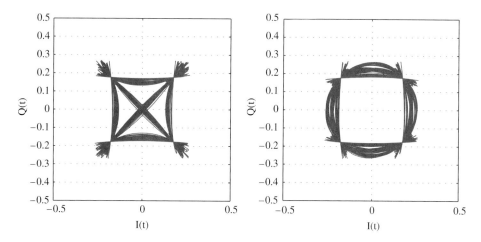

Figure 5.4.4 A comparison of phase trajectories for QPSK (left) and offset QPSK (right) using the SRRC pulse shape with 100% excess bandwidth.

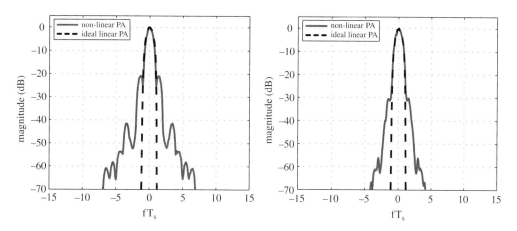

Figure 5.4.5 Power spectral density comparison at the output of the SSPA of Figure 5.4.1 QPSK (left) and offset QPSK (right). In both cases the pulse shape is the SRRC with 100% excess bandwidth.

Generation of an offset QPSK modulated carrier is a straightforward generalization of the QPSK modulator illustrated in Figure 5.4.6. Note that the block diagram is identical to the QPSK modulator of Figure 5.3.10 except for the inclusion of a half-symbol delay on the quadrature path. This half-symbol delay is notional: it could be placed at any point prior to where it is shown in the block diagram. For example, the output of LUT 1 could be clocked half symbol later than the output of LUT 0.

Detection is based on the same structure as the QPSK detector as illustrated in Figure 5.4.7. Observe that the eye diagrams at the outputs of the inphase and quadrature

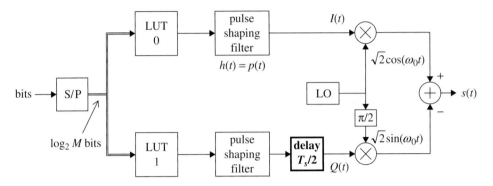

Figure 5.4.6 A notional representation of an offset QPSK modulator using continuous-time processing.

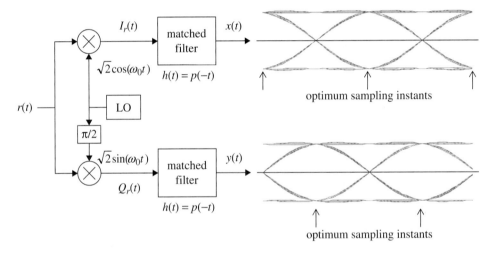

Figure 5.4.7 Detection of offset QPSK using a matched filter detector.

matched filters are offset by one-half symbol time. As a consequence the optimum sampling instants are also offset by one-half symbol time as indicated. The temptation is to delay the inphase component $I_r(t)$ by one-half symbol to realign the optimum sampling instants. In practice, the structure illustrated in Figure 5.4.8 is used. The matched filter outputs are sampled at 2 samples per symbol, time aligned with the optimum sampling instants for $x(t)$ and $y(t)$. Only the samples $x(kT_s)$ and $y(kT_s + T_s/2)$ are used in making symbol decisions. The other samples, $x(kT_s + T_s/2)$ and $y(kT_s)$, are required by the carrier-phase synchronizer and the symbol-timing synchronizer as described in Chapters 7 and 8, respectively.

The effect of carrier-phase offset for offset QPSK is quite different than it is for non-offset QPSK. Recall that the effect of carrier-phase offset for QPSK is a simple rotation of the decision points in the signal space projection, such as that illustrated in Figure 5.3.15

Section 5.5 Multicarrier Modulation

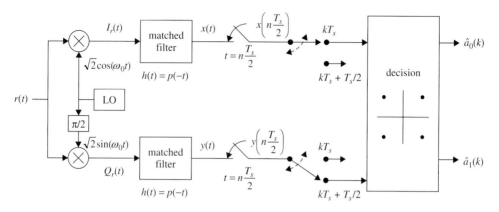

Figure 5.4.8 Typical offset QPSK detector.

for 16-QAM. The offset nature of the inphase and quadrature components in offset QPSK complicates the relationship between the desired samples of the inphase and quadrature components in the presence of a carrier-phase offset. This effect is illustrated in Figure 5.4.9. Observe that something more complicated than a phase rotation occurs. Note that a 90° carrier-phase offset for QPSK produces an I/Q plot that is indistinguishable from the I/Q plot for a 0° carrier-phase offset. (This leads to a potential 90° phase *ambiguity* that is described in Chapter 7.) For offset QPSK, the situation is quite different for a carrier-phase offset of 90°. For this reason, offset QPSK has a potential 180° phase ambiguity.

5.5 MULTICARRIER MODULATION

Signal space concepts can be used to describe a class of modulations known as *multicarrier modulation* or MCM. It will be more convenient to use the complex-valued baseband equivalent to describe this signal. The details of complex valued representations for a band-pass signal are described in Appendix B. The modulated signal $s(t)$ may be expressed as

$$s(t) = \operatorname{Re}\left\{\tilde{s}(t)e^{j2\pi f_c t}\right\} \tag{5.92}$$

The complex equivalent signal $\tilde{s}(t)$ can be thought of as a signal in the span of an N_M-dimensional signal space whose orthonormal basis functions $\phi_0(t), \phi_1(t), \ldots, \phi_{N_M-1}(t)$ are harmonically related complex sinusoids. The k-th basis function is

$$\phi_k(t) = \begin{cases} \dfrac{1}{\sqrt{T_M}} e^{j2\pi k f_M t} & 0 \le t \le T_M \\ 0 & \text{otherwise} \end{cases} \tag{5.93}$$

where f_M is the difference between adjacent harmonics. The frequency f_M is chosen to make the basis functions orthogonal. Orthogonality of the basis set is guaranteed when $\phi_k(t)$ and

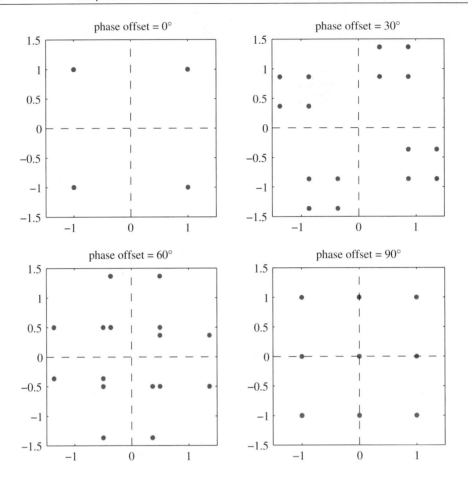

Figure 5.4.9 The effect of carrier-phase offset on the offset QPSK decision points.

$\phi_l(t)$ are orthogonal for $k \neq l$. Thus, f_M must satisfy[9]

$$0 = \int_0^{T_M} \phi_k(t)\phi_l^*(t)dt = \frac{1}{T_M}\int_0^{T_M} e^{j2\pi(k-l)f_M t}dt = \frac{\sin(\pi(k-l)f_M T_M)}{\pi(k-l)f_M T_M} \quad (5.94)$$

for all $l \neq k$. This is true when f_M is an integer multiple of $1/T_M$ and the smallest value of f_M that produces orthogonal functions is

$$f_M = \frac{1}{T_M}. \quad (5.95)$$

---

[9]The definitions for projections involving complex-valued basis functions require the complex conjugate as shown.

Section 5.5 Multicarrier Modulation

Using this relationship, the basis functions may be expressed as

$$\phi_k(t) = \begin{cases} \dfrac{1}{\sqrt{T_M}} e^{j2\pi kt/T_M} & 0 \le t \le T_M \\ 0 & \text{otherwise} \end{cases}. \quad (5.96)$$

Each of the orthonormal basis functions is periodic in t and the period divides T_M. During the interval $0 \le t \le T_M$, the constellation point

$$\mathbf{a}_0(0), \mathbf{a}_1(0), \ldots, \mathbf{a}_{N_M-1}(0)$$

is transmitted using the complex equivalent signal

$$\tilde{s}(t) = \sum_{k=0}^{N_M-1} \mathbf{a}_k(0)\phi_k(t) = \frac{1}{\sqrt{T_M}} \sum_{k=0}^{N_M-1} \mathbf{a}_k(0)e^{j2\pi kt/T_M}. \quad (5.97)$$

In general, a sequence of N_M-dimensional points is transmitted using

$$\tilde{s}(t) = \sum_m \sum_{k=0}^{N_M-1} \mathbf{a}_k(m)\phi_k(t - mT_M) = \frac{1}{\sqrt{T_M}} \sum_m \sum_{k=0}^{N_M-1} \mathbf{a}_k(m)e^{j2\pi k(t-mT_M)/T_M}$$

$$= \frac{1}{\sqrt{T_M}} \sum_m \sum_{k=0}^{N_M-1} \mathbf{a}_k(m)e^{j2\pi kt/T_M} \quad (5.98)$$

where the last line follows from the periodicity of the basis functions. The m-th constellation point

$$\mathbf{a}_0(m), \mathbf{a}_1(m), \ldots, \mathbf{a}_{N_M-1}(m)$$

is a set of N_M complex-valued constants selected by the data. The number bits required to select the N_M complex-constants depends on the number of possibilities allowed for each of the constants. It is frequently the case that the number of possibilities for $\mathbf{a}_k(m)$ is different from the number of possibilities for $\mathbf{a}_l(m)$.

Generation of an MCM modulated carrier is particularly efficient when implemented in discrete time. For simplicity, consider the signal during $0 \le t \le T_M$. Suppose the signal is sampled at T-spaced intervals to produce the sequence

$$s(nT) = \frac{1}{\sqrt{T_M}} \sum_{k=0}^{N_M-1} \mathbf{a}_k(0)e^{j2\pi knT/T_M} \quad (5.99)$$

If the sample rate is equivalent to N_M samples/symbol, then $T = T_M/N_M$ so that

$$s(nT) = \frac{1}{\sqrt{T_M}} \sum_{k=0}^{N_M-1} \mathbf{a}_k(0)e^{j2\pi kn/N_M}. \quad (5.100)$$

This shows that the N_M samples corresponding to the symbol can be produced from the input data by computing a scaled version of the inverse DFT of the sequence

$$\mathbf{a}_0(0), \mathbf{a}_1(0), \ldots, \mathbf{a}_{N_M-1}(0).$$

This remarkable result relies on the fact that using a sample rate equivalent to N_M samples/symbol does not produce aliasing. A cursory inspection of the basis functions suggests that aliasing does indeed occur with this sample rate. Close inspection, however, shows that this is not the case. The starting point is the Fourier transform of the basis functions. The Fourier transform of $\phi_k(t)$ is

$$\Phi_k(f) = \frac{1}{\sqrt{T_M}} \int_0^{T_M} e^{j2\pi kt/T_M} e^{-j2\pi ft} dt = \sqrt{T_M} \frac{\sin\left(\pi\left(f - \frac{k}{T_M}\right)T_M\right)}{\pi\left(f - \frac{k}{T_M}\right)T_M} e^{-j(f - \frac{k}{T_M})T_M}. \quad (5.101)$$

The Fourier transform $\Phi_k(f)$ is a $\sin(X)/X$ function centered at $f = k/T_M$ as illustrated in Figure 5.5.1. The zero-crossing occurs at multiples of $1/T_M$ as shown. Because the frequency spacing between adjacent basis functions is also $1/T_M$, the center of $\Phi_k(f)$ coincides with the zero crossings of all $\Phi_l(f)$ (for $l \neq k$). This characteristic is illustrated in Figure 5.5.2 (a) for the case of $N_M = 4$. Although the bandwidth of each basis function is infinite, something interesting happens when the basis functions are sampled (in time) every $T = T_M/N_M$ seconds. To illustrate the concept, let $N_M = 4$ and suppose the four coefficients $\mathbf{a}_0, \mathbf{a}_1, \mathbf{a}_2, \mathbf{a}_3$ are used to form the signal

$$\tilde{s}(t) = \frac{\mathbf{a}_0}{\sqrt{T_M}} + \frac{\mathbf{a}_1}{\sqrt{T_M}} e^{j2\pi t/T_M} + \frac{\mathbf{a}_2}{\sqrt{T_M}} e^{j2\pi 2t/T_M} + \frac{\mathbf{a}_3}{\sqrt{T_M}} e^{j2\pi 3t/T_M}$$

during the interval $0 \leq t \leq T_M$. Let the Fourier transform of $\tilde{s}(t)$ be $\tilde{S}(f)$ and suppose $\tilde{S}(f)$ is the spectrum illustrated in Figure 5.5.2 (a). Recall that in Chapter 2, an intermediate step

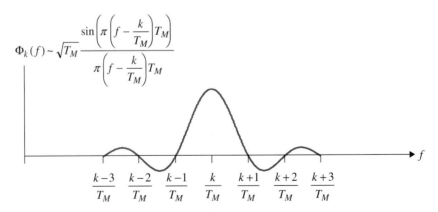

Figure 5.5.1 The Fourier transform of the basis function $\phi_k(t)$.

Section 5.5 Multicarrier Modulation

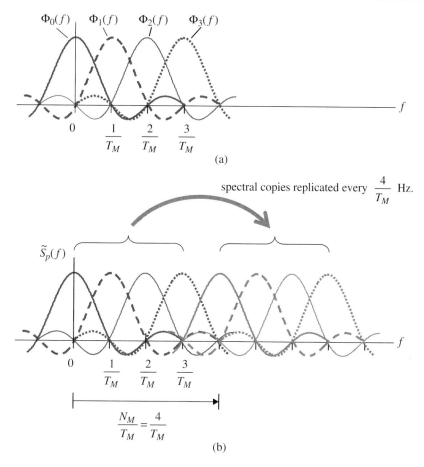

Figure 5.5.2 An example of an MCM signal in the frequency domain for $N_M = 4$: (a) The relationship between the Fourier transforms of the $N_M = 4$ basis functions; (b) the spectrum of the intermediate signal obtained by impulse sampling.

in computing the DTFT of the sampled signal was the signal sampled by an ideal impulse train:

$$\tilde{s}_p(t) = \tilde{s}(t) \times \sum_{n=-\infty}^{\infty} \delta(t - nT) = \sum_{n=-\infty}^{\infty} \tilde{s}(nT)\delta(t - nT).$$

The Fourier transform of $\tilde{s}_p(t)$ is obtained from the Fourier transform of $\tilde{s}(t)$ by a periodic repetition. The Fourier transform $\tilde{S}_p(f)$ is illustrated in Figure 5.5.2 (b). Observe that there is no spectral aliasing at values of f corresponding to multiples of $1/T_M$. It is this property that allows the signal to be constructed from $T = T_M/N_M$-spaced samples.

A block diagram for generating an MCM signal is illustrated in Figure 5.5.3 (a). The input bits are mapped to symbols that are grouped into blocks of four by the serial-to-parallel

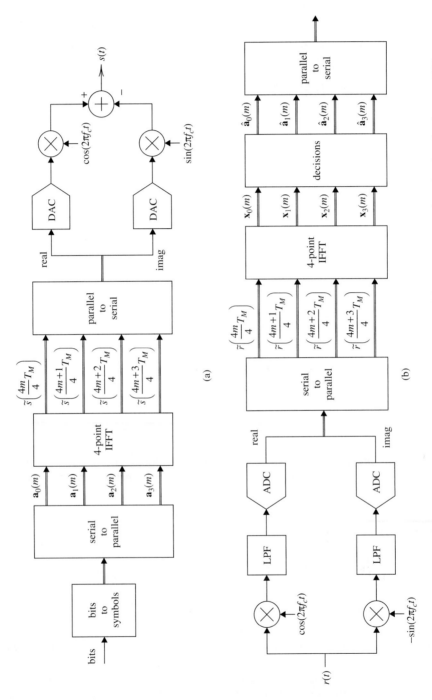

Figure 5.5.3 Generation and detection of an MCM signal for $N_M = 4$: (a) Generation using the 4-point inverse FFT; (b) detection using the 4-point FFT.

Section 5.5 Multicarrier Modulation

converter. The block of symbols is processed by the 4-point inverse FFT block to produce samples of the complex equivalent signal $\tilde{s}(t)$. The discrete-time sequence is converted to a continuous-time signal. The real-valued band-pass signal $s(t)$ is generated using the technique described in Appendix B.

Detection is accomplished using projections onto the signal space spanned by the orthonormal basis functions. The received signal may also be written in the form

$$r(t) = \text{Re}\left\{\tilde{r}(t)e^{j2\pi f_c t}\right\} \tag{5.102}$$

where $\tilde{r}(t)$ is the complex equivalent of real-valued band-pass signal $r(t)$. The demodulator produces $\tilde{r}(t)$ from $r(t)$. The signal $\tilde{r}(t)$ is projected onto the signal space during each interval of T_M seconds. The projection onto the k-basis function is

$$\mathbf{x}_k(m) = \int_0^{T_M} \tilde{r}(t - mT_M)\phi_k^*(t)dt = \frac{1}{\sqrt{T_M}} \int_0^{T_M} \tilde{r}(t - mT_M)e^{-j2\pi kt/T_M}dt. \tag{5.103}$$

When $\tilde{r}(t)$ is sampled every T seconds, the sequence $\tilde{r}(nT)$ may be used to compute the projections as follows:

$$\mathbf{x}_k(m) \approx \frac{1}{\sqrt{N_M}} \sum_{n=0}^{N_M-1} \tilde{r}(nT - mT_M)e^{-j2\pi knT/T_M}. \tag{5.104}$$

As before, when the sample rate is N_M samples/symbol, $T = T_M/N_M$ so that

$$\mathbf{x}_k(m) = \frac{1}{\sqrt{N_M}} \sum_{n=0}^{N_M-1} \tilde{r}\left(\frac{N_M m + n}{N_M}T_M\right) e^{-j2\pi kn/N_M}. \tag{5.105}$$

which is a scaled DFT of the sequence

$$\tilde{r}\left(\frac{N_M m}{N_M}T_M\right), \tilde{r}\left(\frac{N_M m + 1}{N_M}T_M\right), \ldots, \tilde{r}\left(\frac{N_M m + N_M - 1}{N_M}T_M\right).$$

A block diagram of this approach is illustrated in Figure 5.5.3 (b).

MCM has a compelling interpretation in the frequency domain. Each basis function is interpreted as an unmodulated carrier. The k-th unmodulated carrier is amplitude modulated by a constant every T_M seconds. The constant is usually thought of as a point from a QAM constellation. All carriers can be modulated with points from the same constellation (e.g., a QPSK constellation is applied to each carrier) or different constellations can be applied to different carriers (e.g., QPSK can be applied to one carrier and 16-QAM to another). Because there are N_M carriers all amplitude modulation simultaneously, this type of modulation is called *multicarrier modulation*. To emphasize the efficient generation and detection of the modulated carriers, this modulation is also known as *discrete-multitone* modulation or DMT. This is the term used to describe the modulation used for high-speed digital subscriber line or DSL links. To emphasize the orthogonality of the carriers, the name *orthogonal frequency division multiplexing* or OFDM is also used. OFDM is the name used to describe the modulation used for 802.11 wireless networking standards.

As an example, the IEEE 802.11a standard uses OFDM with $N_M = 52$ subcarriers and a symbol time of $T_M = 4\,\mu s$ (i.e., a new OFDM symbol is generated at a rate of $1/T_M = 250,000$ OFDM symbols/s). Of the 52 subcarriers, 48 are used for data. The other four are dedicated *pilot tones* and are used to estimate the conditions of the channel. The data rate experienced by the user is dependent on the channel conditions and available power. The variable data rate is implemented by keeping the symbol time T_M fixed and varying the constellation size and the code rate. The details are summarized in Table 5.5.1. The information should be interpreted as follows: Consider the 16-QAM constellation. Each point in the constellation conveys 4 bits of data. This is indicated by the value in the fourth column: 4 coded bits per subcarrier. When each of the 48 subcarriers is amplitude modulated by a point from the 16-QAM constellation, $48 \times 4 = 192$ bits are contained in the OFDM symbol. This is indicated by the value in the fifth column: 192 coded bits per OFDM symbol. When the rate-3/4 code is used, only 3 out of the 4 bits are user bits. That is, $3/4 \times 192 = 144$ bits in the OFDM symbol are information bits. The other 48 bits represent the overhead required for error correction. This is the information listed in the sixth column: 144 data bits per OFDM symbol. The data rate in the first column is computed as follows:

$$250000\,\frac{\text{OFDM symbols}}{\text{s}} \times 144\,\frac{\text{data bits}}{\text{OFDM symbol}} = 36\,\frac{\text{Mbits}}{\text{s}}.$$

MCM is popular in cases where the pass band of the channel frequency response is not flat. In this case, some frequency components are attenuated more than others. This characteristic causes severe distortion with single carrier modulations, such as QAM described in the previous section. With MCM, each carrier is narrowband and experiences an approximately flat channel pass band in this case. The power-throughput (i.e., bits per

Table 5.5.1 Modulation and coding specifications for OFDM used in the IEEE 802.11a wireless networking standard. This standard uses $N_M = 52$ subcarriers (48 of which are assigned to data) with a symbol time $T_M = 4\,\mu s$

Data Rate (Mbits/s)	Constellation	Code Rate	Coded Bits Per Subcarrier	Coded Bits Per OFDM Symbol	Data Bits Per OFDM Symbol
6	BPSK	1/2	1	48	24
9	BPSK	3/4	1	48	36
12	QPSK	3/4	2	96	48
18	QPSK	3/4	2	96	72
24	16-QAM	1/2	4	192	96
36	16-QAM	3/4	4	192	144
48	64-QAM	2/3	6	288	192
54	64-QAM	3/4	6	288	216

Source: Reproduced from the IEEE 801.11a Standard [80].

5.6 MAXIMUM LIKELIHOOD DETECTION

5.6.1 Introduction

Suppose the detector is presented with a series of numbers, denoted **r**, corresponding to samples of bandlimited waveforms that represent L_0 symbols (or constellation points)

$$\mathbf{a} = \begin{bmatrix} \mathbf{a}(0) & \mathbf{a}(1) & \cdots & \mathbf{a}(L_0 - 1) \end{bmatrix}^T.$$

Detection is the process of estimating the transmitted sequence. The estimate, denoted **â**, is based on observation of the samples **r**.

The true data sequence is not known at the detector.[10] In other words, there exists *uncertainty* about the true data sequence. Uncertainty should not be confused with ignorance. For example, the signal set (or set of constellation points) is known at the detector. Thus, the detector knows that each symbol in the symbol vector **a** has only M possibilities and it knows what those possibilities are.

There are many criteria on which the estimate may be based. The criterion used is the basis for assessing whether or not a candidate data sequence is a better estimate than another candidate data sequence. In the end, the criteria is based on how the uncertainty is modeled. One of the most commonly used models for uncertainty in engineering is probability.

Using probabilities, or probability density functions, to model the uncertainty leads to a detection rule that makes intuitive sense. Let $p(\mathbf{a}|\mathbf{r})$ be the conditional joint probability density function for the data symbols **a** given the observation **r**. Using this model, the best candidate data sequence is the one that maximizes $p(\mathbf{a}|\mathbf{r})$. Expressed mathematically, this estimate is

$$\hat{\mathbf{a}} = \arg\max \left\{ p(\mathbf{a}|\mathbf{r}) \right\}. \tag{5.106}$$

Using Baye's rule, the conditional probability can be rewritten as

$$p(\mathbf{a}|\mathbf{r}) = \frac{p(\mathbf{r}|\mathbf{a})p(\mathbf{a})}{p(\mathbf{r})}. \tag{5.107}$$

Now the estimate may be written as

$$\hat{\mathbf{a}} = \arg\max \left\{ \frac{p(\mathbf{r}|\mathbf{a})p(\mathbf{a})}{p(\mathbf{r})} \right\}$$

$$= \arg\max \left\{ p(\mathbf{r}|\mathbf{a})p(\mathbf{a}) \right\} \tag{5.108}$$

---

[10]If the true data sequence were known, there would be no point in transmitting it. Thus no information would be transmitted. The amount of information transmitted is directly linked to the degree of uncertainty. Think about it.

where the second line follows from the first using the fact that the maximization is with respect to **a**. As a consequence, any nonnegative term in the argument that is not a function of **a** can be omitted.

Conceptually, this maximization can be carried out by listing all M^{L_0} possible length-L_0 M-ary symbol sequences and computing the argument of (5.108) for each one. The candidate sequence that produces the largest value for the right-hand-side of (5.108) is selected as the estimate.

The conditional probability density function in (5.108) can be treated as a function of the data symbols **a**. When treated as a function of the variables to be estimated, this conditional probability is called the *likelihood function* for **a**. It is called the likelihood function because it is a measure of how likely each of the possible symbol sequences are, given the received data samples **r**. The second probability density function in (5.108) is called the *a priori* probability density function. The a priori probability density function, or prior probabilities, quantify the probability that each of the possible candidate symbol sequences was transmitted. If the data source produces certain symbol sequences with greater frequency than others, this term multiplies the likelihoods for these symbol sequences by a value larger than for the other, less likely, symbol sequences.

Often, the a priori probabilities are not known. The system designer has no option than to consider each of the symbol sequences equally likely. Doing so forces

$$p(\mathbf{a}) = \frac{1}{M^{L_0}}.$$

Because $p(\mathbf{a})$ is no longer a function of **a**, it can be omitted from the argument of (5.108) to the decision rule

$$\hat{\mathbf{a}} = \arg\max \left\{ p(\mathbf{r}|\mathbf{a}) \right\}. \tag{5.109}$$

Because the decision rule (5.109) is based solely on the likelihood function, this decision rule is called a *maximum likelihood* decision rule. To implement this rule, the form of the likelihood function $p(\mathbf{r}|\mathbf{a})$ must be known. The form of the likelihood function for linear modulated waveforms is formulated in the next section. Its interpretation in the context of the detectors presented earlier in the chapter is outlined in the last section.

5.6.2 Preliminaries

Let the observation interval be $T_0 = L_0 T_s$ seconds and let the received IF signal be

$$r(t) = s(t) + w(t) \tag{5.110}$$

where

$$s(t) = \sum_{k=0}^{L_0-1} \sum_{d=0}^{K-1} a_d(k) \phi_d(t - kT_s) \tag{5.111}$$

and $w(t)$ is a zero-mean white Gaussian random process with power spectral density $N_0/2$ W/Hz. The signal model (5.111) is the most general formulation based on K orthonormal basis functions. Thus, these results apply to baseband PAM, QAM, and other linear modulation

Section 5.6 Maximum Likelihood Detection

schemes. The symbol value assigned to the d-th basis function has only one subscript. It is understood that

$$a_d(k) \in \{a_{0,d}, a_{1,d}, \ldots, a_{M-1,d}\} \quad \text{for } d = 0, 1, \ldots, K-1. \tag{5.112}$$

The IF signal is sampled every T seconds to produce the sequence

$$r(nT) = s(nT) + w(nT); \quad n = 0, 1, \ldots, NL_0 - 1 \tag{5.113}$$

where N is the ratio of sample rate to symbol rate. For now, it is assumed that the samples are synchronized with the symbol clock so that there are exactly N samples in each symbol interval. This assumption is relaxed in Chapter 7 and 8, where carrier phase synchronization and symbol timing synchronization are explored, respectively. The n-th sample of the sampled signal component may be expressed as

$$s(nT) = \sum_{k=0}^{L_0-1} \sum_{d=0}^{K-1} a_d(k)\phi(nT - kT_s) \tag{5.114}$$

for $n = 0, 1, \ldots, NL_0 - 1$. For notational convenience, the following vectors are defined

$$\mathbf{r} = \begin{bmatrix} r(0) \\ r(T) \\ \vdots \\ r((NL_0-1)T) \end{bmatrix} \quad \mathbf{s} = \begin{bmatrix} s(0) \\ s(T) \\ \vdots \\ s((NL_0-1)T) \end{bmatrix} \quad \mathbf{w} = \begin{bmatrix} w(0) \\ w(T) \\ \vdots \\ w((NL_0-1)T) \end{bmatrix}. \tag{5.115}$$

The vector $\mathbf{w}$ is a sequence of independent and identically distributed Gaussian random variables with zero mean and variance

$$\sigma^2 = \frac{N_0}{2T}. \tag{5.116}$$

The probability density function of $\mathbf{w}$ is

$$p(\mathbf{w}) = \frac{1}{(2\pi\sigma^2)^{L_0 N/2}} \exp\left\{-\frac{1}{2\sigma^2} \sum_{n=0}^{NL_0-1} w^2(nT)\right\}. \tag{5.117}$$

For notational convenience, define the symbol vector $\mathbf{a}$ as

$$\mathbf{a} = \begin{bmatrix} \mathbf{a}(0) & \mathbf{a}(1) & \cdots & \mathbf{a}(L_0-1) \end{bmatrix}^T \tag{5.118}$$

where

$$\mathbf{a}(k) = \begin{bmatrix} a_0(k) \\ \vdots \\ a_{K-1}(k) \end{bmatrix}. \tag{5.119}$$

The vector $\mathbf{a}(k)$ has M possible values:

$$\mathbf{a}(k) \in \left\{ \mathbf{s}_0, \mathbf{s}_1, \ldots, \mathbf{s}_{M-1} \right\} \qquad (5.120)$$

where the $\mathbf{s}_m$ are the vectors defined in (5.14). To emphasize the fact that the $\mathbf{s}$ is a function of $\mathbf{a}$, $\mathbf{s}$ will be expressed as $\mathbf{s}(\mathbf{a})$ and samples of the signal component $s(nT)$ will be expressed as $s(nT; \mathbf{a})$.

5.6.3 Maximum Likelihood Decision Rule

The process of making symbol decisions (formally known as *detection*) can be thought of as a detection problem. The goal is to estimate the KL_0 values in the matrix $\mathbf{a}$ from the samples $r(nT) = s(nT; \mathbf{a}) + w(nT)$. The maximum likelihood estimate is the one that maximizes the logarithm of the conditional probability $p(\mathbf{r}|\mathbf{a})$. Using the probability density function of $\mathbf{w}$ given by (5.117), the conditional probability $p(\mathbf{r}|\mathbf{a})$ is

$$p(\mathbf{r}|\mathbf{a}) = \frac{1}{(2\pi\sigma^2)^{L_0N/2}} \exp\left\{ -\frac{1}{2\sigma^2} \sum_{n=0}^{NL_0-1} |r(nT) - s(nT; \mathbf{a})|^2 \right\}. \qquad (5.121)$$

The log-likelihood function $\Lambda(\mathbf{a})$ is the logarithm of (5.121):

$$\Lambda(\mathbf{a}) = -\frac{L_0 N}{2} \ln(2\pi\sigma^2) - \frac{1}{2\sigma^2} \sum_{n=0}^{NL_0-1} |r(nT) - s(nT; \mathbf{a})|^2$$

$$= -\frac{L_0 N}{2} \ln(2\pi\sigma^2) - \frac{1}{2\sigma^2} \sum_{n=0}^{NL_0-1} \left\{ |r(nT)|^2 - 2r(nT)s(nT; \mathbf{a}) + |s(nT; \mathbf{a})|^2 \right\}. \qquad (5.122)$$

Substituting (5.114) in (5.122) produces

$$\Lambda(\mathbf{a}) =$$

$$-\frac{L_0 N}{2} \ln(2\pi\sigma^2) - \frac{1}{2\sigma^2} \sum_{n=0}^{NL_0-1} |r(nT)|^2 + \frac{1}{\sigma^2} \sum_{k=0}^{L_0-1} \sum_{d=0}^{K-1} a_d(k) \sum_{n=0}^{NL_0-1} r(nT)\phi_d(nT - kT_s)$$

$$-\frac{1}{2\sigma^2} \sum_{k=0}^{L_0-1} \sum_{d=0}^{K-1} \sum_{k'=0}^{L_0-1} \sum_{d'=0}^{K-1} a_d(k) a_{d'}(k') \sum_{n=0}^{NL_0-1} \phi_d(nT - kT_s)\phi_{d'}(nT - k'T_s). \qquad (5.123)$$

The term

$$\sum_{n=0}^{NL_0-1} r(nT)\phi_d(nT - kT_s)$$

Section 5.6 Maximum Likelihood Detection

is the projection of $r(nT)$ onto the d-th basis function and is denoted $1/T \times x_d(kT_s)$. Note that this projection is usually produced by a matched filter as described in Section 5.1.4. The term

$$\sum_{n=0}^{NL_0-1} \phi_d(nT - kT_s)\phi_{d'}(nT - k'T_s)$$

is $1/T \times \delta(k-k')\delta(d-d')$. The $\delta(k-k')$ is a result of the implied assumption that there is no ISI. This assumption is trivially true for basis functions that involve full response pulse shapes and only true for basis functions that involve partial response pulse shapes that satisfy the Nyquist no-ISI condition. The $\delta(d-d')$ term follows from the fact that the basis functions are orthonormal. The last term in (5.123) thus reduces to

$$\frac{1}{2\sigma^2}\sum_{k=0}^{L_0-1}\sum_{d=0}^{K-1}\sum_{k'=0}^{L_0-1}\sum_{d'=0}^{K-1} a_d(k)a_{d'}(k') \sum_{n=0}^{NL_0-1} \phi_d(nT-kT_s)\phi_{d'}(nT-k'T_s) = \frac{1}{N_0}\sum_{k=0}^{L_0-1}\sum_{d=0}^{K-1}|a_d(k)|^2.$$

Putting all this together, the log-likelihood function may be expressed as

$$\Lambda(\mathbf{a}) = -\frac{L_0 N}{2}\ln(2\pi\sigma^2) - \frac{1}{2\sigma^2}\sum_{n=0}^{NL_0-1}|r(nT)|^2 + \frac{2}{N_0}\sum_{k=0}^{L_0-1}\sum_{d=0}^{K-1}a_d(k)x_d(kT_s)$$
$$- \frac{1}{N_0}\sum_{k=0}^{L_0-1}\sum_{d=0}^{K-1}|a_d(k)|^2. \tag{5.124}$$

The maximum likelihood estimate of the L_0 data symbols $\mathbf{a}(0), \mathbf{a}(1), \ldots, \mathbf{a}(L_0-1)$ is the sequence of symbols that maximizes the log-likelihood function. Expressed mathematically, the maximum likelihood estimate $\hat{\mathbf{a}}$ is

$$\hat{\mathbf{a}} = \arg\max_{\mathbf{a}\in\mathcal{S}^{L_0}}\left\{\Lambda(\mathbf{a})\right\}. \tag{5.125}$$

The notation $\mathcal{S}^{L_0}$ means L_0 K-dimensional points drawn from the signal set $\mathcal{S}$. The desired maximization is with respect to the M^{L_0} sequences of constellation points. Two observations should be made at this point:

1. Because each symbol has M distinct values, the maximization is a discrete optimization problem. As such, the usual use of derivatives does not apply. The only generally applicable method is to try all the possibilities for $\mathbf{a}$ to see which one maximizes $\Lambda(\mathbf{a})$.
2. The desired maximization is with respect to $\mathbf{a}$. Hence, terms in (5.124) that are not a function of $\mathbf{a}$ can be ignored.

Applying these two observations produces the following decision rule:

$$\hat{\mathbf{a}} = \arg\max_{\mathbf{a}\in S^{L_0}} \left\{ \frac{1}{N_0}\left[2\sum_{k=0}^{L_0-1}\sum_{d=0}^{K-1} a_d(k)x_d(kT_s) - \sum_{k=0}^{L_0-1}\sum_{d=0}^{K-1} |a_d(k)|^2 \right] \right\}$$

$$= \arg\max_{\mathbf{a}\in S^{L_0}} \left\{ 2\sum_{k=0}^{L_0-1}\sum_{d=0}^{K-1} a_d(k)x_d(kT_s) - \sum_{k=0}^{L_0-1}\sum_{d=0}^{K-1} |a_d(k)|^2 \right\}. \quad (5.126)$$

The maximization (5.126) is equivalent to minimization of the negative of the argument of (5.126):

$$\hat{\mathbf{a}} = \arg\min_{\mathbf{a}\in S^{L_0}} \left\{ \sum_{k=0}^{L_0-1}\sum_{d=0}^{K-1}\left[-2a_d(k)x_d(kT_s) + |a_d(k)|^2 \right] \right\}. \quad (5.127)$$

Adding a positive constant (constant with respect to **a**) to the argument does not change the answer. A convenient positive constant is

$$\sum_{k=0}^{L_0-1}\sum_{d=0}^{K-1} |x_d(kT_s)|^2.$$

Adding this term produces the decision rule

$$\hat{\mathbf{a}} = \arg\min_{\mathbf{a}\in S^{L_0}} \left\{ \sum_{k=0}^{L_0-1}\sum_{d=0}^{K-1}\left[|x_d(k)|^2 - 2a_d(k)x_d(kT_s) + |a_d(k)|^2 \right] \right\} \quad (5.128)$$

$$= \arg\min_{\mathbf{a}\in S^{L_0}} \left\{ \sum_{k=0}^{L_0-1}\sum_{d=0}^{K-1} |x_d(k) - a_d(k)|^2 \right\}. \quad (5.129)$$

Using the vector notation

$$\mathbf{x}(k) = \begin{bmatrix} x(k) \\ x_1(k) \\ \vdots \\ x_{K-1}(k) \end{bmatrix} \quad (5.130)$$

makes more obvious what the interpretation of the decision rule is. Substituting produces the following version for the decision rule:

$$\hat{\mathbf{a}} = \arg\min_{\mathbf{a}\in S^{L_0}} \left\{ \sum_{k=0}^{L_0-1} |\mathbf{x}(k) - \mathbf{a}(k)|^2 \right\}. \quad (5.131)$$

This decision rule has a very nice geometric interpretation: the k-th vector in **a** is the member of S that is closest in Euclidean distance to the vector $\mathbf{x}(k)$. Note that the decisions for $\mathbf{a}(k)$

does not depend on the decisions for any other symbol. As a result, the optimum decision for $\mathbf{a}(k)$ can be made independent of any of the others:

$$\hat{\mathbf{a}}(k) = \arg\min_{\mathbf{a}\in\mathcal{S}} \left\{ |\mathbf{x}(k) - \mathbf{a}(k)|^2 \right\}. \tag{5.132}$$

5.7 NOTES AND REFERENCES

5.7.1 Topics Covered

The geometrical interpretation using signal spaces for characterization and analysis of linear modulations was pioneered by Wozencraft and Jacobs [81]. The matched filter detector was first proposed by North [82] for optimum detection of RADAR returns. An advanced statistical approach to deriving the optimum detector is the Karhunen-Loeve expansion detailed in Refs. [83–85]. Meyr, Oerder, and Polydoros [86] showed that sampled data systems provide sufficient statistics for performing maximum likelihood detection.

Most of the research on signal set design has focused on constellations whose points are restricted to a regular grid (see Campopiano and Glazer [87]) such as the square and cross MQAM constellations. APSK, the placement of points on concentric circles was examined by Lucky and Hancock [88], Weber [89], Salz et al. [90], and Thomas, Weidner, and Durrani [91]. Triangular configurations were described by Kawai et al. [92]. The minimum probability of error constellations under these constraints were identified by Thomas, Weidner, and Durrani [91]. Unconstrained constellations with minimum probability of error were identified by Foschini, Gitlin, and Weinstein [93]. Constellations that maximize mutual information (i.e., come closest to achieving channel capacity) were examined by Blachman [94]. The diagnostic characteristics of the eye diagram were taken from Sousa and Pasupathy [95].

Multicarrier modulation, orthogonal frequency division multiplexing (OFDM), and discrete-multitone have a long history in digital communications with thousands of articles and hundreds of patents analyzing every conceivable issue of such systems. OFDM was developed in the 1960s with notable publications by Saltzberg [96], Chang and Gibbey [97], and Weinstein and Ebert [98]. The application of OFDM to HF radio communications, high-speed modem design [99,100], and mobile communications [101] was investigated in the 1970s and 1980s. In the 1990s and 2000s, OFDM was adopted as the modulation for the popular IEEE 802.11a, 802.11g, and 802.11n wireless networking standards as well as for the digital terrestrial television broadcasting standards in Europe [102]. In asymmetric digital subscriber line (ADSL) applications, OFDM is known as discrete multitone and has been selected as the standard for this application [103,104]. The survey paper by Bingham [105] summarizes the main issues in OFDM design. It is his use of the term "multicarrier modulation" that motivated the heading of Section 5.5. Some recent books also provide a thorough treatment of OFDM. Among them are Prasad [106], Bahai, Saltzberg, and Ergen [107], and Hanzo and Keller [108].

5.7.2 Topics Not Covered

The maximum likelihood detection rule (5.132) that makes each symbol decision independent of the others is actually a consequence of the signal model (5.111), which assumes that the data symbols are not temporally correlated. If the symbols are temporally correlated (e.g., as a result of a code or other formatter), this relationship should be placed in the signal model. The new signal model would then be used in (5.122) to produce an altered version of (5.123) where the dependence of data symbols on one another is quantified. Following the same steps above produces a decision rule where the symbol decisions are temporally related. In other words, the maximum likelihood decision rule estimates a sequence of symbols. In this case, the estimator is a *maximum likelihood sequence estimator*.

A tremendously important class of digital modulation techniques is the class of nonlinear modulations called *continuous phase modulations* or CPMs. CPM signals are of the form

$$s(t) = \cos\left(\omega_0 t + \phi(t)\right)$$

where the phase term $\phi(t)$ is a weighted pulse train formed from the data symbols:

$$\phi(t) = 2\pi h \sum_k a(k) g(t - kT_s).$$

The digital modulation index is h and $g(t)$ is the phase pulse, which is defined as the time integral of a frequency pulse $f(t)$. The classic reference on CPM is Anderson, Aulin, and Sundberg [31]. Chapters 4 and 5 of Proakis [10] also provide a thorough treatment. There are several important, special cases of CPM that should be mentioned: Continuous-phase frequency shift keying (CPFSK) is the special case that uses the NRZ pulse shape for the frequency pulse; minimum shift keying (MSK) uses the NRZ pulse shape for the frequency pulse and has a modulation index of 1/2; Gaussian minimum shift keying (GMSK) uses a low-pass filtered NRZ pulse shape and a modulation index of 1/2 (see Murota and Hirade [109].) The class of CPM modulations is very popular in satellite communications where nonlinear power amplifiers operate in full saturation and bandwidth efficient constant envelope modulations (such as CPM) are preferred.

5.8 EXERCISES

5.1 Let $\mathcal{B} = \{\phi_0(t), \phi_1(t), \ldots, \phi_{K-1}(t)\}$ be a set of K orthonormal functions defined on the interval $T_1 \leq t \leq T_2$ and let $r(t)$ be a signal that is not in Span$\{\mathcal{B}\}$. Let

$$\hat{r}(t) = \sum_{k=0}^{K-1} x_k \phi_k(t)$$

be the approximation to $r(t)$ that is in Span$\{\mathcal{B}\}$ that minimizes the squared error

$$\mathcal{E} = \int_{T_1}^{T_2} |r(t) - \hat{r}(t)|^2 dt.$$

Section 5.8 Exercises

Prove that the coefficients x_k for $k = 0, 1, \ldots, K - 1$ are given by

$$x_k = \int_{T_1}^{T_2} r(t)\phi_k(t)dt.$$

5.2 Let $\mathcal{B}$, $r(t)$, and $\hat{r}(t)$ be as defined in the previous exercise and let $e(t) = r(t) - \hat{r}(t)$. Prove that $e(t)$ is orthogonal to $\hat{r}(t)$.

5.3 Let $\mathcal{B} = \{\phi_0(t), \phi_1(t), \ldots, \phi_{K-1}(t)\}$ be a set of K orthogonal functions defined on the interval $T_1 \leq t \leq T_2$ and let $r(t)$ be a signal that is not in Span$\{\mathcal{B}\}$. Let

$$\hat{r}(t) = \sum_{k=0}^{K-1} x_k \phi_k(t)$$

be the approximation to $r(t)$ that is in Span$\{\mathcal{B}\}$ that minimizes the squared error

$$\mathcal{E} = \int_{T_1}^{T_2} |r(t) - \hat{r}(t)|^2 dt.$$

Derive the expression for the coefficients x_k for $k = 0, 1, \ldots, K - 1$.

5.4 Let $\mathcal{B} = \{\phi_0(t), \phi_1(t), \ldots, \phi_{K-1}(t)\}$ be a set of K functions defined on the interval $T_1 \leq t \leq T_2$ and let $r(t)$ be a signal that is not in Span$\{\mathcal{B}\}$. Note that the functions in $\mathcal{B}$ are not assumed to be orthogonal. Let

$$\hat{r}(t) = \sum_{k=0}^{K-1} x_k \phi_k(t)$$

be the approximation to $r(t)$ that is in Span$\{\mathcal{B}\}$ that minimizes the squared error

$$\mathcal{E} = \int_{T_1}^{T_2} |r(t) - \hat{r}(t)|^2 dt.$$

Derive the expression for the coefficients x_k for $k = 0, 1, \ldots, K - 1$.

5.5 Let $\mathcal{B} = \{\phi_0(t), \phi_1(t), \ldots, \phi_{K-1}(t)\}$ be a set of K orthogonal functions defined on the interval $T_1 \leq t \leq T_2$ and let

$$s(t) = \sum_{k=0}^{K-1} a_k \phi_k(t)$$

be a signal in Span$\{\mathcal{B}\}$. Derive an expression for the energy in $s(t)$ in terms of the coefficients $a_0, a_1, \ldots, a_{K-1}$. What is the geometric interpretation of this result?

5.6 Consider the two basis functions shown below.

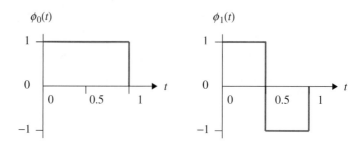

Sketch the waveforms corresponding to the points in the constellation shown below.

5.7 Using the basis functions of Exercise 5.6, sketch the waveforms corresponding to the points in the constellation shown below.

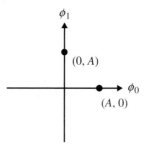

5.8 Using the basis functions of Exercise 5.6, sketch the waveforms corresponding to the points in the constellation shown below.

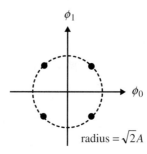

Section 5.8 Exercises

5.9 Using the basis functions of Exercise 5.6, sketch the waveforms corresponding to the points in the constellation shown below.

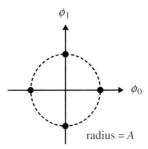

5.10 Using the basis functions of Exercise 5.6, sketch the waveforms corresponding to the points in the constellation shown below.

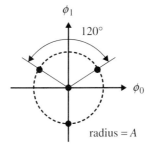

5.11 Using the basis functions of Exercise 5.6, sketch the waveforms corresponding to the points in the constellation shown below.

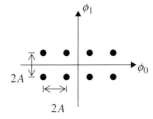

5.12 Sketch the waveform corresponding to the projection of the waveform shown below onto the basis functions of Exercise 5.6.

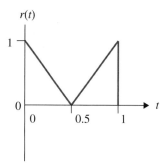

5.13 Sketch the waveform corresponding to the projection of the waveform shown below onto the basis functions of Exercise 5.6.

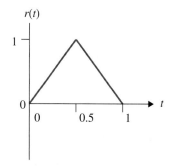

5.14 Sketch the waveform corresponding to the projection of the waveform shown below onto the basis functions of Exercise 5.6.

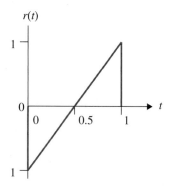

5.15 Sketch the waveform corresponding to the projection of the waveform shown below onto the basis functions of Exercise 5.6.

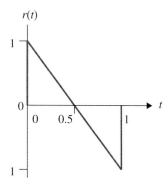

5.16 Consider the signal space spanned by the orthonormal functions of Exercise 5.6.
(a) Sketch the waveform corresponding to the signal space point $\mathbf{s} = [3, 1]$.
(b) Sketch the projection of the waveform shown below onto the signal space.

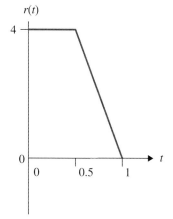

(c) Compare your answers in parts (a) and (b). What conclusions can be made?

5.17 Consider the two basis functions shown below.

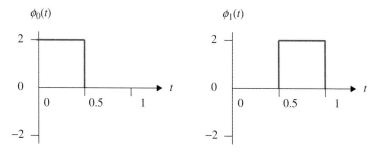

286 Chapter 5 Linear Modulation 1: Modulation, Demodulation, and Detection

Sketch the waveforms corresponding to the points in the constellation shown below.

5.18 Using the basis functions of Exercise 5.17, sketch the waveforms corresponding to the points in the constellation shown below.

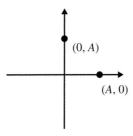

5.19 Using the basis functions of Exercise 5.17, sketch the waveforms corresponding to the points in the constellation shown below.

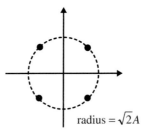

5.20 Using the basis functions of Exercise 5.17, sketch the waveforms corresponding to the points in the constellation shown below.

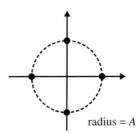

Section 5.8 Exercises

5.21 Using the basis functions of Exercise 5.17, sketch the waveforms corresponding to the points in the constellation shown below.

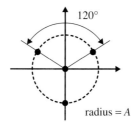

5.22 Using the basis functions of Exercise 5.17, sketch the waveforms corresponding to the points in the constellation shown below.

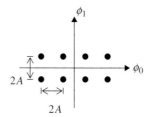

5.23 Given a finite set of M waveforms

$$\{s_0(t), s_1(t), \ldots, s_{M-1}(t)\}$$

defined over the interval $T_1 \leq t \leq T_2$, an orthonormal set of K basis functions

$$\{\phi_0(t), \phi_1(t), \ldots, \phi_{K-1}(t)\}$$

can by constructed by following a procedure known as the *Gram–Schmidt* procedure. The steps are

Step 1 (initialize): Set

$$g_0(t) = s_0(t) \quad \text{and define} \quad \phi_0(t) = \frac{g_0(t)}{\sqrt{\int_{T_1}^{T_2} |g_0(t)|^2 dt}}.$$

Step 2 : Set

$$g_i(t) = s_i(t) - \sum_{j=0}^{i-1} \left[\int_{T_1}^{T_2} s_i(t)\phi_j(t)dt \right] \phi_j(t) \quad \text{and define} \quad \phi_i(t) = \frac{g_i(t)}{\sqrt{\int_{T_1}^{T_2} |g_i(t)|^2 dt}}$$

for $i = 1, 2, \ldots, M-1$.

If one or more of the steps produces $g_i(t) = 0$, omit the normalized version of $g_i(t)$ from the list of basis functions.

Consider the set of $M = 4$ waveforms shown below:

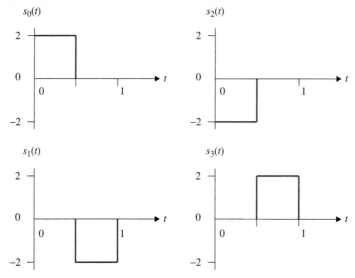

(a) Apply the Gram–Schmidt procedure to obtain a set of K basis functions that span the set of signals. What is the dimension of the signal space that contains these four signals?

(b) Plot the signal constellation corresponding to the signal set.

5.24 Repeat Exercise 5.23, but carry out the Gram–Schmidt procedure in the order $s_3(t)$, $s_2(t), s_1(t), s_0(t)$. If you worked Exercise 5.23, comment on the relationship between the signal constellation obtained in that exercise with the one obtained in this exercise.

5.25 Consider the signal set consisting of $M = 4$ signals shown below:

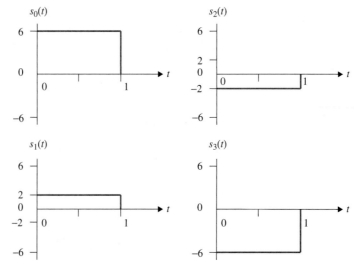

(a) Apply the Gram–Schmidt procedure to obtain a set of K basis functions that span the set of signals. What is the dimension of the signal space that contains these four signals?

(b) Plot the signal constellation corresponding to the signal set.

5.26 Consider the signal set consisting of $M = 4$ signals shown below:

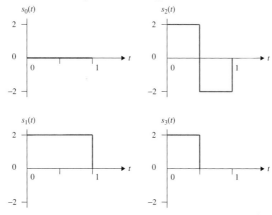

(a) Apply the Gram–Schmidt procedure to obtain a set of K basis functions that span the set of signals. What is the dimension of the signal space that contains these four signals?

(b) Plot the signal constellation corresponding to the signal set.

5.27 Consider the signal set consisting of $M = 4$ signals shown below:

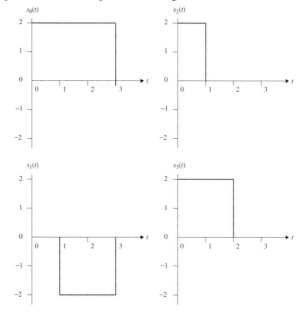

(a) Apply the Gram–Schmidt procedure to obtain a set of K basis functions that span the set of signals. What is the dimension of the signal space that contains these four signals?

(b) Plot the signal constellation corresponding to the signal set.

5.28 Consider the signal set consisting of $M = 3$ signals shown below:

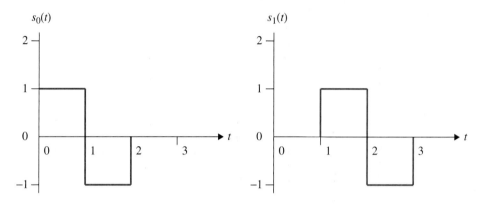

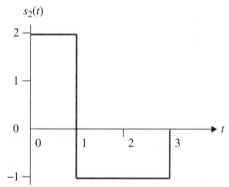

(a) Apply the Gram–Schmidt procedure to obtain a set of K basis functions that span the set of signals. What is the dimension of the signal space that contains these three signals?

(b) Plot the signal constellation corresponding to the signal set.

5.29 Consider the M-ary PAM signal set whose constellation consists of the points

$$-(M-1)A \quad -(M-3)A \quad \cdots -A \quad +A \quad \cdots \quad +(M-3)A \quad +(M-1)A$$

Show that the average energy is

$$E_{avg} = \frac{M^2 - 1}{3} A^2.$$

Section 5.8 Exercises

5.30 Consider the binary PAM signal set with constellation

```
      0           1
◄─────●─────┼─────●─────►
     −A          +A
```

whose average energy is 2.5 mJ.

(a) Sketch a block diagram of the modulator. Be sure to specify the contents of the look-up table.

(b) Assuming the HS pulse shape, sketch the transmitted waveform corresponding to the bit sequence 1 0 0 1 0 1 1 0.

(c) Sketch a block diagram of the matched filter detector.

(d) Determine the estimated symbol sequence

$$\hat{\mathbf{a}} = \hat{a}(0), \hat{a}(1), \hat{a}(2), \hat{a}(3)$$

and corresponding bit sequence for the matched filter outputs

$$\mathbf{x} = \{+0.001, -0.0123, +0.325, +0.071\}.$$

5.31 Consider the 4-ary PAM signal set with constellation

whose average energy is 4.5 mJ.

(a) Sketch a block diagram of the modulator. Be sure to specify the contents of the look-up table.

(b) Assuming the NRZ pulse shape, sketch the transmitted waveform corresponding to the bit sequence 1 0 0 1 0 1 1 0.

(c) Sketch a block diagram of the matched filter detector.

(d) Determine the estimated symbol sequence

$$\hat{\mathbf{a}} = \hat{a}(0), \hat{a}(1), \hat{a}(2), \hat{a}(3)$$

and corresponding bit sequence for the matched filter outputs

$$\mathbf{x} = \{+0.071, +0.055, -0.032, -0.101\}.$$

5.32 Consider the 8-ary PAM signal set with constellation

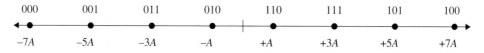

whose average energy is 328.125 mJ.

(a) Sketch a block diagram of the modulator. Be sure to specify the contents of the look-up table.
(b) Assuming the NRZ pulse shape, sketch the transmitted waveform corresponding to the bit sequence 1 0 0 1 0 1 1 1 0.
(c) Sketch a block diagram of the matched filter detector.
(d) Determine the estimated symbol sequence

$$\hat{\mathbf{a}} = \hat{a}(0), \hat{a}(1), \hat{a}(2), \hat{a}(3)$$

and corresponding bit sequence for the matched filter outputs

$$\mathbf{x} = \{+0.601, -0.101, +0.355, -0.777\}.$$

5.33 Consider a binary PAM system using the SRRC pulse shape.
(a) Produce an eye diagram corresponding to 200 randomly generated symbols for a pulse shape with 100% excess bandwidth and $L_p = 6T_s$. Use a sampling rate equivalent to 16 samples/symbol.
(b) Repeat part (a) for a pulse shape with 50% excess bandwidth and $L_p = 12T_s$.
(c) Repeat part (a) for a pulse shape with 25% excess bandwidth and $L_p = 200T_s$.
(d) Compare and contrast the eye diagrams from parts (a)–(c).

5.34 Consider a 4-ary PAM system using the SRRC pulse shape.
(a) Produce an eye diagram corresponding to 200 randomly generated symbols for a pulse shape with 100% excess bandwidth and $L_p = 6T_s$. Use a sampling rate equivalent to 16 samples/symbol.
(b) Repeat part (a) for a pulse shape with 50% excess bandwidth $L_p = 12T_s$.
(c) Repeat part (a) for a pulse shape with 25% excess bandwidth $L_p = 200T_s$.
(d) Compare and contrast the eye diagrams from parts (a)–(c).

5.35 This exercise explores the effect of channel bandwidth on a binary PAM waveform by examining the eye diagram. Consider the PAM system illustrated below

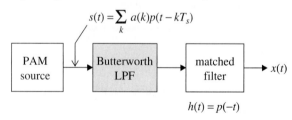

The system uses the NRZ pulse shape and operates at $N = 64$ samples/symbol. The channel is modeled as a third order Butterworth low-pass filter. Create a simulation of this system using MATLAB or some other software simulation tool.
(a) Set the 3-dB cutoff frequency of the Butterworth low-pass filter to $2/T_s$ and plot the eye diagram corresponding to 200 symbols.

(b) Set the 3-dB cutoff frequency of the Butterworth low-pass filter to $1/T_s$ and plot the eye diagram corresponding to 200 symbols.

(c) Set the 3-dB cutoff frequency of the Butterworth low-pass filter to $0.5/T_s$ and plot the eye diagram corresponding to 200 symbols.

(d) Compare and contrast the results obtained in parts (a)–(c).

5.36 This exercise explores the effect of filter order on a low-pass filtered binary PAM system. Consider the PAM system illustrated below

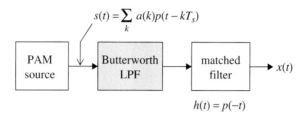

The system uses the NRZ pulse shape and operates at $N = 64$ samples/symbol. The channel is modeled as Butterworth low-pass filter with 3-dB cutoff frequency $0.5/T_s$. Create a simulation of this system using MATLAB or some other software simulation tool.

(a) Set the order of the Butterworth low-pass filter to 1 and plot the eye diagram corresponding to 200 symbols.

(b) Set the order of the Butterworth low-pass filter to 4 and plot the eye diagram corresponding to 200 symbols.

(c) Set the order of the Butterworth low-pass filter to 8 and plot the eye diagram corresponding to 200 symbols.

(d) Compare and contrast the results obtained in parts (a)–(c).

5.37 This exercise investigates the most common method for generating a pair of quadrature sinusoids from a single reference. Consider the system illustrated below.

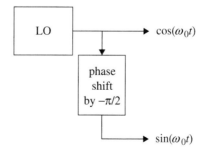

The local oscillator (LO) produces a reference sinusoid that is designated $\cos(\omega_0 t)$. The quadrature sinusoid $\sin(\omega_0 t)$ is generated from the LO by performing a $-\pi/2$ phase shift on the LO output. This produces $\cos(\omega_0 t - \pi/2) = \sin(\omega_0 t)$.

(a) The most straightforward way to produce the $-\pi/2$ phase shift is to delay the input by τ. Delaying the LO output by τ produces

$$\cos(\omega_0(t-\tau)) = \cos(\omega_0 t - \omega_0 \tau).$$

Determine the values of τ required to produce a $-\pi/2$ phase shift.

(b) Now suppose the LO is tunable (i.e., ω_0 can be changed) and that it is desired to use the delay element to produce quadrature sinusoids over the range

$$\omega_0 - \Delta\omega \leq \omega \leq \omega_0 + \Delta\omega.$$

Usually, the delay τ is fixed. In this case, the delay produces a precise $\pi/2$ phase shift only when $\omega = \omega_0$. For the other values of ω, the phase relationship between the two sinusoids is not exactly $\pi/2$. Determine the maximum value of $\Delta\omega$ if the quadrature phase error is to be less than 1°.

5.38 In Exercise 5.37, the use of a delay element to produce a $-\pi/2$ phase shift as part of a system that generated quadrature sinusoids was explored. It was shown that $\cos(\omega_0 t)$ and $\sin(\omega_0 t)$ are produced in a straightforward way. It, however, would be more convenient if $\cos(\omega_0 t)$ and $-\sin(\omega_0 t)$ were produced. An attempt to address this issue is illustrated by the system below.

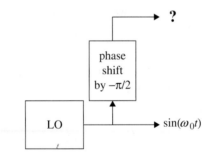

In this case, the LO output is used as the reference for $\sin(\omega_0 t)$ and the delay element is used to impart a $-\pi/2$ phase shift on this element.

(a) Determine the delay τ that produces a $-\pi/2$ phase shift. Express the output of the delay element in terms of the cosine.

(b) Modify the block diagram to show a system that produces $\cos(\omega_0 t)$ and $-\sin(\omega_0 t)$.

5.39 In Exercise 5.37, the use of a delay element to produce a $-\pi/2$ phase shift as part of a system that generated quadrature sinusoids was explored. It was shown that $\cos(\omega_0 t)$ and $\sin(\omega_0 t)$ are produced in a straightforward way. It, however, would be more convenient if $\cos(\omega_0 t)$ and $-\sin(\omega_0 t)$ were produced. An attempt to address this issue is illustrated by the system below.

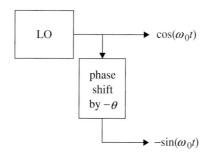

The local oscillator output is used as the reference for $\cos(\omega_0 t)$ and the delay element is used to impart a phase shift of $-\theta$ on the LO output.
(a) Determine the value of θ required to produce $-\sin(\omega_0 t)$ from $\cos(\omega_0 t)$.
(b) Determine the delay τ that produces the phase shift from part (a).
(c) Compare the answer for τ from part (b) to the delay in Exercise 5.37. What conclusions can be drawn?

5.40 Consider the QPSK constellation shown below.

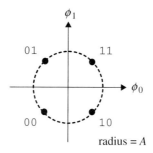

If the pulse shape is NRZ and the average energy is 2 J, complete the following.
(a) Sketch a block diagram of the modulator. Be sure to specify the contents of the look-up tables.
(b) Sketch the modulated waveform corresponding to the data bit sequence 1 1 0 1 0 0 1 1. Use $\omega_0 T_s = 4\pi$.
(c) Sketch a block diagram of the matched filter detector. Be sure to specify the impulse response of the matched filters.
(d) Sketch the decision region boundaries for each point in the constellation.
(e) Determine the estimated symbol sequence

$$\hat{\mathbf{a}} = \begin{bmatrix} \hat{a}_0(0) & \hat{a}_0(1) & \hat{a}_0(2) & \hat{a}_0(3) \\ \hat{a}_1(0) & \hat{a}_1(1) & \hat{a}_1(2) & \hat{a}_1(3) \end{bmatrix}$$

and the corresponding bit sequence for the matched filter outputs

k	0	1	2	3
x(k)	−1.01	+1.02	+1.11	−0.03
y(k)	+1.07	−0.99	+1.00	+0.07

5.41 Consider the 4-ary QAM constellation shown below.

If the pulse shape is HS and the average energy is 0.25 J, complete the following.
(a) Sketch a block diagram of the modulator. Be sure to specify the contents of the look-up tables.
(b) Sketch the modulated waveform corresponding to the data bit sequence 1 1 0 1 0 0 1 1. Use $\omega_0 T_s = 4\pi$.
(c) Sketch a block diagram of the matched filter detector. Be sure to specify the impulse response of the matched filters.
(d) Sketch the decision region boundaries for each point in the constellation.
(e) Determine the estimated symbol sequence

$$\hat{\mathbf{a}} = \hat{a}(0), \hat{a}(1), \hat{a}(2), \hat{a}(3)$$

and the corresponding bit sequence for the matched filter outputs

k	0	1	2	3
x(k)	−0.03	+0.55	+1.53	−2.10

5.42 Consider the 4-ary QAM constellation shown below.

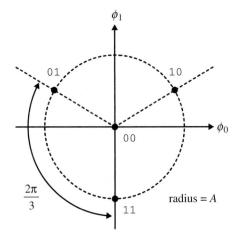

If the pulse shape is NRZ and the average energy is 3 J, complete the following.
(a) Sketch a block diagram of the modulator. Be sure to specify the contents of the look-up tables.

(b) Sketch the modulated waveform corresponding to the data bit sequence 1 1 0 1 0 0 1 1. Use $\omega_0 T_s = 4\pi$.

(c) Sketch a block diagram of the matched filter detector. Be sure to specify the impulse response of the matched filters.

(d) Sketch the decision region boundaries for each point in the constellation.

(e) Determine the estimated symbol sequence

$$\hat{\mathbf{a}} = \begin{bmatrix} \hat{a}_0(0) & \hat{a}_0(1) & \hat{a}_0(2) & \hat{a}_0(3) \\ \hat{a}_1(0) & \hat{a}_1(1) & \hat{a}_1(2) & \hat{a}_1(3) \end{bmatrix}$$

and the corresponding bit sequence for the matched filter outputs

k	0	1	2	3
x(k)	−0.03	−0.01	+0.01	+0.03
y(k)	+1.97	+1.97	+1.97	+1.97

5.43 Consider the 8-ary PSK constellation shown below.

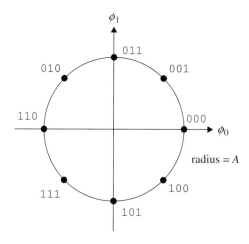

If the pulse shape is NRZ and the average energy is 2 J, complete the following.

(a) Sketch a block diagram of the modulator. Be sure to specify the contents of the look-up tables.

(b) Sketch the modulated waveform corresponding to the data bit sequence 1 1 0 1 0 0 1 1 1 0 1 0. Use $\omega_0 T_s = 4\pi$.

(c) Sketch a block diagram of the matched filter detector. Be sure to specify the impulse response of the matched filters.

(d) Sketch the decision region boundaries for each point in the constellation.

(e) Determine the estimated symbol sequence

$$\hat{\mathbf{a}} = \begin{bmatrix} \hat{a}_0(0) & \hat{a}_0(1) & \hat{a}_0(2) & \hat{a}_0(3) \\ \hat{a}_1(0) & \hat{a}_1(1) & \hat{a}_1(2) & \hat{a}_1(3) \end{bmatrix}$$

and the corresponding bit sequence for the matched filter outputs

k	0	1	2	3
x(k)	−1.30	+1.51	−0.90	+0.65
y(k)	+1.64	−1.49	−0.91	+0.07

5.44 Consider the 8-ary APSK constellation shown below.

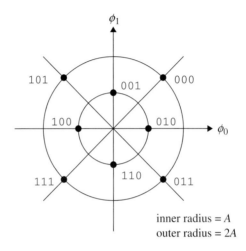

inner radius = A
outer radius = $2A$

If the pulse shape is NRZ and the average energy is 10 J, complete the following.
(a) Sketch a block diagram of the modulator. Be sure to specify the contents of the look-up tables.
(b) Sketch the modulated waveform corresponding to the data bit sequence 1 1 0 1 0 0 1 1 1 0 1 0. Use $\omega_0 T_s = 4\pi$.
(c) Sketch a block diagram of the matched filter detector. Be sure to specify the impulse response of the matched filters.
(d) Sketch the decision region boundaries for each point in the constellation.
(e) Determine the estimated symbol sequence

$$\hat{\mathbf{a}} = \begin{bmatrix} \hat{a}_0(0) & \hat{a}_0(1) & \hat{a}_0(2) & \hat{a}_0(3) \\ \hat{a}_1(0) & \hat{a}_1(1) & \hat{a}_1(2) & \hat{a}_1(3) \end{bmatrix}$$

and the corresponding bit sequence for the matched filter outputs

k	0	1	2	3
x(k)	−2.00	+1.00	+7.20	−1.64
y(k)	−2.01	+1.64	+0.00	+1.25

5.45 Consider the 8-ary APSK constellation shown below.

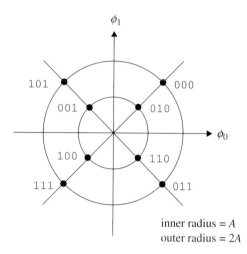

inner radius = A
outer radius = $2A$

If the pulse shape is NRZ and the average energy is 5 J, complete the following.

(a) Sketch a block diagram of the modulator. Be sure to specify the contents of the look-up tables.
(b) Sketch the modulated waveform corresponding to the data bit sequence 1 1 0 1 0 0 1 1 1 0 1 0. Use $\omega_0 T_s = 4\pi$.
(c) Sketch a block diagram of the matched filter detector. Be sure to specify the impulse response of the matched filters.
(d) Sketch the decision region boundaries for each point in the constellation.
(e) Determine the estimated symbol sequence

$$\hat{\mathbf{a}} = \begin{bmatrix} \hat{a}_0(0) & \hat{a}_0(1) & \hat{a}_0(2) & \hat{a}_0(3) \\ \hat{a}_1(0) & \hat{a}_1(1) & \hat{a}_1(2) & \hat{a}_1(3) \end{bmatrix}$$

and the corresponding bit sequence for the matched filter outputs

k	0	1	2	3
$x(k)$	−1.32	+1.51	+0.71	−0.05
$y(k)$	−1.32	−1.51	+1.32	+0.05

5.46 Consider the square 16-ary QAM constellation shown below.

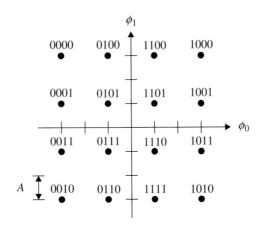

If the pulse shape is HS and the average energy is 0.9 J, complete the following.

(a) Sketch a block diagram of the modulator. Be sure to specify the contents of the look-up tables.
(b) Sketch the modulated waveform corresponding to the data bit sequence 1 1 0 1 0 0 1 1 1 0 1 0 0 1 0 0. Use $\omega_0 T_s = 4\pi$.
(c) Sketch a block diagram of the matched filter detector. Be sure to specify the impulse response of the matched filters.
(d) Sketch the decision region boundaries for each point in the constellation.
(e) Determine the estimated symbol sequence

$$\hat{\mathbf{a}} = \begin{bmatrix} \hat{a}_0(0) & \hat{a}_0(1) & \hat{a}_0(2) & \hat{a}_0(3) \\ \hat{a}_1(0) & \hat{a}_1(1) & \hat{a}_1(2) & \hat{a}_1(3) \end{bmatrix}$$

and the corresponding bit sequence for the matched filter outputs

k	0	1	2	3
x(k)	−0.30	+0.51	−0.90	+0.65
y(k)	+0.64	−0.49	−0.91	+0.07

5.47 Consider the square 16-ary QAM constellation shown below.

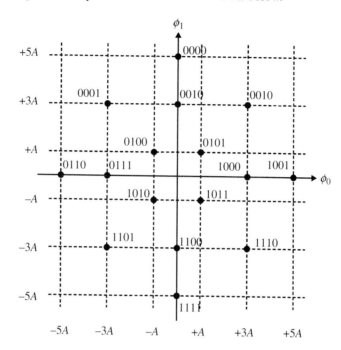

If the pulse shape is HS and the average energy is 1.875 J, complete the following.

(a) Sketch a block diagram of the modulator. Be sure to specify the contents of the look-up tables.

(b) Sketch the modulated waveform corresponding to the data bit sequence 1 1 0 1 0 0 1 1 1 0 1 0 0 1 0 0. Use $\omega_0 T_s = 4\pi$.

(c) Sketch a block diagram of the matched filter detector. Be sure to specify the impulse response of the matched filters.

(d) Sketch the decision region boundaries for each point in the constellation.

(e) Determine the estimated symbol sequence

$$\hat{\mathbf{a}} = \begin{bmatrix} \hat{a}_0(0) & \hat{a}_0(1) & \hat{a}_0(2) & \hat{a}_0(3) \\ \hat{a}_1(0) & \hat{a}_1(1) & \hat{a}_1(2) & \hat{a}_1(3) \end{bmatrix}$$

and the corresponding bit sequence for the matched filter outputs

k	0	1	2	3
x(k)	+0.90	−1.95	+1.90	+0.05
y(k)	+0.54	+0.01	−1.91	−0.94

5.48 Sketch the phase trajectory for QPSK using the RZ pulse shape.

5.49 Sketch the phase trajectory for QPSK using the MAN pulse shape.

5.50 Sketch the phase trajectory for QPSK using the HS pulse shape.

5.51 Sketch the phase trajectory for 16-QAM using the NRZ pulse shape.

5.52 Write a program to generate the phase trajectory for 8-PSK using the SRRC pulse shape with 100% excess bandwidth.

5.53 Compute the average energy for the two APSK constellations shown in Figure 5.3.5. Express your answers in terms of $r_1, r_2,$ and r_3.

5.54 Write a computer program to search for the ratio r_2/r_1 and $\Delta\theta$ that maximize the minimum normalized Euclidean distance between any two points in the 4+12-APSK constellation shown in Figure 5.3.5.

5.55 Write a computer program to search for the ratios r_2/r_1 and r_3/r_1 and the phase angle differences $\Delta\theta_1$ and $\Delta\theta_2$ to maximize the minimum normalized Euclidean distance between any two points in the 4+12+16-APSK constellation shown in Figure 5.3.5.

5.56 Compute the average energy for the two CCITT V.29 constellations shown in Figure 5.3.6.

5.57 The two CCITT V.29 constellations shown in Figure 5.3.6 can also be thought of as APSK constellations.
(a) Show that the 16-ary CCITT V.29 constellation of Figure 5.3.6 (a) is a 4+4+4+4-APSK constellation by determining the values $r_1, r_2, r_3, r_4, \theta_1, \theta_2, \theta_3,$ and θ_4 required to define the points in the signal space.
(b) Show that the 8-ary CCITT V.29 constellation of Figure 5.3.6 (b) is a 4+4-APSK constellation by determining the values of $r_1, r_2, \theta_1,$ and θ_2 required to define the points in the signal space.

5.58 Sketch the phase trajectory for offset QPSK using the RZ pulse shape.

5.59 Sketch the phase trajectory for offset QPSK using the MAN pulse shape.

5.60 Sketch the phase trajectory for offset QPSK using the HS pulse shape.

5.61 Consider the bit-to-symbol mapping for offset QPSK shown below.

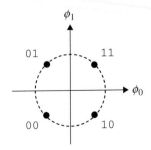

Section 5.8 Exercises

The coherent matched filter detector is illustrated in Figure 5.4.8.

(a) Using the plots for the inphase and quadrature matched filter outputs shown below, sketch the I/Q decision points corresponding to the proper sampling instants.

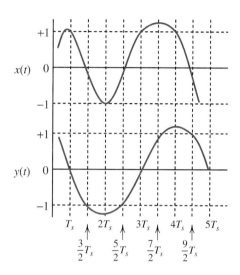

(b) Determine the corresponding transmitted bit sequence.

5.62 Consider an OFDM system with $T_M = 10$ μs/symbol and $N_M = 64$ subcarriers.

(a) Compute the data bit rate if a point from a QPSK constellation is applied to each subcarrier.

(b) Compute the data bit rate if a point from a 16-QAM constellation is applied to each subcarrier.

(c) Compute the data bit rate if a point from a 64-QAM constellation with a rate-2/3 error correcting code is applied to each subcarrier.

5.63 Consider the design of an OFDM system required to transmit 20 Mbits/s. Suppose system limitations impose a $T_M = 5$ μs OFDM symbol time. How many subcarriers are needed if a point from a 16-QAM constellation is applied to each subcarrier?

5.64 Consider the OFDM system, characterized by the $N_M = 8$ subcarriers shown below. Suppose each OFDM symbol is transmitted through a channel whose frequency response, $H(f)$ is shown below. Each received symbol is a distorted version of the transmitted symbol as illustrated by the spectrum $R(f)$ shown below. Explain how the constellations assigned to each subcarrier might be made. What criteria would you use in making your final decision?

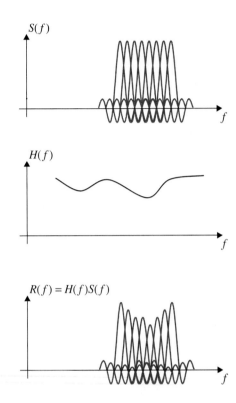

6

Linear Modulation 2: Performance

6.1 PERFORMANCE OF PAM   306
6.2 PERFORMANCE OF QAM   313
6.3 COMPARISONS   325
6.4 LINK BUDGETS   331
6.5 PROJECTING WHITE NOISE ONTO AN ORTHONORMAL BASIS SET   345
6.6 NOTES AND REFERENCES   347
6.7 EXERCISES   348

In the previous chapter, digital modulation was introduced and described in terms of signal spaces. The emphasis of that chapter was the generation of PAM and QAM signals and the recovery of the data from PAM and QAM waveforms. This chapter focuses on the performance of PAM and QAM. Traditionally, performance has been measured using bandwidth and power. The bandwidth is measured from the power spectral density of the modulated waveform. It will be shown that bandwidth is proportional to the bit rate. The constant of proportionality is determined by the pulse shape.

Power is obtained from the probability of bit error and is often expressed in the form of the quantity E_b/N_0 where E_b is the average bit energy and N_0, with units W/Hz, is the power spectral density level of the white noise. This quantity is informally known as the "signal-to-noise ratio." The probability of error analysis is restricted to the important case where additive white Gaussian noise is the only distortion imposed on the received waveform by the channel.

Using these two performance metrics and for a *fixed bit rate*, a given modulation is better than another modulation if it requires less bandwidth and a lower signal-to-noise ratio to achieve a given probability of bit error. In reality, the situation is more complicated. The choice between modulations presents a trade-off between bandwidth and power. For example, a given modulation may require less bandwidth but more power than another modulation. In these cases, other factors such as compatibility with nonlinear power amplifiers or susceptibility to phase jitter become the determining factors in preferring one modulation over another.

This chapter begins with the analysis of PAM. The bandwidth is analyzed in the continuous-time domain because that is where it is important: it measures how much spectrum the modulated waveform requires. The probability of error analysis is performed in the discrete-time domain. It represents a direct application of Gaussian random variables, whose properties were reviewed in Chapter 4. The bandwidth and probability of error for QAM are presented in the following section, where the important concept of a union bound using pairwise error probabilities is introduced. The chapter ends with link budgets. This topic ties system level design issues to both bandwidth and power. It represents a nice application of the theoretical results derived in the first two sections.

6.1 PERFORMANCE OF PAM

6.1.1 Bandwidth

For independent and equally likely symbols, the power spectral density of the PAM pulse train (5.25) is given by

$$P_s(f) = \frac{E_{\text{avg}}}{T_s} |P(f)|^2 \tag{6.1}$$

where $P(f)$ is the continuous-time Fourier transform of the pulse shape $p(t)$ and E_{avg} is the average symbol energy. The power spectral density is determined by the properties of the Fourier transform of the pulse shape. The power spectral density of any full response pulse shapes (e.g., NRZ, RZ, MAN, HS) has infinite support in the frequency domain. Consequently, the absolute bandwidth is infinite for these pulse shapes. In practice, the power spectral densities of these pulse shapes decay with increasing frequency. Thus for a sufficiently large f, $|P(f)|^2$ is "close enough to zero" for the purposes of bandwidth assignments in a wireless communication system. This permits the use of practical measures of bandwidth, such as those described in Appendix A. The bandwidths of the NRZ, RZ, MAN, and HS pulse shapes using these practical measures are summarized in Table A.1.2 in Appendix A. In each case, the bandwidth is of the form

$$\text{BW} = \frac{B}{T_s} = B \times R_s = B \log_2(M) \times R_b \tag{6.2}$$

where R_s is the symbol rate (symbols/second), R_b is the bit rate (bits/second) and B is a constant that depends on the pulse shape and the definition of bandwidth used.

For the ideal SRRC pulse shape, the $|P(f)|^2$ has finite support in the frequency domain. As a consequence, the absolute bandwidth is meaningful and is

$$\text{BW} = \frac{1+\alpha}{2T_s} = \frac{1+\alpha}{2} \times R_s = \frac{1+\alpha}{2} \log_2(M) \times R_b \tag{6.3}$$

where R_s is the symbol rate (symbols/second), R_b is the bit rate (bits/second), and α is the excess bandwidth. Observe that this expression (6.3) is of the same form as (6.2).

6.1.2 Probability of Error

The decision rule (5.16) partitions the signal space into *decision regions* as illustrated for the case of 4-ary baseband PAM in Figure 6.1.1. The probability of error is derived by first computing the conditional probabilities of error and applying the total probability theorem to give the desired result. In mathematical terms, let $s_0, s_1, \ldots, s_{M-1}$ be the M constellation points and let $P(E|a = s_m)$ be the probability of error given that the waveform corresponding to symbol s_m was sent for $m = 0, 1, \ldots, M - 1$. Applying the total probability theorem, the probability of error is

$$P(E) = \sum_{m=0}^{M-1} P(E|a = s_m) P(a = s_m) \quad (6.4)$$

where $P(a = s_m)$ is the probability that symbol s_m was transmitted in the first place. Usually, it is assumed that the transmitted symbols are equally likely (i.e., neither the data nor the transmitter prefers sending one symbol over another). In this case

$$P(a = s_m) = \frac{1}{M} \quad m = 0, 1, \ldots, M - 1 \quad (6.5)$$

and (6.4) reduces to

$$P(E) = \frac{1}{M} \sum_{m=0}^{M-1} P(E|a = s_m). \quad (6.6)$$

All that remains is to compute the conditional probabilities of error.

The procedure for computing the conditional probabilities of error for M-ary PAM is demonstrated for the $M = 4$ case illustrated in Figure 6.1.1. The starting point is the signal space projection (5.47). Assuming $a(k) = +A$, $x(kT_s)$ is a random variable whose probability density function is illustrated in Figure 6.1.2 (a). The conditional probability of error $P(E|a(k) = +A)$ is the probability that $x(kT_s)$ is outside the decision region corresponding to $a(k) = +A$. As it turns out, the easiest way to compute this probability is one

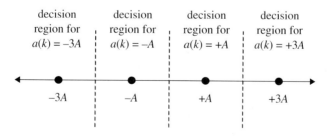

Figure 6.1.1 The four decision regions for 4-ary PAM.

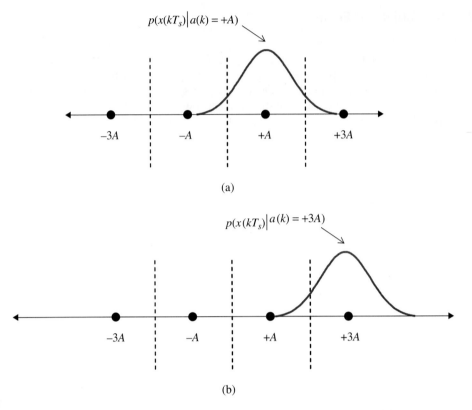

Figure 6.1.2 Probability density function for matched filter output when (a) $a/T = +A$ and (b) $a/T = +3A$.

minus the probability that $x(kT_s)$ is inside the decision region corresponding to $a(k) = +A$. Following this line of reasoning,

$$P(E|a(k) = +A) = 1 - P\left(0 \le x(kT_s) \le \frac{2A}{T}\right)$$

$$= 1 - P\left(0 \le \frac{A}{T} + v(kT_s) \le \frac{2A}{T}\right)$$

$$= 1 - P\left(-\frac{A}{T} \le v(kT_s) \le \frac{A}{T}\right). \tag{6.7}$$

Note that the decision region boundaries have been scaled by T because the signal space projection is also scaled by T. Applying the same procedure for the case $a(k) = -A$ produces

$$P(E|a(k) = -A) = 1 - P\left(-\frac{A}{T} \le v(kT_s) \le \frac{A}{T}\right) \tag{6.8}$$

Section 6.1 Performance of PAM

which is the same as $P(E|a(k) = +A)$ because the decision regions are the same shape. Assuming $a(k) = +3A$, $x(kT_s)$ is a random variable whose probability density function is illustrated in Figure 6.1.2 (b). Following the same procedure as before

$$P(E|a(k) = +3A) = 1 - P\left(\frac{2A}{T} \leq x(kT_s)\right)$$

$$= 1 - P\left(\frac{2A}{T} \leq \frac{3A}{T} + v(kT_s)\right)$$

$$= 1 - P\left(-\frac{A}{T} \leq v(kT_s)\right). \quad (6.9)$$

$P(E|a(k) = +3A)$ is different from $P(E|a(k) = +A)$ because the decision regions have different shapes. The decision region for $a(k) = +3A$ is "open ended" where as the decision region for $a(k) = +A$ is not. Following the same procedure once more, it can be shown that

$$P(E|a(k) = -3A) = 1 - P\left(-\frac{A}{T} \leq v(kT_s)\right) \quad (6.10)$$

which is the same as $P(E|a(k) = +3A)$ because the decision regions are the same shape.

The aforementioned shows that the conditional probabilities of error reduce to a computation involving the random variable $v(kT_s)$ and the constant A. The constant A is related to the received symbol energy. For 4-ary PAM with the four constellation points shown in Figure 6.1.2, the average symbol energy

$$E_{\text{avg}} = 5A^2. \quad (6.11)$$

Once the probability density function of $v(kT_s)$ is known, evaluation of (6.7)–(6.10) is straightforward. The probability of density function of $v(kT_s)$ is derived next.

It was shown in Chapter 4 that when $w(t)$ is white Gaussian random process with zero mean and power spectral density $N_0/2$ W/Hz, T-spaced samples of the ideally bandlimited $w(t)$ are zero-mean Gaussian random variables with autocorrelation function (see Section 4.5.3)

$$E\{w(mT)w(nT)\} = \sigma^2 \delta(m - n) \quad (6.12)$$

where the variance is

$$\sigma^2 = \frac{N_0}{2T}. \quad (6.13)$$

The noise component at the matched filter output, $v(nT)$ is given by

$$v(nT) = \sum_{m=n+\frac{T_1}{T}}^{n+\frac{T_2}{T}} w(mT)p(mT - nT). \quad (6.14)$$

Because $v(nT)$ is a linear combination of the Gaussian random variables $w(mT)$, $v(nT)$ is also a Gaussian random variable. The joint probability density function is determined completely from the mean and auto-covariance of the sequence. The mean of each $v(nT)$ is

$$E\{v(nT)\} = E\left\{\sum_{m=n+\frac{T_1}{T}}^{n+\frac{T_2}{T}} w(mT) p(mT - nT)\right\} = \sum_{m=n+\frac{T_1}{T}}^{n+\frac{T_2}{T}} E\{w(mT)\} p(mT - nT) = 0 \quad (6.15)$$

and the auto-covariance is given by

$$E\{v(nT)v(lT)\} = E\left\{\sum_{m=n+\frac{T_1}{T}}^{n+\frac{T_2}{T}} w(mT) p(mT - nT) \sum_{m'=l+\frac{T_1}{T}}^{l+\frac{T_2}{T}} w(m'T) p(m'T - lT)\right\}$$

$$= \sum_{m=n+\frac{T_1}{T}}^{n+\frac{T_2}{T}} \sum_{m'=l+\frac{T_1}{T}}^{l+\frac{T_2}{T}} E\{w(mT)w(m'T)\} p(mT - nT) p(m'T - lT)$$

$$= \sum_{m=n+\frac{T_1}{T}}^{n+\frac{T_2}{T}} \sum_{m'=l+\frac{T_1}{T}}^{l+\frac{T_2}{T}} \sigma^2 \delta(m - m') p(mT - nT) p(m'T - lT)$$

$$= \sigma^2 \sum_{m=n+\frac{T_1}{T}}^{n+\frac{T_2}{T}} p(mT - nT) p(mT - lT)$$

$$= \frac{\sigma^2}{T} r_p(-(n-l)T). \quad (6.16)$$

The noise component in the signal space projection is a downsampled version of $v(nT)$. Clearly $v(kT_s)$ has zero mean. The correlation is

$$E\{v(kT_s) v(lT_s)\} = \frac{\sigma^2}{T} r_p(-(k-l)T_s) \quad (6.17)$$

$$= \frac{\sigma^2}{T} \delta(k - l) \quad (6.18)$$

where the last line is true only if the pulse shape satisfies the Nyquist no-ISI condition. In summary, the random variables $v(kT_s)$ are independent and identically distributed Gaussian random variables with zero mean and variance

$$\sigma_v^2 = \frac{\sigma^2}{T} = \frac{N_0}{2T^2}. \quad (6.19)$$

Section 6.1 Performance of PAM

Now back to the probability of error for 4-ary PAM. The probabilities (6.7)–(6.10) can be expressed in terms of the area under the upper tail of the standard Gaussian random variable, denoted by the Q-function as described in Chapter 4:

$$P(E|a(k) = +A) = P\left(E \mid a(k) = -\frac{A}{T}\right) = 2Q\left(\frac{A/T}{\sigma_v}\right) = 2Q\left(\sqrt{\frac{2A^2}{N_0}}\right) \quad (6.20)$$

$$P(E|a(k) = +3A) = P\left(E \mid a(k) = -\frac{3A}{T}\right) = Q\left(\frac{A/T}{\sigma_v}\right) = Q\left(\sqrt{\frac{2A^2}{N_0}}\right). \quad (6.21)$$

Note that the last equality follows from the relationship (6.19). Applying the total probability theorem assuming equally likely symbols gives

$$P(E) = \frac{3}{2}Q\left(\sqrt{\frac{2A^2}{N_0}}\right). \quad (6.22)$$

Now using (6.11), the probability of error may be expressed as

$$P(E) = \frac{3}{2}Q\left(\sqrt{\frac{2}{5}\frac{E_{\text{avg}}}{N_0}}\right). \quad (6.23)$$

This shows that when a matched filter detector is used with perfect timing synchronization, the probability of symbol error is a function only of the ratio of the average symbol energy and the power spectral density level of the white noise.

Two normalizations are commonly used to facilitate comparisons between modulations with different signal set sizes. The first is to reduce the symbol-error probability to a bit-error probability. This is due to the notion that the fundamental quantity being transmitted is a bit, no matter how the bits are mapped into symbols. As a consequence, the fundamental measure of reliability is the bit-error rate.[1] The translation from symbol-error probability to bit-error probability follows from the observation that the most likely symbol errors are those involving adjacent symbols. Using a gray code, the bit patterns associated with adjacent symbols differ in only one position. Hence, the most likely symbol errors produce one erroneous bit and $\log_2(M) - 1$ correct bits. Thus

$$P_b = \frac{1}{\log_2 M} P(E). \quad (6.24)$$

---

[1] For a communication system where the data are grouped into frames or packets, many prefer the frame error rate or packet error rate as the fundamental measure of reliability. This follows from the notion that if one or more bits in a packet is in error, then none of the bits in the packet is of use to the end user. See Johannes [110] for a more detailed discussion on this issue.

The translation from symbol-error probability to bit-error probability can be thought of as a scaling of the y-axis in a probability of error versus signal-to-noise ratio plot. The scaling of the x-axis in such a plot involves scaling the signal-to-noise ratio. Because the bit is the fundamental quantity being transmitted, the bit energy is the fundamental unit of energy. The average bit energy, E_b, is related to the average symbol energy, E_{avg}, by

$$E_{\text{avg}} = (\log_2 M) \times E_b. \tag{6.25}$$

Applying these two normalizations to (6.23) produces

$$P_b = \frac{3}{4} Q\left(\sqrt{\frac{4}{5}\frac{E_b}{N_0}}\right). \tag{6.26}$$

Generalizing the aforementioned to M-ary PAM, the probability of symbol error and bit error are

$$P(E) = 2\frac{M-1}{M} Q\left(\sqrt{\frac{6}{M^2-1}\frac{E_{\text{avg}}}{N_0}}\right) \tag{6.27}$$

$$P_b = \frac{2(M-1)}{M \log_2 M} Q\left(\sqrt{6\frac{\log_2 M}{M^2-1}\frac{E_b}{N_0}}\right). \tag{6.28}$$

The probability of bit error (6.28) for $M = 2, 4,$ and 8 is plotted in Figure 6.1.3. A vertical slice through the plot is the line of constant bit energy. The probability of bit error increases as M increases because the average symbol energy must be shared with more bits. A horizontal line through the plot is the line of constant reliability. For a fixed probability of bit error, more bit energy is needed as M increases. For example, at $P_b = 10^{-6}$ binary PAM requires $E_b/N_0 = 10.6$ dB, 4-ary PAM requires $E_b/N_0 = 14.6$ dB, and 8-ary PAM requires $E_b/N_0 = 19.2$ dB. 4-ary PAM is said to be 4 dB worse than binary PAM at $P_b = 10^{-6}$ and 8-ary PAM is 8.6 dB worse than binary PAM at $P_b = 10^{-6}$.

These conclusions, however, ignore bandwidth. 4-ary PAM is two times more bandwidth efficient than binary PAM and 8-ary PAM is three times more bandwidth efficient than binary PAM. This can be seen in one of two ways: assuming the bit rate is constant so that the bandwidth varies with M, or by assuming the bandwidth is constant so that the bit rate varies with M. Either assumption leads to the same conclusion.

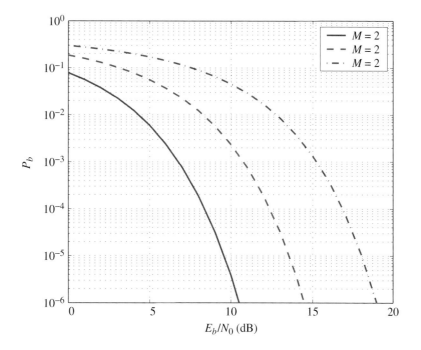

Figure 6.1.3 Probability of bit error versus E_b/N_0 for M-ary PAM for $M = 2, 4,$ and 8.

6.2 PERFORMANCE OF QAM

6.2.1 Bandwidth

For independent and equally likely symbols, the power spectral density of the band-pass QAM signal is given by

$$P_s(f) = \frac{E_{\text{avg}}}{2T_s} |P(f - f_0)|^2 + \frac{E_{\text{avg}}}{2T_s} |P(f + f_0)|^2 \qquad (6.29)$$

where $P(f)$ is the continuous-time Fourier transform of the pulse shape $p(t)$ and E_{avg} is the average symbol energy. Note that this result is a frequency translated version of the power spectral density of a baseband PAM pulse train given by (6.1).

The power spectral density is determined by the properties of the Fourier transform of the pulse shape. The frequency domain properties of many pulse shapes is described in Appendix A. The power spectral density of any full response pulse shape (e.g., NRZ, RZ, MAN, HS) has infinite support in the frequency domain. Consequently, the absolute bandwidth is infinite for these pulse shapes. The power spectral densities of these pulse shapes decay with increasing frequency. Thus, for a sufficiently large f, $|P(f)|^2$ is "close enough to zero" for the purposes of bandwidth assignments in a wireless communication system. This permits the use of practical measures of bandwidth, such as those described in Appendix A.

The bandwidths of the NRZ, RZ, MAN, and HS pulse shapes using these practical measures are summarized in Table A.1.2 in Appendix A. In each case, the bandwidth is of the form B/T_s for some constant B. The bandwidth of the corresponding *band-pass* modulated carrier is twice the bandwidth listed in Table A.1.2. Thus, with reference to the listings in Table A.1.2, the bandwidths of the full response pulses are

$$\text{BW} = \frac{2B}{T_s} = 2B \times R_s = 2B \log_2(M) \times R_b \quad (6.30)$$

where R_s is the symbol rate (symbols/second), R_b is the bit rate (bits/second) and B is a constant that depends on the pulse shape and the definition of bandwidth used.

For the ideal SRRC pulse shape, the $|P(f)|^2$ has finite support in the frequency domain. As a consequence, the absolute bandwidth is meaningful. For the band-pass modulated carrier

$$\text{BW} = 2\frac{1+\alpha}{2T_s} = (1+\alpha) \times R_s = (1+\alpha) \log_2(M) \times R_b \quad (6.31)$$

where R_s is the symbol rate (symbols/second), R_b is the bit rate (bits/second), and α is the excess bandwidth.

6.2.2 Probability of Error

The probability of error calculations follow the same procedure as outlined for baseband PAM. The probability of error is the probability that the 2-dimensional signal space projection given by

$$x(kT_s) = \frac{a_0(k)}{T} + v_0(kT_s)$$
$$y(kT_s) = \frac{a_1(k)}{T} + v_1(kT_s) \quad (6.32)$$

lies in a 2-dimensional decision region other than the one corresponding to the symbol that was transmitted. These calculations require knowledge of the means and covariances of the two random variables $v_0(kT_s)$ and $v_1(kT_s)$. It will be shown that

- $\text{E}\{v_0(kT_s)\} = 0$ and $\text{E}\{v_1(kT_s)\} = 0$.
- The sequence of random variables $v_0(kT_s)$ is uncorrelated for pulse shapes that satisfy the Nyquist no-ISI condition.
- The sequence of random variables $v_1(kT_s)$ is uncorrelated for pulse shapes that satisfy the Nyquist no-ISI condition.
- $v_0(kT_s)$ and $v_1(kT_s)$ are independent Gaussian random variables.

The proof of these properties relies on two important observations. First, both $v_0(kT_s)$ and $v_1(kT_s)$ are functions of the sequence of Gaussian random variables $w(nT)$ whose mean and auto-covariance are

$$\text{E}\{w(nT)\} = 0 \quad (6.33)$$

$$\text{E}\{w(nT)w(n'T)\} = \sigma^2 \delta(n-n') \quad \sigma^2 = \frac{N_0}{2T}. \quad (6.34)$$

Section 6.2 Performance of QAM   **315**

The second important property is that

$$\sum_{n=N_1}^{N_2} y(nT)\cos(\Omega n) = 0 \qquad \sum_{n=N_1}^{N_2} y(nT)\sin(\Omega n) = 0 \qquad (6.35)$$

for a baseband signal $y(nT)$ defined on the interval $N_1 \leq n \leq N_2$.

From the analysis associated with (5.87)–(5.91), the random variables $v_0(kT_s)$ and $v_1(kT_s)$ are defined by

$$v_0(kT_s) = \sum_{l=\frac{T_1}{T}+k\frac{T_s}{T}}^{\frac{T_2}{T}+k\frac{T_s}{T}} w(lT)\sqrt{2}\cos(\Omega_0 l)p(lT - kT_s) \qquad (6.36)$$

$$v_1(kT_s) = -\sum_{l=\frac{T_1}{T}+k\frac{T_s}{T}}^{\frac{T_2}{T}+k\frac{T_s}{T}} w(lT)\sqrt{2}\sin(\Omega_0 l)p(lT - kT_s) \qquad (6.37)$$

where the unit energy pulse shape $p(t)$ is defined over the interval $T_1 \leq t \leq T_2$. Because the $w(lT)$ are Gaussian random variables, $v_0(kT_s)$ and $v_1(kT_s)$ are also Gaussian random variables. The means of $v_0(kT_s)$ and $v_1(kT_s)$ are

$$E\{v_0(kT_s)\} = \sum_{l=\frac{T_1}{T}+k\frac{T_s}{T}}^{\frac{T_2}{T}+k\frac{T_s}{T}} E\{w(lT)\}\sqrt{2}\cos(\Omega_0 l)p(lT - kT_s) = 0$$

$$E\{v_1(kT_s)\} = -\sum_{l=\frac{T_1}{T}+k\frac{T_s}{T}}^{\frac{T_2}{T}+k\frac{T_s}{T}} E\{w(lT)\}\sqrt{2}\sin(\Omega_0 l)p(lT - kT_s) = 0. \qquad (6.38)$$

The auto-covariance of the sequence $v_0(kT_s)$ is given by

$$E\{v_0(kT_s)v_0(k'T_s)\} = 2 \sum_{l=\frac{T_1}{T}+k\frac{T_s}{T}}^{\frac{T_2}{T}+k\frac{T_s}{T}} \sum_{l'=\frac{T_1}{T}+k\frac{T_s}{T}}^{\frac{T_2}{T}+k\frac{T_s}{T}} E\{w(lT)w(l'T)\}$$
$$\times \cos(\Omega_0 l)\cos(\Omega_0 l')p(lT - kT_s)p(l'T - k'T_s) \qquad (6.39)$$

$$= 2 \sum_{l=\frac{T_1}{T}+k\frac{T_s}{T}}^{\frac{T_2}{T}+k\frac{T_s}{T}} \sum_{l'=\frac{T_1}{T}+k\frac{T_s}{T}}^{\frac{T_2}{T}+k\frac{T_s}{T}} \sigma^2 \delta(l - l')\cos(\Omega_0 l)\cos(\Omega_0 l')p(lT - kT_s)p(l'T - k'T_s)$$

$$\qquad (6.40)$$

$$= 2\sigma^2 \sum_{l=\frac{T_1}{T}+k\frac{T_s}{T}}^{\frac{T_2}{T}+k\frac{T_s}{T}} \cos^2(\Omega_0 l) p(lT - kT_s) p(lT - k'T_s) \qquad (6.41)$$

$$\approx \sigma^2 \sum_{l=\frac{T_1}{T}+k\frac{T_s}{T}}^{\frac{T_2}{T}+k\frac{T_s}{T}} p(lT - kT_s) p(lT - k'T_s) \qquad (6.42)$$

$$= \frac{\sigma^2}{T} r_p((k - k')T_s) \qquad (6.43)$$

which is zero except for $k = k'$ when the pulse shape satisfies the Nyquist no-ISI condition. Note that the step from (6.41) to (6.42) holds for $\Omega_0 \neq \pi$. This shows that the sequence $v_0(kT_s)$ is uncorrelated. In the same way, it can be shown that the sequence $v_1(kT_s)$ is also uncorrelated. Finally, to show that $v_0(kT_s)$ and $v_1(kT_s)$ are independent Gaussian random variables, it must be shown that the cross-correlation $\mathrm{E}\{v_0(kT_s)v_1(k'T_s)\} = 0$. Substituting the definitions (6.36) and (6.37) into the cross-correlation and rearranging produces

$$\mathrm{E}\{v_0(kT_s)v_1(k'T_s)\}$$

$$= -2 \sum_{l=\frac{T_1}{T}+k\frac{T_s}{T}}^{\frac{T_2}{T}+k\frac{T_s}{T}} \sum_{l'=\frac{T_1}{T}+k\frac{T_s}{T}}^{\frac{T_2}{T}+k\frac{T_s}{T}} \mathrm{E}\{w(lT)w(l'T)\} \cos(\Omega_0 l) \sin(\Omega_0 l') p(lT - kT_s) p(l'T - k'T_s)$$

$$\qquad (6.44)$$

$$= -2 \sum_{l=\frac{T_1}{T}+k\frac{T_s}{T}}^{\frac{T_2}{T}+k\frac{T_s}{T}} \sum_{l'=\frac{T_1}{T}+k\frac{T_s}{T}}^{\frac{T_2}{T}+k\frac{T_s}{T}} \sigma^2 \delta(l - l') \cos(\Omega_0 l) \sin(\Omega_0 l') p(lT - kT_s) p(l'T - k'T_s) \qquad (6.45)$$

$$= -2\sigma^2 \sum_{l=\frac{T_1}{T}+k\frac{T_s}{T}}^{\frac{T_2}{T}+k\frac{T_s}{T}} \cos(\Omega_0 l) \sin(\Omega_0 l) p(lT - kT_s) p(lT - k'T_s) \qquad (6.46)$$

$$= \sigma^2 \sum_{l=\frac{T_1}{T}+k\frac{T_s}{T}}^{\frac{T_2}{T}+k\frac{T_s}{T}} \sin(2\Omega_0 l) p(lT - kT_s) p(lT - k'T_s) \qquad (6.47)$$

$$= 0 \qquad (6.48)$$

where the last line follows from (6.35) using $\Omega = 2\Omega_0$ and $y(n) = p(nT - kT_s)p(nT - k'T_s)$. Thus, the joint probability density function of the random variables $v_0(kT_s)$ and $v_1(kT_s)$ is given by

Section 6.2 Performance of QAM

$$f_{v_0(kT_s),v_1(kT_s)}(v_0, v_1) = \frac{1}{2\pi \sigma_v^2} \exp\left\{-\frac{v_0^2}{2\sigma_v^2}\right\} \exp\left\{-\frac{v_1^2}{2\sigma_v^2}\right\} \tag{6.49}$$

where

$$\sigma_v^2 = \frac{1}{T}\sigma^2 = \frac{N_0}{2T^2}. \tag{6.50}$$

The probability of error for a signal set is determined by the shape of the decision regions in the signal space for the signals in the set. The 2-dimensional nature of the decision regions makes possible a variety of different decision region shapes that are analyzed in the following examples.

Square MQAM The decision regions for square MQAM constellations are illustrated in Figure 6.2.1 for the case $M = 16$. For notational convenience, all 16 points have been given a label s_m for $m = 0, 1, \ldots, 15$. Each decision region is a square that is either completely enclosed, open on one side, or open on two sides. This regular structure permits straightforward computation of the bit-error rate. The technique is illustrated for 16-QAM.

Assume the symbol s_{12} was sent. Then the matched filter outputs (assuming perfect carrier phase and symbol timing synchronization) are given by (6.32) where $a_0(k) = -3A$ and $a_1(k) = -3A$. The conditional probability of error is

$$P(E|s_{12}) = 1 - P(x(kT_s) < -2A/T, y(kT_s) < -2A/T)$$
$$= 1 - P(-3A/T + v_0(kT_s) < -2A/T, -3A/T + v_1(kT_s) < -2A/T) \tag{6.51}$$
$$= 1 - P(v_0(kT_s) < A/T, v_1(kT_s) < A/T).$$

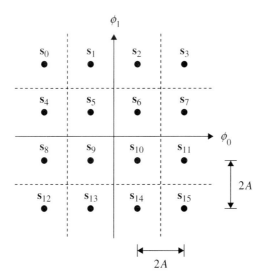

Figure 6.2.1 Decision regions for square 16-QAM.

Because the random variables $v_0(kT_s)$ and $v_1(kT_s)$ are independent Gaussian random variables,

$$P(v_0(kT_s) < A/T, v_1(kT_s) < A/T) = P(v_0(kT_s) < A/T) \times P(v_1(kT_s) < A/T) \quad (6.52)$$

$$= \left[1 - Q\left(\frac{A/T}{\sigma_v}\right)\right] \times \left[1 - Q\left(\frac{A/T}{\sigma_v}\right)\right] \quad (6.53)$$

so that

$$P(E|\mathbf{s}_{12}) = 2Q\left(\frac{A/T}{\sigma_v}\right) - Q^2\left(\frac{A/T}{\sigma_v}\right) \quad (6.54)$$

where σ_v^2 is given by (6.50). Because the decision regions for $\mathbf{s}_0, \mathbf{s}_3,$ and $\mathbf{s}_{14}$ are the same shape as that of the decision region for $\mathbf{s}_{12}$,

$$P(E|\mathbf{s}_0) = P(E|\mathbf{s}_3) = P(E|\mathbf{s}_{14}) = P(E|\mathbf{s}_{12}). \quad (6.55)$$

Now assume the symbol $\mathbf{s}_{13}$ was sent. Then the matched filter outputs (assuming perfect carrier phase and symbol timing synchronization) are given by (6.32) where $a_0(k) = -A$ and $a_1(k) = -3A$. Proceeding as before

$$P(E|\mathbf{s}_{13}) = 1 - P(-2A/T < x(kT_s) < 0, y(kT_s) < -2A/T)$$
$$= 1 - P(-A/T < v_0(kT_s) < A/T)P(v_1(kT_s) < A/T)$$
$$= 1 - \left[1 - 2Q\left(\frac{A/T}{\sigma_v}\right)\right] \times \left[1 - Q\left(\frac{A/T}{\sigma_v}\right)\right] \quad (6.56)$$
$$= 3Q\left(\frac{A/T}{\sigma_v}\right) - 2Q^2\left(\frac{A/T}{\sigma_v}\right).$$

Because the shape of the decision regions for $\mathbf{s}_1, \mathbf{s}_2, \mathbf{s}_4, \mathbf{s}_7, \mathbf{s}_8, \mathbf{s}_{11}, \mathbf{s}_{13},$ and $\mathbf{s}_{14}$ are all the same,

$$P(E|\mathbf{s}_1) = P(E|\mathbf{s}_2) = P(E|\mathbf{s}_4) = P(E|\mathbf{s}_7) = P(E|\mathbf{s}_8) = P(E|\mathbf{s}_{11}) = P(E|\mathbf{s}_{13}) = P(E|\mathbf{s}_{14}). \quad (6.57)$$

Finally assume $\mathbf{s}_9$ was sent. Then the matched filter outputs (assuming perfect carrier phase and symbol timing synchronization) are given by (6.32) where $a_0(k) = -A$ and $a_1(k) = -A$ and the probability of error is

$$P(E|\mathbf{s}_9) = 1 - P(-2A/T < x(kT_s) < 0, -2A/T < y(kT_s) < 0)$$
$$= 1 - P(-A/T < v_0(kT_s) < A/T)P(-A/T < v_1(kT_s) < A/T)$$
$$= 1 - \left[1 - 2Q\left(\frac{A/T}{\sigma_v}\right)\right] \times \left[1 - 2Q\left(\frac{A/T}{\sigma_v}\right)\right] \quad (6.58)$$
$$= 4Q\left(\frac{A/T}{\sigma_v}\right) - 4Q^2\left(\frac{A/T}{\sigma_v}\right).$$

Because the decision regions for $\mathbf{s}_3, \mathbf{s}_6, \mathbf{s}_9,$ and $\mathbf{s}_{10}$ all have the same shape,

$$P(E|\mathbf{s}_5) = P(E|\mathbf{s}_6) = P(E|\mathbf{s}_9) = P(E|\mathbf{s}_{10}). \quad (6.59)$$

Section 6.2 Performance of QAM

Applying the total probability theorem and assuming equally likely symbols, the average probability of error is

$$P(E) = 3Q\left(\frac{A/T}{\sigma_v}\right) - \frac{9}{4}Q^2\left(\frac{A/T}{\sigma_v}\right) \tag{6.60}$$

which is well approximated by

$$P(E) = 3Q\left(\frac{A/T}{\sigma_v}\right). \tag{6.61}$$

Using the substitutions $E_{\text{avg}} = 10A^2$, $E_{\text{avg}} = 4E_b$, and $\sigma_v^2 = N_0/2T^2$ together with the technique for converting symbol-error probability to bit-error probability, the average probabilities of symbol and bit error are

$$P(E) = 3Q\left(\sqrt{\frac{1}{5}\frac{E_{\text{avg}}}{N_0}}\right) \tag{6.62}$$

$$P_b = \frac{3}{4}Q\left(\sqrt{\frac{4}{5}\frac{E_b}{N_0}}\right). \tag{6.63}$$

The probability of bit error for the general case of square MQAM for M an even power of 2 is given in Exercise 6.5.

MPSK The decision regions for MPSK constellations are illustrated in Figure 6.2.2 for the case $M = 8$. For notational convenience, all 8 points have been labeled $\mathbf{s}_m$ for $m = 0, 1, \ldots, 7$. Each decision region is shaped like a pizza slice and therefore each has the same probability of error. The probability of error calculations for MPSK are a little more involved than the calculations for MQAM due to the shape of the decision region. The technique for computing the probability of error is illustrated for 8-PSK.

Assume symbol $\mathbf{s}_0$ was sent. Then the matched filter outputs (assuming perfect carrier phase and symbol timing synchronization) are given by (6.32) where $a_0(k) = A$ and $a_1(k) = 0$. The probability of error is the probability that the point $(x(kT_s), y(kT_s))$ is outside the pie-shaped decision region corresponding to $\mathbf{s}_0$. This computation is best accomplished by using the rectangular-to-polar transformation

$$R = \sqrt{x^2(kT_s) + y^2(kT_s)} \tag{6.64}$$

$$\Theta = \tan^{-1}\left\{\frac{y(kT_s)}{x(kT_s)}\right\} \tag{6.65}$$

and computing the probability that Θ is too large. (R and Θ are random variables because $x(kT_s)$ and $y(kT_s)$ are random variables.) Using standard transformation techniques for random variables, the joint probability density function

$$f_{R,\Theta}(r,\theta) = \frac{r}{2\pi\sigma_v^2}\exp\left\{-\frac{r^2 + A^2/T^2 - 2A/Tr\cos\theta}{2\sigma_v^2}\right\} \tag{6.66}$$

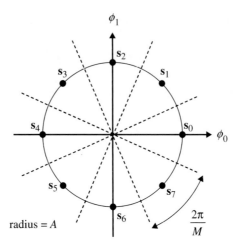

Figure 6.2.2 Decision regions for 8PSK.

where σ_v^2 is given by (6.50). The error event depends only on Θ. As a consequence, the probability density function of Θ is all that is needed. The probability density function of Θ is

$$f_\Theta(\theta) = \int_0^\infty f_{R,\Theta}(r,\theta) dr \qquad (6.67)$$

which, unfortunately, has no closed form. For $A^2/T^2/\sigma_v^2 \gg 1$ and $|\Theta| < \pi/2$, $f_\Theta(\theta)$ is well approximated by

$$f_\Theta(\theta) \approx \sqrt{\frac{A^2/T^2}{2\pi\sigma_v^2}} \cos\theta \exp\left\{-\frac{A^2/T^2}{2\sigma_v^2}\sin^2\theta\right\}. \qquad (6.68)$$

The conditional error probability is thus

$$P(E|\mathbf{s}_0) = 1 - \int_{-\pi/M}^{\pi/M} f_\Theta(\theta) d\theta$$

$$\approx 1 - \int_{-\pi/M}^{\pi/M} \sqrt{\frac{A^2/T}{2\pi\sigma^2}} \cos\theta \exp\left\{-\frac{A^2/T}{2\sigma^2}\sin^2\theta\right\} d\theta$$

$$= 2Q\left(\sqrt{\frac{2A^2}{N_0}\sin^2\left(\frac{\pi}{M}\right)}\right). \qquad (6.69)$$

Because all decision regions have the same shape,

$$P(E|\mathbf{s}_m) = P(E|\mathbf{s}_0) \qquad m = 1, 2, \ldots, M-1. \qquad (6.70)$$

Section 6.2 Performance of QAM

Applying the total probability theorem and using $E_{\text{avg}} = A^2, E_b = (\log_2 M) E_{\text{avg}}$ together with the technique for converting symbol-error probability to bit-error probability, the average probabilities of symbol and bit error are

$$P(E) = 2Q\left(\sqrt{\frac{E_{\text{avg}}}{N_0} 2 \sin^2\left(\frac{\pi}{M}\right)}\right) \qquad (6.71)$$

$$P_b = \frac{2}{\log_2 M} Q\left(\sqrt{\frac{E_b}{N_0} 2(\log_2 M) \sin^2\left(\frac{\pi}{M}\right)}\right). \qquad (6.72)$$

The Union Bound: A Generally Applicable Approach When the decision regions are not rectangular, computing the error probability becomes more complicated as the last section illustrated with MPSK. A simple, but good, approximation follows from a geometric interpretation of error events. Consider the BPSK constellation shown in Figure 6.2.3. The probability density function of the matched filter output assuming the waveform corresponding to $a(k) = -A$ is also shown. The probability of error is

$$P(E) = Q\left(\frac{A/T}{\sigma_v}\right) \qquad (6.73)$$

where σ_v^2, given by (6.50), is the variance of the noise samples at the output of the matched filter. In geometric terms, the probability of error is the probability that the matched filter output exceeds one-half the Euclidean distance between the two symbols. Thus, the probability of error between two symbols is purely a function of how far apart they are in the signal space.

For constellations with more than two points, this concept may still be applied by using *pairwise error probabilities* with the union bound. Consider a constellation with M points labeled $\mathbf{s}_0, \mathbf{s}_1, \ldots, \mathbf{s}_{M-1}$. Assuming $\mathbf{s}_0$ was transmitted, the probability of error is the union of the events $\hat{\mathbf{s}} = \mathbf{s}_n$ for $n \neq 0$. Expressed in mathematical terms, this notion is

$$P(E|\mathbf{s}_0) = P\left([\hat{\mathbf{s}} = \mathbf{s}_1] \cup [\hat{\mathbf{s}} = \mathbf{s}_2] \cup \cdots \cup [\hat{\mathbf{s}} = \mathbf{s}_{M-1}] \,|\, \mathbf{s}_0\right). \qquad (6.74)$$

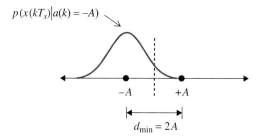

Figure 6.2.3 A geometric interpretation of an error event. The probability of error is the probability that matched filter output exceeds one-half the minimum Euclidean distance between adjacent constellation points.

The probability of a union of events is upper bounded by the sum of the probabilities of the events. Thus the conditional probability (6.74) may be upper bounded as

$$P(E|\mathbf{s}_0) \leq \sum_{n=1}^{M-1} P\left(\hat{\mathbf{s}} = \mathbf{s}_n | \mathbf{s}_0\right). \tag{6.75}$$

Now, assuming the symbols are equally likely and applying the total probability theorem, the average probability of error is bounded by

$$P(E) \leq \frac{1}{M} \sum_{m=0}^{M-1} \sum_{\substack{n=0 \\ n \neq m}}^{M-1} P\left(\hat{\mathbf{s}} = \mathbf{s}_n | \mathbf{s}_m\right). \tag{6.76}$$

Each of the terms in the summation in (6.76) is called a *pairwise error probability* because each one is the probability of error between points $\mathbf{s}_n$ and $\mathbf{s}_m$ when all the other points are ignored. Each pairwise error probability is given by

$$P\left(\hat{\mathbf{s}} = \mathbf{s}_n | \mathbf{s}_m\right) = Q\left(\frac{d_{m,n}/2}{T\sigma_v}\right) = Q\left(\sqrt{\frac{d_{m,n}^2}{2E_{\text{avg}}} \frac{E_{\text{avg}}}{N_0}}\right) \tag{6.77}$$

where $d_{m,n}$ is the Euclidean distance between point $\mathbf{s}_m$ and point $\mathbf{s}_n$.

It is interesting to apply this approach to the 8-PSK and square 16-QAM examined previously. Consider the 8-PSK constellation illustrated in Figure 6.2.2. A good starting point is to list the Euclidean distances between point $\mathbf{s}_0$ of Figure 6.2.2 and the other seven points. These distances are

$$\begin{aligned} d_{0,1} &= d_{0,7} = 2A \sin\left(\tfrac{\pi}{8}\right) \\ d_{0,2} &= d_{0,6} = 2A \sin\left(\tfrac{\pi}{4}\right) \\ d_{0,3} &= d_{0,5} = 2A \sin\left(\tfrac{3\pi}{8}\right) \\ d_{0,4} &= 2A. \end{aligned} \tag{6.78}$$

Using the union bound (6.75) and the pairwise error probability (6.77), the conditional probability of error is bounded by

$$P(E|\mathbf{s}_0) \leq 2Q\left(\sqrt{\frac{2E_{\text{avg}}}{N_0} \sin^2\left(\tfrac{\pi}{8}\right)}\right) + 2Q\left(\sqrt{\frac{2E_{\text{avg}}}{N_0} \sin^2\left(\tfrac{\pi}{4}\right)}\right)$$

$$+ 2Q\left(\sqrt{\frac{2E_{\text{avg}}}{N_0} \sin^2\left(\tfrac{3\pi}{8}\right)}\right) + Q\left(\sqrt{\frac{2E_{\text{avg}}}{N_0}}\right). \tag{6.79}$$

Section 6.2 Performance of QAM

It is easy to show that $P(E|\mathbf{s}_m) = P(E|\mathbf{s}_0)$ for $m = 1, 2, \ldots, M - 1$. Using this observation in (6.76) produces the union bound on the probability of error:

$$P(E) \leq 2Q\left(\sqrt{\frac{2E_{\text{avg}}}{N_0}\sin^2\left(\frac{\pi}{8}\right)}\right) + 2Q\left(\sqrt{\frac{2E_{\text{avg}}}{N_0}\sin^2\left(\frac{\pi}{4}\right)}\right)$$
$$+ 2Q\left(\sqrt{\frac{2E_{\text{avg}}}{N_0}\sin^2\left(\frac{3\pi}{8}\right)}\right) + Q\left(\sqrt{\frac{2E_{\text{avg}}}{N_0}}\right). \quad (6.80)$$

Note that the terms on the right-hand-side of (6.80) are ordered by increasing Euclidean distance. Thus, the terms on the right-hand-side of (6.80) are listed in decreasing order. Sometimes, the term corresponding to the smallest Euclidean distance dominates the expression so that the other terms may be ignored. As it turns out, this is one of those cases. If the last three terms are ignored, the bound (6.80) reduces to the probability of error (6.71).

For the square 16-QAM constellation, it can be shown (see Exercise 6.8) that the union bound for the probability of symbol error is

$$P(E) \leq \frac{48}{16}Q\left(\sqrt{\frac{4}{20}\frac{E_{\text{avg}}}{N_0}}\right) + \frac{36}{16}Q\left(\sqrt{\frac{8}{20}\frac{E_{\text{avg}}}{N_0}}\right) + \frac{32}{16}Q\left(\sqrt{\frac{16}{20}\frac{E_{\text{avg}}}{N_0}}\right)$$
$$+ \frac{48}{16}Q\left(\sqrt{\frac{20}{20}\frac{E_{\text{avg}}}{N_0}}\right) + \frac{16}{16}Q\left(\sqrt{\frac{32}{20}\frac{E_{\text{avg}}}{N_0}}\right) + \frac{16}{16}Q\left(\sqrt{\frac{36}{20}\frac{E_{\text{avg}}}{N_0}}\right)$$
$$+ \frac{24}{16}Q\left(\sqrt{\frac{40}{20}\frac{E_{\text{avg}}}{N_0}}\right) + \frac{16}{16}Q\left(\sqrt{\frac{52}{20}\frac{E_{\text{avg}}}{N_0}}\right) + \frac{4}{16}Q\left(\sqrt{\frac{72}{20}\frac{E_{\text{avg}}}{N_0}}\right). \quad (6.81)$$

Again, the terms on the right-hand-side of (6.81) are ordered by increasing Euclidean distance and hence are listed in decreasing order. This expression is another example where the term corresponding to the minimum Euclidean distance dominates. If the other terms on the right-hand-side of (6.81) are ignored, the resulting expression is identical to the probability of symbol error expression (6.62).

The preceding examples show that for some constellations, a truncated union bound is a good approximation to the probability of symbol error. This is because the most likely errors are those between the closest symbols and the distances to the other symbols are such that those terms can be ignored. Thus, the minimum Euclidean distance is of interest. The minimum Euclidean distance, $d_{\min}$ is defined by

$$d_{\min} = \min_{m \neq n}\{d_{m,n}\} \quad 0 \leq m < M, 0 \leq n < M. \quad (6.82)$$

When the union bound terms corresponding to non-nearest neighbor pairwise symbol-error probabilities can be ignored, the probability of symbol error is well approximated by

$$P(E) \approx N_{\min} Q\left(\sqrt{\frac{d_{\min}^2}{2E_{\text{avg}}} \frac{E_{\text{avg}}}{N_0}}\right) \qquad (6.83)$$

where $N_{\min}$ is the average number of "nearest neighbors" for the constellation.

The real value of the union bound technique is in computing the probability of error for constellations with irregular-shaped decision regions. One such constellation is the 4-ary QAM constellation shown in Figure 6.2.4. The average energy in the constellation

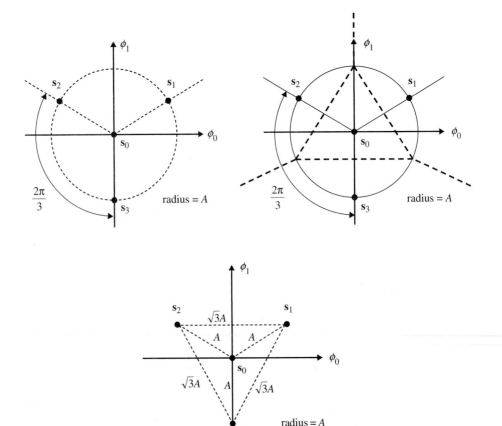

Figure 6.2.4 A 4-ary QAM constellation (top left). Decision regions for the 4-ary QAM constellation (top right). Euclidean distances between the points in the 4-ary QAM constellation (bottom).

is $E_{\text{avg}} = 3A^2/4$ and the Euclidean distances are

$$d_{0,1} = A \quad d_{1,0} = A \quad d_{2,0} = A \quad d_{3,0} = A \quad (6.84)$$

$$d_{0,2} = A \quad d_{1,2} = \sqrt{3}A \quad d_{2,1} = \sqrt{3}A \quad d_{3,1} = \sqrt{3}A \quad (6.85)$$

$$d_{0,3} = A \quad d_{1,3} = \sqrt{3}A \quad d_{2,3} = \sqrt{3}A \quad d_{3,2} = \sqrt{3}A. \quad (6.86)$$

The union bound for the conditions probabilities are

$$P(E|\mathbf{s}_0) \leq 3Q\left(\sqrt{\frac{2}{3}\frac{E_{\text{avg}}}{N_0}}\right) \quad (6.87)$$

$$P(E|\mathbf{s}_1) \leq Q\left(\sqrt{\frac{2}{3}\frac{E_{\text{avg}}}{N_0}}\right) + Q\left(\sqrt{2\frac{E_{\text{avg}}}{N_0}}\right) \quad (6.88)$$

$$P(E|\mathbf{s}_2) = P(E|\mathbf{s}_1) \quad (6.89)$$

$$P(E|\mathbf{s}_3) = P(E|\mathbf{s}_1). \quad (6.90)$$

Using these expressions with the total probability theorem and (6.76), the average probability of symbol error is

$$P(E) \leq \frac{6}{4}Q\left(\sqrt{\frac{2}{3}\frac{E_{\text{avg}}}{N_0}}\right) + \frac{6}{4}Q\left(\sqrt{2\frac{E_{\text{avg}}}{N_0}}\right). \quad (6.91)$$

This approximation technique provides a powerful tool for estimating bit-error rate performance for QAM constellations with irregular decision regions such as the 16-ary and 8-ary CCITT V.29 QAM constellations illustrated in Figure 5.3.6. (See Exercises 6.11 and 6.12.) It should be pointed out that the expression produced from this method are approximations. The approximations are better at high signal-to-noise ratios where the error rate is low.

6.3 COMPARISONS

The bit-error rate performance of the three 4-ary constellations introduced in this chapter is plotted in Figure 6.3.1. Observe that QPSK is the best constellation, the "Y"-shaped QAM is the next best, and 4-PAM is the worst. The signal-to-noise ratio required to achieve $P_b = 10^{-6}$ is summarized below

Constellation	E_b/N_0 (dB)
QPSK	10.6
Y-QAM	12.2
4-PAM	14.4

This data, derived from the plots of Figure 6.3.1, indicates that QPSK is 1.6 dB better than Y-QAM and 3.8 dB better than 4-PAM. The same conclusion could also be estimated by

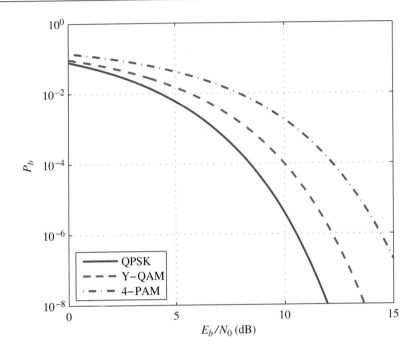

Figure 6.3.1 Probability of bit error versus E_b/N_0 for three 4-ary signal sets described previously.

comparing the multipliers of the ratio E_b/N_0 in the probability of bit-error expressions. For QPSK, the multiplier is 2, for Y-QAM, the multiplier is 4/3, and for 4-PAM, the multiplier is 4/5. Converting the ratios to dB produces

$$10\log_{10}\left(\frac{2}{4/3}\right) = 1.8 \tag{6.92}$$

$$10\log_{10}\left(\frac{2}{4/5}\right) = 4.0. \tag{6.93}$$

This method overestimates the improvement in this case because it does not account for the multiplier outside the Q-function. The QPSK constellation has the best performance among the three constellations because, QPSK packs the points into the space in a more efficient way than the other two.

The bit-error rate performance of four 8-ary constellations is plotted in Figure 6.3.2. From this plot, observe that the signal-to-noise ratio required to achieve $P_b = 10^{-6}$ for each signal set is

Section 6.3 Comparisons

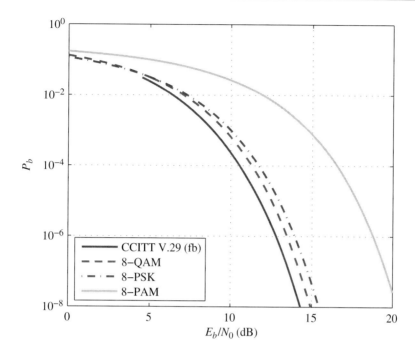

Figure 6.3.2 Probability of bit error versus E_b/N_0 for four 8-ary signal sets described previously.

Constellation	E_b/N_0 (dB)
CCITT V.29 backup	12.8
8 QAM	13.5
8 PSK	14.0
8 PAM	18.8

How much better or worse one constellation is than another can also be estimated using the ratios of the normalized Euclidean distances as described above. The 8-ary QAM constellation used by CCITT V.29 is slightly better than the rectangular 8-QAM constellation. The others have inferior performance because they offer less efficient packing of the 8 constellation points into the signal space.

The bit-error rate performance of four 16-ary constellations is plotted in Figure 6.3.3. The signal-to-noise ratio required to achieve $P_b = 10^{-6}$ for each constellation is summarized below:

Constellation	E_b/N_0 (dB)
16-QAM square	14.4
CCITT V.29	15.3
16 PSK	18.5
16 PAM	23.5

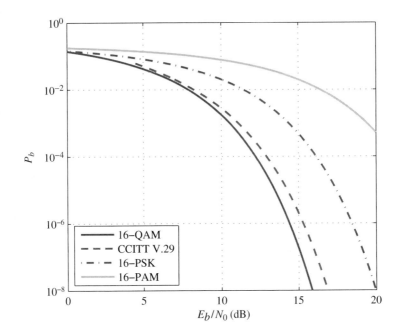

Figure 6.3.3 Probability of bit error versus E_b/N_0 for four 16-ary signal sets described previously.

How much better or worse one constellation is than another can also be estimated using the ratios of the normalized Euclidean distances as described previously. Of the four, the square constellation offers the best bit-error rate performance because it represents the most efficient packing of the four constellations. The others have inferior performance because they offer less efficient packing of the 16 constellation points into the signal space.

Signal set families can also be compared across constellation sizes. A comparison of the bit-error rate performance of square QAM constellations for $M = 4, 16, 64$, and 256 is shown in Figure 6.3.4. A comparison of the bit-error rate performance of MPSK constellations for $M = 4, 8$, and 16 is shown in Figure 6.3.5. One of the challenges in comparing signal sets with different numbers of signals is that bandwidth must also be considered to get a complete picture of how the different constellations relate to one another. The notion of *spectral efficiency* is used to quantify the bandwidth. The spectral efficiency is the ratio of equivalent bit rate to RF bandwidth. It is a normalized bandwidth whose units are bits/second/Hertz.

For example, the bandwidth for M-ary QAM using a square-root raised-cosine pulse shape is

$$B = \frac{1+\alpha}{T_s} = \frac{1+\alpha}{\log_2(M) T_b} = \frac{1+\alpha}{\log_2(M)} R_b \qquad (6.94)$$

Section 6.3 Comparisons

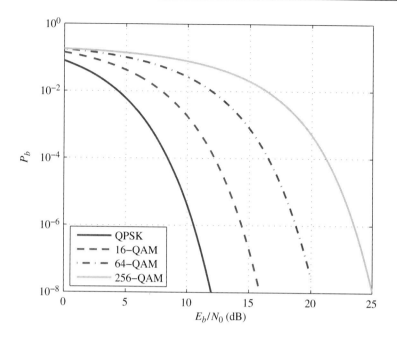

Figure 6.3.4 Probability of bit error versus E_b/N_0 for MPSK for $M = 4, 16, 64,$ and 256.

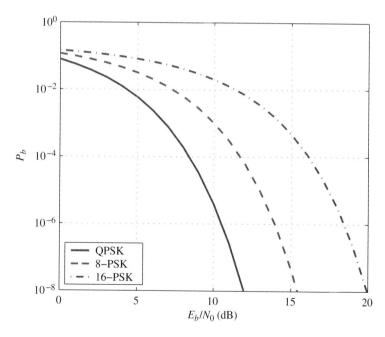

Figure 6.3.5 Probability of bit error versus E_b/N_0 for MPSK for $M = 4, 8,$ and 16.

where $0 \leq \alpha \leq 1$ is the excess bandwidth. Thus, the spectral efficiency for M-ary QAM using a square-root raised-cosine pulse shape is

$$\frac{R_b}{B} = \frac{\log_2(M)}{1+\alpha} \quad \text{(bits/s/Hz)}. \tag{6.95}$$

Constellations with different numbers of points can be compared by plotting the spectral efficiency versus the value of E_b/N_0 required to obtain a specified probability of bit error P_b. Such a plot is shown in Figure 6.3.6 for the MQAM and MPSK constellations for $\alpha = 0.5$ and $P_b = 10^{-6}$. The MQAM constellations are the square and cross constellations illustrated in Figures 5.3.3 and 5.3.4 and the MPSK constellations are drawn from the family of constellations illustrated in Figure 5.3.2. The upper left corner of the plot represents the most desirable operating point: low power requirements and high spectral efficiency. Observe that MQAM provides greater spectral efficiency for a given E_b/N_0 than MPSK. This is a direct consequence of the superior packing efficiency of the MQAM constellations relative to the MPSK constellations. The MQAM point corresponding to $M = 8$ is missing from the plot. The reason for this is explored in Exercise 6.17.

The dashed line is a plot of the channel capacity for a bandlimited channel with additive white Gaussian noise and is included for reference. This channel capacity R_b/B is given by

$$\frac{R_b}{B} = \log_2\left(1 + \frac{R_b}{B}\frac{E_b}{N_0}\right). \tag{6.96}$$

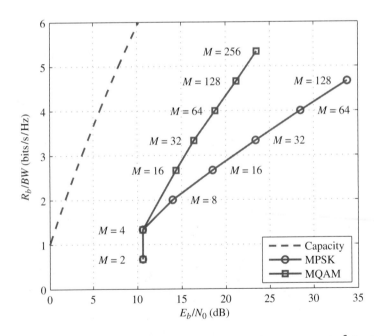

Figure 6.3.6 Spectral efficiency versus E_b/N_0 required to achieve $P_b = 10^{-6}$ for MPSK and MQAM using a square-root raised cosine pulse shape with 50% excess bandwidth.

The channel capacity specifies best spectral efficiency as a function of E_b/N_0 for which reliable communication can be achieved. In other words, it is not possible to operate to the left of the curve and achieve reliable communication. Note that the MQAM constellations are closer to channel capacity than the MPSK constellations. As it turns out, MQAM signal sets do not achieve channel capacity unless error control coding is used.

6.4 LINK BUDGETS

A *link budget* is a powerful computational tool used for wireless network system planning. The link budget expresses E_b/N_0—the fundamental quantity needed to compute the probability of bit error as explained in the previous sections of this chapter—in terms of system parameters such as transmitter power, distance, antenna gains, component noise levels, component losses, etc. The link budget can be used in one of two fundamental ways: Assuming fixed system parameters, the link budget is used to calculate E_b/N_0 and hence, the probability of bit error. Alternatively, given a probability of bit-error goal, the link budget can be used to quantify the various component values required to achieve the corresponding E_b/N_0. The second use is by far the most common use of the link budget and is used to perform system component trade-offs. A closely related use of the link budget is computation of the maximum achievable bit rate for given bandwidth and power resources.

In the following development, an expression for the carrier power to noise power spectral density ratio, usually denoted C/N_0, is derived. This is a fundamental quantity that is common to both digital and analog communication systems. (Fundamental in the sense that it is independent of modulation type, bit rate, and bandwidth.) For digital communication systems, the desired quantity E_b/N_0 is related to C/N_0 by

$$\frac{E_b}{N_0} = \frac{C}{N_0} \times T_b = \frac{C}{N_0} \times \frac{1}{R_b}. \tag{6.97}$$

For an analog communication system operating in a bandwidth of B Hz, the carrier-to-noise ratio C/N is of interest and is given by

$$\frac{C}{N} = \frac{C}{N_0 B}. \tag{6.98}$$

The starting point is the Friis equation that quantifies the received power in terms of well-known system and component parameters. The noise power spectral density level N_0 is also expressed in terms of well-known system component parameters. The end result, C/N_0, is the ratio of the two.

6.4.1 Received Power and the Friis Equation

An isotropic radiator is an RF source that transmits its power uniformly in all spatial directions. Consider an isotropic point source that transmits P_T Watts of RF power such as the one illustrated in Figure 6.4.1 (a). The power radiates in a sphere from the point source. The power flux density at a distance R from the point source is the the ratio of transmitted power P_T to surface area of the sphere:

$$\Phi = \frac{P_T}{4\pi R^2} \quad \frac{W}{m^2}. \tag{6.99}$$

Now replace the isotropic radiator with a real antenna and assume the transmit antenna is located at the origin of a Cartesian coordinate system as shown in Figure 6.4.1 (b). Real antennas are not isotropic because their geometrical configurations define a preference for radiation in certain directions. Consider the point on the sphere with spherical coordinates (R, θ, ϕ) as shown. The transmit antenna gain $G_T(\theta, \phi)$ can be defined as the ratio of the flux density in the direction θ, ϕ to the flux density of an isotropic radiator:

$$G_T(\theta, \phi) = \frac{\text{power flux density at point } (R, \theta, \phi)}{\frac{P_T}{4\pi R^2}}. \tag{6.100}$$

This quantity is usually expressed in decibels with the unit dBi: dB relative to an isotropic radiator. A conceptual visualization of the antenna gain pattern is illustrated in Figure 6.4.1 (b). The antenna gain is used to account for the non-isotropic behavior of real antennas. The power flux density on the surface of the sphere at the point with spherical coordinates (R, θ, ϕ) becomes

$$\Phi(\theta, \phi) = \frac{P_T G_T(\theta, \phi)}{4\pi R^2} \quad \frac{W}{m^2}. \tag{6.101}$$

The received power is computed by integrating the power flux density at the receiver location over the effective area A_{eff} of the receive antenna. The effective area can be thought of as the "capture area" of the receive antenna. Assuming that R is large enough so that the surface area on the sphere is constant over A_{eff}, the received power is the product of the power flux density and the effective area:

$$C = \Phi(\theta, \phi) \times A_{\text{eff}}. \tag{6.102}$$

The effective area is approximately the physical aperture area for large antennas such as reflector antennas, horn antennas, and lens antennas. For smaller antennas such as dipoles and loop antennas, there is no simple relationship between the area of the physical cross-section and the effective area. From electromagnetic theory, it can be shown that the effective area of an antenna is related to the gain[2] of a receive antenna by

$$A_{\text{eff}} = \frac{\lambda^2}{4\pi} G_R(\theta', \phi') \tag{6.103}$$

---

[2]The notion of antenna gain was defined in terms of transmitting antennas, for example, electromagnetic wavefronts emanating from an antenna. It is straightforward to show that the directivity also applies to electromagnetic wavefronts arriving at the antenna so that the concept of antenna gain applies to receive antennas as well.

Section 6.4 Link Budgets   **333**

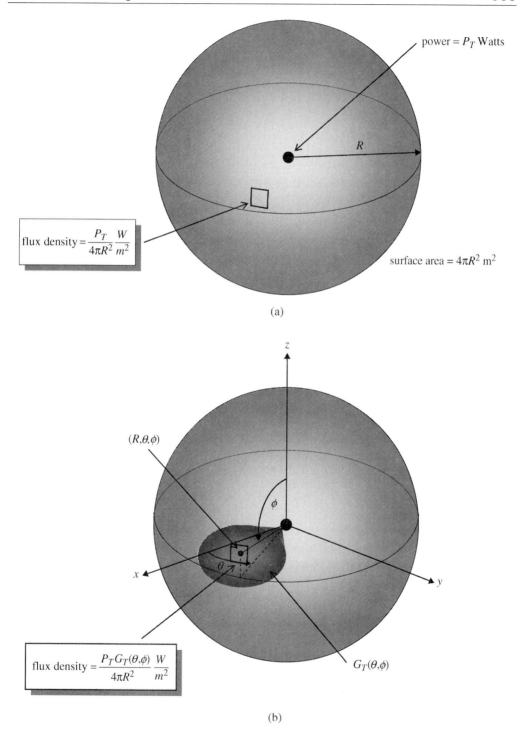

Figure 6.4.1 A conceptual visualization of an isotropic radiator, power flux density, and antenna gain.

where λ is the wavelength of the propagating electromagnetic wavefront and $G_R(\theta', \phi')$ is the receive antenna gain pattern in the direction θ', ϕ' relative to the receive antenna's coordinate system. Using this expression for A_{eff}, the received power may be expressed as

$$C = \frac{G_T(\theta, \phi) G_R(\theta', \phi') \lambda^2}{(4\pi R)^2} P_T. \tag{6.104}$$

This is the Friis radio link formula.

6.4.2 Equivalent Noise Temperature and Noise Figure

In computing link budgets, it is common practice to express the power spectral density of the noise as (see Exercise 4.30)

$$N_0 = kT \tag{6.105}$$

where $k = 1.38 \times 10^{-23}$ J/K is Boltzmann's constant and T is the temperature in Kelvin. But, the temperature of what? The answer is found in examining the thermal noise generated by a simple circuit element: the resistor.

Equivalent Noise Temperature. Consider the resistor shown in Figure 6.4.2 (a). The atoms that constitute the resistor are in motion if the temperature of the resistor is above absolute zero. The motion of each atom creates a very small voltage at the resistor terminals. (The sum of all such voltages is the noise voltage that would be measured at the resistor terminals.) The motions of the atoms are random. Hence, the corresponding terminal voltage is random and has a power spectral density given by (4.71). For frequencies of practical interest in wireless communications, the power spectral density of the noise is well approximated by (6.105). The power in this terminal voltage in a bandwidth B is $P = N_0 B = kTB$ Watts.

The noisy resistor can be replaced by its Thevenin equivalent: a voltage source v_{eq} and an ideal noiseless resistor R as illustrated in Figure 6.4.2 (b). The maximum power that can be delivered by the Thevenin equivalent is to a load with resistance R as illustrated in Figure 6.4.2 (c). The current through the load resistor is $i_{\text{load}} = v_{\text{eq}}/2R$. The power delivered to the load resistor is

$$P_{\text{load}} = E\left\{i_{\text{load}}^2 R\right\} = \frac{1}{4} E\left\{v_{\text{eq}}^2\right\} \tag{6.106}$$

where the expected value is used because the voltage (and corresponding current) are random. Setting P_{load} equal to kTB, the root-mean-square (RMS) value of v_{eq} must be

$$v_{\text{eq,RMS}} = \sqrt{E\left\{v_{\text{eq}}^2\right\}} = \sqrt{4kTB}. \tag{6.107}$$

This concept can be applied to real, noisy components as illustrated in Figure 6.4.3. Consider a real system component with gain G operating over a bandwidth B as illustrated in Figure 6.4.3 (a). The internal components generate thermal noise due to the random motions at the atomic level. As a consequence, the RMS output voltage is nonzero even when there is no signal connected to the input. Let N_{out} represent the power measured

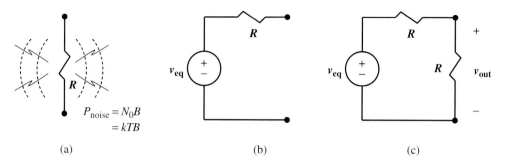

Figure 6.4.2 Noise and real resistors: (a) A noisy resistor; (b) Thevenin equivalent circuit; (c) maximum power transfer.

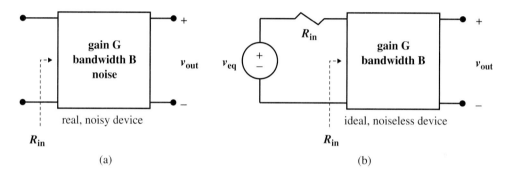

Figure 6.4.3 Noise and real components: (a) A component with gain G and bandwidth B whose internal components generate noise; (b) an ideal, noiseless component with gain G and bandwidth B where the internal noise is replaced by a noise source at the input.

at the output terminals of the real component. To model the noise, the real component is replaced by an ideal, noiseless component with the same gain and operating over the same bandwidth but with a noisy resistor, equal to the input resistance and connected to the input. The Thevenin equivalent circuit of this noisy resistor, connected to the input terminals of the ideal component, is illustrated in Figure 6.4.3 (b). The power delivered to the ideal component by this system is $N_{in} = kTB$. The resulting power measured at the output terminals is $N_{out} = GkTB$. Thus, the temperature of the resistor at the input that produces the equivalent noise power at the output must be

$$T_{eq} = \frac{N_{out}}{kGB}. \tag{6.108}$$

This temperature, called the *equivalent noise temperature* of the component, is the temperature of a noisy resistor, matched to the input impedance, that produces the same noise power at the output terminals (of an ideal, noiseless component) as the internal noise of the actual component. This temperature is related to the physical temperature of the resistor $T_{physical}$ by $T_{eq} = \epsilon T_{physical}$ where ϵ is the emissivity of the resistor. Thus, a complete description of a component includes the gain, bandwidth, and equivalent noise temperature.

Noise Figure. An alternate characterization of the noise added by a component is the *noise figure*, which is the ratio of input signal-to-noise ratio to output signal-to-noise ratio. The relationship between noise figure and equivalent temperature can be derived with the aid of Figure 6.4.4. A component, with gain G, bandwidth B, and equivalent noise temperature T_{eq}, has as inputs a signal (with power S_{in}) and noise (with power N_{in}). The output consists of an amplified version of the input signal and noise plus additional noise added by the component. Let S_{out} be the output signal power and N_{out} be the output noise power. The noise figure is

$$F = \frac{S_{in}/N_{in}}{S_{out}/N_{out}}. \tag{6.109}$$

Using the relationships

$$S_{out} = GS_{in} \tag{6.110}$$

$$N_{out} = GN_{in} + GkT_{eq}B \tag{6.111}$$

the noise figure may be expressed as

$$F = \frac{N_{in} + kT_{eq}B}{N_{in}}. \tag{6.112}$$

The input noise power is usually modeled as the output of a source with equivalent temperature T_0 operating in the same bandwidth B as the component. In this case, $N_{in} = kT_0B$ so that

$$F = 1 + \frac{T_{eq}}{T_0}. \tag{6.113}$$

Observe that $F > 1$ for $T_{eq} > 0$, which is another way of stating that the output signal-to-noise ratio is less than the input signal-to-noise ratio for a device that produces any internal noise.

There is a catch when using the noise figure to quantify the internal noise produced by a component. The noise figure as expressed in (6.113) is a relative measure, that is, it quantifies the noise power relative to a noise source at equivalent temperature T_0. T_0 must be known for (6.113) to be meaningful. In the United States and Western Europe, T_0 has been standardized at 290 K; in Japan, $T_0 = 293$ K. The equivalent temperature may be expressed in terms of the noise figure:

$$T_{eq} = (F - 1)T_0. \tag{6.114}$$

The equivalent temperature and noise figure for noisy components in cascade is explored in Exercises 6.25–6.28.

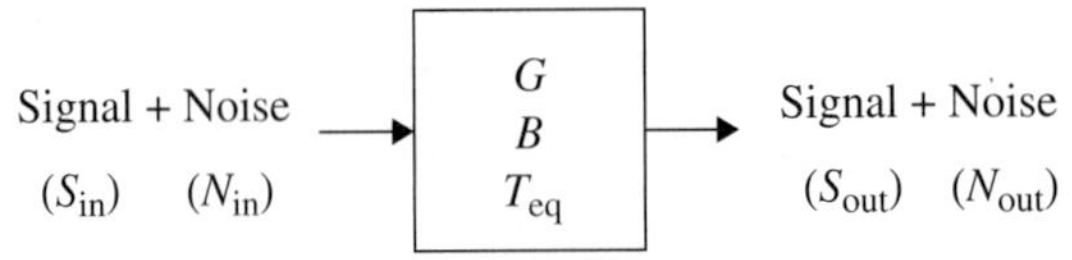

Figure 6.4.4 A system with gain G, bandwidth B, and equivalent noise temperature T_{eq}. The input consists of a signal (with power S_{in}) and noise (with power N_{in}). The output consists of a signal (with power S_{out}) and noise (with power N_{out}).

Measuring Noise Temperature and Noise Figure. In practice, the equivalent noise temperature and corresponding noise figure of components are measured in a laboratory setting. There are two commonly used measurement methods: the "Y method" and the "gain method." In its most common form, the Y method measures the equivalent noise temperature of a component (or system). In contrast, the gain is designed to measure the noise figure directly. Each method has its strengths and weaknesses.

Y Method The Y method uses two noise sources and a power meter to provide the data required to compute the equivalent noise temperature as illustrated in Figure 6.4.5. First, a calibrated noise source with equivalent noise temperature T_{cold} is connected to the component input. In Figure 6.4.5, this is accomplished by placing the input switch in position A. The corresponding output power, N_{cold}, is measured. Next, a different noise source, with equivalent noise temperature T_{hot}, is connected to the input by placing the switch in Figure 6.4.5 in position B. The corresponding output power, N_{hot}, is measured. The ratio

$$Y = \frac{N_{\text{hot}}}{N_{\text{cold}}} \qquad (6.115)$$

is used to compute T_{eq}. Using the basic principles that lead to (6.111), Y may be expressed as

$$Y = \frac{k\left(T_{\text{hot}} + T_{\text{eq}}\right) BG}{k\left(T_{\text{cold}} + T_{\text{eq}}\right) BG}. \qquad (6.116)$$

Solving for T_{eq} produces

$$T_{\text{eq}} = \frac{T_{\text{hot}} - Y T_{\text{cold}}}{Y - 1}. \qquad (6.117)$$

This method has the advantage that the gain and bandwidth of the component do not have to be known. On the other hand, the equivalent temperatures of the two noise sources must be known. Care must be taken in selecting the two temperatures: T_{cold} should be much less than T_{hot}. Otherwise, Y will be close to unity, which makes the denominator of (6.117)

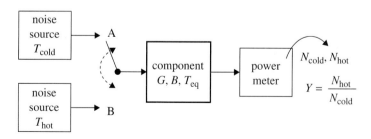

Figure 6.4.5 Block diagram representation of the Y method used for measuring the equivalent noise temperature T_{eq}.

very small (this can lead to numerical accuracy problems). When possible, the input is set to zero. In this case, $T_{\text{cold}} = 0$ and (6.117) simplifies to

$$T_{\text{eq}} = \frac{T_{\text{hot}}}{Y - 1}. \tag{6.118}$$

Calibrated noise sources are available for performing the measurements required for the Y method. When the system antenna is included in the test—especially if the antenna is a highly directional antenna—it is common practice to point the antenna at sources with known equivalent temperatures. For satellite communications and radio astronomy, celestial sources are often used.

It was just such a test that Arno Penzias and Robert Wilson were conducting in 1962 when they inadvertently discovered the background radiation that proved the "big bang"[3] theory of the origin of the universe. In 1960, Bell Labs launched the ECHO communications satellite. Because ECHO was designed to reflect signals transmitted to it, the transmitted signal had to be very powerful and the receive antenna had to have a very large gain. Bell Labs built a large horn antenna (illustrated in Figure 6.4.6) in Holmdel, New Jersey, to meet these requirements. In 1962, TELSTAR, equipped with a receiver, a mixer, and a transmitter (*transponder* for short), was launched. Communications with TELSTAR could be accomplished with a smaller,

Figure 6.4.6 The Holmdel, New Jersey, antenna used by Penzias and Wilson.

---

[3]The theory of a sudden expansion at a single instant in the past as the basis for creation of the universe was proposed by Georges Lemaître, a Roman Catholic abbot in Belgium in 1927. Lemaître did not use the term "big bang" to describe the theory. He actually used the language "a day that would not have had a yesterday." The name was first used by Fred Hoyle, "this hot big bang," in comments made in 1948 ridiculing the idea.

Section 6.4 Link Budgets

more manageable ground antenna. As a result, the Holmdel antenna became available to radio astronomers. Penzias and Wilson, both radio astronomers, carried out a series of careful measurements to determine the equivalent noise temperature of the receiver electronics because the signals they were looking for were very weak. In the process, they noticed a discrepancy of 2.7 K in their noise temperature measurements. What was further puzzling was that this discrepancy was present no matter which celestial noise source was used (i.e., it appeared to be the same no matter what direction the antenna was pointed). Remarkably, the existence of low-level background radiation as residue from the "big bang" had been theorized by Robert Dicke at Princeton University years before. Eventually, the observations made by Penzias and Wilson recognized as experimental proof of Dicke's theories. Penzias and Wilson received the Nobel Prize in physics in 1978.

Gain Method The starting point for the gain method is the basic definition of noise figure (6.109), which may be expressed as

$$F = \frac{N_{\text{out}}}{GN_{\text{in}}}. \tag{6.119}$$

Expressing the input noise power as $N_{\text{in}} = kT_0B$, this expression becomes

$$F = \frac{N_{\text{out}}}{kT_0BG} = \frac{N_{\text{out}}}{4.002 \times 10^{-21}BG} \tag{6.120}$$

where $T_0 = 290$ K has been used. This method requires a power meter (to measure N_{out}) and knowledge of the component gain G and bandwidth B as illustrated in Figure 6.4.7 (a). The bandwidth requirement may be removed to produce an alternate form for (6.120):

$$F = \frac{N_{0,\text{out}}}{kT_0G} = \frac{N_{0,\text{out}}}{4.002 \times 10^{-21}G} \tag{6.121}$$

where $N_{0,\text{out}}$ is the power spectral density of the output noise. The power spectral density of the output noise can be measured with a spectrum analyzer as illustrated in Figure 6.4.7 (b). Care must be taken when applying this method to a system with low gain and low noise figure. In this case, the noise floor of the spectrum analyzer may be higher than the power spectral density of noise output of the component.

6.4.3 The Link Budget Equation

The desired quantity C/N_0 is the ratio of the received power, given by the Friis equation (6.104), and N_0, given by (6.105) with T_{eq} used in place of T. This ratio is usually expressed by arranging the terms as

$$\frac{C}{N_0} = P_T G_T(\theta,\phi) \times \left(\frac{\lambda}{4\pi R}\right)^2 \times \frac{G_R(\theta',\phi')}{T_{\text{eq}}} \times \frac{1}{k}. \tag{6.122}$$

The equation is usually organized this way to collect terms with a common theme. The product $P_T G_T(\theta,\phi)$ is the *effective isotropic radiated power*, or EIRP, and is a function solely of the transmitter. The second term

$$L_p = \left(\frac{\lambda}{4\pi R}\right)^2 \tag{6.123}$$

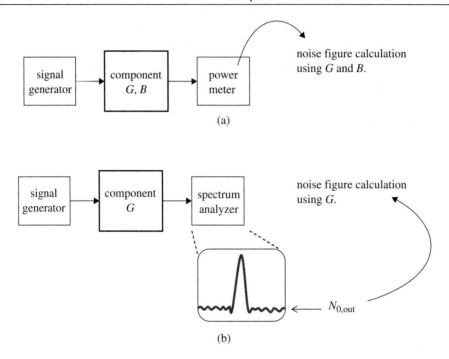

Figure 6.4.7 Two alternative forms for measuring the noise figure using the gain method.

is called the *spreading loss* and represents the "inverse square law" component of free-space propagation. The third term $G_R(\theta',\phi')/T_{eq}$ is solely a function of the receiver. The final term is Boltzmann's constant. For historical reasons, the computations are carried out using decibels (dB). Each of the terms is converted to decibels to convert the product into a sum:

$$\left[\frac{C}{N_0}\right]_{dB} = \left[P_T G_T(\theta,\phi)\right]_{dB} + \left[\left(\frac{\lambda}{4\pi R}\right)^2\right]_{dB} + \left[\frac{G_R(\theta',\phi')}{T_{eq}}\right]_{dB} - [k]_{dB} \qquad (6.124)$$

where

$$[X]_{dB} = 10\log_{10}(X). \qquad (6.125)$$

It is straightforward to include other systems losses, such as atmospheric attenuation, polarization mismatch loss, waveguide attenuation, connector losses, etc., in the calculation. One way to accomplish this is to lump all these losses into a single loss term L and include it

Section 6.4 Link Budgets

in the link equation:[4]

$$\left[\frac{C}{N_0}\right]_{\text{dB}} = \left[P_T G_T(\theta,\phi)\right]_{\text{dB}} + \left[\left(\frac{\lambda}{4\pi R}\right)^2\right]_{\text{dB}} + \left[\frac{G_R(\theta',\phi')}{T_{\text{eq}}}\right]_{\text{dB}} - [k]_{\text{dB}} + [L]_{\text{dB}}. \quad (6.126)$$

The following examples illustrate the use of link budget concept.

Space Exploration Link Budget. This example examines the sometimes overlooked space communication link. The link budget can be used in planning future missions to explore the solar system. An interesting example from the earliest beginnings of the US space program is illustrated in Figure 6.4.8. The purpose of these calculations was to show that soon-to-be realized advances in electronics would increase the distance over which a communication link could be maintained. The prediction was that by 1960, communication between Earth and Mars or Venus would be possible. The interesting feature of the table is the distance of the space-to-earth link is increased by making improvements to both the (space-borne) transmitter and the (earth-based) receiver. In space, the transmitter power is increased whereas on earth, the noise temperature of the receiver electronics is decreased. Note that *both* improvements were required to achieve the goal.

The gain of the earth-based antenna was computed using the formula for the boresight gain of a parabolic reflector (or "dish") antenna:

$$G = \eta \left(\frac{\pi d}{\lambda}\right)^2 \quad (6.127)$$

where d is the antenna diameter, λ is the wavelength, and η is the *illumination efficiency*. The value $\eta = 1$ means that *all* of the incident electromagnetic energy is reflected to the focal point of the parabola where the antenna feed is located. Most parabolic reflector antennas achieve $50\% \leq \eta \leq 60\%$.

Satellite Link Budget. This is an example of downlink from a satellite in geostationary orbit. The application is a digital video broadcast (DVB) signal transmitted from the TELSTAR 402R satellite in geostationary orbit at 89°W longitude to a small antenna mounted on a housetop in Chicago, Illinois. The modulation is QPSK. The link budget summarized in Table 6.4.1 is used to compute the link margin. *Link margin* is the extra power available over that needed to satisfy system performance requirements. In the link budget, the receive antenna gain is computed using the formula for the boresight gain of a parabolic reflector antenna (6.127). The equivalent temperature of the receiver is computed using

$$T_{\text{eq}} = T_a + (F-1)T_0 \quad (6.128)$$

where T_a is the antenna noise temperature referenced to the input of the receiver. The required E_b/N_0 may seem low, but includes the bit-error rate improvement due to forward

---

[4]The notation adopted in (6.126) assumes that L is a number less than unity, so that $[L]_{\text{dB}} < 0$. Often, losses are specified in decibels as positive numbers, such as "2 dB rain attenuation" or "0.5 dB waveguide loss." Care must be taken to properly interpret the information that is provided. All losses reduce the received power available and should be incorporated into a link budget accordingly.

Table I
Summary of Illustrative Communication System Characteristics for Space Program

Characteristics	1958	1959	1960	Remarks
A. Space-to-Earth Path				
1. Space-to-earth frequency	1,000-2,000mc	1,000-2,000mc	1,000-2,000mc	Maximum S/N ratio: best compromise between tracking accuracy and angle acquisition. Early use of solar energy.
2. Vehicle transmitter power	0.1 watt	1 watt	10 watt	
3. Vehicle antenna gain	6 db	10 db	18 db	
4. Ground tracking stations	2	4	4	Three 85' tracking antennas in world net. One 8' tracking antenna at launch site. Gain at 1,000 mc 46 db
5. Ground antenna	85' diam.	85' diam.	85' diam.	
6. Beam width of ground antenna	0.8°	0.8°	0.8°	
7. Angle tracking accuracy	1 - 3 mils	1 - 3 mils	.1 - .3 mils	Use radio stars for calibration and compute for correcting angle data.
8. Ground receiver bandwidth	60 cps	25 cps	10 cps	Using oscillators with increased stability
9. Ground receiver noise temp.	2000°K	1000°K	400°K	Using low temp. solid state techniques.
10. Ground receiver sensitivity	-148 DBM	-155 DBM	-165 DBM	
11. Space-to-earth range for S/N = 10 DB	350,000 mi.	3,500,000 mi.	50,000,000 mi.	
B. Earth-to-Space Path				
1. Earth-to-space frequency	none	1,000 - 2,000mc	1000 - 2000 mc	
2. Ground transmitter power	none	10 KW	10 KW	
3. Ground transmitter stations	–	1	2	
4. Ground transmitter antenna	none	85'	85'	Additional 85' dishes required.
5. Doppler velocity	one way	two way	two way	
6. Range tracking accuracy	–	100 miles	100 miles	
7. Vehicle receiver BW	–	100 cps	100 cps	
8. Vehicle receiver noise temp.	–	30,000°K	3,000°K	
Vehicle receiver sensitivity	–	-134 DBM	-144 DBM	
Earth-to-space range for S/N = 10 DB	–	10,000,000 mi.	50,000,000 mi.	

Figure 6.4.8 Link budget calculations compiled in April 1958 by engineers at the Jet Propulsion Laboratory in Pasadena, California. Reproduced from Craig Waff [111].

Section 6.4 Link Budgets

Table 6.4.1 Digital video broadcasting (DVB) link budget

Transmitter		
EIRP		51 dBW
Propagation		
Path distance	37792.1 km	
Carrier frequency	12.5 GHz	
L_p		−205.93 dB
Receiver		
Antenna diameter	0.46 m	
Antenna efficiency	0.60	
Antenna noise temperature	30 K	
LNB noise figure	1.0 dB	
G/T		13.16 dB^{-1}
Boltzmann's constant		−228.6 dB W/Hz-K
Atmospheric absorption	0.12 dB	
Tropospheric scintillation fading	0.06 dB	
Attenuation due to precipitation	0.67 dB	
Pointing and polarization loss	0.1 dB	
L		−0.95 dB
C/N_0		85.88 dB/Hz
Bit rate	40 Mbits/s	
E_b/N_0		9.86 dB
Required E_b/N_0	7.1 dB	
Link Margin		2.76 dB

error correcting codes that are applied to the signal. The link margin is 2.76 dB, that is, the link provides a C/N_0 that is 2.76 dB larger than the minimum value required to meet the performance specifications. Link margin is an important part of wireless network design and should be included in all links. Sometimes, the margins are specified by the requirement. In other applications, the link margin is a feature that can be used as a selling point for the system.

Cellular Telephony Link Budget. In this example, the goal is to determine the cell radius for a UMTS[5] cellular system in a suburban setting. Because of the limited power

---

[5]The universal mobile telecommunications system (UMTS) is one of the third generation (3G) cellular telephony standards. In the United States, first generation (1G) cellular telephony was based on narrowband FM, an analog modulation. The transition to digital modulation occurred with the adoption of second generation (2G) systems. There are two, incompatible 2G standards. One is based on the European GSM system which uses Gaussian minimum shift keying (GMSK). The other, defined in the IS-95 standard, uses direct sequence spread spectrum with QPSK to create a code division multiple access (CDMA) system. There are competing technologies for 3G systems as well. UMTS is based on wideband CDMA; freedom of mobile multimedia access (FOMA), developed by Japanese mobile phone operator NTT DoCoMo, is also based on wideband CDMA; CDMA 2000, that offers backward compatibility with IS-95; and time division synchronous code division multiple access (TD-SCDMA), developed in China and based on direct sequence spread spectrum in combination with both code division and time division multiple access. The appeal of 3G cellular telephony is increased bandwidth to support multimedia, videoconferencing, internet, etc. Given the author's experience with those who drive while using 2G cell phones, he is skeptical about the safety and utility these added features.

Table 6.4.2 A simple link budget for a UMTS uplink in a suburban environment

Transmitter		
Max power	125 mW	
Antenna gain	0 dBi	
EIRP		−9 dBW
Propagation		
L_p		? dB
Receiver		
Antenna gain	17 dBi	
Noise figure	4 dB	
G/T		−9.4 dB/K
Boltzmann's constant		−228.6 dB W/Hz-K
Body loss	2 dB	
Cable and connector losses	3 dB	
Interference margin	4 dB	
Log-normal shadowing margin	7 dB	
L		−19 dB
Required E_b/N_0 for speech	5 dB	
Bit rate	12.2 kbits/s	
C/N_0		45.9 dB/Hz

available in a battery-powered handset, the uplink (mobile the base station) is the limiting case. In mobile telephony, the propagation environment is complicated by the presence of people, structures, plants, and automobiles that produce reflected copies of the propagating electromagnetic wavefront and severe attenuation. The presence of reflected copies of the electromagnetic wavefront produces a phenomenon known as *multipath propagation*. When reflected copies arrive out-of-phase with each other, the resulting destructive interference produces what is known as a *multipath fade*. Multipath fading severely reduces the distance over which reliable communication can be maintained. As a consequence cellular system designers rarely use (6.123) for L_p in their link budget calculations. The form of the link budget is still (6.126), but they use a propagation model that accounts for the attenuation due to multipath propagation instead. One of the most popular models is the Okamura–Hata propagation model [112]. For the system parameters of interest, the loss an electromagnetic wavefront experiences on propagation of a distance R km in this environment is

$$[L_p]_{\text{dB}} = -137.4 - 35.2 \log_{10}(R). \tag{6.129}$$

In other words, (6.129) is used in place of (6.123) in (6.126). The link budget is summarized in Table 6.4.2. Using these numbers in (6.126), the propagation loss is $L_p = -145.3$ dB. Using this value in (6.129) gives $R = 1.7$ km. Note that the concept of margin (extra power needed to overcome an impairment) is incorporated into the link budget as a loss.

6.5 PROJECTING WHITE NOISE ONTO AN ORTHONORMAL BASIS SET

The analysis of the noise samples in the matched filter detectors for baseband PAM and band-pass QAM are special cases of a more general result: the projection of white noise onto an orthonormal basis set. The most efficient way to analyze this case is to apply the general result of Section 4.3. To do so, the projection process needs to reformulated in vector-matrix notation.

Let $\phi_0(t), \phi_1(t), \ldots, \phi_{K-1}(t)$ be a set of K orthonormal functions defined over the interval $T_1 \leq t \leq T_2$. Suppose these functions are sampled at T-spaced intervals. The corresponding sampled versions are $\phi_0(nT), \phi_1(nT), \ldots, \phi_{K-1}(nT)$ for $N_1 \leq n \leq N_2$. A consequence of the orthonormality of the continuous-time functions is that the sampled versions are orthogonal. Using the relation

$$\int_{T_1}^{T_2} \phi_i(t)\phi_j(t)dt \approx T \sum_{N_1}^{N_2} \phi_i(nT)\phi_j(nT), \tag{6.130}$$

it is straightforward to show that

$$\sum_{n=N_1}^{N_2} \phi_i(nT)\phi_j(nT) = \begin{cases} \dfrac{1}{T} & i=j \\ 0 & i \neq j \end{cases}. \tag{6.131}$$

Let the noise samples be $w(nT)$ for $N_1 \leq n \leq N_2$. Each noise sample is a Gaussian random variable with zero mean and variance $\sigma^2 = N_0/2T$. Further, the noise samples are uncorrelated.

The process is illustrated in Figure 6.5.1. The noise samples $w(nT)$ are projected onto the function $\phi_k(nT)$ (for $0 \leq k < K$) to produce the sample v_k given by

$$v_k = \sum_{n=N_1}^{N_2} \phi_k(nT)w(nT). \tag{6.132}$$

To facilitate the vector-matrix formulation for these operations, the $L = N_2 - N_1 + 1$ samples of each of the K orthonormal functions may be organized into an $L \times K$ matrix

$$\mathbf{A} = \begin{bmatrix} \phi_0(N_1 T) & \phi_1(N_1 T) & \cdots & \phi_{K-1}(N_1 T) \\ \vdots & \vdots & & \vdots \\ \phi_0(N_2 T) & \phi_1(N_2 T) & \cdots & \phi_{K-1}(N_2 T) \end{bmatrix} \tag{6.133}$$

The noise samples are organized into the $L \times 1$ vector $\mathbf{w}$ and the $K \times 1$ vector $\mathbf{v}$:

$$\mathbf{w} = \begin{bmatrix} w(N_1 T) \\ \vdots \\ w(N_2 T) \end{bmatrix} \qquad \mathbf{v} = \begin{bmatrix} v_0 \\ \vdots \\ v_{K-1} \end{bmatrix}. \tag{6.134}$$

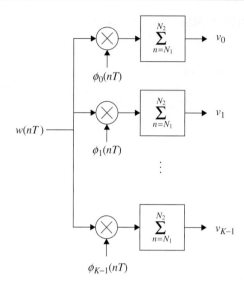

Figure 6.5.1 The projection of noise onto *K* orthonormal functions.

The probability density function of the Gaussian random vector **w** is

$$p(\mathbf{w}) = \frac{1}{(2\pi\sigma^2)^{L/2}} \exp\left\{-\frac{1}{2\sigma^2}\mathbf{w}^T\mathbf{w}\right\}. \tag{6.135}$$

Note that this is a special case of the probability density function for a multivariate Gaussian random vector given by (4.48) where $\boldsymbol{\mu} = \mathbf{0}$ (an $L \times 1$ vector of zeros) and $\mathbf{M} = \sigma^2\mathbf{I}$.

The relationship between **v** and **w** is

$$\mathbf{v} = \mathbf{A}^T\mathbf{w}. \tag{6.136}$$

The important observation is that this is a linear operator acting on the Gaussian random vector **w**. As a consequence, **v** is also a Gaussian random vector whose probability density function is of the form

$$p(\mathbf{v}) = \frac{1}{(2\pi)^{K/2}|\mathbf{M_v}|^{1/2}} \exp\left\{-\frac{1}{2}(\mathbf{v}-\boldsymbol{\mu_v})^T\mathbf{M_v}^{-1}(\mathbf{v}-\boldsymbol{\mu_v})\right\} \tag{6.137}$$

where the mean vector and covariance matrix are

$$\boldsymbol{\mu_v} = E\{\mathbf{v}\} \tag{6.138}$$

$$\mathbf{M_v} = E\left\{\mathbf{vv}^T\right\}. \tag{6.139}$$

Substituting the relationship (6.136) produces

$$\boldsymbol{\mu_v} = E\left\{\mathbf{A}^T\mathbf{w}\right\} = \mathbf{A}^T E\{\mathbf{w}\} = \mathbf{0} \tag{6.140}$$

$$\mathbf{M_v} = E\left\{\mathbf{A}^T\mathbf{w}\mathbf{w}^T\mathbf{A}\right\} = \mathbf{A}^T E\left\{\mathbf{w}\mathbf{w}^T\right\}\mathbf{A} = \sigma^2\mathbf{A}^T\mathbf{A}. \tag{6.141}$$

Because the functions are orthonormal, the matrix inner product is

$$\mathbf{A}^T\mathbf{A} = \frac{1}{T}\mathbf{I}. \tag{6.142}$$

Consequently, the covariance matrix $\mathbf{M_v}$ is

$$\mathbf{M_v} = \frac{\sigma^2}{T}\mathbf{I} = \sigma_v^2\mathbf{I}. \tag{6.143}$$

The interpretation is that the vector of samples in $\mathbf{v}$ are uncorrelated, and hence independent, zero-mean Gaussian random variables. Observe that this result is very general and applies to baseband and band-pass modulations based on orthonormal projections. This result is also related to the unitary operators explored in Exercise 4.20.

6.6 NOTES AND REFERENCES

6.6.1 Topics Covered

The geometrical interpretation using signal spaces for characterization and analysis of linear modulations was pioneered by Wozencraft and Jacobs [81]. The geometric approach used in calculating the probability of error was also described. A nice tutorial overview emphasizing the comparisons of different modulations has been published by Sklar [113,114].

The derivation of bandwidth for linear modulations based on an analysis of cyclostationary random processes is developed by Proakis [10].

The characterization of thermal noise as the temperature of an equivalent blackbody radio is the classical technique for modeling noise generated in microwave devices. There are many excellent textbooks in this field such as Pozar [115], Gonzalez [116], and Rohde and Newkirk [117]. Other types of noise generated by electronic communications circuitry and design methods to reduce these effects are described by Lee [118]. Link budgets are a standard tool for determining the capacity of a line-of-sight radio link. This technique is covered in much more detail in textbooks on satellite communications such as those by Gagliardi [119], Pratt and Bostian [120], Pritchard, Suyderhoud, and Nelson [121], and Richharia [122].

6.6.2 Topics Not Covered

The maximum likelihood detection rule (5.132) that makes each symbol decision independent of the others is actually a consequence of the signal model (5.111) which assumes that the data symbols are not temporally correlated. If the symbols are temporally correlated (e.g., as a result of a code or other formatter), this relationship should be placed in the signal model. The new signal model would then be used in (5.122) to produce an altered version of (5.123) where the dependence of data symbols on one another is quantified. Following the same steps above produces a decision rule where the symbol decisions are temporally related. In other words, the maximum likelihood decision rule estimates a sequence of symbols. In this case, the estimator is a *maximum likelihood sequence estimator*.

6.7 EXERCISES

6.1 Show that the probability of bit error for binary PAM in the AWGN environment is

$$P_b = Q\left(\sqrt{\frac{2E_b}{N_0}}\right).$$

6.2 Show that the probability of bit error for 8-ary PAM in the AWGN environment is

$$P_b = \frac{7}{12}Q\left(\sqrt{\frac{2}{7}\frac{E_b}{N_0}}\right).$$

6.3 Show that the probability of bit error for M-ary PAM in the AWGN environment is

$$P_b = \frac{2(M-1)}{M\log_2 M}Q\left(\sqrt{6\frac{\log_2 M}{M^2-1}\frac{E_b}{N_0}}\right).$$

6.4 Starting with probability of symbol error given by (6.62), show that the probability of bit error for square 16-QAM in the AWGN environment is

$$P_b = \frac{3}{4}Q\left(\sqrt{\frac{4}{5}\frac{E_b}{N_0}}\right).$$

6.5 Show that the probability of bit error for square M-ary QAM (M is an even power of 2) in the AWGN environment is

$$P_b = \frac{4}{\log_2 M}\frac{\sqrt{M}-1}{\sqrt{M}}Q\left(\sqrt{\frac{3(\log_2 M)}{M-1}\frac{E_b}{N_0}}\right).$$

6.6 QPSK can be thought of either as a square QAM constellation with four points or as 4-ary PSK.
  (a) Using the technique for analyzing square QAM constellations outlined in Section 6.2, derive the probability of bit error for QPSK.
  (b) Apply the general result for MPSK to the special case $M = 4$ to derive an expression for the probability of bit error for QPSK.
  (c) Compare the answers obtained in parts (a) and (b).

6.7 Derive the probability of bit error for the 8-ary QAM constellation shown below.

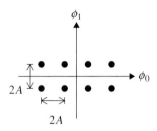

6.8 This exercise explores the use of the union bound to approximate the probability of error for the square 16-QAM constellation shown in Figure 6.2.1.
   (a) Compute the list of Euclidean distances from constellation point s_0 to the 15 other points. Note that this list of differences is the same for points s_3, s_{12}, and s_{15}. Use this list of distances to compute the conditional probabilities $P(E|s_m)$ for $m = 0, 3, 12, 15$.
   (b) Compute the list of Euclidean distances from constellation point s_1 to the 15 other points. Note that this list of differences is the same for points s_2, s_4, s_7, s_8, s_{11}, s_{13}, and s_{14}. Use these list of distances to compute the conditional probabilities $P(E|s_m)$ for $m = 1, 2, 4, 7, 8, 11, 13, 14$.
   (c) Compute the list of Euclidean distances from constellation point s_5 to the 15 other points. Note that this list of differences is the same for points s_6, s_9, and s_{10}. Use this list of distances to compute the conditional probabilities $P(E|s_m)$ for $m = 5, 6, 9, 10$.
   (d) Applying your answers from the previous parts to (6.76), show that the union bound on the average probability of symbol error is given by (6.81).
   (e) Evaluate each term on the right-hand-side of (6.81) for $E_{\text{avg}}/N_0 = 10$ dB. Compare the values and assess how small they are relative to each other. What do you conclude from this comparison?

6.9 Use the union bound to upper bound the probability of bit error for the 4+12-APSK constellation shown in Figure 5.3.5 using $r_1 = A, r_2 = 2A, \theta_1 = \pi/4, \theta_2 = 0$.

6.10 Use the union bound to upper bound the probability of bit error for the 4+12+32-APSK constellation shown in Figure 5.3.5 using $r_1 = A, r_2 = 2A, r_3 = 3A, \theta_1 = \pi/6$, and $\theta_2 = 0$.

6.11 Use the union bound to upper bound the probability of bit error for the CCITT V.29 16-ary QAM constellation shown in Figure 5.3.6 (a).

6.12 Use the union bound to upper bound the probability of bit error for the CCITT V.29 backup 8-ary QAM constellation shown in Figure 5.3.6 (b).

6.13 Use the union bound to upper bound the probability of bit error for the 32-ary QAM cross constellation shown below.

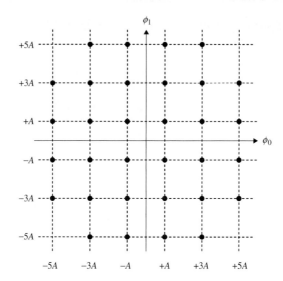

6.14 Use the union bound to upper bound the probability of bit error for the 128-ary QAM cross constellation shown below.

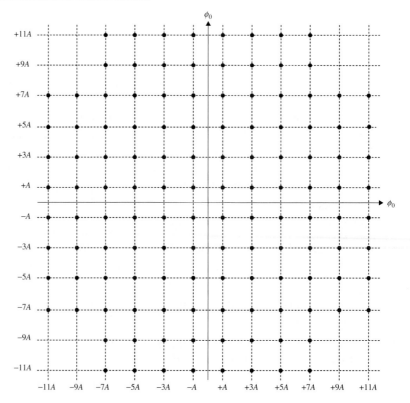

Section 6.7 Exercises

6.15 Consider the two 16-QAM constellations shown below.

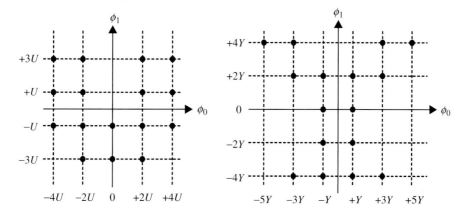

Which constellation has the lower probability of bit error?

6.16 Consider the three 8-QAM constellations shown below.

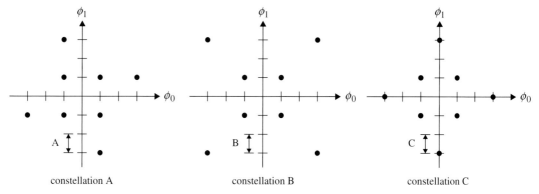

(a) Which constellation has the lowest probability of bit error?
(b) Which constellation has the highest probability of bit error?

6.17 The spectral efficiency versus E_b/N_0 for the MQAM family shown in Figure 6.3.6 displays a definite pattern for $M \geq 4$. This suggests there should be an 8-ary MQAM constellation that achieves $P_b = 10^{-6}$ at $E_b/N_0 \approx 12.5$ dB. Unfortunately, the rectangular constellation shown in Figure 5.3.4 achieves $P_b = 10^{-6}$ at $E_b/N_0 = 13.5$ dB. Perhaps another constellation achieves $P_b = 10^{-6}$ at the desired E_b/N_0. Using the union bound, compute the probability of bit error for the three constellations in Exercise 6.16 and determine which constellation comes closest to the desired operating point.

6.18 Add the operating points for M-ary PAM for $M = 2, 4, 8, 16, 32, 64, 128$ to the plot of Figure 6.3.6. Assume band-pass operation so that the spectral efficiency is given by (6.95). Use $\alpha = 0.5$ and $P_b = 10^{-6}$.

6.19 Determine the spectral efficiency versus E_b/N_0 operating point for the two CCITT V.29 constellations illustrated in Figure 5.3.6 for $P_b = 10^{-6}$. Assume a square-root raised-cosine pulse shape with 50% excess bandwidth. Plot these operating points along with the MQAM and MPSK points of Figure 6.3.6. How well do the CCITT V.29 constellations compare?

6.20 Determine the spectral efficiency versus E_b/N_0 operating point for the two APSK constellations illustrated in Figure 5.3.5 for $P_b = 10^{-6}$. Assume a square-root raised-cosine pulse shape with 50% excess bandwidth. Plot these operating points along with the MQAM and MPSK points of Figure 6.3.6. How well do the APSK constellations compare?

6.21 Determine the spectral efficiency versus E_b/N_0 operating point for the Y-QAM constellation illustrated in Figure 6.2.1 for $P_b = 10^{-6}$. Assume a square-root raised-cosine pulse shape with 50% excess bandwidth. Plot these operating points along with the MQAM and MPSK points of Figure 6.3.6. How well does the Y-QAM constellation compare?

6.22 The exercise explores the effect of the choice of P_b on how close a constellation comes to achieving capacity.
- Reproduce the plot of Figure 6.3.6 except use $P_b = 10^{-5}$.
- Reproduce the plot of Figure 6.3.6 except use $P_b = 10^{-8}$.
- Compare the two plots. What conclusions can you draw about the effect of P_b on how close a constellation comes to achieving capacity.

6.23 This exercise explores the effect of the excess bandwidth on how close a constellation comes to achieving capacity.
- Reproduce the plot of Figure 6.3.6 except use $\alpha = 1$.
- Reproduce the plot of Figure 6.3.6 except use $\alpha = 0$.
- Compare the two plots. What conclusions can you draw about the effect of excess bandwidth on how close a constellation comes to achieving capacity.

6.24 Derive (6.114) from (6.113).

6.25 Consider a cascade of two components shown below.

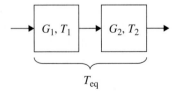

The gain of the first component is G_1 and the gain of the second component is G_2. The equivalent noise temperature of the first component is T_1 and the equivalent temperature of the second component is T_2. Assume both components operate over the same bandwidth B. Show that the equivalent temperature of the cascaded system is

$$T_{eq} = T_1 + \frac{T_2}{G_1}. \tag{6.144}$$

6.26 Consider a cascade of two components shown below.

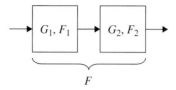

The gain of the first component is G_1 and the gain of the second component is G_2. The noise figure of the first component is F_1 and the noise figure of the second component is F_2. Assume both components operate over the same bandwidth B. Show that the noise figure of the cascaded system is

$$F = F_1 + \frac{F_2 - 1}{G_1}. \tag{6.145}$$

6.27 The first two components of a typical RF front end are a band-pass filter and a low-noise amplifier. The band-pass filter is usually a passive circuit. Thus, it is a lossy component (in other words, it has a gain less than one). Undergraduate linear system theory teaches that the order of the cascade of ideal, noiseless components is not important (i.e., either ordering of the two in cascade performs the same mathematical operation). This problem explores the consequences of ordering when the internal noise produced by each one is taken into account.

(a) Consider the cascade of the band-pass filter followed by the low-noise amplifier illustrated below. Determine the noise figure of this ordering.

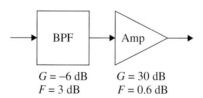

(b) Now consider the cascade of the same two components in reverse order, as illustrated below. Determine the noise figure of this ordering.

(c) Which of the two orderings has the best noise figure? Based on the result of Exercise 6.26, could you have predicted this result? Explain.

(d) Might there be a situation where the ordering with the worse noise figure is preferred? If so, give an example.

6.28 Consider the cascade of three noisy components illustrated below.

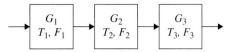

(a) Show that the equivalent noise temperature of the cascaded system is

$$T_{eq} = T_1 + \frac{T_2}{G_1} + \frac{T_3}{G_1 G_2}. \qquad (6.146)$$

(b) Show that the noise figure of the cascaded system is

$$F = F_1 + \frac{F_2 - 1}{G_1} + \frac{F_3 - 1}{G_1 G_2}. \qquad (6.147)$$

6.29 A variation on the Y method for computing equivalent noise temperature is the use of the excess noise ratio (ENR) that is defined as

$$\text{ENR} = \frac{T_{\text{hot}} - 290}{290}. \qquad (6.148)$$

(a) Express the equivalent noise temperature in terms of ENR and Y.

(b) Express the noise figure in terms of ENR and Y.

6.30 The noise figure of a receiver system is to be measured using the Y method. When a calibrated noise source with equivalent noise temperature 1268 K is connected to the input, the power spectral density of the output is -87 dBm/Hz. A different noise source with equivalent noise temperature 290 K is connected to the input and the power spectral density of the output is -90 dBm/Hz. Determine the noise figure of the receiver system.

6.31 A low noise amplifier (LNA) used in cellular telephony has a gain of 15 dB, operates over 500 MHz of bandwidth, and has a noise figure of 0.9 dB.

(a) If the gain method is used to measure the noise figure, what power spectral density level must be measured at the output of the LNA if the input is a noise source with an equivalent noise temperature of 290 K?

(b) If the Y method is to be used with two sources with equivalent noise temperatures $T_{\text{hot}} = 1000$ K and $T_{\text{cold}} = 290$ K, what power levels must be measured at the LNA output terminals?

(c) Based on the measurement capabilities of the spectrum analyzers and RF power meters currently available, which of these methods do you think is the most practical?

6.32 The noise figure of a mixer is to be measured. Using a noise source with equivalent noise temperature 290 K, the output noise power is measured as −83.5 dBm. Using a noise source with equivalent noise temperature 2468 K, the output noise power is measured as −80.8 dBm.
(a) Determine the noise figure of the mixer.
(b) Based on your answer in part (a) and the results of Exercise 6.26, where would you place the mixer in the RF front end: before the LNA or after?

6.33 This problem explores the relationship between power and bandwidth using the link budget. Sometimes power is the limiting factor in determining the maximum achievable bit rate. Such links (or channels) are called *power limited channels*. Sometimes bandwidth is the limiting factor in determining the maximum achievable bit rate. In this case, the link (or channel) is called a *bandwidth limited channel*. Consider a band-pass communications link with a bandwidth of 1.5 MHz and with an available $C/N_0 = 82$ dB Hz. The desired bit-error rate is 10^{-6}.
(a) If the modulation is 16-PSK using the SRRC pulse shape with $\alpha = 0.5$, What is the maximum achievable bit rate on the link? Is this a power limited or bandwidth limited channel?
(b) If the modulation is square 16-QAM using the SRRC pulse shape with $\alpha = 0.5$, what is the maximum achievable bit rate on this link? Is this a power limited or bandwidth limited channel?

6.34 Consider a communications link that achieves a C/N_0 of 70 dB W/Hz with a bandwidth of 375 kHz. The modulation is 16-QAM using the SRRC pulse shape with 50% excess bandwidth. If the bit-error rate is not to exceed 10^{-5}, determine the maximum achievable bit rate.

6.35 Consider the downlink from the Mars Orbital Surveyor operating at X-band (8400 MHz).

Transmitter (Mars Orbital Surveyor)
    EIRP   50 dBi

Receiver (Ground Station)
    Antenna diameter   70 m
    Antenna efficiency   60 %
    Equivalent noise temperature   50 K
    Atmospheric losses   2 dB
    Polarization losses   1 dB
    System losses   2 dB

    Modulation   QPSK
    Required bit-error rate 10^{-6}

The mean distance between the Sun and Earth is 149.6×10^6 km and the mean distance between the Sun and Mars is 227.9×10^6 km. Determine the maximum and minimum achievable bits rates on this link. Express your answer in kbits/second. (Note: bandwidth is not an issue in this application).

6.36 Consider a "point-to-point" microwave link illustrated below. (Such links were the key component in the telephone company's long distance network before fiber-optic cables were installed.) Both antenna gains are 20 dB and the transmit antenna power is 10 W. The modulation is 51.84 Mbits/s 256 QAM with a carrier frequency of 4 GHz. Atmospheric losses are 2 dB and other incidental losses are 2 dB. A pigeon in the line-of-sight path causes an additional 2 dB loss. The receiver has an equivalent noise temperature of 400 K and an implementation loss of 1 dB. How far away can the two towers be if the bit-error rate is not to exceed 10^{-8}? Include the pigeon. (Hint: see Exercise 6.5.)

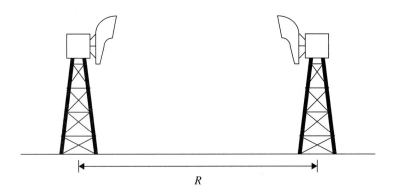

R

6.37 Consider a communications link between a satellite in geostationary orbit and a ground equipped with a parabolic reflector antenna with illumination efficiency $\eta = 55\%$. The satellite EIRP is 40 dBW and the satellite-to-ground distance is 40,000 km. The modulation is 20 Mbit/s QPSK with a carrier frequency of 12 GHz. Assume atmospheric, rain, polarization, and other losses are 2 dB and that the QPSK detector has an implementation loss of 1.4 dB (i.e., it requires 1.4 dB more in E_b/N_0 than predicted by theory). The receiver has an equivalent noise temperature of 300 K. Determine the diameter of the ground station antenna required to achieve a bit-error rate of 10^{-6}.

6.38 In many applications, a direct link between the source and destination is not possible. In these circumstances, a *repeater* is used. The repeater is situated so that it is able to communicate with both the source and the destination. There are two basic forms a repeater can take. The first is a simple amplifier (coupled with a frequency translation for implementation reasons). A link based on a simple amplifier repeater is sometimes called a *bent-pipe* link. An example of a bent-pipe link is illustrated below.

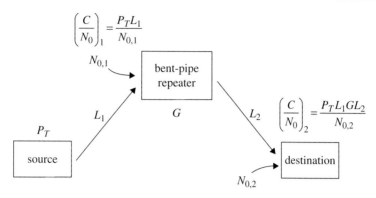

The source, equipped with a transmitter with P_T Watts, transmits the signal to the repeater through a path with loss L_1. The repeater receives the signal from the source together with additive noise whose power spectral density is $N_{0,1}$ W/Hz. An expression for C/N_0 on this first link is shown. The repeater is equipped with a power amplifier that produces a gain G. It amplifies the received signal (and noise) with a gain G and transmits the amplified signal to the destination through a link with loss L_2. An expression for C/N_0 on this second link is also shown.

(a) Show that the expression for the composite C/N_0 on this link is

$$\frac{C}{N_0} = \frac{P_T L_1 G L_2}{G L_2 N_{0,1} + N_{0,2}}.$$

(b) Using the definitions for C/N_0 on both links, show that the composite C/N_0 may be expressed as

$$\frac{C}{N_0} = \frac{1}{\left(\dfrac{C}{N_0}\right)_1^{-1} + \left(\dfrac{C}{N_0}\right)_2^{-1}}.$$

The other kind of repeater is the *regenerative* repeater. The regenerative repeater demodulates the received signal to produce an estimate of the original bit stream. The bits are then remodulated and sent to the destination. For purposes of analysis, a link based on the regenerative repeater is shown below.

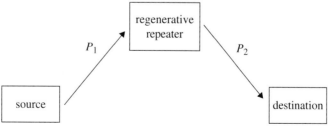

The link from the source to the regenerative repeater is characterized by a probability of bit error P_1 and the link from the regenerative repeater is characterized a probability of bit error P_2.

(c) Show that the probability of correct reception at the destination is

$$P_c = (1 - P_1)(1 - P_2).$$

(d) Show that the overall probability of bit error is

$$P_b \approx P_1 + P_2.$$

(e) Suppose a 1 Mbit/s QPSK link has $(C/N_0)_1 = 72$ dB W/Hz and $(C/N_0)_2 = 70$ dB W/Hz. Which type of repeater provides the lowest composite bit-error rate: the bent-pipe repeater or the regenerative repeater?

7

Carrier Phase Synchronization

7.1 BASIC PROBLEM FORMULATION   360
7.2 CARRIER PHASE SYNCHRONIZATION FOR QPSK   365
7.3 CARRIER PHASE SYNCHRONIZATION FOR BPSK   375
7.4 CARRIER PHASE SYNCHRONIZATION FOR MQAM   381
7.5 CARRIER PHASE SYNCHRONIZATION FOR OFFSET QPSK   382
7.6 CARRIER PHASE SYNCHRONIZATION FOR BPSK AND QPSK USING CONTINUOUS-TIME TECHNIQUES   391
7.7 PHASE AMBIGUITY RESOLUTION   394
7.8 MAXIMUM LIKELIHOOD PHASE ESTIMATION   409
7.9 NOTES AND REFERENCES   421
7.10 EXERCISES   423

As explained in Section 5.3, a carrier phase error causes a rotation in the signal space projections. If the rotation is large enough, the signal space projections for each possible symbol lie in the wrong decision region. As a consequence, decision errors occur even with perfect symbol timing synchronization and in the absence of additive noise. Estimating the carrier phase is the role of *carrier phase synchronization*.[1] Conceptually, carrier phase synchronization is the process of forcing the local oscillators in the detector to oscillate in both phase and frequency with the carrier oscillator used at the transmitter.

The synchronizers presented in this chapter are all based on the phase-locked loop, or PLL. The fundamentals of PLL operation and analysis are reviewed, in detail, in Appendix C. There, the analysis is applied to the simple case of tracking the phase and frequency of a simple sinusoid. Unfortunately, for carrier phase synchronization in band-pass communications, a sinusoid at the desired phase and frequency is rarely available. In QPSK, for example, the received waveform possesses 90° phase shifts due to the data. These phase shifts are in addition to the unknown carrier phase. If a QPSK waveform were input directly into a PLL designed to track the phase of a simple sinusoid, the PLL would try to track the phase shifts due to the

---

[1]The word *synchronization* comes from *Chronos* ($\chi\rho\acute{o}\nu o\varsigma$), the Greek god of time, and *syn*, a prefix meaning "the same." "To synchronize" thus means to cause one thing to occur or operate with exact coincidence in time or rate as another thing. As applied to digital communications, it usually means the process of causing one oscillator to oscillate with the same frequency and phase as another oscillator.

data and probably never lock. Thus, the carrier phase synchronization PLL must *remove* the phase shifts due to the data and track the remaining phase. This task can be accomplished by proper design of the phase detector.

Because the phase detector is responsible for removing the effects of the modulation on the underlying unmodulated carrier, the focus of this chapter is on the design and analysis of the phase detector. It is important to keep in mind that the overall PLL still has the structure illustrated in Figure C.1.1 for continuous-time PLLs or Figure C.2.1 (a) for discrete-time PLLs.

In this chapter, the traditional order of presenting continuous-time systems followed by the discrete-time counterpart is reversed. This is done because it is easier to understand the operation of carrier phase synchronization in terms of a rotation in two-space using discrete-time signal processing.

7.1 BASIC PROBLEM FORMULATION

The received band-pass MQAM waveform may be represented as

$$r(t) = G_a \sum_k \left\{ a_0(k)p(t - kT_s)\sqrt{2}\cos(\omega_0 t + \theta) - a_1(k)p(t - kT_s)\sqrt{2}\sin(\omega_0 t + \theta) \right\} + w(t) \tag{7.1}$$

where $a_0(k)$ and $a_1(k)$ are the inphase and quadrature components of the k-th symbol, $p(t)$ is the unit-energy pulse shape with support on $-L_p T_s \le t \le L_p T_s$, T_s is the symbol time, ω_0 is the center frequency in radians/second, θ is the unknown carrier phase offset, and $w(t)$ is the additive white Gaussian noise. The constant G_a represents all the amplitude gains and losses through the antennas, propagation medium, amplifiers, mixers, filters, and other RF components. The received signal is sampled at a rate $F_s = 1/T$ samples/s. The n-th sample of the received signal is

$$r(nT) = G_a \sum_k a_0(k)p(nT - kT_s)\sqrt{2}\cos(\Omega_0 n + \theta) - a_1(k)p(nT - kT_s)\sqrt{2}\sin(\Omega_0 n + \theta) + w(nT) \tag{7.2}$$

where $\Omega_0 = \omega_0 T$ radians/sample. There are two approaches incorporating estimates of the carrier phase θ into the processing. The first approach adjusts the phase of the quadrature sinusoids used to translate the received samples from band-pass to I/Q baseband. This approach is described in Section 7.1.1. The second approach differs from the first in that it uses quadrature sinusoids with a fixed frequency and phase to perform the translation from band-pass to I/Q baseband. Carrier phase offset compensation is performed by rotating the downsampled matched filter outputs to remove rotation due to carrier phase offset. This approach is described in Section 7.1.2. These approaches are mathematically equivalent, but produce PLL-based carrier phase synchronizers with different structures.

7.1.1 Approach 1

The first approach is illustrated in Figure 7.1.1. This approach mimics the approach used with continuous-time processing described in Section 7.6. The received signal is downconverted

Section 7.1 Basic Problem Formulation

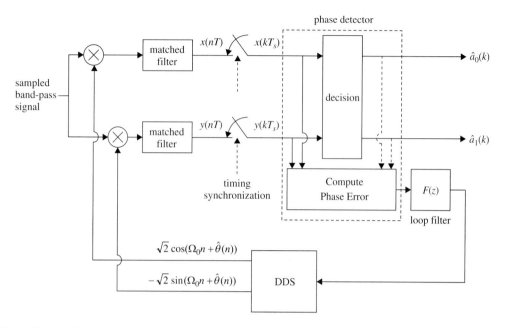

Figure 7.1.1 Carrier phase synchronization using phase adjusted quadrature sinusoids.

using quadrature sinusoids $\sqrt{2}\cos(\Omega_0 n + \hat{\theta}(n))$ and $-\sqrt{2}\sin(\Omega_0 n + \hat{\theta}(n))$ produced by a DDS. The multiplication produces inphase and quadrature components which, neglecting the double frequency terms, may be expressed as

$$I(nT) = G_a \sum_k \left\{ a_0(k) p(nT - kT_s) \cos(\theta - \hat{\theta}(n)) - a_1(k) p(nT - kT_s) \sin(\theta - \hat{\theta}(n)) \right\}$$
$$+ w_I(nT)$$
$$Q(nT) = G_a \sum_k \left\{ a_0(k) p(nT - kT_s) \sin(\theta - \hat{\theta}(n)) + a_1(k) p(nT - kT_s) \cos(\theta - \hat{\theta}(n)) \right\}$$
$$+ w_Q(nT). \tag{7.3}$$

The inphase and quadrature components are filtered by the matched filter, whose impulse response is $p(-nT)$, to produce the inphase and quadrature matched filter outputs

$$x(nT) = \frac{G_a}{T} \sum_k \left\{ \left[a_0(k) \cos(\theta - \hat{\theta}(n)) - a_1(k) \sin(\theta - \hat{\theta}(n)) \right] r_p(nT - kT_s) \right\} + v_I(nT)$$
$$y(nT) = \frac{G_a}{T} \sum_k \left\{ \left[a_0(k) \sin(\theta - \hat{\theta}(n)) + a_1(k) \cos(\theta - \hat{\theta}(n)) \right] r_p(nT - kT_s) \right\} + v_Q(nT).$$
$$\tag{7.4}$$

where $r_p(u)$ is the autocorrelation function of the pulse shape given by

$$r_p(u) = \int_{-L_p T_s}^{L_p T_s} p(t)p(t-u)dt, \quad (7.5)$$

$v_I(nT) = p(-nT) * w_I(nT)$, and $v_Q(nT) = p(-nT) * w_Q(nT)$. Assuming perfect timing synchronization, $x(nT)$ and $y(nT)$ are sampled at $n = kT_s/T = kN$ to produce the signal space projection corresponding to the k-th symbol. When the pulse shape satisfies the Nyquist No-ISI condition, $r_p(0) = 1$ and $r_p(mT_s) = 0$ for $m \neq 0$ so that

$$\begin{aligned} x(kT_s) &= K\left[a_0(k)\cos(\theta - \hat{\theta}(kN)) - a_1(k)\sin(\theta - \hat{\theta}(kN))\right] + v_I(kT_s) \\ y(kT_s) &= K\left[a_0(k)\sin(\theta - \hat{\theta}(kN)) + a_1(k)\cos(\theta - \hat{\theta}(kN))\right] + v_Q(kT_s) \end{aligned} \quad (7.6)$$

where $K = G_a/T$. The effect of the uncompensated carrier phase offset $\theta - \hat{\theta}(kN)$ is more evident when (7.6) is organized into a matrix equation:

$$\begin{bmatrix} x(kT_s) \\ y(kT_s) \end{bmatrix} = K \underbrace{\begin{bmatrix} \cos(\theta - \hat{\theta}(kN)) & -\sin(\theta - \hat{\theta}(kN)) \\ \sin(\theta - \hat{\theta}(kN)) & \cos(\theta - \hat{\theta}(kN)) \end{bmatrix}}_{\text{rotation matrix}} \begin{bmatrix} a_0(k) \\ a_1(k) \end{bmatrix} + \begin{bmatrix} v_I(kT_s) \\ v_Q(kT_s) \end{bmatrix}. \quad (7.7)$$

This shows that the point $(x(kT_s), y(kT_s))$ is a rotated version of the point $(a_0(k), a_1(k))$. The angle of rotation is the uncompensated phase error $\theta - \hat{\theta}(kN)$ in the counterclockwise direction. The rotated point is scaled by K and perturbed from its rotation position by the additive noise.

The carrier phase PLL is formed by the closed-loop path created by the phase detector, loop filter, and DDS as illustrated in Figure 7.1.1. The PLL locks when the error signal is forced to zero. The error signal is generated by the "Compute Phase Error" block using the signal space projections and the data symbols (or the data symbol estimates) as described later. Because the error signal is proportional to the uncompensated phase error $\theta - \hat{\theta}(kN)$, the PLL locks when $\hat{\theta}(kN)$ is equal to θ.

This approach is conceptually straightforward. Note, however, that it is a multi-rate system. The phase detector operates at 1 sample/symbol whereas the DDS operates at N samples/symbol. As such, an upsample operation must be inserted somewhere in between the "Compute Phase Error" block output and the DDS input.

7.1.2 Approach 2

The second approach is illustrated in Figure 7.1.2. In this approach, fixed frequency quadrature sinusoids are used for the downconversion operation and phase compensation is performed at the output of the matched filter. The phase compensation is realized by the "CCW Rotation" block as described below.

Section 7.1 Basic Problem Formulation

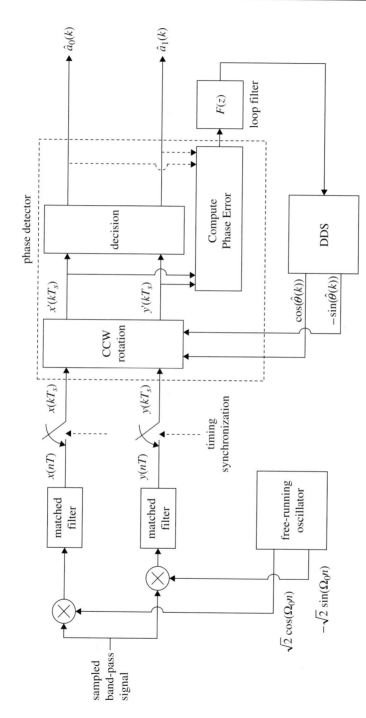

Figure 7.1.2 Carrier phase synchronization using a post-matched filter de-rotation operation.

Assuming the quadrature sinusoids used for downconversion are $\sqrt{2}\cos(\Omega_0 n)$ and $-\sqrt{2}\sin(\Omega_0 n)$, the signal space projections are computed following the same steps used in the development from (7.3) to (7.6). The signal space projections are given by

$$\begin{aligned} x(kT_s) &= K\left[a_0(k)\cos(\theta) - a_1(k)\sin(\theta)\right] + v_I(kT_s) \\ y(kT_s) &= K\left[a_0(k)\sin(\theta) + a_1(k)\cos(\theta)\right] + v_Q(kT_s) \end{aligned} \qquad (7.8)$$

or, in matrix form, by

$$\begin{bmatrix} x(kT_s) \\ y(kT_s) \end{bmatrix} = K \begin{bmatrix} \cos(\theta) & -\sin(\theta) \\ \sin(\theta) & \cos(\theta) \end{bmatrix} \begin{bmatrix} a_0(k) \\ a_1(k) \end{bmatrix} + \begin{bmatrix} v_I(kT_s) \\ v_Q(kT_s) \end{bmatrix}. \qquad (7.9)$$

The sampled matched filter outputs $(x(kT_s), y(kT_s))$ are de-rotated by the estimated carrier phase offset $\hat{\theta}(k)$. The DDS provides carrier phase estimate as $\cos(\hat{\theta}(k))$ and $-\sin(\hat{\theta}(k))$. The point $(x(kT_s), y(kT_s))$ is rotated by $-\hat{\theta}(k)$ to form the de-rotated signal space projection

$$\begin{bmatrix} x'(kT_s) \\ y'(kT_s) \end{bmatrix} = \begin{bmatrix} \cos(\hat{\theta}(k)) & \sin(\hat{\theta}(k)) \\ -\sin(\hat{\theta}(k)) & \cos(\hat{\theta}(k)) \end{bmatrix} \begin{bmatrix} x(kT_s) \\ y(kT_s) \end{bmatrix}. \qquad (7.10)$$

This relationship defines the operation of the "CCW Rotation" block in Figure 7.1.2.

Using (7.9), the de-rotated signal space projection may be expressed as

$$\begin{aligned} \begin{bmatrix} x'(kT_s) \\ y'(kT_s) \end{bmatrix} &= K \begin{bmatrix} \cos(\hat{\theta}(k)) & \sin(\hat{\theta}(k)) \\ -\sin(\hat{\theta}(k)) & \cos(\hat{\theta}(k)) \end{bmatrix} \begin{bmatrix} \cos(\theta) & -\sin(\theta) \\ \sin(\theta) & \cos(\theta) \end{bmatrix} \begin{bmatrix} a_0(k) \\ a_1(k) \end{bmatrix} \\ &\quad + \begin{bmatrix} \cos(\hat{\theta}(k)) & \sin(\hat{\theta}(k)) \\ -\sin(\hat{\theta}(k)) & \cos(\hat{\theta}(k)) \end{bmatrix} \begin{bmatrix} v_I(kT_s) \\ v_Q(kT_s) \end{bmatrix} \end{aligned} \qquad (7.11)$$

$$= K \begin{bmatrix} \cos(\theta - \hat{\theta}(k)) & -\sin(\theta - \hat{\theta}(k)) \\ \sin(\theta - \hat{\theta}(k)) & \cos(\theta - \hat{\theta}(k)) \end{bmatrix} \begin{bmatrix} a_0(k) \\ a_1(k) \end{bmatrix} + \begin{bmatrix} v'_I(kT_s) \\ v'_Q(kT_s) \end{bmatrix}. \qquad (7.12)$$

The expression (7.12) is identical to (7.7) with the exception that the phase estimate $\hat{\theta}$ in (7.7) is sampled at $1/T$ samples/second whereas $\hat{\theta}$ in (7.12) is sampled at $1/T_s$ samples/second. This shows that the signal space projections $(x(kT_s), y(kT_s))$ are rotated versions of the constellation point $(a_0(k), a_1(k))$ where the rotation angle is the phase error $\theta - \hat{\theta}(k)$.

The carrier phase PLL is formed by the closed-loop path created by the phase detector, loop filter, and DDS as illustrated in Figure 7.1.2. The PLL locks when the error signal is forced to zero. The error signal is generated by the "Compute Phase Error" block using the de-rotated signal space projections and the data symbols (or the data symbol estimates) as described later. The error signal is proportional to the uncompensated phase error $\theta - \hat{\theta}(k)$. Thus, the loop locks when $\hat{\theta}(k)$ is equal to θ.

Because the sampled matched filter outputs form a discrete-time sequence, this approach is a purely discrete-time approach. In addition, the PLL operates at 1 sample/symbol. This is in contrast with the PLL used in the first approach, which must operate at N samples/symbol.

In both approaches, the dashed line represents an optional connection between the symbol estimates $\hat{a}_0(k)$ and $\hat{a}_1(k)$ and the phase error detector represented by the "Compute

Section 7.2 Carrier Phase Synchronization for QPSK

Phase Error" block. When the phase error detector uses the symbol estimates to compute the phase error, the resulting PLL is called a *decision-directed* loop. Alternatively, the phase error may be computed using knowledge of the transmitted data symbols. Usually, the known data takes the form of a predefined data sequence, known as a *training sequence*, that is inserted at the beginning of the transmission for the purposes of phase acquisition. This approach is commonly used for packetized data links where the training sequence forms part of the packet header or preamble. A carrier phase PLL that uses known data is often called a *data aided* PLL.

Most PLLs operate in the decision-directed mode. This being the case, decision-directed PLLs are emphasized in the following sections. Application of the ideas presented in the following sections to the data-aided case is straightforward.

7.2 CARRIER PHASE SYNCHRONIZATION FOR QPSK

Carrier phase synchronization for QPSK begins with the development of a phase error detector based on heuristic reasoning founded in the basic principles of what is to be accomplished. Next, the maximum likelihood phase error detector is introduced. Its relationship to the heuristic phase error detector is emphasized. In both cases, the general approach outlined in Figure 7.1.2 is the basis for the development.

7.2.1 A Heuristic Phase Error Detector

Consider the QPSK carrier phase synchronizer illustrated in Figure 7.1.2. The inphase and quadrature matched filter outputs, $x(kT_s)$ and $y(kT_s)$, are rotated by $-\hat{\theta}(k)$ in an attempt to align the signal space projection $(x'(kT_s), y'(kT_s))$ with one of the four constellation points at $(\pm A, \pm A)$. In the absence of noise, the relationship between signal space projection $(x'(kT_s), y'(kT_s))$ and the transmitted data symbol $(a_0(k), a_1(k))$ is

$$\begin{bmatrix} x'(kT_s) \\ y'(kT_s) \end{bmatrix} = K \begin{bmatrix} \cos(\theta - \hat{\theta}(k)) & -\sin(\theta - \hat{\theta}(k)) \\ \sin(\theta - \hat{\theta}(k)) & \cos(\theta - \hat{\theta}(k)) \end{bmatrix} \begin{bmatrix} a_0(k) \\ a_1(k) \end{bmatrix}. \tag{7.13}$$

Assume, for the moment, that the phase error detector knows the data symbols. The phase error is extracted from the point $(x'(kT_s), y'(kT_s))$ by computing the residual phase difference between $(x'(kT_s), y'(kT_s))$ and the transmitted constellation point $(a_0(k), a_1(k))$.

Computation of the phase error is easily understood in geometric terms. Consider the scenario shown in Figure 7.2.1. The phase angle of the de-rotated matched filter outputs is

$$\theta_r(k) = \tan^{-1}\left\{\frac{y'(kT_s)}{x'(kT_s)}\right\} \tag{7.14}$$

and the phase angle of the transmitted constellation point is

$$\theta_d(k) = \tan^{-1}\left\{\frac{a_1(k)}{a_0(k)}\right\}. \tag{7.15}$$

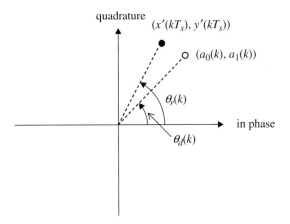

Figure 7.2.1 Geometric representation of the phase error computation for QPSK.

The *phase error detector output* for the k-th symbol is thus

$$e(k) = \theta_r(k) - \theta_d(k)$$
$$= \tan^{-1}\left\{\frac{y'(kT_s)}{x'(kT_s)}\right\} - \tan^{-1}\left\{\frac{a_1(k)}{a_0(k)}\right\}. \tag{7.16}$$

The error signal (7.16) requires knowledge of the transmitted symbols. When the actual transmitted data symbols are unknown (either the training sequence has passed or there was no training sequence provided) a *decision-directed* carrier phase synchronizer can be used. A decision-directed carrier phase synchronizer uses the data decisions to compute the phase error. The phase error is extracted from the point $(x'(kT_s), y'(kT_s))$ by computing the residual phase difference between $(x'(kT_s), y'(kT_s))$ and the nearest constellation point $(\hat{a}_0(k), \hat{a}_1(k))$. Thus, the decision-directed carrier phase synchronizer replaces $a_0(k)$ and $a_1(k)$ in (7.16) with the decisions $\hat{a}_0(k)$ and $\hat{a}_1(k)$:

$$e(k) = \tan^{-1}\left\{\frac{y'(kT_s)}{x'(kT_s)}\right\} - \tan^{-1}\left\{\frac{\hat{a}_1(k)}{\hat{a}_0(k)}\right\}. \tag{7.17}$$

For QPSK, the decisions are

$$\hat{a}_0(k) = A \times \text{sgn}\left\{x'(kT_s)\right\}$$
$$\hat{a}_1(k) = A \times \text{sgn}\left\{y'(kT_s)\right\}. \tag{7.18}$$

This approach is illustrated in Figure 7.2.2.

Section 7.2 Carrier Phase Synchronization for QPSK

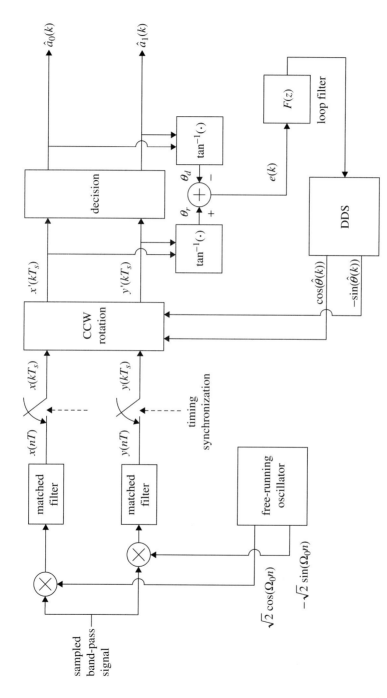

Figure 7.2.2 Block diagram of the QPSK carrier phase PLL using an error signal based on (7.17).

Both the data-aided phase error detector (7.16) and the decision-directed phase error detector (7.17) express the phase error as a function of the rotated signal space projections. The phase error detector output may also be expressed in terms of the phase error θ_e and the data symbols. A plot of $e(k)$ versus θ_e is called an S-curve and is usually denoted $g(\theta_e)$. The function $g(\theta_e)$ is the nonlinear phase error detector function that was assumed known in the general PLL analysis in Appendix C.

For the data-aided case, the S-curve is derived from (7.16) by expressing the de-rotated signal space projections in terms of θ_e and substituting into (7.16). The de-rotated signal space projections may be expressed in terms of the phase error and the data symbols as

$$\begin{bmatrix} x'(kT_s) \\ y'(kT_s) \end{bmatrix} = K \begin{bmatrix} \cos(\theta_e) & -\sin(\theta_e) \\ \sin(\theta_e) & \cos(\theta_e) \end{bmatrix} \begin{bmatrix} a_0(k) \\ a_1(k) \end{bmatrix}. \tag{7.19}$$

Substituting this expression for $x'(kT_s)$ and $y'(kT_s)$ in (7.16) produces

$$g(\theta_e, a_0(k), a_1(k)) = \tan^{-1}\left\{ \frac{a_0(k)\sin(\theta_e) + a_1(k)\cos(\theta_e)}{a_0(k)\cos(\theta_e) - a_1(k)\sin(\theta_e)} \right\} - \tan^{-1}\left\{ \frac{a_1(k)}{a_0(k)} \right\}. \tag{7.20}$$

The average S-curve, denoted $\overline{g}(\theta_e)$, is obtained from $g(\theta_e, a_0(k), a_1(k))$ by averaging over the four possible symbols $(a_0(k), a_1(k)) \in \{\pm A, \pm A\}$. After a little algebra, the result is

$$\overline{g}(\theta_e) = \theta_e. \tag{7.21}$$

The average S-curve (7.21) is plotted in Figure 7.2.3 (a) where it is seen that this phase detector is an ideal linear phase detector with $K_p = 1$.

The average S-curve for the decision-directed phase error detector (7.17) is computed in the same way. The average S-curve for the decision-directed phase detector is

$$\overline{g}(\theta_e) = \begin{cases} \theta_e + \pi & -\pi \leq \theta_e < -\dfrac{3\pi}{4} \\ \theta_e + \dfrac{\pi}{2} & -\dfrac{3\pi}{4} < \theta_e < -\dfrac{\pi}{4} \\ \theta_e & -\dfrac{\pi}{4} < \theta_e < \dfrac{\pi}{4} \\ \theta_e - \dfrac{\pi}{2} & \dfrac{\pi}{4} < \theta_e < \dfrac{3\pi}{4} \\ \theta_e - \pi & \dfrac{3\pi}{4} < \theta_e \leq \pi \end{cases} \tag{7.22}$$

and is plotted in Figure 7.2.3 (b). Note that the slope of the S-curve at $\theta_e = 0$ is 1 so that $K_p = 1$ for this phase detector.

Comparing the S-curves for the data-aided and decision-directed phase detectors in Figure 7.2.3 reveals some interesting differences. The S-curve for the decision-directed loop crosses zero at

$$\theta_e = -\frac{3\pi}{4}, -\frac{\pi}{2}, -\frac{\pi}{4}, 0, \frac{\pi}{4}, \frac{\pi}{2}, \frac{3\pi}{4}, \pi.$$

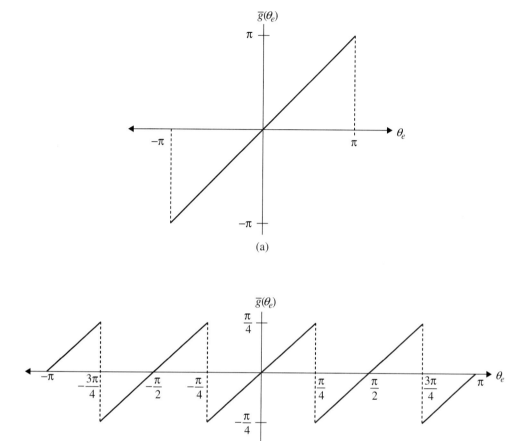

Figure 7.2.3 S-curve for (a) the data-aided phase detector (7.21) and (b) the decision-directed phase detector (7.22).

Because a PLL locks at $\theta_e = 0$, the question arises: which of these zero crossings represent a stable lock point? As it turns out, only those values of θ_e where $g(\theta_e)$ passes through zero with a positive slope are stable lock points. Thus, the stable lock points are

$$\theta_e = -\frac{\pi}{2}, 0, \frac{\pi}{2}, \pi.$$

As a consequence, the QPSK carrier phase PLL could lock in phase with true carrier phase, ±90° out of phase with the true carrier phase, or 180° out of phase with the true carrier phase. This PLL possesses what is called a $\pi/2$ *phase ambiguity*.[2] The phase ambiguity is a

---

[2]An S-curve with L stable lock points produces a phase-locked loop with a $2\pi/L$ phase ambiguity.

by-product of removing the data-induced phase shifts from the received signal. The QPSK constellation has a $\pi/2$ rotational symmetry; a $\pi/2$ phase ambiguity is to be expected.

7.2.2 The Maximum Likelihood Phase Error Detector

The heuristic phase error detectors (7.16) and (7.17) were based on a geometric interpretation of the effects of uncompensated carrier phase offset. Although the phase detector is linear (a good property), it requires two four-quadrant arctangent operations (a bad property). The maximum likelihood (ML) phase error detector selects a phase estimate $\hat{\theta}$ that minimizes the energy in the phase error $\theta_e = \theta - \hat{\theta}$. The derivation of the ML phase estimator from basic principles is outlined in Section 7.8.

An alternate derivation of the ML phase error detector starts with the heuristic phase error detectors (7.16) and (7.17) and uses the sine of the phase error rather than the actual phase error. Taking the sine of (7.16) and applying the identity $\sin(A - B) = \sin A \cos B - \cos A \sin B$ produces

$$\sin(\theta_r(k) - \theta_d(k)) = \sin(\theta_r(k)) \cos(\theta_d(k)) - \cos(\theta_r(k)) \sin(\theta_d(k)) \quad (7.23)$$

$$= \frac{y'(kT_s)a_0(k) - x'(kT_s)a_1(k)}{\sqrt{x'^2(kT_s) + y'^2(kT_s)}\sqrt{a_0^2(k) + a_1^2(k)}}. \quad (7.24)$$

To avoid the division suggested by (7.24), the numerator alone can be used as the error signal while the denominator terms are absorbed into the phase detector gain K_p. Thus, the ML phase error for the data-aided case is

$$e(k) = y'(kT_s)a_0(k) - x'(kT_s)a_1(k). \quad (7.25)$$

The decision-directed version of the ML phase error detector is obtained by replacing $a_0(k)$ and $a_1(k)$ in (7.25) with $\hat{a}_0(k)$ and $\hat{a}_1(k)$:

$$e(k) = y'(kT_s)\hat{a}_0(k) - x'(kT_s)\hat{a}_1(k). \quad (7.26)$$

For QPSK, the decisions are

$$\hat{a}_0(k) = A \times \text{sgn}\left\{x'(kT_s)\right\}$$
$$\hat{a}_1(k) = A \times \text{sgn}\left\{y'(kT_s)\right\}. \quad (7.27)$$

A block diagram of a QPSK carrier phase PLL using the ML phase error detector (7.26) is shown in Figure 7.2.4.

The S-curves for the data-aided phase error detector (7.25) and the decision-directed phase error detector (7.26) are obtained using (7.19) to express $x'(kT_s)$ and $y'(kT_s)$ in terms of θ_e, $a_0(k)$, and $a_1(k)$. Substituting for $x'(kT_s)$ and $y'(kT_s)$ in (7.25) produces the S-curve for

Section 7.2 Carrier Phase Synchronization for QPSK

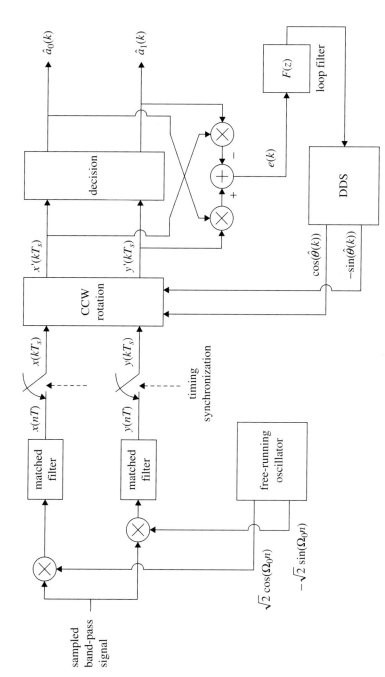

Figure 7.2.4 Block diagram of the QPSK carrier phase PLL using an error signal based on (7.26).

the data-aided phase error detector:

$$g(\theta_e, a_0(k), a_1(k)) = K\Big(a_0(k)\sin(\theta_e) + a_1(k)\cos(\theta_e)\Big)a_0(k)$$
$$- K\Big(a_0(k)\cos(\theta_e) - a_1(k)\sin(\theta_e)\Big)a_1(k) \tag{7.28}$$
$$= K\Big(a_0^2(k) + a_1^2(k)\Big)\sin(\theta_e). \tag{7.29}$$

Averaging over the four possible symbols $(a_0(k), a_1(k)) \in \{\pm A, \pm A\}$ produces

$$\bar{g}(\theta_e) = 2KA^2 \sin\theta_e. \tag{7.30}$$

A plot of (7.30) is illustrated in Figure 7.2.5 (a). Note that for $\theta_e \approx 0$, $\bar{g}(\theta_e) \approx 2KA^2\theta_e$ from which $K_p = 2KA^2$. The average S-curve for the decision-directed phase error detector is

$$\bar{g}(\theta_e) = \begin{cases} -2KA^2 \sin(\theta_e) & -\pi \leq \theta_e < -\dfrac{3\pi}{4} \\ 2KA^2 \cos(\theta_e) & -\dfrac{3\pi}{4} < \theta_e < -\dfrac{\pi}{4} \\ 2KA^2 \sin(\theta_e) & -\dfrac{\pi}{4} < \theta_e < \dfrac{\pi}{4} \\ -2KA^2 \cos(\theta_e) & \dfrac{\pi}{4} < \theta_e < \dfrac{3\pi}{4} \\ -2KA^2 \sin(\theta_e) & \dfrac{3\pi}{4} < \theta_e \leq \pi \end{cases}. \tag{7.31}$$

The S-curve is plotted in Figure 7.2.5 (b). Note, again, that for $\theta_e \approx 0$, $\bar{g}(\theta_e) \approx 2KA^2\theta_e$ from which $K_p = 2KA^2$.

Observe that the gain of both phase error detectors is proportional to KA^2. As a consequence, K_p depends on the amplitude of the received signal. The received signal amplitude must be either estimated or fixed at a known value before proper operation of the PLL can be obtained. If the received signal amplitude K is estimated, the loop filter constants must be recomputed each time K changes. Because this is inconvenient, most systems using a PLL based on (7.26) adjust the amplitude of the received signal to a predetermined value using an *automatic gain control* or AGC subsystem, such as the ones described in Section 9.5. In this case, K and A are fixed at a known value. This, in turn, fixes K_p which allows the loop filter constants to be computed to achieve the desired PLL response.

The S-curve for the data-aided phase error detector crosses zero with a positive slope only at $\theta_e = 0$. Hence, this phase error detector does not posses an ambiguity. The S-curve for the decision-directed phase error detector crosses zero with a positive slope at

$$\theta_e = -\frac{\pi}{2}, 0, \frac{\pi}{2}, \pi$$

and thus has four stable lock points resulting in a $\pi/2$ phase ambiguity.

Section 7.2 Carrier Phase Synchronization for QPSK

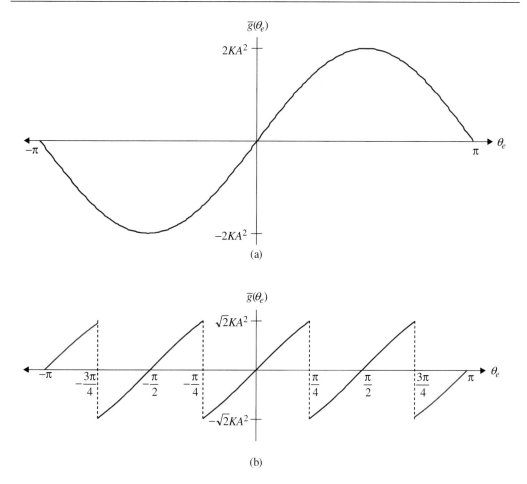

Figure 7.2.5 S-curve for (a) the data-aided phase detector (7.30) and (b) the decision-directed phase detector (7.31).

Sometimes, the decision-directed phase error detector (7.26), uses

$$\hat{a}_0(k) = \text{sgn}\left\{x'(kT_s)\right\}$$
$$\hat{a}_1(k) = \text{sgn}\left\{y'(kT_s)\right\}$$
(7.32)

instead of (7.27). In this case, the S-curve is given by (7.31) except that the constant multiplying $\sin(\theta_e)$ and $\cos(\theta_e)$ is $2KA$. As a consequence $K_p = 2KA$. Other than the scaling, the PLL based on (7.32) is identical to the PLL based on (7.27).

7.2.3 Examples

As an example of loop design using the ML decision-directed phase error detector, suppose system requirements call for a QPSK carrier phase synchronizer PLL with an equivalent loop bandwidth of 2% of the symbol rate and a damping factor equal to $1/\sqrt{2}$. For this and the following example, $KA^2 = 1$ for simplicity. Using a discrete-time proportional-plus-integrator loop filter, $\zeta = 1/\sqrt{2}$ and $B_n T_s = B_n T = 0.02$ together with $K_p = 2$ and $K_0 = 1$ in (C.61), the loop filter constants are

$$K_1 = 2.6 \times 10^{-2} \tag{7.33}$$

$$K_2 = 6.9 \times 10^{-4}. \tag{7.34}$$

The phase estimate $\hat{\theta}(k)$ and phase error $e(k)$ for a sequence of 250 randomly generated QPSK symbols is illustrated in Figure 7.2.6. The carrier phase offset is a $\pi/4$ step. Observe that $\hat{\theta}(k)$ settles to $\pi/4$ after about 200 symbols and that the phase error settles to zero at the same time. The nature of the transient response is controlled by the loop filter constants that are determined by the damping factor and loop bandwidth. This loop is slightly underdamped and exhibits an overshoot in response to the phase step input in the carrier.

Another popular architecture for QPSK carrier phase synchronization is based on the general architecture illustrated in Figure 7.1.1. In this system, the DDS is designed to

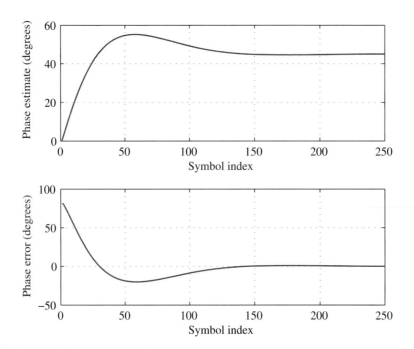

Figure 7.2.6 Phase estimate $\hat{\theta}(k)$ (top) and corresponding phase error $e(k)$ (bottom) for the first QPSK carrier phase PLL example.

operate at the band-pass center frequency Ω_0 rads/sample. Carrier phase compensation is incorporated into the DDS used to generate the quadrature sinusoids used for the downconversion from band-pass to I/Q baseband. As such, there is no need for the phase rotation block. Any of the four phase error detectors (7.16), (7.17), (7.25), and (7.26) described above can be used in the "Compute Phase Error" block where the sampled matched filter outputs $x(kT_s)$ and $y(kT_s)$ are used in place of $x'(kT_s)$ and $y'(kT_s)$. Note that the DDS operates at N samples/symbol while the phase error estimate is updated once per symbol. As a consequence, a rate conversion is required. As an example, consider the block diagram in Figure 7.2.7. In this example, the upsample block is placed in between the phase detector and the loop filter. In this arrangement, the loop filter also functions as an interpolation filter. (See Section 3.2.3 for a description of the role of interpolation filters in an upsampling operation.) As a consequence, the phase detector and loop filter operate at one sample/symbol while the rest of the loop (the DDS and matched filters) operate at N samples/symbol. The filter constants should be computed assuming operation at 1 sample/symbol with a DDS constant $K_0 = N$. This is because, from the loop filter point of view, the DDS increments N times for each step in the loop filter. Because the matched filter is included in the closed-loop path, a small equivalent loop bandwidth is required for stable operation when the loop filter constants are based on a second-order system.

Returning to the design example requiring a QPSK carrier phase PLL with an equivalent loop bandwidth 2% of the symbol rate, assume the band-pass sample rate is $N = 16$ samples/symbol. For the QPSK carrier phase PLL illustrated in Figure 7.2.7, $K_p = 2$ and $K_0 = 16$. Using $\zeta = 1/\sqrt{2}$ and $B_n T_s = 16 B_n T_s = 0.02$ in (C.61), the loop filter constants are

$$K_1 = 1.7 \times 10^{-3} \tag{7.35}$$

$$K_2 = 2.8 \times 10^{-6}. \tag{7.36}$$

7.3 CARRIER PHASE SYNCHRONIZATION FOR BPSK

Either of the two general architectures illustrated in Figures 7.1.1 and 7.1.2 can be used for BPSK carrier phase synchronization. As with QPSK, both approaches use the same phase error detector. In fact, the only difference between carrier phase synchronization for BPSK and QPSK lies in the phase error detector. The architecture of Figure 7.1.2 is examined in this section. Applications to the architecture of Figure 7.1.1 are straightforward.

For BPSK, the received band-pass signal is

$$r(t) = G_a \sum_k a_0(k) p(t - kT_s) \sqrt{2} \cos(\omega_0 t + \theta) + w(t) \tag{7.37}$$

where $a_0(k) \in \{-A, +A\}$ is the k-th data symbol, $p(t)$ is the unit-energy pulse shape with support on $-L_p T_s \leq t \leq L_p T_s$, T_s is the symbol time, ω_0 is the center frequency in

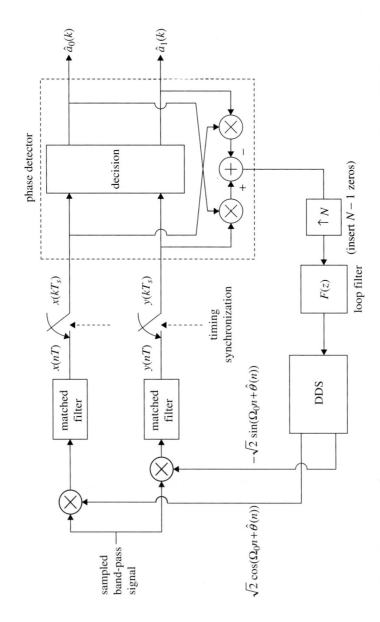

Figure 7.2.7 QPSK carrier phase synchronization based on the general architecture shown in Figure (7.1.1) and using the phase error (7.26).

Section 7.3 Carrier Phase Synchronization for BPSK

radians/second, θ is the unknown carrier phase offset, and $w(t)$ is the additive white Gaussian noise. The constant G_a represents all the amplitude gains and losses through the antennas, propagation medium, amplifiers, mixers, filters, and other RF components. The received signal is sampled at a rate $F_s = 1/T$ samples/s. The n-th sample of the received signal is

$$r(nT) = G_a \sum_k a_0(k) p(nT - kT_s) \sqrt{2} \cos(\Omega_0 n + \theta) + w(nT) \quad (7.38)$$

where $\Omega_0 = \omega_0 T$ radians/sample. Following the analysis of Section 7.1.2, sampled matched filter outputs corresponding to the k-th symbol may be expressed as

$$\begin{aligned} x(kT_s) &= K a_0(k) \cos(\theta) + v_I(kT_s) \\ y(kT_s) &= K a_0(k) \sin(\theta) + v_Q(kT_s) \end{aligned} \quad (7.39)$$

and the de-rotated matched filter outputs as

$$\begin{aligned} x'(kT_s) &= K a_0(k) \cos(\theta - \hat{\theta}(k)) + v_I(kT_s) \\ y'(kT_s) &= K a_0(k) \sin(\theta - \hat{\theta}(k)) + v_Q(kT_s). \end{aligned} \quad (7.40)$$

The BPSK phase error detector based on heuristic reasoning uses the error signal

$$e(k) = \tan^{-1}\left\{\frac{y'(kT_s)}{x'(kT_s)}\right\} - \tan^{-1}\left\{\frac{0}{a_0(k)}\right\} \quad (7.41)$$

for the data-aided phase error detector, and

$$e(k) = \tan^{-1}\left\{\frac{y'(kT_s)}{x'(kT_s)}\right\} - \tan^{-1}\left\{\frac{0}{\hat{a}_0(k)}\right\} \quad (7.42)$$

for the decision-directed phase error detector. The estimate of k-th data symbol is given by

$$\hat{a}_0(k) = A \times \text{sgn}\left\{x'(kT_s)\right\}. \quad (7.43)$$

The numerator of the second term in (7.41) and (7.42) is zero because the BPSK constellation is a 1-dimensional constellation: There is no data on the quadrature carrier component when the receiver is operating in phase coherence with the transmitter. Note that the second term on the right-hand side of (7.42) is 0 when $x'(kT_s) > 0$ and π when $x'(kT_s) < 0$. It should also be observed that (7.41) follows from (7.16) and (7.42) follows from (7.17) when $a_1(k) = 0$. When the architecture of Figure 7.1.1 is used, $x(kT_s)$ and $y(kT_s)$ are used in place of $x'(kT_s)$ and $y'(kT_s)$ in (7.41) and (7.42).

The S-curve for the heuristic phase error detector (7.41) is

$$\bar{g}(\theta_e) = \theta_e \quad (7.44)$$

while the S-curve for the phase detector (7.42) is

$$\bar{g}(\theta_e) = \begin{cases} \theta_e + \pi & -\pi < \theta_e < -\dfrac{\pi}{2} \\ \theta_e & -\dfrac{\pi}{2} < \theta_e < \dfrac{\pi}{2} \\ \theta_e - \pi & \dfrac{\pi}{2} < \theta_e < \pi \end{cases}. \qquad (7.45)$$

The S-curves (7.44) and (7.45) are plotted in Figure 7.3.1 (a) and (b), respectively. Note that the S-curve for the decision-directed phase error detector possesses two stable lock points at $\theta_e = 0$ and $\theta_e = \pi$ and therefore has a π-phase ambiguity equal to the rotational symmetry of the BPSK constellation. Both phase error detectors have unity slope at $\theta_e = 0$. Thus $K_p = 1$.

The maximum likelihood (ML) phase error detectors are

$$e(k) = y'(kT_s)a_0(k) \qquad (7.46)$$

for the data-aided phase detector and

$$e(k) = y'(kT_s)\hat{a}_0(k) \qquad (7.47)$$

for the decision-directed phase detector where the estimate of the k-th symbol is given by (7.43). Note that (7.46) follows from (7.25) and (7.47) follows from (7.26) when $a_1(k) = 0$. An example of the BPSK carrier phase PLL architecture based on the general architecture of Figure 7.1.2 and using the error signal (7.47) is illustrated in Figure 7.3.2. As before, if the architecture of Figure 7.1.1 is used, $x(kT_s)$ and $y(kT_s)$ are used in place of $x'(kT_s)$ and $y'(kT_s)$ in (7.46) and (7.47).

The S-curve for the ML data-aided phase error detector (7.46) is

$$\bar{g}(\theta_e) = KA^2 \sin\theta_e \qquad (7.48)$$

while the S-curve for the ML decision-directed phase error detector (7.47) is

$$\bar{g}(\theta_e) = \begin{cases} -KA^2 \sin\theta_e & -\pi < \theta_e < -\dfrac{\pi}{2} \\ KA^2 \sin\theta_e & -\dfrac{\pi}{2} < \theta_e < \dfrac{\pi}{2} \\ -KA^2 \sin\theta_e & \dfrac{\pi}{2} < \theta_e < \pi \end{cases}. \qquad (7.49)$$

The S-curves (7.48) and (7.49) are plotted in Figure 7.3.3 (a) and (b), respectively. Observe that the S-curve for the ML decision-directed phase detector possesses two stable lock points at $\theta_e = 0$ and $\theta_e = \pi$ and therefore has a π-phase ambiguity. Both S-curves are approximated by $KA^2 \sin\theta_e \approx KA^2\theta_e$ for $\theta_e \approx 0$ and therefore have $K_p = KA^2$. As before,

Section 7.3 Carrier Phase Synchronization for BPSK

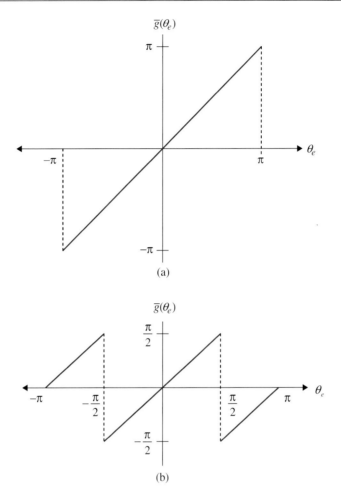

Figure 7.3.1 S-curves for (a) the data-aided BPSK phase detector (7.44) and (b) the decision-directed BPSK phase detector (7.45).

the phase error detector gains are proportional to the amplitude of the received signal. The undesirable characteristic may be addressed by estimating K or using automatic gain control (see Section 9.5) to set the received signal amplitude to a fixed value.

At first, it may seem odd that a carrier phase synchronizer for BPSK requires the quadrature matched filter outputs. The quadrature component is needed to compute the phase rotation at the matched filter outputs. Because there is no information on the quadrature component of the carrier for BPSK, the BPSK carrier phase PLL locks when the residual quadrature component goes to zero. Observe that all of the BPSK phase detectors go to zero when $y'(kT_s)$ is zero.

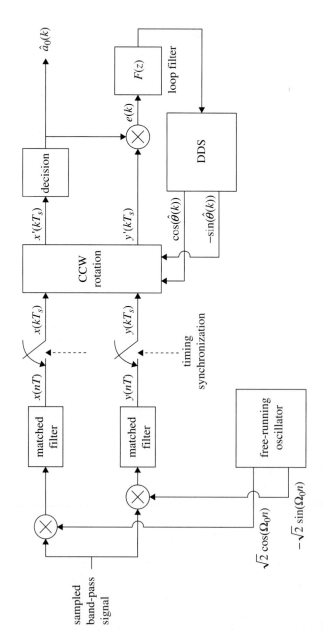

Figure 7.3.2 BPSK carrier phase synchronization system based on the general architecture of Figure (7.1.2) and using the phase error signal (7.47).

Section 7.4 Carrier Phase Synchronization for MQAM

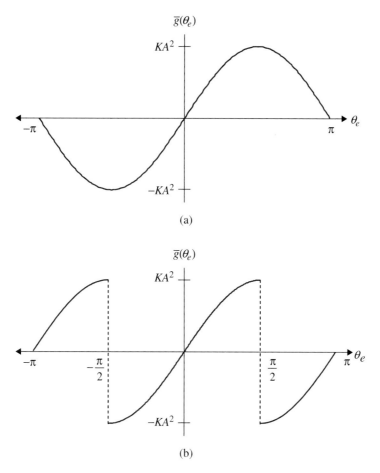

Figure 7.3.3 S-curves for (a) the data-aided BPSK phase detector (7.48) and (b) the decision-directed BPSK phase detector (7.49).

7.4 CARRIER PHASE SYNCHRONIZATION FOR MQAM

Because MQAM is a straightforward generalization of QPSK, the carrier phase PLL for MQAM is also a straightforward generalization of the carrier phase PLL for QPSK. The carrier phase synchronization for the general case of M-ary QAM can be based on either of the architectures illustrated in Figures 7.1.1 and 7.1.2. In fact, the QPSK carrier phase PLL illustrated in Figure 7.2.2 can be used for MQAM in general with appropriate changes to the "decision block." This is demonstrated for the case illustrated in Figure 7.1.2.

The received band-pass signal is given by (7.1) whose samples are (7.2). Following the analysis of Section 7.1.2, sampled matched filter outputs corresponding to the k-th symbol may be expressed as

$$x(kT_s) = K\left[a_0(k)\cos(\theta) - a_1(k)\sin(\theta)\right] + v_I(kT_s)$$
$$y(kT_s) = K\left[a_0(k)\sin(\theta) + a_1(k)\cos(\theta)\right] + v_Q(kT_s) \quad (7.50)$$

and the de-rotated matched filter outputs as

$$x'(kT_s) = K\left[a_0(k)\cos(\theta - \hat{\theta}(k)) - a_1(k)\sin(\theta - \hat{\theta}(k))\right] + v_I(kT_s)$$
$$y'(kT_s) = K\left[a_0(k)\sin(\theta - \hat{\theta}(k)) + a_1(k)\cos(\theta - \hat{\theta}(k))\right] + v_Q(kT_s). \quad (7.51)$$

The data-aided and decision-directed phase error detectors based on the heuristic arguments are

$$e(k) = \tan^{-1}\left\{\frac{y'(kT_s)}{x'(kT_s)}\right\} - \tan^{-1}\left\{\frac{a_1(k)}{a_0(k)}\right\} \quad (7.52)$$

$$e(k) = \tan^{-1}\left\{\frac{y'(kT_s)}{x'(kT_s)}\right\} - \tan^{-1}\left\{\frac{\hat{a}_1(k)}{\hat{a}_0(k)}\right\}, \quad (7.53)$$

respectively. The ML data-aided phase error detector is

$$e(k) = y'(kT_s)a_0(k) - x'(kT_s)a_1(k). \quad (7.54)$$

The ML decision-directed phase error detector is obtained by replacing $a_0(k)$ and $a_1(k)$ in (7.54) with the decisions $\hat{a}_0(k)$ and $\hat{a}_1(k)$:

$$e(k) = y'(kT_s)\hat{a}_0(k) - x'(kT_s)\hat{a}_1(k). \quad (7.55)$$

Note that when the architecture of Figure 7.1.1 is used, $x(kT_s)$ and $y(kT_s)$ are used in place of $x'(kT_s)$ and $y'(kT_s)$ in the phase error detector.

Even though the block diagrams of the MQAM carrier phase PLLs are identical to the QPSK carrier phase PLLs, the properties of S-curves are strongly dependent on the constellation. For example, consider the S-curves for the 8-PSK, square 16-QAM, and CCITT V.29 16-QAM constellations using the phase detector (7.55) plotted in Figures 7.4.1–7.4.3, respectively. Note that although the S-curves are different, they do have a few features in common. First, each S-curve crosses zero at $\theta_e = 0$ with a positive slope thereby indicating that $\theta_e = 0$ is a stable lock point for the PLL. Second, each S-curve is approximately linear for $\theta_e \approx 0$. Finally, each S-curve possess multiple stable lock points. The number of stable lock points and the values of θ_e where they occur is determined by the rotational symmetry of the constellation.

7.5 CARRIER PHASE SYNCHRONIZATION FOR OFFSET QPSK

Either of the carrier phase PLL basic structures introduced in Figures 7.1.1 and 7.1.2 may be applied to carrier phase synchronization using offset QPSK. The application requires one

Section 7.5 Carrier Phase Synchronization for Offset QPSK

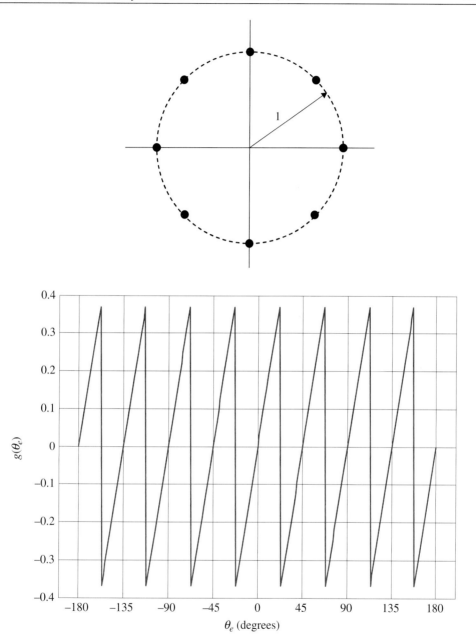

Figure 7.4.1 S-curve for the square 8-PSK constellation using the phase detector (7.55).

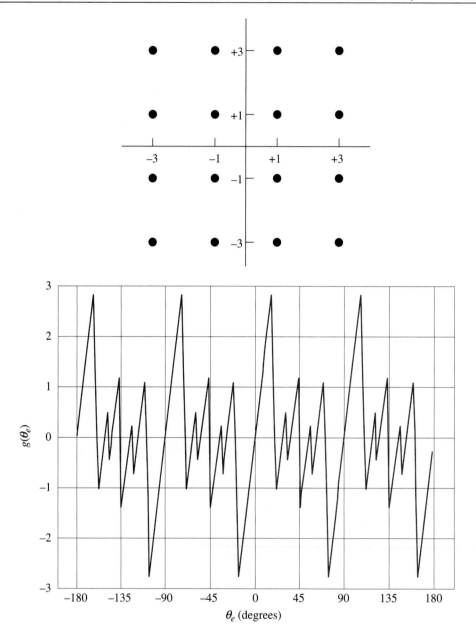

Figure 7.4.2 S-curve for the square 16-QAM constellation using the phase detector (7.55).

Section 7.5 Carrier Phase Synchronization for Offset QPSK

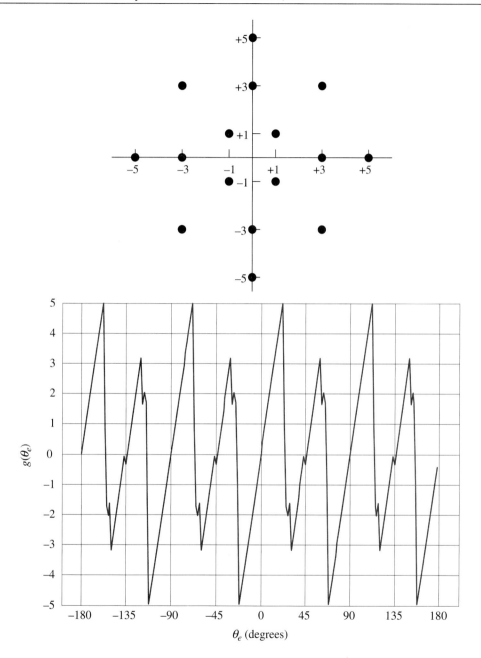

Figure 7.4.3 S-curve for the CCITT V.29 16-QAM constellation using the phase detector (7.55).

important modification: The matched filters are sampled at 2 samples/symbol because the quadrature component of the transmitted signal is delayed by one-half a symbol period.

The offset QPSK carrier phase PLL counterpart to Figure 7.1.2 is illustrated in Figure 7.5.1. Samples of the bandlimited band-pass signal are downconverted using a free running oscillator. The resulting inphase and quadrature signals, $I(nT)$ and $Q(nT)$, respectively, are filtered by matched filters. The inphase and quadrature matched filter outputs, $x(nT)$ and $y(nT)$, respectively, are sampled at 2 samples/symbol with perfect timing synchronization. For convenience, the samples are indexed by the symbol index k. The optimum sampling instants for the inphase component are kT_s whereas the optimum sampling instants for the quadrature component are $kT_s + T_s/2$ for $k = 0, 1, \ldots$. The samples $\ldots, x(kT_s), x(kT_s + T_s/2), \ldots$ and $\ldots, y(kT_s), y(kT_s + T_s/2), \ldots$ are rotated by an angle $-\hat{\theta}(k)$ to produce $\ldots, x'(kT_s), x'(kT_s + T_s/2), \ldots$ and $\ldots, y'(kT_s), y'(kT_s + T_s/2), \ldots$. Using the index notation just described, $x'(kT_s)$ and $y'(kT_s + T_s/2)$ are used for detection. As shown below, $x'(kT_s + T_s/2)$ and $y'(kT_s)$ are used to compute the carrier phase error signal.

Let the samples of the received band-pass signal be given by

$$r(nT) = G_a \sum_m a_0(m) p(nT - mT_s) \cos(\Omega_0 n + \theta)$$
$$- G_a \sum_m a_1(m) p(nT - mT_s - T_s/2) \sin(\Omega_0 n + \theta) + w(nT) \quad (7.56)$$

where $1/T$ is the sample rate, $a_0(m) \in \{-A, +A\}$ and $a_1(m) \in \{-A, +A\}$ are the information symbols, $p(nT)$ is a unit-energy pulse shape with support on the interval $-L_p T_s/T < n < L_p T_s/T$, Ω_0 is the center frequency in radians/sample, θ is the unknown carrier phase offset, and G_a represents the cumulative effect of the gains and losses from the transmitter through the propagation medium and through the receiver up to the ADC. Neglecting noise, the matched filter outputs may be expressed as

$$x(nT) = K \sum_m \left\{ a_0(m) r_p(nT - mT_s) \cos(\theta) - a_1(m) r_p(nT - mT_s - T_s/2) \sin(\theta) \right\} \quad (7.57)$$

$$y(nT) = K \sum_m \left\{ a_0(m) r_p(nT - mT_s) \sin(\theta) + a_1(m) r_p(nT - mT_s - T_s/2) \cos(\theta) \right\} \quad (7.58)$$

where $r_p(u)$ is the autocorrelation function of the pulse shape given by (7.5) and $K = G_a/T$. After some algebra and the application of basic trigonometric identities, the rotated matched filter outputs, sampled at $n = kT_s/T$ and $n = (k + 1/2)T_s/T$, can be shown to be

$$x'(kT_s) = K a_0(k) \cos(\theta - \hat{\theta}(k)) - K \sum_{m \neq k} a_1(m) r_p((k - m - 1/2)T_s) \sin(\theta - \hat{\theta}(k)) \quad (7.59)$$

$$y'(kT_s) = K a_0(k) \sin(\theta - \hat{\theta}(k)) - K \sum_{m \neq k} a_1(m) r_p((k - m - 1/2)T_s) \cos(\theta - \hat{\theta}(k)) \quad (7.60)$$

Section 7.5 Carrier Phase Synchronization for Offset QPSK

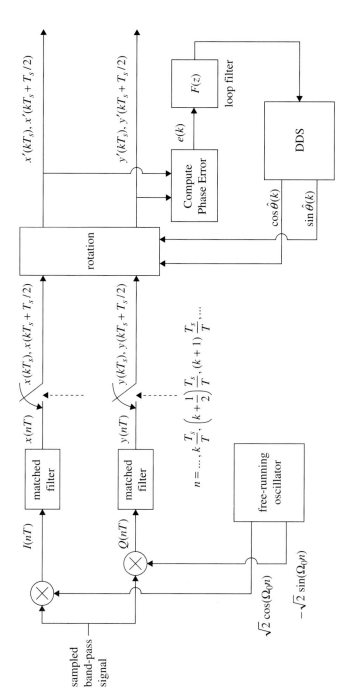

Figure 7.5.1 Carrier phase synchronization PLL for offset QPSK based on a free-running quadrature LO and a phase rotation. Compare with the carrier phase synchronization system for non-offset MQAM illustrated in Figure 7.1.2.

$$x'((k+1/2)T_s) = -Ka_1(k)\sin(\theta - \hat{\theta}(k)) + K\sum_{m\neq k} a_1(m)r_p((k-m+1/2)T_s)\cos(\theta - \hat{\theta}(k))$$
(7.61)

$$y'((k+1/2)T_s) = Ka_1(k)\cos(\theta - \hat{\theta}(k)) + K\sum_{m\neq k} a_1(m)r_p((k-m+1/2)T_s)\sin(\theta - \hat{\theta}(k))$$
(7.62)

where $r_p(0) = 1$ and $r_p((k-m)T_s) = 0$ for $m \neq k$ is assumed. The terms $y'(kT_s)$ and $x'(kT_s + T_s/2)$ contain the product of a single symbol and the sine of the phase error $\theta - \hat{\theta}(k)$. Knowledge of the symbol or the symbol estimate can by used to provide the correct sign to the sine of the phase error. The data-aided phase error is thus

$$e(k) = a_0(k)y'(kT_s) - a_1(k)x'((k+1/2)T_s).$$
(7.63)

The decision-directed phase error is obtained from (7.63) by substituting data symbol estimates $\hat{a}_0(k)$ and $\hat{a}_1(k)$ for $a_0(k)$ and $a_1(k)$, respectively:

$$e(k) = \hat{a}_0(k)y'(kT_s) - \hat{a}_1(k)x'((k+1/2)T_s).$$
(7.64)

The data symbol estimates are

$$\hat{a}_0(k) = A \times \text{sgn}\left\{x'(kT_s)\right\}$$
$$\hat{a}_1(k) = A \times \text{sgn}\left\{y'((k+1/2)T_s)\right\}.$$
(7.65)

The S-curve for the data-aided phase error detector may be computed by substituting (7.60) and (7.61) for $y'(kT_s)$ and $x'((k+1/2)T_s)$, respectively, in (7.63). Using $\theta_e = \theta - \hat{\theta}$, $a_0^2(k) = A^2$, and $a_1^2(k) = A^2$, the S-curve is

$$g(\theta_e) = 2KA^2 \sin\theta_e + 2KA^2 \sum_{m\neq k} [a_0(k)a_1(m) - a_1(k)a_0(m)]$$
$$\times r_p((k-m)T_s + T_s/2)\cos(\theta_e).$$
(7.66)

The S-curve is thus the familiar sine of the phase error plus a second term that represents the self-noise of this phase error detector. The self-noise has an average value of zero if $a_0(m)$ and $a_1(m)$ are uncorrelated. The S-curve for both the data-aided error signal (7.63) and the decision-directed error signal (7.64) are plotted in Figure 7.5.2 for the case $KA^2 = 1$. Note that the decision-directed error detector does not have a stable lock point at $\theta_e = \pm\pi/2$ as its non-offset counterpart did. Because stable lock points are those where the S-curve crosses zero with positive slope, the data-aided phase error detector has one stable lock point at $\theta_e = 0$ whereas the decision-directed phase error detector has two stable lock points (at $\theta_e = 0$ and $\theta_e = \pi$) and hence a π-phase ambiguity.

Section 7.5 Carrier Phase Synchronization for Offset QPSK

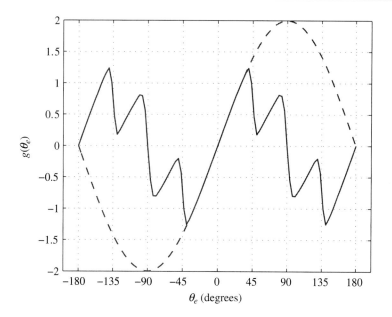

Figure 7.5.2 S-curves for the data-aided OQPSK phase error detector (dashed line) and the decision-directed OQPSK phase error detector (solid line) for $E_b/N_0 = 20$ dB.

Carrier phase synchronization for offset QPSK can also be accomplished using a tunable DDS at band pass as illustrated in Figure 7.5.3. Using the same notation as before, samples of the inphase and quadrature matched filter outputs may be expressed as (neglecting noise)

$$x(nT) = K \sum_m a_0(m) r_p(nT - mT_s) \cos(\theta - \hat{\theta}(k))$$

$$- K \sum_m a_1(m) r_p(nT - mT_s - T_s/2) \sin(\theta - \hat{\theta}(k)) \quad (7.67)$$

$$y(nT) = K \sum_m a_0(m) r_p(nT - mT_s) \sin(\theta - \hat{\theta}(k))$$

$$+ K \sum_m a_1(m) r_p(nT - mT_s - T_s/2) \cos(\theta - \hat{\theta}(k)). \quad (7.68)$$

Sampling the matched filter outputs at $n = kT_s/T$ and $n = (k+1/2)T_s/T$ produces

$$x(kT_s) = K a_0(k) \cos(\theta - \hat{\theta}(k)) - K \sum_m a_1(m) r_p((k - m - 1/2)T_s) \sin(\theta - \hat{\theta}(k))$$

$$(7.69)$$

$$y(kT_s) = K a_0(k) \sin(\theta - \hat{\theta}(k)) + K \sum_m a_1(m) r_p((k - m - 1/2)T_s) \cos(\theta - \hat{\theta}(k))$$

$$(7.70)$$

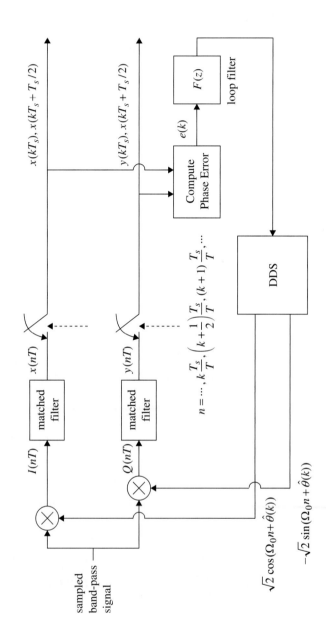

Figure 7.5.3 Carrier phase synchronization PLL for offset QPSK based on a band-pass DDS. Compare with the carrier phase synchronization system for non-offset MQAM illustrated in Figure 7.1.1.

$$x((k+1/2)T_s) = -Ka_1(k)\sin(\theta - \hat{\theta}(k)) + K\sum_m a_1(m)r_p((k-m+1/2)T_s)\cos(\theta - \hat{\theta}(k))$$
(7.71)

$$y((k+1/2)T_s) = Ka_1(k)\cos(\theta - \hat{\theta}(k)) + K\sum_m a_1(m)r_p((k-m+1/2)T_s)\sin(\theta - \hat{\theta}(k))$$
(7.72)

which are identical to (7.59)–(7.62) except that phase compensation is performed at bandpass through the DDS instead of after the matched filters. Using the same line of reasoning as before, the data-aided error signal is

$$e(k) = a_0(k)y(kT_s) - a_1(k)x((k+1/2)T_s) \quad (7.73)$$

and the decision-directed phase error is

$$e(k) = \hat{a}_0(k)y(kT_s) - \hat{a}_1(k)x((k+1/2)T_s) \quad (7.74)$$

where the data decisions are given by (7.65).

7.6 CARRIER PHASE SYNCHRONIZATION FOR BPSK AND QPSK USING CONTINUOUS-TIME TECHNIQUES

Carrier phase PLLs using continuous-time processing almost always use the architecture illustrated in Figure 7.1.1 due to the difficulty of constructing a baseband VCO required by the architecture of Figure 7.1.2. Figure 7.6.1 illustrates a hybrid architecture using both continuous-time and discrete-time processing for QPSK carrier phase synchronization. The phase detector is a discrete-time processor that updates the carrier phase offset once per symbol using the sequence of matched filter outputs. Any of the data-aided or decision-directed phase error signals for QPSK or BPSK introduced above can be used here. The sequence of phase errors is converted to a continuous-time signal by the digital-to-analog converter. The converted signal forms the input to the continuous-time loop filter $F(s)$ that drives a continuous-time VCO. The sinusoidal output of the VCO is split into quadrature sinusoids using a phase shifter (usually a delay element). The quadrature sinusoids are used to downconvert the band-pass signal and to separate the band-pass signal into its quadrature components. The quadrature components are matched filtered and sampled to produce the signal space projection used by the phase detector to update the phase error.

A purely "analog" solution to QPSK carrier phase synchronization is illustrated in Figure 7.6.2. This structure is called a *Costas loop*. The phase error computation involves the difference between the cross products of the baseband inphase and quadrature signals and their signs. This structure—which results from a recursive solution to maximum likelihood phase estimation as shown in Section 7.8—is reminiscent of the ML decision-directed phase error given by (7.26) and illustrated in Figure 7.2.4. (Note that the sign on the error signal is switched in Figures 7.2.4 and 7.6.2. This is due to the use of $-\sin(\cdot)$ for the quadrature

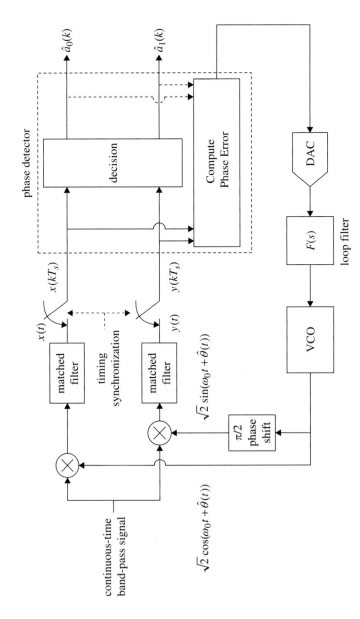

Figure 7.6.1 A hybrid continuous-time/discrete-time carrier phase PLL for QPSK.

Section 7.6 Carrier Phase Synchronization for BPSK and QPSK

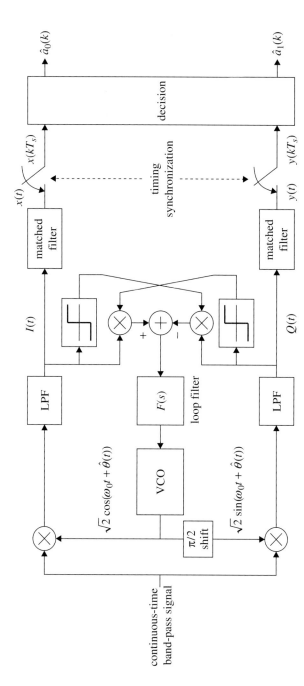

Figure 7.6.2 A QPSK carrier phase PLL based on the Costas loop.

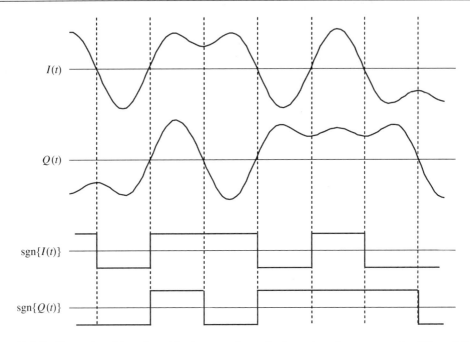

Figure 7.6.3 Example of baseband inphase and quadrature signal components, $I(t)$ and $Q(t)$, respectively, and their signs used by the QPSK Costas loop shown in 7.6.2.

component in Figure 7.2.4 and $\sin(\cdot)$ for the quadrature component in Figure 7.6.2.) To see how this works, consider the plot of the baseband inphase and quadrature components plotted in Figure 7.6.3. During each symbol interval, the signs of the baseband inphase and quadrature components could be used as decisions from which the sine of the phase error can be continuously updated.

The Costas loop for BPSK carrier phase synchronization is illustrated in Figure 7.6.4. Note that this structure is reminiscent of the discrete-time BPSK carrier phase PLL based on the phase error given by (7.47) where the sign of the baseband inphase component plays the role of the decision. As with QPSK, a sign reversal (not shown) is present due to the use of $-\sin(\cdot)$ in discrete-time carrier phase PLLs using a quadrature DDS and $\sin(\cdot)$ in Figure 7.6.4.

7.7 PHASE AMBIGUITY RESOLUTION

Decision-directed carrier phase synchronizers for MPSK possess a *phase ambiguity*. As a consequence, the carrier phase PLL can lock to the unmodulated carrier with a phase offset. The phase ambiguity is due to the rotational symmetry of the constellation. There are two commonly used methods to resolve this phase ambiguity: the unique word method and differential encoding.

Section 7.7 Phase Ambiguity Resolution

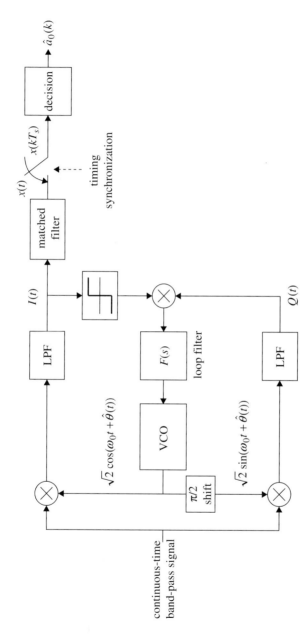

Figure 7.6.4 A BPSK carrier phase PLL based on the Costas loop.

7.7.1 Unique Word

One of the ways to resolve phase ambiguity is to insert a pattern of known symbols (or "unique word"—UW) in the data stream. After carrier phase lock, the detector operates on the decisions $\hat{a}_0(k)$ and $\hat{a}_1(k)$ searching for the possible rotations of the UW due to phase ambiguity. When the UW (or its rotated version) is found, the phase ambiguity is known and can be resolved. This method is best illustrated by an example.

UW with BPSK. The decision-directed carrier phase PLL for BPSK exhibits a 180° phase ambiguity. As a consequence, the PLL can lock with in phase with the carrier or 180° out of phase with the carrier. Thus, the phase ambiguity resolution process must search for two versions of the UW: UW and −UW as illustrated in Figure 7.7.1. If UW is found, the carrier phase PLL has locked in phase with the carrier and the decisions $\hat{a}_0(k)$ are the correct ones. If −UW is found then the carrier phase PLL has locked 180° out of phase with the carrier and the decisions are really the negatives of what they should be. In this case, the phase ambiguity correction can either change the signs on $\hat{a}_0(k)$ or use a different symbol-to-bit map.

As an example, suppose UW = +1 −1 +1 +1 and that the bit-to-symbol map is indicated on the constellation below.

```
      bit = 0            bit = 1
         ●                  ●
      ─────────────┼─────────────
       a₀ = −1          a₁ = +1
```

Suppose the symbol estimates are

k	0	1	2	3	4	5	6	7
$\hat{a}_0(k)$	+1	−1	+1	+1	−1	+1	−1	−1

where the first four symbol estimates correspond to the unique word. After the forth symbol is clocked in, the output of the "find UW" block in Figure 7.7.1 is asserted (the "find−UW"

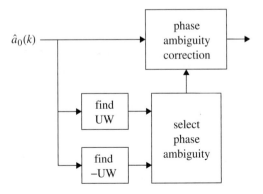

Figure 7.7.1 A block diagram illustrating the use of the unique word for phase ambiguity resolution for BPSK.

Section 7.7 Phase Ambiguity Resolution

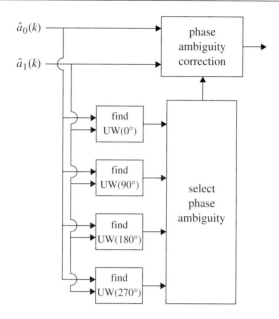

Figure 7.7.2 A block diagram illustrating the use of the unique word for phase ambiguity resolution for QPSK.

block output is not asserted). This indicates that the phase ambiguity is 0° and the output bit sequence is 0 1 0 0. Now suppose the symbol estimates are

k	0	1	2	3	4	5	6	7
$\hat{a}_0(k)$	−1	+1	−1	−1	+1	−1	+1	−1

After the forth symbol is clocked in, the output of the "find −UW" block is asserted. This indicates that the phase ambiguity is 180° so that the output bits should be 0 1 0 1.

UW with QPSK. The decision-directed carrier phase PLL for BPSK exhibits a 90° phase ambiguity. As a consequence, the PLL can lock in phase with the carrier, 90° out of phase with the carrier, 180° out of phase with the carrier, or 270° out of phase with the carrier. Thus, the phase ambiguity resolution process must search for four versions of the UW which, for convenience, are denoted UW(0°), UW(90°), UW(180°), and UW(270°), as illustrated in Figure 7.7.2. If UW(0°) is found, carrier phase PLL has locked in phase with the carrier and the decisions $\hat{a}_0(k)$ and $\hat{a}_1(k)$ are the correct ones. If UW(90°) is found, then the carrier phase PLL has locked 90° out of phase with the carrier and a correction must be made. This correction can be made by rotating the estimates $\hat{a}_0(k)$ and $\hat{a}_1(k)$ and using the original bit-to-symbol map, or by simply using a different bit-to-symbol map. The process proceeds in a similar way if UW(180°) or UW(270°) are found.

Consider the following example where the bit-to-symbol map is

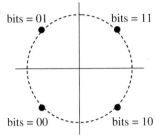

and the unique word is[3] 1 0 1 1 0 1 0 0. Suppose the symbol estimates are

k	0	1	2	3	4	5	6	7
$\hat{a}_0(k)$	−1	+1	+1	−1	+1	+1	−1	−1
$\hat{a}_1(k)$	−1	−1	+1	+1	−1	+1	+1	−1

where the first four symbols correspond to the UW. When this symbol estimate sequence is presented to the system of Figure 7.7.2, the "find UW(270°)" output is asserted at $k=3$. This indicates that the carrier phase PLL has locked 270° out of phase with the carrier. The output bit sequence could be determined in one of two ways: The remaining symbol estimates could be rotated by 90° and the bit-to-symbol map illustrated above applied, or the bit-to-symbol map could be altered so that the bit patterns above apply to the constellation points 90° clockwise. Either way, a compensation for the phase ambiguity can be used to produce the correct output bit sequence 1 1 0 1 0 0 1 0.

In both examples, the "find UW" function was applied to the symbol estimates $\hat{a}_0(k)$ (and $\hat{a}_1(k)$ in the case of QPSK). The "find UW" function could also be applied to the corresponding bit decision in a very straightforward way. Using either method performs the equivalent task and produces the same result.

7.7.2 Differential Encoding

A second method used to overcome the phase ambiguity is the use of differential encoding. The MPSK modulation presented in Chapter 5 mapped the data to the phase of the carrier. Phase shifts in the carrier were incidental to the changes in the data. In differential encoding, the data are mapped to the *phase shifts* of the modulated carrier. Because phase shifts (or transitions) are preserved when the constellation is rotated by one of the phase ambiguities, it is still possible to recover the data.

To understand differential encoding when applied to MPSK, a slightly different representation of the modulated waveform must be used. The starting point is the traditional representation of an MPSK modulated carrier:

$$s(t) = \sum_k a_0(k)p(t-kT_s)\sqrt{2}\cos(\omega_0 t) - a_1(k)p(t-kT_s)\sqrt{2}\sin(\omega_0 t) \qquad (7.75)$$

---

[3]Just to illustrate the variations available, the UW is defined using an 8-bit bit pattern. The equivalent symbol pattern is $(+1,-1),(+1,+1),(-1,+1),(-1,-1)$.

where $a_0(k)$ and $a_1(k)$ represent the k-th constellation point, T_s is the symbol time, and $p(t)$ is a unit-energy pulse shape. Using the relationship

$$X \cos(\phi) + Y \sin(\phi) = \sqrt{X^2 + Y^2} \cos\left(\phi - \tan^{-1}\left\{\frac{Y}{X}\right\}\right), \qquad (7.76)$$

the MPSK modulated carrier may also be expressed as

$$s(t) = \sum_k A p(t - kT_s) \cos(\omega_0 t + \theta_k) \qquad (7.77)$$

where

$$A = \sqrt{2}\sqrt{a_0^2(k) + a_1^2(k)} \qquad (7.78)$$

$$\theta_k = \tan^{-1}\left\{\frac{a_1(k)}{a_0(k)}\right\}. \qquad (7.79)$$

The interpretation of (7.77) is that the data symbols $a_0(k)$ and $a_1(k)$ select the phase of the carrier. For BPSK, the two possibilities for θ_k are 0 and π. For QPSK, the four possibilities for θ_k are $\pi/4, 3\pi/4, 5\pi/4$, and $7\pi/4$.

When differential encoding is used, the phase *shift* is determined by the data. For example, in BPSK, there are two possible phase shifts: 0 and π. In QPSK, there are four possible phase shifts: $0, \pi/2, \pi$, and $3\pi/2$. If the shift corresponding to the k-th data symbol is $\Delta\theta_k$, then the carrier phase corresponding to the k-th data symbol is

$$\theta_k = \theta_{k-1} + \Delta\theta_k. \qquad (7.80)$$

Thus, the MPSK modulated carrier may be expressed as

$$s(t) = \sum_k A p(t - kT_s) \cos\left(\omega_0 t + \theta_{k-1} + \Delta\theta_k\right). \qquad (7.81)$$

The most common way to do differential encoding is to encode the bits prior to the bit-to-waveform mapping and use the encoded bits to perform this mapping. This process usually involves Boolean operations on the bits.

Let $b_0, b_1, \ldots$ be the input bit sequence. This bit sequence is encoded to produce a sequence $\delta_0, \delta_1, \ldots$. The bits are segmented into nonoverlapping blocks of $L = \log_2 M$ bits. The bits in the k-th block are $b_{Lk}, b_{Lk+1}, \ldots, b_{Lk+L-1}$ for $k = 0, 1, \ldots$. Each block of input bits produces a block of encoded bits. The k-th encoded bits $\delta_{Lk}, \delta_{Lk+1}, \ldots, \delta_{Lk+L-1}$ is a function of the k-th block of input bits and the $(k-1)$-th block of encoded bits. The k-th block of encoded bits is used to choose the k-th symbol from the look-up tables. The k-th symbol is used to form the modulated waveform. This process is illustrated in Figure 7.7.3.

A traditional MPSK demodulator/detector can be used as illustrated in Figure 7.7.4. The k-th block of bits output from the decision block are estimates of the k-th block of encoded bits, denoted $\hat{\delta}_{Lk}, \hat{\delta}_{Lk+1}, \ldots, \hat{\delta}_{Lk+L-1}$ as shown. The differential decoder operates on the current block of $\log_2(M)$ coded bit estimates and the previous block of $\log_2(M)$ coded bit estimates.

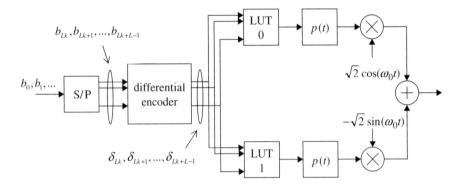

Figure 7.7.3 A block diagram of an MPSK modulator with a differential encoder. Note that the input bits b_k are encoded to produce differentially encoded bits δ_k. The differentially encoded bits are used to select the constellation points from the look-up tables.

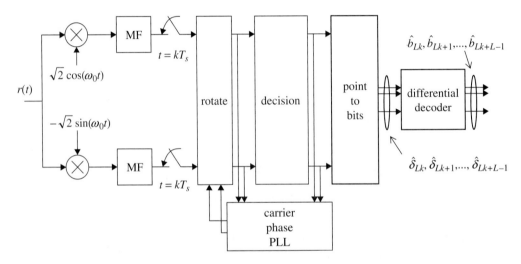

Figure 7.7.4 A block diagram of an MPSK detector with a differential decoder. Note that the detector decisions are estimates of the differentially encoded bits $\hat{\delta}_k$. The estimates of these encoded bits are used by the differential decoder to produce estimates of the original bits.

BPSK Example. As an example, consider BPSK with the following specifications:

$\delta = 0$: $-A$ $\delta = 1$: $+A$

b_k	$\Delta\theta_k$
0	π
1	0

The differential encoder is a function of the encoded bit from the previous interval and the current input bit. The encoded bit from the previous interval is required because the input bit defines a phase shift (i.e., the shift from the previous phase). The new phase is thus a function

Section 7.7 Phase Ambiguity Resolution

of the previous phase. Using the definitions of input bits b_k and differentially encoded bits δ_k, the following truth table may be constructed:

b_k	δ_{k-1}	δ_k
0	0	1
0	1	0
1	0	0
1	1	1

In the first two rows, δ_k is the complement of δ_{k-1} because the input is 0, which forces a phase shift of π. In the second two rows, δ_k and δ_{k-1} are the same because the input bit is 1, which forces a phase shift of 0. A block diagram of the differential encoder is illustrated in Figure 7.7.5. The differential encoder may be implemented as a four-row look-up table or as the Boolean "exclusive NOR" function as shown.

An example of encoding the bit sequence 1 0 1 1 0 1 0 0 is shown below. Note that the assumption $\delta_{-1} = 0$ was used.

k	0	1	2	3	4	5	6	7
b_k	1	0	1	1	0	1	0	0
δ_k	0	1	1	1	0	0	1	0

The differential decoder is placed after the decision block. The differential decoding rule is derived from the definitions of the relationship between input bits and the phase shift and the relationship between the encoded bits and the constellation points. For this case, the differential decoding rule is

$\hat{\delta}_{k-1}$	$\hat{\delta}_k$	$\hat{b}_k$
0	0	1
0	1	0
1	0	0
1	1	1

A block diagram of the differential decoder is illustrated in Figure 7.7.6. Note that the decoder can be thought of as a look-up table or the equivalent Boolean function (which, in this case, is the exclusive NOR function).

Applying the decoding rule to the previously encoded bit sequence reveals some interesting properties. First consider the case where the carrier phase ambiguity is 0. In this case, $\hat{\delta}_k = \delta_k$ as shown below. Applying the differential decoding rule to $\hat{\delta}_k$ produces the

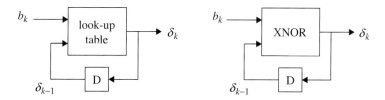

Figure 7.7.5 Two equivalent versions of the differential encoder for the BPSK example.

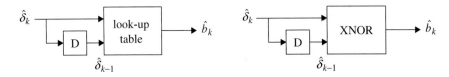

Figure 7.7.6 Two equivalent versions of the differential decoder for BPSK.

bit sequence $\hat{b}_k$ as shown. (Note that the assumption $\hat{\delta}_{-1} = 0$ was used.) The bit sequence is identical to the original bit sequence.

k	0	1	2	3	4	5	6	7
b_k	1	0	1	1	0	1	0	0
δ_k	0	1	1	1	0	0	1	0
$\hat{\delta}_k$	0	1	1	1	0	0	1	0
$\hat{b}_k$	1	0	1	1	0	1	0	0

Now consider the case where the carrier phase ambiguity is π. In the case, $\hat{\delta}_k$ is the complement of δ_k as shown below. Applying the differential decoding rule to $\hat{\delta}_k$ produces the bit sequence shown in the fourth row. Note that the assumption $\hat{\delta}_{-1} = 0$ was used. A consequence of this assumption is that the first bit is wrong. All the rest of the bits, however, are correct.

k	0	1	2	3	4	5	6	7
b_k	1	0	1	1	0	1	0	0
δ_k	0	1	1	1	0	0	1	0
$\hat{\delta}_k$	1	0	0	0	1	1	0	1
$\hat{b}_k$	0	0	1	1	0	1	0	0

This demonstrates that the π phase ambiguity for BPSK is transparent to the differential encoding and decoding operations. There is the possibility, however, that the first bit may be decoded in error. This is usually not a problem in real systems because the frame synchronization is not acquired until after carrier and symbol timing synchronization have been achieved. In this case, the erroneous first bit is but a distant memory to the system.

QPSK Example. Consider QPSK with the following specifications:

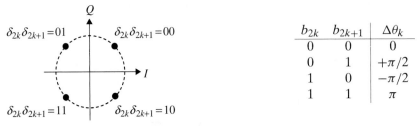

b_{2k}	b_{2k+1}	$\Delta\theta_k$
0	0	0
0	1	$+\pi/2$
1	0	$-\pi/2$
1	1	π

The differential encoder processes blocks of two bits at a time, b_{2k} and b_{2k+1} for $k = 0, 1, \ldots$ and produces two encoded bits, δ_{2k} and δ_{2k+1}. Because the encoding process is differential,

Section 7.7 Phase Ambiguity Resolution

the encoder also needs to know the previous two encoded bits δ_{2k-2} and δ_{2k-1}. Using the definitions relating the input bits to phase shifts and the relationship between encoded bits and constellation points, the following truth table may be constructed:

b_{2k}	b_{2k+1}	δ_{2k-2}	δ_{2k-1}	δ_{2k}	δ_{2k+1}
0	0	0	0	0	0
0	0	0	1	0	1
0	0	1	0	1	0
0	0	1	1	1	1
0	1	0	0	0	1
0	1	0	1	1	1
0	1	1	0	0	0
0	1	1	1	1	0
1	0	0	0	1	0
1	0	0	1	0	0
1	0	1	0	1	1
1	0	1	1	0	1
1	1	0	0	1	1
1	1	0	1	1	0
1	1	1	0	0	1
1	1	1	1	0	0

A block diagram showing the differential encoder for this example is illustrated in Figure 7.7.7. Note that the encoding function may be implemented by either a look-up table or by combinational logic.

As an example, consider the bit sequence 1 0 1 1 0 1 0 0 1 1 1 0 1 0 0 1. The bit sequence is organized into two-bit blocks as shown in the table below. Applying the encoding rule defined above (assuming $\delta_{-2} = 0$ and $\delta_{-1} = 0$) to the input bit sequences in the second and third columns of the table produces the differentially encoded bit sequence in the fourth and fifth columns.

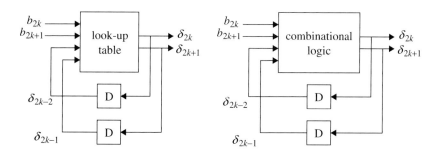

Figure 7.7.7 Two equivalent versions of the differential encoder for the QPSK example.

k	b_{2k}	b_{2k+1}	δ_{2k}	δ_{2k+1}
0	1	0	1	0
1	1	1	0	1
2	0	1	1	1
3	0	0	1	1
4	1	1	0	0
5	1	0	1	0
6	1	0	1	1
7	0	1	1	0

The differential decoder is placed after the decision block in the detector as illustrated in Figure 7.7.4. The differential decoding rule is derived from the definitions of the relationship between input bits and the phase shift and the relationship between the encoded bits and the constellation points. For this case, the differential decoding rule is shown in the table below. Note that the fifth column lists, as an intermediate result, the phase shift corresponding to the pairs of encoded bits. This is included to help you see how the table was constructed.

$\hat{\delta}_{2k-2}$	$\hat{\delta}_{2k-1}$	$\hat{\delta}_{2k}$	$\hat{\delta}_{2k+1}$	$\Delta\theta_k$	$\hat{b}_{2k}$	$\hat{b}_{2k+1}$
0	0	0	0	0	0	0
0	0	0	1	$+\pi/2$	0	1
0	0	1	0	$-\pi/2$	1	0
0	0	1	1	π	1	1
0	1	0	0	$-\pi/2$	1	0
0	1	0	1	0	0	0
0	1	1	0	π	1	1
0	1	1	1	$+\pi/2$	0	1
1	0	0	0	$+\pi/2$	0	1
1	0	0	1	π	1	1
1	0	1	0	0	0	0
1	0	1	1	$-\pi/2$	1	0
1	1	0	0	π	1	1
1	1	0	1	$-\pi/2$	1	0
1	1	1	0	$+\pi/2$	0	1
1	1	1	1	0	0	0

A block diagram of the differential decoder is illustrated in Figure 7.7.8. Note that the decoder can be thought of as a look-up table or the equivalent Boolean function.

Consider the application of this decoding rule to the encoded bit sequence presented above. First suppose the carrier phase ambiguity is 0. In this case, $\hat{\delta}_{2k} = \delta_{2k}$ and $\hat{\delta}_{2k+1} = \delta_{2k+1}$ as shown in the table below. Applying the decoding rule with the assumption $\delta_{-2} = 0$ and $\delta_{-1} = 0$ produces the bit estimates in the right-most two columns. Observe that this is the correct bit sequence.

Section 7.7 Phase Ambiguity Resolution

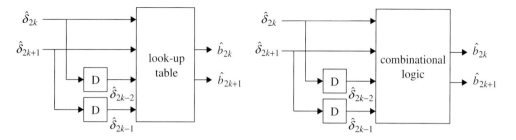

Figure 7.7.8 Two equivalent versions of the differential decoder for QPSK.

k	b_{2k}	b_{2k+1}	δ_{2k}	δ_{2k+1}	$\hat{\delta}_{2k}$	$\hat{\delta}_{2k+1}$	$\hat{b}_{2k}$	$\hat{b}_{2k+1}$
0	1	0	1	0	1	0	1	0
1	1	1	0	1	0	1	1	1
2	0	1	1	1	1	1	0	1
3	0	0	1	1	1	1	0	0
4	1	1	0	0	0	0	1	1
5	1	0	1	0	1	0	1	0
6	1	0	1	1	1	1	1	0
7	0	1	1	0	1	0	0	1

Now consider the case where the phase ambiguity is $\pi/2$. In this case, the detected sequence of encoded bits is as shown in the table below. Applying the decoding rule produces the bit sequence shown in the right-most two columns in the table below. Note that $\delta_{-2} = 0$ and $\delta_{-1} = 0$ was assumed and that this assumption resulted in an erroneous first bit. Again, this is not a problem in a real system using a decision-directed carrier phase PLL because this erroneous bit occurs prior to the time when the carrier phase synchronization PLL achieves lock.

k	b_k	b_{k+1}	δ_k	δ_{k+1}	$\hat{\delta}_k$	$\hat{\delta}_{k+1}$	$\hat{b}_k$	$\hat{b}_{k+1}$
0	1	0	1	0	0	0	0	0
1	1	1	0	1	1	1	1	1
2	0	1	1	1	1	0	0	1
3	0	0	1	1	1	0	0	0
4	1	1	0	0	0	1	1	1
5	1	0	1	0	0	0	1	0
6	1	0	1	1	1	0	1	0
7	0	1	1	0	0	0	0	1

Offset QPSK Example. The decision-directed carrier phase synchronization PLL for offset QPSK possesses a 180° phase ambiguity as shown by the S-curve in Figure 7.5.2. The carrier phase ambiguity situation for offset QPSK, however, is somewhat more complicated due to a phenomenon known as *delay axis ambiguity*. The delay axis ambiguity is a consequence of the offset way in which the inphase and quadrature matched filter outputs

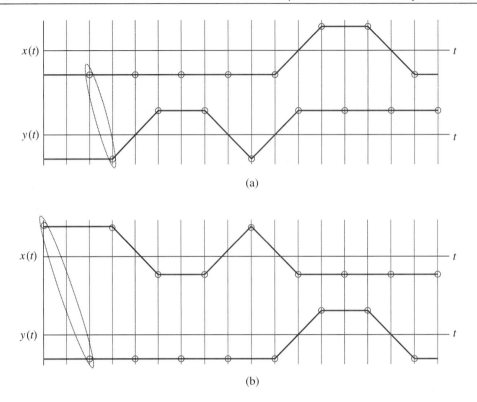

Figure 7.7.9 The relationship between the matched filter outputs, carrier phase, and timing for offset QPSK: (a) The matched filter outputs resulting with a phase ambiguity of 0. (b) The matched filter outputs corresponding to a phase shift of $\pi/2$. The consequence of the phase shift is a reversal of the half-symbol delay relationship between the inphase and quadrature matched filter outputs.

are sampled and paired. The effect is illustrated in Figure 7.7.9. Suppose the inphase and quadrature matched filter outputs for no phase ambiguity are given by Figure 7.7.9 (a). The optimum sampling instants are identified by the circles. Symbol decisions are based on the pairings indicated by the ellipse. *A sample from the inphase matched filter output is always paired with the sample from the quadrature matched filter output one-half symbol time later.* It is this staggered sample-paring that leads to the delay axis ambiguity. Now suppose the input to the matched filters is rotated by $\pi/2$. The effect of this rotation is that the negative of the quadrature matched filter output becomes the inphase matched filter output while the inphase matched filter output becomes the quadrature matched filter output. This effect is shown in Figure 7.7.9 (b). The staggered samples of the two matched filter outputs are paired as shown by the ellipse. Observe that the original relationship between the staggered matched filter output samples has been replaced by a new relationship.

Section 7.7 Phase Ambiguity Resolution

A differential encoding rule for offset QPSK must account for both the 180° phase ambiguity from the carrier phase synchronization PLL *and* the delay axis ambiguity. One such encoding rule that operates on the k-th block of input bits b_{2k}, b_{2k+1} to produce the block of differentially encoded bits $\delta_{2k}, \delta_{2k+1}$ is

$$\delta_{2k} = b_{2k} \oplus \overline{\delta}_{2k-1}$$
$$\delta_{2k+1} = b_{2k+1} \oplus \delta_{2k} \tag{7.82}$$

where $\oplus$ is the Boolean "exclusive or" operation and $\overline{\delta}_{2k-1}$ is the logical complement of δ_{2k-1}. As an example, consider the application of the differential encoding rule (7.82) to the bit sequence 1 0 1 1 0 1 0 0 1 1 1 0 1 0 0 1. The bit sequence is organized into blocks of two bits as illustrated in the table below. The resulting encoded bit sequence assuming $\delta_{-1} = 0$ is also shown.

k	b_{2k}	b_{2k+1}	δ_{2k}	δ_{2k+1}
0	1	0	0	0
1	1	1	0	1
2	0	1	0	1
3	0	0	0	0
4	1	1	0	1
5	1	0	1	1
6	1	0	1	1
7	0	1	0	1

The differential decoding rule is obtained by applying Boolean identities

$$A \oplus A = 0$$
$$A \oplus 0 = A$$

to solve (7.82) for the input bits b_{2k} and b_{2k+1}. The result is

$$\hat{b}_{2k} = \hat{\delta}_{2k} \oplus \overline{\delta}_{2k-1}$$
$$\hat{b}_{2k+1} = \hat{\delta}_{2k+1} \oplus \delta_{2k} \tag{7.83}$$

where again $\oplus$ is the Boolean "exclusive or" operation and $\overline{\delta}_{2k-1}$ is the logical complement of δ_{2k-1}.

An example of differential decoding and its ability to compensate for the carrier phase ambiguity and the delay axis ambiguity requires a known (differentially encoded) bit-to-symbol mapping. Suppose the mapping is the one shown in the constellation below.

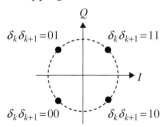

First consider the case where the phase ambiguity is 0. In this case, $\hat{\delta}_{2k} = \delta_{2k}$ and $\hat{\delta}_{2k+1} = \delta_{2k+1}$. Organizing the bit decisions in blocks of two and applying the rule (7.83) with $\hat{\delta}_{-1} = 0$ produces the table shown below.

k	b_{2k}	b_{2k+1}	δ_{2k}	δ_{2k+1}	$\hat{\delta}_{2k}$	$\hat{\delta}_{2k+1}$	$\hat{b}_{2k}$	$\hat{b}_{2k+1}$
0	1	0	0	0	0	0	1	0
1	1	1	0	1	0	1	1	1
2	0	1	0	1	0	1	0	1
3	0	0	0	0	0	0	0	0
4	1	1	0	1	0	1	1	1
5	1	0	1	1	1	1	1	0
6	1	0	1	1	1	1	1	0
7	0	1	0	1	0	1	0	1

Observe that the original bit sequence is recovered perfectly. Now assume the phase ambiguity is π. In this case, the constellation is simply rotated by 180° so that $\hat{\delta}_{2k} = \bar{\delta}_{2k}$ and $\hat{\delta}_{2k+1} = \bar{\delta}_{2k+1}$. Organizing the bit decisions in blocks of two and applying the rule (7.83) with $\hat{\delta}_{-1} = 0$ produces the table shown below.

k	b_{2k}	b_{2k+1}	δ_{2k}	δ_{2k+1}	$\hat{\delta}_{2k}$	$\hat{\delta}_{2k+1}$	$\hat{b}_{2k}$	$\hat{b}_{2k+1}$
0	1	0	0	0	1	1	0	0
1	1	1	0	1	1	0	1	1
2	0	1	0	1	1	0	0	1
3	0	0	0	0	1	1	0	0
4	1	1	0	1	1	0	1	1
5	1	0	1	1	0	0	1	0
6	1	0	1	1	0	0	1	0
7	0	1	0	1	1	0	0	1

Note the first bit is incorrect, but the remaining bits are correct. As before, this is not a problem because first erroneous bit occurs before the carrier phase synchronization PLL has achieved lock.

Finally, one of the two possible axis ambiguity cases is considered. Returning to Figure 7.7.9, the matched filter outputs corresponding to the running example with no phase ambiguity are plotted in Figure 7.7.9 (a). Observe that proper pairing of the two matched filter outputs at the indicated sampling instants produces the sequence of encoded bit estimates (the $\hat{\delta}$'s) in the two examples above. The matched filter outputs shown in Figure 7.7.9 (b) are the result of a $\pi/2$ phase shift at the input to the matched filters. Using the staggered pairings as shown, the estimates for the differentially encoded bits listed in the table below is produced. Applying the decoding rule (7.83) with $\hat{\delta}_{-1} = 0$ produces the bit estimates also listed below.

k	b_{2k}	b_{2k+1}	δ_{2k}	δ_{2k+1}	$\hat{\delta}_{2k}$	$\hat{\delta}_{2k+1}$	$\hat{b}_{2k}$	$\hat{b}_{2k+1}$
0	1	0	0	0	1	0	0	1
1	1	1	0	1	0	0	1	0
2	0	1	0	1	0	0	1	0
3	0	0	0	0	1	0	0	1
4	1	1	0	1	0	1	1	1
5	1	0	1	1	0	1	0	1
6	1	0	1	1	0	0	0	0
7	0	1	0	1	0	0	1	0

At first glance it appears that the estimated bit sequence is incorrect. This is not quite the case. The axis ambiguity and associated timing offset produced a delayed version of the correct bit sequence. This can be seen by writing the transmitted bit sequence and the estimated sequence in rows with the estimated sequence delayed one bit:

transmitted bits   1 0 1 1 0 1 0 0 1 1 1 0 1 0 0 1 –
estimated bits   – 0 1 1 0 1 0 0 1 1 1 0 1 0 0 1 0

Observe that the list of estimated bits in the table is missing the first bit in the transmitted sequence but that an extra bit is produced. This is usually not a problem because carrier phase and symbol timing synchronization are usually followed by a frame synchronizer. The frame synchronizer searches the bit estimates for a known pattern that identifies where information and data bits occur in the received bit stream.

7.8 MAXIMUM LIKELIHOOD PHASE ESTIMATION

Maximum likelihood estimation uses conditional probabilities as a measure of "how likely" a parameter is given noisy observations. This technique was applied in Chapter 5 to derive the optimum (in the maximum likelihood sense) structure for detectors. The problem was cast as an estimation problem where the information symbols were the unknown quantity. Maximum likelihood estimation can also be applied to synchronization. In this case, the carrier phase offset is the unknown to be estimated. The technique is demonstrated for QPSK. Extensions to other 2-dimensional signal sets and other D-dimensional signal sets are straightforward.

7.8.1 Preliminaries

Let the observation interval be $T_0 = L_0 T_s$ seconds and let the received band-pass signal be

$$r(t) = s(t) + w(t) \quad (7.84)$$

where

$$s(t) = \sum_{k=0}^{L_0} a_0(k) p(t - kT_s) \sqrt{2} \cos(\omega_0 t + \theta) - a_1(k) p(t - kT_s) \sqrt{2} \sin(\omega_0 t + \theta) \quad (7.85)$$

where $a_0(k) \in \{-1,+1\}$ and $a_1(k) \in \{-1,+1\}$ are the inphase and quadrature components of the k-th symbol, $p(t)$ is unit-energy pulse shape with support on $-L_p T_s \leq t \leq L_p T_s$, T_s is the symbol time, θ is the unknown carrier phase to be estimated, and $w(t)$ is a zero-mean white Gaussian random process with power spectral density $N_0/2$ W/Hz. The band-pass signal is sampled every T seconds to produce the sequence

$$r(nT) = s(nT) + w(nT); \quad n = 0, 1, \ldots, NL_0 - 1. \quad (7.86)$$

The sampled signal component may be expressed as

$$s(nT) = \sum_{k=0}^{L_0-1} a_0(k) p(nT - kT_s) \sqrt{2} \cos(\Omega_0 n + \theta)$$

$$- a_1(k) p(nT - kT_s) \sqrt{2} \sin(\Omega_0 n + \theta) \quad (7.87)$$

for $n = 0, 1, \ldots, NL_0 - 1$. Note that perfect symbol timing synchronization is assumed. For convenience, the following vectors are defined

$$\mathbf{r} = \begin{bmatrix} r(0) \\ r(T) \\ \vdots \\ r((NL_0-1)T) \end{bmatrix} \quad \mathbf{s} = \begin{bmatrix} s(0) \\ s(T) \\ \vdots \\ s((NL_0-1)T) \end{bmatrix} \quad \mathbf{w} = \begin{bmatrix} w(0) \\ w(T) \\ \vdots \\ w((NL_0-1)T) \end{bmatrix}. \quad (7.88)$$

The vector $\mathbf{w}$ is a sequence of independent and identically distributed Gaussian random variables with zero mean and variance

$$\sigma^2 = \frac{N_0}{2T}. \quad (7.89)$$

The probability density function of $\mathbf{w}$ is

$$p(\mathbf{w}) = \frac{1}{(2\pi\sigma^2)^{L_0 N/2}} \exp\left\{-\frac{1}{2\sigma^2} \sum_{n=0}^{NL_0-1} w^2(nT)\right\}. \quad (7.90)$$

For notational convenience, define the symbol vector $\mathbf{a}$ as

$$\mathbf{a} = \begin{bmatrix} \mathbf{a}(0) & \mathbf{a}(1) & \cdots & \mathbf{a}(L_0-1) \end{bmatrix}^T \quad (7.91)$$

where

$$\mathbf{a}(k) = \begin{bmatrix} a_0(k) \\ a_1(k) \end{bmatrix}. \quad (7.92)$$

To emphasize the fact that the $\mathbf{s}$ is a function of $\mathbf{a}$ and θ, $\mathbf{s}$ is expressed as $\mathbf{s}(\mathbf{a}, \theta)$ and samples of the signal component $s(nT)$ are expressed as $s(nT; \mathbf{a}, \theta)$.

Carrier phase synchronization can be thought of as an estimation problem. The goal is to estimate the parameter θ from the samples $r(nT) = s(nT; \mathbf{a}, \theta) + w(nT)$. The maximum

Section 7.8 Maximum Likelihood Phase Estimation

likelihood estimate is the one that maximizes the logarithm of the conditional probability $p(\mathbf{r}|\mathbf{a},\theta)$. Using the probability density function of $\mathbf{w}$ given by (7.90), the conditional probability $p(\mathbf{r}|\mathbf{a},\theta)$ is

$$p(\mathbf{r}|\mathbf{a},\theta) = \frac{1}{(2\pi\sigma^2)^{L_0 N/2}} \exp\left\{-\frac{1}{2\sigma^2} \sum_{n=0}^{NL_0-1} [r(nT) - s(nT;\mathbf{a},\theta)]^2\right\}. \tag{7.93}$$

The log-likelihood function $\Lambda(\mathbf{a},\theta)$ is the logarithm of (7.93):

$$\Lambda(\mathbf{a},\theta) = -\frac{L_0 N}{2}\ln(2\pi\sigma^2) - \frac{1}{2\sigma^2}\sum_{n=0}^{NL_0-1}[r(nT) - s(nT;\mathbf{a},\theta)]^2. \tag{7.94}$$

Later it will be convenient to express the cross product sum as

$$\sum_{n=0}^{NL_0-1} r(nT)s(nT;\mathbf{a},\theta) = \sum_{k=0}^{L_0-1} a_0(k) \sum_{n=(k-L_p)N}^{(k+L_p)N} r(nT)p(nT-kT_s)\sqrt{2}\cos(\Omega_0 n + \theta)$$

$$- \sum_{k=0}^{L_0-1} a_1(k) \sum_{n=(k-L_p)N}^{(k+L_p)N} r(nT)p(nT-kT_s)\sqrt{2}\sin(\Omega_0 n + \theta). \tag{7.95}$$

Two approaches will be taken to obtain the maximum likelihood estimator for θ. The first approach assumes $\mathbf{a}$ is known.[4] In this case, the estimator for θ is a function of the data symbols. This estimator is called a *data-aided estimator*.

The second approach does not assume $\mathbf{a}$ is known. In this case, the dependence on $\mathbf{a}$ is removed by assuming the symbol sequence $\mathbf{a}$ is random and using the total probability theorem to obtain the average probability density function $p(\mathbf{r}|\theta)$. The maximum likelihood estimate maximizes the logarithm of $p(\mathbf{r}|\theta)$. Such an estimator is called a *non-data-aided estimator* or *blind estimator*. It will be shown that the high signal-to-noise ratio approximation of the non-data-aided estimator is the decision-directed estimator introduced in Section 7.2.

The conditional probability density function $p(\mathbf{r}|\theta)$ is obtained from the conditional probability density function $p(\mathbf{r}|\mathbf{a},\theta)$ using the total probability theorem:

$$p(\mathbf{r}|\theta) = \int p(\mathbf{r}|\mathbf{a},\theta)p(\mathbf{a})d\mathbf{a} \tag{7.96}$$

where $p(\mathbf{a})$ is the probability density function of the symbol sequence $\mathbf{a}$. The most commonly used probability density function for the data sequence assumes the symbols are independent and equally likely. Independence implies

$$p(\mathbf{a}) = \prod_{k=0}^{L_0-1} p(\mathbf{a}(k)) \tag{7.97}$$

---

[4]For packetized burst mode communication systems with a known preamble or header, the L_0 data symbols are known and should be used for synchronization.

while equally likely implies

$$p(\mathbf{a}(k)) = \frac{1}{4}\delta(a_0(k)-1)\delta(a_1(k)-1) + \frac{1}{4}\delta(a_0(k)-1)\delta(a_1(k)+1)$$
$$+ \frac{1}{4}\delta(a_0(k)+1)\delta(a_1(k)-1) + \frac{1}{4}\delta(a_0(k)+1)\delta(a_1(k)+1). \quad (7.98)$$

Thus,

$$p(\mathbf{r}|\theta) = \int p(\mathbf{r}|\mathbf{a},\theta)p(\mathbf{a})d\mathbf{a} \quad (7.99)$$

$$= \prod_{k=0}^{L_0-1} \int p(\mathbf{r}|\mathbf{a}(k),\theta)p(\mathbf{a}(k))d\mathbf{a}(k) \quad (7.100)$$

$$= \prod_{k=0}^{L_0-1} \left\{ \frac{1}{4}p(\mathbf{r}|\mathbf{a}(k)=[1,1],\theta) + \frac{1}{4}p(\mathbf{r}|\mathbf{a}(k)=[1,-1],\theta) \right.$$
$$\left. + \frac{1}{4}p(\mathbf{r}|\mathbf{a}(k)=[-1,1],\theta) + \frac{1}{4}p(\mathbf{r}|\mathbf{a}(k)=[-1,-1],\theta) \right\}. \quad (7.101)$$

By writing (7.93) as

$$p(\mathbf{r}|\mathbf{a},\theta) = \prod_{k=0}^{L_0-1} \frac{1}{(2\pi\sigma^2)^{N/2}} \exp\left\{ -\frac{1}{2\sigma^2} \sum_{n=N(k-L_p)}^{N(k+L_p)} [r(nT) - s(nT;\mathbf{a},\theta)]^2 \right\} \quad (7.102)$$

and using the substitution

$$s(nT;\mathbf{a}(k),\theta) = a_0(k)p(nT - kT_s)\sqrt{2}\cos(\Omega_0 n + \theta)$$
$$- a_1(k)p(nT - kT_s)\sqrt{2}\sin(\Omega_0 n + \theta) \quad (7.103)$$

each term in (7.101) may be expressed as

$$p(\mathbf{r}|\mathbf{a}(k)=[1,1],\theta) = \prod_{k=0}^{L_0-1} \frac{1}{(2\pi\sigma^2)^{N/2}} \exp\left\{ -\frac{1}{2\sigma^2} \sum_{n=N(k-L_p)}^{N(k+L_p)} r^2(nT) + p^2(nT - kT_s) \right\}$$

$$\times \exp\left\{ \frac{1}{\sigma^2} \sum_{n=N(k-L_p)}^{N(k+L_p)} r(nT)p(nT - kT_s)\sqrt{2}\cos(\Omega_0 n + \theta) \right\}$$

$$\times \exp\left\{ -\frac{1}{\sigma^2} \sum_{n=N(k-L_p)}^{N(k+L_p)} r(nT)p(nT - kT_s)\sqrt{2}\sin(\Omega_0 n + \theta) \right\} \quad (7.104)$$

Section 7.8 Maximum Likelihood Phase Estimation

$$p(\mathbf{r}|\mathbf{a}(k) = [1, -1], \theta) = \prod_{k=0}^{L_0-1} \frac{1}{(2\pi\sigma^2)^{N/2}} \exp\left\{-\frac{1}{2\sigma^2} \sum_{n=N(k-L_p)}^{N(k+L_p)} r^2(nT) + p^2(nT - kT_s)\right\}$$

$$\times \exp\left\{\frac{1}{\sigma^2} \sum_{n=N(k-L_p)}^{N(k+L_p)} r(nT)p(nT - kT_s)\sqrt{2}\cos(\Omega_0 n + \theta)\right\}$$

$$\times \exp\left\{\frac{1}{\sigma^2} \sum_{n=N(k-L_p)}^{N(k+L_p)} r(nT)p(nT - kT_s)\sqrt{2}\sin(\Omega_0 n + \theta)\right\} \quad (7.105)$$

$$p(\mathbf{r}|\mathbf{a}(k) = [-1, 1], \theta) = \prod_{k=0}^{L_0-1} \frac{1}{(2\pi\sigma^2)^{N/2}} \exp\left\{-\frac{1}{2\sigma^2} \sum_{n=N(k-L_p)}^{N(k+L_p)} r^2(nT) + p^2(nT - kT_s)\right\}$$

$$\times \exp\left\{-\frac{1}{\sigma^2} \sum_{n=N(k-L_p)}^{N(k+L_p)} r(nT)p(nT - kT_s)\sqrt{2}\cos(\Omega_0 n + \theta)\right\}$$

$$\times \exp\left\{-\frac{1}{\sigma^2} \sum_{n=N(k-L_p)}^{N(k+L_p)} r(nT)p(nT - kT_s)\sqrt{2}\sin(\Omega_0 n + \theta)\right\}$$

$$(7.106)$$

$$p(\mathbf{r}|\mathbf{a}(k) = [-1, -1], \theta) = \prod_{k=0}^{L_0-1} \frac{1}{(2\pi\sigma^2)^{N/2}} \exp\left\{-\frac{1}{2\sigma^2} \sum_{n=N(k-L_p)}^{N(k+L_p)} r^2(nT) + p^2(nT - kT_s)\right\}$$

$$\times \exp\left\{-\frac{1}{\sigma^2} \sum_{n=N(k-L_p)}^{N(k+L_p)} r(nT)p(nT - kT_s)\sqrt{2}\cos(\Omega_0 n + \theta)\right\}$$

$$\times \exp\left\{\frac{1}{\sigma^2} \sum_{n=N(k-L_p)}^{N(k+L_p)} r(nT)p(nT - kT_s)\sqrt{2}\sin(\Omega_0 n + \theta)\right\}. \quad (7.107)$$

Substituting (7.104)–(7.107) into (7.101) and collecting similar terms produces

$$p(\mathbf{r}|\theta) = \frac{1}{4} \prod_{k=0}^{L_0-1} \frac{1}{(2\pi\sigma^2)^{N/2}} \exp\left\{-\frac{1}{2\sigma^2} \sum_{n=N(k-L_p)}^{N(k+L_p)} r^2(nT) + p^2(nT - kT_s)\right\}$$

$$\times \left(\exp\left\{\frac{1}{\sigma^2} \sum_{n=N(k-L_p)}^{N(k+L_p)} r(nT)p(nT - kT_s)\sqrt{2}\cos(\Omega_0 n + \theta)\right\}\right.$$

$$+ \exp\left\{-\frac{1}{\sigma^2}\sum_{n=N(k-L_p)}^{N(k+L_p)} r(nT)p(nT-kT_s)\sqrt{2}\cos(\Omega_0 n + \theta)\right\}\right)$$

$$\times \left(\exp\left\{\frac{1}{\sigma^2}\sum_{n=N(k-L_p)}^{N(k+L_p)} r(nT)p(nT-kT_s)\sqrt{2}\sin(\Omega_0 n + \theta)\right\}\right.$$

$$\left.+ \exp\left\{-\frac{1}{\sigma^2}\sum_{n=N(k-L_p)}^{N(k+L_p)} r(nT)p(nT-kT_s)\sqrt{2}\sin(\Omega_0 n + \theta)\right\}\right). \quad (7.108)$$

Applying the identity

$$\frac{e^x + e^{-x}}{2} = \cosh(x) \quad (7.109)$$

to (7.108) produces

$$p(\mathbf{r}|\theta) = \prod_{k=0}^{L_0-1} \frac{1}{(2\pi\sigma^2)^{N/2}} \exp\left\{-\frac{1}{2\sigma^2}\sum_{n=N(k-L_p)}^{N(k+L_p)} r^2(nT) + p^2(nT-kT_s)\right\}$$

$$\times \cosh\left(\frac{1}{\sigma^2}\sum_{n=N(k-L_p)}^{N(k+L_p)} r(nT)p(nT-kT_s)\sqrt{2}\cos(\Omega_0 n + \theta)\right)$$

$$\times \cosh\left(\frac{1}{\sigma^2}\sum_{n=N(k-L_p)}^{N(k+L_p)} r(nT)p(nT-kT_s)\sqrt{2}\sin(\Omega_0 n + \theta)\right). \quad (7.110)$$

The average log-likelihood function is

$$\overline{\Lambda}(\theta) = -\frac{NL_0}{2}\ln(2\pi\sigma^2) - \frac{1}{2\sigma^2}\sum_{k=0}^{L_0-1}\sum_{N(k-L_p)}^{N(k+L_p)} r^2(nT) + p^2(nT-kT_s)$$

$$+ \sum_{k=0}^{L_0-1} \ln\cosh\left(\frac{1}{\sigma^2}\sum_{n=N(k-L_p)}^{N(k+L_p)} r(nT)p(nT-kT_s)\sqrt{2}\cos(\Omega_0 n + \theta)\right)$$

$$+ \sum_{k=0}^{L_0-1} \ln\cosh\left(\frac{1}{\sigma^2}\sum_{n=N(k-L_p)}^{N(k+L_p)} r(nT)p(nT-kT_s)\sqrt{2}\sin(\Omega_0 n + \theta)\right). \quad (7.111)$$

7.8.2 Carrier Phase Estimation

Known Symbol Sequence. For the case where the data symbols are known, the maximum likelihood estimate $\hat{\theta}$ is the value of θ that maximizes the log-likelihood function

Section 7.8 Maximum Likelihood Phase Estimation

$\Lambda(\mathbf{a}, \theta)$ given by (7.94). This estimate is the value of θ that forces the partial derivative of $\Lambda(\mathbf{a}, \theta)$ with respect to θ to be zero. The partial derivative of $\Lambda(\mathbf{a}, \theta)$ is

$$\frac{\partial}{\partial \theta} \Lambda(\mathbf{a}, \theta) = -\frac{1}{2\sigma^2} \frac{\partial}{\partial \theta} \sum_{n=0}^{NL_0-1} [r(nT) - s(nT; \mathbf{a}, \theta)]^2 \quad (7.112)$$

$$= -\frac{1}{2\sigma^2} \frac{\partial}{\partial \theta} \sum_{n=0}^{NL_0-1} \left[r^2(nT) - 2r(nT)s(nT; \mathbf{a}, \theta) + s^2(nT; \mathbf{a}, \theta) \right]. \quad (7.113)$$

The partial derivatives of the first and third terms are zero because the energy in the received signal and the energy in a QPSK waveform are the same for all phase rotations. All that remains is the middle term. Substituting (7.87) for $s(nT; \mathbf{a}, \theta)$, interchanging the order of summations, and computing the derivative yields

$$\frac{\partial}{\partial \theta} \Lambda(\mathbf{a}, \theta) = -\frac{1}{\sigma^2} \sum_{k=0}^{L_0-1} a_0(k) \sum_{n=(k-L_p)N}^{(k+L_p)N} r(nT)p(nT - kT_s)\sqrt{2} \sin(\Omega_0 n + \theta)$$

$$- \sum_{k=0}^{L_0-1} a_1(k) \sum_{n=(k-L_p)N}^{(k+L_p)N} r(nT)p(nT - kT_s)\sqrt{2} \cos(\Omega_0 n + \theta). \quad (7.114)$$

The inner sums of the first and second terms may be interpreted as the outputs at $n = kT_s/T$ of a filter matched to the pulse shape $p(nT)$. The inputs to the matched filter are the results of a coherent translation of $r(nT)$ from band pass to baseband as illustrated in Figure 7.1.1. Using

$$x(kT_s; \theta) = \sum_{n=N(k-L_p)}^{N(k+L_p)} r(nT)p(nT - kT_s)\sqrt{2} \cos(\Omega_0 n + \theta) \quad (7.115)$$

$$y(kT_s; \theta) = -\sum_{n=N(k-L_p)}^{N(k+L_p)} r(nT)p(nT - kT_s)\sqrt{2} \sin(\Omega_0 n + \theta) \quad (7.116)$$

to represent these quantities, the partial derivative may be expressed in the more compact form

$$\frac{\partial}{\partial \theta} \Lambda(\mathbf{a}, \theta) = \frac{1}{\sigma^2} \sum_{k=0}^{L_0-1} a_0(k)y(kT_s; \theta) - \sum_{k=0}^{L_0-1} a_1(k)x(kT_s; \theta). \quad (7.117)$$

The maximum likelihood estimate $\hat{\theta}$ is the value of θ that forces the right-hand side of (7.117) to zero:

$$0 = \sum_{k=0}^{L_0-1} a_0(k)y(kT_s; \hat{\theta}) - a_1(k)x(kT_s; \hat{\theta}). \quad (7.118)$$

This equation may be solved iteratively. A value for θ is chosen and used to compute the right-hand side of (7.118). The estimate for θ is increased (if the computation is negative)

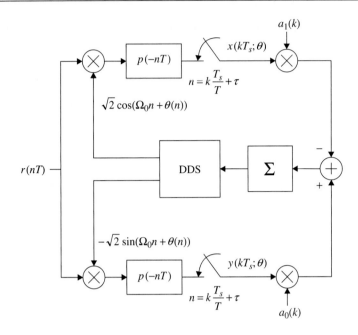

Figure 7.8.1 Block diagram of the maximum likelihood QPSK phase estimator based on the form (7.118).

or decreased (if the computation is positive) until θ satisfies (7.118). A block diagram of a system which finds the maximum likelihood estimate iteratively is illustrated in Figure 7.8.1. Note that it is a PLL structure that uses the right-hand side of (7.118) as the error signal. The summation block plays the role of the loop filter (recall that the loop filter contains an integrator). Compare this block diagram with the QPSK carrier phase PLL shown in Figure 7.2.7. If the symbol decisions in Figure 7.2.7 are replaced by the true symbols in the error detector, then the two systems are equivalent.

An alternate expression for the data-aided ML estimate is obtained from (7.114) using the identities

$$\cos(A + B) = \cos A \cos B - \sin A \sin B$$
$$\sin(A + B) = \sin A \cos B + \cos A \sin B.$$

Equation (7.114) may be expressed as

$$\frac{\partial}{\partial \theta} \Lambda(\mathbf{a}, \theta) = -\sum_{k=0}^{L_0-1} a_0(k) \sum_{n=(k-L_p)N}^{(k+L_p)N} r(nT) p(nT - kT_s) \sqrt{2} \sin(\Omega_0 n) \sin \theta$$

$$- \sum_{k=0}^{L_0-1} a_0(k) \sum_{n=(k-L_p)N}^{(k+L_p)N} r(nT) p(nT - kT_s) \sqrt{2} \cos(\Omega_0 n) \sin \theta$$

Section 7.8 Maximum Likelihood Phase Estimation

$$-\sum_{k=0}^{L_0-1} a_1(k) \sum_{n=(k-L_p)N}^{(k+L_p)N} r(nT)p(nT - kT_s)\sqrt{2}\cos(\Omega_0 n)\cos\theta$$

$$+\sum_{k=0}^{L_0-1} a_1(k) \sum_{n=(k-L_p)N}^{(k+L_p)N} r(nT)p(nT - kT_s)\sqrt{2}\sin(\Omega_0 n)\sin\theta. \quad (7.119)$$

Using

$$x(kT_s) = \sum_{n=(k-L_p)N}^{(k+L_p)N} r(nT)p(nT - kT_s)\sqrt{2}\cos(\Omega_0 n) \quad (7.120)$$

$$y(kT_s) = -\sum_{n=(k-L_p)N}^{(k+L_p)N} r(nT)p(nT - kT_s)\sqrt{2}\sin(\Omega_0 n) \quad (7.121)$$

Equation (7.119) may be expressed as

$$\frac{\partial}{\partial \theta}\Lambda(\mathbf{a},\theta) = \sum_{k=0}^{L_0-1} a_0(k)\left[y(kT_s)\cos\theta - x(kT_s)\sin\theta\right]$$

$$-\sum_{k=0}^{L_0-1} a_1(k)\left[x(kT_s)\cos\theta + y(kT_s)\sin\theta\right]. \quad (7.122)$$

Here, the quantities $x(kT_s)$ and $y(kT_s)$ may be interpreted as the outputs of filters, matched to the pulse shape $p(nT)$, at the sample instants $n = kT/T_s$. The inputs to the matched filter are the results of a noncoherent translation of $r(nT)$ from band pass to baseband as illustrated in Figure 7.1.2. The terms in the square brackets are the equations for the rotation of the point $(x(kT_s), y(kT_s))$ by an angle $-\theta$. Following the notation introduced in Section 7.2, let $(x'(kT_s; \theta), y'(kT_s; \theta))$ represent the rotated point (θ is included to emphasize the dependence on θ) so that

$$\begin{bmatrix} x'(kT_s;\theta) \\ y'(kT_s;\theta) \end{bmatrix} = \begin{bmatrix} \cos\theta & \sin\theta \\ -\sin\theta & \cos\theta \end{bmatrix} \begin{bmatrix} x(kT_s) \\ y(kT_s) \end{bmatrix}. \quad (7.123)$$

Thus, (7.122) may be expressed as

$$\frac{\partial}{\partial \theta}\Lambda(\mathbf{a},\theta) = \sum_{k=0}^{L_0-1} a_0(k)y'(kT_s;\theta) - a_1(k)x'(kT_s;\theta) \quad (7.124)$$

so that the alternate expression for the ML phase estimate is

$$0 = \sum_{k=0}^{L_0-1} a_0(k)y'(kT_s;\hat\theta) - a_1(k)x'(kT_s;\hat\theta). \quad (7.125)$$

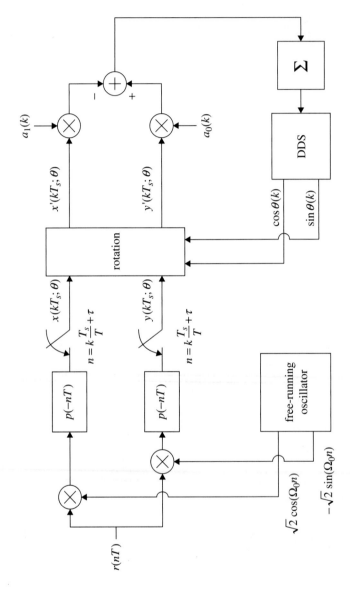

Figure 7.8.2 Block diagram of the maximum likelihood QPSK phase estimator based on the form (7.125).

Note that two forms for the ML estimator (7.118) and (7.125) are identical. The difference is where carrier phase compensation occurs.

The solution to the alternate formulation may be found iteratively or solved directly in closed form. It is easiest to visualize the iterative solution using the form (7.125). A block diagram illustrating the iterative solution to (7.125) is shown in Figure 7.8.2. This is a PLL structure where the right-hand side of (7.125) is the error signal. The solution shown in Figure 7.8.2 is almost identical to that shown in Figure 7.2.4.

The closed form solution starts with the form (7.122). Setting (7.122) to zero and solving for θ results in a closed form solution for the ML phase estimate. Grouping the terms which have the cosine in common and grouping the terms that have the sine in common and solving produces

$$\frac{\sin\hat{\theta}}{\cos\hat{\theta}} = \frac{\sum_{k=0}^{L_0-1} a_0(k)y(kT_s) - a_1(k)x(kT_s)}{\sum_{k=0}^{L_0-1} a_0(k)x(kT_s) + a_1(k)y(kT_s)} \qquad (7.126)$$

from which the maximum likelihood phase estimate is

$$\hat{\theta} = \tan^{-1}\left\{\frac{\sum_{k=0}^{L_0-1} a_0(k)y(kT_s) - a_1(k)x(kT_s)}{\sum_{k=0}^{L_0-1} a_0(k)x(kT_s) + a_1(k)y(kT_s)}\right\}. \qquad (7.127)$$

This solution is useful for packetized communications links where the carrier phase offset θ will remain constant over the duration of the data packet. Such detectors typically use block processing in place of iterative processing.

Unknown Symbol Sequence. When the symbol sequence is unknown, the ML phase estimate is the one that maximizes the average log-likelihood function $\overline{\Lambda}(\theta)$ given by (7.111). The partial derivative of $\overline{\Lambda}(\theta)$ is

$$\frac{\partial}{\partial\theta}\overline{\Lambda}(\theta) = -\sum_{k=0}^{L_0-1} \tanh\left(\frac{1}{\sigma^2}\sum_{n=N(k-L_p)}^{N(k+L_p)} r(nT)p(nT-kT_s)\sqrt{2}\cos(\Omega_0 n + \theta)\right)$$

$$\times \left[\frac{1}{\sigma^2}\sum_{n=N(k-L_p)}^{N(k+L_p)} r(nT)p(nT-kT_s)\sqrt{2}\sin(\Omega_0 n + \theta)\right]$$

$$+ \sum_{k=0}^{L_0-1} \tanh\left(\frac{1}{\sigma^2} \sum_{n=N(k-L_p)}^{N(k+L_p)} r(nT)p(nT - kT_s)\sqrt{2}\sin(\Omega_0 n + \theta)\right)$$

$$\times \left[\frac{1}{\sigma^2} \sum_{n=N(k-L_p)}^{N(k+L_p)} r(nT)p(nT - kT_s)\sqrt{2}\cos(\Omega_0 n + \theta)\right]. \quad (7.128)$$

As before, the inner sums over the index n may be interpreted as the outputs of filter matched to the pulse shape. Using the relationships (7.115) and (7.116) for the inphase and quadrature matched filter outputs, respectively, (7.128) may be expressed in the more compact form

$$\frac{\partial}{\partial \theta}\overline{\Lambda}(\theta) = \sum_{k=0}^{L_0-1} \tanh\left(\frac{1}{\sigma^2}x(kT_s;\theta)\right)\frac{1}{\sigma^2}y(kT_s;\theta)$$

$$- \tanh\left(\frac{1}{\sigma^2}y(kT_s;\theta)\right)\frac{1}{\sigma^2}x(kT_s;\theta). \quad (7.129)$$

The ML phase estimate is the value of θ that forces (7.129) to zero:

$$0 = \sum_{k=0}^{L_0-1} \tanh\left(\frac{1}{\sigma^2}x(kT_s;\hat{\theta})\right)\frac{1}{\sigma^2}y(kT_s;\hat{\theta}) - \tanh\left(\frac{1}{\sigma^2}y(kT_s;\hat{\theta})\right)\frac{1}{\sigma^2}x(kT_s;\hat{\theta}). \quad (7.130)$$

A block diagram outlining an iterative approach to finding $\hat{\theta}$ based on (7.130) is shown in Figure 7.8.3. This is a PLL structure where the right-hand side of (7.130) is the error signal. This complexity of this structure is often reduced by replacing the hyperbolic tangent with an approximation. A plot of $\tanh(X)$ versus X is shown in Figure 7.8.4. This plot shows that the hyperbolic tangent is well approximated by

$$\tanh(X) \approx \begin{cases} X & |X| < 0.3 \\ \operatorname{sgn}\{X\} & |X| > 3 \end{cases}. \quad (7.131)$$

Thus, the form of the approximation is determined by the magnitude of the argument. Equation (7.130) shows that the magnitude of the argument is proportional to the reciprocal of the noise variance σ^2. The magnitude of σ^2 relative to the magnitudes of $a_0(k)$ and $a_1(k)$ is determined by the signal-to-noise ratio. When the signal-to-noise ratio is large, the input to the hyperbolic tangent is large. Using the large signal approximation for the hyperbolic tangent, the error signal (input to the summation block) is, ignoring the scaling constant σ^2,

$$e(k) \approx \operatorname{sgn}\{x(kT_s;\theta)\}y(kT_s;\theta) - \operatorname{sgn}\{y(kT_s;\theta)\}x(kT_s;\theta). \quad (7.132)$$

Section 7.9 Notes and References

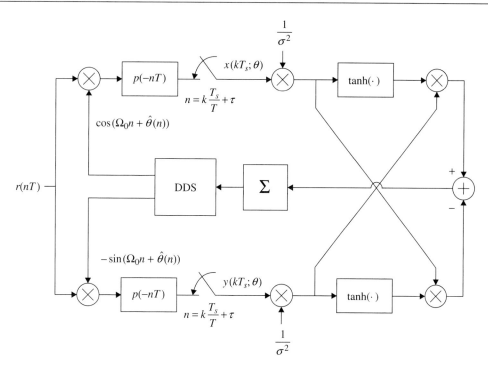

Figure 7.8.3 Block diagram of the maximum likelihood QPSK phase estimator based on the form (7.130).

Note that for QPSK, the symbol decisions are

$$\hat{a}_0(k) = \text{sgn}\left\{x(kT_s;\theta)\right\}$$
$$\hat{a}_1(k) = \text{sgn}\left\{y(kT_s;\theta)\right\}.$$

Thus, error signal may also be expressed as

$$e(k) \approx \hat{a}_0(k)y(kT_s;\theta) - \hat{a}_1(k)x(kT_s;\theta). \tag{7.133}$$

The large signal-to-noise ratio approximation for the non-data-aided ML phase estimator may be realized as a decision-directed PLL, such as those described in Section 7.2.2.

7.9 NOTES AND REFERENCES

7.9.1 Topics Covered

As with timing synchronization, much of the work on carrier phase synchronization in the early years of digital communications was devoted to adhoc techniques that later

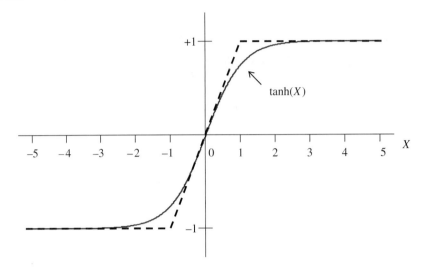

Figure 7.8.4 Plot of tanh(X) versus X illustrating the accuracy of the approximation (7.131).

turned out to be approximations to maximum likelihood estimation. The material in this chapter relies heavily on the text by Mengali and D'Andrea [123]. The emphasis of this chapter has been on close-loop techniques based on the discrete-time PLL. PLL-based carrier phase synchronizers operate in the non-data-aided mode most of the time. The basic formulation for the non-data-aided carrier phase synchronizing PLLs illustrated in Figures 7.1.1 and 7.1.2 are generalized versions of the Costas loop described by Proakis, Drouilhet, and Price [124], Viterbi [125], Natali and Walbesser [126], Gardner [127], and Linsey and Simon [128]. The performance of the Costas loop has been thoroughly analyzed over the years: see Kobayashi [129], Lindsey [130], Simon and Smith [131], Falconer [132], Mengali [133], Falconer and Salz [134], Simon [135], Meyers and Franks [136], Leclerc and Vandamme [137], and Moridi and Sari [138]. The tutorial by Franks [139] surveys continuous-time solutions for carrier phase synchronization.

The examples for differential encoding were limited to BPSK, QPSK, and offset QPSK. Differential encoding can also be applied to MPSK in a straightforward generalization. The application of differential encoding for MQAM is discussed by Weber [140]. The differential encoding and decoding rules for offset QPSK are the ones specified in the 2004 edition of the IRIG-106 Standard [141]. The adoption of this rule was motivated by its inclusion in the text by Feher [142]. This rule was motivated by the more general work of Weber [140] who interpreted offset I/Q modulations as forms of continuous-phase frequency shift keying. In this case, the differential encoding applies to the sign of the frequency shift. The derivation of the differential encoding/decoding equations for offset QPSK presented in this chapter is detailed by Rice [143].

7.9.2 Topics Not Covered

There are several aspects of synchronization that were not covered in this chapter. The first is carrier phase synchronization in burst-mode communications. Usually, a sequence of known data symbols is included in each burst transmission and the phase synchronizer estimates the carrier phase offset and applies this estimate to de-rotate all data corresponding to the burst. The maximum likelihood estimator is given by (7.127). Non-data-aided approaches are also possible. An approach suitable for MPSK was described by Viterbi and Viterbi [144]. This technique was generalized to MQAM by Moeneclaey and de Jonghe [145].

The carrier frequency offset problem is a very important issue. The phase locked loops described in this chapter are capable of compensating for some degree of frequency offset because a small frequency offset creates a slow time-variation on the carrier phase. The maximum frequency offset with which a PLL can obtain lock is called the *pull-in range* and is approximated by (C.41) in Appendix C. In the presence of a nonzero frequency offset, first order PLLs lock with a constant phase error that is proportional to the ratio of the frequency offset to the open loop gain. Second-order loops lock with zero phase error. The frequency acquisition of a carrier phase PLL in a digital communications system has been investigated by Frazier and Page [146], Messerschmitt [147], Mengali [148], and Cahn [149]. When the anticipated frequency offset is too large to be handled effectively by a phase locked loop, other techniques must be applied. The most common type of closed-loop frequency offset compensation is the use of a frequency locked loop to perform automatic frequency control. Frequency locked loops are described by Natali [150,151], Gardner [152], Sari and Moridi [153], Meyr and Ascheid [154], D'Andrea and Mengali [155], and Karam, Daffara, and Sari [156]. In burst mode communications (where the carrier phase offset is estimated using a block of training symbols) there is no tracking of the phase error through the burst. As a consequence, a separate frequency offset estimator must be used. The most popular data-aided frequency estimators are by Kay [157], Fitz [158,159], and Luise and Regiannini [160]. Non-data-aided techniques suitable for use in burst mode communication include those by Chuang and Sollenberger [161] and Classen, Meyr, and Sehier [162]. Uncompensated frequency offsets on the order of 10%–20% of the symbol rate can produce intersymbol interference in digital receivers [163].

False lock and cycle slips are important properties of carrier phase synchronizers that were not mentioned in this chapter. See the Notes and References Section of Appendix C for references on these issues.

Differential encoding, which was described in the context of phase ambiguity resolution, also permits a noncoherent detection technique known as *differential detection*. The detection algorithm along with its performance analysis is discussed in more advanced textbooks such as Ref. [10].

7.10 EXERCISES

7.1 Derive the expression given by (7.21) for the average S-curve for the heuristic QPSK data-aided phase error detector based on the error signal (7.16).

7.2 Derive the expression given by (7.22) for the average S-curve for the heuristic QPSK decision-directed phase error detector based on the error signal (7.17).

7.3 Show that the sine of the phase error for the heuristic QPSK phase error detector is given by (7.24).

7.4 Derive the expression given by (7.30) for the average S-curve for the ML QPSK data-aided phase error detector based on the error signal (7.25).

7.5 Derive the expression given by (7.31) for the average S-curve for the ML QPSK decision-directed phase error detector based on the error signal (7.26).

7.6 This exercise explores the performance of carrier phase synchronization for QPSK.
  (a) Compare the S-curves for the data-aided phase error detector (7.21) and the decision-directed phase error detector (7.22) for the heuristic phase error detector. How are they the same? How are they different?
  (b) Compare the S-curves for the data-aided phase error detector (7.30) and the decision-directed phase error detector (7.31) for the ML phase error detector. How are they the same? How are they different?
  (c) Compare the S-curves for the heuristic data-aided phase error detector (7.21) and the ML data-aided phase error detector (7.30). How are they the same? How are they different?
  (d) Compare the S-curve for the heuristic decision-directed phase error detector (7.22) and the ML decision-directed phase error detector (7.31). How are they the same? How are they different?

7.7 Derive the expression given by (7.45) for the average S-curve for the heuristic BPSK data-aided phase error detector based on the error signal (7.41).

7.8 Derive the expression given by (7.45) for the average S-curve for the heuristic BPSK decision-directed phase error detector based on the error signal (7.42).

7.9 Derive the expression given by (7.48) for the average S-curve for the ML BPSK data-aided phase error detector based on the error signal (7.46).

7.10 Derive the expression given by (7.49) for the average S-curve for the ML BPSK decision-directed phase error detector based on the error signal (7.47).

7.11 This exercise explores the performance of carrier phase synchronization for BPSK.
  (a) Compare the S-curves for the data-aided phase error detector (7.44) and the decision-directed phase error detector (7.45) for the heuristic phase error detector. How are they the same? How are they different?
  (b) Compare the S-curves for the data-aided phase error detector (7.48) and the decision-directed phase error detector (7.49) for the ML phase error detector. How are they the same? How are they different?

Section 7.10 Exercises

(c) Compare the S-curves for the heuristic data-aided phase error detector (7.44) and the ML data-aided phase error detector (7.48). How are they the same? How are they different?

(d) Compare the S-curve for the heuristic decision-directed phase error detector (7.45) and the ML decision-directed phase error detector (7.49). How are they the same? How are they different?

7.12 This exercise explores S-curves for the Y constellation.
(a) Derive the average S-curve for the Y constellation for a phase error detector based on an error signal of the form (7.16).
(b) Derive the average S-curve for the Y constellation for a phase error detector based on an error signal of the form (7.17).
(c) Derive the average S-curve for the Y constellation for a phase error detector based on an error signal of the form (7.25).
(d) Derive the average S-curve for the Y constellation for a phase error detector based on an error signal of the form (7.26).

7.13 This exercise explores S-curves for the 8-PSK constellation.
(a) Derive the average S-curve for the 8-PSK constellation for a phase error detector based on an error signal of the form (7.16).
(b) Derive the average S-curve for the 8-PSK constellation for a phase error detector based on an error signal of the form (7.17).
(c) Derive the average S-curve for the 8-PSK constellation for a phase error detector based on an error signal of the form (7.25).
(d) Derive the average S-curve for the 8-PSK constellation for a phase error detector based on an error signal of the form (7.26).

7.14 This exercise explores S-curves for the 16-QAM constellation.
(a) Derive the average S-curve for the 16-QAM constellation for a phase error detector based on an error signal of the form (7.16).
(b) Derive the average S-curve for the 16-QAM constellation for a phase error detector based on an error signal of the form (7.17).
(c) Derive the average S-curve for the 16-QAM constellation for a phase error detector based on an error signal of the form (7.25).
(d) Derive the average S-curve for the 16-QAM constellation for a phase error detector based on an error signal of the form (7.26).

7.15 The optimum coherent QPSK detector (and corresponding constellation) shown below is used to produce the inphase and quadrature matched filter outputs plotted below, where the optimum sampling instants are indicated by the circles:

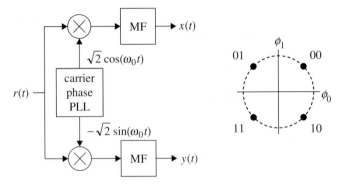

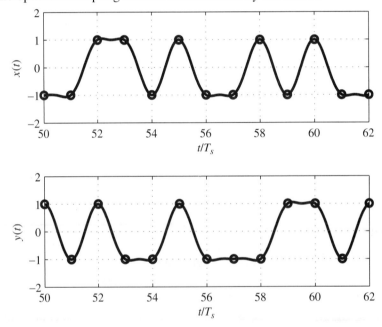

Suppose the system uses the unique word method for carrier phase ambiguity resolution where the data are organized into a frame as follows:

8 bits	18 bits
UW	DATA

UW = 10001101

In this case, the labels on the constellation refer to the true data bits. If the first four matched filter outputs correspond to the unique word, determine the phase ambiguity and the 18 data bits.

7.16 The optimum coherent QPSK detector (and corresponding constellation) shown below

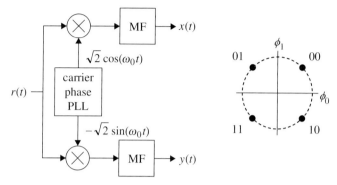

is used to produce the inphase and quadrature matched filter outputs plotted below, where the optimum sampling instants are indicated by the circles:

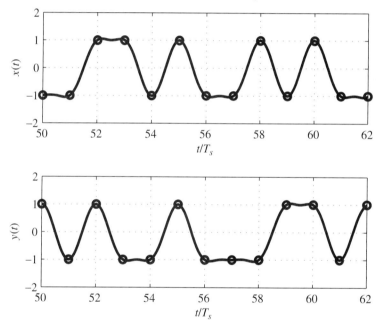

Suppose the QPSK system uses differential encoding for carrier phase ambiguity resolution. The differential encoding rule is shown in the table below.

bits	phase shift
00	180°
01	−90°
10	0°
11	+90°

In this case, all of the 13 matched filter outputs correspond to data and the labels on the constellation refer to differentially encoded bits. Determine the 24 data bits corresponding to the last 12 matched filter outputs.

7.17 The optimum coherent QPSK detector (and corresponding constellation) shown below

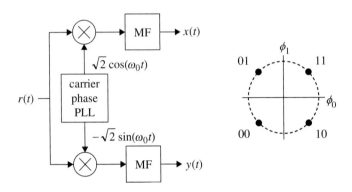

is used to produce the inphase and quadrature matched filter outputs plotted below, where the optimum sampling instants are indicated by the circles:

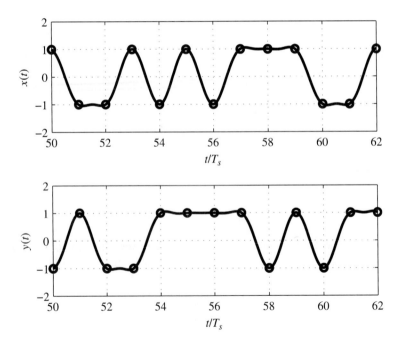

Suppose the system uses the unique word method for carrier phase ambiguity resolution where the data are organized into a frame as follows:

UW = 01101101

In this case, the labels on the constellation refer to the true data bits. If the first four matched filter outputs correspond to the unique word, determine the phase ambiguity and the 18 data bits.

7.18 The optimum coherent QPSK detector (and corresponding constellation) shown below

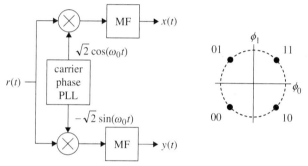

is used to produce the inphase and quadrature matched filter outputs plotted below, where the optimum sampling instants are indicated by the circles:

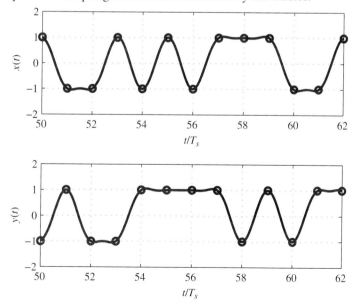

Suppose the QPSK system uses differential encoding for carrier phase ambiguity resolution. The differential encoding rule is shown in the table below.

bits	phase shift
00	$-90°$
01	$0°$
10	$180°$
11	$+90°$

In this case, all of the 13 matched filter outputs correspond to data and the labels on the constellation refer to differentially encoded bits. Determine the 24 data bits corresponding to the last 12 matched filter outputs.

7.19 The optimum coherent BPSK detector (and corresponding constellation) shown below

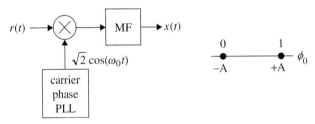

is used to produce the matched filter outputs plotted below, where the optimum sampling instants are indicated by circles:

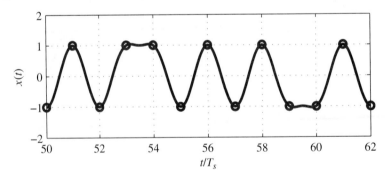

Suppose the system uses the unique word method for carrier phase ambiguity resolution where the data are organized into a frame as follows:

5 bits	18 bits
UW	DATA

UW = 10100

Section 7.10 Exercises

In this case, the labels on the constellation refer to the true data bits. If the first five matched filter outputs correspond to the unique word, determine the phase ambiguity and the 8 data bits.

7.20 The optimum coherent BPSK detector (and corresponding constellation) shown below

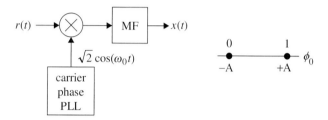

is used to produce the matched filter outputs plotted below, where the optimum sampling instants are indicated by circles:

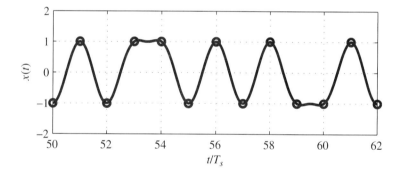

Suppose the system uses differential encoding for carrier phase ambiguity resolution. The differential encoding rule is shown in the table below:

bits	phase shift
0	0°
1	180°

In this case, all of the 13 matched filter outputs correspond to data and the labels on the constellation refer to the differentially encoded bits. Determine the 12 bits corresponding to the last 12 matched filter outputs.

7.21 Consider a differentially encoded BPSK system defined by the constellation shown below.

Suppose the matched filter output (of the optimum coherent BPSK detector) is plotted below where the circles represent the optimum sampling instants.

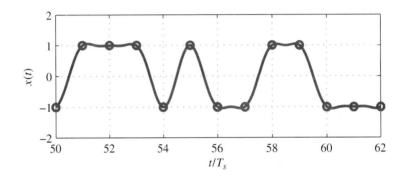

There are two possible bit sequences, depending on which of the two possible differential encoding rules were used. Determine the two possible bit sequences and indicate which differential encoding rule corresponds to which bit sequence.

7.22 The optimum coherent offset QPSK detector (and corresponding constellation) shown below

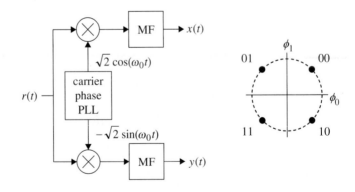

is used to produce the inphase and quadrature matched filter outputs plotted below, where the optimum sampling instants are indicated by the circles:

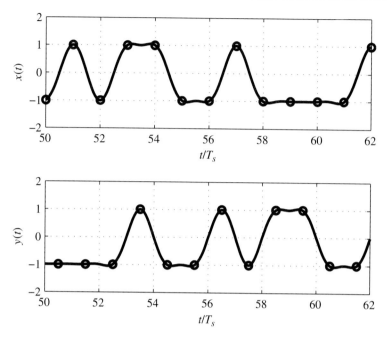

Suppose the system uses differential encoding for carrier phase and delay axis ambiguity resolution. In this case, all of the matched filter outputs correspond to data and the labels on the constellation refer to differentially encoded bits. Using the differential encoding rule (7.82) and the corresponding differential decoding rule (7.83), determine the 22 data bits corresponding to the last 11 pairs of matched filter outputs.

8

Symbol Timing Synchronization

8.1 BASIC PROBLEM FORMULATION 436
8.2 CONTINUOUS-TIME TECHNIQUES FOR M-ARY PAM 438
8.3 CONTINUOUS-TIME TECHNIQUES FOR MQAM 443
8.4 DISCRETE-TIME TECHNIQUES FOR M-ARY PAM 445
8.5 DISCRETE-TIME TECHNIQUES FOR MQAM 494
8.6 DISCRETE-TIME TECHNIQUES FOR OFFSET QPSK 497
8.7 DEALING WITH TRANSITION DENSITY: A PRACTICAL CONSIDERATION 501
8.8 MAXIMUM LIKELIHOOD ESTIMATION 503
8.9 NOTES AND REFERENCES 513
8.10 EXERCISES 515

Conceptually, symbol timing synchronization[1] is the process of estimating a clock signal that is aligned in both phase and frequency with the clock used to generate the data at the transmitter. Because it is not efficient to allocate spectrum to transmit a separate clock signal from the transmitter to the receiver for the purposes of timing synchronization, the data clock must be extracted from the noisy received waveforms that carry the data. For matched filter detectors, the clock signal is used to identify the instants when the matched filter output should be sampled.

The effect of a timing error for QPSK is illustrated in Figure 8.1. The optimum sampling instant for the inphase and quadrature matched filter outputs corresponds to the center of the eye diagrams. A timing error can be thought of as sampling the eye diagram at an instant that is not at the maximum average eye opening in the center of the eye diagram. The effects of this error are shown by the signal space projections. These signal space projections correspond to the values of the eye diagram at the timing instants shown. Even though perfect phase synchronization is used and no noise is added, the signal space projections are scattered; many are closer to their decision boundaries than they would be otherwise. As the timing error gets larger, the effect is more dramatic as illustrated.

The form of the symbol timing synchronizer is quite different for continuous-time and discrete-time systems and is perhaps one of the biggest differences between the two

---

[1]The origins of the word *synchronize* are explained in the introduction of Chapter 7.

Chapter 8 Symbol Timing Synchronization

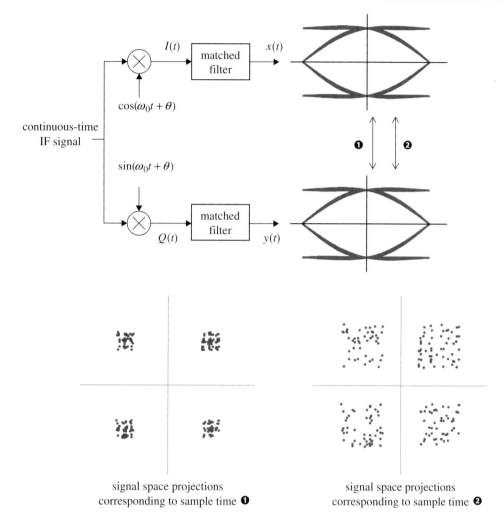

Figure 8.1 The effect of timing error on QPSK using a square-root raised-cosine pulse shape with 50% excess bandwidth. Perfect carrier phase synchronization is assumed.

implementations. Continuous-time techniques are reviewed in Section 8.2. Discrete-time techniques for symbol timing synchronization are covered in detail in Section 8.4. In both cases, symbol timing synchronization for M-ary PAM is developed. Extensions to MQAM are described in Sections 8.3 and 8.5 for continuous-time and discrete-time systems, respectively. The synchronizers presented in this chapter are all based on the phase-locked loop, or PLL. The fundamentals of PLL operation and analysis are reviewed, in detail, in Appendix C.

436   Chapter 8   Symbol Timing Synchronization

8.1 BASIC PROBLEM FORMULATION

Figure 8.1.1 shows a block diagram of the basic architecture for symbol timing synchronization for M-ary PAM using continuous-time techniques. Let the received M-ary PAM signal be

$$r(t) = G_a \sum_k a(k)p(t - kT_s - \tau) + w(t) \tag{8.1}$$

where $a(k) \in \{-(M-1), -(M-3), \ldots, -1, 1, \ldots, M-3, M-1\}$ is the k-th PAM symbol; T_s is the symbol time; τ is the unknown timing delay; $p(t)$ is a unit-energy pulse shape with support on the interval $-L_p T_s \leq t \leq L_p T_s$; G_a is the composite of all the amplitude gains and losses from the transmitter, through the transmission medium, to the receiver; and $w(t)$ is the additive white Gaussian noise. The received signal is passed through a matched filter whose impulse response is $p(-t)$. The output of the matched filter $x(t)$ may be expressed as

$$x(t) = G_a \sum_k a(k)r_p(t - kT_s - \tau) + v(t) \tag{8.2}$$

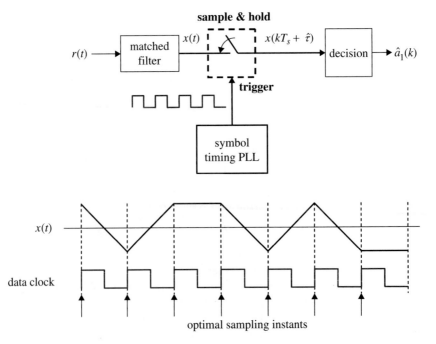

Figure 8.1.1 Block diagram for a binary PAM detector showing the role of the symbol timing PLL and the relationship between the matched filter output and the symbol timing PLL output.

Section 8.1 Basic Problem Formulation

where $r_p(u)$ is the autocorrelation function of the pulse shape defined by

$$r_p(u) = \int_{-L_p T_s}^{L_p T_s} p(t)p(t-u)dt \qquad (8.3)$$

and $v(t) = w(t) * p(-t)$ represents the noise at the output of the matched filter. Ideally, the matched filter output should be sampled at $t = kT_s + \tau$ for detection. This is easy if τ is known. When τ is not known, it must be estimated. This is the role of the symbol timing synchronizer.

Using the estimate $\hat{\tau}$ provided by symbol timing synchronizer, the matched filter output at $t = kT_s + \hat{\tau}$ is

$$x(kT_s + \hat{\tau}) = G_a a(k) r_p(\hat{\tau} - \tau) + G_a \sum_{m \neq k} a(m) r_p((k-m)T_s + \hat{\tau} - \tau). \qquad (8.4)$$

In the following, it will be convenient to express this output in terms of the timing error $\tau_e = \tau - \hat{\tau}$:

$$x(kT_s + \hat{\tau}) = G_a a(k) r_p(-\tau_e) + G_a \sum_{m \neq k} a(m) r_p((k-m)T_s - \tau_e). \qquad (8.5)$$

Note that when the pulse shape satisfies the Nyquist condition for no ISI, the second term is zero for $\tau_e = 0$.

In a continuous-time detector, the goal of symbol timing is to produce a clock signal aligned with the data transitions as illustrated in Figure 8.1.1 for binary PAM. In this example, the rising edge of the clock is aligned with the symbol transitions and is used to trigger the sample-and-hold operation at the matched filter output. The clock signal is produced by a symbol timing PLL.

The symbol timing PLL in Figure 8.1.1 consists of a timing error detector (TED), loop filter, and voltage controlled clock (VCC) arranged as shown in Figure 8.1.2. The TED computes the phase error between the VCC output and the clock signal embedded in the matched filter outputs. This timing error is filtered and used to adjust the phase of the VCC output to align the clock edges with the symbol boundaries.

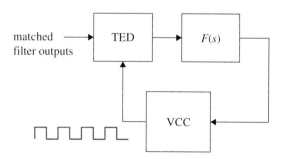

Figure 8.1.2 The three basic components of an continuous-time symbol timing PLL.

When cast in terms of the basic PLL structure of Figure C.1.1, the matched filter output plays the role of the input sinusoid, the TED plays the role of the phase detector (i.e., it detects the difference in *phase* between the data clock embedded in the matched filter output and the VCC output), and the VCC plays the role of the voltage controlled oscillator (VCO). Because the matched filter output does not look like a sinusoid (or pure clock, in this case), the TED must extract the clock information from the matched filter output. For this reason, TED design is the component that requires the most ingenuity and creativity.

8.2 CONTINUOUS-TIME TECHNIQUES FOR M-ARY PAM

The operation of the TED is best understood using the eye diagram. Figure 8.2.1 illustrates this concept for binary PAM. Observe that the optimum sampling instant coincides with the time instant of maximum average eye opening. The time instant of maximum eye opening

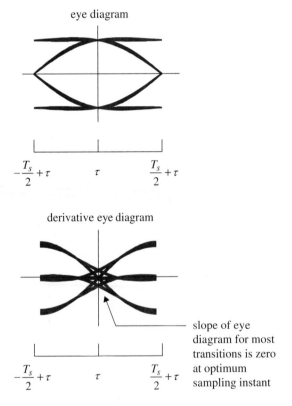

Figure 8.2.1 Eye diagram demonstrating that the optimum sampling instant and the instant of maximum average eye opening coincide with the instant where the slope of the average eye is 0.

occurs at the time instant when the average slope of the eye diagram is zero. The nonzero slope at $t = \tau$ are points in trajectories corresponding to no data sign transition followed by a data sign transition or a data sign transition followed by no data sign transition. This feature reinforces the fact that symbol timing synchronizers rely on data sign transitions to produce a proper timing error signal.

The slope of the eye diagram can be used to generate a timing error. Figure 8.2.2 demonstrates that the sign of the slope of the eye must be qualified to provide the correct timing error. Figure 8.2.2 (a) illustrates the case where the current sampling instant is early (i.e., $\hat{\tau}(k) < \tau$, which means $\tau_e(k) > 0$) and the data symbol $a(k) = +1$. The next sampling instant $\hat{\tau}(k+1)$ should be greater than $\hat{\tau}(k)$. This is accomplished by increasing the period of the VCC output. The slope of the matched filter output at $t = \hat{\tau}(k)$ is positive and can be used as the error signal. Note that this applies only to the part of the eye corresponding to a transition from $a(k-1) = -1$ to $a(k) = +1$. The approximately horizontal portion of

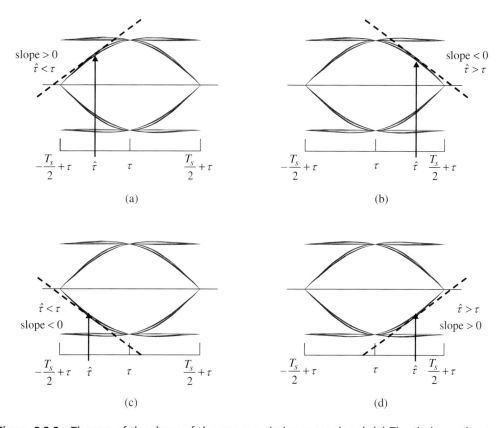

Figure 8.2.2 The use of the slope of the eye as a timing error signal. (a) The timing estimate is early and the corresponding symbol is positive. (b) The timing estimate is late and the corresponding symbol is positive. (c) The timing estimate is early and the corresponding symbol is negative. (d) The timing estimate is late and the corresponding symbol is negative.

the eye diagram just above the end of the arrow corresponds to the case where there is no data transition (i.e., $a(k-1) = +1$). The slope of the eye diagram is very small along this trajectory. As a consequence, little timing error information is provided in the absence of a data transition. This property will be more obvious in the examples to come.

Figure 8.2.2 (b) demonstrates the case where the current sampling instant is late (i.e., $\hat{\tau}(k) > \tau$ or $\tau_e < 0$) and the data symbol $a(k) = +1$. In this case, the period of the VCC should be decreased to force $\hat{\tau}(k+1) < \hat{\tau}(k)$. The slope of the eye at $t = \hat{\tau}(k)$ along the trajectory from $a(k) = +1$ to $a(k+1) = -1$ is negative and thus indicates the proper adjustment to the VCC period. As before, the trajectory from $a(k) = +1$ to $a(k+1) = +1$ has a very small slope and thus provides little or no timing error information.

In the preceding two cases, the data symbol $a(k) = +1$. Figures 8.2.2 (c) and (d) demonstrate what happens when $a(k) = -1$. In Figure 8.2.2 (c), the current sampling instant is early (i.e., $\hat{\tau}(k) < \tau$ or $\tau_e > 0$) and the period of the VCC should be increased to force $\hat{\tau}(k+1) > \hat{\tau}(k)$. Unfortunately, the slope of the eye corresponding to the transition from $a(k-1) = +1$ to $a(k) = -1$ is negative at $t = \hat{\tau}(k)$ and therefore does not indicate the proper adjustment to the VCC period. This is because the data symbol $a(k)$ is negative. If the slope of the eye $t = \hat{\tau}(k)$ is altered by the sign of the data symbol, then a signal that provides the proper adjustment to the VCC period is obtained. Similarly, Figure 8.2.2 (d) shows the case where the current sampling instant is late (i.e., $\hat{\tau}(k) > \tau$ or $\tau_e < 0$), which requires that $\hat{\tau}(k+1) < \hat{\tau}(k)$ and a decrease in the VCC period. Because the slope of the eye corresponding to the transition from $a(k) = -1$ to $a(k+1) = +1$ at $t = \hat{\tau}(k)$ is positive, the product of the slope and the sign of $a(k)$ provides the proper signal for adjusting the VCC period.

The aforementioned observations suggest a timing error signal of the form

$$e(k) = a(k)\dot{x}\left(kT_s + \hat{\tau}(k)\right) \tag{8.6}$$

for the data-aided case, and

$$e(k) = \hat{a}(k)\dot{x}\left(kT_s + \hat{\tau}(k)\right) \tag{8.7}$$

for the decision-directed case where $\dot{x}(t)$ is the time derivative of the matched filter output. For binary PAM, the symbol estimate is given by

$$\hat{a}(k) = \text{sgn}\left\{x(kT_s + \hat{\tau})\right\}. \tag{8.8}$$

As it turns out, this error signal follows from the maximum likelihood estimate for timing offset as outlined in Section 8.8. A block diagram of a symbol timing PLL based on (8.7) is illustrated in Figure 8.2.3 for binary PAM. Generation of the error signal requires a differentiator connected to the matched filter output. The output of the differentiator is sampled at $t = kT_s + \hat{\tau}(k)$ to provide the slope of the eye at $\hat{\tau}(k)$. The sign of this slope is qualified by the sign of the data by multiplying by the sign of the corresponding sampled matched filter output to form the error signal (8.7).

Figure 8.2.3 also plots an example of the received waveform $r(t)$ (neglecting noise), the corresponding matched filter output $x(t)$, the derivative of the matched filter output $\dot{x}(t)$,

Section 8.2 Continuous-Time Techniques for M-ary PAM

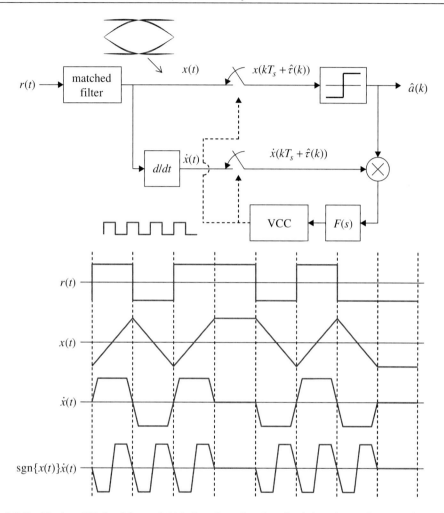

Figure 8.2.3 Timing PLL for binary PAM showing the detail of the phase detector based on (8.7).

and the product $\text{sgn}\{x(t)\}\dot{x}(t)$ whose samples are used as the timing error signal. Observe that the timing error signal $\text{sgn}\{x(t)\}\dot{x}(t)$ is zero at the optimum sampling time.[2] In the absence of data transitions, the timing error signal is zero throughout the entire symbol interval because the derivative of the matched filter output is zero. This demonstrates that data transitions are necessary to provide sufficient timing error information to obtain symbol

---

[2]The timing error signal $\text{sgn}\{x(t)\}\dot{x}(t)$ is also zero halfway between the optimum sampling times. This zero crossing is not a stable lock point.

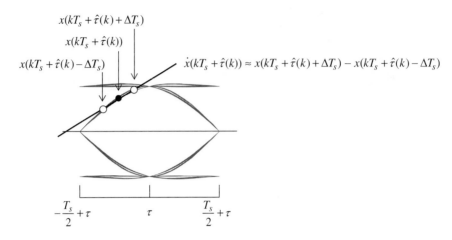

Figure 8.2.4 Illustration of using a difference to approximate the derivative of the eye diagram.

timing synchronization. If too many consecutive symbols are the same, the symbol timing PLL could drift out of lock.

Often, it is desirable to reduce the complexity of the TED by approximating the derivative operation with a difference as illustrated in Figure 8.2.4 for binary PAM and the case $\hat{\tau}(k) < \tau$ and $a(k) = 1$. Because

$$\dot{x}(t_0) = \lim_{\Delta \to 0} \frac{x(t_0 + \Delta) - x(t_0 - \Delta)}{2\Delta}, \tag{8.9}$$

the derivative of the matched filter output may by approximated using the difference as shown. Qualifying the sign of the difference by the data produces a TED known as an *early–late gate detector*. The data-aided version of the early–late gate timing error is

$$e(k) = a(k)\left[x\left(kT_s + \hat{\tau}(k) + \Delta T_s\right) - x\left(kT_s + \hat{\tau}(k) - \Delta T_s\right)\right] \tag{8.10}$$

and the decision-directed version is

$$e(k) = \hat{a}(k)\left[x\left(kT_s + \hat{\tau}(k) + \Delta T_s\right) - x\left(kT_s + \hat{\tau}(k) - \Delta T_s\right)\right]. \tag{8.11}$$

For binary PAM, a popular form of the early–late gate error signal that does not rely directly on the decisions is

$$e(k) = \left|x\left(kT_s + \hat{\tau}(k) + \Delta T_s\right)\right| - \left|x\left(kT_s + \hat{\tau}(k) - \Delta T_s\right)\right|. \tag{8.12}$$

A block diagram of a binary PAM timing PLL based on the early–late gate (8.12) is illustrated in Figure 8.2.5. Also shown are the baseband received signal $r(t)$ (neglecting noise), the corresponding matched filter output $x(t)$, and the signal $|x(t + \Delta T_s)| - |x(t - \Delta T_s)|$ for $\Delta = 1/4$. Observe that the timing error signal $|x(t + \Delta T_s)| - |x(t - \Delta T_s)|$ is zero at the optimum

Section 8.3 Continuous-Time Techniques for MQAM

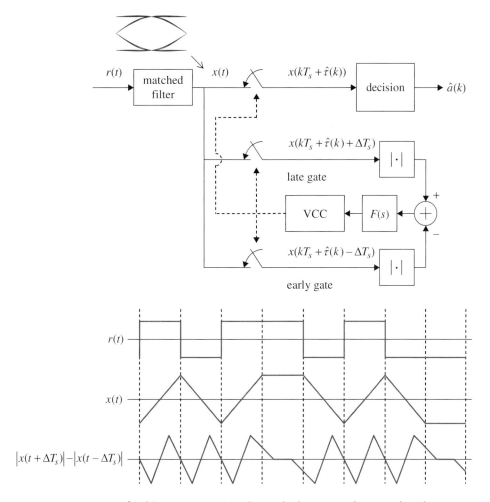

Figure 8.2.5 Timing PLL for binary PAM using the early–late gate detector (8.12).

sampling instants when there is a data transition. When there is no data transition, the timing error signal is not zero at the optimum sampling instant. This nonzero value is called *self noise* and can limit the accuracy of the timing PLL if the density of data transitions is not sufficiently high.

8.3 CONTINUOUS-TIME TECHNIQUES FOR MQAM

The general form for the non-offset MQAM symbol timing PLL is shown in Figure 8.3.1 for the case of QPSK. Assuming perfect carrier phase synchronization and neglecting the double

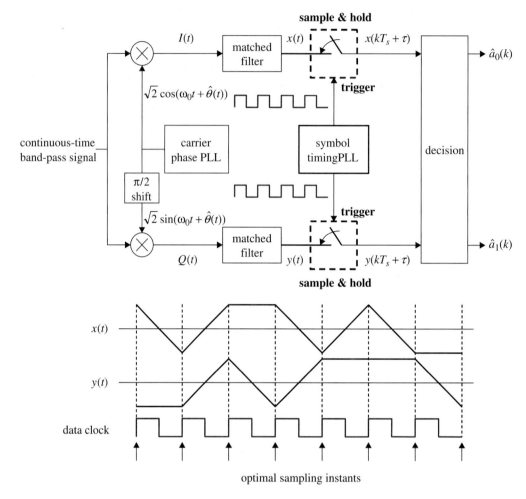

Figure 8.3.1 Block diagram for a QPSK detector showing the role of the symbol timing PLL and the relationship between the matched filter output and the symbol timing PLL output.

frequency terms and noise, the inphase and quadrature components are

$$I(t) = G_a \sum_m a_0(m) p(t - mT_s - \tau) \tag{8.13}$$

$$Q(t) = G_a \sum_m a_1(m) p(t - mT_s - \tau) \tag{8.14}$$

Section 8.4 Discrete-Time Techniques for M-ary PAM **445**

where τ is the unknown delay to be estimated by the symbol timing synchronizer and G_a is the amplitude of the received signal. The matched filter outputs are

$$x(t) = G_a \sum_m a_0(m) r_p(t - mT_s - \tau) \qquad (8.15)$$

$$y(t) = G_a \sum_m a_1(m) r_p(t - mT_s - \tau) \qquad (8.16)$$

where $r_p(u)$ is the autocorrelation function of the pulse shape defined by (8.3). Both these equations are of the same form as (8.2), the sampled matched filter output for M-ary PAM. Thus, timing error information can be derived from both the inphase and quadrature components in parallel. Extending the notions developed in Section 8.2, the data-aided timing error signal is

$$e(k) = a_0(k)\dot{x}(kT_s) + a_1(k)\dot{y}(kT_s) \qquad (8.17)$$

and the decision-directed timing error signal is

$$e(k) = \hat{a}_0(k)\dot{x}(kT_s) + \hat{a}_1(k)\dot{y}(kT_s) \qquad (8.18)$$

where $\dot{x}(t)$ and $\dot{y}(t)$ are the time derivatives of the inphase and quadrature matched filter outputs, respectively. A popular form for the QPSK decision-directed error signal results from a straightforward extension of (8.18):

$$e(k) = \operatorname{sgn}\{a_0(k)\}\dot{x}(kT_s) + \operatorname{sgn}\{a_1(k)\}\dot{y}(kT_s). \qquad (8.19)$$

The derivative operation can be replaced by an early-late gate structure on both the inphase and quadrature components to form another popular QPSK error signal

$$e(k) = |x(kT_s + \hat{\tau}(k) + \Delta T_s)| - |x(kT_s + \hat{\tau}(k) + \Delta T_s)| \\ + |y(kT_s + \hat{\tau}(k) + \Delta T_s)| - |y(kT_s + \hat{\tau}(k) + \Delta T_s)|. \qquad (8.20)$$

A QPSK symbol timing PLL based on (8.19) is shown in Figure 8.3.2.

8.4 DISCRETE-TIME TECHNIQUES FOR M-ARY PAM

When the matched filter is implemented as a discrete-time filter, an analog-to-digital converter (ADC) preceding the matched filter is required. An ADC produces T-spaced samples of (8.1) at a rate N samples/symbol. The n-th sample of this waveform may be represented by

$$r(nT) = G_a \sum_m a(m) p(nT - mT_s - \tau) + w(nT) \qquad (8.21)$$

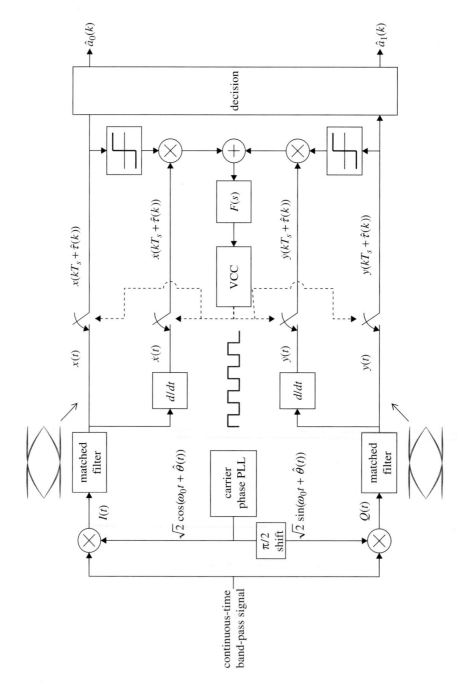

Figure 8.3.2 QPSK symbol timing PLL showing the detail of the phase detector based on (8.19).

Section 8.4 Discrete-Time Techniques for M-ary PAM

where $a(k) \in \{-(M-1)A, -(M-3)A, \ldots, -A, +A, \ldots, +(M-3)A, +(M-1)A\}$ is the k-th symbol; T_s is the symbol time; τ is the unknown timing delay; $p(nT)$ are samples of $p(t)$, the bandlimited unit-energy pulse shape with support on the interval $-L_p T_s \leq t \leq L_p T_s$; G_a is the composite of all the amplitude gains and losses from the transmitter, through the transmission medium, to the receiver; and $w(nT)$ are samples of the bandlimited thermal noise. It is assumed the data symbols are uncorrelated:

$$E\{a(k)a(m)\} = E_{\text{avg}}\delta(m-k) \tag{8.22}$$

where E_{avg} is the average symbol energy. The received signal is processed by a discrete-time matched filter whose impulse response consists of samples of the time reversed pulse shape waveform. The matched filter output is

$$x(nT) = \frac{G_a}{T} \sum_m a(m) r_p(nT - mT_s - \tau) + v(nT) \tag{8.23}$$

where $r_p(u)$ is the autocorrelation function of the pulse shape given by (8.3) and $v(nT) = p(-nT) * w(nT)$ is the component of the matched filter output due to the noise.

The goal of symbol timing synchronization is to produce N samples at the matched filter outputs during each symbol interval such that one of the samples is aligned with the maximum eye opening. There are two basic approaches to the problem. The first approach, illustrated in Figure 8.4.1, uses the timing error to adjust the phase of the voltage controlled clock (VCC)

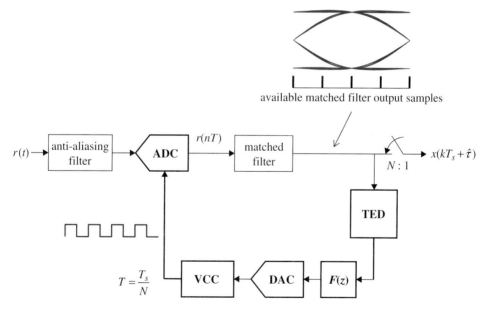

Figure 8.4.1 A hybrid continuous-time/discrete-time approach to symbol timing synchronization for sampled-data detectors.

that triggers the ADC. As a result, the samples of $r(t)$ are aligned with the symbol boundaries and the optimum eye opening as shown. This approach has the advantage that it produces samples that are aligned in both phase and frequency with the data clock (i.e., T and T_s are commensurate). There are four disadvantages to this approach.

1. First, a feedback path to the continuous-time part of the system is required. The hardware overhead of transferring from the digital to analog domains via a multi-bit output bus, a data control line, a multi-bit DAC, and an analog filter to supply the control voltage to the VCC has the potential to complicate the analog front-end design.
2. Second, the transport delay of the matched filter now resides in the feedback path of the timing control loop. This significantly reduces the response time of the timing recovery loop.
3. Third, higher levels of phase noise (and hence, timing jitter) are contributed by the VCC relative to the phase noise contributed by fixed-frequency sampling clocks. See Section 9.1.1 for a discussion on the effects of timing jitter on ADCs.
4. Fourth, this technique does not allow the ADC to be applied directly to the band-pass signal if the band-pass signal contains multiplexed signals whose symbol clocks are derived from independent sources. In software defined radios, the goal is to "push the ADC to the antenna." To meet this goal, demultiplexing and channel selection must be performed using digital signal processing on asynchronous samples of $r(t)$.

The second approach, illustrated in Figure 8.4.2, addresses these issues by sampling the received signal $r(t)$ at a fixed rate $1/T$ that is asynchronous with the symbol rate $1/T_s$. The time delay τ is estimated solely from the samples $x(nT)$, the asynchronous samples at the output of the matched filter. This approach produces samples that are not aligned with the symbol boundaries as shown by the eye diagram at the output of the matched filter. The role of symbol timing synchronization is to "move" the samples to the desired time instants. Another name for "moving" samples in time is *interpolation*. Because the timing synchronizer has to adapt to an unknown time delay, the interpolator must be adaptive. When working properly, the interpolator produces matched filter outputs that are aligned with the symbol boundaries and the optimum sampling instant as illustrated by the eye diagram at the output of the interpolator in Figure 8.4.2.

The major disadvantage to this approach is *interpolation jitter* that occurs when $T_i \neq NT$. In this case, an interpolant is output every N samples, on average. But, due to the condition $T_i \neq NT$, the fractional timing error accumulates and eventually becomes unity. When this occurs, an interpolant is output $N-1$ samples or $N+1$ after the previous interpolant samples to make up the difference. (Which one it is depends on the sign of the accumulating fractional timing error.) This interpolation jitter is especially problematic if the data bits must be retransmitted over a synchronous link to some other destination.

The asynchronous sampling approach shown in Figure 8.4.2 is the more common approach used for timing synchronization in sampled-data detectors. Often, timing synchronization is performed using a discrete-time PLL composed of three basic units: the TED, loop filter, and the interpolation control. Cast in terms of the general PLL described in Appendix C, the interpolator and TED combination plays the role of the phase detector and the interpolator

Section 8.4 Discrete-Time Techniques for M-ary PAM

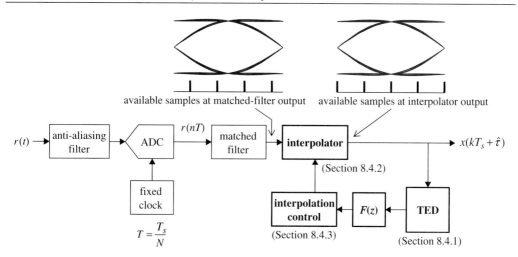

Figure 8.4.2 A discrete-time approach to symbol timing synchronization for sampled-data detectors.

control plays the role of the DDS, or oscillator. Timing error detectors are described in Section 8.4.1, interpolation in Section 8.4.2, and interpolation control in Section 8.4.3.

8.4.1 Timing Error Detectors

In general, the TEDs produce an error signal once every symbol based on the current timing estimate and using matched filter input, $r(nT)$, and the matched filter output $x(nT)$. In other words, the discrete-time error signal is updated at the symbol rate.

Assume an ideal interpolator is available that computes the interpolant $x(kT_s + \hat{\tau})$ using a timing delay estimate $\hat{\tau}$ and the outputs of the matched filter. The interpolant may be expressed as

$$x(kT_s + \hat{\tau}) = K \sum_m a(m) r_p \left((k-m)T_s + \hat{\tau} - \tau \right) + v(kT_s + \hat{\tau}) \tag{8.24}$$

$$= K \sum_m a(m) r_p \left((k-m)T_s - \tau_e \right) + v(kT_s + \hat{\tau}) \tag{8.25}$$

where $\tau_e = \tau - \hat{\tau}$ is the timing error and $K = G_a/T$. The TED produces a signal that is a function of the timing error τ_e in the same way the phase detector in the carrier phase PLL produced a signal that was a function of the phase error. The output of the TED, $e(kT_s)$, is a function of the interpolated matched filter outputs and the data symbols (or their estimates). The characteristics of the TED are described by the S-curve $g(\tau_e)$.

Maximum Likelihood Timing Error Detector. The maximum likelihood timing error detector (MLTED) is derived in Section 8.8 and uses the sign-corrected slope of the eye diagram for the error signal as described in Section 8.2 and illustrated in Figure 8.2.2. The error signal for the data-aided TED is

$$e(k) = a(k)\dot{x}(kT_s + \hat{\tau}) \tag{8.26}$$

while the error signal for the decision-directed TED is

$$e(k) = \hat{a}(k)\dot{x}(kT_s + \hat{\tau}) \tag{8.27}$$

where $\dot{x}(kT_s + \hat{\tau})$ is the time derivative of the matched filter output at $t = kT_s + \hat{\tau}$. For binary PAM, the data symbol estimates are

$$\hat{a}(k) = A \times \text{sgn}\left\{x(kT_s + \hat{\tau})\right\}. \tag{8.28}$$

As an alternative, the sign of the matched filter output can also be used.

$$\hat{a}(k) = \text{sgn}\left\{x(kT_s + \hat{\tau})\right\}. \tag{8.29}$$

The effect of this minor simplification is a simple scaling of the S-curve. The TED gain, K_p is also scaled by the same value.

For nonbinary PAM, the data symbol estimates are used in the obvious way. A commonly used simplification is to use (8.29) in place of the true data symbol estimates in the nonbinary case. The performance of this approach is acceptable in most applications. When (8.29) is used, the TED gain will be smaller than that derived below. Care must be taken to ensure that the proper value of K_p is used in computing the loop filter constants.

The S-curve for the MLTED is obtained by computing the expected value of the error signal (8.26) using (8.25) for $x(kT_s + \hat{\tau})$ and the property (8.22):

$$g(\tau_e) = \text{E}\left\{a(k)\frac{d}{dt}x(kT_s + \tau)\right\}$$

$$= \text{E}\left\{a(k)\frac{d}{dt}K\sum_m a(m)r_p\left((k-m)T_s - \tau_e\right)\right\}$$

$$= KE_{\text{avg}}\dot{r}_p(-\tau_e) \tag{8.30}$$

where the last line follows from (8.22), and $\dot{r}_p(-\tau_e)$ is the time derivative of the pulse shape autocorrelation function evaluated at $-\tau_e$. The S-curve for the decision-directed MLTED is obtained by assuming $\hat{a}(k) = a(k)$ and proceeding as outlined above. As long as the decisions are correct, the S-curve for the decision-directed MLTED is identical to the S-curve for the data-aided MLTED. This is illustrated in Figure 8.4.3, which is a plot of the S-curve for the square-root raised-cosine pulse shape with 50% excess bandwidth. The S-curves for both the data-aided detector and the decision-directed detector are identical for $|\tau_e/T_s| < 0.35$. This is because $\hat{a}(k) = a(k)$. When $|\tau_e/T_s| > 0.35$ the S-curve for the decision-directed detector departs from the S-curve for the data-aided detector due to decision errors. Note that when (8.29) is used, E_{avg} in (8.30) is replaced by a different value that is a function of A and M.

Section 8.4 Discrete-Time Techniques for M-ary PAM

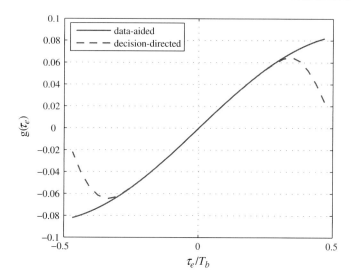

Figure 8.4.3 S-curves for the data-aided MLTED (solid line) and the decision-directed MLTED (dashed line). These are simulation results for binary PAM using a square-root raised-cosine pulse shape with 50% excess bandwidth and $KE_{\text{avg}} = 1$. The signal-to-noise ratio is $E_b/N_0 = 20$ dB. The derivative matched filter was obtained by computing the first central difference of a unit-energy matched filter at $N = 32$ samples/symbol.

The detector gain K_p is the slope of $g(\tau_e)$ at $\tau_e = 0$. Equation (8.30) shows that K_p is a function of the received signal amplitude K, the average symbol energy E_{avg}, and the slope of $\dot{r}_p(\tau_e)$ at $\tau_e = 0$. Because $r_p(t)$ and its derivatives are a function of the pulse shape $p(t)$, K_p is also a function of the excess bandwidth when $p(t)$ is the square-root raised-cosine pulse shape. This dependence is plotted in Figure 8.4.4.

The fact that K_p is a function of the received signal level presents a challenge in achieving the desired PLL performance. The desired loop performance is determined by the loop filter constants K_1 and K_2 (for a proportional-plus-integrator loop filter). These constants are a function of the loop bandwidth, K_0, and K_p as explained in Appendix C — see (C.61). If K is not known, then K_p is not known and it is impossible to properly compute the loop filter constants to achieve the desired loop performance. Conceptually, there are two ways to address this issue. The first is to estimate the received signal level K and use the estimated value to compute the loop filter constants. In dynamic environments (such as mobile communications), the received signal amplitude changes. In this case, the loop filter constants must be recomputed every time K changes significantly. For many reasons, this is an undesirable mode of operation. The other option is to dynamically adjust the amplitude of the received signal to produce a signal whose amplitude is fixed at a predetermined value. A system that performs this function is called an *automatic gain control* or AGC and is described in Section 9.5. In this case, K is fixed at a known value so that the loop filter constants need only be computed once.

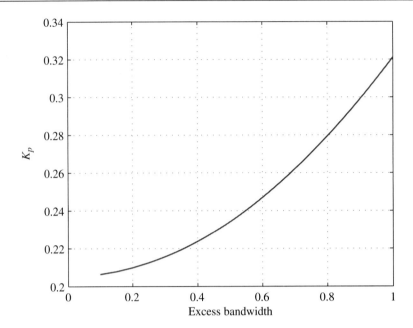

Figure 8.4.4 Phase detector gain, K_p, of the MLTED as a function of excess bandwidth for the square-root raised-cosine pulse shape and binary PAM with $KE_{avg} = 1$. The derivative of $r_p(\cdot)$ was obtained by computing the first central difference of a unit-energy raised-cosine response sampled at $N = 32$ samples/symbol.

Computing samples of the time derivative of a sampled waveform can be performed using the discrete-time differentiator developed in Section 3.3.3. Denoting the impulse response of the differentiating filter as $d(nT)$, samples of the derivative of $x(nT)$ may be expressed in one of two forms as

$$\dot{x}(nT) = x(nT) * d(nT) \qquad (8.31)$$
$$= (r(nT) * p(-nT)) * d(nT)$$
$$= r(nT) * (p(-nT) * d(nT))$$
$$= r(nT) * \dot{p}(-nT) \qquad (8.32)$$

where the third line follows from the second line by the associative property of convolution. The two expressions (8.31) and (8.32) suggest two alternate discrete-time systems for producing the desired samples. The system in Figure 8.4.5 (a) illustrates the discrete-time processing defined by (8.31). The system in Figure 8.4.5 (b) illustrated the discrete-time processing defined by (8.32), which uses a filter whose impulse response consists of samples of the time derivative of $p(-t)$.

A complete detector requires both matched filter outputs and derivative matched filter outputs. Thus, the use of either approach to compute the derivative matched filter requires

Section 8.4 Discrete-Time Techniques for M-ary PAM

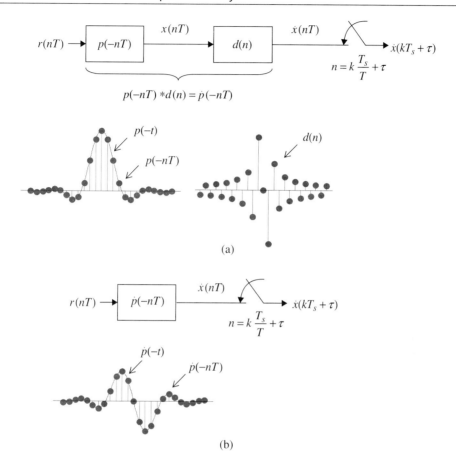

Figure 8.4.5 Two approaches for computing samples of the derivative of the matched filter output.

two filters. The approach illustrated in Figure 8.4.5 (a), however, uses the two filters in series, whereas the approach illustrated in Figure 8.4.5 (b) uses two filters operating on the same input samples in parallel. This second approach can be important in a delay-sensitive application such as a phase-locked loop.

In either case, the sample rate used to produce the samples of $r(t)$ or $x(t)$ must satisfy the Nyquist sampling theorem. Samples of the time derivative of the $x(t)$ cannot be obtained from undersampled signals that have been distorted by aliasing. Thus the MLTED is, in general, a multirate discrete-time system. The input sample rate is N samples/symbol whereas the output sample rate is 1 sample/symbol.

Early–Late Timing Error Detector. The early–late timing error detector (ELTED) uses time differences, as described in Section 8.2, to approximate the derivative required by the

MLTED. The data-aided early–late error signal is of the form

$$e(k) = a(k)\left[x\left(kT_s + \hat{\tau} + \Delta T_s\right) - x\left(kT_s + \hat{\tau} - \Delta T_s\right)\right]$$

where ΔT_s is usually selected to be a value conveniently supplied by the sample rate. Because a sample rate of 2 samples/symbol is commonly used, $\Delta = 1/2$ is a popular choice. When sampled at 2 samples/symbol, the matched filter outputs may be indexed using the symbol index k as

$$\ldots, x((k-1)T_s - \tau), x((k-1/2)T_s - \tau), x(kT_s - \tau), x((k+1/2)T_s - \tau), x((k+1)T_s - \tau), \ldots$$

The early–late timing error for a sample rate of 2 samples/symbol is

$$e(k) = a(k)\left[x\left((k+1/2)T_s + \hat{\tau}\right) - x\left((k-1/2)T_s + \hat{\tau}\right)\right] \quad (8.33)$$

for the data-aided detector, and

$$e(k) = \hat{a}(k)\left[x\left((k+1/2)T_s + \hat{\tau}\right) - x\left((k-1/2)T_s + \hat{\tau}\right)\right] \quad (8.34)$$

for the decision-directed detector. For binary PAM, the data symbol estimates are defined by (8.28). The approximation (8.29) can also be used for both binary PAM and M-ary PAM.

The S-curve for the ELTED is obtained by computing the expected value of the error signal (8.33) using (8.25) for $x((k+1/2)T_s + \hat{\tau})$ and $x((k-1/2)T_s + \hat{\tau})$ along with the property (8.22):

$$g(\tau_e) = \mathrm{E}\left\{a(k)\left[x((k+1/2)T_s + \tau) - x((k-1/2)T_s + \tau)\right]\right\}$$

$$= \mathrm{E}\left\{a(k)\left[K\sum_m a(m)r_p\left((k-m+1/2)T_s - \tau_e\right)\right.\right.$$

$$\left.\left. -K\sum_{m'} a(m')r_p\left((k-m'-1/2)T_s - \tau_e\right)\right]\right\}$$

$$= KE_{\mathrm{avg}}\left[r_p(T_s/2 - \tau_e) - r_p(-T_s/2 - \tau_e)\right] \quad (8.35)$$

where the last line follows from (8.22). The S-curve is thus an approximation to the derivative of $r_p(t)$ at $t = -\tau_e$ using values of $r_p(t)$ one-half a symbol time before and after $-\tau_e$. (Compare the S-curve for the ELTED with the S-curve for the MLTED.) As an autocorrelation function, $r_p(-\tau_e)$ is symmetric about $\tau_e = 0$. Consequently, the difference $r_p(T_s/2 - \tau_e) - r_p(-T_s/2 - \tau_e)$ is zero at $\tau_e = 0$. The S-curve for the decision-directed MLTED is obtained by assuming $\hat{a}(k) = a(k)$ and proceeding as outlined above. As long as the decisions are correct, the S-curve for the decision-directed MLTED is identical to the S-curve for the data-aided MLTED as illustrated in Figure 8.4.6 for the square-root raised-cosine pulse shape with 50% excess bandwidth. Again, the S-curves for both the data-aided detector and the decision-directed detector are identical for $|\tau_e/T_s| < 0.35$. This is because $\hat{a}(k) = a(k)$. When $|\tau_e/T_s| > 0.35$

Section 8.4 Discrete-Time Techniques for M-ary PAM

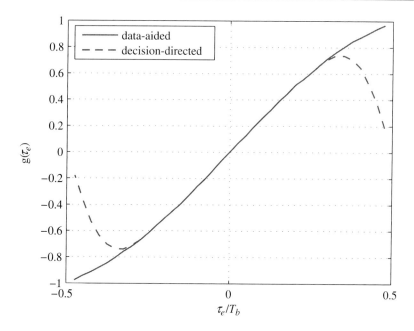

Figure 8.4.6 S-curves for the data-aided early–late TED (solid line) and the decision-directed early–late TED (dashed line). These are simulation results for binary PAM using a square-root raised-cosine pulse shape with 50% excess bandwidth and $KE_{avg} = 1$. The signal-to-noise ratio is $E_b/N_0 = 20$ dB.

the S-curve for the decision-directed detector departs from the S-curve for the data-aided detector due to decision errors. When the true data symbol estimates are used, the S-curve is proportional to E_{avg}. When the approximation (8.29) is used, E_{avg} is replaced by a different constant of proportionality.

The gain of the ELTED, K_p, is the slope of $g(0)$ and is proportional to the received signal amplitude K and the average symbol energy E_{avg}, and a function of $r_p(t)$. Because $r_p(t)$ is a function of the pulse shape $p(t)$, K_p is also a function of the excess bandwidth when the pulse shape is the square-root raised-cosine pulse shape. This dependence is plotted in Figure 8.4.7. The dependence of K_p on K presents the same challenges to timing synchronizers based on the ELTED as was the case for timing synchronizers based on the MLTED. Estimates of K or automatic gain control can be applied as described in the context of the MLTED.

Zero-Crossing Timing Error Detector. The zero-crossing timing error detector (ZCTED) is based on finding the zero crossings in the eye diagram. It operates at 2 samples/symbol and provides zero error when every other sample is time-aligned with a zero crossing in the matched filter output. (The other samples are time-aligned with the optimum sampling instants corresponding to the maximum average eye opening.) For convenience, assume that the matched filter outputs are available at a rate of 2 samples/symbol and are

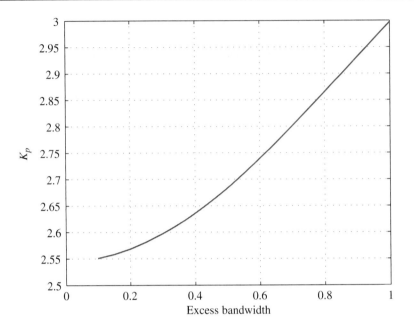

Figure 8.4.7 Phase detector gain, K_p, of the early–late TED as a function of excess bandwidth for the square-root raised-cosine pulse shape and binary PAM with $KE_{avg} = 1$.

indexed using the symbol index k:

$$\ldots, x((k-1)T_s - \tau), x((k-1/2)T_s - \tau), x(kT_s - \tau), x((k+1/2)T_s - \tau), x((k+1)T_s - \tau), \ldots$$

The timing error signal is

$$e(k) = x((k-1/2)T_s + \hat{\tau})\left[a(k-1) - a(k)\right] \qquad (8.36)$$

for the case of data-aided symbol timing synchronization and

$$e(k) = x((k-1/2)T_s + \hat{\tau})\left[\hat{a}(k-1) - \hat{a}(k)\right] \qquad (8.37)$$

for decision-directed symbol timing synchronization.

A graphical interpretation of the ZCTED is shown in Figure 8.4.8. The sign of the error is controlled by the difference $a(k-1) - a(k)$ or $\hat{a}(k-1) - \hat{a}(k)$. Figure 8.4.8 (a) illustrates the case where the timing error τ_e is positive and the data transition is positive-to-negative. The timing error is $e(k) = x((k-1/2)T_s + \hat{\tau})[a(k-1) - a(k)] = 2Ax((k-1/2)T_s + \hat{\tau}) > 0$ and provides an error signal with the correct sign. In Figure 8.4.8 (b), the timing error τ_e is negative. The error signal is $e(k) = x((k-1/2)T_s + \hat{\tau})[a(k-1) - a(k)] = 2Ax((k-1/2)T_s + \hat{\tau}) < 0$ and provides an error signal with the correct sign. When the data transition is negative-to-positive, the sign of $x((k-1/2)T_s + \hat{\tau})$ is wrong but is corrected by $a(k-1) - a(k) = -2A$. Note that when there is no data transition, $a(k-1) - a(k) = 0$ and no timing error information is provided.

Section 8.4 Discrete-Time Techniques for M-ary PAM

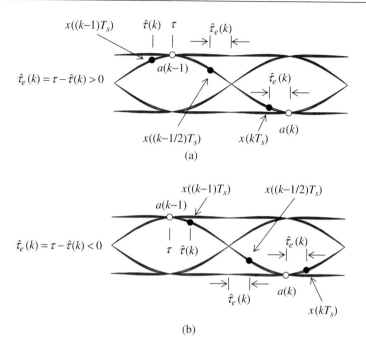

Figure 8.4.8 Example showing the operation of the zero-crossing detector for positive-to-negative data transitions: (a) The timing estimate is early; (b) the timing estimate is late.

For binary PAM, the symbol decisions are

$$\hat{a}(k-1) = A \times \text{sgn}\left\{x((k-1)T_s + \hat{\tau})\right\}$$
$$\hat{a}(k) = A \times \text{sgn}\left\{x(kT_s + \hat{\tau})\right\}. \tag{8.38}$$

As before, an alternative is to use

$$\hat{a}(k-1) = \text{sgn}\left\{x((k-1)T_s + \hat{\tau})\right\}$$
$$\hat{a}(k) = \text{sgn}\left\{x(kT_s + \hat{\tau})\right\} \tag{8.39}$$

for both binary PAM and nonbinary PAM. The approximation (8.39) provides the proper sign on the estimate of the timing error, but not the correct amplitude. In most applications, the incorrect amplitude does not appear to impact performance in any significant way.

The S-curve for the ZCTED may be obtained by computing the expected value of $e(k)$ and using an expression of the form (8.25) for $x((k-1/2)T_s + \hat{\tau})$:

$$g(\tau_e) = \mathrm{E}\left\{x((k-1/2)T_s - \tau_e)[a(k-1) - a(k)]\right\} \tag{8.40}$$

$$= \mathrm{E}\left\{K\sum_m a(m)r_p\left((k-m)T_s - T_s/2 - \tau_e\right)\left[a(k-1) - a(k)\right]\right\} \tag{8.41}$$

$$= K E_{\text{avg}}\left[r_p\left(T_s/2 - \tau_e\right) - r_p\left(-T_s/2 - \tau_e\right)\right] \tag{8.42}$$

where the last line follows from (8.22). The S-curve is thus an estimate of the slope of $r_p(-\tau_e)$ using values of $r_p(t)$ one-half a symbol time before and after $-\tau_e$. As an autocorrelation function, $r_p(t)$ is symmetric about $t = 0$. Consequently, $r_p(T_s/2 - \tau_e) - r_p(-T_s/2 - \tau_e) = 0$ at $\tau_e = 0$ which, in turn, forces the S-curve to be zero at $\tau_e = 0$. The S-curve for the decision-directed ZCTED is identical when $\hat{a}(k-1) = a(k-1)$ and $\hat{a}(k) = a(k)$. Note that the S-curve for the ZCTED given by (8.42) is identical to the S-curve for the ELTED given by (8.35) (see Figure 8.4.6). The ZCTED performance, however, is superior to that of the ELTED because the ELTED suffers from a higher degree of self-noise than the ZCTED.

The gain of the ZCTED, K_p, is the slope of $g(0)$ and is proportional to the received signal amplitude K and the average symbol energy E_{avg}, and a function of $r_p(t)$. Because $r_p(t)$ is a function of the pulse shape $p(t)$, K_p is also a function of the excess bandwidth when the pulse shape is the square-root raised-cosine pulse shape. This dependence is identical to that for the ELTED (see Figure 8.4.7). The dependence of K_p on K presents the same challenges to timing synchronizers based on the ZCTED as was the case for timing synchronizers based on the MLTED. Estimates of K or automatic gain control can be applied as described in the context of the MLTED.

Gardner Timing Error Detector. Like the ZCTED, the Gardner timing error detector (GTED) is based on finding the zero crossing in the eye diagram. It operates at 2 samples/symbol and was developed to operate with BPSK and QPSK (and, by extension, binary baseband PAM). Assume the matched filter outputs are available at 2 samples/symbol and that they are indexed using the symbol index k:

$$\ldots, x((k-1)T_s - \tau), x((k-1/2)T_s - \tau), x(kT_s - \tau), x((k+1/2)T_s - \tau), x((k+1)T_s - \tau), \ldots$$

The GTED uses the error signal

$$e(k) = x\left((k-1/2)T_s + \hat{\tau}\right)\left[x\left((k-1)T_s + \hat{\tau}\right) - x\left(kT_s + \hat{\tau}\right)\right]. \tag{8.43}$$

An important observation is that the GTED is purely a non-data-aided TED. The graphical representation of GTED operation is very similar to that for the ZCTED illustrated in Figure 8.4.8 except that the matched filter outputs $x((k-1)T_s + \hat{\tau})$ and $x(kT_s + \hat{\tau})$ are used in place of the symbol decisions $\hat{a}(k-1)$ and $\hat{a}(k)$. Gardner showed that this TED is *rotationally invariant*, that is, for a fixed $\hat{\tau}$, $e(k)$ is independent of any carrier phase rotation.

Section 8.4 Discrete-Time Techniques for M-ary PAM

This property makes the GTED ideally suited for achieving timing synchronization before carrier phase synchronization in systems using BPSK and QPSK.

The S-curve for the GTED is obtained by expressing $e(k)$ in terms of τ_e and computing the expected value:

$$g(\tau_e) = E\left\{x((k-1/2)T_s - \tau_e)\left[x((k-1)T_s - \tau_e) - x(kT_s - \tau_e)\right]\right\}$$

$$= E\left\{K\sum_m a(m)r_p((k-m-1/2)T_s - \tau_e)\left[K\sum_{m'} a(m')r_p((k-m'-1)T_s - \tau_e)\right.\right.$$

$$\left.\left. - K\sum_{m''} a(m'')r_p((k-m'')T_s - \tau_e)\right]\right\}$$

$$= K^2 E_{\text{avg}} \sum_m r_p((m-1/2)T_s - \tau_e)\left[r_p((m-1)T_s - \tau_e) - r_p(mT_s - \tau_e)\right]. \quad (8.44)$$

The last line follows from (8.22) with k set to 0. For the SRRC pulse shape with excess bandwidth α it can be shown that (8.44) may also be expressed as

$$g(\tau_e) = \frac{4K^2 E_{\text{avg}}}{T_s} C(\alpha) \sin\left(2\pi \frac{\tau_e}{T_s}\right) \quad (8.45)$$

where the constant $C(\alpha)$ is a function of the excess bandwidth given by

$$C(\alpha) = \frac{1}{4\pi\left(1 - \frac{\alpha^2}{4}\right)} \sin\left(\frac{\pi\alpha}{2}\right). \quad (8.46)$$

A plot of the S-curve for the GTED is illustrated in Figure 8.4.9. From this expression, the gain of the GTED is

$$K_p = \frac{4K^2 E_{\text{avg}}}{T_s} \times \frac{1}{4\pi\left(1 - \frac{\alpha^2}{4}\right)} \sin\left(\frac{\pi\alpha}{2}\right). \quad (8.47)$$

As with the ELTED and the ZCTED, the gain of the GTED diminishes as the the excess bandwidth decreases.

Mueller and Müller Timing Error Detector. The Mueller and Müller timing error detector (MMTED) operates on the matched filter outputs sampled at 1 sample/symbol. The symbol timing error signal is

$$e(k) = a(k-1)x(kT_s + \hat{\tau}) - a(k)x((k-1)T_s + \hat{\tau}) \quad (8.48)$$

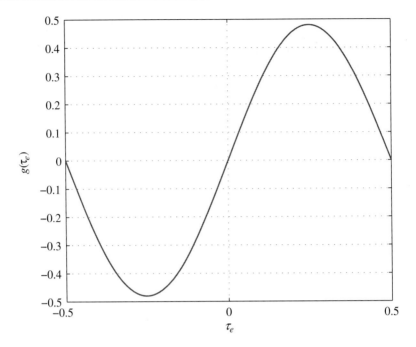

Figure 8.4.9 S-Curve for the Gardner timing error detector for binary PAM using the SRRC pulse shape with 50% excess bandwidth and with $K^2 E_{avg}/T_s = 1$.

for data-aided symbol timing synchronization and

$$e(k) = \hat{a}(k-1)x(kT_s + \hat{\tau}) - \hat{a}(k)x((k-1)T_s + \hat{\tau}) \tag{8.49}$$

for decision-directed symbol timing synchronization. An interpretation of MMTED operation may be obtained through the expression for the S-curve. As before, the S-curve is obtained by computing the expected value of $e(k)$ using (8.25) for $x(kT_s - \tau_e)$ and $x((k-1)T_s - \tau_e)$. The S-curve for the data-aided timing error detector is

$$g(\tau_e) = \mathrm{E}\left\{a(k-1)x(kT_s - \tau_e) - a(k)x((k-1)T_s - \tau_e)\right\}$$

$$= \mathrm{E}\left\{a(k-1)K\sum_m a(m)r_p\left((k-m)T_s - \tau_e\right) - a(k)K\sum_{m'} a(m')r_p\left((k-1-m')T_s - \tau_e\right)\right\}$$

$$= KE_{avg}\left[r_p\left(T_s - \tau_e\right) - r_p\left(-T_s - \tau_e\right)\right]. \tag{8.50}$$

The S-curve is thus an estimate of the slope of $r_p(\tau_e)$ using values of $r_p(t)$ a symbol time before and after $-\tau_e$. Because $r_p(t)$ is an autocorrelation function, it is symmetric about $\tau_e = 0$. As a consequence, the S-curve is zero at $\tau_e = 0$. The S-curve for the decision-directed MMTED

Section 8.4 Discrete-Time Techniques for M-ary PAM

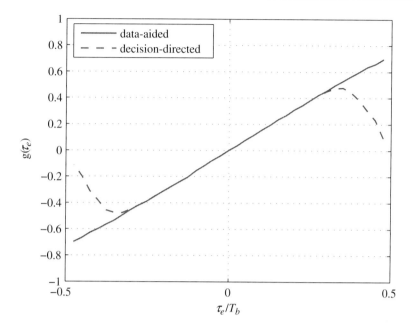

Figure 8.4.10 S-curves for the data-aided Mueller and Müller timing error detector (solid line) and the decision-directed Mueller and Müller timing error detector (dashed line). These are simulation results for binary PAM using a square-root raised-cosine pulse shape with 50% excess bandwidth and $KE_{\text{avg}} = 1$. The signal-to-noise ratio is $E_b/N_0 = 20$ dB.

is identical when $\hat{a}(k-1) = a(k-1)$ and $\hat{a}(k) = a(k)$ as shown in Figure 8.4.10 for the square-root raised-cosine pulse shape with 50% excess bandwidth. When $|\tau_e/T_s| < 0.35$, the symbol decisions are correct and the two S-curves are identical. When $|\tau_e/T_s| > 0.35$, some of the symbol decisions are incorrect and reduce the MMTED gain as indicated by the departure of the S-curve for the decision-directed MMTED from the S-curve for the data-aided MMTED.

The phase detector gain, K_p, is a function of the pulse shape which, for the square-root raised-cosine pulse shape, is a function of the excess bandwidth as shown in Figure 8.4.11.

The gain of the MMTED, K_p, is the slope of $g(0)$ and is proportional to the received signal amplitude K and the average symbol energy E_{avg} and is a function of $r_p(t)$. Because $r_p(t)$ is a function of the pulse shape $p(t)$, K_p is also a function of the excess bandwidth when the pulse shape is the square-root raised-cosine pulse shape. This dependence is plotted in Figure 8.4.11. The dependence of K_p on K presents the same challenges to timing synchronizers based on the MMTED as was the case for timing synchronizers based on the MLTED. Estimates of K or automatic gain control can be applied as described in the context of the MLTED.

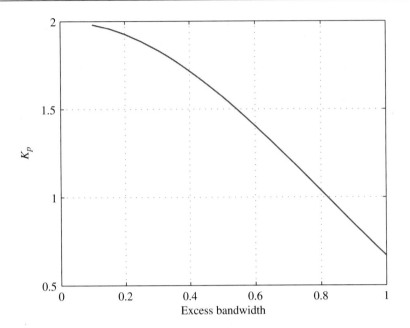

Figure 8.4.11 Phase detector gain, K_p, of the Mueller and Müller detector as a function of excess bandwidth for the square-root raised-cosine pulse shape and binary PAM with $KE_{avg} = 1$.

8.4.2 Interpolation

The commonly used terms to describe interpolation are illustrated by the diagram in Figure 8.4.12. T-spaced samples of the bandlimited continuous-time signal $x(t)$ are available and denoted

$$\ldots, x((n-1)T), x(nT), x((n+1)T), x((n+2)T), \ldots .$$

The desired sample is a sample of $x(t)$ at $t = kT_I$ and is called the k-th *interpolant*. The process used to compute $x(kT_I)$ from the available samples is called *interpolation*. When the k-th interpolant is between samples $x(nT)$ and $(x(n+1)T)$, the sample index n is called the k-th *basepoint index* and is denoted $m(k)$. The time instant kT_I is some fraction of a sample time greater than $m(k)T$. This fraction is called the k-th fractional interval and is denoted $\mu(k)$. The k-th fractional interval satisfies $0 \leq \mu(k) < 1$ and is defined by $\mu(k)T = kT_I - m(k)T$.

The fundamental equation for interpolation may be derived by considering a fictitious system involving continuous-time processing illustrated in Figure 8.4.13. The samples $x(nT)$ ($n = 0, 1, \ldots$) are converted to a weighted impulse train

$$x_a(t) = \sum_n x(nT)\delta(t - nT) \qquad (8.51)$$

Section 8.4 Discrete-Time Techniques for M-ary PAM

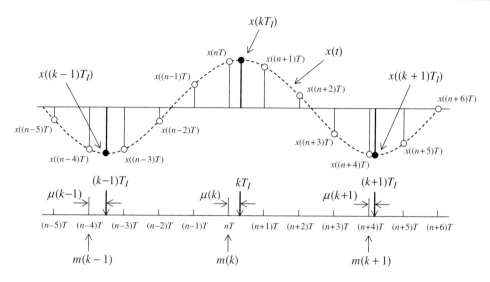

Figure 8.4.12 Illustration of the relationships between the interpolation interval T_I, the sample time T, the basepoint indexes, and fractional intervals.

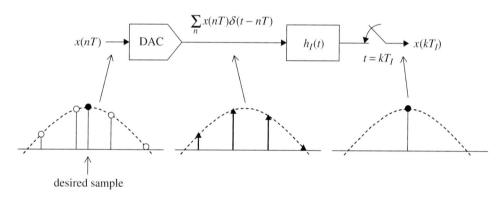

Figure 8.4.13 Fictitious system using continuous-time processing for performing interpolation.

by the digital-to-analog converter (DAC). The impulse train is filtered by an interpolating filter with impulse response $h_I(t)$ to produce the continuous-time output $x(t)$. The continuous-time signal $x(t)$ may be expressed as

$$x(t) = \sum_n x(nT) h_I(t - nT). \qquad (8.52)$$

To produce the desired interpolants, $x(t)$ is resampled at intervals[3] kT_I ($k = 0, 1, \ldots$). The k-th interpolant is (8.52) evaluated at $t = kT_I$ and may be expressed as

$$x(kT_I) = \sum_n x(nT) h_I(kT_i - nT). \tag{8.53}$$

The index n indexes the signal samples. The convolution sum (8.53) may be reexpressed using a filter index i. Using $m(k) = \lfloor kT_I/T \rfloor$ and $\mu(k) = kT_I/T - m(k)$, the filter index is $i = m(k) - n$. Using the filter index, equation (8.53) may be expressed as

$$x(kT_I) = \sum_i x\left((m(k) - i)T\right) h_I\left((i + \mu(k))T\right). \tag{8.54}$$

Equation (8.54) will serve as the fundamental equation for interpolation and shows that the desired interpolant can be obtained by computing a weighted sum of the available samples. The optimum interpolation filter is an ideal low-pass filter whose impulse response is

$$h_I(t) = \frac{\sin(\pi t/T)}{\pi t/T}. \tag{8.55}$$

Given a fractional interval μ, the ideal impulse response is sampled at $t = iT - \mu T$ to produce the filter coefficients required by (8.54).

The role of the interpolation control block in Figure 8.4.2 is to provide the interpolator with the basepoint index and fractional interval for each desired interpolant.

For asynchronous sampling, the sample clock is independent of data clock used by the transmitter. As a consequence, the sampling instants are not synchronized to the symbol periods. The sample rate and symbol rate are *incommensurate* and the sample times never coincide exactly with the desired interpolant times. When the symbol timing PLL is in lock and the interpolants are desired once per symbol, $T_I = T_s$. The behavior of the fractional interval $\mu(k)$ as a function of k depends on the relationship between the sample clock period T and the symbol period T_s as follows:

- When T_s is incommensurate with NT, $\mu(k)$ is irrational and changes for each k for infinite precision or progresses through a finite set of values, never repeating exactly for finite precision.
- When $T_s \approx NT$, $\mu(k)$ changes very slowly for infinite precision or remains constant for many k for finite precision.
- When T_s is commensurate with NT, but not equal, $\mu(k)$ cyclically progresses through a finite set of values.

The ideal interpolation filter is IIR. Thus, its use poses an often unacceptable computational burden—especially when the fractional interval changes. For this reason, FIR filters that

---

[3] If $T_I = T$, then the process produces one interpolant for each sample. This is the strict definition of *interpolation*. When $T_I \neq T$, the sample rate of the output is different than the sample rate of the input. This process is known as *resampling* or *rate conversion*. In digital communication applications, $T_I > T$ is the case typically encountered because T is the reciprocal of the sample rate at the input to the matched filter and T_I is the reciprocal of the symbol rate.

Section 8.4 Discrete-Time Techniques for M-ary PAM

approximate the ideal interpolation filter are preferred in digital communication applications. A popular class of FIR interpolating filters are piecewise polynomial filters discussed below. Another alternative is to massively upsample the matched filter input, match filter at the high sample rate, then downsample the matched filter output with the appropriately chosen sample offset to obtain the desired interpolant. This approach leads to a polyphase-filterbank interpolator.

Piecewise Polynomial Interpolation. The underlying continuous-time waveform $x(t)$ is approximated by a polynomial in t of the form

$$x(t) \approx c_p t^p + c_{p-1} t^{p-1} + \cdots + c_1 t + c_0. \tag{8.56}$$

The polynomial coefficients are determined by the $p + 1$ sample values surrounding the basepoint index. Once the coefficient values are known, the interpolant at $t = kT_I = (m(k) + \mu(k))T$ is obtained using

$$x(kT_I) \approx c_p (kT_I)^p + c_{p-1} (kT_I)^{p-1} + \cdots + c_1 (kT_I) + c_0. \tag{8.57}$$

Three special cases, $p = 1, 2,$ and 3 are of interest and are illustrated in Figure 8.4.14. When $p = 1$, the first degree polynomial

$$x(t) \approx c_1 t + c_0 \tag{8.58}$$

is used to approximate the underlying continuous-time waveform. The desired interpolants are computed from

$$x((m(k) + \mu(k))T) = c_1((m(k) + \mu(k))T) + c_0. \tag{8.59}$$

The coefficients c_1 and c_0 are determined by the available samples and satisfy the equation

$$\begin{bmatrix} x(m(k)T) \\ x((m(k)+1)T) \end{bmatrix} = \begin{bmatrix} m(k)T & 1 \\ (m(k)+1)T & 1 \end{bmatrix} \begin{bmatrix} c_1 \\ c_0 \end{bmatrix}. \tag{8.60}$$

Solving the above for c_1 and c_0 and substituting into (8.59) produces

$$x((m(k) + \mu(k))T) = \mu(k)x((m(k)+1)T) + (1 - \mu(k))x(m(k)T) \tag{8.61}$$

which is the familiar linear interpolator.

Four observations are important. The first is that the interpolant is a linear combination of the available samples. As a consequence, the interpolant can be thought of as the output of a filter with coefficients suggested by (8.61):

$$x((m(k) + \mu(k))T) = \sum_{i=-1}^{0} h_1(i) x((m(k) - i)T) \tag{8.62}$$

where

$$\begin{aligned} h_1(-1) &= \mu(k) \\ h_1(0) &= 1 - \mu(k). \end{aligned} \tag{8.63}$$

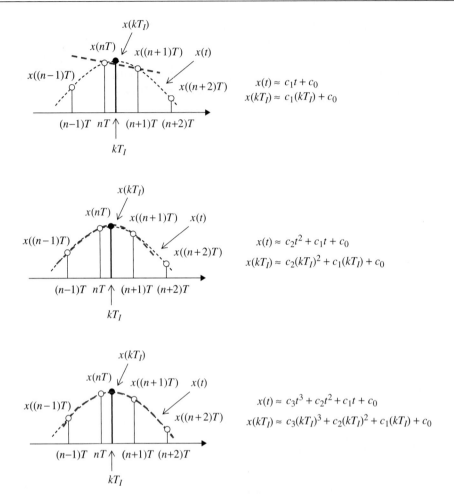

Figure 8.4.14 Three special cases of polynomial interpolation: (top) linear interpolation, (middle) quadratic interpolation, and (bottom) cubic interpolation.

The second important observation is that the equivalent filter coefficients are a function only of the fractional interval and not a function of the basepoint index. The basepoint index defines *which* set of samples should be used to compute the interpolant. The third observation is that the interpolating filter is linear phase FIR filter, which is an extremely important property for digital communications. To see that this filter is linear phase, note that the coefficients are symmetric about the center point of the filter that is defined by $\mu(k) = 1/2$. In other words, $h((m + 1/2)T) = h((-m + 1/2)T)$ for $m = 0, 1, 2, \ldots$. This is a result of using an even number of samples to compute an interpolant that is between the middle two. The final observation is that the sum of the coefficients is unity and is therefore independent of $\mu(k)$.

Section 8.4 Discrete-Time Techniques for M-ary PAM

As a consequence, the interpolating filter does not alter the amplitude of the underlying continuous-time waveform in the process of producing the interpolant.

The second observation is an attractive feature because any finite precision computing device would eventually overflow as $m(k)$ increased. The third property requires the use of an even number of samples by the interpolator. An even number of samples is needed to define an odd-degree approximating polynomial. For this reason, odd-degree approximating polynomials are popular. The next highest odd-degree polynomial is $p = 3$. In this case

$$x(t) \approx c_3 t^3 + c_2 t^2 + c_1 t + c_0 \qquad (8.64)$$

is used to approximate the underlying continuous-time waveform. The desired interpolants are computed from

$$x((m(k) + \mu(k))T) = c_3((m(k) + \mu(k))T)^3 + c_2((m(k) + \mu(k))T)^2 \\ + c_1((m(k) + \mu(k))T) + c_0. \qquad (8.65)$$

The coefficients c_3, c_2, c_1, and c_0 are defined by

$$\begin{bmatrix} x((m(k)-1)T) \\ x(m(k)T) \\ x((m(k)+1)T) \\ x((m(k)+2)T) \end{bmatrix} = \begin{bmatrix} ((m(k)-1)T)^3 & ((m(k)-1)T)^2 & (m(k)-1)T & 1 \\ (m(k)T)^3 & (m(k)T)^2 & m(k)T & 1 \\ ((m(k)+1)T)^3 & ((m(k)+1)T)^2 & (m(k)+1)T & 1 \\ ((m(k)+2)T)^3 & ((m(k)+2)T)^2 & (m(k)+2)T & 1 \end{bmatrix} \begin{bmatrix} c_3 \\ c_2 \\ c_1 \\ c_0 \end{bmatrix}. \qquad (8.66)$$

Solving the above for c_3, c_2, c_1, and c_0 and substituting into (8.65) produces

$$x((m(k) + \mu(k))T) = \left(\frac{\mu(k)^3}{6} - \frac{\mu(k)}{6} \right) x((m(k) + 2)T)$$

$$- \left(\frac{\mu(k)^3}{2} - \frac{\mu(k)^2}{2} - \mu(k) \right) x((m(k) + 1)T)$$

$$+ \left(\frac{\mu(k)^3}{2} - \mu(k)^2 - \frac{\mu(k)}{2} + 1 \right) x(m(k)T)$$

$$- \left(\frac{\mu(k)^3}{6} - \frac{\mu(k)^2}{2} + \frac{\mu(k)}{3} \right) x((m(k) - 1)T) \qquad (8.67)$$

which is called a cubic interpolator. When interpreted as a filter, the cubic interpolator output is of the form

$$x((m(k) + \mu(k))T) = \sum_{i=-2}^{1} h_3(i) x((m(k) - i)T) \qquad (8.68)$$

where the filter coefficients are

$$h_3(-2) = \frac{\mu(k)^3}{6} - \frac{\mu(k)}{6}$$
$$h_3(-1) = -\frac{\mu(k)^3}{2} + \frac{\mu(k)^2}{2} + \mu(k)$$
$$h_3(0) = \frac{\mu(k)^3}{2} - \mu(k)^2 - \frac{\mu(k)}{2} + 1$$
$$h_3(1) = -\frac{\mu(k)^3}{6} + \frac{\mu(k)^2}{2} - \frac{\mu(k)}{3}.$$
(8.69)

Finally, for the case $p = 2$, using the approximation

$$x(t) \approx c_2 t^2 + c_1 t + c_0 \tag{8.70}$$

to approximate the underlying continuous-time waveform and

$$x((m(k) + \mu(k))T) = c_2((m(k) + \mu(k))T)^2 + c_1((m(k) + \mu(k))T) + c_0 \tag{8.71}$$

to compute the desired interpolant requires the use of three samples. Because the number of samples is odd, the desired interpolant is not in between the middle two and the resulting filter will not be symmetric with respect to $\mu(k) = 1/2$. The desire to use four points introduces a wrinkle that is explored in Exercise 8.7 where it is shown that the desired interpolant can be thought of as the output of a filter of the form

$$x((m(k) + \mu(k))T) = \sum_{i=-2}^{1} h_2(i) x((m(k) - i)T) \tag{8.72}$$

where the filter coefficients are

$$h_2(-2) = \alpha \mu(k)^2 - \alpha \mu(k)$$
$$h_2(-1) = -\alpha \mu(k)^2 + (1+\alpha)\mu(k)$$
$$h_2(0) = -\alpha \mu(k)^2 - (1-\alpha)\mu(k) + 1$$
$$h_2(1) = \alpha \mu(k)^2 - \alpha \mu(k)$$
(8.73)

and α is a free parameter required to account for the additional degree of freedom introduced by using four points. Simulation results have shown that $\alpha = 0.43$ is the optimal value for binary PAM using the root raised-cosine pulse shape with 100% excess bandwidth. Erup, Gardner, and Harris showed that using $\alpha = 0.5$ substantially reduces the complexity of the hardware and results in a performance loss less than 0.1 dB [165].

Using a piecewise polynomial interpolator to produce the desired interpolant results in a computation of the form

$$x((m(k) + \mu(k))T) = \sum_{i=-I_1}^{I_2} h_p(i; \mu(k)) x((m(k) - i)T) \tag{8.74}$$

where the filter coefficients are given by (8.63), (8.73), and (8.69) for $p = 1, 2$, and 3, respectively. Comparing (8.74) with the fundamental interpolation equation (8.54) shows that the filter coefficients $h_p(i; \mu(k))$ play the role of approximating the samples of the ideal interpolation filter $h_I((i - \mu(k))T)$. Plots of $h_1(i; \mu(k))$, $h_2(i; \mu(k))$, and $h_3(i; \mu(k))$ are shown in Figure 8.4.15. Observe that as p increases, $h_p(i; \mu(k))$ approximates (8.55) with greater and greater accuracy. In fact, in the limit $p \to \infty$, $h_p(i; \mu(k))$ approaches (8.55).

Because the filter coefficients suggested by the filter structure defined by (8.63), (8.69), and (8.73) are a function of the variable $\mu(k)$, a hardware implementation requires two-input multipliers with two variable quantities. The complexity can be reduced by formulating the problem in terms of two-input multipliers where one of the inputs is fixed. To do this, observe that each filter coefficient $h_p(i; \mu(k))$ in (8.74) may be written as a polynomial in $\mu(k)$. Let

$$h_p(i; \mu(k)) = \sum_{l=0}^{p} b_l(i) \mu(k)^l \qquad (8.75)$$

represent the polynomial. Substituting (8.75) into (8.74) and rearranging produces

$$x((m(k) + \mu(k))T) = \sum_{l=0}^{p} \mu(k)^l \underbrace{\sum_{i=I_1}^{I_2} b_l(i) x((m(k) - i)T)}_{v(l)}. \qquad (8.76)$$

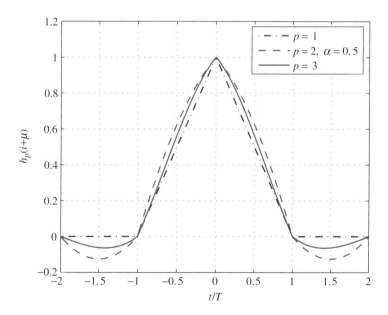

Figure 8.4.15 Plot of the filter impulse responses resulting from piecewise polynomial interpolation.

The inner sum looks like a filter equation where the input data samples $x((m(k) - i)T)$ pass through a filter with impulse response $b_l(i)$. The $b_l(i)$ are independent of $\mu(k)$. Thus, this filter has fixed coefficients and an efficient implementation. Computing (8.76) by nested evaluation produces an expression of the form

$$x((m(k) + \mu(k))T) = (v(2)\mu(k) + v(1))\mu(k) + v(0) \tag{8.77}$$

for piecewise parabolic interpolation and

$$x((m(k) + \mu(k))T) = ((v(3)\mu(k) + v(2))\mu(k) + v(1))\mu(k) + v(0) \tag{8.78}$$

for cubic interpolation. Mapping these expressions to hardware results in an efficient filter structure called the *Farrow structure* illustrated in Figure 8.4.16. The *Farrow coefficients* for the Farrow structure are listed in Tables 8.4.1 and 8.4.2. Note that when $\alpha = 1/2$ for the piecewise parabolic interpolator, all of the filter coefficients but one become 0, 1, or $\pm 1/2$. The resulting filter structure is elegantly simple.

Polyphase Filterbank Interpolation. An alternate approach to interpolation is to upsample the matched filter output by a factor Q and then downsample with the appropriate offset to produce a sample close to the desired interpolant. How close the sample is to the desired interpolant is controlled by the upsample factor Q. A conceptual block diagram of this process is shown in Figure 8.4.17 (a) for the case of binary PAM. (Generalizations to M-ary PAM are straightforward.) The input to the matched filter consists of samples the received signal $r(nT)$ sampled at N samples/symbol (i.e., $T_s = NT$). The impulse response $h(nT)$ of the matched filter consists of T-spaced samples of a time-reversed version of the

Table 8.4.1 Farrow coefficients $b_l(i)$ for the piecewise parabolic interpolator

i	$b_2(i)$	$b_1(i)$	$b_0(i)$
-2	α	$-\alpha$	0
-1	$-\alpha$	$1 + \alpha$	0
0	$-\alpha$	$\alpha - 1$	1
1	α	$-\alpha$	0

Table 8.4.2 Farrow coefficients $b_l(i)$ for the cubic interpolator

i	$b_3(i)$	$b_2(i)$	$b_1(i)$	$b_0(i)$
-2	$\frac{1}{6}$	0	$-\frac{1}{6}$	0
-1	$-\frac{1}{2}$	$\frac{1}{2}$	1	0
0	$\frac{1}{2}$	-1	$-\frac{1}{2}$	1
1	$-\frac{1}{6}$	$\frac{1}{2}$	$-\frac{1}{3}$	0

Section 8.4 Discrete-Time Techniques for M-ary PAM

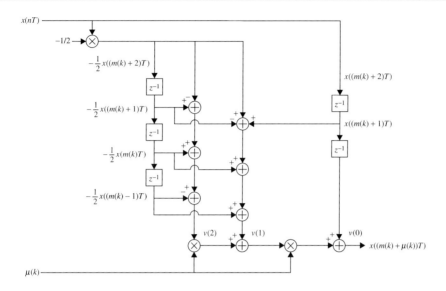

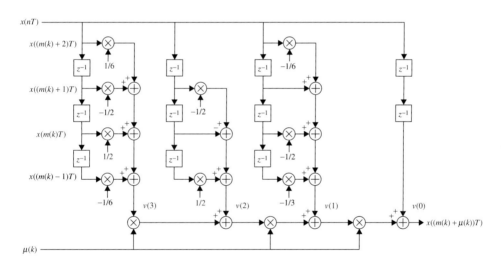

Figure 8.4.16 Farrow interpolator structures for the piecewise parabolic with $\alpha = 1/2$ (top) and cubic (bottom) interpolators.

pulse shape $p(t)$: $h(nT) = p(-nT)$. The matched filter output, $x(nT)$ is upsampled by inserting $Q - 1$ zeros between each sample. An interpolating low-pass filter is used to produce samples of the matched filter output at a rate of NQ samples/symbol. This signal is denoted $x(nT/Q)$. The matched filter output with the desired delay is obtained by downsampling $x(nT/Q)$ with the proper offset.

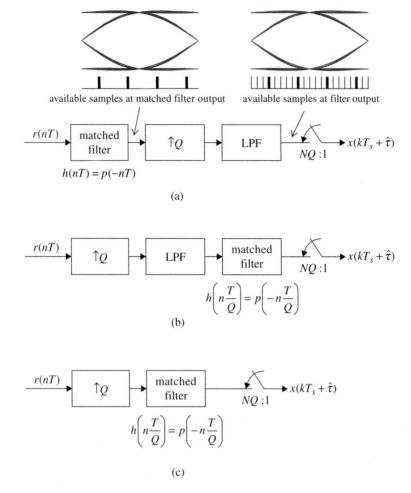

Figure 8.4.17 An upsample approach to interpolation: (a) Upsample and interpolation applied to the matched filter output; (b) upsample and interpolation applied the matched filter input; and (c) using the matched filter for both interpolation and shaping.

The upsample-and-interpolate operation can be applied to the matched filter input instead of the output as illustrated in Figure 8.4.17 (b). The received signal is upsampled by inserting $Q - 1$ zeros between each sample. The upsampled signal is low-pass filtered to produce $r(nT/Q)$, which consists of samples of the received signal at the high sample rate. In this case, the impulse response of the matched filter consists of T/Q-spaced samples of $p(-t)$. The desired matched filter output is obtained by downsampling the matched filter outputs at the high sample rate, $x(nT/Q)$, with the proper offset.

Both the interpolating filter and matched filter are low-pass filters. Thus, it is not necessary to filter twice. The low-pass interpolating filter may be removed as shown in Figure 8.4.17 (c).

Section 8.4 Discrete-Time Techniques for M-ary PAM   **473**

The key difference here is that the matched filter is performing two functions: interpolation and shaping. In other words, the matched filter outputs at the high sample rate, $x(nT/Q)$, are not identical to an upsampled version of the input $r(nT/Q)$.

Assuming the matched filter corresponds to a pulse shape with support on the interval $-L_p T_s \le t \le L_p T_s$, the discrete-time matched filter operating at QN samples/symbol is a length-$2QNL_p + 1$ FIR filter with indexes corresponding to $-QNL_p \le l \le QNL_p$. Thus, the matched filter outputs at the high sample may be expressed as

$$x\left(n\frac{T}{Q}\right) = \sum_{l=-QNL_p}^{QNL_p} r\left((n-l)\frac{T}{Q}\right) h\left(l\frac{T}{Q}\right). \tag{8.79}$$

The sequence $x(nT/Q)$ may be downsampled by Q to produce a sequence at N samples/symbol where every N-th sample is as close to $x(kT_s + \tau)$ as the resolution allows. The polyphase decomposition is due to the fact that not all of the multiplications defined by (8.79) are required. Because

$$r\left(n\frac{T}{Q}\right) = \begin{cases} r(nT) & n = 0, \pm Q, \pm 2Q, \ldots \\ 0 & \text{otherwise,} \end{cases} \tag{8.80}$$

only every Q-th value of $r(nT/M)$ in the FIR matched filter is nonzero. At a time instant during the high sample rate, these nonzero values coincide with the filter coefficients

$$\ldots, h(-2QT), h(-QT), h(0), h(QT), h(2QT), \ldots$$

and the filter output may be expressed as

$$\sum_{i=-NL_p}^{NL_p} r((n-i)T) h(iT) = x(nT). \tag{8.81}$$

At the next time instant, the nonzero values of $r(nT/Q)$ coincide with the filter coefficients

$$\ldots, h(-2QT+1), h(-QT+1), h(1), h(QT+1), h(2QT+1), \ldots$$

so that the filter output may be expressed as

$$\sum_{i=-NL_p}^{NL_p} r((n-i)T) h\left(\left(i+\frac{1}{Q}\right)T\right) = x\left(\left(n-\frac{1}{Q}\right)T\right). \tag{8.82}$$

At the q-th time instant, the nonzero values of $r(nT/Q)$ coincide with the filter coefficients

$$\ldots, h(-2QT+q), h(-QT+q), h(q), h(QT+q), h(2QT+q), \ldots$$

so that the filter output may be expressed as

$$\sum_{i=-NL_p}^{NL_p} r((n-i)T) h\left(\left(i+\frac{q}{Q}\right)T\right) = x\left(\left(n-\frac{q}{Q}\right)T\right). \tag{8.83}$$

This characteristic is illustrated in Figure 8.4.18 where a parallel bank of Q filters operating at the low sample rate $1/T$ is shown. Each filter in the filterbank is a downsampled version of the matched filter, except with a different index offset. The impulse response for $h_q(nT)$ is

$$h_q(nT) = h\left(nT + \frac{q}{Q}T\right) \quad \text{for } q = 0, 1, \ldots, Q-1. \tag{8.84}$$

The data samples $r(nT)$ form the input to all the filters in the filterbank simultaneously. The desired phase shift of the output is selected by connecting the output to the appropriate filter in the filterbank.

To see that the output of the q-th filter in the polyphase filter bank given by (8.83) does indeed produce the desired result given by (8.53), assume for the moment that T_I/T in (8.53) is sufficiently close to 1 so that $m(k) = n$. Then (8.54) becomes

$$x(kT_I) = \sum_i x((k-i)T) h_I((i+\mu(k))T). \tag{8.85}$$

Because the polyphase filterbank implementation uses the matched filter as the interpolation filter, the input data sequence $r(nT_s)$ in (8.83) plays the role of the matched filter output $x(nT)$ in (8.54) and the matched filter $h(nT)$ in (8.83) plays the role of the interpolation filter in (8.54). The comparison shows that the ratio of the polyphase filter stage index q to the number of

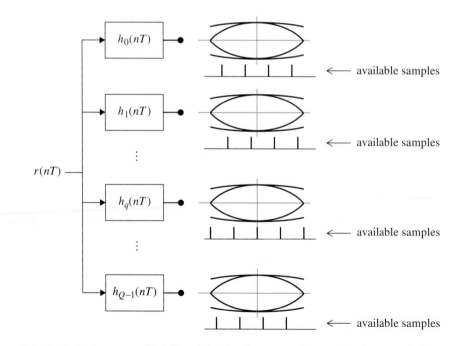

Figure 8.4.18 Polyphase matched filter filterbank outputs illustrating how each filter in the filterbank produces an output sequence with a different delay.

filterbank stages Q plays the same role as the fractional interval $\mu(k)$ in the interpolation filter. In this way, the polyphase filterbank implements the interpolation defined by (8.54) with a quantized fractional interval. The degree of quantization is controlled by the number of polyphase filter stages in the filterbank. The observations regarding the behavior of $\mu(k)$ above apply to the filter stage index q for the cases where T and T_s are not commensurate.

8.4.3 Interpolation Control

The purpose of the interpolator control block in Figure 8.4.2 is to provide the interpolator with the k-th basepoint index $m(k)$ and the k-th fractional interval $\mu(k)$. The basepoint index is usually not computed explicitly but rather identified by a signal often called a *strobe*. Two commonly used methods for interpolation control are a counter-based method and a recursive method.

Modulo-1 Counter Interpolation Control. For the case where interpolants are required for every N samples, interpolation control can be accomplished using a modulo-1 counter designed to underflow every N samples where the underflows are aligned with the basepoint indexes. A block diagram of this approach is shown in Figure 8.4.19. The T-spaced samples of the matched filter input are clocked into the matched filter with the same clock used to update the counter. A decrementing modulo-1 counter is shown here as it simplifies the computation of the fractional interval. An incrementing modulo-1 counter could also be used and is explored in Exercise 8.11.

The counter decrements by $1/N$ on average so that underflows occur every N samples on average. The loop filter output $v(n)$ adjusts the amount by which the counter decrements. This is done to align the underflows with the sample times of the desired interpolant. When operating properly, the modulo-1 counter underflows occur a clock period after the desired interpolant as illustrated in Figure 8.4.20. The underflow condition is indicated by a strobe and is used by the interpolator to identify the basepoint index.

The fractional interval may be computed directly from the contents of the modulo-1 counter on underflow. In general, the counter value satisfies the recursion

$$\eta(n+1) = \Big(\eta(n) - W(n)\Big) \mod 1 \qquad (8.86)$$

where $W(n) = 1/N + v(n)$ is the counter input and is the current estimate of the ratio T_i/T. When the decrementing counter underflows, the index n is the basepoint index $m(k)$. Incorporating the modulo-1 reduction produces

$$\eta\Big(m(k)+1\Big) = 1 + \eta\Big(m(k)\Big) - W\Big(m(k)\Big). \qquad (8.87)$$

As illustrated in Figure 8.4.20, the counter values $\eta(m(k))$ and $1 - \eta(m(k)+1)$ form similar triangles. This observation leads to the relationship

$$\frac{\mu\Big(m(k)\Big)}{\eta\Big(m(k)\Big)} = \frac{1 - \mu\Big(m(k)\Big)}{1 - \eta\Big(m(k)+1\Big)}. \qquad (8.88)$$

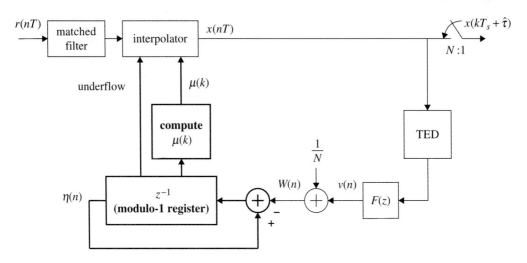

Figure 8.4.19 Modulo-1 counter for interpolation control in a baseband PAM system. The basepoint index is identified by the underflow strobe and the fractional interval updated using the counter contents on underflow.

Solving for $\mu\big(m(k)\big)$ produces

$$\mu\big(m(k)\big) = \frac{\eta\big(m(k)\big)}{1 - \eta\big(m(k)+1\big) + \eta\big(m(k)\big)} = \frac{\eta\big(m(k)\big)}{W\big(m(k)\big)}. \qquad (8.89)$$

The underflow period (in samples) of the decrementing modulo-1 counter is

$$\frac{1}{W(n)} = \frac{1}{\frac{1}{N} + v(n)} \qquad (8.90)$$

$$= \frac{N}{1 + Nv(n)}. \qquad (8.91)$$

When in lock, $v(n)$ is zero on average and the decrementing modulo-1 counter underflow period is N samples on average. During acquisition, $v(n)$ adjusts the underflow period to align the underflow events with the symbol boundaries as described above.

The decrementing modulo-1 counter plays the same role in this closed loop system that the DDS played in the general PLL of Figure C.2.1 (a). Note that because the counter is a *decrementing counter*, the DDS gain is $K_0 = -1$. The modulo-1 operation of the counter corresponds to the modulo-2π operation of the DDS.

Section 8.4 Discrete-Time Techniques for M-ary PAM

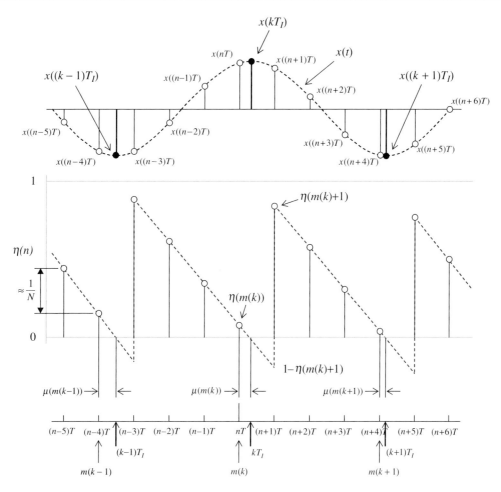

Figure 8.4.20 Illustration of the relationship between the available samples, the desired interpolants, and the modulo-1 counter contents.

Recursive Interpolation Control. The relationship for recursive interpolation control can be obtained by writing the expressions for two successive interpolation instants as

$$kT_I = (m(k) + \mu(k))T$$
$$(k+1)T_I = (m(k+1) + \mu(k+1))T \quad (8.92)$$

and subtracting the two to obtain the recursion

$$m(k+1) = m(k) + \frac{T_I}{T} + \mu(k) - \mu(k+1). \quad (8.93)$$

Because $m(k)$ and $m(k+1)$ are integers, the fractional part of the right-hand side of (8.93) must be zero from which the recursion for the fractional interval is obtained:

$$\mu(k+1) = \left(\mu(k) + \frac{T_I}{T}\right) \bmod 1. \tag{8.94}$$

The condition $0 \le \mu(k+1) < 1$ means the relationship

$$m(k+1) + \mu(k+1) = m(k) + \frac{T_I}{T} + \mu(k) < m(k+2) \tag{8.95}$$

must hold. The recursion on the sample count increment is thus

$$m(k+1) - m(k) = \left\lfloor \frac{T_I}{T} + \mu(k) \right\rfloor. \tag{8.96}$$

The sample count increment is a more useful quantity than the actual basepoint index because any finite-precision counter used to compute and/or store $m(k)$ would eventually overflow. As was the case with the counter-based control, the ratio T_I/T required by (8.94) and (8.96) is estimated by $W(n) = 1/N + v(n)$ where $v(n)$ is the output of the loop filter.

8.4.4 Examples

Two examples are provided to put all the pieces together. Both examples use binary PAM as the modulation. The first uses the MLTED and operates at 16 samples/symbol. The second uses the ZCTED and operates at 2 samples/symbol.

Binary PAM with MLTED. This example illustrates the use of the MLTED, the decrementing modulo-1 counters for interpolator control, and a linear interpolator for performing symbol timing synchronization for binary PAM. A block diagram is illustrated in Figure 8.4.21. The pulse shape is the square-root raised-cosine with 50% excess bandwidth. The received signal is sampled at a rate equivalent to $N = 16$ samples/s. Because $r(t)$ is eight times oversampled, a linear interpolator is adequate. Note that this system is different from the one suggested by the system in Figures 8.4.2 and 8.4.19 in that the interpolator precedes the matched filter. This was done to illustrate that the interpolator may be placed at either location in the processing chain.

Samples of the received signal are filtered by a discrete-time matched filter and derivative matched filter in parallel. The outputs are downsampled to 1 sample/symbol as directed by the controller. The timing error signal is formed as prescribed by the decision-directed MLTED (8.27). In this implementation, the loop filter and decrementing modulo-1 counter operate at the high sample rate of 16 samples/symbol. As a consequence, the error signal, which is updated at 1 sample/symbol, must be upsampled. The upsampling is performed by inserting zeros in between the error signal updates. The error signal is filtered by a discrete-time proportional-plus-integrator loop filter. The loop filter output forms the input to a decrementing modulo-1 counter. The counter controls the interpolation process as described in Section 8.4.3. Because the interpolator is not performing a sample rate change, there is no

Section 8.4 Discrete-Time Techniques for M-ary PAM

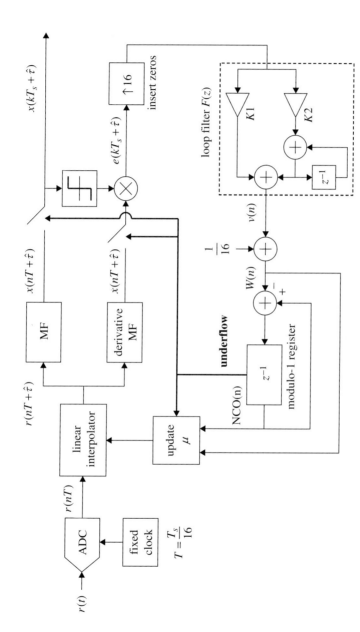

Figure 8.4.21 Binary PAM symbol timing synchronization system based on the MLTED using a linear interpolator and an proportional-plus-integrator loop filter.

need to provide basepoint index information. The interpolator produces one interpolant for each input sample.

The timing synchronization system can also be described as a computer program. The challenge with this approach is that the timing synchronization system is a parallel system while a computer program is a sequential representation. This is a common problem in system modeling: simulating an inherently parallel system on a sequential processor. A common method for generating the sequential representation is to write a program loop where each pass through the loop represents a clock cycle in the digital system. Within the loop, the parallel arithmetic (combinatorial) expressions are evaluated in topological order. Next the registered values (memory) are updated.

The code segment listed below is written using a MATLAB style syntax and consists of a `for` loop iterating on the samples of the received signal. The structure of the `for` loop follows the convention of updating the arithmetic (or combinatorial) quantities first and the registered values (or memory) last. The variable names used in the code segment are the same as those used in Figure 8.4.21 with the following additions:

- `rI`   a scalar representing the interpolant $r(nT + \hat{\tau})$.
- `mf`   a row vector consisting of samples of the matched filter impulse response
- `dmf`   a row vector consisting of samples of the derivative matched filter impulse response
- `rIBuff` a column vector of interpolator outputs used by the matched filter and derivative matched filter
- `xx`   a vector holding the desired matched filter outputs $x(kT_s + \hat{\tau})$ for $k = 0, 1, \ldots$.

The code segment is not written in the most efficient manner, but rather to explain the sequence of operations for proper PLL operation.

```
for n=2:length(r)

    % evaluate arithmetic expressions in topological order

    CNT = CNT_next;
    mu = mu_next;
    rI = mu*r(n) + (1 - mu)*r(n-1);
    x = mf*[rI; rIBuff];
    xdot = dmf*[rI; rIBuff];
    if underflow == 1
        e = sign(x)*xdot;
        xx(k) = x;                  % update output
        k = k + 1;
    else
        e = 0;
    end
    vp = K1*e;                      % proportional component of loop filter
    vi = vi + K2*e;                 % integrator component of loop filter
    v = vp + vi;                    % loop filter output
    W = 1/N + v;                    % counter control word

    % update registers
```

```
        CNT_next = CNT - W;            % update counter value for next cycle
        if CNT_next < 0                % test to see if underflow has occurred
            CNT_next = 1 + CNT_next;   % reduce counter value modulo-1 if underflow
            underflow = 1;             % set underflow flag
            mu_next = CNT/W;           % update mu
        else
            underflow = 0;
            mu_next = mu;
        end
        rIBuff = [rI; rIBuff(1:end-1)];
end
```

As an example, consider a symbol timing PLL with performance requirements $B_n T_s = 0.005$ and $\zeta = 1/\sqrt{2}$. Figure 8.4.4 gives the phase detector gain $K_p = 0.235$. Using $N = 16$, the loop constants given by (C.61) are

$$K_1 K_p K_0 = 8.3299 \times 10^{-4}$$

$$K_2 K_p K_0 = 3.4708 \times 10^{-7}.$$

Finally, solving for K_1 and K_2 using $K_p = 0.235$ and $K_0 = -1$ (to account for the fact that the modulo-1 counter is a *decrementing* counter) gives

$$K_1 = -3.5446 \times 10^{-3}$$

$$K_2 = -1.4769 \times 10^{-6}.$$

A plot of the timing error signal $e(k)$ and the fractional interval $\mu(k)$ are illustrated in Figure 8.4.22 for 1000 random symbols. The plot of $\mu(k)$ shows that the loop locks after about 400 symbols at the steady-state value $\mu = 0.5$. The plot of $\mu(k)$ looks "noisy." This is due to the self-noise produced by the TED.

Although the interpolator does not require basepoint index information from the controller, the rate change at the matched filter and derivative matched filter outputs does require basepoint index information. During acquisition, the PLL has to find the right basepoint index for the desired matched filter output. This search is indicated by the "ramping" effect observed in the plot of μ during the first 100 symbols. Each time μ touches zero, it wraps to $\mu = 1$ and reduces the interval between the current basepoint index and the next basepoint index by 1.

A practical variation of this design is illustrated in Figure 8.4.23. In this example, interpolation is moved to the output side of the matched filter and derivative matched filter. This placement requires two interpolators operating in parallel as shown. In this architecture, the two interpolators are required to perform a sample rate conversion. Hence the underflow strobe from the controller is required to provide basepoint index information to the interpolators. Relative to the architecture illustrated in Figure 8.4.21, this architecture has the disadvantage that two interpolators are required. But, it has the advantage that the matched filter and derivative matched filters are not in the closed loop path.

As before, the received signal is sampled at a rate equivalent to 16 samples/symbol to produce the samples $r(nT)$. These samples are filtered by a matched filter and derivative

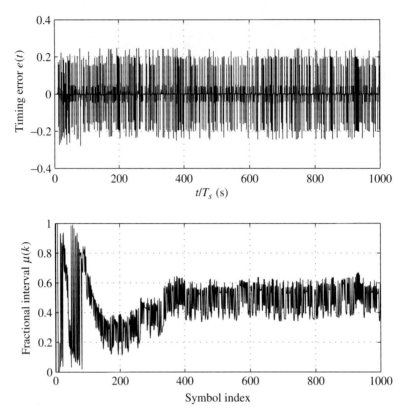

Figure 8.4.22 Timing error signal and fractional interpolation interval for the symbol timing synchronization system illustrated in Figure 8.4.21.

matched filter operating at 16 samples/symbol to produce the outputs $x(nT)$ and $\dot{x}(nT)$. These outputs form the inputs to two linear interpolators also operating in parallel. The interpolators produce one interpolant per symbol as directed by the controller. The controller provides both the basepoint index (via the underflow strobe) and the fractional interval. The two interpolator outputs $x(kT_s + \tau)$ and $\dot{x}(kT_s + \tau)$ are used to compute the timing error signal $e(k)$ given by (8.27). The error signal is upsampled by 16 to match the operating rate of the loop filter and controller.

An equivalent description using a MATLAB style code segment is shown below. The code segment uses the same variable names as Figure 8.4.23 with the following additions:

>   xI   a scalar representing the interpolant $x\left(kT_s + \hat{\tau}\right)$.
>   xdotI   a scalar representing the interpolant $\dot{x}\left(kT_s + \hat{\tau}\right)$.
>   xx   a vector holding the matched filter outputs $x\left(kT_s + \hat{\tau}\right)$ for
> $k = 0, 1, \ldots$.

Section 8.4 Discrete-Time Techniques for M-ary PAM

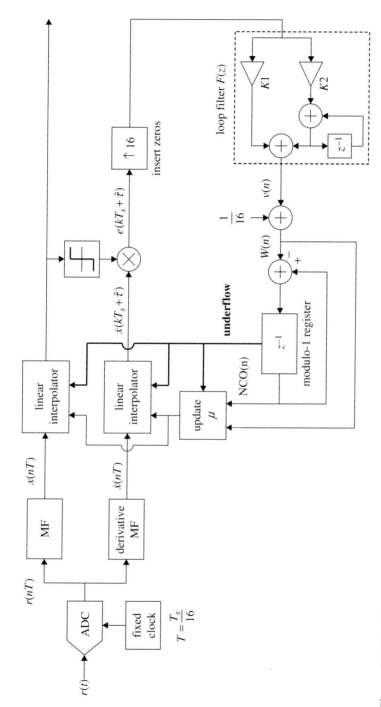

Figure 8.4.23 Binary PAM symbol timing synchronization system based on the MLTED using a linear interpolator and a proportional-plus-integrator loop filter.

The code segment consists of a `for` loop that iterates on the matched filter and derivative matched filter output samples. The code segment is not written in the most efficient manner, but rather to explain the sequence of operations for proper PLL operation.

```
for n=2:length(x)

    % evaluate arithmetic expressions in topological order

    CNT = CNT_next;
    mu = mu_next;
    if underflow == 1
        xI = mu*x(n) + (1 - mu)*x(n-1);
        xdotI = mu*xdot(n) + (1 - mu)*xdot(n-1);
        e = sign(xI)*xdotI;
        xx(k) = xI;
        k = k + 1;
    else
        e = 0;
    end
    vp = K1*e;                    % proportional component of loop filter
    vi = vi + K2*e;               % integrator component of loop filter
    v = vp + vi;                  % loop filter output
    W = 1/N + v;                  % counter control word

    % update registers

    CNT_next = CNT - W;           % update counter value for next cycle
    if CNT_next < 0               % test to see if underflow has occurred
        CNT_next = 1 + CNT_next;  % reduce counter value modulo-1 if underflow
        underflow = 1;            % set underflow flag
        mu_next = CNT/W;          % update mu
    else
        underflow = 0;
        mu_next = mu;
    end
end
```

An example of the phase error and fractional interval are plotted in Figure 8.4.24 for 1000 random symbols. The loop filter constants are identical to those used previously. As before, the timing PLL locks after about 400 symbols. The shape of the fractional interval plot is quite similar to the fractional interval plot in Figure 8.4.22. Differences are due to the placement of the matched filter and derivative matched filter. In Figure 8.4.22, the matched filter and derivative matched filters are in the closed loop path whereas in Figure 8.4.24 they are not.

Binary PAM with ZCTED. This example illustrates the use of the ZCTED along with the decrementing modulo-1 counter for interpolator control and the piecewise parabolic interpolator to perform symbol timing synchronization for binary PAM. A block diagram is illustrated in Figure 8.4.25. The pulse shape is the square-root raised-cosine with 50% excess bandwidth. The received signal is sampled at a rate equivalent to $N = 2$ samples/symbol. Samples of the received signal are filtered by a discrete-time matched filter operating at 2 samples/symbol. The matched filter outputs $x(nT)$ are used by the piece-wise parabolic interpolator to compute the interpolants $x(nT + \hat{\tau})$. These interpolants form the input to the zero crossing detector described in Section 8.4.1 and given by (8.37). The timing error

Section 8.4 Discrete-Time Techniques for M-ary PAM

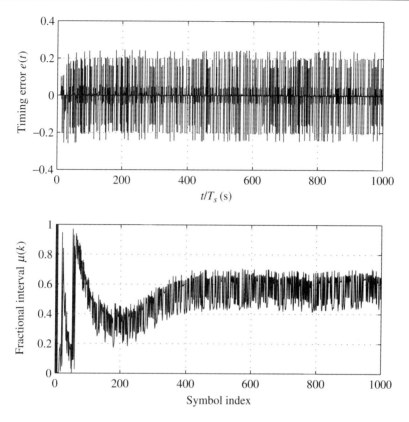

Figure 8.4.24 Timing error signal and fractional interpolation interval for the symbol timing synchronization system illustrated in Figure 8.4.23.

signal is updated at 1 sample/symbol. Because the loop filter and controller operate at $N = 2$ samples/symbol, the timing error signal is upsampled by inserting a zero(s) in between the updates. The upsampled timing error signal is filtered by the proportional-plus-integrator loop filter. The loop filter output forms the input to a decrementing modulo-1 counter. The counter controls the interpolation process as described in Section 8.4.3.

A code segment modeling the system is listed below. It is written using a MATLAB style syntax and consists of a `for` loop iterating on the samples of the matched filter output. The structure of the `for` loop follows the convention of updating the arithmetic (or combinatorial) quantities first and the registered values (or memory) last. The variable names used in the code segment are the same as those used in Figure 8.4.21 with the following additions:

    `TEDBuff`   a 2 × 1 column vector of interpolator outputs used by the TED
    `xx`   a vector holding the matched filter outputs $x\left(kT_s + \hat{\tau}\right)$ for $k = 0, 1, \ldots$.

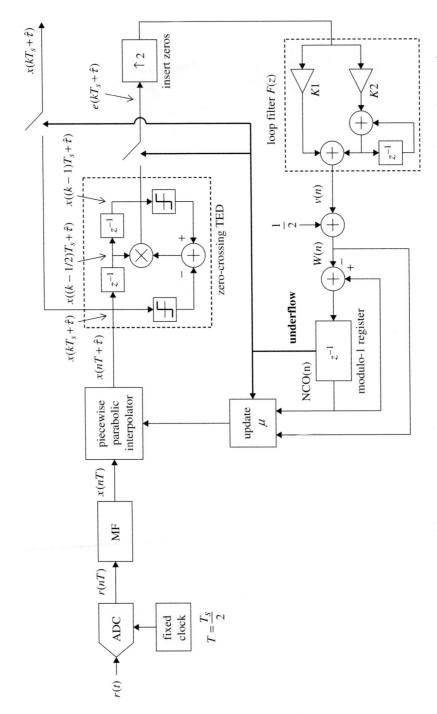

Figure 8.4.25 Binary PAM symbol timing synchronization system based on the ZCTED using a linear interpolator and an proportional-plus-integrator loop filter.

Section 8.4 Discrete-Time Techniques for M-ary PAM

```
for n=2:length(x)-2

    % evaluate arithmetic expressions in topological order

    CNT = CNT_next;
    mu = mu_next;
    v2 = 1/2*[1, -1, -1, 1]*x(n+2:-1:n-1);   % Farrow structure for the
    v1 = 1/2*[-1, 3, -1, -1]*x(n+2:-1:n-1);  % piecewise parabolic
    v0 = x(n);                                % interpolator
    xI = (mu*v2 + v1)*mu + v0;                % interpolator output
    if underflow == 1
        e = TEDBuff(1) * (sign(TEDBuff(2)) - sign(xI));
        xx(k) = xI;
        k = k + 1;
    else
        e = 0;
    end
    vp = K1*e;                 % proportional component of loop filter
    vi = vi + K2*e;            % integrator component of loop filter
    v = vp + vi;               % loop filter output
    W = 1/N + v;               % NCO control word

    % update registers

    CNT_next = CNT - W;        % update counter value for next cycle
    if CNT_next < 0            % test to see if underflow has occurred
        CNT_next = 1 + CNT_next;  % reduce counter value modulo-1 if underflow
        underflow = 1;         % set underflow flag
        mu_next = CNT/W;       % update mu
    else
        underflow = 0;
        mu_next = mu;
    end
    TEDBuff = [xI; TEDBuff(1)];
end
```

As an example, consider a symbol timing PLL with performance requirements $B_n T_s = 0.005$ and $\zeta = 1/\sqrt{2}$. Figure 8.4.7 gives the phase detector gain $K_p = 2.7$. Using $N = 2$, the loop constants given by (C.61) are

$$K_1 K_p K_0 = 6.6445 \times 10^{-3}$$

$$K_2 K_p K_0 = 2.2148 \times 10^{-5}.$$

Finally, solving for K_1 and K_2 using $K_p = 2.7$ and $K_0 = -1$ (to account for the fact that the controller is a *decrementing* modulo-1 counter) gives

$$K_1 = -2.4609 \times 10^{-3}$$

$$K_2 = -8.2030 \times 10^{-6}.$$

A plot of the timing error signal $e(k)$ and the fractional interval $\mu(k)$ are illustrated in Figure 8.4.26 for 1000 random symbols. The plot of $\mu(k)$ shows that the loop locks after about 200 symbols at the steady-state value $\mu = 0.5$. Because the ZCTED does not produce

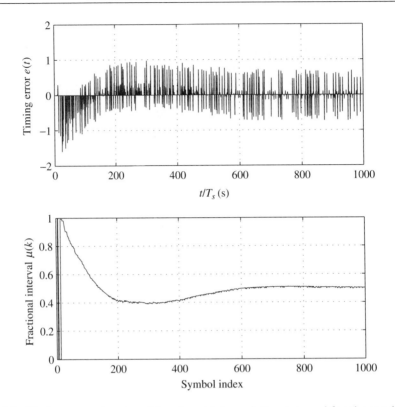

Figure 8.4.26 Timing error signal and fractional interpolation interval for the symbol timing synchronization system illustrated in Figure 8.4.26.

any self-noise, the plot of μ has a much "cleaner" look than the plot of μ for the MLTED in Figure 8.4.22.

The code segment listed above relies on a noncausal interpolation filter. The piecewise parabolic interpolation filter operates on $x((n+2)T), x((n+1)T), x(nT)$, and $x((n-1)T)$ to produce an interpolant in between $x((n+1)T)$ and $x(nT)$. Thus, the relationships between the NCO contents, the available samples, and the desired interpolants illustrated in Figure 8.4.20 apply. In most practical applications, a *causal* operation is required. Consequently, a causal interpolation filter is required. The length-4 piecewise parabolic interpolation filter can operate as a causal filter with a two-sample delay. That is, it operates on $x(nT), x((n-1)T)$, $x((n-2)T)$, and $x((n-3)T)$ and produces an interpolant between $x((n-1)T)$ and $x((n-2)T)$. This requires careful interpretation of the interpolator output.

The relationships between the available samples, the desired interpolants, the NCO contents, and the strobe assuming a causal length-4 interpolation filter is illustrated in Figure 8.4.27. Suppose the matched filter output sample at time nT is clocked in and the strobe is high. The strobe is high because the counter underflowed after processing the previous matched filter output $x((n-1)T)$. As a consequence, the k-th basepoint index is $(n-2)T$

Section 8.4 Discrete-Time Techniques for M-ary PAM

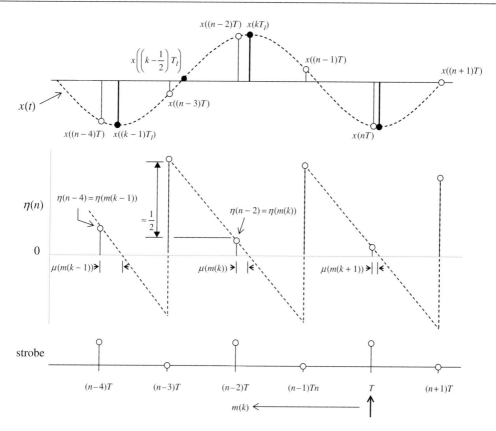

Figure 8.4.27 A detailed diagram illustrating the relationship between the available samples, the NCO contents, the strobe, and the desired interpolants. The focus of the discussion is on the matched filter output at time nT when the strobe is high.

as shown. Because the k-th basepoint index is $(n-2)T$, the desired interpolant is between $x((n-2)T)$ and $x((n-1)T)$. Thus, the interpolation filter output corresponding to the inputs $x(nT), x((n-1)T), x((n-2)T)$, and $x((n-3)T)$ is the desired interpolant $x(kT_i)$. The ZCTED uses this output, $x(kT_i)$, together with the two previous interpolator outputs $x\left(\left(k-\frac{1}{2}\right)T_i\right)$ and $x((k-1)T_i)$ to compute the error signal. A segment of code based on these principles is illustrated below.

```
for n=4:length(x)

    % evaluate arithmetic expressions in topological order

    CNT = CNT_next;
    mu  = mu_next;
    v2  = 1/2*[1, -1, -1, 1]*x(n:-1:n-3);   % Farrow structure for the
    v1  = 1/2*[-1, 3, -1, -1]*x(n:-1:n-3);  % piecewise parabolic
    v0  = x(n-2);                           % interpolator
```

```
        xI = (mu*v2 + v1)*mu + v0;          % interpolator output
        if underflow == 1
            e = TEDBuff(1) * (sign(TEDBuff(2)) - sign(xI));
            xx(k) = xI;
            k = k + 1;
        else
            e = 0;
        end
        vp = K1*e;                           % proportional component of loop filter
        vi = vi + K2*e;                      % integrator component of loop filter
        v = vp + vi;                         % loop filter output
        W = 1/N + v;                         % NCO control word

        % update registers

        CNT_next = CNT - W;                  % update counter value for next cycle
        if CNT_next < 0                      % test to see if underflow has occurred
            CNT_next = 1 + CNT_next;         % reduce counter value modulo-1 if underflow
            underflow = 1;                   % set underflow flag
            mu_next = CNT/W;                 % update mu
        else
            underflow = 0;
            mu_next = mu;
        end
        TEDBuff = [xI; TEDBuff(1:end-1)];
end
```

The code listings above (both the noncausal and causal versions) do not work for the case of sample clock frequency offset. That is, for the case $T \neq T_s/2$, the code must be modified to account for the cases when an interpolant is required during two consecutive clock cycles ($T > T_s/2$) or for the case when two clock cycles occur between consecutive interpolants ($T < T_s/2$).

The case $T > T_s/2$ is illustrated in Figure 8.4.28. The desired samples appear to "slide to the left" because the samples are spaced slightly further apart than $T_s/2$. Most of the time, a desired matched filter interpolant is produced for every two available matched filter samples. Because $T > T_s/2$, a residual timing error accumulates. As the residual timing error accumulates, the fractional interval $\mu(k)$ decreases with time as shown. Eventually, the accumulated residual timing error exceeds a sample period. This coincides with $\mu(k)$ decreasing to 0 and wrapping around to 1. When this occurs, desired matched filter interpolants occur one sample apart instead of the normal two. As shown, when this occurs, one of the samples needed by the ZCTED is never produced. This missing sample must be inserted or "stuffed" into the ZCTED registers to ensure proper operation after the "wrap around."

The case $T < T_s/2$ is illustrated in Figure 8.4.29. In this case, the desired samples appear to "slide to the right" because the samples are spaced slightly closer together than $T_s/2$. Most of the time, a desired matched filter interpolant is produced for every two available matched filter samples. Because $T < T_s/2$, a residual timing error accumulates. As the residual timing error accumulates, the fractional interval $\mu(k)$ increases with time as shown. Eventually, the accumulated residual timing error exceeds a sample period. This coincides with $\mu(k)$ exceeding 1 and wrapping around to 0. When this occurs, the desired matched filter interpolants are spaced three samples apart instead of the normal two. As a consequence, the

Section 8.4 Discrete-Time Techniques for M-ary PAM

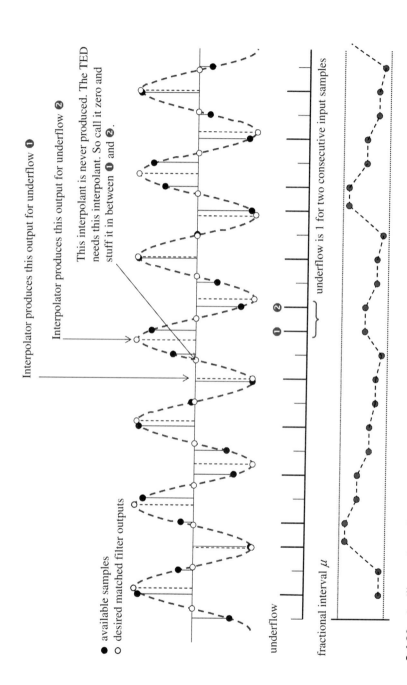

Figure 8.4.28 An illustration of the relationship between the available matched filter output samples, the desired interpolants, the underflow from the NCO interpolation controller, and the fractional interval for the case where the sample clock frequency is slightly slower than 2 samples/symbol (i.e., $T > T_s/2$). A causal piecewise parabolic interpolation filter is assumed.

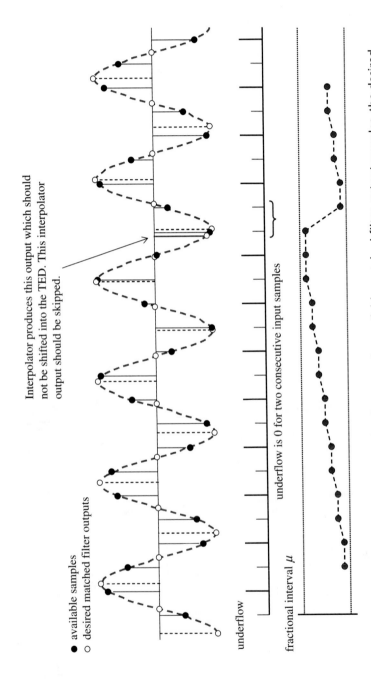

Figure 8.4.29 An illustration of the relationship between the available matched filter output samples, the desired interpolants, the underflow from the NCO interpolation controller, and the fractional interval for the case where the sample clock frequency is slightly faster than 2 samples/symbol (i.e., $T < T_s/2$). A causal piecewise parabolic interpolation filter is assumed.

interpolator produces an extra sample that should be ignored, or "skipped" by the ZCTED. This is accomplished by not shifting the ZCTED registers after the "wrap around."

A modified segment of code to account for this condition is shown below. A new variable `old_underflow` is introduced. This variable, together with `underflow`, is used to determine whether normal operation, "stuffing," or "skipping" should occur. Again, the code is not written in the most efficient manner, but rather to provide a description of the subtleties associated with proper operation of the ZCTED.

```
for n=2:length(x)-2

    % evaluate arithmetic expressions in topological order

    CNT = CNT_next;
    mu = mu_next;
    v2 = 1/2*[1, -1, -1, 1]*x(n:-1:n-3);       % Farrow structure for the
    v1 = 1/2*[-1, 3, -1, -1]*x(n:-1:n-3);      % piecewise parabolic
    v0 = x(n-2);                                % interpolator
    xI = (mu*v2 + v1)*mu + v0;                  % interpolator output
    if underflow == 1                           % update output
        xx(k) = xI;
        k = k + 1;
    end
    if underflow == 1 & old_underflow == 0
        e = TEDBuff(1) * (sign(TEDBuff(2)) - sign(xI));
    else
        e = 0;
    end
    vp = K1*e;                                  % proportional component of loop filter
    vi = vi + K2*e;                             % integrator component of loop filter
    v = vp + vi;                                % loop filter output
    W = 1/N + v;                                % NCO control word

    % update registers

    if underflow == 0 & old_underflow == 0
        TEDBuff = TEDBuff;                      % skip current sample
    elseif underflow == 0 & old_underflow == 1
        TEDBuff = [xI; TEDBuff(1)];             % normal operation
    elseif underflow == 1 & old_underflow == 0
        TEDBuff = [xI; TEDBuff(1)];             % normal operation
    elseif underflow == 1 & old_underflow == 1
        TEDBuff = [xI; 0];                      % stuff missing sample
    end
    CNT_next = CNT - W;                         % update counter value for next cycle
    if CNT_next < 0                             % test to see if underflow has occurred
        CNT_next = 1 + CNT_next;                % reduce counter value modulo-1 if underflow
        old_underflow = underflow;
        underflow = 1;                          % set underflow flag
        mu_next = CNT/W;                        % update mu
    else
        old_underflow = underflow;
        underflow = 0;
        mu_next = mu;
    end
end
```

As this code segment illustrates, the "upsample by 2" function inserted between the TED and the loop filter is only an abstraction. The upsample operation is performed by inserting zeros between the timing error updates. Most of the time 1 zero is inserted. But sometimes no zeros are inserted; sometimes 2 zeros are inserted.

As an example of operation for the case where the sample clock frequency is slightly higher than 2 samples/symbol (i.e., $T < T_s/2$), suppose the samples $r(nT)$ were obtained where T satisfied

$$T = \frac{T_s}{2 + \frac{1}{400}}$$

or, what is equivalent

$$\text{sample rate} = \left(2 + \frac{1}{400}\right) \times \text{symbol rate}.$$

The sampling clock frequency is 1/400 of the symbol rate faster than 2 samples/symbol. The error signal and fractional interval for the same timing PLL considered previously are plotted in Figure 8.4.30. As expected, the fractional interval ramps from 0 to 1 and rolls over every 400 symbol times. This is because the frequency error in the sample clock is 1/400 of the symbol rate. The error signal indicates that the timing PLL locks after about 200 symbols. This case is the symbol timing PLL equivalent of a phase ramp input for the generic PLL reviewed in Section C.1.1 of Appendix C.

8.5 DISCRETE-TIME TECHNIQUES FOR MQAM

Let the received band-pass MQAM signal be

$$r(t) = G_a \sum_k a_0(k) p(t - kT_s - \tau) \sqrt{2} \cos(\omega_0 t + \theta)$$

$$- G_a \sum_k a_1(k) p(t - kT_s - \tau) \sqrt{2} \sin(\omega_0 t + \theta) + w(t) \quad (8.97)$$

where $a_0(k)$ and $a_1(k)$ are the inphase and quadrature components of the k-th symbol; $p(t)$ is a unit-energy pulse shape with support on the interval $-L_p T_s \leq t \leq L_p T_s$; T_s is the symbol time; τ is the unknown timing delay to be estimated; ω_0 is the center frequency in radians/second; θ is the carrier phase offset; and $w(t)$ is the additive white Gaussian noise. The constant G_a represents all the amplitude gains and losses through the antennas, propagation medium, amplifiers, mixers, filters, and other RF components. ADC placement is an important system-level consideration that requires some discussion at this point.

There are two locations where the ADC is commonly placed as illustrated in Figure 8.5.1. Figure 8.5.1 (a) shows a configuration commonly referred to as "band-pass sampling." The ADC samples the bandlimited signal $r(t)$ every T_{BP} seconds where the sampling rate

Section 8.5 Discrete-Time Techniques for MQAM

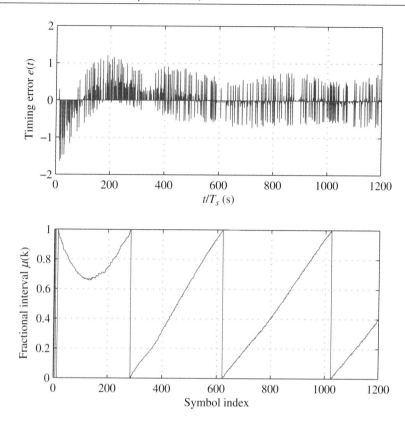

Figure 8.4.30 Timing error signal and fractional interpolation interval for the symbol timing synchronization system illustrated in Figure 8.4.25 for the case where the sample clock is slightly faster than 2 samples/symbol.

satisfies the Nyquist rate condition for the band-pass signal. These samples are mixed by quadrature discrete-time sinusoids to produce samples of the baseband inphase and quadrature components $I(nT_{BP})$ and $Q(nT_{BP})$. $I(nT_{BP})$ and $Q(nT_{BP})$ are filtered by the discrete-time matched filters with impulse response $h(nT_{BP}) = p(-nT_{BP})$. The desire is to produce N_{BP} samples of the inphase and quadrature matched filter outputs during each symbol such that one of the samples on both the inphase and quadrature components are aligned with the maximum average eye opening.

The second commonly used option for ADC placement is shown in Figure 8.5.1 (b). The band-pass signal $r(t)$ is mixed to baseband using continuous-time quadrature sinusoids and low-pass filtered to produce the inphase and quadrature baseband components $I(t)$ and $Q(t)$. $I(t)$ and $Q(t)$ and are sampled by a pair of ADCs (or a dual-channel ADC) to produce samples of the inphase and quadrature baseband components $I(nT_{BB})$ and $Q(nT_{BB})$, respectively. $I(nT_{BB})$ and $Q(nT_{BB})$ are filtered by the discrete-time matched filters with impulse response $h(nT_{BB}) = p(-nT_{BB})$. As before, the desire is to produce N_{BB} samples of the inphase and

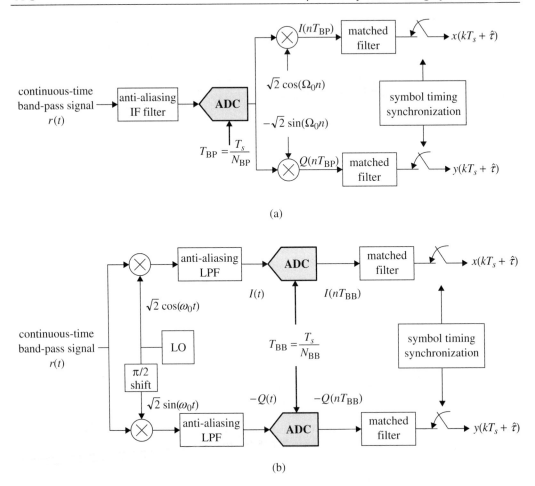

Figure 8.5.1 Two commonly used options for ADC placement: (a) Band-pass sampling; (b) Dual-channel baseband sampling.

quadrature matched filter outputs during each symbol such that one of the samples on both the inphase and quadrature components are aligned with the maximum average eye opening.

Which of the two approaches is preferred depends on many factors including the symbol rate and the center frequency (which determine the required sample rate), cost, performance requirements, the availability of good analog band-pass filters for channel selection and/or adjacent channel rejection, etc. These issues are discussed in detail in Chapter 10.

It is not important which of the two approaches is used for the purposes of describing symbol timing synchronization using discrete-time techniques. In either case, the matched filter inputs are the samples of $I(t)$ and $Q(t)$. These samples are denoted $I(nT)$ and $Q(nT)$, respectively; whether $T = T_s/N_{BP}$ or $T = T_s/N_{BB}$ is not important as long as it is known. $I(nT)$ and $Q(nT)$ are of the same form as $r(nT)$ in Section 8.4. TEDs operate on both $I(nT)$

and $Q(nT)$ in the same way they operated on $r(nT)$ in Section 8.4. Let $x(nT)$ and $y(nT)$ be the outputs of matched filters operating on $I(nT)$ and $Q(nT)$, respectively. Applying the data-aided TEDs outlined in Section 8.4 to QAM produces the following:

MLTED: $\quad e(k) = a_0(k)\dot{x}\left(kT_s + \hat{\tau}\right) + a_1(k)\dot{y}\left(kT_s + \hat{\tau}\right)$ (8.98)

ELTED: $\quad e(k) = a_0(k)\left[x\left((k+1/2)T_s + \hat{\tau}\right) - x\left((k-1/2)T_s + \hat{\tau}\right)\right]$

$\quad\quad\quad\quad + a_1(k)\left[y\left((k+1/2)T_s + \hat{\tau}\right) - y\left((k-1/2)T_s + \hat{\tau}\right)\right]$ (8.99)

ZCTED: $\quad e(k) = x\left((k-1/2)T_s + \hat{\tau}\right)[a_0(k-1) - a_0(k)]$

$\quad\quad\quad\quad + y\left((k-1/2)T_s + \hat{\tau}\right)[a_1(k-1) - a_1(k)]$ (8.100)

GTED: $\quad e(k) = x\left((k-1/2)T_s + \hat{\tau}\right)\left[x\left((k-1)T_s + \hat{\tau}\right) - x\left(kT_s + \hat{\tau}\right)\right]$

$\quad\quad\quad\quad + y\left((k-1/2)T_s + \hat{\tau}\right)\left[y\left((k-1)T_s + \hat{\tau}\right) - y\left(kT_s + \hat{\tau}\right)\right]$ (8.101)

MMTED: $\quad e(k) = a_0(k-1)x\left(kT_s + \hat{\tau}\right) - a_0(k)x\left((k-1)T_s + \hat{\tau}\right)$

$\quad\quad\quad\quad + a_1(k-1)y\left(kT_s + \hat{\tau}\right) - a_1(k)y\left((k-1)T_s + \hat{\tau}\right).$ (8.102)

This shows that a TED is applied to the matched filter outputs corresponding to the inphase and quadrature components of the received signal. The outputs of the two TEDs are summed to form the error signal. The error signal is filtered by the loop filter and drives the interpolation control. The decision-directed versions of the TEDs are obtained in the obvious way. The general structure for MQAM symbol timing synchronization with band-pass sampling is illustrated in Figure 8.5.2.

8.6 DISCRETE-TIME TECHNIQUES FOR OFFSET QPSK

Assuming band-pass sampling and perfect phase synchronization, the discrete-time band-pass offset QPSK signal may be represented as

$$r(nT) = G_a \sum_m a_0(m) p(nT - mT_s - \tau)\sqrt{2}\cos(\Omega_0 n + \theta)$$

$$- G_a \sum_m a_1(m) p(nT - mT_s - T_s/2\tau)\sqrt{2}\sin(\Omega_0 n + \theta) \quad (8.103)$$

where $1/T$ is the sample rate; $a_0(m) \in \{-A, +A\}$ and $a_1(m) \in \{-A, +A\}$ are the information symbols; $p(nT)$ is a unit-energy pulse shape with support on the interval $-L_p T_s/T < n < L_p T_s/T$; Ω_0 is the center frequency in radians/sample; G_a is the cumulative amplitude gain from the transmitter, through the propagation medium and through the receiver; and τ is the

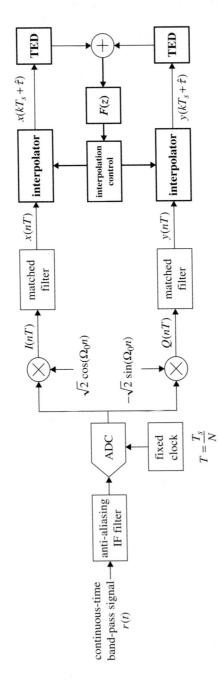

Figure 8.5.2 General structure for symbol timing synchronization for MQAM using band-pass sampling.

Section 8.6 Discrete-Time Techniques for Offset QPSK

unknown symbol timing offset. Assuming perfect carrier phase synchronization, $\theta = 0$ and the matched filter outputs may be expressed as

$$x(nT) = K \sum_m a_0(m) r_p(nT - mT_s - \tau) \tag{8.104}$$

$$y(nT) = K \sum_m a_1(m) r_p(nT - mT_s - T_s/2 - \tau) \tag{8.105}$$

where $r_p(u)$ is the autocorrelation function of the pulse shape given by (8.3).

The relationship between the two eye patterns formed by $x(nT)$ and $y(nT)$ is illustrated in Figure 8.6.1. The maximum average eye opening on $y(nT)$ is delayed from the maximum average eye opening on $x(nT)$ by $T_s/2$. The inphase matched filter output $x(nT)$ should be sampled at

$$n = k\frac{T_s}{T} + \tau \tag{8.106}$$

while the quadrature matched filter output $y(nT)$ should be sampled at

$$n = \left(k + \frac{1}{2}\right)\frac{T_s}{T} + \tau \tag{8.107}$$

for $k = 0, 1, \ldots$.

Following the same line of reasoning as before, the slope of eye patterns can be used as a timing error signal. Because the eye patterns are delayed $T_s/2$ from each other, this method must be modified. The maximum likelihood data-aided TED uses the error signal

$$e(k) = a_0(k)\dot{x}\left(kT_s + \hat{\tau}(k)\right) + a_1(k)\dot{y}\left(\left(k + \frac{1}{2}\right)T_s + \hat{\tau}(k)\right) \tag{8.108}$$

where $\dot{x}(kT_s + \hat{\tau}(k))$ is the time derivative of $x(t)$ evaluated at $t = kT_s + \hat{\tau}(k)$ and $\dot{y}((k + 1/2)T_s + \hat{\tau}(k))$ is the time derivative of $y(t)$ evaluated at $t = (k + 1/2)T_s + \hat{\tau}(k)$. The slopes of the matched filter outputs at time instants offset by half a symbol period are combined to form the error signal. The decision-directed maximum likelihood TED uses the error signal

$$e(k) = \hat{a}_0(k)\dot{x}\left(kT_s + \hat{\tau}(k)\right) + \hat{a}_1(k)\dot{y}\left(\left(k + \frac{1}{2}\right)T_s + \hat{\tau}(k)\right) \tag{8.109}$$

where the data symbol estimates are

$$\begin{aligned}\hat{a}_0(k) &= A \times \text{sgn}\left\{x\left(kT_s + \hat{\tau}(k)\right)\right\} \\ \hat{a}_1(k) &= A \times \text{sgn}\left\{y\left(\left(k + \frac{1}{2}\right)T_s + \hat{\tau}(k)\right)\right\}.\end{aligned} \tag{8.110}$$

The time derivative may be computed using the techniques described in Section 8.4.1 and illustrated in Figure 8.4.5. The early–late techniques, described in Section 8.4.1 can be used to approximate the derivatives with the appropriate modifications suggested by (8.108) and (8.109). The zero crossing detector can also be applied to $x(nT)$ and $y(nT)$ with appropriate delays.

500 Chapter 8 Symbol Timing Synchronization

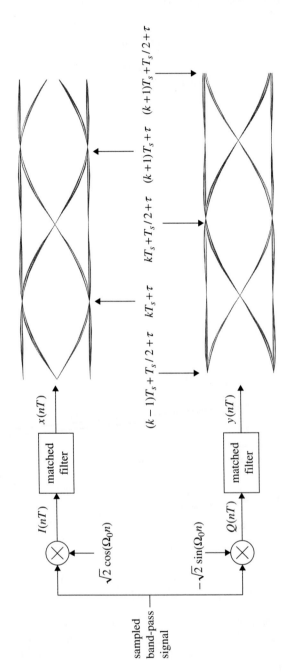

Figure 8.6.1 Eye diagrams of the inphase and quadrature matched filter outputs for offset QPSK showing the relationship between the maximum average eye openings.

8.7 DEALING WITH TRANSITION DENSITY: A PRACTICAL CONSIDERATION

The TEDs described in Section 8.4.1 all require data transitions to produce a non-zero timing error. The MLTED and ELTED are based on the time derivative (or its approximate) of the matched filter output. In the absence of data transitions, the matched filter output is almost a constant whose time derivative approaches zero for any timing error. The ZCTED and GTED are based on zero crossings in the matched filter output. Because zero crossings in the matched filter output only occur with a data transition, these TEDs also require data transitions to produce a nonzero timing error.

When used in a PLL with a relatively small loop bandwidth, the averaging performed by the loop filter permits reasonably good performance in the absence of data transitions over a few symbol periods. The absence of data transitions over a sufficiently long period, however, will not provide enough timing error information for the PLL to stay in lock. For this reason, long sequences of consecutive ones or zeros are undesirable.

The presence or absence of data transitions is often measured by the *transition density*. For purely random data, the transition density is approximately[4] 50%. The S-curves for the TEDs described in Section 8.4.1 were all based on a transition density of 50%. Data sequences with smaller transition densities reduce the TED gain K_p while data sequences with higher transition densities increase the TED gain. In general, the TED gain varies as the square root of the transition density.

In many circumstances, there is little control over the data bits generated by the user. Thus, there is little control over the transition density. There are two approaches to this problem. The first is to use a pulse shape that guarantees a data transition. The MAN pulse shape, described in Appendix A, is such a pulse shape. It possesses a transition at the midpoint of the symbol interval. Thus, even in the presence of a long sequence of consecutive ones or zeros, the matched filter output contains a data transition every symbol interval. Unfortunately, the bandwidth of the MAN pulse shape is enormous. As a consequence, this approach is not feasible for most wireless communication links.

The second approach applies some bit-level processing to the data bits produced by the data source. The goal of the bit-level processing is to produce a "more random" bit sequence. Error control coding and encryption, when used, perform this function well. Another approach is to use a simple shift register system to perform pseudorandom scrambling. The goal of pseudorandom scrambling is to produce "random looking" bit sequences in response to long sequences of consecutive ones or zeros. These systems are required in links that do not use error control coding or encryption, and are often desirable in links that use error control coding without encryption.

As an example of pseudorandom scrambling, consider the length-7 shift register system specified for this purpose in the IEEE 802.11a standard. This system, illustrated in Figure 8.7.1, is called a *scrambler* in the standard. This system produces an output bit stream based on the input bit stream and the contents of the shift register. As each input bit is clocked in, the shift register is also clocked to shift its contents to the right. The contents of the fourth and

---
[4] Note that an alternating 1 0 1 0 ... pattern with binary PAM produces a transition density of 100%. As this is not a very useful information sequence, real data rarely has a transition density close to 100%.

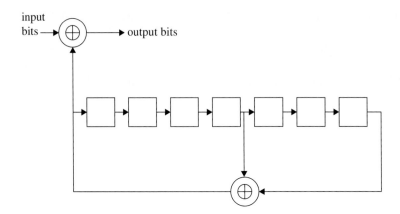

Figure 8.7.1 Pseudorandom scrambler used to ensure data transitions in the IEEE 802.11a wireless networking standard. (Reproduced from [80].)

seventh element in the shift register are combined using the XOR (exclusive OR) operation and tied to the input of the shift register and to the input of another XOR that computes the XOR with the input bit. The shift register must be initialized to some nonzero state (i.e., at least one of the shift register elements must be 1) for desired operation. The same system is used as a descrambler as well.

To see that this is so, suppose the input bit stream is

$$1\ 0\ 1\ 1\ 0\ 1\ 0\ 0\ 1\ 1\ 1\ 0\ 1\ 0\ 0\ 1$$

and that the initial state of the shift register is all zeros except for the right-most register. In this case, the scrambled output is

$$0\ 0\ 1\ 1\ 1\ 1\ 0\ 1\ 0\ 1\ 1\ 0\ 0\ 0\ 1\ 0.$$

When the scrambled bit sequence above is applied to the input of the shift register with the same initial state, the output is the unscrambled input. The real benefit is what the scrambler does to the all-zeros sequence. With the initial state of the shift register set to all zeros except for the right-most register, the input sequence

$$0\ 0\ 0\ 0\ 0\ 0\ 0\ 0\ 0\ 0\ 0\ 0\ 0\ 0\ 0\ 0$$

produces the scrambled output

$$1\ 0\ 0\ 0\ 1\ 0\ 0\ 1\ 1\ 0\ 0\ 0\ 1\ 0\ 1\ 1.$$

The all-zeros input produces no symbol transitions. The scrambled output is sufficiently "random" to produce an acceptable number of symbol transitions. Note that the scrambled bit sequence applied to the same shift register system with the same initial state produces the all-zeros output.

Another example of the use of pseudorandom scrambling used to introduce data transitions is the linear feedback shift register system from the IRIG-106 standard illustrated in Figure 8.7.2. In this standard, this type of bit-level processor is called a *randomizer*. The IRIG-106 randomizer is illustrated in Figure 8.7.2 (a). It is based on a length-15 shift register with feedback. The inverse operation, called a *derandomizer*, is based on the same length-15 shift register, but with different connections between the input bits and the shift register contents.

An alternative approach is to use substitution codes that limit the runs of consecutive ones or zeros. An example of this approach was applied to the digital audio standard for the compact disk. In this case, each block of 8 bits is replaced by a block of 14 bits that limits the number of consecutive ones or zeros to 11.

8.8 MAXIMUM LIKELIHOOD ESTIMATION

Maximum likelihood estimation uses conditional probabilities as a measure of "how likely" a parameter is given noisy observations. This technique was applied in Chapter 5 to derive the optimum (in the maximum likelihood sense) structure for detectors. The problem was cast as an estimation problem where the information symbols were the unknown quantity. Maximum likelihood estimation can also be applied to synchronization. In this case, the timing delay is the unknown quantity that needs to be estimated. The technique is demonstrated for QPSK. Extensions to other signal sets are straightforward.

8.8.1 Preliminaries

Let the observation interval be $T_0 = L_0 T_s$ seconds and let the received band-pass signal be

$$r(t) = s(t) + w(t) \tag{8.111}$$

where

$$s(t) = \sum_{k=0}^{L_0} a_0(k) p(t - kT_s - \tau) \sqrt{2} \cos(\omega_0 t) - a_1(k) p(t - kT_s - \tau) \sqrt{2} \sin(\omega_0 t) \tag{8.112}$$

where $a_0(k) \in \{-1, +1\}$ and $a_1(k) \in \{-1, +1\}$ are the inphase and quadrature components of the k-th symbol; $p(t)$ is a unit-energy pulse shape with support on $-L_p T_s \leq t \leq L_p T_s$; T_s is the symbol time; τ is the unknown delay to be estimated; and $w(t)$ is a zero-mean white Gaussian random process with power spectral density $N_0/2$ W/Hz. The band-pass signal is sampled every T seconds to produce the sequence

$$r(nT) = s(nT) + w(nT); \quad n = 0, 1, \ldots, NL_0 - 1. \tag{8.113}$$

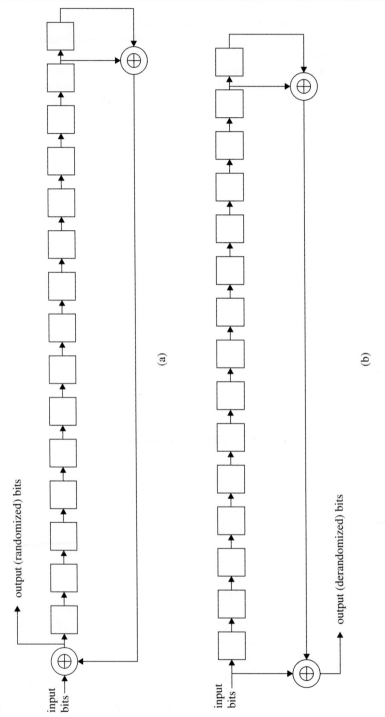

Figure 8.7.2 Pseudorandom scrambler used to ensure data transitions in the IRIG-106 standard: (a) The randomizer; (b) The derandomizer. (Reproduced from [141].)

Section 8.8 Maximum Likelihood Estimation

The sampled signal component may be expressed as

$$s(nT) = \sum_{k=0}^{L_0-1} a_0(k) p(nT - kT_s - \tau) \sqrt{2} \cos(\Omega_0 n)$$
$$- a_1(k) p(nT - kT_s - \tau) \sqrt{2} \sin(\Omega_0 n) \qquad (8.114)$$

for $n = 0, 1, \ldots, NL_0 - 1$. Note that the carrier phase offset is assumed known. For notational simplicity, $\theta = 0$ is used. For convenience, the following vectors are defined

$$\mathbf{r} = \begin{bmatrix} r(0) \\ r(T) \\ \vdots \\ r((NL_0-1)T) \end{bmatrix} \quad \mathbf{s} = \begin{bmatrix} s(0) \\ s(T) \\ \vdots \\ s((NL_0-1)T) \end{bmatrix} \quad \mathbf{w} = \begin{bmatrix} w(0) \\ w(T) \\ \vdots \\ w((NL_0-1)T) \end{bmatrix}. \qquad (8.115)$$

The vector $\mathbf{w}$ is a sequence of independent and identically distributed Gaussian random variables with zero mean and variance

$$\sigma^2 = \frac{N_0}{2T}. \qquad (8.116)$$

The probability density function of $\mathbf{w}$ is

$$p(\mathbf{w}) = \frac{1}{(2\pi\sigma^2)^{L_0 N/2}} \exp\left\{ -\frac{1}{2\sigma^2} \sum_{n=0}^{NL_0-1} w^2(nT) \right\}. \qquad (8.117)$$

For notational convenience, define the symbol vector $\mathbf{a}$ as

$$\mathbf{a} = \begin{bmatrix} \mathbf{a}(0) & \mathbf{a}(1) & \cdots & \mathbf{a}(L_0-1) \end{bmatrix}^T \qquad (8.118)$$

where

$$\mathbf{a}(k) = \begin{bmatrix} a_0(k) \\ a_1(k) \end{bmatrix}. \qquad (8.119)$$

To emphasize that $\mathbf{s}$ is a function of $\mathbf{a}$ and τ, $\mathbf{s}$ will be expressed as $\mathbf{s}(\mathbf{a}, \tau)$ and samples of the signal component $s(nT)$ will be expressed as $s(nT; \mathbf{a}, \tau)$.

Symbol timing synchronization can be thought of as an estimation problem. The goal is to estimate the parameter τ from the samples $r(nT) = s(nT; \mathbf{a}, \tau) + w(nT)$. The maximum likelihood estimate is the one that maximizes the logarithm of the conditional probability $p(\mathbf{r}|\mathbf{a}, \tau)$. Using the probability density function of $\mathbf{w}$ given by (8.117), the conditional probability $p(\mathbf{r}|\mathbf{a}, \tau)$ is

$$p(\mathbf{r}|\mathbf{a}, \tau) = \frac{1}{(2\pi\sigma^2)^{L_0 N/2}} \exp\left\{ -\frac{1}{2\sigma^2} \sum_{n=0}^{NL_0-1} [r(nT) - s(nT; \mathbf{a}, \tau)]^2 \right\}. \qquad (8.120)$$

The log-likelihood function $\Lambda(\mathbf{a}, \tau)$ is the logarithm of (8.120):

$$\Lambda(\mathbf{a}, \tau) = -\frac{L_0 N}{2} \ln(2\pi\sigma^2) - \frac{1}{2\sigma^2} \sum_{n=0}^{NL_0-1} [r(nT) - s(nT; \mathbf{a}, \tau)]^2$$

$$= -\frac{L_0 N}{2} \ln(2\pi\sigma^2)$$

$$- \frac{1}{2\sigma^2} \sum_{n=0}^{NL_0-1} \left\{ r^2(nT) - 2r(nT)s(nT; \mathbf{a}, \tau) + s^2(nT; \mathbf{a}, \tau) \right\}. \qquad (8.121)$$

Later it will be convenient to express the cross product sum as

$$\sum_{n=0}^{NL_0-1} r(nT)s(nT; \mathbf{a}, \tau) = \sum_{k=0}^{L_0-1} a_0(k) \sum_{n=(k-L_p)N}^{(k+L_p)N} r(nT)p(nT - kT_s - \tau)\sqrt{2}\cos(\Omega_0 n)$$

$$- \sum_{k=0}^{L_0-1} a_1(k) \sum_{n=(k-L_p)N}^{(k+L_p)N} r(nT)p(nT - kT_s - \tau)\sqrt{2}\sin(\Omega_0 n). \qquad (8.122)$$

Two approaches will be taken to obtain the maximum likelihood estimator for τ. The first approach assumes $\mathbf{a}$ is known.[5] In this case, the estimator for τ is a function of the data symbols. The second approach does not assume $\mathbf{a}$ is known. In this case, the dependence on $\mathbf{a}$ is removed by assuming the symbol sequence $\mathbf{a}$ is random and using the total probability theorem to obtain the conditional probability density function $p(\mathbf{r}|\tau)$. The maximum likelihood estimate maximizes the logarithm of $p(\mathbf{r}|\tau)$.

The conditional probability density function $p(\mathbf{r}|\tau)$ is related to the conditional probability density function $p(\mathbf{r}|\mathbf{a}, \tau)$ by the total probability theorem:

$$p(\mathbf{r}|\tau) = \int p(\mathbf{r}|\mathbf{a}, \tau) p(\mathbf{a}) d\mathbf{a} \qquad (8.123)$$

where $p(\mathbf{a})$ is the probability density function of the symbol sequence $\mathbf{a}$. The most commonly used probability density function for the data sequence assumes the symbols are independent and equally likely. Independence implies

$$p(\mathbf{a}) = \prod_{k=0}^{L_0-1} p(\mathbf{a}(k)) \qquad (8.124)$$

---

[5] For packetized burst mode communication systems with a known preamble or header, the L_0 data symbols are known and should be used for synchronization.

Section 8.8 Maximum Likelihood Estimation   507

while equally likely implies

$$p(\mathbf{a}(k)) = \frac{1}{4}\delta(a_0(k)-1)\delta(a_1(k)-1) + \frac{1}{4}\delta(a_0(k)-1)\delta(a_1(k)+1)$$
$$+ \frac{1}{4}\delta(a_0(k)+1)\delta(a_1(k)-1) + \frac{1}{4}\delta(a_0(k)+1)\delta(a_1(k)+1). \quad (8.125)$$

Thus,

$$p(\mathbf{r}|\tau) = \int p(\mathbf{r}|\mathbf{a},\tau)p(\mathbf{a})d\mathbf{a} \quad (8.126)$$

$$= \prod_{k=0}^{L_0-1} \int p(\mathbf{r}|\mathbf{a}(k),\tau)p(\mathbf{a}(k))d\mathbf{a}(k) \quad (8.127)$$

$$= \prod_{k=0}^{L_0-1} \left\{ \frac{1}{4}p(\mathbf{r}|\mathbf{a}(k)=[1,1],\tau) + \frac{1}{4}p(\mathbf{r}|\mathbf{a}(k)=[1,-1],\tau) \right.$$
$$\left. + \frac{1}{4}p(\mathbf{r}|\mathbf{a}(k)=[-1,1],\tau) + \frac{1}{4}p(\mathbf{r}|\mathbf{a}(k)=[-1,-1],\tau) \right\}. \quad (8.128)$$

By writing (8.120) as

$$p(\mathbf{r}|\mathbf{a},\tau) = \prod_{k=0}^{L_0-1} \frac{1}{(2\pi\sigma^2)^{N/2}} \exp\left\{ -\frac{1}{2\sigma^2} \sum_{n=N(k-L_p)}^{N(k+L_p)} [r(nT) - s(nT;\mathbf{a},\tau)]^2 \right\} \quad (8.129)$$

and using the substitution

$$s(nT;\mathbf{a}(k),\tau) = a_1(k)p(nT-kT_s-\tau)\sqrt{2}\cos(\Omega_0 n) - a_2(k)p(nT-kT_s-\tau)\sqrt{2}\sin(\Omega_0 n)$$
$$(8.130)$$

each term in (8.128) may be expressed as

$$p(\mathbf{r}|\mathbf{a}(k)=[1,1],\tau) = \prod_{k=0}^{L_0-1} \frac{1}{(2\pi\sigma^2)^{N/2}} \exp\left\{ -\frac{1}{2\sigma^2} \sum_{n=N(k-L_p)}^{N(k+L_p)} r^2(nT) + p^2(nT-kT_s-\tau) \right\}$$

$$\times \exp\left\{ \frac{1}{\sigma^2} \sum_{n=N(k-L_p)}^{N(k+L_p)} r(nT)p(nT-kT_s-\tau)\sqrt{2}\cos(\Omega_0 n) \right\}$$

$$\times \exp\left\{ -\frac{1}{\sigma^2} \sum_{n=N(k-L_p)}^{N(k+L_p)} r(nT)p(nT-kT_s-\tau)\sqrt{2}\sin(\Omega_0 n) \right\} \quad (8.131)$$

$$p(\mathbf{r}|\mathbf{a}(k) = [1, -1], \tau) = \prod_{k=0}^{L_0-1} \frac{1}{(2\pi\sigma^2)^{N/2}} \exp\left\{-\frac{1}{2\sigma^2} \sum_{n=N(k-L_p)}^{N(k+L_p)} r^2(nT) + p^2(nT - kT_s - \tau)\right\}$$

$$\times \exp\left\{\frac{1}{\sigma^2} \sum_{n=N(k-L_p)}^{N(k+L_p)} r(nT)p(nT - kT_s - \tau)\sqrt{2}\cos(\Omega_0 n)\right\}$$

$$\times \exp\left\{\frac{1}{\sigma^2} \sum_{n=N(k-L_p)}^{N(k+L_p)} r(nT)p(nT - kT_s - \tau)\sqrt{2}\sin(\Omega_0 n)\right\} \quad (8.132)$$

$$p(\mathbf{r}|\mathbf{a}(k) = [-1, 1], \tau) = \prod_{k=0}^{L_0-1} \frac{1}{(2\pi\sigma^2)^{N/2}} \exp\left\{-\frac{1}{2\sigma^2} \sum_{n=N(k-L_p)}^{N(k+L_p)} r^2(nT) + p^2(nT - kT_s - \tau)\right\}$$

$$\times \exp\left\{-\frac{1}{\sigma^2} \sum_{n=N(k-L_p)}^{N(k+L_p)} r(nT)p(nT - kT_s - \tau)\sqrt{2}\cos(\Omega_0 n)\right\}$$

$$\times \exp\left\{-\frac{1}{\sigma^2} \sum_{n=N(k-L_p)}^{N(k+L_p)} r(nT)p(nT - kT_s - \tau)\sqrt{2}\sin(\Omega_0 n)\right\} \quad (8.133)$$

$$p(\mathbf{r}|\mathbf{a}(k) = [-1, -1], \tau) = \prod_{k=0}^{L_0-1} \frac{1}{(2\pi\sigma^2)^{N/2}} \exp\left\{-\frac{1}{2\sigma^2} \sum_{n=N(k-L_p)}^{N(k+L_p)} r^2(nT) + p^2(nT - kT_s - \tau)\right\}$$

$$\times \exp\left\{-\frac{1}{\sigma^2} \sum_{n=N(k-L_p)}^{N(k+L_p)} r(nT)p(nT - kT_s - \tau)\sqrt{2}\cos(\Omega_0 n)\right\}$$

$$\times \exp\left\{\frac{1}{\sigma^2} \sum_{n=N(k-L_p)}^{N(k+L_p)} r(nT)p(nT - kT_s - \tau)\sqrt{2}\sin(\Omega_0 n)\right\}. \quad (8.134)$$

Substituting (8.131) – (8.134) into (8.128) and collecting similar terms produces

$$p(\mathbf{r}|\tau) = \frac{1}{4} \prod_{k=0}^{L_0-1} \frac{1}{(2\pi\sigma^2)^{N/2}} \exp\left\{-\frac{1}{2\sigma^2} \sum_{n=N(k-L_p)}^{N(k+L_p)} r^2(nT) + p^2(nT - kT_s - \tau)\right\}$$

$$\times \left(\exp\left\{\frac{1}{\sigma^2} \sum_{n=N(k-L_p)}^{N(k+L_p)} r(nT)p(nT - kT_s - \tau)\sqrt{2}\cos(\Omega_0 n)\right\}\right.$$

Section 8.8 Maximum Likelihood Estimation

$$+ \exp\left\{-\frac{1}{\sigma^2} \sum_{n=N(k-L_p)}^{N(k+L_p)} r(nT)p(nT - kT_s - \tau)\sqrt{2}\cos(\Omega_0 n)\right\}\right)$$

$$\times \left(\exp\left\{\frac{1}{\sigma^2} \sum_{n=N(k-L_p)}^{N(k+L_p)} r(nT)p(nT - kT_s - \tau)\sqrt{2}\sin(\Omega_0 n)\right\}\right.$$

$$+ \exp\left\{-\frac{1}{\sigma^2} \sum_{n=N(k-L_p)}^{N(k+L_p)} r(nT)p(nT - kT_s - \tau)\sqrt{2}\sin(\Omega_0 n)\right\}\right). \quad (8.135)$$

Applying the identity

$$\frac{e^x + e^{-x}}{2} = \cosh(x) \quad (8.136)$$

to (8.135) produces

$$p(\mathbf{r}|\tau) = \prod_{k=0}^{L_0-1} \frac{1}{(2\pi\sigma^2)^{N/2}} \exp\left\{-\frac{1}{2\sigma^2} \sum_{n=N(k-L_p)}^{N(k+L_p)} r^2(nT) + p^2(nT - kT_s - \tau)\right\}$$

$$\times \cosh\left(\frac{1}{\sigma^2} \sum_{n=N(k-L_p)}^{N(k+L_p)} r(nT)p(nT - kT_s - \tau)\sqrt{2}\cos(\Omega_0 n)\right)$$

$$\times \cosh\left(\frac{1}{\sigma^2} \sum_{n=N(k-L_p)}^{N(k+L_p)} r(nT)p(nT - kT_s - \tau)\sqrt{2}\sin(\Omega_0 n)\right). \quad (8.137)$$

The average log-likelihood function is

$$\overline{\Lambda}(\tau) = -\frac{NL_0}{2}\ln(2\pi\sigma^2) - \frac{1}{2\sigma^2}\sum_{k=0}^{L_0-1}\sum_{N(k-L_p)}^{N(k+L_p)} r^2(nT) + p^2(nT - kT_s - \tau)$$

$$+ \sum_{k=0}^{L_0-1} \ln\cosh\left(\frac{1}{\sigma^2} \sum_{n=N(k-L_p)}^{N(k+L_p)} r(nT)p(nT - kT_s - \tau)\sqrt{2}\cos(\Omega_0 n)\right)$$

$$+ \sum_{k=0}^{L_0-1} \ln\cosh\left(\frac{1}{\sigma^2} \sum_{n=N(k-L_p)}^{N(k+L_p)} r(nT)p(nT - kT_s - \tau)\sqrt{2}\sin(\Omega_0 n)\right). \quad (8.138)$$

8.8.2 Symbol Timing Estimation

Known Symbol Sequence. For the case of known symbols, the maximum likelihood timing estimate is the value of τ that maximizes the log-likelihood function $\Lambda(\mathbf{a}, \tau)$ given by (8.121). The partial derivative of $\Lambda(\mathbf{a}, \tau)$ is

$$\frac{\partial}{\partial \tau} \Lambda(\mathbf{a}, \tau) = -\frac{1}{2\sigma^2} \frac{\partial}{\partial \tau} \sum_{n=0}^{NL_0-1} [r(nT) - s(nT; \mathbf{a}, \tau)]^2 \tag{8.139}$$

$$= -\frac{1}{2\sigma^2} \frac{\partial}{\partial \tau} \sum_{n=0}^{NL_0-1} \left[r^2(nT) - 2r(nT)s(nT; \mathbf{a}, \tau) + s^2(nT; \mathbf{a}, \tau) \right]. \tag{8.140}$$

The partial derivative of the first term is zero because the energy in the received signal does not depend on the timing offset. The partial derivative of the third term is approximately zero as there is a weak dependence on τ. For QPSK, this approximation is quite good and shall be carried through with the remainder of this development. As was the case with carrier phase estimation, all that remains is the middle term. Substituting (8.114) for $s(nT; \mathbf{a}, \tau)$ and interchanging the order of summations produces

$$\frac{\partial}{\partial \tau} \Lambda(\mathbf{a}, \tau) = \frac{1}{\sigma^2} \frac{\partial}{\partial \tau} \sum_{k=0}^{L_0-1} a_0(k) \sum_{n=(k-L_p)N}^{(k+L_p)N} r(nT)p(nT - kT_s - \tau)\sqrt{2}\cos(\Omega_0 n)$$

$$- \frac{1}{\sigma^2} \frac{\partial}{\partial \tau} \sum_{k=0}^{L_0-1} a_1(k) \sum_{n=(k-L_p)N}^{(k+L_p)N} r(nT)p(nT - kT_s - \tau)\sqrt{2}\sin(\Omega_0 n). \tag{8.141}$$

Recognizing the inner summations as matched filter outputs and using the identities

$$x(kT_s + \tau) = \sum_{n=N(k-L_p)}^{N(k+L_p)} r(nT)p(nT - kT_s - \tau)\sqrt{2}\cos(\Omega_0 n) \tag{8.142}$$

$$y(kT_s + \tau) = -\sum_{n=N(k-L_p)}^{N(k+L_p)} r(nT)p(nT - kT_s - \tau)\sqrt{2}\sin(\Omega_0 n), \tag{8.143}$$

the partial derivative of the log-likelihood function (8.141) may be expressed as

$$\frac{\partial}{\partial \tau} \Lambda(\mathbf{a}, \tau) = \frac{1}{\sigma^2} \frac{\partial}{\partial \tau} \sum_{k=0}^{L_0-1} a_0(k)x(kT_s + \tau) + a_1(k)y(kT_s + \tau) \tag{8.144}$$

$$= \frac{1}{\sigma^2} \sum_{k=0}^{L_0-1} a_0(k)\dot{x}(kT_s + \tau) + a_1(k)\dot{y}(kT_s + \tau) \tag{8.145}$$

Section 8.8 Maximum Likelihood Estimation

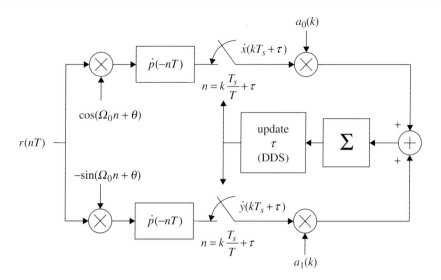

Figure 8.8.1 Block diagram of the maximum likelihood QPSK timing estimator based on (8.146).

where $\dot{x}(kT_s + \tau)$ and $\dot{y}(kT_s + \tau)$ are samples of the time derivatives of the inphase and quadrature matched filter outputs, respectively. These time derivatives may be computed from samples of the matched filter inputs using a filter whose impulse response consists of samples of the time derivative of the pulse shape as illustrated in Figure 8.4.5 in Section 8.4.1.

The maximum likelihood timing estimate $\hat{\tau}$ is the value of τ that forces (8.145) to zero:

$$0 = \sum_{k=0}^{L_0-1} a_0(k)\dot{x}(kT_s + \hat{\tau}) + a_1(k)\dot{y}(kT_s + \hat{\tau}). \tag{8.146}$$

Unlike the maximum likelihood carrier phase estimate, there is no closed form solution for $\hat{\tau}$. A block diagram illustrating an iterative method for finding $\hat{\tau}$ is shown in Figure 8.8.1. The solution is a PLL structure where the right-hand side of (8.146) is the error signal.

Unknown Symbol Sequence. For the case of unknown data symbols, the maximum likelihood timing estimate is the value of τ that maximizes the average log-likelihood function $\overline{\Lambda}(\tau)$ given by (8.138). The partial derivative of $\overline{\Lambda}(\tau)$ is

$$\frac{\partial}{\partial \tau}\overline{\Lambda}(\tau) = \sum_{k=0}^{L_0-1} \tanh\left(\frac{1}{\sigma^2}\sum_{n=N(k-L_p)}^{N(k+L_p)} r(nT)p(nT - kT_s - \tau)\sqrt{2}\cos(\Omega_0 n)\right)$$

$$\times \frac{\partial}{\partial \tau}\left[\frac{1}{\sigma^2}\sum_{n=N(k-L_p)}^{N(k+L_p)} r(nT)p(nT - kT_s - \tau)\sqrt{2}\cos(\Omega_0 n)\right]$$

$$+ \sum_{k=0}^{L_0-1} \tanh\left(\frac{1}{\sigma^2} \sum_{n=N(k-L_p)}^{N(k+L_p)} r(nT)p(nT - kT_s - \tau)\sqrt{2}\sin(\Omega_0 n)\right)$$

$$\times \frac{\partial}{\partial \tau}\left[\frac{1}{\sigma^2} \sum_{n=N(k-L_p)}^{N(k+L_p)} r(nT)p(nT - kT_s - \tau)\sqrt{2}\sin(\Omega_0 n)\right]. \qquad (8.147)$$

Recognizing the inner summations on the index n as matched filter outputs, the relationships (8.142) and (8.143) may be used to express (8.147) in the more compact form

$$\frac{\partial}{\partial \tau}\overline{\Lambda}(\tau) = \sum_{k=0}^{L_0-1} \tanh\left(\frac{1}{\sigma^2}x(kT_s + \tau)\right) \frac{\partial}{\partial \tau}\left[\frac{1}{\sigma^2}x(kT_s + \tau)\right]$$

$$+ \sum_{k=0}^{L_0-1} \tanh\left(\frac{1}{\sigma^2}y(kT_s + \tau)\right) \frac{\partial}{\partial \tau}\left[\frac{1}{\sigma^2}y(kT_s + \tau)\right]. \qquad (8.148)$$

Denoting the time derivatives of the inphase and quadrature matched filter outputs by $\dot{x}(kT_s + \tau)$ and $\dot{y}(kT_s + \tau)$, respectively, the maximum likelihood timing estimate $\hat{\tau}$ satisfies

$$0 = \sum_{k=0}^{L_0-1} \tanh\left(\frac{1}{\sigma^2}x(kT_s + \hat{\tau})\right) \frac{1}{\sigma^2}\dot{x}(kT_s + \hat{\tau})$$

$$+ \sum_{k=0}^{L_0-1} \tanh\left(\frac{1}{\sigma^2}y(kT_s + \hat{\tau})\right) \frac{1}{\sigma^2}\dot{y}(kT_s + \hat{\tau}). \qquad (8.149)$$

A block diagram outlining an iterative method for finding $\hat{\tau}$ is shown in Figure 8.8.2. The basic structure is that of a PLL that uses the right-hand side of (8.149) as the error signal. The computational burden can be reduced by replacing the hyperbolic tangent with its small signal or large signal approximation. A plot of the hyperbolic tangent is given in Figure 7.8.4. From this plot the approximations given by (7.131) may be used. For example, when the signal-to-noise ratio is high, $\tanh(X) \approx \text{sgn}\{X\}$ so that the error signal (input to the summation block in Figure 8.8.2) is well approximated by

$$e(k) \approx \text{sgn}\{x(kT_s + \tau)\}\dot{x}(kT_s + \tau) + \text{sgn}\{y(kT_s + \tau)\}\dot{y}(kT_s + \tau). \qquad (8.150)$$

Observe that for QPSK

$$\hat{a}_0(k) = \text{sgn}\{x(kT_s + \tau)\}$$
$$\hat{a}_1(k) = \text{sgn}\{y(kT_s + \tau)\}$$

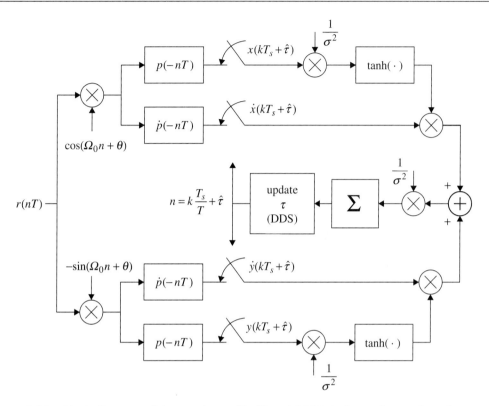

Figure 8.8.2 Block diagram of the maximum likelihood QPSK timing estimator based on (8.149).

so that the large signal-to-noise ratio approximation may also be expressed as

$$e(k) \approx \hat{a}_0(k)\dot{x}(kT_s + \tau) + \hat{a}_1(k)\dot{y}(kT_s + \tau). \tag{8.151}$$

This shows that the large signal-to-noise ratio approximation for the non-data-aided ML timing estimator may be formulated as a decision-directed PLL. Compare this with the QPSK timing PLL illustrated in Figure 8.3.2.

8.9 NOTES AND REFERENCES

8.9.1 Topics Covered

The emphasis of this chapter is on PLL-based symbol timing estimation. In the early years of digital communications, the PLL-based approach was by far the most common approach. The tutorial by Franks [139] surveys continuous-time solutions for symbol timing synchronization. As digital communication systems were developed, symbol timing synchronization subsystems were characterized by adhoc techniques that later were shown to be approximations to maximum likelihood estimation. The form of the maximum likelihood TED given by (8.26) was published by Kobayashi [129], Meyers and Franks [136], and Gardner [165]. A more general form of the maximum likelihood TED was developed by Gardner [165] and Bergmans and Wong-Lam [166]. The early–late TED (8.33) was described by Lindsey and Simon [128]. The zero–crossing TED (8.36) was suggested by Gardner [165] and described by Mengali and D'Andrea [123]. The Gardner TED was defined and analyzed by Gardner in Ref. [167]. The Mueller and Müller TED was described by Mueller and Müller in Ref. [168].

General discussions on interpolation are found in most textbooks on discrete-time signal processing. A nice tutorial overview was published by Shafer and Rabiner [169]. The treatment of interpolation in this chapter draws heavily on the seminal papers by Gardner and his colleagues at the European Space Agency [164,170] and the wonderful text by Mengali and D'Andrea [123]. The use of polyphase filterbanks for symbol timing synchronization was described by harris and Rice [171]. The use of a decrementing counter to control the interpolations and the closed loop operation was described by Gardner [170]. The recursive interpolation control described in the chapter is due to Moenecleay and described in Gardner [170]. Many textbooks provide a more in-depth look at timing synchronization from a more theoretical point of view [123,154,172]. A more detailed discussion of interpolation jitter and ways to overcome it are offered by Gardner [170] and Qin, Wang, Zeng, and Xiong [173].

A discussion of transition density and its effects on the performance of symbol timing synchronizers is in Waggener [174]. The discussion of the EFM technique used in compact disk digital audio is described by Peek [175].

8.9.2 Topics Not Covered

An important characteristic of TEDs in closed loop symbol timing estimators is *self noise*. This was only briefly mentioned in the chapter. See Franks and Bubrouski [176], Moeneclaey [177], D'Andrea, Mengali, and Moro [178], Gardner [167], D'Andrea and Luise [179,180], Cowley [181], and Mengali and D'Andrea [123] for a more thorough treatment of self-noise for many commonly used TEDs.

This chapter focused primarily on the application of FIR filters in symbol timing synchronization. The application of recursive IIR filters has also been investigated by Dick and harris [182] and harris [183].

In addition to closed loop symbol timing synchronizers based on a PLL, "feed-forward" symbol timing synchronization has also been investigated. Most often, feed-forward timing synchronization is used in packetized burst communication systems (such as GSM) and relies on a sequence of known symbols called *training symbols* or *pilot symbols* embedded in each packet. The data-aided maximum likelihood estimate is the value of τ that maximizes (8.121) or, what is equivalent, the value of τ that satisfies (8.146). As there is no closed form solution in these cases, some form of search must be performed. A common approach is to quantize the time axis to Q parts per symbol and compute (8.121) at these quantized values and select the value that produces the maximum. Non-data aided solutions are also possible. The most popular is the estimator described by Oerder and Meyr [184].

8.10 EXERCISES

8.1 Derive the S-curve for the data-aided MLTED given by (8.30) based on the error signal (8.26).

8.2 Derive the S-curve for the data-aided ELTED given by (8.35) based on the error signal (8.33).

8.3 Derive the S-curve for the data-aided ZCTED given by (8.42) based on the error signal (8.36).

8.4 Derive the S-curve for the data-aided MMTED given by (8.50) based on the error signal (8.48).

8.5 Derive the linear interpolator filter (8.61) from (8.59) and (8.60).

8.6 Derive the cubic interpolator filter (8.67) from (8.65) and (8.66).

8.7 This exercise steps through the derivation of the piecewise parabolic interpolator (8.72).

(a) Using the second-order polynomial approximation

$$x(t) = c_2 t^2 + c_1 t + c_0$$

express $x((m + \mu)T)$ as a polynomial in μ. The answer should be of the form

$$x((m + \mu)T) = b_2 \mu^2 + b_1 \mu + b_0$$

where the b's are functions of the c's, m, and T.

(b) Using the boundary conditions $x(mT)$ and $x((m+1)T)$, solve for b_0 and b_1 and show that $x((m + \mu)T)$ may be expressed as

$$x((m + \mu)T) = c_2 T^2 \left(\mu^2 - \mu \right) + \mu x((m+1)T) + (1 - \mu) x(mT)$$

This result shows that $x((m + \mu)T)$ is a linear combination of $x((m+1)T)$ and $x(mT)$ plus another term. If c_2 is also a linear combination of $x((m+1)T)$ and $x(mT)$, then $x((m+\mu)T)$ can be regarded as the output of a filter with inputs $x(mT)$ and $x((m+1)T)$. Part (c) shows that c_2 must be a function of more than $x(mT)$ and $x((m+1)T)$ in order to produce a piecewise parabolic interpolator. In part (d), a piecewise parabolic interpolator of the form given by (8.72) is derived.

(c) Suppose c_2 is a linear combination of $x((m+1)T)$ and $x(mT)$, that is,

$$c_2 = A_{-1} x((m+1)T) + A_0 x(mT).$$

Substitute the above relationship into the expression in part (b) and express $x((m + \mu)T)$ as a linear combination of $x(mT)$ and $x((m+1)T)$:

$$x((m + \mu)T) = B_{-1} x((m+1)T) + B_0 x(mT).$$

There are two unknowns in the resulting equation: A_{-1} and A_0. The linear phase and unity gain constraints provide two conditions that can be used to solve for the unknowns. The linear phase constraint means the coefficients are symmetric about the center of the filter. Because the center of the filter corresponds to $\mu = 1/2$, this constraint imposes the relationship $B_{-1} = B_0$ when $\mu = 1/2$. The unity gain constraint means $B_{-1} + B_0 = 1$. Show that the application of these two constraints requires $A_{-1} = A_0 = 0$ so that a linear interpolator is the only interpolator that satisfies all the constraints for this case.

(d) Because an even number of filter taps are required, suppose c_2 is a linear combination of $x((m+2)T), x((m+1)T), x(mT)$ and $x((m-1)T)$, that is,

$$c_2 = A_{-2}x((m+2)T) + A_{-1}x((m+1)T) + A_0 x(mT) + A_1 x((m-1)T).$$

Substitute the above relationship into the expression in part (b) and express $x((m+\mu)T)$ as a linear combination of $x((m+2)T), x((m+1)T), x(mT)$ and $x((m-1)T)$ of the form

$$x((m+\mu)T) = B_{-2}x((m+2)T) + B_{-1}x((m+1)T) + B_0 x(mT) + B_1 x((m-1)T).$$

There are four unknowns in the resulting expression: A_{-2}, A_{-1}, A_0, and A_1. The linear phase and unity gain constraints provide three equations that the four unknowns must satisfy. The linear phase constraint imposes the condition $B_{-1} = B_0$ and $B_{-2} = B_1$ when $\mu = 1/2$. The unity gain constraint imposes the condition $B_{-2} + B_{-1} + B_0 + B_1 = 1$. One more equation is needed to solve for the four unknowns. This remaining condition is provided by setting $A_{-2}T^2 = \alpha$ where α is a free parameter. Show that using these conditions to solve for A_{-2}, A_{-1}, A_0, and $A_1, x((m+\mu)T)$ may be expressed as

$$x((m+\mu)T) = \left[\alpha\mu^2 - \alpha\mu\right]x((m+2)T) + \left[-\alpha\mu^2 + (\alpha+1)\mu\right]x((m+1)T)$$
$$+ \left[-\alpha\mu^2 + (\alpha-1)\mu + 1\right]x(mT) + \left[\alpha\mu^2 - \alpha\mu\right]x((m-1)T).$$

8.8 Derive the Farrow filter structure for the linear interpolator.
  (a) Produce a table similar to Table 8.4.1.
  (b) Sketch a block diagram of the resulting Farrow filter similar to those shown in Figure 8.4.16.

8.9 Do the following for the piecewise parabolic interpolator:
  (a) Derive the Farrow coefficients for the piecewise parabolic interpolator listed in Table 8.4.1.
  (b) Sketch a block diagram of the Farrow filter similar to that shown in Figure 8.4.16 for the general piecewise parabolic interpolator.
  (c) Show that when $\alpha = 1/2$, the answer in part (b) reduces to the structure shown in Figure 8.4.16.

Section 8.10 Exercises

8.10 Derive the Farrow coefficients for the cubic interpolator listed in Table 8.4.2.

8.11 This exercise derives the formula for the fractional interpolation interval when an incrementing modulo-1 counter is used for interpolation control. In general, the value of an incrementing modulo-1 counter is

$$\eta(n+1) = \Big(\eta(n) + W(n)\Big) \mod 1$$

where $W(n) = 1/N + v(n)$ is the control value and $v(n)$ is the loop filter output. When the counter overflows, the previous index is the k-th basepoint index $m(k)$ as illustrated below.

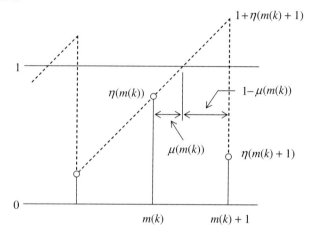

(a) Show that on overflow, the expression for the modulo-1 counter value is

$$\eta\Big(m(k)+1\Big) = \eta\Big(m(k)\Big) + W\Big(m(k)\Big) - 1.$$

Solve this expression for $W\Big(m(k)\Big)$.

(b) Show that the similar triangles formed on overflow give the relationship

$$\frac{\mu\Big(m(k)\Big)}{1 - \eta\Big(m(k)\Big)} = \frac{1 - \mu\Big(m(k)\Big)}{\eta\Big(m(k)+1\Big)}.$$

(c) Solve the relationship in part (b) for $\mu\Big(m(k)\Big)$. Compare the answer with (8.89), the expression for $\mu\Big(m(k)\Big)$, using a decrementing modulo-1 counter.

8.12 Derive the maximum likelihood carrier phase estimator for BPSK assuming a known bit sequence and known timing.

8.13 Derive the maximum likelihood carrier phase estimator for BPSK assuming an unknown bit sequence and known timing.

8.14 Derive the maximum likelihood carrier phase estimator for BPSK assuming and unknown bit sequence and unknown timing.

8.15 Show that the data-aided carrier phase error signal (7.73) follows from the maximum likelihood carrier phase estimator for offset QPSK assuming a known symbol sequence and known symbol timing.

8.16 Derive the maximum likelihood bit timing estimator for BPSK assuming a known bit sequence and known carrier phase.

8.17 Derive the maximum likelihood bit timing estimator for BPSK assuming an unknown bit sequence and known carrier phase.

8.18 Derive the maximum likelihood bit timing estimator for BPSK assuming an unknown bit sequence and unknown carrier phase.

8.19 Derive the maximum likelihood symbol timing estimator for offset QPSK assuming a known symbol sequence and known carrier phase.

8.20 Derive the maximum likelihood symbol timing estimator for offset QPSK assuming an unknown symbol sequence and known carrier phase.

9

System Components

9.1 THE CONTINUOUS-TIME DISCRETE-TIME INTERFACE   519
9.2 DISCRETE-TIME OSCILLATORS   537
9.3 RESAMPLING FILTERS   555
9.4 CoRDiC: COORDINATE ROTATION DIGITAL COMPUTER   578
9.5 AUTOMATIC GAIN CONTROL   588
9.6 NOTES AND REFERENCES   593
9.7 EXERCISES   597

There are many important components required to implement a discrete-time demodulator and detector. These include the analog-to-digital and digital-to-analog converters that interface the continuous-time and discrete-time domains; discrete-time oscillators for generating discrete-time cosines and sines required for frequency translations and discrete-time PLLs; structures for performing sample rate changes; structures for computing transcendental functions such as cosine, sine, tangent, and arctangent (the CoRDiC algorithm); and automatic gain control. The capabilities and limitations of these components determine how their use impacts the performance of the larger system. This chapter characterizes these components in terms that are useful to system designers.

9.1 THE CONTINUOUS-TIME DISCRETE-TIME INTERFACE

The continuous-time discrete-time interface is what makes discrete-time processing possible in wireless communications. A demodulator and detector rely on an analog-to-digital converter (ADC) to produce samples of the received continuous-time waveform suitable for discrete-time processing. The ADC is not ideal and introduces distortion in the signal processing path. This distortion is characterized in Section 9.1.1. ADC placement is another design decision that is also summarized.

Transmitters that use discrete-time processing rely on a digital-to-analog converter (DAC) to produce a waveform, suitable for the transmission medium, from the samples produced by the discrete-time processor. Again, the DAC is not ideal and introduces distortion. The nonideal behavior is characterized in Section 9.1.2. As before, DAC placement is an important design consideration and is summarized in Section 9.1.2.

9.1.1 Analog-to-Digital Converter

An analog-to-digital converter (ADC) produces a sampled-time version of a bandlimited continuous-time signal. The operation of an ideal ADC is illustrated in Figure 9.1.1. The sample period is T s/sample and the sample rate is $F_s = 1/T$ samples/s. The output of the ADC is a discrete-time sequence $x(nT)$. This sequence consists of T-spaced samples of the bandlimited continuous-time signal $x(t)$. For an ideal ADC, $x(nT)$ is exactly $x(t)$ at $t = nT$, that is, the ADC has infinite precision in amplitude. A real ADC, however, has finite precision amplitudes and small errors in the sampling instants. Both of these effects limit the performance of an ADC and are explored in the next two sections.

Quantization Effects. A conceptual model of an ADC with finite precision amplitudes is illustrated in Figure 9.1.2 (a). An ideal sample-and-hold circuit produces $x(nT)$ from $x(t)$. The sample-and-hold output $x(nT)$ is mapped to a discrete amplitude (from a finite number of possible amplitudes) to produce the quantized output $x_q(nT)$. The result of sample-and-hold and quantizer for the case illustrated in Figure 9.1.1 is shown in Figure 9.1.3.

A uniform quantizer with b bits partitions the dynamic range $2X_m$ (see Figure 9.1.3) into 2^b discrete levels. The width of each level is

$$\Delta = \frac{2X_m}{2^b} = \frac{X_m}{2^{b-1}}. \tag{9.1}$$

An alternate way of viewing the effect of the quantizer is the stair-step input/output plot of Figure 9.1.4 (a). The quantized output can be expressed as

$$x_q(nT) = k\Delta X_m \quad \text{when} \quad \left(k\Delta - \frac{\Delta}{2}\right)X_m < x(nT) \leq \left(k\Delta + \frac{\Delta}{2}\right)X_m. \tag{9.2}$$

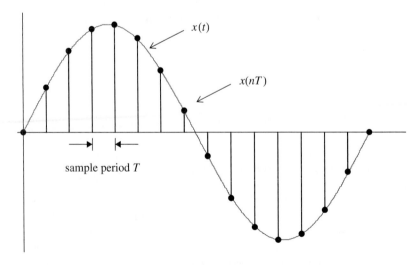

Figure 9.1.1 Sampling a continuous-time signal using an ideal ADC.

Section 9.1 The Continuous-Time Discrete-Time Interface

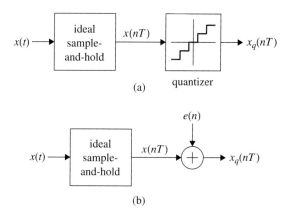

Figure 9.1.2 A model for a finite precision ADC: (a) An ideal sample-and-hold followed by a quantizer; (b) an ideal sample-and-hold followed by additive quantization noise.

The quantization error is

$$e(n) = x(nT) - x_q(nT). \tag{9.3}$$

When

$$-X_m - \frac{\Delta}{2} < x(nT) \leq X_m - \frac{\Delta}{2} \tag{9.4}$$

the quantization error is bounded by

$$-\frac{\Delta}{2} \leq e(n) < \frac{\Delta}{2} \tag{9.5}$$

as illustrated in Figure 9.1.4 (b). When the input signal exceeds the dynamic range of the ADC, that is

$$x(nT) < -X_m - \frac{\Delta}{2} \quad \text{or} \quad X_m - \frac{\Delta}{2} < x(nT), \tag{9.6}$$

then the error is not bounded by (9.5) and the input signal is "clipped" by the quantizer. The term "clip" is used in the sense that all values of $x(nT)$ that are greater than $X_m - \Delta/2$ are mapped to the same quantized amplitude ($X_m - \Delta/2$) and cannot be larger. A similar observation is made for $x(nT) < -X_m - \Delta/2$. This is why the magnitude of the error depicted in Figure 9.1.4 (b) grows large for $x > X_m - \Delta/2$ and $x < -X_m - \Delta/2$. Clipping introduces severe distortion and should be avoided in normal operation.

The error signal $e(n)$ is a function of the amplitude of $x(nT)$. When $x(t)$ is a deterministic test signal (such as a sinusoid), $e(n)$ may be computed directly using the equation for $x(nT)$ and the properties of the quantizer. When $x(t)$ is random, $e(n)$ is still a function of the amplitude of $x(nT)$ but is also random. In this case, the effects of quantization are quantified using the signal-to-quantization-noise ratio SNR_q, which is the ratio of signal power to quantization

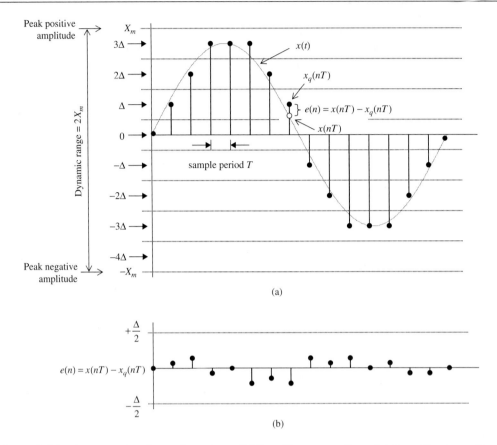

Figure 9.1.3 An example of samples using a 3-bit uniform quantizer: (a) Quantized amplitude samples and their relationship to the continuous-time signal; (b) the corresponding quantization error.

noise power (usually expressed in decibels). Using (9.3), the quantized sampled may be expressed as

$$x_q(nT) = x(nT) - e(n) \qquad (9.7)$$

which suggests the model illustrated in Figure 9.1.2 (b). This model treats the quantization process as one that *adds* quantization noise to the ideal samples $x(nT)$. In the absence of clipping and for input signals with sufficient amplitude variations, the statistical properties of the quantization noise are well approximated by the following assumptions:

1. The mean and variance of $e(n)$ do not depend on n.
2. The probability density function of $e(n)$ is uniform on the interval $-\frac{\Delta}{2} \leq e(n) < \frac{\Delta}{2}$.
3. Sample $e(n)$ is uncorrelated with $e(n')$ for $n \neq n'$.
4. The error sequence $e(n)$ is uncorrelated with the sequence $x(nT)$.

Section 9.1 The Continuous-Time Discrete-Time Interface

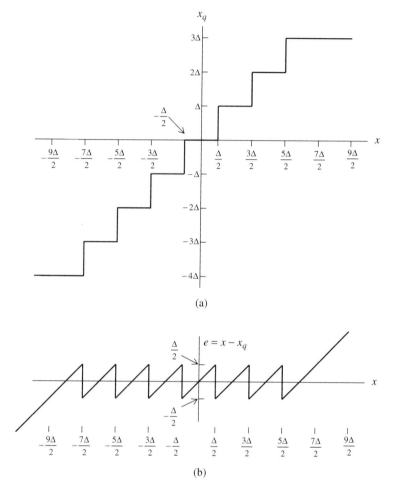

Figure 9.1.4 (a) An alternate representation of a uniform quantizer as a memoryless input/output mapper. (b) The corresponding quantization error.

A consequence of the fourth assumption is that the signal-to-quantization-noise ratio may be expressed as

$$\text{SNR}_q = 10 \log_{10} \left(\frac{\sigma_x^2}{\sigma_e^2} \right) \tag{9.8}$$

where σ_x^2 is the variance of the sequence $x(nT)$ and σ_e^2 is the variance of the additive quantization noise that may be expressed as

$$\sigma_e^2 = E\{|e(n) - \mu_e|^2\} = E\{|e(n)|^2\} - \mu_e^2 \tag{9.9}$$

where μ_e is the expected value of $e(n)$. Using the first and second assumptions,

$$\mu_e = \int_{-\frac{\Delta}{2}}^{\frac{\Delta}{2}} e \frac{1}{\Delta} de = 0 \tag{9.10}$$

$$E\{|e(n)|^2\} = \int_{-\frac{\Delta}{2}}^{\frac{\Delta}{2}} e^2 \frac{1}{\Delta} de = \frac{\Delta^2}{12} \tag{9.11}$$

so that

$$\sigma_e^2 = \frac{\Delta^2}{12} = \frac{1}{3} \frac{X_m^2}{2^{2b}} \tag{9.12}$$

where the last step follows from the definition of Δ in (9.1). Substituting (9.12) into (9.8) gives

$$\text{SNR}_q = 4.8 + 6.02b - 20 \log_{10}\left(\frac{X_m}{\sigma_x}\right). \tag{9.13}$$

Two important observations regarding (9.13) should be made.

1. Each bit of quantization precision contributes approximately 6 dB to the signal-to-quantization-noise ratio. In other words, each doubling of the number of quantization levels increases SNR_q by 6 dB.
2. The last term in (9.13) involves the dynamic range of the ADC (defined by X_m) and the RMS value σ_x of the quantizer input. To ensure that no clipping occurs, the peak amplitude of $x(nT)$, X_{peak}, should be less than or equal to X_m. The signal-to-quantization-noise ratio (9.13) may be expressed in terms of X_{peak} as

$$\text{SNR}_q = 4.8 + 6.02b - 20 \log_{10}\left(\frac{X_m}{X_{\text{peak}}}\right) - 20 \log_{10}\left(\frac{X_{\text{peak}}}{\sigma_x}\right). \tag{9.14}$$

The last term is known as the *peak-to-average ratio* of the input sequence $x(nT)$ and is an inherent property of the signal being quantized (i.e., it cannot be changed by scaling). For example, the peak-to-average ratio of a constant 0 dB; the peak-to-average ratio of a real-valued sinusoid is 3 dB (see Exercise 9.2 for the peak-to-average ratio of a complex-valued exponential); and the peak-to-average ratio of binary PAM with the SRRC pulse shape is a function of the excess bandwidth as illustrated in Figure 9.1.5 and varies from 3.5 to 7 dB for excess bandwidths greater than 10%. Peak-to-average ratios for some representative modulations are summarized in Table 9.1.1. For a fixed number of conversion bits b, signals with large peak-to-average ratios have a lower signal-to-quantization-ratio than signals with small peak-to-average ratios. The next to the last term in (9.14) is the ratio of full-scale amplitude to peak amplitude and represents the performance penalty for not operating at full scale. This penalty is controlled by scaling the input signal. This ratio must always be greater than or equal to

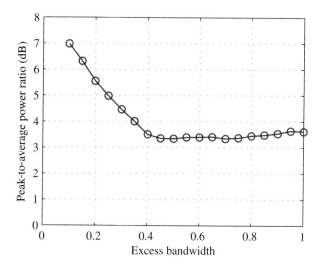

Figure 9.1.5 The peak-to-average ratio for binary PAM with the SRRC pulse shape as a function of excess bandwidth.

Table 9.1.1 Peak-to-average power ratios for different QAM constellations using the SRRC pulse shape with excess bandwidth $0 < \alpha \leq 1$

Constellation	Peak-to-Average Ratio (dB)	
	$\alpha = 0.25$	$\alpha = 0.50$
MPSK	4.7	3.2
Square 16-QAM	6.6	5.4
Square 64-QAM	6.8	6.2
Square 256-QAM	7.0	6.5
Cross 32-QAM	6.1	5.2
Cross 128 QAM	6.3	5.7
CCITT V.29 16-QAM	6.7	5.5

1 to avoid clipping. To maximize SNR_q, $x(nT)$ should be scaled to produce X_m/X_{peak} as close to 1 as possible. Any calibration errors, however, could result in clipping, which violates (9.5) and renders the approximation (9.14) invalid. Scaling $x(nT)$ below full scale reduces the risk of clipping, but also reduces the signal-to-quantization-noise ratio.

The impact of these observations in the frequency domain is illustrated in Figure 9.1.6 for the case where

$$x(n) = \exp\left\{j\frac{205}{512}\pi n\right\}. \tag{9.15}$$

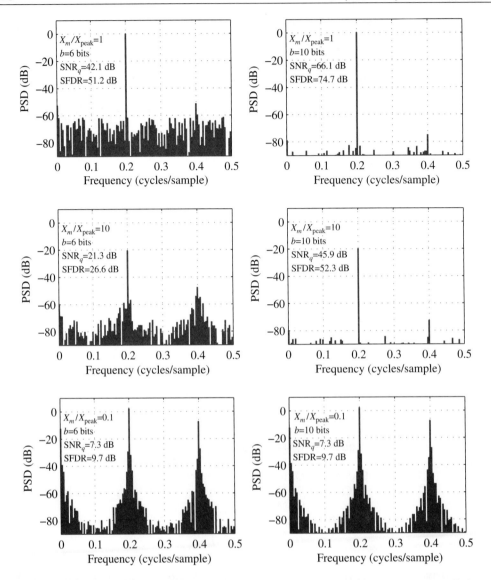

Figure 9.1.6 Quantization effects for a complex exponential using a 6-bit quantizer (left column) and a 10-bit quantizer (right column) for $X_m/X_{peak} = 1$ (top row), $X_m/X_{peak} = 10$ (middle row), and $X_m/X_{peak} = 0.1$ (bottom row).

Section 9.1 The Continuous-Time Discrete-Time Interface

A complex exponential is commonly used to assess the performance of an ADC because the DTFT is a single impulse; thus, the distortion effects are easy to observe. The power spectral densities in each plot were computed using a length-1024 FFT and averaged using the Welch periodogram method with no window and no overlap. Two figures of merit are used to assess the impact of the impairments: the signal-to-quantization-noise ratio SNR_q and the spurious-free dynamic range SFDR. As predicted by (9.14), the signal-to-quantization-noise ratio increases as the number of bits increases and as the ratio X_m/X_{peak} is closer to 1. Severe distortion occurs when the signal amplitude exceeds X_m as observed in the two plots on the bottom row. This demonstrates why "clipping" should be avoided. The SFDR is also indicated for each plot. The SFDR is the ratio of the spectral line corresponding to the desired signal and the maximum spectral line due to distortion. When there is no clipping, the SFDR improves as the number of bits increases and as the X_m/X_{peak} is closer to 1.

Clock-Jitter Effects. Small variations in the sample clock period limit the effective number of conversion bits in an ADC. These small variations are called *clock jitter*. Sampling clock jitter produces small amplitude errors as illustrated in Figure 9.1.7. The relationship between sample rate, clock jitter, and amplitude error may be derived by using a sinusoid as the input signal to the ADC. Let $x(t) = X_{\text{peak}} \cos(\omega_0 t)$ be the signal to be sampled. The sampling period is T and the sampling instant error due to clock jitter is δT. The amplitude error $\Delta x(n)$ due to clock jitter at sampling instant $t = nT$ is

$$
\begin{aligned}
\Delta x(n) &= X_{\text{peak}} \cos(\omega_0 nT) - X_{\text{peak}} \cos\left(\omega_0(nT + \delta T)\right) \\
&= X_{\text{peak}} \cos(\omega_0 nT) - X_{\text{peak}} \cos(\omega_0 \delta T)\cos(\omega_0 nT) + X_{\text{peak}} \sin(\omega_0 \delta T)\sin(\omega_0 nT) \\
&\approx X_{\text{peak}} \cos(\omega_0 nT) - X_{\text{peak}} \cos(\omega_0 nT) + X_{\text{peak}} \omega_0 \delta T \sin(\omega_0 nT) \\
&= X_{\text{peak}} \omega_0 \delta T \sin(\omega_0 nT)
\end{aligned}
\tag{9.16}
$$

where the approximation assumes $\omega_0 \delta T$ is small and uses the small argument approximations for cosine and sine. Because $-1 \leq \sin(\theta) \leq +1$, the amplitude error is bounded by

$$-X_{\text{peak}} \omega_0 \delta T \leq \Delta x(n) \leq X_{\text{peak}} \omega_0 \delta T. \tag{9.17}$$

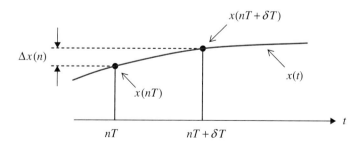

Figure 9.1.7 The relationship between sampling clock jitter and amplitude error.

To relate the amplitude error to the number of bits used by the quantizer, the amplitude error bound due to quantization given by (9.5) is applied as an upper bound to (9.17):

$$X_{peak}\omega_0 \delta T < \frac{\Delta}{2} = \frac{X_m}{2^b}. \qquad (9.18)$$

Using $\omega_0 = 2\pi f_0$ and the fact that f_0 must be less than one-half the sample rate F_s, the relationship between the number of bits b and the sampling clock jitter is obtained:

$$2^b < \frac{X_m}{X_{peak}} \frac{1}{\pi F_s \delta T} \quad \text{or} \quad b < \log_2\left(\frac{X_m}{X_{peak}}\right) - \log_2(F_s \delta T) - 1.6516. \qquad (9.19)$$

Thus, the effective number of conversion bits for an ADC is a function of the peak amplitude to dynamic range ratio and the clock jitter to sample period ratio. A signal whose peak amplitude is full scale ($X_m/X_{peak} = 1$) has a higher penalty than a signal whose peak amplitude is less than full scale. A plot of (9.19) is illustrated in Figure 9.1.8 where the effective number of conversion bits in the ADC is plotted as a function of sample rate for sample clock jitters of 1, 10, and 100 ps. This plot shows the maximum number of bits of precision as a function of normalized peak amplitude, sample rate, and sample clock jitter. For a fixed sample clock jitter, every doubling of the sample rate decreases the effective number of bits by one. For

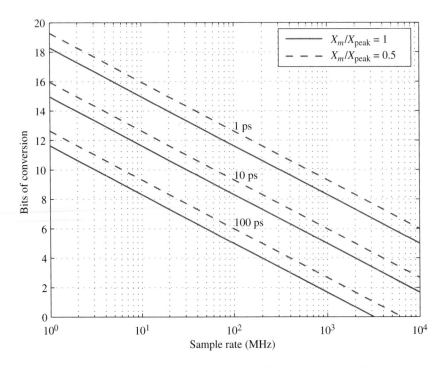

Figure 9.1.8 The effect of sample clock jitter on the effective number of bits of conversion.

example, an ADC operating at 10 Msamples/s with a 1-ps clock jitter can produce samples of a full-scale sinusoid with at most 15 bits of precision. Increasing the sample rate by 2^4 to 160 Msamples/s, a 1-ps clock jitter produces samples of a full-scale sinusoid with at most 11 bits of precision. (If the full 15 bits of conversion precision were to be maintained, the ADC operating at 160 Msamples/s would need a clock jitter less than 0.06 ps.) Current technology supports sample clock jitters between 1 and 5 ps RMS and improvements in this parameter have been very slow. Figure 9.1.8 shows that in the foreseeable future, ADCs operating above 1 Gsample/s will have a conversion precision no greater than 7 or 8 bits.

9.1.2 Digital-to-Analog Converter

Conversion of Baseband Signals. A conceptualization of the signal processing involved in the discrete-time to continuous-time conversion process is illustrated in Figure 9.1.9. The system is driven by a clock with period T whose reciprocal is the conversion rate. In its most basic form, the DAC uses the input samples $x(nT)$ to form a continuous-time stair-step signal $x_{SH}(t)$ that is smoothed by a low-pass reconstruction filter with impulse response $h_R(t)$. To determine the requirements of the reconstruction filter and quantify the distortion introduced by this process, a mathematically equivalent operation to the sample-and-hold operation is examined. The stair-step signal $x_{SH}(t)$ can be generated by passing an impulse train

$$x_p(t) = \sum_n x(nT)\delta(t - nT) \tag{9.20}$$

through a filter with impulse response

$$g(t) = \begin{cases} 1 & 0 \leq t \leq T \\ 0 & \text{otherwise} \end{cases} \tag{9.21}$$

as illustrated in the lower portion of Figure 9.1.9.

The relationships between these signals and their transforms is illustrated in Figure 9.1.10. The input sample sequence $x(nT)$ with DTFT $X_c\left(e^{j\Omega}\right)$ is shown in Figure 9.1.10 (a). Following the analysis in Section 2.6, the Fourier transform of the impulse train $x_p(t)$ is given by

$$X_p(j\omega) = \frac{1}{T}\sum_k X_c\left(j\left(\omega - k\frac{2\pi}{T}\right)\right). \tag{9.22}$$

This spectrum is illustrated in Figure 9.1.10 (b) and consists of periodic replicas of $X_c(j\omega)$ whose spacing is the sample rate. Also shown is the Fourier transform of the sample-and-hold filter

$$G(j\omega) = T\frac{\sin\left(\frac{\omega T}{2}\right)}{\frac{\omega T}{2}}e^{-j\omega T/2}. \tag{9.23}$$

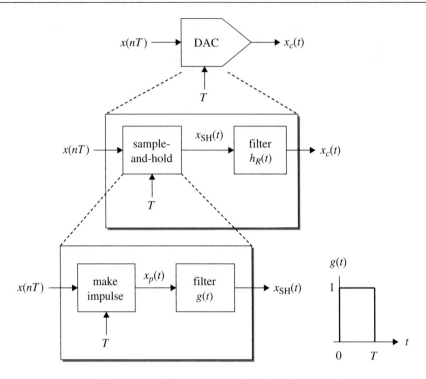

Figure 9.1.9 A conceptual diagram of the steps involved in discrete-time to continuous-time conversion.

The stair-step signal and its Fourier transform are

$$x_{SH}(t) = x_p(t) * g(t) = \sum_n x(nT)g(t - nT) \quad (9.24)$$

$$X_{SH}(j\omega) = X_p(j\omega)\, G(j\omega) \quad (9.25)$$

$$= \sum_k X_c\left(j\left(\omega - k\frac{2\pi}{T}\right)\right) \frac{\sin\left(\frac{\omega T}{2}\right)}{\frac{\omega T}{2}} e^{-j\omega T/2} \quad (9.26)$$

and are illustrated in Figure 9.1.10 (c). Observe that the effect of the sample-and-hold operation is to suppress the copies of $X_p(j\omega)$ that are not centered at $\omega = 0$. This suppression, however, is not complete. A low-pass filter, called a reconstruction filter, is used to remove these artifacts, thereby producing a smooth signal as illustrated in Figure 9.1.10 (d).

The requirements for the reconstruction can be determined by close examination of the spectral plot in Figure 9.1.10 (c). To avoid distortion, the bandwidth must be W_c radians/s. The filter must also attenuate the residual spectral copy centered at $\omega = 2\pi/T$ radians/s. As such,

Section 9.1 The Continuous-Time Discrete-Time Interface   **531**

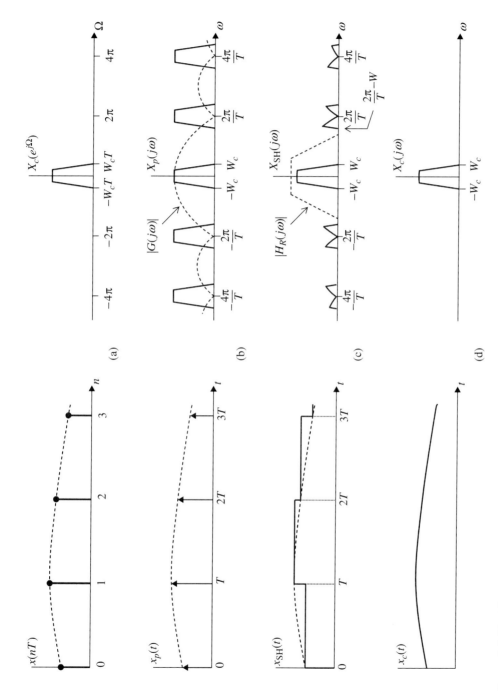

Figure 9.1.10 Illustrations of the signal in Figure 9.1.9: (a) The discrete-time sequence and its DTFT; (b) the impulse train and its Fourier transform; (c) the output of the sample-and-hold circuit and its Fourier transform; (d) the output of the reconstruction filter.

the filter must achieve the desired attenuation at $\omega = 2\pi/T - W_c$ radians/s. The transition band for the filter is thus $2\pi/T - W_c - W_c = 2\pi/T - 2W_c$ radians/s. When the sample rate is not much larger than twice the bandwidth, the bandwidth available for the transition band is small as illustrated in Figure 9.1.11 (a). In this case, a complicated high-order filter is required to achieve the desired attenuation over the small transition band. Such filters can be problematic to implement. If the sample rate is increased, the transition band increases, thereby relaxing the transition band requirements on the reconstruction filter as illustrated in Figure 9.1.11 (b). In this way, system complexity can be moved between the discrete-time portion and the continuous-time portion. Simple continuous-time filters can be used if a high clock rate can be used. For this reason, DACs are usually operated at as high a clock rate as possible.

The sample-and-hold operation also distorts the magnitude of the spectral copy at $\omega = 0$. This distortion results from the fact that $|G(j\omega)|$ is not constant in the vicinity of $\omega = 0$. In the example shown in Figure 9.1.10, this distortion manifests itself as a "rounding" of the spectrum. For a fixed bandwidth W_c, the severity of this distortion diminishes as the sample rate increases because $|G(j\omega)|$ appears increasingly flat over a decreasing region of ω centered

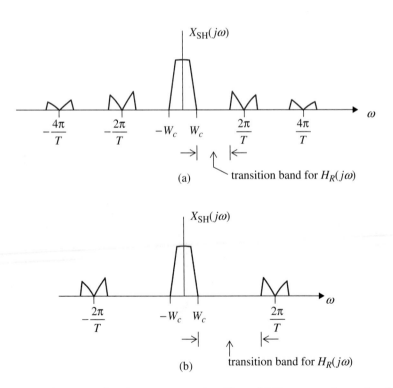

Figure 9.1.11 Requirements for the reconstruction filter: (a) The case where the signal is not oversampled; (b) the case where the signal is oversampled.

about 0. Compensation for this distortion can be performed in continuous-time processing by replacing the low-pass reconstruction filter with a low-pass filter with frequency response

$$H_R(j\omega) = \begin{cases} \dfrac{1}{G(j\omega)} & |\omega| \leq W_c \\ 0 & \text{otherwise} \end{cases}. \quad (9.27)$$

Instead of having a flat pass band, this filter exhibits some "peaking" in the pass band. There are two primary drawbacks to this approach.

- A realization of (9.27) significantly complicates the design of the reconstruction filter. (See the Section 9.6 for references to continuous-time designs.)
- In some applications, it is desirable to operate the DAC at several different clock rates. The filter with frequency response (9.27) cannot be used with multiple sample rates. Thus a different filter must be used for each clock rate.

As a consequence, compensation is often performed in discrete-time processing prior to discrete-time to continuous-time conversion.

Discrete-time compensation takes the form of a discrete-time filter that precedes the DAC as illustrated in Figure 9.1.12. The filter "pre-distorts" the discrete-time sequence so that the signal after conversion to continuous-time is not distorted. To derive the discrete-time compensation, an ideally bandlimited version of the continuous-time distortion $G(j\omega)$ is projected into discrete-time. Let

$$\tilde{G}(j\omega) = \begin{cases} G(j\omega) & -\dfrac{\pi}{T} \leq \omega \leq \dfrac{\pi}{T} \\ 0 & \text{otherwise} \end{cases} \quad (9.28)$$

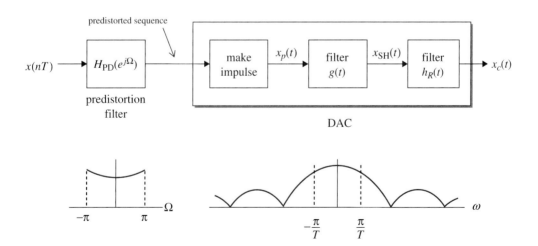

Figure 9.1.12 A predistortion filter to compensate for the DAC distortion.

be the ideally bandlimited version of $G(j\omega)$. Applying the notions developed in Section 2.6.2, the equivalent bandlimited discrete-time DAC distortion is, neglecting the linear phase term,

$$G_d\left(e^{j\Omega}\right) = \frac{1}{T}\tilde{G}\left(j\frac{\Omega}{T}\right) = \frac{\sin\left(\frac{\Omega}{2}\right)}{\frac{\Omega}{2}} \quad -\pi \leq \Omega \leq \pi. \tag{9.29}$$

The ideal discrete-time predistortion filter, $H_{\text{PD}}\left(e^{j\Omega}\right)$ equalizes the DAC distortion:

$$G_d\left(e^{j\Omega}\right) H_{\text{PD}}\left(e^{j\Omega}\right) = 1 \tag{9.30}$$

so that

$$H_{\text{PD}}\left(e^{j\Omega}\right) = \frac{1}{G_d(e^{j\Omega})} = \frac{\frac{\Omega}{2}}{\sin\left(\frac{\Omega}{2}\right)}. \tag{9.31}$$

There are several methods for producing FIR filters that approximate (9.31). A common approach is to use the Parks–McClellan equiripple design algorithm to produce linear phase FIR filter approximations. Examples of filters designed using the Parks–McClellan algorithm with a maximum pass-band error of 0.01 dB are illustrated in Figure 9.1.13. The frequency responses for filters of lengths 5, 11, 21, 51, and 101 are plotted in Figure 9.1.13 (a). $|G_d(e^{j\Omega})|$ is also plotted for a reference. The plot of $|G_d(e^{j\Omega}) H_{\text{PD}}(e^{j\Omega})|$ shown in Figure 9.1.13 (b) demonstrates how well the predistortion filters satisfy the goal (9.30). Note that as the filter length increases, the usable bandwidth also increases: The length-5 filter provides adequate predistortion over 10% of the sample rate and the length-101 filter provides adequate predistortion over 48% of the sample rate. If the maximum allowable pass-band ripple is increased, the usable bandwidths for each filter length increase. The information in Figure 9.1.13 can be used to quantify the trade-off between clock rate and filter length for the predistortion filter. Other approaches are summarized in Section 9.6. An interesting approach using a length-3 FIR filter is explored in Exercise 9.10.

Conversion of Band-pass Signals. Discrete-time modulators often produce band-pass signals that are to be converted to continuous-time. The mathematical description of the conversion for band-pass signals is identical to that for baseband signals and given by (9.20)–(9.26), except that the picture is different. An example of the spectra for $X(e^{j\Omega})$, $X_p(j\omega)$, and $X_{\text{SH}}(j\omega)$ for a band-pass signal is illustrated in Figure 9.1.14. The spectrum $X_p(j\omega)$ consists of periodic replicas of $X_c(j\omega)$ as shown in Figure 9.1.14 (b). Observe that the spectral copies are centered at frequencies determined by T and ω_0 and that there are no spectral copies centered at multiples of $2\pi/T$, in general. This characteristic introduces possibilities for the reconstruction filters that do not exist with conversion of baseband signals. These possibilities are illustrated in Figures 9.1.14 (c) and (d). Figure 9.1.14 (c) illustrates the case where the reconstruction filter is a band-pass filter centered at ω_0 rad/s. Another option, illustrated in Figure 9.1.14 (d), uses a band-pass filter centered at $2\pi/T + \omega_0$ rad/s as the

Section 9.1 The Continuous-Time Discrete-Time Interface 535

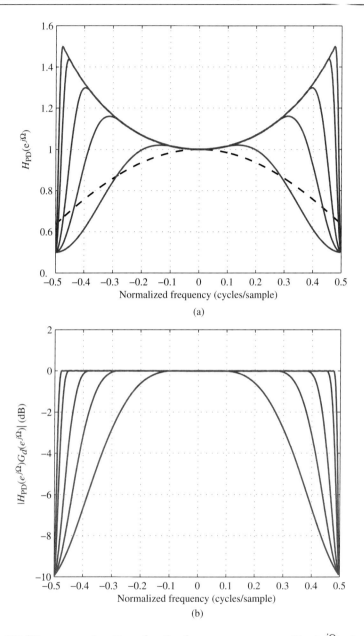

Figure 9.1.13 FIR filter approximations for the frequency response $H_{PD}(e^{j\Omega})$ given by (9.31). The filters were designed using the Parks–McClellan algorithm for equiripple FIR approximations with a maximum allowed pass-band error of 0.01 dB. (a) $H_{PD}(e^{j\Omega})$ for filter lengths 5, 11, 21, 51, and 101. (b) $|G_d(e^{j\Omega})H_{PD}(e^{j\Omega})|$ for the same filters in (a). The usable bandwidth increases with filter length.

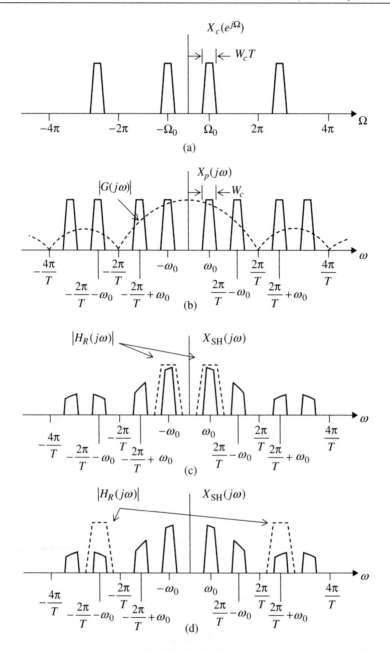

Figure 9.1.14 The spectra of band-pass signals in discrete-time to continuous-time conversion: (a) the DTFT of the discrete-time band-pass sequence; (b) the Fourier transform of the impulse train; (c) the Fourier transform of the output of the sample-and-hold circuit showing one option for the reconstruction filter; (d) the same as (c) except showing a different option for the reconstruction filter.

reconstruction filter. This example shows that it is possible to realize frequency translations "for free" by proper selection of the reconstruction filter. The possible frequency translations are determined by the center frequency of the band-pass signal and the sample rate used for conversion to continuous-time.

Each possible frequency translation is accompanied by an amplitude that is determined by $|G(j\omega)|$. The power of the continuous-time signal becomes smaller as the center frequency increases. In addition, $|G(j\omega)|$ distorts the magnitude of the continuous-time waveform. This distortion is illustrated by the sloped magnitudes of the spectral copies in Figures 9.1.14 (c) and (d). The compensation techniques described for conversion of baseband signals can be used when the desired spectral copy resides in the main Nyquist zone: $-\pi/T \leq \omega \leq \pi/T$. Care must be taken to ensure that the usable bandwidth of the predistortion filter includes the frequencies occupied by the band-pass discrete-time signal. When one of the other spectral copies are desired, compensation is usually performed at baseband using an FIR low-pass predistortion filter. In this case, the predistortion filter is designed by projecting an appropriately bandlimited version of $G(j\omega)$, for ω in the region of the desired spectral copy, to baseband and applying one of the standard FIR filter design techniques such as those outlined in Section 3.3.2. Because the frequency response of the filter is not symmetric about $\Omega = 0$, the filter coefficients are complex values. The real part of the complex filter is applied to $I(nT)$ and the imaginary part of the complex filter is applied to $Q(nT)$ as described by Harris in [185].

9.2 DISCRETE-TIME OSCILLATORS

Discrete-time oscillators play an important role in both modulators and demodulators in sampled data systems. Oscillators are used to perform frequency translations and phase rotations as discussed in Chapters 5, 6, and 7 and are frequently used with multipliers to construct the discrete-time equivalent of the continuous-time mixer. Although multirate processing can be used to perform many of the frequency translations required in discrete-time modulators and demodulators, discrete-time oscillators are still required for carrier phase synchronization, compensation for carrier frequency offsets, and for those cases where multirate processing is not convenient.

In principle, samples of the cosine and sine functions operating at the desired frequency could be computed, stored in memory, and read out to the multiplier as needed. This approach is simple, but does not provide any easy method for adjusting the frequency (other than re-storing recomputed samples in the memory). Oscillators with an adjustable frequency are necessary for carrier phase synchronization (as outlined in Chapter 7) and desired when operation at different channels in a frequency division multiplexed system is required. For this reason, this section focuses on discrete-time methods for producing samples of a sinusoid with an adjustable frequency.

There are two approaches to designing discrete-time oscillators. The first is the design of an LTI system whose impulse response is a discrete-time sinusoid at the desired frequency. When the input is set to a discrete-time impulse, the resulting output is the desired sinusoid. The LTI system takes the form of an IIR filter. Adjustable frequency is realized by changing one or more of the feedback coefficients. Finite precision number representations and arithmetic can lead to stability problems with this approach. For this reason, the other approach, known as a direct digital synthesizer (DDS) is commonly used. The DDS consists of a phase accumulation register and a look-up table where samples of the cosine are stored. Finite precision arithmetic leads to phase noise and limits frequency resolution of the DDS, but it is stable.

9.2.1 Discrete Oscillators Based on LTI Systems

A filter with impulse response

$$h(n) = \cos(\Omega_0 n)\, u(n) = \frac{1}{2} e^{j\Omega_0 n} u(n) + \frac{1}{2} e^{-j\Omega_0 n} u(n) \tag{9.32}$$

has z-transform

$$H(z) = \frac{1/2}{1 - e^{j\Omega_0} z^{-1}} + \frac{1/2}{1 - e^{-j\Omega_0} z^{-1}} = \frac{1 - \cos(\Omega_0)\, z^{-1}}{1 - 2\cos(\Omega_0)\, z^{-1} + z^{-2}}. \tag{9.33}$$

The direct-form realization of this system is illustrated in Figure 9.2.1. The oscillation frequency is changed by adjusting the filter coefficient $\cos(\Omega_0)$. The system has two poles on the unit circle at complex conjugate positions $z = e^{\pm j\Omega_0}$. As a consequence, the system is only marginally stable and is very sensitive to changes in the pole locations due to finite precision numerical representations and arithmetic.

This is demonstrated by plots of possible pole locations for 4-bit and 6-bit precision shown in Figure 9.2.2. Using 4-bit quantization, the filter coefficient $\cos(\Omega_0)$ can assume 16 values

$$-1, -\frac{7}{8}, -\frac{6}{8}, -\frac{5}{8}, -\frac{4}{8}, -\frac{3}{8}, -\frac{2}{8}, -\frac{1}{8}, 0, +\frac{1}{8}, +\frac{2}{8}, +\frac{3}{8}, +\frac{4}{8}, +\frac{5}{8}, +\frac{6}{8}, +\frac{7}{8}.$$

For 6-bit quantization, the filter coefficients assume the corresponding multiples of $1/32$. Two observations are important:

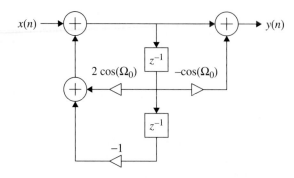

Figure 9.2.1 Direct-form realization of an IIR filter whose impulse response is $\cos(\Omega_0 n)\, u(n)$.

Section 9.2 Discrete-Time Oscillators

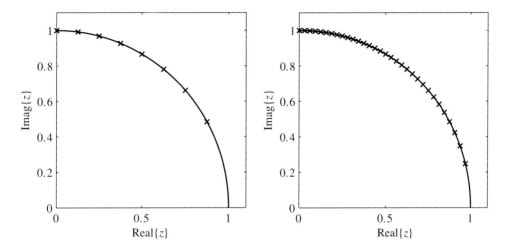

Figure 9.2.2 Possible pole locations for the direct-form realization (9.33) using 4-bit quantization (left) and 6-bit quantization (right).

1. Pole locations are limited to a finite set of possibilities. If the desired pole location is not in the set, then it is not possible to realize the desired filter response.
2. The pole locations are sparse near the real axis. This makes it impossible to realize low-frequency oscillators.

The second observation is a consequence of the interaction between closely spaced poles in a finite precision system as described by Kaiser in Chapter 7 of Ref. [64]. One of the ways to overcome this limitation is to realize a system with just one pole at $z = e^{j\Omega_0}$, compute the complex-valued output, and retain only the real part. The direct-form realization of this system is illustrated in Figure 9.2.3. The impulse response of this system is $h(n) = e^{j\Omega_0 n}u(n)$. When the input is $x(n) = \delta(n)$, the output is $y(n) = e^{j\Omega_0 n}u(n)$. Because $e^{j\Omega_0 n} = e^{j\Omega_0}e^{j\Omega_0(n-1)}$, the output follows the simple recursion $y(n) = e^{j\Omega_0}y(n-1)$. By expressing $y(n) = y_r(n) + jy_i(n)$, the recursions for the real and imaginary parts of the output are

$$y_r(n) = \cos(\Omega_0)y_r(n-1) - \sin(\Omega_0)y_i(n-1)$$
$$y_i(n) = \sin(\Omega_0)y_r(n-1) + \cos(\Omega_0)y_i(n-1). \quad (9.34)$$

A system that realizes (9.34) is illustrated in Figure 9.2.4. (An interesting connection between this system and continuous-time oscillators is explored in Exercise 9.12.) Observe that the filter coefficients to be quantized are $\cos(\Omega_0)$ and $\sin(\Omega_0)$, which are the real and imaginary parts of the desired pole location. The pole locations, using quantized filter coefficients, for this structure are illustrated in Figure 9.2.5.

Although this system eliminates the interaction of the complex conjugate poles, it still does not allow placement of a pole at an arbitrary location on the unit circle. This is because the finite precision arithmetic does not always permit $y_r^2(n) + y_i^2(n) = 1$ for all possible values of $y_r(n)$ and $y_i(n)$. One way to address this shortcoming is to move all the multiplication

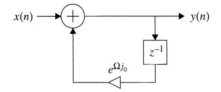

Figure 9.2.3 Direct-form realization of an IIR filter with one pole at $z = e^{j\Omega_0}$.

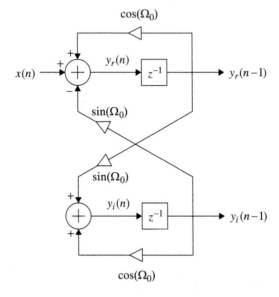

Figure 9.2.4 An alternate realization of the direct-form realization of an IIR filter with one pole at $z = e^{j\Omega_0}$ illustrated in Figure 9.2.3.

operations to the input of the register as illustrated in Figure 9.2.6 (a). The multiplier A is allowed to vary and is selected to force

$$A^2 y_r^2(n) + A^2 y_i^2(n) = 1. \tag{9.35}$$

The adaptation for A is derived as follows: Suppose

$$y_r^2(n) + y_i^2(n) = 1 + \epsilon. \tag{9.36}$$

Substituting (9.36) into (9.35) and solving for A produces

$$A = \frac{1}{\sqrt{1+\epsilon}}. \tag{9.37}$$

Section 9.2 Discrete-Time Oscillators

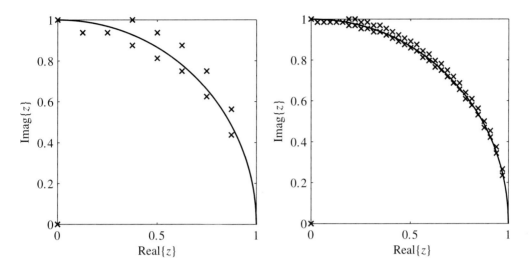

Figure 9.2.5 Possible pole locations for the system that realizes (9.34) for 4-bit quantization (left) and 6-bit quantization (right).

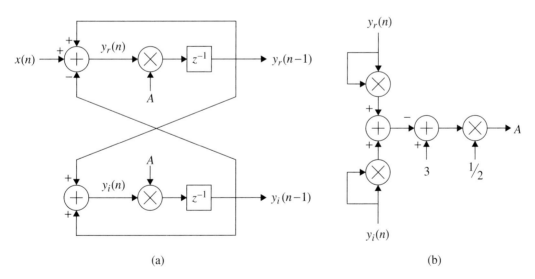

Figure 9.2.6 Recursive system for producing a sinusoid. (a) The system motivated by the system of Figure 9.2.3. The multiplier A is selected to force $A^2 y_r^2(n) + A^2 y_i^2(n) = 1$. (b) A system for computing A from $y_r(n)$ and $y_i(n)$ as suggested by (9.38).

Assuming ϵ is small, A may be approximated as

$$A \approx \frac{1}{1+\frac{\epsilon}{2}} \approx 1 - \frac{\epsilon}{2}. \tag{9.38}$$

A system for computing the desired value A from $y_r(n)$ and $y_i(n)$ using the relationship (9.36) is illustrated in Figure 9.2.6 (b).

9.2.2 Direct Digital Synthesizer

The DDS method is the preferred method for producing a reference frequency whenever extremely precise frequency resolution and fast clocking speed are required. The operating principles of the DDS are best understood by reexamining the fundamental definitions of frequency and phase and applying those definitions to discrete-time sinusoids.

Fundamental Relationships. Consider the continuous-time sinusoid

$$\cos\left(\omega_0 t + \phi(t)\right) \tag{9.39}$$

with time-varying "phase" $\phi(t)$. To determine the frequency of this sinusoid, the fundamental relationship between "phase" and "frequency" must be understood. This fundamental relationship is captured by the following definitions:

$$\text{instantaneous phase} = \omega_0 t + \phi(t) \quad \text{rad} \tag{9.40}$$

$$\text{instantaneous excess phase} = \phi(t) \quad \text{rad} \tag{9.41}$$

$$\text{instantaneous frequency} = \omega_0 + \frac{d}{dt}\phi(t) \quad \text{rad/s} \tag{9.42}$$

$$\text{instantaneous excess frequency} = \frac{d}{dt}\phi(t) \quad \text{rad/s}. \tag{9.43}$$

These definitions encapsulate the notion that "frequency is the time derivative of the phase." In other words, the slope of the instantaneous phase is the instantaneous frequency. These definitions also express the notion that sinusoids are often defined as operating about a desired operating frequency (ω_0 rad/s in this case) and that excursions from this operating point might be of interest. The relationships between phase, excess phase, frequency, and excess frequency are illustrated in Figure 9.2.7. In this figure, and for much of the development that follows, the instantaneous phase is denoted

$$\theta(t) = \omega_0 t + \phi(t). \tag{9.44}$$

Observe that although the instantaneous frequency can be discontinuous, the instantaneous phase is continuous.

These definitions are the basis for the input/output relationship for a continuous-time voltage controlled oscillator (VCO) as illustrated in Figure 9.2.8. A VCO is a continuous-time component with input $x(t)$ and sinusoidal output $y(t)$. The instantaneous excess frequency of

Section 9.2 Discrete-Time Oscillators

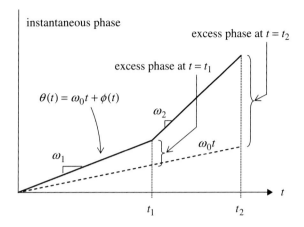

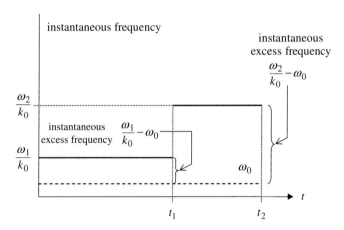

Figure 9.2.7 An illustration of the relationship between phase, excess phase, frequency, and excess frequency.

$y(t)$ is proportional to the input $x(t)$. The constant of proportionality k_0 is often called the *VCO gain* and has units radians/volt (or radians/amp if the device is current-driven). The relationship between instantaneous excess phase $\phi(t)$ and the input $x(t)$ is

$$\phi(t) = k_0 \int_{-\infty}^{t} x(\tau)d\tau. \tag{9.45}$$

A conceptual block diagram representing the signal processing operations of a VCO is illustrated in Figure 9.2.8 (b). The input is scaled by k_0 and biased by the constant ω_0. The sum is integrated to form the input to a function that computes the cosine. (A VCO is not really built this way—this is just a conceptualization.) The instantaneous frequency of the output

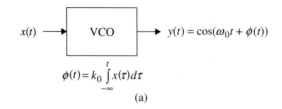

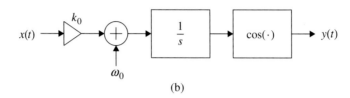

Figure 9.2.8 The voltage controlled oscillator (VCO): (a) Block diagram showing the input/output relationship; (b) a conceptual block diagram emphasizing the signal processing.

is adjusted by changing the input $x(t)$. When $x(t) = 0$, the instantaneous frequency of $y(t)$ is ω_0 rad/s. When $x(t) > 0$, the instantaneous frequency of $y(t)$ is greater than ω_0 rad/s (i.e., the instantaneous excess frequency is positive) and when $x(t) < 0$, the instantaneous frequency of $y(t)$ is less than ω_0 rad/s (i.e., the instantaneous excess frequency is negative). The amount by which the instantaneous frequency changes is controlled by the amplitude of $x(t)$ and the VCO gain k_0.

Now consider T-spaced samples of the input represented by the sequence $x(nT)$. The desire is to compute samples of the output given by

$$y(nT) = \cos(\omega_0 nT + \phi(nT)). \tag{9.46}$$

Applying the backward difference rule for approximating continuous-time integration outlined in Section 3.3.3, $\phi(nT)$ may be approximated from the samples $x(nT)$ as

$$\phi(nT) = k_0 T \sum_{k=-\infty}^{n-1} x(kT). \tag{9.47}$$

This leads to the discrete-time version of the VCO illustrated in Figure 9.2.9 (a). The instantaneous frequency of the discrete-time sinusoid is changed by adjusting the amplitude of the samples $x(nT)$ in the same way that $x(t)$ controlled the instantaneous frequency of the VCO output.

A block diagram that realizes (9.47) is illustrated in Figure 9.2.9 (b). The input samples are scaled by the constant $K_0 = k_0 T$ rad/unit amplitude and biased by the constant $\Omega_0 = \omega_0 T$ rad/sample. The sum forms the input to a recursive system consisting of a register with unity feedback. The register stores the running sum of its input. Because the sum that is stored is

Section 9.2 Discrete-Time Oscillators

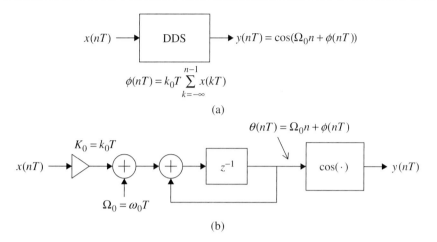

Figure 9.2.9 The direct digital synthesizer (DDS): (a) Block diagram showing the input/output relationship; (b) a conceptual block diagram emphasizing the signal processing.

the instantaneous phase of the discrete-time sinusoid, the register is often called the *phase accumulation register*. A function that computes the cosine (or sine) of the contents of the phase accumulation register completes the processing chain to create the output

$$y(nT) = \cos\left(\Omega_0 n + K_0 \sum_{k=-\infty}^{n-1} x(kT)\right). \tag{9.48}$$

This architecture is the basis for the direct digital frequency synthesizer, or DDS for short. The diagram of Figure 9.2.9 (b) is an idealized system. In real systems, finite precision arithmetic impacts the performance of this architecture. These issues are addressed in the next section.

Finite Precision Arithmetic and Implementation Issues. Finite precision numerical representations impact the DDS in three basic ways: the phase accumulator, the way in which samples of cosine and sine are computed, and the precision with which the samples of cosine and sine are computed. Although, in principle, the samples of the cosine and sine could be computed using a numerical subroutine, this approach is much too slow for most applications. The most common approach is to pre-compute samples of the cosine (or sine) and store them in a look-up table or LUT. The number of samples stored in the LUT is determined by the number of LUT address lines and is related to the number of bits used to represent the phase accumulator contents. The number of bits used to represent each entry in the LUT is independent of the number of entries.

The phase accumulator in the DDS defines the instantaneous phase for the next sample of the sinusoidal output. When a LUT is used to store samples of the cosine and sine, the phase accumulator contents form the address to the LUT. This architecture is illustrated in Figure 9.2.10 (a). The most convenient method for coupling the instantaneous phase with

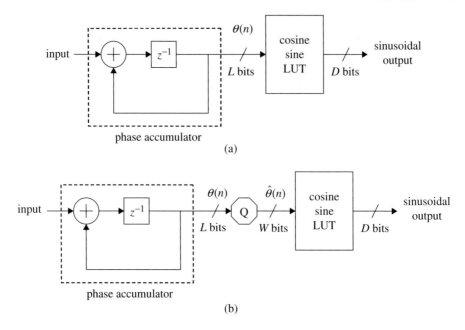

Figure 9.2.10 The DDS with finite precision numerical representations: (a) An L-bit phase accumulator and cosine/sine LUT with L address bits; (b) a practical alternative using "phase truncation" with an L-bit phase accumulator and a cosine/sine LUT with $W < L$ address bits.

the LUT address is to use the two's complement representation for the phase accumulator contents.

An L-bit two's complement integer consists of L bits of the form

$$b_0 b_1 \cdots b_{L-1}.$$

The corresponding decimal value depends on where the radix point (i.e., the separation between positive and negative powers of 2) is assumed to be. If the radix point is to the right of b_{L-1}, then decimal equivalents are integers given by

$$-b_0 2^{L-1} + b_1 2^{L-2} + \cdots + b_{L-1} 2^0.$$

Observe that the left-most bit plays the role of a "sign bit." Moving the radix point to the left (just to the right of the "sign bit") produces an equivalent decimal value

$$-b_0 2^0 + b_1 2^{-1} + \cdots + b_{L-1} 2^{-(L-1)}.$$

All of these values are multiples of $2^{-(L-1)}$ between -1 and $1 - 2^{L-1}$. The attractive feature of the two's complement method for quantizing the interval $[-1, +1)$ is the way in which the values "wrap around" on overflow. As an example, consider the eight possible values of a three-bit two's complement word and the corresponding decimal equivalents illustrated

Section 9.2 Discrete-Time Oscillators

in Figure 9.2.11 (a). The most significant bit (MSB) plays the role of the "sign bit" and the possible values are multiples of 1/4. Starting with 000 and incrementing the least significant bit (LSB) each time produces the series of values plotted in Figure 9.2.11 (b). Observe that incrementing the LSB for the two's complement representation of 3/4 produces the two's complement representation of -1. In this way "+1" wraps around to -1 and vice versa. When the decimal equivalent is interpreted as the coefficient of π in $\cos(x\pi)$, the two's complement numbers quantize the interval $[-\pi, +\pi)$ and perform the modulo-2π reduction automatically. If the phase accumulator of Figure 9.2.10 (a) is three bits wide, then the cosine LUT interprets the phase accumulator contents as three address lines that access the eight entries summarized in Figure 9.2.12.

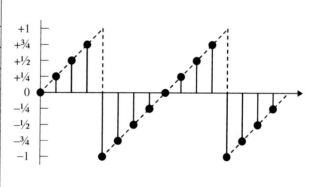

Binary Pattern	Interpretation	Decimal Equivalent
000	0.00	0
001	0.01	+1/4
010	0.10	+1/2
011	0.11	+3/4
100	1.00	-1
101	1.01	$-3/4$
110	1.10	$-1/2$
111	1.11	$-1/4$

Figure 9.2.11 Three-bit two's complement representation: (left) The eight possible values along with the decimal equivalent values when the radix point is positioned as shown; (right) the sequence of decimal equivalent values associated with overflow as the two's complement value is incremented.

Address	LUT Contents	
000	$\cos(0)$	1
001	$\cos(\pi/4)$	0.7071
010	$\cos(\pi/2)$	0
011	$\cos(3\pi/4)$	-0.7071
100	$\cos(-\pi)$	-1
101	$\cos(-3\pi/4)$	-0.7071
110	$\cos(-\pi/2)$	0
111	$\cos(-\pi/4)$	0.7071

Figure 9.2.12 The LUT contents for each address corresponding to the list of two's complement values in Figure 9.2.11.

The resolution of the phase increment is a function of the number of bits used to represent the phase accumulator contents. Let L represent the number of bits used to represent the phase accumulator contents. L is determined by the resolution and maximum frequency requirements of the DDS. Suppose the sample rate is $F_s = 1/T$ samples/s and that the DDS is capable of producing samples of a sinusoid with maximum frequency $f_{\max}$ cycles/s with resolution $f_{\min}$. The sampling theorem imposes the limit $f_{\max} = F_s/2$ while the resolution is determined by the sinusoid with the longest period. Because an L-bit two's complement word has 2^L discrete values, the longest period is 2^L and the smallest frequency is $f_{\min} = F_s/2^L$. The ratio of the two defines the number of bits needed to represent the contents of the accumulation register:

$$2^{L-1} = \frac{f_{\max}}{f_{\min}}. \qquad (9.49)$$

As an example, suppose a DDS with $f_{\max} = 200$ MHz and 0.1 Hz resolution is required. Then

$$2^{L-1} = \frac{200 \times 10^6}{10^{-1}} = 2 \times 10^9 \quad \rightarrow \quad L = 32 \quad \text{bits}. \qquad (9.50)$$

If the architecture of Figure 9.2.10 (a) is used, a 32-bit accumulation register coupled to a LUT with 32 address lines (and $2^{32} = 4{,}294{,}967{,}296$ entries) is required. Implementing such a large LUT, especially if high speed operation at 400 MHz is required, can be challenge.

There are two approaches that have proven useful in reducing the size of the LUT.

1. Phase Accumulator Truncation: In this approach, the output of the phase accumulator (after feedback) is truncated by removing the B least significant bits. The truncated $W = (L - B)$-bit word is then used to address the LUT contents as illustrated in Figure 9.2.10 (b). This approach has the advantage of requiring a smaller LUT, but introduces quantization error in the instantaneous phase. The effect of phase accumulator truncation is the presence of spurious lines in the spectrum of the LUT output. An example of this for a complex exponential operating at a normalized frequency of 11/500 cycles/sample is illustrated in Figure 9.2.13. The phase accumulator uses $L = 28$ bits and the cosine/sine LUT uses $W = 8$ bits and stores $2^8 = 256$ samples of the cosine uniformly spaced on the interval $[-\pi, \pi)$. Each LUT entry has $D = 16$ bits of precision. The large spectral line at $11/500 = 0.022$ represents the complex exponential. The other lines, called "spurs," are a form of phase noise produced by this technique. The normalized frequency corresponds to a phase increment of $11\pi/250$. Quantifying this phase increment exactly requires an angular resolution of $\pi/250$ in the LUT. Because the LUT uses $W = 8$ address bits, the angular resolution is $2\pi/256 = \pi/128$. As a consequence, true phase is only approximated and this approximation is the cause of the spurs observed in Figure 9.2.13. Three methods for reducing these spurs are outlined in the next section.
2. Exploiting Cosine/Sine Symmetries: Another useful approach is to exploit the inherit symmetries in the cosine and sine functions. A clue that this might be possible was demonstrated in Figure 9.2.12 where the eight LUT entries require only three unique values (and their negatives). The symmetries of interest are those centered about

Section 9.2 Discrete-Time Oscillators

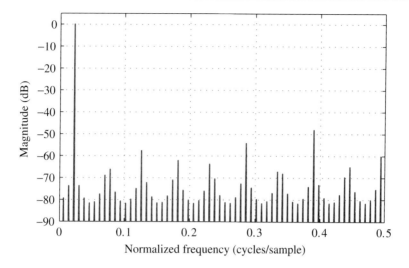

Figure 9.2.13 The frequency-domain effects of phase accumulator truncation for a complex exponential. The normalized frequency of the complex exponential is 11/500 cycles/sample. The phase accumulator uses $L = 28$ bits, the LUT uses $W = 8$ bits, and the LUT contents are quantized to 16 bits. A length-1500 FFT was applied to 1500 samples to produce this realization.

$\theta = \pi/2$ and $\theta = \pi$:

$$\begin{aligned} \cos\left(\frac{\pi}{2} - \theta\right) &= -\cos\left(\frac{\pi}{2} + \theta\right) & 0 \leq \theta \leq \frac{\pi}{2} \\ \cos(\theta) &= \cos(-\theta) & -\pi \leq \theta \leq \pi \end{aligned} \quad (9.51)$$

$$\begin{aligned} \sin\left(\frac{\pi}{2} - \theta\right) &= \sin\left(\frac{\pi}{2} + \theta\right) & 0 \leq \theta \leq \frac{\pi}{2} \\ \sin(\theta) &= -\sin(-\theta) & -\pi \leq \theta \leq \pi. \end{aligned} \quad (9.52)$$

Values of the $\sin(\theta)$ for $0 \leq \theta \leq \pi/2$ may be used to compute $\sin(\theta)$ for $0 \leq \theta \leq \pi$ by using θ modulo $\pi/2$ and the absolute value of θ. This is easily realized in hardware by truncating the MSB of the phase (i.e., the "sign bit") and using the second MSB to determine the quadrant. A block diagram illustrating the application of these symmetries to reduce the size of the LUT by four is illustrated in Figure 9.2.14 for the sine function.

Note that these techniques can be used separately or together.

Dealing with Spurious Spectral Lines. The spurious spectral lines are a product of the phase error

$$\delta\theta(n) = \theta(n) - \hat{\theta}(n). \quad (9.53)$$

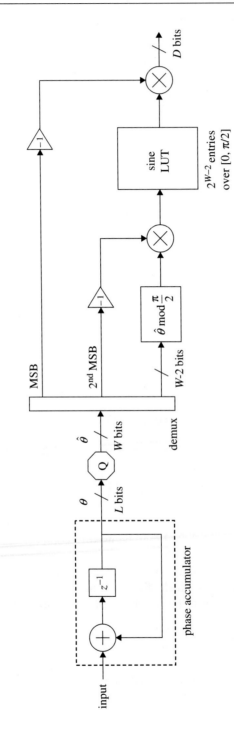

Figure 9.2.14 Exploiting the trigonometric symmetries (9.51) and (9.52) to reduce the size of the cosine/sine LUT. (Adapted from Ref. [186].)

Section 9.2 Discrete-Time Oscillators

For the case of a complex exponential, the output may be expressed as

$$e^{j\theta(n)} = e^{j\hat{\theta}(n)} e^{j\delta\theta(n)} = e^{j\hat{\theta}(n)} \left[\cos\left(\delta\theta(n)\right) + j\sin\left(\delta\theta(n)\right)\right] \approx e^{j\hat{\theta}(n)} [1 + j\delta\theta(n)] \quad (9.54)$$

where the last approximation follows from small argument approximations for cosine and sine. When the input is a constant phase increment, the phase error $\delta\theta(n)$ is not random and is a periodic ramp. As such, its spectrum is discrete and is the source of the spurious spectral lines observed in Figure 9.2.13. The spectral locations of these lines are determined by the period of the phase error ramp. The period is a function of the phase resolution of the LUT $(2\pi/2^W)$ and the desired phase increment which is a multiple of $2\pi/2^L$. The amplitudes are determined by the Fourier series coefficients of the periodic phase error ramp. The largest amplitude is 2^{-W}. For the DDS example considered above, the phase was quantized to $W = 8$ bits so that the peak spectral line is -48 dB at 0.39 cycles/sample as shown in Figure 9.2.13. The largest of the spurious lines is the limiting factor in many applications. The amplitude of this line is quantified by the *spurious-free dynamic range* which is the ratio of the amplitude of the spectral line corresponding to the desired sinusoidal output to the amplitude of the largest spurious spectral lines. (The spurious-free dynamic range for the oscillator illustrated in Figure 9.2.13 is 48 dB.) Quantization effects can also add to the amplitude of the spurious spectral lines. This produces a trade-off between the number of entries the LUT most store (2^W) and the number of bits D used to represent each stored value in the LUT. Figure 9.2.15 illustrates the trade-off between W and D.

There are three basic methods for reducing the peak amplitudes of the spurious spectral lines due to phase accumulator truncation.

1. **Error Feedforward:** The error feedforward method exploits the fact that the phase error is known and can be used to adjust the output of the LUT in compensation. Starting with (9.54) and applying Euler's identity to the complex exponential, the desired output may be expressed as

$$e^{j\theta(n)} = \cos\left(\hat{\theta}(n)\right) - \delta\theta(n)\sin\left(\hat{\theta}(n)\right) + j\left[\sin\left(\hat{\theta}(n)\right) + \delta\theta(n)\cos\left(\hat{\theta}(n)\right)\right]. \quad (9.55)$$

   This system may be realized in hardware as illustrated in Figure 9.2.16 (a). Applying this system to the example illustrated in Figure 9.2.13 produces the spectrum shown in Figure 9.2.17. Observe that the only spur visible in the 90 dB dynamic range shown is the one whose amplitude is -86 dB at 0.39 cycles/sample. The potential drawback of this approach is the requirement to perform two multiplications and two additions with D bits of precision at the sample rate.

2. **Dithering:** The primary reason the amplitudes of the spurious spectral lines are so large is the periodic structure of the phase error $\delta\theta(n)$. The periodicity tends to concentrate the energy in the phase error at certain frequencies. Dithering disrupts the periodic structure of the phase error to produce a more "random" effect. The result is that the phase error energy becomes more distributed across the spectrum. Dithering is realized by adding a small error term to the phase accumulator output just prior to

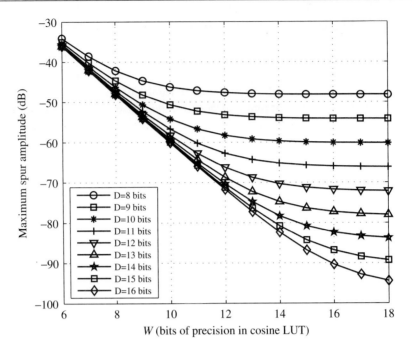

Figure 9.2.15 Peak amplitude of the spurious spectral line as a function of cosine/sine LUT size 2^W and LUT quantization D bits. Quantization to D bits is assumed to contribute 1 LSB to the magnitude of the peak spurious spectral line. (Adapted from Ref. [186].)

the quantizer as illustrated in Figure 9.2.16 (b). The amplitude of the dithering signal should be small (on the order of 2^{-W}) and random, preferably drawn from a probability density function with tails. The performance of this technique applied to the example of Figure 9.2.13 is illustrated by the spectrum in Figure 9.2.18. Observe that the spectral lines have been removed and that a "noise floor" is now present. The peak amplitude of this noise floor is −60 dB. The improvement is 12 dB, or about 2 bits, which can be interpreted to mean the system with dithering is performing as if it were using a LUT with $W = 10$ bits and no corrective measures. The dithering signal may be generated by an independent random-number generator as shown in Figure 9.2.16 (b) and used in the simulations. A commonly used alternative is to use the system clock signal, an appropriate number of address lines from a system memory element that is not a part of the DDS, or some other set of bits in the system that is not a direct part of the DDS. Another interesting source is to use the phase error from the previous instantaneous phase value. This is a special case of the error feedback technique that is summarized next.

3. **Error Feedback:** The error feedback technique treats the phase truncation from L bits to W bits as an additive noise process and attempts to subtract the error (that the quantizer will add) prior to quantization. Since the phase error $\delta\theta(n)$ is only known

Section 9.2 Discrete-Time Oscillators

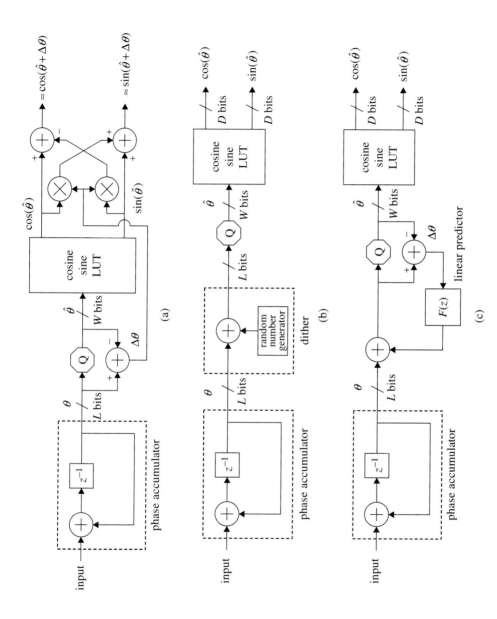

Figure 9.2.16 Three ways to fix the spurs in a DDS using phase accumulator truncation: (a) Error feedfoward; (b) dithering; (c) error feedback using a linear predictor.

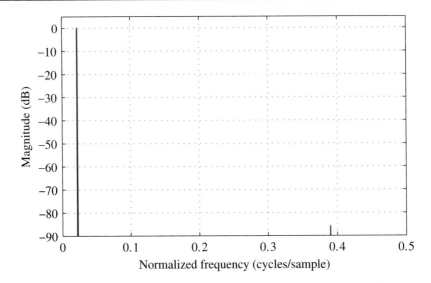

Figure 9.2.17 Performance of the error feedfoward technique to compensate for phase accumulator truncation for a sinusoid at 11/500 cycles/sample. $L = 28$, $W = 8$, $D = 16$. All of the compensation arithmetic at the output of the LUT were performed using values quantized to $D = 16$ bits. A length-1500 FFT was used to compute the DFT corresponding to 1500 samples. Compare with Figure 9.2.13.

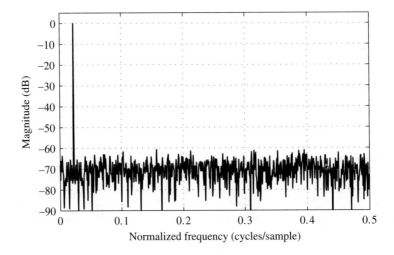

Figure 9.2.18 Performance of the dithering technique to compensate for phase accumulator truncation for a sinusoid at 11/500 cycles/sample. $L = 28$, $W = 8$, $D = 16$. The dithering sequence was generated using a Gaussian random number generator with zero mean and standard deviation $2^{-(L-B+1)}$. A length-1500 FFT was used to compute the DFT corresponding to 1500 samples. Compare with Figure 9.2.13.

Section 9.3 Resampling Filters

after quantization, the error feedback mechanism must predict $\delta\theta(n)$ from previous values. This prediction is performed by an FIR filter, called a *linear predictor*, that computes an estimate of $\delta\theta(n)$ by computing a linear combination of the previous values of $\delta\theta(n)$. The basic structure of this approach is illustrated in Figure 9.2.16 (c). The simplest linear predictor is the one-step linear predictor represented by the prediction filter $F(z) = z^{-1}$. The performance of the one-step linear predictor applied to the example of Figure 9.2.13 is illustrated in Figure 9.2.19 (a). The peak spurious spectral line has been reduced to -54 dB and occurs at 0.382 cycles/sample. Observe that the spurious spectral lines are small in the immediate vicinity of the desired signal but are much larger for frequencies that are not close to that of the desired signal. This system can be thought of as using $\delta\theta(n-1)$ as the dithering signal at time n. This behavior, often called "noise shaping," is characteristic of error feedback using a linear predictor. By using longer filters, the noise shaping can be generalized. In Figure 9.2.19 (b), the prediction filter $F(z)$ was designed to suppress the quantization noise over 50% of the bandwidth. The result is that the spurious lines are smaller than -76 dB around the desired signal but approach -50 dB only at frequencies above 0.25 cycles/sample. In Figure 9.2.19 (c) the prediction filter was designed to suppress the quantization noise over 25% of the bandwidth. The result in this case is that the spurious spectral lines are less than -88 dB for frequencies less than 0.125 cycles/sample, but can be as high as -50 dB for higher frequencies.

9.3 RESAMPLING FILTERS

Resampling filters are required in most practical modulator and demodulator implementations. In modulators, pulse shaping is applied to the data symbols at a clock rate that is four to eight times the symbol rate. If a band-pass signal is to be created in discrete-time, the sample rate must be significantly higher. The primary function of resampling in modulators is to increase the sampling rate. When the desired sample rate is a multiple of the symbol rate, the upsampling process is straightforward. However, there are many applications where resampling by a ratio P/Q is required. For example, a modulator may support a number of different symbol rates. An adjustable clock could be used to clock the DAC to guarantee DAC operation at an integer number of samples per symbol. But, clock-jitter limitations on DAC performance often require the DAC to operate with a fixed clock since a fixed clock has much lower jitter than an adjustable clock. As a consequence, the amount by which the discrete-time signal must be upsampled is not always fixed at an integer.

In demodulators, samples of the received waveform are produced by the ADC. If the ADC is allowed to operate with a fixed clock, clock-jitter performance is improved and system integration is greatly simplified. Usually, the ADC sample rate is a function of the carrier frequency ω_0 as explained in Section 10.1.3. Since ω_0 is rarely a function of the symbol rate, some form of resampling is required between the ADC interface and the matched filters. The primary function of resampling in demodulators is the reduction of the sampling rate.

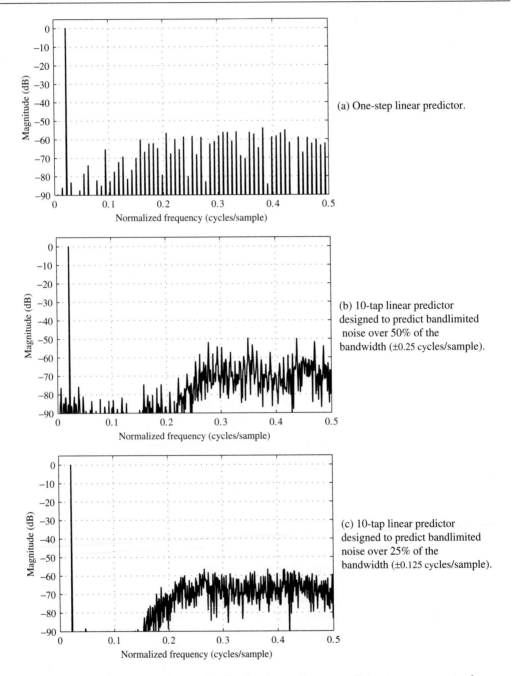

Figure 9.2.19 Performance of error feedback using a linear predictor to compensate for phase accumulator truncation for a sinusoid at 11/500 cycles/sample. $L = 28$, $W = 8$, $D = 16$. A length-1500 FFT was used to compute the DFT corresponding to 1500 samples. Compare with Figure 9.2.13.

Section 9.3 Resampling Filters

This section summarizes three popular classes of resampling filters: CIC and Hogenauer filters, half-band filters, and polyphase filterbanks. Because the emphasis of the text is on demodulators and detectors, the treatment deals mostly with sample rate reduction. Sample rate increases are mentioned where the application is not obvious.

9.3.1 CIC and Hogenauer Filters

The *cascade-integrator-comb* (CIC) filter is a multiply-free low-pass filter. It is not a particularly good filter, but the ability to perform even modest filtering without multiplications is attractive at high sample rates. When used as a resampling filter, the CIC is called a Hogenauer filter and is a popular method for changing the sample rate by an integer.

The starting point is the simple all-ones FIR filter (sometimes called a "boxcar" filter):

$$h(n) = \begin{cases} 1 & 0 \leq n < M \\ 0 & \text{otherwise} \end{cases}. \tag{9.56}$$

The z-transform and DTFT of this filter are

$$H(z) = \frac{1 - z^{-m}}{1 - z^{-1}} \tag{9.57}$$

$$H\left(e^{j\Omega}\right) = \frac{\sin\left(\frac{M\Omega}{2}\right)}{\sin\left(\frac{\Omega}{2}\right)} e^{-j\Omega\left(\frac{M-1}{2}\right)}. \tag{9.58}$$

The filter has M zeros equally spaced around the unit circle and one pole at $z = 1$. The pole at $z = 1$ exactly cancels the zero at $z = 1$ to produce a pole-zero plot such as the one illustrated in Figure 9.3.1 (a) for the case $M = 10$. The corresponding DTFT is plotted in Figure 9.3.1 (b) which shows that this filter is not a particularly good low-pass filter because the peak stop-band attenuation is only 13 dB below the peak pass-band gain. This situation can be remedied somewhat by cascading boxcar filters. The DTFTs of $h(n)$, $h(n) * h(n)$, $h(n) * h(n) * h(n)$, and $h(n) * h(n) * h(n) * h(n)$ are plotted in Figure 9.3.2 (a)–(d), respectively. Observe that with each additional stage, the stop-band attenuation increases by 13 dB and the pass-band gain increases. The null-to-null bandwidth of the filter remains the same but the steepness (or spectral distortion) increases. As a result, the usable bandwidth of a K-stage filter is limited to about 25% of the main lobe width.

Rewriting the z-transform (9.57) as

$$H(z) = \underbrace{\frac{1}{1 - z^{-1}}}_{I(z)} \times \underbrace{1 - z^{-M}}_{C(z)} \tag{9.59}$$

shows that each stage can be treated as the cascade of two filters: an integrator $I(z)$ and a "comb" filter $C(z)$. The integrator has a single pole at $z = 1$ and is a zero-delay accumulator

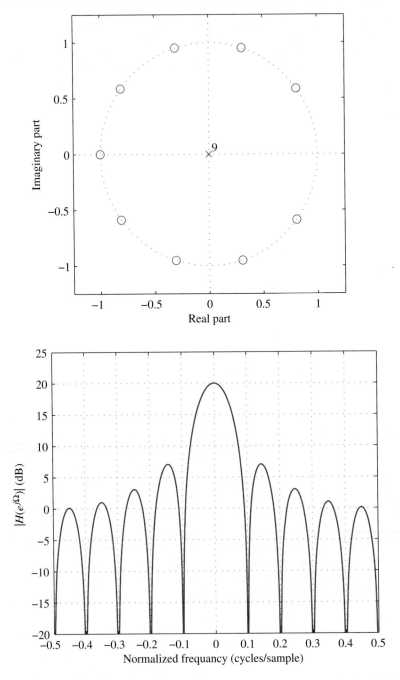

Figure 9.3.1 Frequency domain representations of the length-10 boxcar filter (9.56): (a) The z-transform (9.57); (b) the DTFT (9.58).

Section 9.3 Resampling Filters

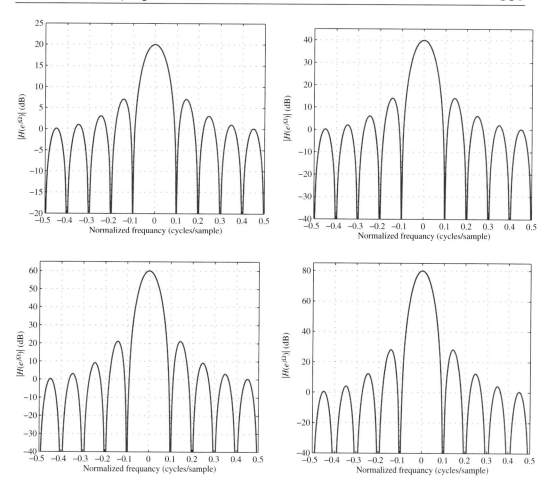

Figure 9.3.2 (a) The DTFTs of the length-10 boxcar filter (9.56), (b) the cascade of two length-10 boxcar filters, (c) the cascade of three length-10 boxcar filters, (d) and the cascade of four length-10 boxcar filters (9.56).

that approximates integration as described in Section 3.3.3. The comb filter has M zeros equally spaced around the unit circle. As a consequence, the corresponding DTFT,

$$C\left(e^{j\Omega}\right) = j2\sin\left(\frac{M\Omega}{2}\right)e^{-j\frac{M\Omega}{2}} \qquad (9.60)$$

has M nulls in the frequency response as illustrated in Figure 9.3.3 for the case $M = 10$. These nulls eliminate the spectral content at equally spaced intervals in the same way a comb parts hair at equally spaced intervals. Hence the name "comb" filter. When the boxcar filter

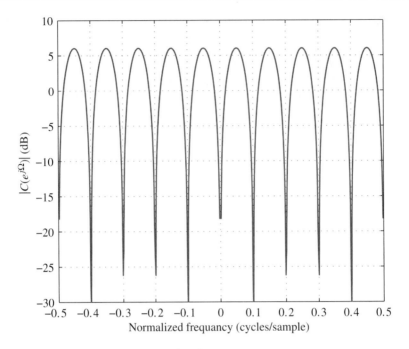

Figure 9.3.3 The DTFT of the comb filter for the case $M = 10$.

is realized as two filters in cascade as suggested by (9.60), the filter is called a "cascade-integrator-comb" filter.

It is important to note that, in general, treating the numerator and denominator of a filter's transfer frequency response as separate filters, especially where pole-zero cancellations exist, is a bad idea. Care must be taken to account for finite precision effects that would not exist if the filter were realized as a single unit. For the CIC filter, if the two's complement representation is used and if the word width of the accumulator is sufficiently large, then the CIC filter is stable and performs the intended low-pass filter function. Each CIC filter used in cascade increases the gain of the filter and thus increases the required word length. For this reason, practical CIC filters are limited to $K = 3$ or 4 stages.

The Hogenauer filter is a CIC filter with upsampling or downsampling incorporated into the circuit. The sample rate change is by a factor M, the length of the underlying boxcar filter. The number of stages K controls the attenuation of the spectral content folded into the pass band after downsampling. The conceptual steps in generating the Hogenauer filter for downsampling by M is illustrated in Figure 9.3.4. Figure 9.3.4 (a) shows a K-stage CIC filter preceding an M-to-1 downsample operation. The K integrators and K comb filters that make up the CIC filter are separated as shown in Figure 9.3.4 (b). The comb filters are placed next to the downsample operation so that the downsample operation can be moved inside the filter using the Noble identities. Application of the Noble identities results in the system illustrated in Figure 9.3.4 (c). The comb filters are now discrete-time differentiators and operate at the

Section 9.3 Resampling Filters

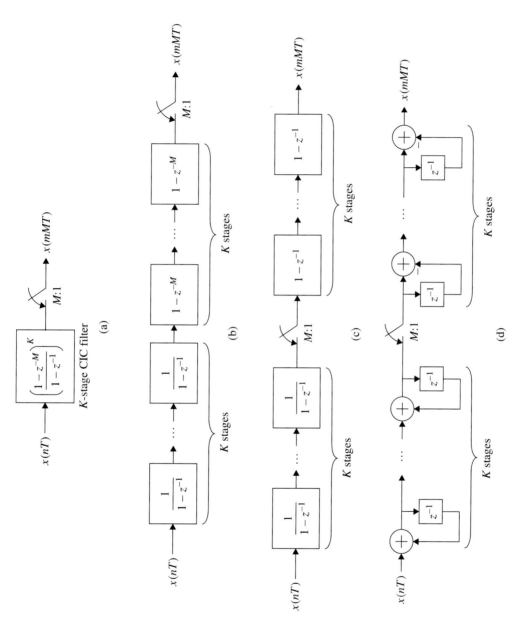

Figure 9.3.4 The conceptual steps in the development of the Hogenauer downsampling filter from the CIC filter.

lower sample rate. The block diagram illustrated in Figure 9.3.4 (d) shows that the filtering is accomplished using only additions. The conceptual procedure applies to upsampling filters except that the comb filters are placed at the input, next to the zero-insertion operation. Using the Noble identities, the zero-insertion operation can be moved to the output side of the comb filters.

The pass-band of the Hogenauer filter distorts the resulting downsampled signal by imposing a $\sin(M\Omega/2)/\sin(\Omega/2)$ shape on the magnitude of the spectrum. A correction filter is usually applied *after* the Hogenauer downsampling filter to compensate for this distortion. (The filter is placed here so that it can operate at the lower sample rate.) The design of this filter is very similar to that of the predistortion filter used to compensate for sample-and-hold distortion in a DAC as described in Section 9.1.2 except that the desired frequency response is $\sin(\Omega/2)/\sin(M\Omega/2)$ instead of $(\Omega/2)/\sin(\Omega/2)$. For a Hogenauer upsampling filter, this correction is performed by a predistortion filter placed *before* the filter. This placement allows the predistortion filter to operate at the lower sample rate.

9.3.2 Half-Band Filters

A half-band filter has odd symmetry about $\Omega = \pi/2$ as illustrated in Figure 9.3.5 (a) for the case of a half-band low-pass filter. For a zero-phase, real-valued, FIR half-band filter, this symmetry may be expressed as

$$H\left(e^{j\Omega}\right) = C - H\left(e^{j\Omega+\pi}\right) \tag{9.61}$$

for a real-valued positive constant C as illustrated in Figure 9.3.5 (b). The filter coefficients $h(n)$ may be expressed as

$$\begin{aligned} h(n) &= \frac{1}{2\pi} \int_{-\pi}^{\pi} H\left(e^{j\Omega}\right) e^{j\Omega n} d\Omega \\ &= \frac{1}{2\pi} \int_{-\pi}^{\pi} \left[C - H\left(e^{j\Omega+\pi}\right)\right] e^{j\Omega n} d\Omega \\ &= C\delta(n) - \frac{1}{2\pi} \int_{0}^{2\pi} (-1)^n H\left(e^{j\theta}\right) e^{j\theta n} d\theta \\ &= C\delta(n) - (-1)^n h(n) \end{aligned} \tag{9.62}$$

from which the fundamental relationship

$$h(n)\left[1 + (-1)^n\right] = C\delta(n) \tag{9.63}$$

follows. This relationship imposes constraints on $h(n)$ for even values of n. In particular

$$\begin{aligned} h(0) &= C/2 \\ h(n) &= 0 \quad \text{for } n \neq 0 \text{ and even.} \end{aligned} \tag{9.64}$$

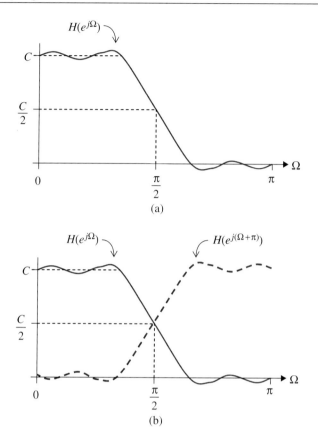

Figure 9.3.5 (a) The frequency response of a half-band low-pass filter. (b) The frequency response of the half-band low-pass filter and a frequency-shifted version illustrating the symmetry about $\Omega = \pi/2$.

As an example, consider the ideal half-band low-pass filter illustrated in Figure 9.3.6 (a). The impulse response is

$$h(n) = \frac{1}{2\pi} \int_{-\pi/2}^{\pi/2} e^{j\Omega n} d\Omega = \frac{1}{2} \frac{\sin\left(\frac{n\pi}{2}\right)}{\frac{n\pi}{2}} \qquad (9.65)$$

and is plotted in Figure 9.3.6 (b) for $-6 \leq n \leq 6$. Observe that, as predicted by (9.64), $h(0) = 1/2$ and the nonzero even-indexed values are zero.

Because almost half of the filter coefficients are zero, half-band filters are particularly attractive for use with a sample rate change by 2. This fact is illustrated by using the filter of the previous example to reduce the sample rate by 2. The most efficient method for performing the downsampling is to use a polyphase filterbank with two subfilters as illustrated

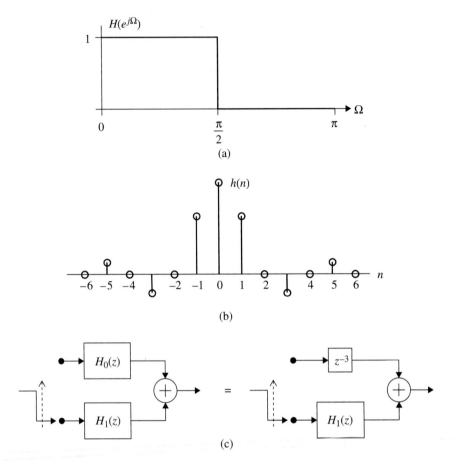

Figure 9.3.6 An example of the ideal low-pass half-band filter: (a) Frequency response; (b) corresponding impulse response for $-6 \leq n \leq 6$; (c) the use of the half-band filter in a downsample-by-2 resampler based on a polyphase filterbank.

in Figure 9.3.6 (c). The polyphase partition is applied to a causal version of the half-band filter. For example, a causal version of the filter of Figure 9.3.6 (b) is produced by delaying the filter coefficients by 6. In this case, the filter coefficient with index 6 is the only nonzero even-indexed coefficient. As a result, the subfilter in the top branch of the polyphase partition is all zeros except for the coefficient with index 3. Scaling the coefficients by 2 makes the value of the only nonzero coefficient in the top branch equal to 1. This scaling eliminates the need for multiplication in the top branch, thus simplifying the filterbank. The resulting filterbank is shown on the right-hand side of Figure 9.3.6 (c).

Half-band FIR filters may be designed by applying the windowing technique described in Chapter 3 to the impulse response (9.65) or by using any of the other filter approximation techniques, such as the Parks–McClellan algorithm. In order to use the Noble identity to create the simple form for the polyphase partition (such as the one illustrated in Figure 9.3.6 (c)), the filter should have a "center tap." This requires the length of the filter to be odd, that is, of the form $2L + 1$. The lone nonzero even-indexed value is in the center of the response when L is even. This is equivalent to requiring the end points of the prototype half-band filter to be zero. If a filter design produces a filter with nonzero end points (i.e., L is odd), then zeros can be appended to each end of the impulse response without increasing the complexity of the filter.

9.3.3 Arbitrary Resampling Using Polyphase Filterbanks

The Hogenauer filter structure described in Section 9.3.1 can be used to perform a downsample by an integer and the half-band filter of Section 9.3.2 can be used to perform a downsample by 2. It is almost always the case that downsampling by a factor other than an integer is required in the demodulator. A polyphase filterbank can be used to change the sample rate by a factor P/Q. When $P < Q$, the filter performs a downsample-by-Q/P operation. Likewise, when $P > Q$, the filter performs an upsample-by-P/Q operation. As this terminology can be confusing to those new to the area, the language "resample by P/Q" will be used. The variable r will be used to denote the "resample rate." When the resample rate is rational, the resample rate is of the form $r = P/Q$. If P is too large, or the exact sample is rate is not known until run time, the resample rate $r = P/Q$ is only an approximation. In some instances, r may be irrational. As before, the emphasis will be placed on sample rate reductions which are common in the demodulator.

Resampling by a Rational Factor. Suppose the input signal $x(nT)$, with sampling rate $1/T$ and bandwidth W, is to be resampled to a new rate that may be expressed as P/Q times the input rate. That is, the samples $x(kTQ/P)$ are to be produced from the samples $x(nT)$. This may be accomplished using the cascaded system illustrated in Figure 9.3.7 (a). The input signal is upsampled by P and filtered to produce the intermediate signal $x(mT/P)$ with sample rate P/T. The intermediate signal is downsampled by Q by selecting every Q-th sample.

Direct application of this approach requires the intermediate filter to operate at P times the input sample rate. A polyphase decomposition of the resampling filter can be used to perform the upsample/filter operation as described in Section 3.2.3. Each input sample produces P output samples in parallel, thus allowing the resampling filter to operate at the input rate. The polyphase implementation is illustrated in Figure 9.3.7 (b). The output commutator strides through the polyphase filterbank outputs Q subfilters at a time, since only every Q-th output is desired. Each time the output commutator index exceeds P, it "rolls over" modulo-P and a new input sample is clocked into the filterbank. When $P < Q$, multiple inputs are clocked into the filter in between each output sample. When $P > Q$, multiple outputs are taken for each input clocked into the filter.

The diagram in Figure 9.3.7 (b) suggests the need for P subfilters. Usually, hardware implementations are based on one filter with P sets of filter coefficients stored in memory as illustrated in Figure 9.3.7 (c). This figure applies to the case where the system reduces the sample rate (i.e., $P < Q$). The system operates at the input sample rate by continuously

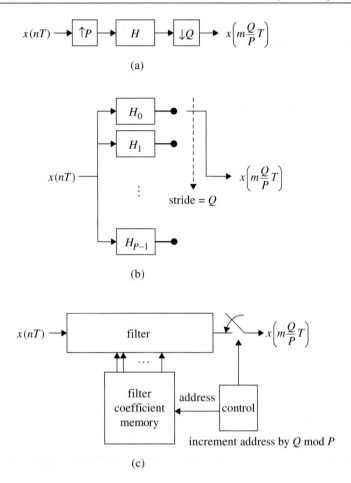

Figure 9.3.7 Resampling by P/Q: (a) A conceptual approach using an upsample-by-P and filter operations followed by a downsample-by-Q operation; (b) a conceptual polyphase filterbank implementation; (c) an efficient implementation of the system in (b) based on a single filter and memory to store the subfilter coefficients.

clocking input samples into the filter and only computing the desired samples. On average, the filter produces Q outputs for every P inputs.

As an example, consider a 12.5 Msymbol/s system that presents to the ADC an IF signal centered at 70 MHz. Based on the principles discussed in Section 10.1.3, a desirable sample rate is $93\frac{1}{3}$ Msamples/s. Thus, the ADC presents to the discrete-time processor a signal sampled at

$$\frac{93\frac{1}{3}}{12.5} = \frac{112}{15} \quad \text{samples/symbol}.$$

Section 9.3 Resampling Filters

This signal must be resampled to 2 samples/symbol for matched filtering, timing synchronization, and carrier phase synchronization. The resample rate is thus $r = 15/56$. The system of Figures 9.3.7 (b) and (c) can be used with $P = 15$ and $Q = 56$.

Resampling by an Arbitrary Ratio. There are several cases when the direct approach described above is not practical.

- When P is large, the number of subfilters (of Figure 9.3.7 (b)) or the required storage (of Figure 9.3.7 (c)) may be prohibitive. In this case, the desired resampling rate r may be approximated by a ratio P/Q of smaller integers.
- In some applications, the resample rate r has to be chosen "on the fly." In this case, the number of subfilters in the polyphase filterbank cannot be fixed before-hand during the design cycle. Using the direct approach of Figure 9.3.7 (c), the system would need sufficient resources to recompute the filter coefficients and store them in a memory that is capable of holding a dynamically assignable number of rows. These requirements can present substantial challenges to the design of a real-time system.
- The relationship between the input and output sample rates may not be rational. A time-varying input or output sample rate is an example.

Simple modifications to the polyphase filterbank described above can be used to handle these cases. The modification is best described by returning to the example of designing a resampler to change the sample rate by $r = 15/56$. Suppose system constraints limit the number of subfilters in the polyphase filterbank to $P = 5$. As such, the resampling ratio may be expressed as

$$r = \frac{5}{18\,2/3}.$$

This ratio requires the stride of the output commutator to be $18\ 2/3$. To see how this works, consider the first six multiples of the stride

$$
\begin{aligned}
18\ 2/3 &= 3 \times 5 + 3\ 2/3 \\
37\ 1/3 &= 7 \times 5 + 2\ 1/3 \\
56 &= 11 \times 5 + 1 \\
74\ 2/3 &= 14 \times 5 + 4\ 2/3 \\
93\ 1/3 &= 18 \times 5 + 3\ 1/3 \\
112 &= 22 \times 5 + 2
\end{aligned}
$$

Each multiple is expressed in the form $q \times 5 + \rho$, which is the result after dividing by $P = 5$. The quotient q defines how many input samples should be clocked into the filterbank to compute the desired output sample and ρ is used to define the required polyphase filter index.

Because ρ is not always an integer, an approximation is required. The first approximation, called "nearest neighbor," approximates the output by rounding the output index pointer ρ to the nearest integer. The result is an integer that specifies which subfilter output contains the desired output sample. In this example, the first output is taken from subfilter 4 after 3 input samples are clocked into the filterbank. The next output is taken from subfilter 2 after 7 input samples are clocked into the filterbank. A block diagram of this polyphase resampler is illustrated in Figure 9.3.8 (a). An example of the relationship between the input index, the intermediate index, and the desired output sample is illustrated in Figure 9.3.9.

The second approximation uses linear interpolation. The illustration in the top part of Figure 9.3.9 shows that the integer part of ρ is the index (counting samples at the intermediate

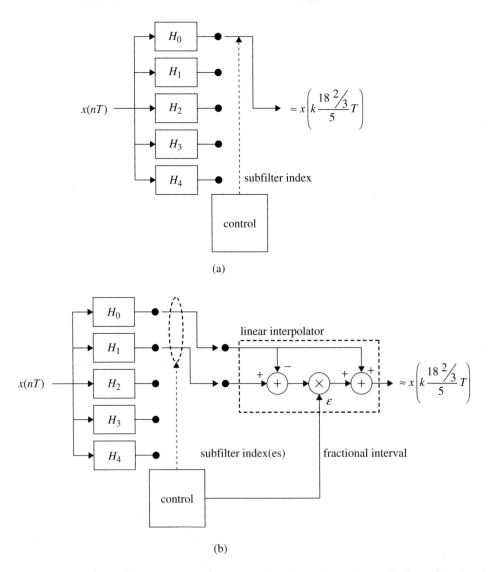

Figure 9.3.8 Polyphase filter structures for resampling by 112/15 using polyphase filterbank with $P = 5$ subfilters: (a) The nearest neighbor approximation; (b) the linear interpolator approximation. The block diagrams suggest the need for P filters, but an actual hardware realization requires one filter for (a) and possibly two filters for (b) with P sets of coefficients.

Section 9.3 Resampling Filters

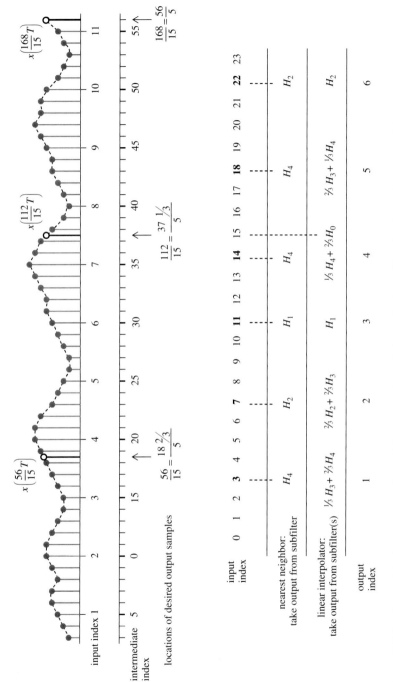

Figure 9.3.9 A graphical representation of the operation of the polyphase filterbanks with $P = 5$ subfilters, illustrated in Figure 9.3.8 to perform resampling by 112/15. The relationship between the input samples, the intermediate samples, and the output samples for the nearest neighbor and linear interpolator approximations are shown.

sample rate reduced modulo-5) of the sample just to the left of the desired output. The next higher indexed subfilter in the polyphase filterbank outputs the sample just to the right of the desired output. The fractional part of ρ specifies the fractional interval between these two outputs. A linear interpolator can be used to approximate the desired output value from the two available filterbank outputs. Using the interpolation terminology introduced in Chapter 8, the integer part of ρ is the base-point index and the fractional part of ρ is the fractional interval. Returning to the example, the first output is interpolated between subfilters 3 and 4 using a fractional interval of $2/3$. The next output is interpolated between subfilters 2 and 3 using a fractional interval of $1/3$. Observe that the next output coincides exactly with the output of subfilter 1 and no interpolation is needed. These relationships are depicted graphically in Figure 9.3.9. A block diagram of the polyphase resampler using linear interpolation is illustrated in Figure 9.3.8 (b).

To generalize, let n be the input index and k the output index. The k-th output is determined by expressing the k-th multiple of the stride as

$$k \times \text{stride} = q \times P + \rho.$$

The controller keeps track of q to determine when an output sample should be produced. Because the stride is not an integer, ρ is not, in general, an integer. The quotient q specifies the input index n, and ρ is used to specify the polyphase subfilter index(es). For the nearest neighbor approximation, ρ rounded to the nearest integer to specify polyphase subfilter index. For the linear interpolation approximation, the integer part of ρ and this integer plus one specify the two polyphase filter outputs involved in the interpolation and the fractional part of ρ defines the fractional interpolation interval.

The number of subfilters P that is required to produce an acceptable approximation depends on the bandwidth of the input signal and a performance metric that defines what is acceptable. Two commonly used performance metrics are the ratio of signal power to distortion power and the ratio of average signal level to peak distortion level.

Signal-to-Distortion Power Ratio Let W_d be the bandwidth of the input signal $x(nT)$. (Note that W_d is relative to the input sample rate $1/T$.) For the purposes of analysis, assume that the DTFT of the input signal $x(nT)$ is

$$X\left(e^{j\Omega}\right) = \begin{cases} A & -W_d \leq \Omega \leq W_d \\ 0 & \text{otherwise} \end{cases}. \tag{9.66}$$

The P-stage polyphase filterbank produces samples of the intermediate signal with sample rate P/T. The DTFT of the intermediate signal $x(mT/P)$ is

$$X\left(e^{j\Omega}\right) = \begin{cases} A & -\frac{W_d}{P} \leq \Omega \leq \frac{W_d}{P} \\ 0 & \text{otherwise} \end{cases}. \tag{9.67}$$

Using Parseval's theorem, the power of this signal is

$$S^2 = \frac{1}{2\pi} \int_{-W_d/P}^{W_d/P} A^2 d\Omega = \frac{A^2 W}{P\pi}. \tag{9.68}$$

Section 9.3 Resampling Filters

Referring to the example illustrated in the top part of Figure 9.3.9, suppose the desired signal is

$$x\left((m+\epsilon)\frac{T}{P}\right)$$

for some integer m that indexes the intermediate sample time T/P and some fractional interval ϵ. The nearest neighbor approximation produces

$$x\left(m\frac{T}{P}\right).$$

In this case, the fractional interval is sampling time error and satisfies $-1/2 \leq \epsilon \leq 1/2$. The error is

$$x\left((m+\epsilon)\frac{T}{P}\right) - x\left(m\frac{T}{P}\right)$$

and the DTFT of the error is

$$X\left(e^{j\Omega}\right)e^{j\Omega\epsilon} - X\left(e^{j\Omega}\right).$$

The error represents distortion introduced by using the approximation. The power of this distortion may be computed using Parseval's theorem:

$$D^2 = \frac{1}{2\pi}\int_{-W_d/P}^{W_d/P}\left|X\left(e^{j\Omega}\right)e^{j\Omega\epsilon} - X\left(e^{j\Omega}\right)\right|^2 d\Omega = \frac{A^2}{2\pi}\int_{-W_d/P}^{W_d/P}\left|e^{j\Omega\epsilon} - 1\right|^2 d\Omega. \quad (9.69)$$

The Taylor series of e^{ju} about 0 is

$$e^{ju} = 1 + ju - \frac{u^2}{2!} - j\frac{u^3}{3!} + \frac{u^4}{4!} + \cdots \quad (9.70)$$

Using the first two terms in the Taylor series for $e^{j\Omega\epsilon}$ produces

$$D^2 \approx \frac{A^2\epsilon^2}{2\pi}\int_{-W_d/P}^{W_d/P}\Omega^2 d\Omega = \frac{A^2\epsilon^2 W_d^3}{3\pi Q^3} \leq \frac{A^2 W_d^3}{12\pi P^3} \quad (9.71)$$

where the inequality follows from the bound on ϵ. The signal-to-distortion ratio for the nearest neighbor approximation is thus

$$\frac{S^2}{D^2} \geq 12\frac{P^2}{W_d^2}. \quad (9.72)$$

The signal-to-distortion ratio, in amplitude, is the square root of (9.72). This shows that the amplitude signal-to-distortion ratio is proportional to P or, what is equivalent, inversely proportional to the bandwidth of the intermediate signal. For this reason, the nearest neighbor approximation is sometimes called a "first-order approximation."

The linear interpolation approximation produces

$$(1-\epsilon)x\left(m\frac{T}{P}\right) + \epsilon x\left((m+1)\frac{T}{P}\right).$$

In this case, the fractional interval ϵ is the fractional interpolation interval and satisfies $0 \leq \epsilon < 1$. Proceeding as before, the error is

$$x\left((m+\epsilon)\frac{T}{P}\right) - (1-\epsilon)x\left(m\frac{T}{P}\right) - \epsilon x\left((m+1)\frac{T}{P}\right)$$

and the DTFT of the error is

$$X\left(e^{j\Omega}\right)e^{j\Omega\epsilon} - (1-\epsilon)X\left(e^{j\Omega}\right) - \epsilon X\left(e^{j\Omega}\right)e^{j\Omega}.$$

The distortion power is

$$D^2 = \frac{1}{2\pi}\int_{-W_d/P}^{W_d/P} \left|X\left(e^{j\Omega}\right)\left[e^{j\Omega\epsilon} - (1-\epsilon) - \epsilon e^{j\Omega}\right]\right|^2 d\Omega \tag{9.73}$$

$$= \frac{A^2}{2\pi}\int_{-W_d/P}^{W_d/P} \left|e^{j\Omega\epsilon} - (1-\epsilon) - \epsilon e^{j\Omega}\right|^2 d\Omega. \tag{9.74}$$

Using the first three terms in the Taylor series for the complex exponentials produces

$$D^2 \approx \frac{A^2}{2\pi}\int_{-W_d/P}^{W_d/P} \left|\left(1 + j\Omega\epsilon - \frac{\Omega^2\epsilon^2}{2}\right) - (1-\epsilon) - \epsilon\left(1 + j\Omega - \frac{\Omega^2}{2}\right)\right|^2 d\Omega \tag{9.75}$$

$$= \frac{A^2(1-\epsilon)^2\epsilon^2}{20\pi}\frac{W_d^5}{P^5}. \tag{9.76}$$

Because $0 \leq \epsilon \leq 1$, the term $(1-\epsilon)\epsilon$ is bounded by $1/4$, so that

$$D^2 \leq \frac{A^2}{320\pi}\frac{W_d^5}{P^5}. \tag{9.77}$$

The signal-to-distortion ratio is

$$\frac{S^2}{D^2} \geq 320\frac{P^4}{W_d^4}. \tag{9.78}$$

The signal-to-distortion ratio in amplitude, which is the square root of (9.78), is proportional to P^2 or, what is equivalent, inversely proportional to the square of the bandwidth of the intermediate signal. For this reason, the linear interpolation approximation is sometimes called a "second-order approximation."

Section 9.3 Resampling Filters

For both the nearest neighbor and linear interpolator approximations, the signal-to-distortion ratio increases as P increases and as W_d decreases. This shows that increasing the number of subfilters improves the performance of the approximations and that the more oversampled the input signal is, the better the approximations are. The linear interpolator approximation requires significantly fewer subfilters than the nearest neighbor approximation. The difference between the two becomes more dramatic as the sample rate of the input signal approaches the Nyquist sampling rate. These features are illustrated in Figure 9.3.10. In practice, the nearest neighbor approximation is able to produce a signal-to-distortion ratio on the order of 50–60 dB whereas the linear interpolator approximation is able to produce a signal-to-distortion ratio almost as low as desired.

Signal Amplitude to Peak Distortion Level Ratio An alternate approach is to compute the ratio of the average signal level and the peak distortion level. The peak distortion for the two approximations is best quantified in the frequency domain. The analysis proceeds as follows: A fictitious continuous-time signal $\tilde{x}(t)$ is constructed from the signal at the intermediate sample rate P/T. This signal is constructed to place the approximate values at the desired sampling instants. In the frequency domain, the effect of this approximation is clearly visible and is used to quantify the peak distortion. As before, the DTFT of the input signal is assumed to be (9.66).

A continuous-time signal that represents the nearest neighbor approximation is one that is produced by converting the samples at the intermediate sample rate to a continuous-time impulse train and processing the impulse train by a filter with impulse response shown

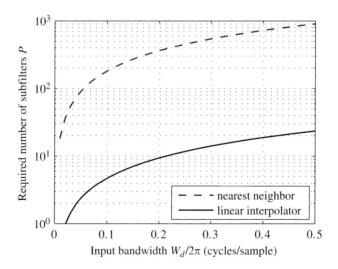

Figure 9.3.10 A plot of the required number of subfilters P as a function of the input signal bandwidth W for a fixed signal-to-distortion ratio $S/D = 60$ dB for the nearest neighbor approximation (dashed line) and the linear interpolator approximation (solid line).

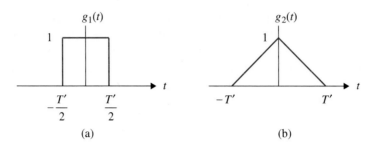

Figure 9.3.11 Continuous-time filters used to create a fictitious continuous-time signal to represent the two approximations used for arbitrary sample rate changes using a polyphase filterbank: (a) The filter used for the nearest neighbor approximation; (b) the filter used for the linear interpolator approximation.

in Figure 9.3.11 (a) where $T' = T/P$ is the sample time of the intermediate signal. The continuous-time signal is

$$\tilde{x}(t) = \sum_m x(mT')\delta(t - mT') * g_1(t) = \sum_m x(mT')g_1(t - mT'). \quad (9.79)$$

This signal is almost identical to the stair-step signal produced by the sample-and-hold characteristic of a DAC summarized in Section 9.1.2. Sampling $\tilde{x}(t)$ at $t = mT/P + \epsilon$ produces the nearest neighbor approximation. The distortion in $\tilde{x}(t)$ is quantified in the frequency domain. The Fourier transform of $\tilde{x}(t)$ is the product of the Fourier transform of the impulse train (which consists of periodically spaced replicas of $X(j\omega)$) and the Fourier transform of the filter

$$G_1(j\omega) = T' \frac{\sin\left(\frac{\omega T'}{2}\right)}{\frac{\omega T'}{2}}. \quad (9.80)$$

An illustration of $\tilde{X}(j\omega)$ is shown in Figure 9.3.12 (a). The bandwidth of $X(j\omega)$ is $W_c = W_d/T'$ rad/s. The distortion is represented by the residual copies $X(j\omega)$ centered at multiples of $\omega = 2\pi/T'$. The highest distortion level occurs at $\omega = 2\pi/T' - W_c$ as shown. The peak distortion is

$$D_{\text{peak}} = \left| AG_1\left(j\left(\frac{2\pi}{T'} - W_c\right)\right)\right| = AT' \frac{\sin\left(\left(\frac{2\pi}{T'} - W_c\right)\frac{T'}{2}\right)}{\left(\frac{2\pi}{T'} - W_c\right)\frac{T'}{2}}. \quad (9.81)$$

Using the first term in the Taylor series expansion

$$\frac{\sin(au)}{(au)} = -\frac{a}{\pi}\left(u - \frac{\pi}{a}\right) + \frac{a^2}{\pi^2}\left(u - \frac{\pi}{a}\right)^2 + \frac{a^3}{\pi^3}\left(\frac{\pi^2}{6} - 1\right)\left(u - \frac{\pi}{a}\right)^3 + \cdots, \quad (9.82)$$

Section 9.3 Resampling Filters

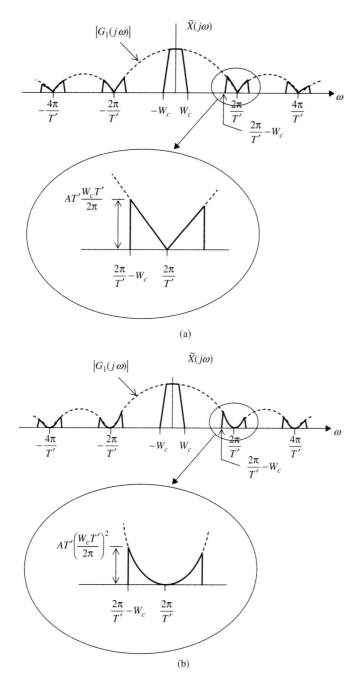

Figure 9.3.12 An illustration of the peak distortion for arbitrary resampling using a polyphase filterbank with P subfilters using (a) the nearest neighbor approximation and (b) the linear interpolator approximation.

the peak distortion is approximately

$$D_{\text{peak}} \approx AT' \frac{W_c T'}{2\pi} \qquad (9.83)$$

as shown in Figure 9.3.12 (a). The average signal magnitude is $S = AG_1(0) = AT'$. The signal-to-peak-distortion ratio is

$$\frac{S}{D_{\text{peak}}} = \frac{2\pi}{W_c T'} = \frac{2\pi P}{W_c T} = 2\pi \frac{P}{W_d}. \qquad (9.84)$$

Comparing (9.84) with the square root of (9.72) shows that both predict the signal-to-distortion ratio in amplitude to be inversely proportional to the intermediate signal bandwidth W_d/P. The constant of proportionality is approximately 1.8 times larger for the peak distortion criterion.

The ratio S/D_{peak} given by (9.84) can also be interpreted as the usable dynamic range due to distortion. Quantization also limits the usable dynamic range as described in Section 9.1.1 for ADCs. Finite precision arithmetic based on b-bits of precision has a usable dynamic range 2^b. An alternate method for selecting the number of subfilters is to select P to make the usable dynamic range due to the approximation larger than the usable dynamic range due to quantization. This makes quantization the limiting factor, rather than the error introduced by the approximation. Following this line of reasoning produces

$$2\pi \frac{P}{W_d} > 2P > 2^b \quad \rightarrow \quad P > 2^{b-1} \qquad (9.85)$$

where the fact that $W_d < \pi$ has been used to simplify the bound.

A continuous-time signal that represents the linear interpolator approximation is one that is produced by converting the samples at the intermediate sample rate to a continuous-time impulse train and processing the impulse train by a filter with a triangular impulse response $g_2(t)$ illustrated in Figure 9.3.11 (b). The continuous-time signal is

$$\tilde{x}(t) = \sum_m x(mT')\delta(t - mT') * g_2(t) = \sum_m x(mT')g_2(t - mT') \qquad (9.86)$$

and is characterized by a straight line connecting each of the impulse weights. Sampling $\tilde{x}(t)$ at $t = mT/P + \epsilon$ produces the linear interpolator approximation. The Fourier transform of $\tilde{x}(t)$ is the product of the Fourier transform of the impulse train (which consists of periodically spaced replicas of $X(j\omega)$) and the Fourier transform of the filter

$$G_2(j\omega) = T' \frac{\sin^2\left(\frac{\omega T'}{2}\right)}{\left(\frac{\omega T'}{2}\right)^2} \qquad (9.87)$$

Section 9.3 Resampling Filters

as illustrated in Figure 9.3.12 (b). As before, the bandwidth of $X(j\omega)$ is $W_c = W_d/T'$ rad/s. The distortion is represented by the residual copies $X(j\omega)$ centered at multiples of $\omega = 2\pi/T'$. The highest distortion level occurs at $\omega = 2\pi/T' - W_c$ as shown. The peak distortion is

$$D_{\text{peak}} = \left| AG_2\left(j\left(\frac{2\pi}{T'} - W_c\right)\right)\right| = AT' \frac{\sin^2\left(\left(\frac{2\pi}{T'} - W_c\right)\frac{T'}{2}\right)}{\left[\left(\frac{2\pi}{T'} - W_c\right)\frac{T'}{2}\right]^2}. \tag{9.88}$$

Using the first term in the Taylor series expansion

$$\frac{\sin^2(au)}{(au)^2} = \frac{a^2}{\pi^2}\left(u - \frac{\pi}{a}\right)^2 - 2\frac{a^3}{\pi^3}\left(u - \frac{\pi}{a}\right)^3 + \frac{a^4}{\pi^4}\left(3 - \frac{\pi^2}{3a^2}\right)\left(u - \frac{\pi}{a}\right)^4 + \cdots, \tag{9.89}$$

the peak distortion is approximately

$$D_{\text{peak}} \approx AT'\left(\frac{W_c T'}{2\pi}\right)^2 \tag{9.90}$$

as shown in Figure 9.3.12 (b). Using $S = AG_2(0) = AT'$, the signal-to-peak-distortion ratio is

$$\frac{S}{D_{\text{peak}}} = \left(\frac{2\pi}{W_c T'}\right)^2 = \left(\frac{2\pi P}{W_c T}\right)^2 = (2\pi)^2 \left(\frac{P}{W_d}\right)^2. \tag{9.91}$$

Comparison of this result with the square root of (9.78) shows that both criteria predict that the signal-to-distortion ratio in amplitude is proportional to P^2 and inversely proportional to the square of the bandwidth of the intermediate signal. (The constant of proportionality is 2.2 times larger using the peak-distortion criterion.)

Following the same dynamic range analysis developed for the nearest neighbor approximation, the relationship is

$$(2\pi)^2 \left(\frac{P}{W_d}\right)^2 > 4P^2 > 2^b \quad \rightarrow \quad P > 2^{\frac{b-2}{2}}. \tag{9.92}$$

The reduced distortion offered by the linear interpolation approximation manifests itself as the increased attenuation of the spectral copies of $X(j\omega)$ produced by the $\sin^2(u)/u^2$ characteristic compared to the $\sin(u)/u$ characteristic. Just as predicted by the signal-to-distortion ratio criterion, the peak-distortion criterion shows that performing linear interpolation substantially reduces the number of required subfilters.

The above analysis has performed one "slight of hand" that should be addressed. The distortion was quantified using the intermediate signal. But it is the distortion in the downsampled signal that is of interest. Sampling $\tilde{x}(t)$ produces shifted copies of $\tilde{X}(j\omega)$ replicated at the sampling rate. As a result, the residual copies of $X(j\omega)$ not completely eliminated by $G_1(j\omega)$ or $G_2(j\omega)$ are aliased into the first Nyquist zone. Because different copies may overlap, the peak distortion may be larger than that predicted by the above analysis if the copies add constructively. This does not appear to happen in practice. As a result, the bounds for D_{peak} are useful.

9.4 CoRDiC: COORDINATE ROTATION DIGITAL COMPUTER

Many of the functions required for wireless communications involve multiplications by a pair of quadrature sinusoids. This multiplication can be interpreted as performing a rotation in the Cartesian coordinate system. The DDS, described in Section 9.2.2, is an effective method for performing this function, but it requires either the ability to compute the cosine and sine of an arbitrary angle or the ability to store the cosine and sine at a sufficiently dense sampling of the interval $[0, 2\pi)$. A low complexity method for computing the cosine and sine for an arbitrary angle is desirable when memory limitations prohibit the use of large cosine/sine LUTs.

The Coordinate Rotation Digital Computer (CoRDiC) is a method for rotating a point in the Cartesian coordinate system by using only shifts, additions, and subtractions. Not only can it be used in place of the cosine/sine LUT in a DDS, but it can also replace the cosine/sine multipliers as well. When the algorithm is operated in the reverse direction, it can compute the square root and arctangent functions as well. CoRDiC is an attractive alternative for computing trigonometric functions and for performing polar to rectangular conversions.

9.4.1 Rotations: Moving on a Circle

Rotating the Cartesian point (x_0, y_0) by an angle θ in the positive (counterclockwise) direction is equivalent to multiplying the complex number $\tilde{x}_0 = x_0 + jy_0$ by the complex exponential $e^{j\theta}$. Now, suppose θ can be approximated as the sum of N well-chosen angles:

$$\theta \approx \sum_{n=0}^{N-1} \theta_n. \tag{9.93}$$

The desired rotation may be expressed as

$$\tilde{x}_N = \tilde{x}_0 \exp\left\{j \sum_{n=0}^{N-1} \theta_n\right\} = \tilde{x}_0 \prod_{n=0}^{N-1} \exp\{j\theta_n\}. \tag{9.94}$$

Separating the real and imaginary parts produces the matrix equation

$$\begin{bmatrix} x_N \\ y_N \end{bmatrix} = \begin{bmatrix} \cos\theta_{N-1} & \sin\theta_{N-1} \\ \sin\theta_{N-1} & \cos\theta_{N-1} \end{bmatrix} \cdots \begin{bmatrix} \cos\theta_1 & \sin\theta_1 \\ \sin\theta_1 & \cos\theta_1 \end{bmatrix} \begin{bmatrix} \cos\theta_0 & \sin\theta_0 \\ \sin\theta_0 & \cos\theta_0 \end{bmatrix} \begin{bmatrix} x_0 \\ y_0 \end{bmatrix}. \tag{9.95}$$

Because each matrix is a rotation matrix, this shows that the rotation by θ can be accomplished using a series of rotations by the angles θ_0, θ_1, and so on.

This series of rotations is illustrated in Figure 9.4.1 for $N = 4$. Figure 9.4.1 (a) illustrates the use of these rotations to rotate an initial point (x_0, y_0) by an angle θ. When used in this way, the CoRDiC algorithm is operating in the "forward" direction (or rotation mode). CoRDiC can also operate in the opposite or "reverse" direction (or vector mode) as shown in Figure 9.4.1 (b). The series of rotations is performed until the y-component is zero. This action rotates the initial point (x_0, y_0) to the x-axis. The x-value is the magnitude of the initial

Section 9.4 CoRDiC: Coordinate Rotation Digital Computer

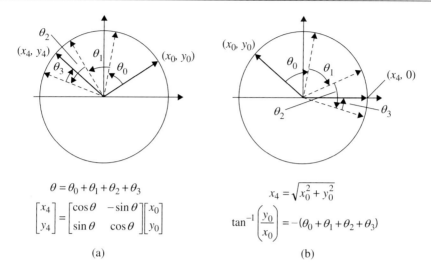

Figure 9.4.1 A graphical representation of the CoRDiC algorithm. (a) Operation in the "forward direction." The final point is a rotation of the initial point by θ. (b) Operation in the "reverse direction." The final point is on the x-axis and is the magnitude of the initial point. The angles required to align the initial point with the x-axis are the negative of the angle of the initial point.

point $\sqrt{x_0^2 + y_0^2}$. The sum of the angles required to achieve this is the negative of the angle of the initial point. This shows that CoRDiC, operating in the reverse direction, can compute the square root and arctangent operations.

The best way to describe the CoRDiC algorithm is as a recursion. Defining the intermediate rotations as

$$\begin{bmatrix} \cos\theta_{N-1} & \sin\theta_{N-1} \\ \sin\theta_{N-1} & \cos\theta_{N-1} \end{bmatrix} \cdots \begin{bmatrix} \cos\theta_1 & \sin\theta_1 \\ \sin\theta_1 & \cos\theta_1 \end{bmatrix} \underbrace{\begin{bmatrix} \cos\theta_0 & \sin\theta_0 \\ \sin\theta_0 & \cos\theta_0 \end{bmatrix} \begin{bmatrix} x_0 \\ y_0 \end{bmatrix}}_{\begin{bmatrix} x_1 & y_1 \end{bmatrix}^T} \quad (9.96)$$

$$\underbrace{}_{\begin{bmatrix} x_2 & y_2 \end{bmatrix}^T}$$

$$\underbrace{}_{\begin{bmatrix} x_N & y_N \end{bmatrix}^T}$$

defines the fundamental CoRDiC recursion for vector rotation:

$$\begin{bmatrix} x_{k+1} \\ y_{k+1} \end{bmatrix} = \begin{bmatrix} \cos\theta_k & -\sin\theta_k \\ \sin\theta_k & \cos\theta_k \end{bmatrix} \begin{bmatrix} x_k \\ y_k \end{bmatrix} \quad \text{for} \quad k = 0, 1, \ldots, N-1. \quad (9.97)$$

Factoring out the cosine produces

$$\begin{bmatrix} x_{k+1} \\ y_{k+1} \end{bmatrix} = \cos\theta_k \begin{bmatrix} 1 & -\tan\theta_k \\ \tan\theta_k & 1 \end{bmatrix} \begin{bmatrix} x_k \\ y_k \end{bmatrix} \quad (9.98)$$

which motivates the definition of the CoRDiC recursion for pseudo vector rotation

$$\begin{bmatrix} x'_{k+1} \\ y'_{k+1} \end{bmatrix} = \begin{bmatrix} 1 & -\tan\theta_k \\ \tan\theta_k & 1 \end{bmatrix} \begin{bmatrix} x'_k \\ y'_k \end{bmatrix}. \tag{9.99}$$

Observe that (9.99) differs from (9.98) only by a constant scale factor. Each step in the recursion requires two multiplications and two additions. The multiplications can be eliminated by choosing the set of angles $\{\theta_0, \theta_1, \ldots, \theta_{N-1}\}$ to be the angles whose tangents are powers of $1/2$, that is,

$$\tan\theta_k = \pm\frac{1}{2^k}. \tag{9.100}$$

The first eight such angles are listed in second column of Table 9.4.1.

The plus and minus signs are used to indicate that the signs of these angles are changed as necessary to direct the sum toward the desired angle. An example using the first eight angles to approximate $50°$ is

$$50° \approx 45.00° + 26.56° - 14.04° - 7.13° - 3.58° + 1.79° + 0.90° + 0.45° = 49.96°.$$

The cumulative angle is plotted in Figure 9.4.2.

This example shows that the difference between the running sum and the desired angle can be used to determine the signs for the angles θ_n. Let $\delta_n \in \{-1, +1\}$ be the sign for θ_n. Then the cumulative sum at the k-th step is $\sum_{n=0}^{k-1} \delta_n \theta_n$. When operating in the forward direction, the condition $\sum_{n=0}^{k-1} \delta_n \theta_n < \theta$ implies that the next rotation should be in the positive (CCW) direction by θ_k (i.e., $\delta_k = +1$). Otherwise, the next rotation should be in the negative (CW) direction by θ_k. This concept is realized by defining the error

$$e_k = \theta - \sum_{n=0}^{k-1} \delta_n \theta_n \tag{9.101}$$

Table 9.4.1 A CoRDiC example: Rotating the point $(1, 0)$ by $50°$

k	θ_k (degrees)	$\sum_{n=0}^{k} \delta_{n-1}\theta_n$ (degrees)	e_k (degrees)	δ_k	x'_k	y'_k
0	45.00	45.00	+5.00	+1	+1.00	+1.00
1	26.56	71.56	−12.57	−1	+0.50	+1.50
2	14.04	57.53	−7.53	−1	+0.88	+1.38
3	7.13	50.50	−0.40	−1	+1.05	+1.27
4	3.58	46.83	+3.17	+1	+1.13	+1.20
5	1.79	48.62	+1.38	+1	+1.09	+1.24
6	0.90	49.51	+0.49	+1	+1.07	+1.25
7	0.45	49.96	+0.04	+1	+1.06	+1.26

Section 9.4 CoRDiC: Coordinate Rotation Digital Computer

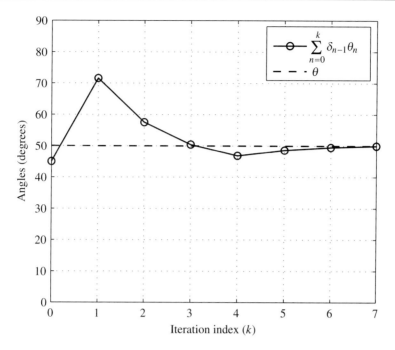

Figure 9.4.2 A plot of cumulative sum/difference of the first eight angles whose arctangents are a power of 1/2 to approximate 50°.

and using the sign of e_k to determine the direction of rotation by θ_k. This is accomplished using $\delta_k = \text{sgn}\{e_k\}$. To use this with the other recursions, a recursive relationship for e_k is needed. The desired recursion is obtained by rewriting the expression for e_{k+1} in terms of e_k:

$$e_{k+1} = \theta - \sum_{n=0}^{k} \delta_n \theta_n = \theta - \sum_{n=0}^{k-1} \delta_n \theta_n - \delta_k \theta_k = e_k - \delta_k \theta_k. \quad (9.102)$$

All the elements for the CoRDiC recursion are now in place. Performing the matrix arithmetic of (9.99) and incorporating the recursion (9.102), the set of recursions that define the CoRDiC algorithm in the forward direction is

$$\delta_k = \text{sgn}\{e_k\}$$
$$x'_{k+1} = x'_k - \delta_k \frac{y'_k}{2^k}$$
$$y'_{k+1} = \delta_k \frac{x'_k}{2^k} + y'_k \quad (9.103)$$
$$e_{k+1} = e_k - \delta_k \theta_k$$

for $k = 0, 1, \ldots, N-1$. The recursion is initialized by setting $x'_0 = x_0$, $y'_0 = y_0$, and $e_0 = \theta$. At the end of the recursion the point (x'_N, y'_N) is obtained. The range of convergence is approximately $-99.88° \le \theta \le 00.88°$.

Repeated application of (9.98) shows that the relationship between the desired point (x_N, y_N) and the point (x'_N, y'_N) is

$$\begin{bmatrix} x_N \\ y_N \end{bmatrix} = K \begin{bmatrix} x'_N \\ y'_N \end{bmatrix} \tag{9.104}$$

where

$$K = \prod_{k=0}^{N-1} \cos \theta_k = \prod_{k=0}^{N-1} \frac{1}{\sqrt{1 + \frac{1}{4^k}}}. \tag{9.105}$$

This constant is used to adjust the magnitude of the result to compensate for the factoring performed in (9.98). Note that the constant is only a function of the number of iterations N. If the number of iterations is fixed, K may be precomputed and stored for use at the end of the iteration.

As an example, consider rotating the point $(1,0)$ by $50°$ in the positive (CCW) direction using $N = 8$ iterations. The values of δ_k, x'_k, y'_k, and e_k using the recursions (9.103) are tabulated in Table 9.4.1. The cumulative angle, tabulated in the third column of Table 9.4.1, is plotted in Figure 9.4.2. The final point (x'_8, y'_8) is scaled by

$$K = \prod_{k=0}^{7} \frac{1}{\sqrt{1 + 4^{-k}}} = 0.6073 \tag{9.106}$$

to produce the desired result $(x_8, y_8) = (0.6433, 0.7656)$. It is easy to confirm that this point has unity magnitude and angle $49.96°$.

The set of recursions that define the CoRDiC algorithm in the reverse direction is

$$\begin{aligned} \delta_k &= -\text{sgn}\{e_k\} \\ x'_{k+1} &= x'_k - \delta_k \frac{y'_k}{2^k} \\ y'_{k+1} &= \delta_k \frac{x'_k}{2^k} + y'_k \\ e_{k+1} &= e_k - \delta_k \theta_k \end{aligned} \tag{9.107}$$

for $k = 0, 1, \ldots, N-1$. The recursion is initialized by setting $x'_0 = x_0$, $y'_0 = y_0$, and $e_0 = 0$. At the end of the recursion, the point $(x'_N, 0)$ is obtained. The desired results are

$$\begin{aligned} \sqrt{x_0^2 + y_0^2} &= K x'_N \\ \tan^{-1}\left(\frac{y_0}{x_0}\right) &= -\sum_{n=0}^{N-1} \delta_n \theta_n. \end{aligned} \tag{9.108}$$

Section 9.4 CoRDiC: Coordinate Rotation Digital Computer

The recursive formulations (9.103) and (9.107) are useful for implementation in a programmable processor. For hardware realizations, the recursive formulations may also be used to form a serial processor, such as the one illustrated in Figure 9.4.3 for the forward direction. The serial processor consists of a unit to perform the pseudo rotation (9.107) together with a unit to track the error signal. The block diagram shows two multipliers and a tangent block. Note that these are not necessary because the tangent of θ_k is known in advance (it is 2^{-k}) and the multipliers may be replaced by right-shifts (shifting a binary word to the right by k positions is equivalent to multiplying by 2^{-k}), a pipelined approach such as the one illustrated in Figure 9.4.4. Again, the tangent operation and multipliers are not necessary.

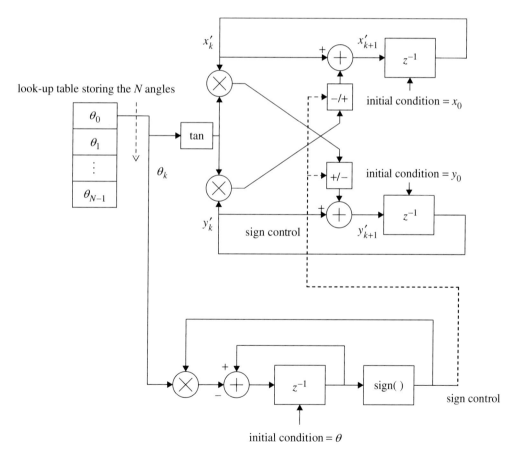

Figure 9.4.3 A serial realization of the forward recursion (9.103).

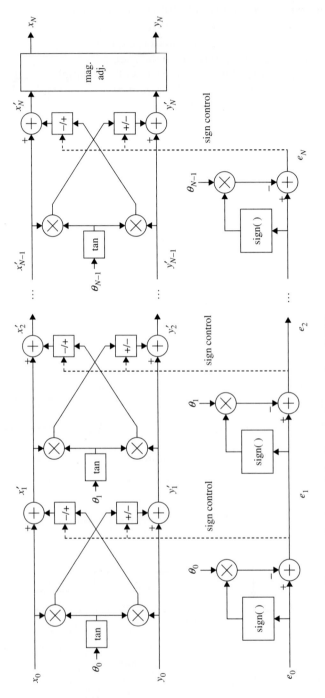

Figure 9.4.4 A pipelined realization of the forward recursion (9.103).

9.4.2 Moving Along Other Shapes

The CoRDiC recursions (9.103) and (9.107) translate a point from an initial position to a final position using a series of moves that require only shifts and additions. The savings in computational complexity result from choosing translations that require multiplications by powers of $1/2$. This notion has been extended to translations along contours other than the circle. The most straightforward extension involves translations along the vertical line $y = x_0$, such as those illustrated in Figure 9.4.5.

Figure 9.4.5 (a) illustrates the operation of CoRDiC in the forward direction. When operating in the forward direction, the algorithm moves an initial point on the x-axis, $(x_0, 0)$ to the point (x_0, y_4) using a series of steps that are scaled versions of x_0. The final ordinate y_4 is a multiple of x_0 as shown. Thus, this algorithm can be used to perform multiplications using a series of shifts and additions.

The more interesting case corresponds to operation in the reverse direction as illustrated in Figure 9.4.5 (b). Again, translations are along the line $y = x_0$. The starting point is (x_0, y_0) and the ending point is $(x_0, 0)$. Each translation along the line $y = x_0$ is designed to move the point to the x-axis. The sum of these translations is the constant of proportionality between x_0 and y_0. Thus, this algorithm performs division using a series of shifts and additions.

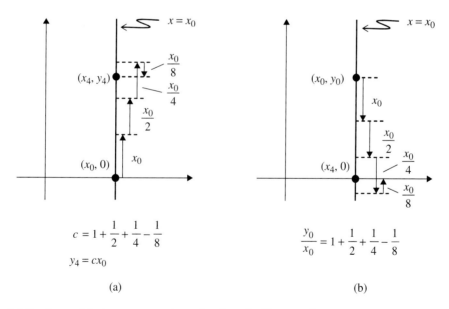

Figure 9.4.5 A graphical representation of a CoRDiC-like translation along the line $y = x_0$. (a) Operation in the forward direction produces the point (x_4, y_4) where y_4 is a multiple of x_0. This algorithm is used to perform multiplication. (b) Operation in the reverse direction produces the point $(x_4, 0)$. The translations required to achieve this form the constant of proportionality between y_0 and x_0. This algorithm is used to perform division.

The development of the recursions that define translations along the line $y = x_0$ follows that of the previous section. In the forward direction, the ultimate goal is to move from the initial point (x_0, y_0) to the point (x, y). Because the translation is along the line $y = x_0, x = x_0$ and y may be expressed as $y = cx_0$. The matrix equation that defines this relationship is

$$\begin{bmatrix} x \\ y \end{bmatrix} = \begin{bmatrix} 1 & 0 \\ c & 0 \end{bmatrix} \begin{bmatrix} x_0 \\ y_0 \end{bmatrix}. \tag{9.109}$$

Now suppose the constant c may be expressed as sums and differences of powers of $1/2$:

$$c \approx \sum_{n=0}^{N-1} \delta_n 2^{-n} \tag{9.110}$$

where $\delta_n \in \{-1, +1\}$ indicates whether 2^{-n} should be added or subtracted. The final point y may be expressed as

$$y = cx_0 \approx x_0 \sum_{n=0}^{N-1} \delta_n 2^{-n}. \tag{9.111}$$

A recursion may be defined by expanding the sum as follows:

$$\underbrace{\underbrace{\underbrace{x_0 + \delta_1 2^{-1} x_0}_{y_1} + \delta_2 2^{-2} x_0}_{y_2} + \cdots + \delta_{N-1} 2^{-(N-1)} x_0}_{y_N}. \tag{9.112}$$

This provides the recursions for x_k and y_k:

$$\begin{aligned} x_{k+1} &= x_k \\ y_{k+1} &= y_k + \delta_k 2^{-k} x_k \end{aligned} \tag{9.113}$$

for $k = 0, 1, \ldots, N-1$.

The sign for each power of $1/2$ is determined by the error of the cumulative sum. The error for step k is

$$e_k = c - \sum_{n=0}^{k-1} \delta_n 2^{-n} \tag{9.114}$$

and the recursion for e_{k+1} is

$$e_{k+1} = c - \sum_{n=0}^{k} \delta_n 2^{-n} = c - \sum_{n=0}^{k-1} \delta_n 2^{-n} - \delta_k 2^{-k} = e_k - \delta_k 2^{-k} \tag{9.115}$$

for $k = 0, 1, \ldots, N-1$. The sign indicator δ_k is the sign of the error e_k.

Section 9.4 CoRDiC: Coordinate Rotation Digital Computer

Putting this all together, the recursion for the forward direction is

$$\delta_k = \operatorname{sgn}\{e_k\}$$
$$x_{k+1} = x_k$$
$$y_{k+1} = y_k + \delta_k 2^{-k} x_k \quad (9.116)$$
$$e_{k+1} = e_k - \delta_k 2^{-k}$$

for $k = 0, 1, \ldots, N-1$. The recursion is initialized by setting x_0 to one of the multiplicands, $y_0 = 0$, and e_0 to the other multiplicand. The recursion produces $y_N = y_0 + e_0 \times x_0 = e_0 \times x_0$. The range of convergence is $-2 < e_0 < 2$.

To perform division, the recursion in the reverse direction is used:

$$\delta_k = -\operatorname{sgn}\{e_k\}$$
$$x_{k+1} = x_k$$
$$y_{k+1} = y_k + \delta_k 2^{-k} x_k \quad (9.117)$$
$$e_{k+1} = e_k - \delta_k 2^{-k}$$

for $k = 0, 1, \ldots, N-1$. The recursion is initialized by setting x_0 to the divisor, y_0 to the dividend, and $e_0 = 0$. When completed, the recursion produces $e_N = e_0 + y_0/x_0 = y_0/x_0$. The range of convergence is $-2 < x_0/y_0 < 2$.

An example of the forward recursion computing the product $0.8 \times 0.2 = 0.16$ using $N = 10$ iterations is summarized in Table 9.4.2.

Table 9.4.2 An example using CoRDiC-like translations to perform multiplication in $N = 10$ iterations.

k	δ_k	x_k	y_k	e_k
0	+1	0.8	0.0000	+0.20000
1	−1	0.8	0.8000	−0.80000
2	−1	0.8	0.4000	−0.30000
3	−1	0.8	0.2000	−0.05000
4	+1	0.8	0.1000	+0.07500
5	+1	0.8	0.1500	+0.01250
6	−1	0.8	0.1750	−0.01875
7	−1	0.8	0.1625	−0.00313
8	+1	0.8	0.1563	+0.00469
9	+1	0.8	0.1594	+0.00078
10	−	0.8	0.1609	−0.00117

9.5 AUTOMATIC GAIN CONTROL

Signal level is important in discrete-time processing because many of the detection functions that need to be performed require precise knowledge of the signal amplitude. Examples include the following:

- The decision regions for constellations such as square and cross M-QAM (for $M > 4$), M-APSK, and CCITT V.29 are a function of the average symbol energy of the received signal. The average symbol energy is, in turn, a function of the average signal level. An erroneous assumption about the received signal level produces incorrect decision-region boundaries. The result is a high symbol error rate, even when no noise is present.
- In synchronizers based on phase locked loops, the gains (the K_p's) of the phase error detectors (7.21)–(7.22), (7.40)–(7.41), (7.48)–(7.49), and (7.45)–(7.46) described in Chapter 7 and the timing error detectors (8.26)–(8.27), (8.33)–(8.34), (8.36)–(8.37), and (8.48)–(8.49) described in Chapter 8 are dependent on the average symbol energy (and, hence, on the received signal level). The loop bandwidth is set by the gains K_1 and K_2 of the loop filter. Because K_1 and K_2 are a function of K_p, an erroneous assumption about the received signal level (which leads to an erroneous assumption for K_p) produces incorrect values for K_1 and K_2. This, in turn, produces a loop with a loop bandwidth different from the desired loop bandwidth.

If the average level of the received signal is not known in advance, then the received signal level must be estimated and used to either dynamically adjust the decision region boundaries and loop filter constants or to dynamically scale the amplitude of the input signal. In general, the second option is easier to accomplish and is the most widely used in practice.

In most cases, the exact signal level is not known in advance and will vary over time. Let the sampled band-pass signal be

$$r(nT) = I_r(nT) \cos(\Omega_0 n) - Q_r(nT) \sin(\Omega_0 n) \quad (9.118)$$

which can be written as

$$= A_r(nT) \cos(\Omega_0 n + \theta_r(nT)) \quad (9.119)$$

where

$$A_r(nT) = \sqrt{I_r^2(nT) + Q_r^2(nT)} \quad (9.120)$$

$$\theta_r(nT) = \tan^{-1}\left\{\frac{Q_r(nT)}{I_r(nT)}\right\}. \quad (9.121)$$

To track variations in $A_r(nT)$ due to changes in the channel attenuation, automatic gain control (AGC) is most often based on a feedback algorithm that tracks variations in $A_r(nT)$ with respect to a reference level R. The amplitude error is averaged to reduce the effects of noise and to average over symbols with different energies (such as with square 16-QAM).

Section 9.5 Automatic Gain Control

When operating properly, the AGC loop forces the amplitude error to zero thereby producing a signal with the desired amplitude.

There are several options for AGC placement. One example is illustrated in Figure 9.5.1 (a) where the AGC operates on the sampled band-pass signal. The closed-loop processing is illustrated in Figure 9.5.1 (b). The amplitude error $e_A(n)$, given by

$$e_A(n) = R - A(n)A_r(nT) \tag{9.122}$$

is scaled and averaged. The constant R is the reference level and is the amplitude to which the average amplitude of the band-pass signal will be scaled when AGC converges. The scaled and averaged error signal is used to update the AGC gain $A(n)$. A potential difficulty with this approach is the process for extracting $A_r(nT)$ from $r(nT)$. This is explored in Exercise 9.26. Another alternative is to place the AGC after the matched filters as illustrated in Figure 9.5.2 (a). The corresponding closed-loop system is shown in Figure 9.5.2 (b). The AGC processor

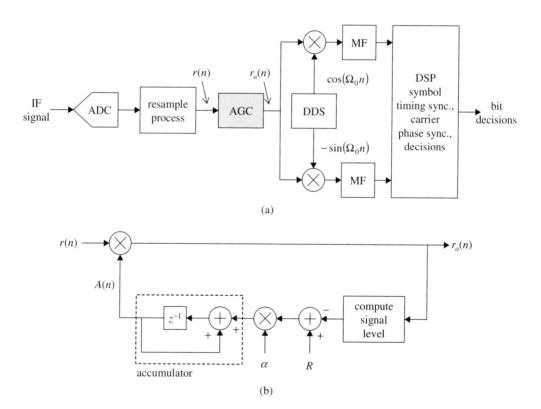

Figure 9.5.1 AGC operating on the band-pass signal.

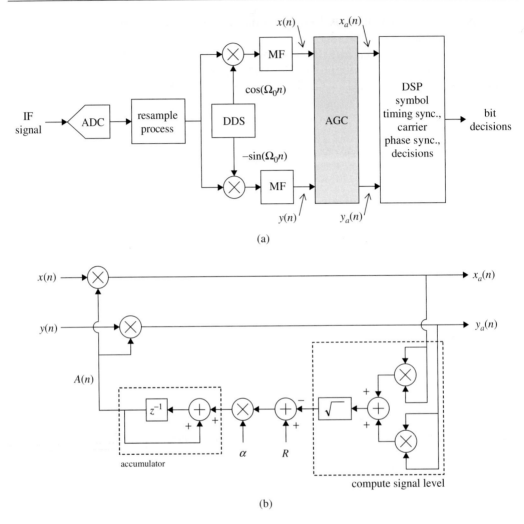

Figure 9.5.2 AGC operating on the baseband signal.

operates on the quadrature matched filter outputs $x(nT)$ and $y(nT)$ to compute the amplitude after the matched filter outputs using

$$A_{\mathrm{MF}}(nT) = \sqrt{x^2(nT) + y^2(nT)}. \tag{9.123}$$

(Note that due to the amplitude scaling performed by the matched filters, the amplitude $A_{\mathrm{MF}}(nT)$ will be different than the amplitude of the band-pass signal $A_r(nT)$.) Other possibilities for AGC placement exist. In general, the AGC is placed either at band-pass or baseband. When operating on the band-pass signal, the block diagrams of Figure 9.5.1

Section 9.5 Automatic Gain Control

apply to AGC placement at any point in the resampling process where $r(nT)$ is still a bandpass signal. When operating at baseband, the block diagrams of Figure 9.5.2 apply to AGC placement at any point where the low-pass quadrature components of $r(nT)$ (or filtered versions of them) are available. The only constraints on AGC placement are the clock rate at which the algorithm is to operate and that the scaling should occur before the symbol timing and carrier phase synchronizing PLLs.

Close examination of the closed-loop processing of Figures 9.5.1 (b) and 9.5.2 (b) show that this portion of the processing is the same, except for the nature of the signal amplitude estimator. The amplitude update equation is

$$A(n+1) = A(n) + \alpha \left[R - |A(n)A_r(nT)| \right] = A(n) \left[1 - \alpha |A_r(nT)| \right] + \alpha R. \tag{9.124}$$

The amplitude estimator is a nonlinear operation and as such makes (9.124) a nonlinear equation. Insight into the behavior of this algorithm is gained by examining the case where $A_r(nT)$ is a step. Suppose $A_r(nT) = cu(n)$ for a constant $c > 0$. Then

$$A(n+1) = A(n) \left[1 - \alpha c \right] + \alpha R, \quad n \geq 0 \tag{9.125}$$

which is a linear difference equation that can be solved for $A(n)$:

$$A(n) = \frac{R}{c} \left[1 - (1 - \alpha c)^n \right] u(n). \tag{9.126}$$

The steady-state value for $A(n)$ is R/c, which is the desired value. Because this is a first-order system, the transient behavior is characterized by the time constant n_0, which is the value of n for which $A(n)$ achieves $(1 - e^{-1})$ times its steady-state value[1]:

$$A(n_0) = \left(1 - \frac{1}{e}\right) \frac{R}{c}. \tag{9.127}$$

Substituting (9.126) for $A(n)$ in (9.127) and solving for n_0 yields (see Exercise 9.23)

$$n_0 \approx \frac{1}{\alpha c}. \tag{9.128}$$

The time constant is a function of both α and the input signal level c. As a consequence, convergence time of the AGC is long when the signal level is low, and short when the signal level is long.

---

[1] The notion of a time constant has its origins in first-order continuous-time systems. For such a system, the transient portion of the step response is of the form $e^{-t/\tau}$ where τ is the time constant. When $t = \tau$, the transient portion of the step response has decayed to e^{-1} times its initial value. Or, put another way, the step response has achieved $(1 - e^{-1})$ times its steady-state value. This latter definition is carried over to the discrete-time case even though the transient portion of the step response is not usually expressed as e^{-n/n_0}.

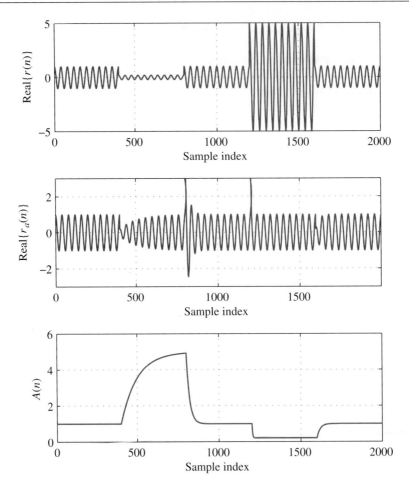

Figure 9.5.3 An example of the linear AGC loop performance operating on $r(n) = M(n)e^{j\Omega_0 n}$: (a) The real part of $r(n)$; (b) the real part of $r_a(n) = A(n)r(n)$; (c) the AGC output $A(n)$.

This behavior is illustrated by the simple example shown in Figure 9.5.3. In this case, the input signal is $r(n) = A_r(n)e^{j\Omega_0 n}$ where $\Omega_0 = \pi/20$ and

$$A_r(n) = \begin{cases} 1 & 0 \leq n < 400 \\ 1/5 & 400 \leq n < 800 \\ 1 & 800 \leq n < 1200 \\ 5 & 1200 \leq n < 1600 \\ 1 & 1600 \leq n < 2000 \end{cases}.$$

as illustrated by the plot of Re $\{r(n)\}$ in Figure 9.5.3 (a). A plot of Re $\{r_a(n)\}$ for $R = 1$ and $\alpha = 0.05$ is shown in Figure 9.5.3 (b) and the corresponding $A(n)$ is plotted in Figure 9.5.3 (c).

The plots in Figures 9.5.3 (b) and (c) for $400 \leq n < 800$ (where $A_r(n) = 1/5$) show that the time constant is very large. The AGC is just reaching the desired steady-state value of 5 at the end of the interval. Contrast this characteristic with the behavior for $1200 \leq n < 1600$ (where $A_r(n) = 5$). The time constant is very short so that the AGC reaches its desired steady-state value of 1/5 almost instantaneously.

This behavior is undesirable in a real system where the signal processing following the AGC must wait until the AGC has converged to function properly. This difficulty is overcome by operating the AGC on the logarithm of the signal level using the close-loop system illustrated in Figure 9.5.4 for the case where the AGC operates at baseband. (The application of this closed-loop algorithm to the band-pass signal is straightforward.) In this case, the amplitude update equation is

$$\log\left(A(n+1)\right) = \log\left(A(n)\right) + \alpha \left[\log(R) - \log\left(|A(n)A_r(nT)|\right)\right]$$
$$= \log\left(A(n)\right)[1-\alpha] - \alpha \log\left(\frac{|A_r(nT)|}{R}\right) \quad (9.129)$$

which is a nonlinear equation in $\log(A(n))$. Following the same steps as before, the update equation is examined for the case $A_r(n) = cu(n)$ for a constant $c > 0$. In this case,

$$\log\left(A(n+1)\right) = \log\left(A(n)\right)[1-\alpha] - \alpha \log\left(\frac{c}{R}\right) \quad n \geq 0 \quad (9.130)$$

which is a linear difference equation for $\log\{A(n)\}$. The solution is (see Exercise 9.24)

$$\log\left(A(n)\right) = -\log\left(\frac{c}{R}\right)\left[1 - (1-\alpha)^n\right]u(n). \quad (9.131)$$

The steady-state value for $\log(A(n))$ is $\log(R/c)$, which is the desired value. The time constant is obtained by solving

$$\log\left(A(n_0)\right) = -\left(1 - \frac{1}{e}\right)\log\left(\frac{c}{R}\right) \quad (9.132)$$

for n_0. Following the steps outlined in Exercise 9.25, the time constant is

$$n_0 \approx \frac{1}{\alpha} \quad (9.133)$$

which is independent of the signal amplitude c. The performance of the log-domain AGC on the same complex exponential signal described above is illustrated in Figure 9.5.5. The plot of $A(n)$ in Figure 9.5.5 (c) makes it clear that the time constant is the same for all changes in the signal amplitude.

9.6 NOTES AND REFERENCES

9.6.1 Topics Covered

The development of the quantization effects for ADCs and DACs presented here was strongly influenced by the development in Ref. [41]. Continuous-time techniques for DAC distortion

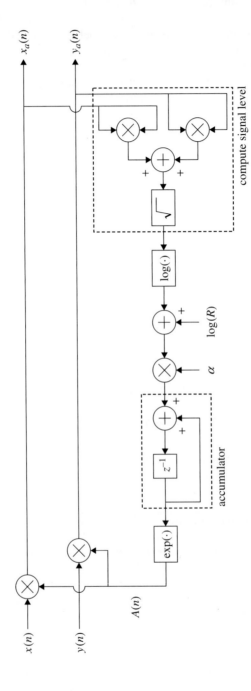

Figure 9.5.4 Log-based AGC operating at baseband.

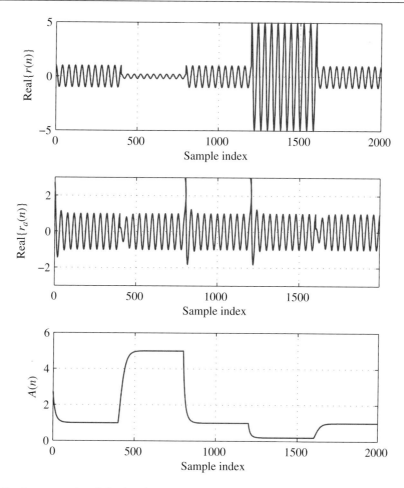

Figure 9.5.5 An example of the log-based AGC loop performance operating on $r(n) = M(n)e^{j\Omega_0 n}$: (a) The real part of $r(n)$; (b) the real part of $r_a(n) = A(n)r(n)$; (c) the AGC output $A(n)$.

compensation have been investigated by Bermudez, Filho, and Seara in Refs. [187–190]. The discrete-time techniques for DAC distortion compensation presented in above were based on the Parks–McClellan equiripple FIR filter design algorithm. The Parks–McClellan algorithm is described in textbooks on discrete-time signal processing such as Ref. [41]. The original paper on the algorithm is Ref. [61]. On the basis of these principles, Samueli developed a length-11 filter that does not require any multiplications [191–193] and was successfully used in the digital modulator described in Ref. [194]. A length-3 predistortion filter, based on the principles outlined in Exercise 9.10, was developed by Henriques and Franca [195]. Predistortion filter design based on the Taylor series approximation of $[\sin(M\Omega/2)/\sin(\Omega/2)]^K$ is described by Harris in Chapter 13 of Ref. [53]. This technique is developed in the context of a compensation for the CIC filter distortion but can be easily applied to develop predistortion filters to compensate for $\sin(\Omega/2)/(\Omega/2)$ DAC distortion. Predistortion filter design for discrete-time to continuous-time conversion of band-pass signals is described by Harris in Ref. [185].

The DDS architecture based on periodically overflowing two's complement phase accumulator coupled with a cosine/sine LUT was first presented by Tierney, Rader, and Gold [196]. Reduction of the cosine/sine LUT using the symmetries of the cosine and sine functions was first presented by Sunderland et al. [197] and thoroughly examined by Nicholas, Samueli, and Kim [186]. The spectral analysis of a real DDS output was first performed by Mehrgardt [198] and later generalized by Nicholas and Samueli [199] who investigated the impact of phase truncation, and Kroupa [200] who extended the analysis to include quantization effects of the DAC. Garvey and Babitch [201] presented an exact spectral analysis of the spectral spurs and provided experimental measurements of the largest spurs as a function of the number of bits used to address the cosine/sine LUT and the number of DAC bits used and confirmed the approximations developed in [199]. Dithering the phase at the cosine/sine LUT input to reduce the amplitude of the spurs was introduced by Jasper [202] and examined by Flanagan and Zimmerman [203]. The error feedforward technique and the application of linear prediction to this problem were presented by Harris and Knight [204].

The use of CIC filters for resampling was first described by Hogenauer [205]. The effects of finite precision arithmetic were analyzed by Hogenauer [205] and Chu and Burrus [206]. A more detailed explanation of the Hogenauer filter can be found in Chapter 11 of Ref. [53]. Half-band filters are described in many DSP texts such as Refs. [54], [55]. A particularly nice tutorial is in Chapter 8 of Harris [53]. The use of the Parks–McClellan algorithm for designing half-band filters is described by Vaidyanathan and Ngyuen [207] and Harris [53]. The polyphase filterbank resampler is described in many textbooks on discrete-time signal processing and multirate processing such as Refs. [53–55,208]. Only nearest neighbor and linear interpolator approximations were discussed. Higher order interpolation can also be used as discussed by Ramstad [209]. The higher order interpolators are more accurate, thus requiring fewer subfilters, but require more computations to produce each output sample. The signal-to-distortion power ratio analysis followed closely the analysis presented in Refs. [55,208]. The signal-to-peak-distortion amplitude ratio analysis followed that of Harris in Ref. [53]. An alternate derivation is summarized by Ramstad [209].

The CoRDiC algorithm was first published by Volder [210] as a method for performing rectangular-to-polar and polar-to-rectangular conversions. CoRDic and CoRDiC-like algorithms were presented in a unified context by Walther [211] and later by Ahmed, Delosme, and Morf [212]. Convergence of CoRDiC was proved by Muller [213]. CoRDiC is not only popular in signal processing for communications, but is also used in handheld calculators [214]. An intriguing new algorithm, called the "BKM" algorithm, published by Bajard, Kla, and Muller [215], generalizes the CoRDiC algorithm in a way that eliminates the need for the scale factor K for the trigonometric functions.

Automatic gain control using continuous-time processing has been investigated by Oliver [216], Victor and Brockman [217], Banta [218], Simpson and Tranter [219], and Ohlson [220] with a nice overview by Mercy [221]. An analysis of AGC in the presence of noise was carried by Schachter and Bergstein [222] and summarized in Chapter 7 of Meyr [223]. The discrete-time version of automatic gain control is analyzed by Morgan [224] and in Chapters 6 and 7 of Frerking [5]. The algorithm presented in this chapter is independent of the modulation (i.e. it is a "blind") algorithm. A decision-directed version of AGC for linear modulations was examined by Weber [225].

9.6.2 Topics Not Covered

The Farrow filter structure, introduced by Farrow [226], is another important class of resampling filter. The Farrow structure was introduced in Chapter 8 in the context of interpolation filters. A more general form provides an efficient means for performing resampling, especially when the sample rate is increased. Because this chapter focused primarily on demodulators that use resampling filters that reduce the sample rate, the Farrow filter was not covered. A nice treatment of the application of Farrow filters to QAM modulators is offered by Harris in Ref. [53].

The resampling filters presented in this chapter were all based on FIR filters. IIR filters have also been investigated as resampling filters by Harris, d'Oreye de Lantremange, and Constantinides [227] and Harris, Gurantz, and Tzukerman [228].

9.7 EXERCISES

9.1 Let $x(t) = X \cos(\omega_0 t + \theta)$.
   (a) Compute X_{peak} using
   $$X_{\text{peak}} = \max_t \{|x(t)|\}.$$
   (b) Compute σ_x using
   $$\sigma_x^2 = \frac{1}{T_0} \int_{T_0} |x(t)|^2 dt$$
   where $T_0 = 2\pi/\omega_0$.
   (c) Compute the peak-to-average ratio
   $$20 \log_{10} \left(\frac{X_{\text{peak}}}{\sigma_x} \right).$$

9.2 Let $x(t) = X \exp\{j\omega_0 t\}$.
   (a) Compute X_{peak} using
   $$X_{\text{peak}} = \max_t \{|x(t)|\}.$$
   (b) Compute σ_x using
   $$\sigma_x^2 = \frac{1}{T_0} \int_{T_0} |x(t)|^2 dt$$
   where $T_0 = 2\pi/\omega_0$.

(c) Compute the peak-to-average ratio

$$20 \log_{10} \left(\frac{X_{\text{peak}}}{\sigma_x} \right).$$

9.3 Consider an audio signal that is sampled at 48 ksamples/s and is to be quantized using a uniform quantizer with b bits. The goal is to determine the number of bits necessary to achieve a signal-to-quantization noise ratio in the range 90–96 dB as required for CD-quality audio. Assume the probability density function is Gaussian with zero mean and variance σ_x^2. (It really is not, but we will use the assumption anyway.) The peak amplitude is infinite, but this value occurs with probability 0.

(a) Assume $X_{\text{peak}} = 4\sigma_x$.
  i. What is the probability that $|x(nT)|$ exceeds X_{peak}?
  ii. Assuming full-scale operation, derive the expression for the signal-to-quantization-noise ratio as a function of b.
  iii. Plot SNR_q as a function of b. What value of b gives a signal-to-quantization-noise ratio in the range 90–96 dB?

(b) Assume $X_{\text{peak}} = 3\sigma_x$.
  i. What is the probability that $|x(nT)|$ exceeds X_{peak}?
  ii. Assuming full-scale operation, derive the expression for the signal-to-quantization-noise ratio as a function of b.
  iii. Plot SNR_q as a function of b. What value of b gives a signal-to-quantization-noise ratio in the range 90–96 dB?

(c) Determine the value of X_{peak} such that the probability that $|x(nT)|$ exceeds X_{peak} is less than 0.1%.
  i. Assuming full-scale operation, derive the expression for the signal-to-quantization-noise ratio as a function of b.
  ii. Plot SNR_q as a function of b. What value of b gives a signal-to-quantization-noise ratio in the range 90–96 dB?

9.4 Consider the six cases for quantizing a complex exponential illustrated in Figure 9.1.6. Use (9.14) to compute the signal-to-quantization-noise ratio SNR_q for each case and compare your answer with the measured value indicated in the plot. Are the answers the same or different? If different, explain why.

9.5 Consider a QPSK signal using the SRRC pulse shape with 50% excess bandwidth centered at ω_0 rad/s whose RMS amplitude is 10 dBm. This signal is to be sampled by an ADC operating at 100 Msamples/s using a clock whose jitter is 4 ps RMS.

(a) The dynamic range of the ADC is ± 3.5 V. An amplifier is inserted before the ADC to achieve full-scale operation without clipping. What is the gain of this amplifier?

(b) What is the maximum achievable signal-to-quantization-noise ratio for this system? Assume full-scale operation.

9.6 Repeat Exercise 9.5, except use 8-PSK in place of QPSK.

9.7 Repeat Exercise 9.5, except use 16-QAM in place of QPSK.

Section 9.7 Exercises

9.8 Repeat Exercise 9.5, except use 16-ary CCITT V.29 in place of QPSK.

9.9 Repeat Exercise 9.5, except use 256-QAM in place of QPSK.

9.10 This exercise explores the use of a length-3 linear phase FIR filter as a predistortion filter to compensate for DAC distortion. The filter coefficients are

$$h_{PD}(n) = \begin{cases} -\dfrac{\alpha}{2} & n = -1 \\ 1 + \alpha & n = 0 \\ -\dfrac{\alpha}{2} & n = +1 \\ 0 & \text{otherwise} \end{cases}.$$

(a) Compute $H_{PD}(e^{j\Omega})$, the DTFT of $h_{PD}(n)$.
(b) Plot $|H_{PD}(e^{j\Omega})|$ and $|G_d(e^{j\Omega})|$ on the same set of axes.
(c) Plot $|H_{PD}(e^{j\Omega})| |G_d(e^{j\Omega})|$.
(d) On the basis of your results in part (c), what value of α is best? What is the usable bandwidth of this predistortion filter? Compare this filter with the longer filters described in Figure 9.1.13.

9.11 The impulse response of the discrete-time system of Figure 9.2.1 is $h(n) = \cos(\Omega_0 n) u(n)$. What is the impulse response of the discrete-time system shown below? Does it suffer from the same quantization effects demonstrated in Figure 9.2.2? Why or why not?

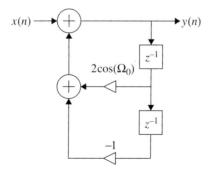

9.12 This exercise explores an interesting connection between the discrete-time oscillator of Figure 9.2.4 and a commonly used continuous-time oscillator structure.
   (a) Consider the linear, continuous-time system shown below. Determine the impulse response of this system.

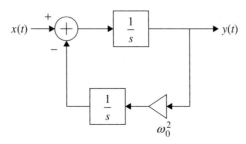

(b) Show that the discrete-time oscillator of Figure 9.2.4 can also be drawn as the system shown below.

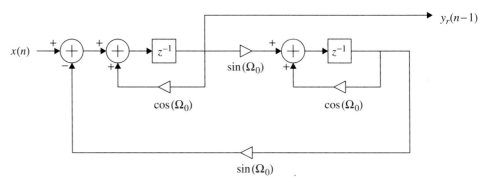

(c) Comment on the similarities between the oscillators in part (a) and part (b).

9.13 Sketch a block diagram, similar to that of Figure 9.2.14, of a DDS that exploits the quarter-wave symmetries of the cosine to reduce the size of the cosine LUT by 4.

9.14 Consider a LUT with W address bits that stores samples of $\cos(\theta)$ for 2^W evenly spaced samples of θ in the interval $-\pi \leq \theta < \pi$. Assume two's complement representation. Suppose the LUT is a memory device capable of producing two outputs (in parallel) for a given address. The address of the first output is the input address and the address of the second output is the input address plus a fixed offset. Devise an addressing scheme to produce cosine and sine in parallel from such a memory device.

9.15 Repeat Exercise 9.14, except to produce cosine and $-$sine terms in parallel.

9.16 Most of the detectors outlined in Chapters 5 and 7 relied on a DDS that produced quadrature sinusoids of the form $\cos(\Omega_0 n)$ and $-\sin(\Omega_0 n)$. This exercise explores the use of feedforward error correction (to compensate for phase accumulator truncation) with DDSs of this form.
  (a) Derive the equation, similar to (9.55), that defines the relationship between $e^{-j\theta(n)}$ and $\hat{\theta}(n), \delta\theta(n)$.
  (b) Sketch a block diagram, similar to that of Figure 9.2.16 (a), that realizes feedforward error correction.

9.17 Sketch a series of four block diagrams, similar to those of Figure 9.3.4, for an upsampling Hogenauer filter.

9.18 Show that if the spectral raised-cosine pulse shape described by (A.27) is thought of as an impulse response, the corresponding filter is half-band filter.

9.19 Let θ_k be an angle whose tangent is 2^{-k}. Show that

$$\cos \theta_k = \frac{1}{\sqrt{1 + 4^{-k}}}.$$

Hint: construct the triangle

and apply the definitions of tangent and cosine.

9.20 Construct a table, similar to that of Table 9.4.1, for CoRDiC operating in the reverse direction with the initial point $(3, 4)$. Use eight iterations. In the reverse direction, CoRDiC performs a rectangular to polar conversion. Is your answer what you expect?

9.21 Construct a table, similar to that of Table 9.4.2, for the CoRDiC-like algorithm that performs translations along a line in the reverse direction. The desire is to compute the quotient $1/1.5$. Is your answer what you expect?

9.22 This exercise explores the derivation of (9.126) from (9.125).
(a) Show that the z-transform of (9.125) is

$$z\mathcal{A}(z) = \mathcal{A}(z)[1 - \alpha c] + \frac{\alpha R}{1 - z^{-1}}$$

where $\mathcal{A}(z)$ is the z-transform of $A(n)$.
(b) Solve the result of part (a) for $\mathcal{A}(z)$.
(c) Compute the inverse z-transform of the answer in part (b). Your answer should be in terms of $u(n - 1)$. Evaluate your answer at $n = 0$ and use the result to express $A(n)$ in terms of $u(n)$.

9.23 This exercise explores the development of the time constant given by (9.127) for the linear AGC.
(a) Show that substituting (9.126) for $A(n)$ in (9.127) and simplifying produces

$$n_0 \ln(1 - \alpha c) = -1.$$

(b) The power series representation for $\ln(1 + x)$ is

$$\ln(1 + x) = x - \frac{x^2}{2} + \frac{x^3}{3} - \frac{x^4}{4} + \cdots, \quad \text{for } |x| < 1.$$

Assuming $\alpha c \ll 1$, the power series can be truncated to retain only the first term. Derive the expression for n_0 using this approximation.

9.24 This exercise explores the derivation the log-based AGC amplitude given by (9.131).

(a) Using
$$l(n) = \log\left(A(n)\right),$$
show that (9.130) may be written as
$$l(n+1) = l(n)\left[1-\alpha\right] - \alpha \log\left(\frac{c}{R}\right).$$

(b) Show that the z-transform of the result in part (a) may be expressed as
$$L(z) = -\log\left(\frac{c}{R}\right) \frac{\alpha z}{\left(z - (1-\alpha)\right)\left(z - 1\right)}$$
where $L(z)$ is the z-transform of $l(n)$.

(c) Show that the inverse z-transform of the result in part (b) is
$$l(n) = -\log\left(\frac{c}{R}\right)\left[u(n-1) - (1-\alpha)(1-\alpha)^{n-1} u(n-1)\right].$$
By evaluating this expression at $n = 0$, show that an equivalent form for this expression is
$$l(n) = -\log\left(\frac{c}{R}\right)\left[1 - (1-\alpha)^n\right] u(n).$$

9.25 This exercise explores the derivation of the time constant for the log-based AGC.

(a) Using the result of Exercise 9.24 (c), show that
$$l(n_0) = -\left(1 - \frac{1}{e}\right)$$
simplifies to
$$n_0 \ln(1-\alpha) = -1.$$

(b) Using the power series representation for $\ln(1+x)$ described in Exercise 9.23, show that
$$n_0 \approx \frac{1}{\alpha}.$$

9.26 This exercise explores the development of a discrete-time envelope detector for use with the AGC for a discrete-time band-pass signal of the form (9.119).

(a) Show that the band-pass signal (9.119) may be expressed in the form
$$x(n) = I_c(nT) e^{j\Omega_0 n} + j I_c^*(nT) e^{-j\Omega_0 n}.$$
Derive an expression for $I_c(nT)$ in terms of $A_r(nT)$ and $\theta_r(nT)$.

(b) A band-pass filter, such as the one shown below, may be used to isolate the first term in the alternate form above.

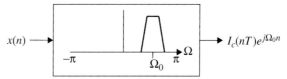

Describe the properties of this band-pass filter.

(c) Show that
$$A_r(nT) = |I_c(nT)e^{j\Omega_0 n}|.$$

(d) Sketch a block diagram of the "compute signal level" block for use in an AGC operating on a real-valued band-pass discrete-time signal.

9.27 The "compute signal level" subsystem described in the previous exercise is a special case of the Hilbert transform. The Hilbert transform is described by a filter whose frequency response is

$$H\left(e^{j\Omega}\right) = \begin{cases} -j & 0 < \Omega < \pi \\ j & -\pi < \Omega < 0 \end{cases}.$$

(a) Derive the output of the Hilbert transform filter when the band-pass signal (9.119) is the input.

(b) Show how to use the answer in part (a), together with the band-pass signal (9.119), to produce signals from which the signal level can be calculated.

(c) Sketch a block diagram of this "compute signal level" subsystem.

(d) Show that the subsystem described in Exercise 9.26 is a special case of this subsystem.

# 10

System Design

10.1 ADVANCED DISCRETE-TIME ARCHITECTURES   604
10.2 CHANNELIZATION   637
10.3 NOTES AND REFERENCES   658
10.4 EXERCISES   662

This chapter applies much of the concepts in the previous chapters to the larger design issues in digital communications. The chapter begins with a discussion of system design for discrete-time modulators and demodulators. Important considerations such as DAC and ADC placement, sample rate selection, the application of multirate processing, and some issues with "low-IF" architectures are summarized. The evolution of demodulators in response to advances in the performance of discrete-time components is also discussed.

The chapter concludes with the important issue of channelizers. The analysis of demodulators and detectors in this and most other textbooks assumes that only the modulated carrier of interest is presented to the receiver. As this is rarely the case in practice, a technique for selecting the desired frequency band, or "channel" is required. The class of techniques that perform this important function are called channelizers. Channelizers in continuous-time processing are almost always based on the superheterodyne architecture pioneered by Howard Armstrong in the 1910s. In discrete-time processing, channelizers are very different. The most efficient channelizers are based on filterbanks and exploit the frequency translations performed in multirate processing. Continuous-time channelizers and discrete-time channelizers are described side-by-side to emphasize their similarities and differences.

10.1 ADVANCED DISCRETE-TIME ARCHITECTURES

The discrete-time QAM modulators and demodulators described in Chapter 5 were discrete-time equivalents of their continuous-time counterparts. These architectures are important because they show *what* functions need to be performed. *How* these functions are performed is the subject of this section. The primary focus is on DAC and ADC placement and the use of multirate processing to perform frequency translations and to parallelize the operations.

Section 10.1 Advanced Discrete-Time Architectures

The complex notation for QAM, developed in Appendix B, is used here because it makes more obvious the trigonometric identities the multirate approach depends on.

10.1.1 Discrete-Time Architectures for QAM Modulators

DAC Placement. The first important issue is DAC placement. The two most common options are illustrated in Figure 10.1.1. The system of Figure 10.1.1 (a) converts the discrete-time pulse trains $I(nT)$ and $Q(nT)$ to continuous-time pulse trains $I(t)$ and $Q(t)$ by a pair of DACs. The QAM waveform is formed in continuous-time using quadrature mixers and a summer as shown.

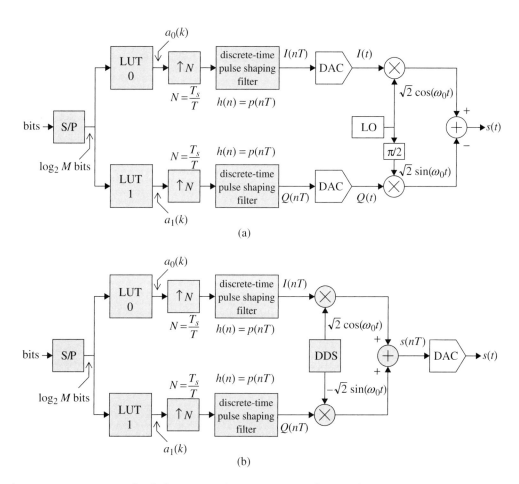

Figure 10.1.1 Two methods for generating a QAM signal using discrete-time processing: (a) Produce quadrature baseband signals in discrete-time, convert to continuous-time, translate to IF using continuous-time processing. (b) Produce a discrete-time band-pass signal at IF and convert to continuous-time at IF.

The advantage of this approach is that the DACs (and hence the discrete-time processing) can operate at a relatively low sample rate (usually two to eight times the symbol rate) because the discrete-time pulse trains are baseband sequences. The disadvantage of this approach is that it can be difficult, using continuous-time hardware, to produce perfectly balanced mixers with equal gains. Further, if the transmitter is to be located in a harsh environment, producing a pair of sinusoids exactly in quadrature can also be a challenge. The "I/Q-imbalance" effects introduce distortion into the transmitted waveforms that increase the bandwidth and degrade the bit error rate performance. Simple examples of this distortion are examined in Exercises 10.1 and 10.2.

The second approach, illustrated in Figure 10.1.1 (b), alleviates the disadvantages by forming the intermediate frequency (IF) signal in discrete-time processing. The DAC is moved to the IF so that the quadrature sinusoids are generated in discrete time and the multiplications required for frequency translations are performed in discrete time. In discrete-time processing, the sampled sinusoids are numbers so that proper amplitude and phase balance can be guaranteed (to the degree that finite precision allows). The gain imbalance issue associated with the mixers does not exist in discrete time because the mixers are replaced by arithmetic multiplications. These advantages are obtained at the cost of sampling rate. The discrete-time processing must operate at a much higher sample rate so that the first Nyquist zone has enough space to accommodate a band-pass signal. This cost can be reduced somewhat by trading clock rate for area as described next.

Multirate Processing. As explained in Appendix B, a sampled QAM modulated carrier may be represented using the complex-valued signal

$$\tilde{s}(nT) = [I(nT) + jQ(nT)] \sqrt{2} e^{j\Omega_0 n} \tag{10.1}$$

where

$$I(nT) = \sum_k a_0(k) p(nT - kT_s)$$
$$Q(nT) = \sum_k a_1(k) p(nT - kT_s) \tag{10.2}$$

for a unit-energy pulse shape $p(nT)$ and where the k-th constellation point is $(a_0(k), a_1(k))$. Using $\tilde{s}(nT)$, the real-valued, band-pass sampled QAM signal may be expressed as

$$s(nT) = \text{Re}\left\{ [I(nT) + jQ(nT)] \sqrt{2} e^{j\Omega_0 n} \right\} \tag{10.3}$$

$$= I(nT)\sqrt{2} \cos(\Omega_0 n) - Q(nT)\sqrt{2} \sin(\Omega_0 n). \tag{10.4}$$

The QAM modulator is equivalent to the complex-valued system shown in Figure 10.1.2 (a). The complex-valued symbol sequence $\tilde{a}(k) = a_0(k) + ja_1(k)$ is the output of look-up tables (not shown for clarity). The upsample-by-N operation inserts $N-1$ zeros to produce the sequence

$$\tilde{a}\left(\frac{n}{N}\right) = \sum_m \left[a_0\left(\frac{m}{N}\right) + ja_1\left(\frac{m}{N}\right) \right] \delta(n - mN) \tag{10.5}$$

Section 10.1 Advanced Discrete-Time Architectures

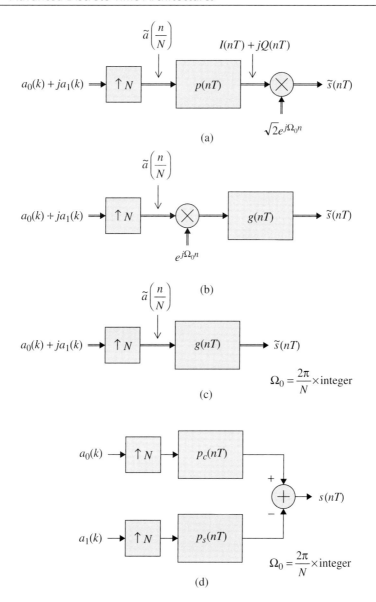

Figure 10.1.2 Creation of a discrete-time band-pass QAM signal: (a) QAM modulator using complex notation; (b) equivalent QAM modulator interchanging the order of frequency translation and shaping; (c) equivalent QAM modulator when $\Omega_0 = $ integer $\times 2\pi/N$; (d) efficient system for producing the real-valued band-pass QAM signal.

where $a_0\left(\frac{m}{N}\right)$ is understood to mean

$$a_0\left(\frac{m}{N}\right) = \begin{cases} a_0\left(\frac{m}{N}\right) & m = \text{integer} \times N \\ 0 & \text{otherwise} \end{cases}. \tag{10.6}$$

The corresponding understanding applies to $a_1\left(\frac{m}{N}\right)$. Applying the results of Section 3.2.3, the DTFT of $\tilde{a}(n)$ consists of N copies of the compressed spectrum of $a_0(k) + ja_1(k)$ as illustrated in Figure 10.1.3 (a). The system of Figure 10.1.1 (b) isolated the spectral copy at baseband using a low-pass filter whose impulse response was equal to the sampled pulse shape. This operation is illustrated in Figure 10.1.3 (b). The low-pass copy was translated to the desired center frequency using multiplication by a complex exponential.

It is inefficient to keep the baseband copy and perform multiplications to shift the baseband signal up to a frequency that was occupied by a spectral copy that was rejected in the first place. When $\Omega_0 = \text{integer} \times 2\pi/N$, a more efficient procedure rejects the baseband component and keeps the desired band-pass component as illustrated in Figure 10.1.3 (c).

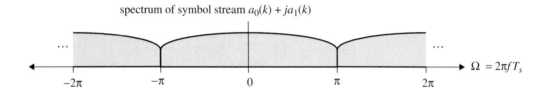

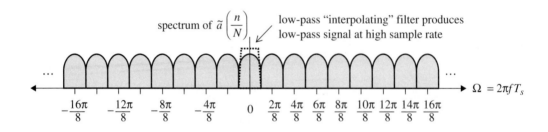

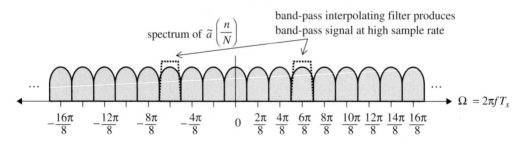

Figure 10.1.3 A graphical representation of the spectra of the signals in Figure 10.1.2 for the special case of upsampling by $N = 8$.

Section 10.1 Advanced Discrete-Time Architectures

To do this, a band-pass version of the pulse shaping filter is required. This pulse shaping filter, with impulse response $g(nT)$, is obtained from the pulse shape by

$$g(nT) = p(nT)\sqrt{2}e^{j\Omega_0 n}. \tag{10.7}$$

Observe that $p(nT)$ is a real-valued filter, whereas $g(nT)$ is a complex-valued filter. To show that this can be done, the mathematical expressions for these operations are now derived.

Applying the results of Section 2.7, the mixer may be moved to the input of the pulse shaping filter as illustrated in Figure 10.1.2 (b). This shows that frequency translation can be applied first to the complex-valued symbol impulse train. Pulse shaping then occurs at band-pass using a band-pass pulse shaping filter with impulse response given by (10.7). The band-pass impulse train can be expressed as

$$\tilde{a}\left(\frac{n}{N}\right)e^{j\Omega_0 n} = \sum_m \left[a_0\left(\frac{m}{N}\right) + ja_1\left(\frac{m}{N}\right)\right]\delta(n - Nm)e^{j\Omega_0 n}$$

$$= \sum_m \left[a_0\left(\frac{m}{N}\right) + ja_1\left(\frac{m}{N}\right)\right]\delta(n - Nm)e^{j\Omega_0 Nm}. \tag{10.8}$$

When $\Omega_0 = 2\pi/N \times$ integer, $e^{j\Omega_0 Nm} = 1$ so that

$$\tilde{a}\left(\frac{n}{N}\right)e^{j\Omega_0 n} = \sum_m \left[a_0\left(\frac{m}{N}\right) + ja_1\left(\frac{m}{N}\right)\right]\delta(n - Nm) = \tilde{a}\left(\frac{n}{N}\right) \tag{10.9}$$

and the system of Figure 10.1.2 (c) results. The impulse response of the band-pass pulse shaping filter may be expressed as

$$g(nT) = p(nT)\sqrt{2}e^{j\Omega_0 n}$$

$$= \underbrace{p(nT)\sqrt{2}\cos(\Omega_0 n)}_{p_c(nT)} + j\underbrace{p(nT)\sqrt{2}\sin(\Omega_0 n)}_{p_s(nT)}. \tag{10.10}$$

The complex-valued band-pass signal may be expressed as

$$\tilde{s}(nT) = \tilde{a}\left(\frac{n}{N}\right) * g(nT). \tag{10.11}$$

Expressing $g(nT)$ as $g(nT) = p_c(nT) + jp_s(nT)$ and retaining only the real part produces an expression for the real-valued band-pass signal $s(nT)$:

$$s(nT) = \sum_m a_0\left(\frac{m}{N}\right)\delta(n - Nm) * p_c(nT) - \sum_m a_1\left(\frac{m}{N}\right)\delta(n - Nm) * p_s(nT). \tag{10.12}$$

The discrete-time system that performs these operations is illustrated in Figure 10.1.2 (d). The shaping filters are frequency translated versions of the baseband pulse shaping filters. The shaping filters extract the desired spectral copy from all those made available by the insertion of $N - 1$ zeros by the upsampler. Note that this system has eliminated the multiplications required by the frequency translation because frequency translation occurs for free due to the upsampling operation. The filters, however, must run at a high clock rate.

The high clock rate can be exchanged for area by using polyphase partitions of the two filters $p_c(nT)$ and $p_s(nT)$. Direct application of the polyphase partition described in Section 3.2 produces a system with two polyphase filterbanks operating in parallel: one for the inphase component and the other for the quadrature component. (See Exercise 10.6.)

Additional insight is gained by performing a polyphase partition of the complex-valued band-pass filter $g(nT)$ in the system of Figure 10.1.2 (c). The z-transform of $g(nT)$ may be expressed as

$$
\begin{aligned}
G(z) = \quad & g(0) & + \quad & g(N)z^{-N} & + \cdots \\
+ \quad & g(1)z^{-1} & + \quad & g(N+1)z^{-(N+1)} & + \cdots \\
+ \quad & g(2)z^{-2} & + \quad & g(N+2)z^{-(N+2)} & + \cdots \\
& \vdots \\
+ \quad & g(N-1)z^{-(N-1)} & + \quad & g(2N-1)z^{-(2N-1)} & + \cdots
\end{aligned}
\tag{10.13}
$$

Using the relationship (10.7), $G(z)$ may be reexpressed as

$$
\begin{aligned}
G(z) = \quad & \sqrt{2}p(0) & + \quad & \sqrt{2}p(N)e^{j\Omega_0 N}z^{-N} & + \cdots \\
+ \quad & \sqrt{2}p(1)e^{j\Omega_0}z^{-1} & + \quad & \sqrt{2}p(N+1)e^{j\Omega_0(N+1)}z^{-(N+1)} & + \cdots \\
+ \quad & \sqrt{2}p(2)e^{j\Omega_0 2}z^{-2} & + \quad & \sqrt{2}p(N+2)e^{j\Omega_0(N+2)}z^{-(N+2)} & + \cdots \\
& \vdots \\
+ \quad & \sqrt{2}p(N-1)e^{j\Omega_0(N-1)}z^{-(N-1)} & + \quad & \sqrt{2}p(2N-1)e^{j\Omega_0(2N-1)}z^{-(2N-1)} & + \cdots
\end{aligned}
\tag{10.14}
$$

When $\Omega_0 = \text{integer} \times 2\pi/N$, this expression may be simplified to

$$
\begin{aligned}
G(z) = \quad & \sqrt{2}p(0) & + \quad & \sqrt{2}p(N)z^{-N} & + \cdots \\
+ \quad & \sqrt{2}p(1)e^{j\Omega_0}z^{-1} & + \quad & \sqrt{2}p(N+1)e^{j\Omega_0}z^{-(N+1)} & + \cdots \\
+ \quad & \sqrt{2}p(2)e^{j\Omega_0 2}z^{-2} & + \quad & \sqrt{2}p(N+2)e^{j\Omega_0 2}z^{-(N+2)} & + \cdots \\
& \vdots \\
+ \quad & \sqrt{2}p(N-1)e^{j\Omega_0(N-1)}z^{-(N-1)} & + \quad & \sqrt{2}p(2N-1)e^{j\Omega_0(N-1)}z^{-(2N-1)} & + \cdots
\end{aligned}
\tag{10.15}
$$

Each row is the z-transform of a subfilter in the polyphase partition of $G(z)$. The z-transform of the m-th subfilter is

$$G_m(z^N) = z^{-m}\sqrt{2}e^{j\Omega_0 m}\left[p(m) + p(m+N)z^{-N} + p(m+2N)z^{-2N} + \cdots\right] \tag{10.16}$$

for $m = 0, 1, \ldots, N-1$. The sum in the square brackets is the z-transform of the m-th subfilter in the polyphase partition of $p(nT)$. Using $P_m(z^N)$ to represent the z-transform of the m-th subfilter, $G_m(z^N)$ may be expressed as

$$G_m(z^N) = z^{-m}\sqrt{2}e^{j\Omega_0 m}P_m(z^N). \tag{10.17}$$

Section 10.1 Advanced Discrete-Time Architectures   **611**

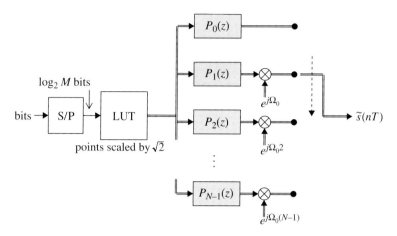

Figure 10.1.4 Creation of a discrete-time band-pass QAM signal using a polyphase filterbank.

This shows that the polyphase partition of the complex-valued filter $G(z)$ is based on the polyphase partition of the real-valued filter $P(z)$. As a consequence, the input can be processed by a polyphase filterbank consisting of real-valued subfilters followed by multiplications by a complex-valued constant. Following the procedure, outlined in Section 3.2, for applying a polyphase filterbank to the upsample-filter operation, a block diagram for the QAM modulator, based on the polyphase filterbank, is illustrated in Figure 10.1.4. The advantage of this approach is that the filterbank operates at the symbol rate. Hence, all multipliers run at the symbol rate. The high clock rate data is produced by the commutator operating at the outputs of the two filterbanks. The system can be simplified a little more by recognizing that only the real part of $\tilde{s}(nT)$ needs to be computed. The simplifications are explored in Exercise 10.7.

10.1.2 Discrete-Time Architectures for QAM Demodulators

As explained in Section 5.3, the received real-valued, band-pass QAM signal may be represented as

$$r(t) = I_r(t) \cos(\omega_0 t) - Q_r(t) \sin(\omega_0 t). \qquad (10.18)$$

The fundamental role of the discrete-time demodulator is to produce samples of $I_r(t)$ and $Q_r(t)$ for processing by the discrete-time matched filters. The matched-filter outputs form, after appropriate timing adjustments and phase rotations, the signal space projections used for detection. The most useful interpretation of extracting $I_r(t)$ and $Q_r(t)$ from the received band-pass IF signal is as a complex frequency translation as described in Appendix B. The ADC can be placed anywhere in the processing chain to produce the desired samples. Two options are discussed first: I/Q sampling and IF sampling.

ADC Placement. There are two basic options for ADC placement as illustrated in Figure 10.1.5. The first option, outlined in Figure 10.1.5 (a), performs the quadrature mixing

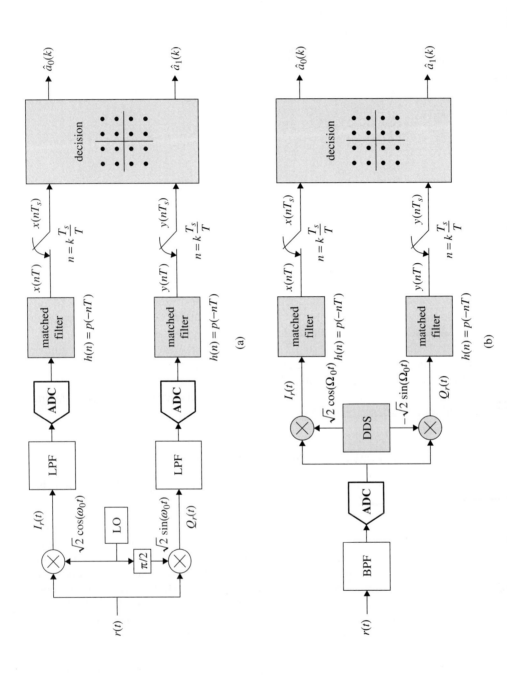

Figure 10.1.5 Two options for ADC placement in a discrete-time QAM detector: (a) I/Q baseband sampling using two ADCs performing baseband sampling; (b) IF sampling using a single ADC performing band-pass sampling.

Section 10.1 Advanced Discrete-Time Architectures

in the continuous-time domain using a pair of mixers and quadrature sinusoids generated from a single source called a "local oscillator" or LO. The outputs of the mixers are low-pass filtered to remove the double-frequency components and sampled by a pair of ADCs. The ADC outputs are processed by discrete-time matched filters and processed to produce the desired signal space projections. The attractive feature of this approach is that the ADCs can operate at a relatively low clock rate, usually two to four times the symbol rate as determined by the pulse shape. In general, higher precision quantization is available with ADCs at lower sample rates. Consequently, higher dynamic range is available.

The disadvantages of this approach are linked to the manufacturing tolerances that limit the accuracy of the parameter values of the components used to construct the continuous-time circuits.

- *DC Offsets*: This problem is related to imperfections in the mixers used to perform the quadrature downconversion. The isolation between the two input ports of a mixer is not perfect. As a consequence, the LO sinusoid "leaks" into the other input port. As a result, the "leaked" version of the sinusoid is multiplied by the LO sinusoid and this difference appears at the mixer output. For example, suppose $A_1 \cos(\omega_0 t)$ is the LO sinusoid (connected to the LO port on the mixer) and $A_2 \cos(\omega_0 t + \theta)$ is the leaked version (that appears on the input port of the mixer). In addition to multiplying the received signal by the LO, the mixer also produces the product of the LO and leaked sinusoids. The product includes the DC term $A_1 A_2/2 \cos(\theta)$ and a double frequency term $A_1 A_2/2 \cos(2\omega_0 t + \theta)$. The DC term is added to the desired signal component resulting in a type of distortion called "DC Offset." If the product $A_1 A_2$ is large enough, the signal plus DC offset can saturate the electronic circuits that follow. In addition, the DC offset moves the signal space projections closer to (or across!) decision region boundaries thus producing degradation in bit error rate performance.
- *LO Leakage*: Another problem associated with LO leakage is the fact that the "leaked" version of the LO sinusoid can work its way back to the receive antenna and be radiated from this antenna. Because a zero-IF receiver uses an LO that is equal to the carrier frequency, this is equivalent to radiating an unmodulated carrier in the band of interest for other receivers. This interference degrades the performance of other receivers that are close enough to receive this unintended radiated energy.
- *I/Q Imbalance*: To preserve the proper amplitude and phase relationship between the inphase and quadrature components of the received signal, the gains through each mixer-filter-ADC branch must be identical and the quadrature sinusoids produced by the LO and phase shifter must have precisely the same amplitude and be exactly 90° out of phase. Manufacturing tolerances make this almost impossible to achieve. The resulting distortion alters the positions of the signal space projections used by the detector. This effect is explored in Exercises 10.3 and 10.4.
- *Even Order Distortion*: Intermodulation interference results when two band-pass signals, centered at different carrier frequencies, are amplified by a nonlinear device. All amplifiers that operate at RF exhibit some degree of nonlinear behavior. Usually, the nonlinear behavior is modeled in amplitude (voltage in this case) using a polynomial

of the form
$$v_{\text{out}} = a_1 v_{\text{in}} + a_2 v_{\text{in}}^2 + a_3 v_{\text{in}}^3 + a_4 v_{\text{in}}^4 + \cdots$$

In other receiver architectures—the architecture of Figure 10.1.5 (b) for example—only the distortion resulting from odd-order nonlinearities impacts performance. This defines which frequencies can cause interference. In the architecture of Figure 10.1.5 (a), even-order non-linearities also cause distortion. This increases the number of candidate frequency components that can cause intermodulation interference. Limiting this interference imposes stricter linearity requirements on the RF amplifiers that precede the quadrature downconversion.

These challenges are (mostly) alleviated by the basic architecture outlined in Figure 10.1.5 (b). Here the IF signal is sampled by a single ADC operating on an appropriately bandlimited band-pass signal. The generation of quadrature sinusoids and the corresponding mixing operations occur entirely in the discrete-time domain. The discrete-time quadrature sinusoids are a list of numbers. It is a trivial matter to ensure equal gain and $\pi/2$ phase difference (at least to the degree permitted by finite precision). The gain differences of continuous-time mixers is also not an issue because the continuous-time mixers are replaced by arithmetic multipliers. As a consequence, the I/Q imbalance problem is nonexistent. The gain difference issue with the ADCs in not an issue because only one ADC is required.

These desirable features are obtained at a cost: the ADC must operate at a much higher clock rate than the ADCs in the I/Q baseband sampling architecture of Figure 10.1.5 (a). As a consequence, lower precision quantization is available, which can impact dynamic range. A second consequence is that much of the signal processing must operate at a higher clock rate. This tends to increase the cost of discrete-time hardware.

Which option is the best will depend on what is most important in the design. In applications where dynamic range is the most important or where high clock-rate operation is not possible (due to cost, hardware limitations, or power consumption), the I/Q baseband sampling architecture tends to be preferred. Discrete-time techniques to compensate for the I/Q balance issues are often applied in these cases. In most other applications, the IF sampling architecture is preferred. In these systems, continuous-time automatic gain control (AGC) is important to help overcome the dynamic range limitations. When implemented properly, either architecture produces a demodulator with excellent performance. A good system designer must be able to move back and forth between the continuous-time and discrete-time domains to perform the cost/performance trade-off analysis for each option for a given set of constraints.

Sampling Rates. The sample rate for the pair of ADCs in the I/Q baseband sampling architecture is determined by the bandwidth of $I_r(t)$ and $Q_r(t)$. The bandwidth is a function of the pulse shape and the symbol rate as outlined in Chapter 6. For example, consider an $R_s = 5$ Msymbol/s QAM signal that occupies $B = 6$ MHz centered at an intermediate frequency $f_0 = 70$ MHz. After mixing to I/Q baseband and eliminating the double-frequency terms with the continuous-time low-pass filter, $I_r(t)$ and $Q_r(t)$ are baseband signals whose bandwidth is 3 MHz. The sampling theorem requires the sample rate to be at least $F_s = 6$ Msamples/s. In practice, the sample rate is 10%–20% above this rate (i.e., $F_s = 6.6$ Msamples/s to

$F_s = 7.2$ Msamples/s) to allow for a transition band in the subsequent discrete-time filtering. These sampling rates require a resampling filter in discrete-time processing to produce samples at a multiple of the symbol rate suitable for use with the matched filter and timing error detectors. The applicable resampling filters outlined in Section 9.3 may be used to perform the resampling. The need for resampling filters is eliminated if the sample rate is chosen to be a multiple of the symbol rate. Sampling at 10 Msamples/s produces a discrete-time system operating at $N = 2$ samples/symbol whereas sampling at 20 Msamples/s produces a discrete-time system operating at $N = 4$ samples/symbol.

There are also several options for the sample rate when IF sampling is used. The IF sample rate could be selected to satisfy the sampling theorem, that is, the sample rate is 10%–20% higher than twice the highest frequency of the IF signal. Returning to the example, the highest frequency of the IF signal is 73 MHz. The sampling theorem requires a sample rate in excess of $F_s = 146$ Msamples/s and practical considerations require a sample rate in the range 160.6–175.2 Msamples/s. Note that this is substantially higher than the sample rates required for I/Q baseband sampling. These sample rates also require resampling filters to produce sampled baseband sequences $I_r(nT)$ and $Q_r(nT)$ at an integer multiple of the symbol rate for the matched filter and the timing error detector. The resampling filters become very simple, (or may not even be required) when the sample rate is an integer multiple of the symbol rate. Sample rates that both satisfy the sampling theorem and are a multiple of the symbol rate are $F_s = 175, 180,$ or 185 Msamples/s (for $N = 35, 36,$ or 37 samples/symbol, respectively).

The sample rate may also be selected to be a multiple of the intermediate frequency. This has advantages in the subsequent quadrature downconversion processing. For example, if the IF sample rate is four times the intermediate frequency, that is, $F_s = 4f_0$ samples/s, then

$$\Omega_0 = 2\pi f_0 T = \frac{2\pi f_0}{F_s} = \frac{\pi}{2}$$

so that the discrete-time sinusoids $\cos(\Omega_0 n) = \cos(n\pi/2)$ and $\sin(\Omega_0 n) = \sin(n\pi/2)$ assume only three trivial values:

n	0	1	2	3	4	5	...
$\cos(n\pi/2)$	1	0	-1	0	1	0	...
$\sin(n\pi/2)$	0	1	0	-1	0	1	...

Thus, the multiplications by $\cos(\Omega_0 n)$ and $\sin(\Omega_0 n)$ do not require any multiplications. The frequency translation is performed by simple sign-alternations as indicated above. Returning to the example, setting $F_s = 280$ Msamples/s produces a band-pass discrete-time sequence centered at $\Omega_0 = \pi/2$.

Band-pass subsampling can also be used to reduce the required sample rate. In this case, the sample rate is selected to alias the intermediate frequency to a lower frequency such that the aliased copies of the positive-frequency spectrum and the negative-frequency spectrum do not overlap. A very common choice is to chose the sample frequency so as to force f_0 to alias to the quarter-sample-rate frequency. When the discrete-time band-pass signal is centered at the quarter-sample-rate frequency, $\Omega_0 = \pi/2$ so that the quadrature sinusoids assume the $0, \pm 1$ values as described above and the quadrature frequency translation is trivial. The intermediate frequency f_0 aliases to the quarter-sample-rate frequency when F_s satisfies

$$kF_s \pm \frac{1}{4}F_s = f_0 \tag{10.19}$$

for an integer k. Note that the solution corresponding to $k = 0$ is the sample rate discussed in the previous paragraph. Commonly used IFs are listed in Section 10.2.1. For example, sample rates that alias $f_0 = 70$ MHz to the quarter-sample-rate frequency are summarized in Table 10.1.1. Returning to the example, the information in Table 10.1.1 shows that there are 12 sample rates, ranging from $12\frac{4}{23}$ to 280 Msamples/s that simultaneously satisfy the bandpass version of the sampling theorem and (10.19). Note that two of the sample rates, $F_s = 280$ Msamples/s and $F_s = 40$ Msamples/s are integer multiples of the symbol rate.

Multirate Processing. The properties of multirate processing can be used to reduce the complexity of a QAM demodulator. In this section, four cases are examined to illustrate the concept. In the first case, the IF sample rate aliases intermediate frequency to the quarter sample rate and is an integer multiple of the symbol rate. The next three cases examine the scenario where the IF sample rate is not a multiple of the symbol rate and a resampling filter is required. These scenarios are more commonly encountered in practice. These three cases differ in the relationship between the intermediate frequency and the sample rate. The second case assumes the intermediate frequency aliases to the quarter sample rate. In the third case, the intermediate frequency aliases to a simple rational multiple of the IF sample rate, that is, $\Omega_0 = Q/D \times 2\pi$. (Note that the second case is a special case of this case where $Q = 1$ and $D = 4$.) In the fourth case, the IF sample rate is arbitrary.

Case 1: The IF Sample Rate Aliases the Signal to the Quarter Sample Rate and is a Multiple of the Symbol Rate

As outlined in Appendix B, the QAM demodulator/detector is equivalent to the complex-valued system illustrated Figure 10.1.6 (a). Let

$$r(nT) = I_r(nT)\sqrt{2}\cos(\Omega_0 n) - Q_r(nT)\sqrt{2}\sin(\Omega_0 n) \tag{10.20}$$

represent the sampled IF signal where $\Omega_0 = \pi/2$. The product resulting from the complex spectral shift may be expressed as (see Exercise B.18)

$$r(nT)\sqrt{2}e^{j\Omega_0 n} = I_r(nT) + jQ_r(nT) + [I_r(nT) - jQ_r(nT)]e^{-j2\Omega_0 n}. \tag{10.21}$$

The DTFT of the first term is centered about $\Omega = 0$ and is the complex baseband representation of $r(nT)$. The second term is centered about $\Omega = -2\Omega_0 \mod 2\pi$ and represents the "double frequency" products that result from the multiplication of sinusoids. The matched filter removes the "double frequency" terms and produces the complex-valued baseband output $x(nT) + jy(nT)$ from the baseband term. The matched filter output $x(nT) + jy(nT)$ can be downsampled by D because its bandwidth has been reduced by the

Section 10.1 Advanced Discrete-Time Architectures

Table 10.1.1 Sample rates that alias the IF to one-quarter the sample rate for a 70 MHz IF. Also shown are the maximum usable bandwidths as determined by the sampling theorem.

Sample Rate (Msamples/s)	Maximum Bandwidth (MHz)	Sample Rate (Msamples/s)	Maximum Bandwidth (MHz)
280	140.0000	$4\frac{4}{9}$	2.2222
$93\frac{1}{3}$	46.6667	$4\frac{4}{13}$	2.1538
56	28.0000	$4\frac{12}{67}$	2.0896
40	20.0000	$4\frac{4}{69}$	2.0290
$31\frac{1}{9}$	15.5556	$3\frac{67}{71}$	1.9718
$25\frac{5}{11}$	12.7273	$3\frac{61}{73}$	1.9178
$21\frac{7}{13}$	10.7692	$3\frac{11}{15}$	1.8667
$18\frac{2}{3}$	9.3333	$3\frac{7}{11}$	1.8182
$16\frac{8}{17}$	8.2353	$3\frac{43}{79}$	1.7722
$14\frac{14}{19}$	7.3684	$3\frac{37}{81}$	1.7284
$13\frac{1}{3}$	6.6667	$3\frac{31}{83}$	1.6867
$12\frac{4}{23}$	6.0870	$3\frac{5}{17}$	1.6471
$11\frac{1}{5}$	5.6000	$3\frac{19}{87}$	1.6092
$10\frac{10}{27}$	5.1852	$3\frac{13}{89}$	1.5730
$9\frac{19}{29}$	4.8276	$3\frac{1}{13}$	1.5385
$8\frac{16}{33}$	4.2424	$3\frac{1}{93}$	1.5054
8	4.0000	$2\frac{18}{19}$	1.4737
$7\frac{21}{37}$	3.7838	$2\frac{86}{97}$	1.4433
$7\frac{7}{39}$	3.5897	$2\frac{82}{99}$	1.4141
$6\frac{34}{41}$	3.4146	$2\frac{78}{101}$	1.3861
$6\frac{22}{43}$	3.2558	$2\frac{74}{103}$	1.3592
$6\frac{2}{9}$	3.1111	$2\frac{2}{3}$	1.3333
$5\frac{45}{47}$	2.9787	$2\frac{66}{107}$	1.3084
$5\frac{5}{7}$	2.8571	$2\frac{62}{109}$	1.2844
$5\frac{25}{51}$	2.7451	$2\frac{58}{111}$	1.2613
$5\frac{15}{53}$	2.6415	$2\frac{54}{113}$	1.2389
$5\frac{1}{11}$	2.5455	$2\frac{10}{23}$	1.2174
$4\frac{52}{57}$	2.4561	$2\frac{46}{117}$	1.1966
$4\frac{44}{59}$	2.3729	$2\frac{6}{17}$	1.1765
$4\frac{36}{61}$	2.2951	$2\frac{38}{121}$	1.1570

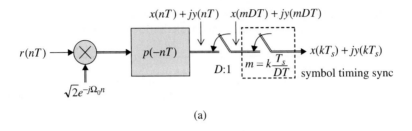

(a)

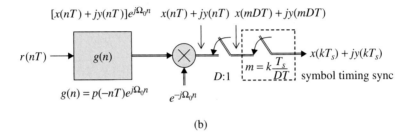

(b)

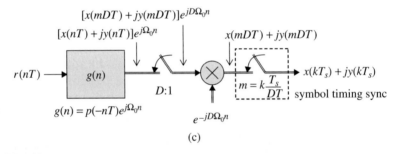

(c)

Figure 10.1.6 Development of multirate processing for the QAM demodulator: (a) The QAM demodulator using complex notation where the sampling rate is able to support a downsample by D; (b) exchanging the order of filtering and multiplying by the complex exponentials described in Section 2.7; (c) sliding the downsample operation through the complex exponential multiplier to the filter output.

matched filter. The resulting downsampled signal can be expressed as $x(mDT) + jy(mDT)$. The symbol timing synchronizer, operating on $x(mDT) + jy(mDT)$, produces a sequence at 1 sample/symbol from which the desired signal space projections are produced after rotation by the carrier phase offset (not shown). It will become clear in a moment how downsampling from T_s/T samples/symbol to 1 sample/symbol can be accomplished efficiently in two steps.

Applying the results outlined in Section 2.7, the heterodyne-filter operations used to compute the matched filter output may be reversed by replacing the matched filter (a low-pass

Section 10.1 Advanced Discrete-Time Architectures

filter) with the complex-valued band-pass filter

$$g(n) = p(-nT)\sqrt{2}e^{j\Omega_0 n} \quad (10.22)$$

and by performing the complex heterodyne on the output of this filter. The system that performs these operations in this order is illustrated in Figure 10.1.6 (b). The output of the complex-valued band-pass matched filter can be expressed as (see Exercise 10.9)

$$[x(nT) + jy(nT)]e^{j\Omega_0 n}. \quad (10.23)$$

The complex heterodyne removes the $e^{j\Omega_0 n}$ term to produce the complex baseband matched filter outputs as shown. In Figure 10.1.6 (c), the downsample operation is moved to the other side of the complex heterodyne operation and placed at the output of the band-pass filter. As a consequence, the complex exponential oscillates at a frequency $D\Omega_0$ rad/sample.

A frequency domain interpretation of the processing performed by the three systems of Figure 10.1.6 is illustrated in Figure 10.1.7. The spectra of the signals in Figure 10.1.6 (a) are shown in Figure 10.1.7 (a). This system is characterized by a one-sided frequency translation that moves the positive frequency component of the received signal to baseband, a low-pass filter operation (using a matched filter) that isolates the baseband component, and a downsample operation that expands the spectrum of the matched filter output relative to the sample rate. Figure 10.1.7 (b) illustrates the spectra of the signals in Figure 10.1.6 (b). Here the frequency translation and filtering operations have been interchanged. The filtering is performed using one-sided band-pass filter centered at Ω_0 rad/sample. This filter isolates the positive frequency component of the received signal. The filter output is translated to baseband where it is then downsampled. The spectra of the signals in Figure 10.1.6 (c) are shown in Figure 10.1.7 (c). The positive frequency component is isolated using a one-sided band-pass filter. The resulting signal, centered at Ω_0 rad/sample, is downsampled by D. As a result, the bandwidth increases and the center frequency aliases to $D\Omega_0 \mod (2\pi)$ rad/sample. The final frequency translation moves this signal to baseband as shown.

Note that if

$$D\Omega_0 \mod (2\pi) = 0, \quad (10.24)$$

then the downsample operation aliases the band-pass filter output directly to baseband and the final frequency translation is not required. The advantage of using complex-valued signal processing is now evident: The spectrum is one-sided and on the positive frequency axis so that there is no negative frequency component that could also alias to baseband and cancel out components of the desired signal. When $\Omega_0 = \pi/2$, the condition (10.24) is met whenever D is an integer multiple of 4 (or even integer multiple of 2). Another convenient value for the final frequency translation occurs when

$$D\Omega_0 \mod (2\pi) = \pi. \quad (10.25)$$

In this case, the final frequency translation is accomplished using multiplications by

$$e^{jD\Omega_0 m} = e^{j\pi m} = (-1)^m. \quad (10.26)$$

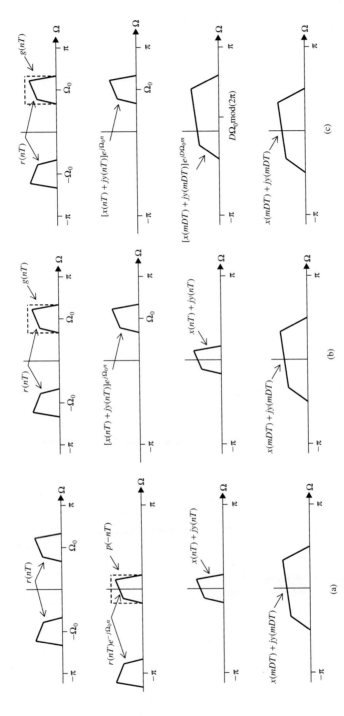

Figure 10.1.7 A frequency domain interpretation of the processing outlined in Figure 10.1.6: (a) The spectra of the signals in Figure 10.1.6 (a); (b) the spectra of the signals in Figure 10.1.6 (b); (c) the spectra of the signals in Figure 10.1.6 (c).

No multiplications are required since multiplying a sequence by $(-1)^m$ is equivalent to changing the sign on the odd-indexed samples. When $\Omega_0 = \pi/2$, the condition (10.25) is met when D is an odd integer multiple of 2. The two most commonly used choices for D are $D = 2$ and $D = 4$ and are explored below.

The structure of Figure 10.1.6 (c) begins with the filter-downsample operations. An efficient method for performing this sequence of tasks using a polyphase filterbank was described in Section 3.2.2. A direct application of polyphase filterbank processing results in the system illustrated in Figure 10.1.8 (a).

Incorporating the polyphase filter structure with the important special case $D = 2$ produces an efficient receiver implementation. The two-stage polyphase representation of the filter $G(z)$ is

$$G(z) = G_0(z^2) + z^{-1}G_1(z^2) \tag{10.27}$$

where

$$\begin{aligned}
G_0(z^2) &= \sqrt{2}p(0) + \sqrt{2}p(-2)e^{j2\Omega_0}z^{-2} + \sqrt{2}p(-4)e^{j4\Omega_0}z^{-4} + \cdots \\
&= \sqrt{2}\left[p(0) - p(-2)z^{-2} + p(-4)z^{-4} + \cdots\right] \\
G_1(z^2) &= \sqrt{2}p(-1)e^{j\Omega_0} + \sqrt{2}p(-3)e^{j3\Omega_0}z^{-2} + \sqrt{2}p(-5)e^{j5\Omega_0}z^{-4} + \cdots \\
&= j\sqrt{2}\left[p(-1) - p(-3)z^{-2} + p(-5)z^{-4} - \cdots\right].
\end{aligned} \tag{10.28}$$

Applying the Noble identities, the downsample-by-2 operation is moved to the input side of the filters. The resulting filters, $G_0(z)$ and $G_1(z)$, now operate at half the IF sample rate. The filter $G_0(z)$ is purely real and the filter $G_1(z)$ is purely imaginary. Because the input to the filterbank is real, the output of filter $G_0(z)$ is purely real. This output is the desired inphase signal $x(2mT)$. Likewise, the output of the filter $G_1(z)$ is purely imaginary and it is the desired quadrature output $y(2mT)$. Note that the factor j appearing in the polyphase decomposition (10.28) serves as a placeholder to identify the imaginary portion of the output.

Each subfilter is almost a polyphase partition of the matched filter $P(1/z)$. The difference is the alternating signs of the filter coefficients. Consider the downsampled input to the filter $G_0(z)$. The downsample operation has aliased the center frequency of the input from the quarter sample rate (relative to the high sample rate) to the half sample rate (relative to the low sample rate). Interpreting the sign alternations as a heterodyne by $e^{j\pi m}$, the filter $G_0(z)$ may be interpreted as a band-pass filter centered at the half sample rate. The same signal processing could be performed by heterodyning the downsampled input signal with $e^{j\pi m} = (-1)^m$ and using the low-pass version of the filter. The same line of reasoning applies to the input to the filter $G_1(z)$.

Applying these principles produces the system illustrated in Figure 10.1.8 (b). Observe that the combination of sign alternations and downsampling has removed the need to perform any multiplications by the complex exponential. The commutator operating on the input signal performs the downsample-by-2 operation by switching the even- and odd-indexed samples between the two subfilters. Multiplication by $(-1)^m$ is performed by alternating the sign of each sample in both of the resulting parallel sequences. This translates the center

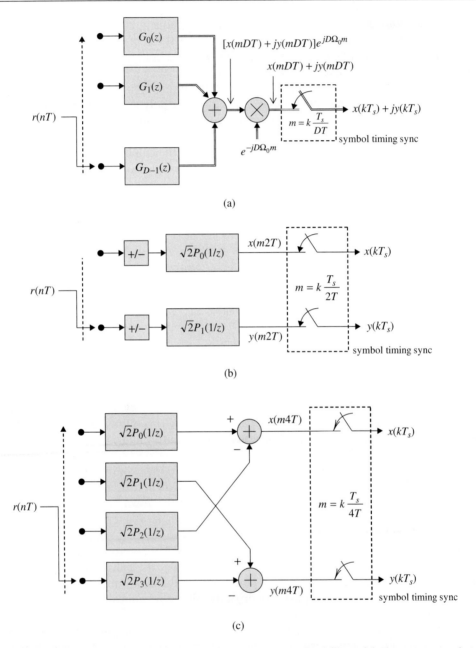

Figure 10.1.8 A QAM demodulator using a polyphase matched filter: (a) The structure that follows by direct application of polyphase processing to the system of Figure 10.1.6 (c); (b) the special case where the discrete-time IF frequency aliases to the quarter sample rate and a downsample by $D = 2$ is performed; (c) the special case where the discrete-time IF frequency aliases to the quarter sample rate and a downsample by $D = 4$ is performed.

frequency of each of the sequences from the half sample rate to baseband. The real-valued subfilters are used to produce $x(2mT)$ and $y(2mT)$ as shown.

If the IF sampling rate is high enough to permit a downsample by $D = 4$, the sign alternations can be eliminated. For $D = 4$, the four-stage polyphase partition of $G(z)$ is

$$G(z) = G_0(z^4) + z^{-1}G_1(z^4) + z^{-2}G_2(z^4) + z^{-3}G_3(z^4) \qquad (10.29)$$

where

$$\begin{aligned}
G_0(z^4) &= \sqrt{2}p(0) + \sqrt{2}p(-4)e^{j4\Omega_0}z^{-4} + \sqrt{2}p(-8)e^{j8\Omega_0}z^{-8} + \cdots \\
&= \sqrt{2}\left[p(0) + p(-4)z^{-4} + p(-8)z^{-8} + \cdots\right] \\
&= \sqrt{2}P_0(z^{-4}) \\
G_1(z^4) &= \sqrt{2}p(-1)e^{j\Omega_0} + \sqrt{2}p(-5)e^{j5\Omega_0}z^{-4} + \sqrt{2}p(-9)e^{j9\Omega_0}z^{-8} + \cdots \\
&= j\sqrt{2}\left[p(-1) + p(-5)z^{-4} + p(-9)z^{-8} + \cdots\right] \\
&= j\sqrt{2}P_1(z^{-4}) \\
G_2(z^4) &= \sqrt{2}p(-2)e^{j2\Omega_0} + \sqrt{2}p(-6)e^{j6\Omega_0}z^{-4} + \sqrt{2}p(-10)e^{j10\Omega_0}z^{-8} + \cdots \qquad (10.30)\\
&= -\sqrt{2}\left[p(-2) + p(-6)z^{-4} + p(-10)z^{-8} + \cdots\right] \\
&= -\sqrt{2}P_2(z^{-4}) \\
G_3(z^4) &= \sqrt{2}p(-3)e^{j3\Omega_0} + \sqrt{2}p(-7)e^{j7\Omega_0}z^{-4} + \sqrt{2}p(-11)e^{j11\Omega_0}z^{-8} + \cdots \\
&= -j\sqrt{2}\left[p(-3) + p(-7)z^{-4} + p(-11)z^{-8} + \cdots\right] \\
&= -j\sqrt{2}P_3(z^{-4}).
\end{aligned}$$

The Noble identities are applied to move the downsample-by-4 operation to the input side of these filters. Observe that the filters $G_0(z)$ and $G_2(z)$ are purely real whereas the filters $G_1(z)$ and $G_3(z)$ are purely imaginary. Because the input is real, the combination of the outputs of filters $G_0(z)$ and $G_2(z)$ is $x(4mT)$ whereas the combination of the outputs of filters $G_1(z)$ and $G_3(z)$ is $y(4mT)$. The system that exploits this property is shown in Figure 10.1.8 (c). The commutator downsamples the input signal by 4 so that at each switch position output, a version of the IF signal sampled at one-quarter the IF sampling rate is present. The downsampling aliases the signal components to baseband with four times the bandwidth relative to the output sampling rate. These baseband signals are filtered by the four subfilters in the polyphase decomposition of the matched filter and combined to produce the desired inphase and quadrature outputs at 1/4 the IF sample rate. Note that no multiplications by complex exponentials are required to produce the frequency translation. The frequency translations occur "for free" as a result of the multirate processing applied at the input.

Case 2: The IF Sample Rate Aliases the Signal to the Quarter Sample Rate But is Not a Multiple of the Symbol Rate

The architectures described in Figure 10.1.8 assume that the sample rate that produces $\Omega_0 = \pi/2$ is also an integer multiple of the symbol rate. Because the symbol rate is only indirectly related to the intermediate frequency, the sample rates that alias the IF to the quarter sample rate are rarely an integer multiple of the symbol rate. In most cases, the matched filter must operate at a sample rate that is an integer multiple of the symbol rate. As a consequence, a sample rate change must be performed.

The mathematical development is based on the principles outlined in Sections 2.7 and 3.2 and proceeds as outlined in Figure 10.1.9. The direct approach, shown in Figure 10.1.9 (a), performs a complex heterodyne to translate the real-valued IF signal $r(nT)$ to complex baseband. The signal is resampled in two steps. The first step is a downsample-by-D operation that will be combined with the heterodyne operation in a moment. The second step is a resampling filter that produces samples at an equivalent rate of N samples/symbol in preparation for matched filtering. Methods for performing this step are outlined in Section 9.3.

Let the sampled IF signal be

$$r(nT) = I_r(nT)\sqrt{2}\cos(\Omega_0 n) - Q_r(nT)\sqrt{2}\sin(\Omega_0 n). \tag{10.31}$$

Then the result of the complex heterodyne is (see Exercise B.18)

$$r(nT)\sqrt{2}e^{j\Omega_0 n} = I_r(nT) + jQ_r(nT) + [I_r(nT) - jQ_r(nT)]e^{j2\Omega_0 n}. \tag{10.32}$$

The ideal low-pass filter $h(n)$ removes the "double frequency" term and retains the complex-valued baseband output $I_r(nT) + jQ_r(nT)$. This signal can be downsampled by D because its bandwidth has been reduced by the low-pass filter. The resulting downsampled signal, $I_r(mDT) + jQ_r(mDT)$, is resampled to produce $I_r(lT_s/N) + jQ_r(lT_s/N)$, which is sampled at a rate equivalent to N samples/symbol. This signal is processed by the matched filter operating at N samples/symbol and downsampled to produce the samples used to compute the desired signal space projections.

Following the analysis outlined in Section 2.7, the order of the heterodyne-filter operations can be reversed to produce the filter-heterodyne system shown in Figure 10.1.9 (b). The band-pass filter $g(n)$ is generated from the low-pass filter $h(n)$ by heterodyning the coefficients of the low-pass filter:

$$g(n) = h(n)\sqrt{2}e^{j\Omega_0 n}. \tag{10.33}$$

The downsample and complex heterodyne operations are swapped to produce the system in Figure 10.1.9 (c). The complex heterodyne operates on the downsampled signal and performs a residual frequency translation of $D\Omega_0 \mod (2\pi)$ rad/sample.

The frequency domain interpretation of the processing outlined in Figure 10.1.9 is illustrated in Figure 10.1.10. The spectra of the signals associated with the processing represented by the diagram in Figure 10.1.9 (a) are shown in Figure 10.1.10 (a). The first heterodyne performs a one-sided frequency translation that moves the positive frequency

Section 10.1 Advanced Discrete-Time Architectures

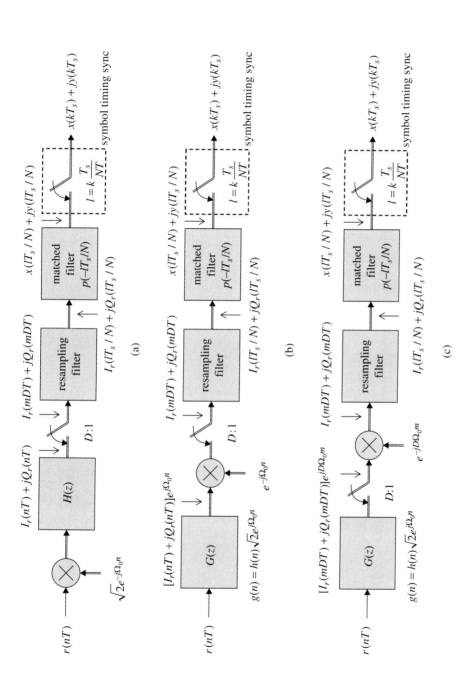

Figure 10.1.9 Development of multirate processing for the QAM demodulator using both a resampling filter and a matched filter: (a) The QAM demodulator (with complex-valued notation) based on a heterodyne and a matched filter prior to the matched filter; (b) exchanging the order of filtering and multiplying by the complex exponentials described in Section 2.7; (c) sliding the downsample operation through the complex exponential multiplier to the filter output.

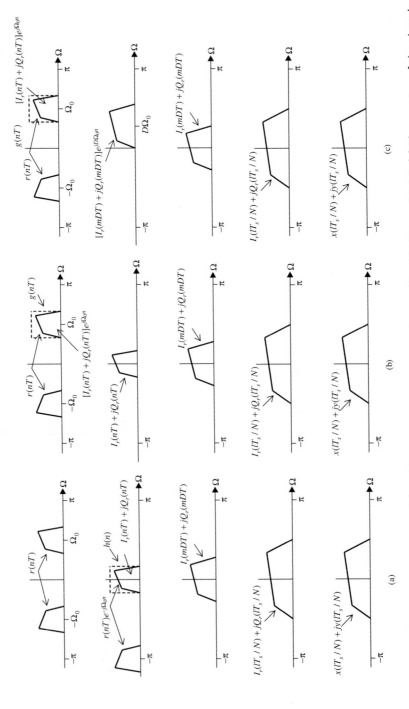

Figure 10.1.10 A frequency domain interpretation of the processing outlined in Figure 10.1.9: (a) The spectra of the signals in Figure 10.1.9 (a); (b) the spectra of the signals in Figure 10.1.9 (b); (c) the spectra of the signals in Figure 10.1.9 (c).

component of the received signal to baseband. The low-pass filter isolates the baseband component. The baseband component is then resampled in two steps: a simple downsample by D and a resampling to produce a signal whose sample rate is an integer multiple of the symbol rate. Each resampling step increases the bandwidth of the signal as shown. The signal is processed by the matched filter to produce the desired signal. The spectra shown in Figure 10.1.10 (b) represent the signals in associated with the processing illustrated in Figure 10.1.9 (b). The positive frequency component of the received signal is isolated using the band-pass filter and translated to baseband using the complex heterodyne operation. The resulting baseband signal is resampled and processed by the matched filter as described before. The spectra of the signals associated with the processing represented by the diagram in Figure 10.1.9 (c) are shown in Figure 10.1.10 (c). As before, the positive frequency component of the received signal is isolated by the band-pass filter. In this case, the resampling is applied to the positive frequency component. The downsample by D increases the bandwidth of the signal and aliases the center frequency to $D\Omega_0 \mod (2\pi)$ rad/sample as shown. The resulting signal is translated to baseband using a complex heterodyne operating at a lower sample rate. The baseband signal is resampled to an integer multiple of the symbol rate and matched filtered as shown.

As in the previous case, $D = 2$ and $D = 4$ are particularly attractive choices for $\Omega_0 = \pi/2$. When coupled with multirate processing, an efficient demodulator is produced. Following the same analysis in the development from Figure 10.1.6 (c) through Figure 10.1.8 (a) to Figure 10.1.8 (b), the systems of Figure 10.1.11 can be produced. The system of Figure 10.1.11 (a) is based on a $D = 2$ stage polyphase decomposition of the band-pass filter $g(n)$:

$$G(z) = G_0(z^2) + z^{-1}G_1(z^2). \tag{10.34}$$

The two polyphase subfilters are

$$\begin{aligned}
G_0(z^2) &= \sqrt{2}h(0) + \sqrt{2}h(2)e^{j2\Omega_0}z^{-2} + \sqrt{2}h(4)e^{j4\Omega_0}z^{-4} + \cdots \\
&= \sqrt{2}\left[h(0) - h(2)z^{-2} + h(4)z^{-4} - \cdots\right] \\
G_1(z^2) &= \sqrt{2}h(1)e^{j\Omega_0} + \sqrt{2}h(3)e^{j3\Omega_0}z^{-2} + \sqrt{2}h(5)e^{j5\Omega_0}z^{-4} + \cdots \\
&= j\sqrt{2}\left[h(1) - h(3)z^{-2} + h(5)z^{-4} - \cdots\right].
\end{aligned} \tag{10.35}$$

The Noble identities are applied to move the downsample-by-2 operation to the input side of the filterbank. The filterbank consists of two filters $G_0(z)$ and $G_1(z)$. Both subfilters contain the sign alternations that may be interpreted as a frequency translation by the half sample rate. As discussed in the previous section, the sign alternations may be applied to the commutated input data sequences and processed by a filterbank whose subfilters, $H_0(z)$ and $H_1(z)$, are based on the two-stage polyphase decomposition of the low-pass filter $H(z)$. The resulting system is illustrated in Figure 10.1.11 (a). This system has the advantage that the first polyphase filterbank can operate at half the IF sample rate and that no complex-valued arithmetic is required.

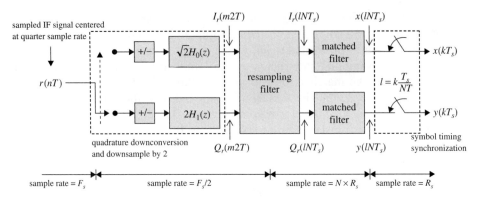

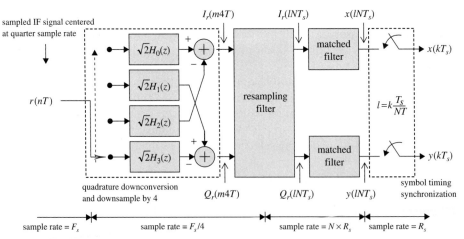

Figure 10.1.11 A QAM demodulator based on a polyphase filter implementation of the system in Figure 10.1.9 (c) for the case where the sampled IF signal is centered at the quarter sample rate: (a) Performing the quadrature downconversion using a downsample by $D = 2$. (b) performing the quadrature downconversion using a downsample by $D = 4$.

The system of Figure 10.1.11 (b) is based on the $D = 4$ stage polyphase decomposition of the band-pass filter $g(n)$:

$$G(z) = G_0(z^4) + z^{-1}G_1(z^4) + z^{-2}G_2(z^4) + z^{-3}G_3(z^4). \tag{10.36}$$

The four polyphase subfilters are

$$G_0(z^4) = \sqrt{2}h(0) + \sqrt{2}h(4)e^{j4\Omega_0}z^{-4} + \sqrt{2}h(8)e^{j8\Omega_0}z^{-8} + \cdots$$
$$= \sqrt{2}\left[h(0) + h(4)z^{-4} + h(8)z^{-8} + \cdots\right]$$
$$= \sqrt{2}H_0(z^4)$$

Section 10.1 Advanced Discrete-Time Architectures

$$\begin{aligned}
G_1(z^4) &= \sqrt{2}h(1)e^{j\Omega_0} + \sqrt{2}h(5)e^{j5\Omega_0}z^{-4} + \sqrt{2}h(9)e^{j9\Omega_0}z^{-8} + \cdots \\
&= j\sqrt{2}\left[h(1) + h(5)z^{-4} + h(9)z^{-8} + \cdots\right] \\
&= j\sqrt{2}H_1(z^4) \\
G_2(z^4) &= \sqrt{2}h(2)e^{j2\Omega_0} + \sqrt{2}h(6)e^{j6\Omega_0}z^{-4} + \sqrt{2}h(10)e^{j10\Omega_0}z^{-8} + \cdots \quad (10.37)\\
&= -\sqrt{2}\left[h(2) + h(6)z^{-4} + h(10)z^{-8} + \cdots\right] \\
&= -\sqrt{2}H_2(z^4) \\
G_3(z^4) &= \sqrt{2}h(3)e^{j3\Omega_0} + \sqrt{2}h(7)e^{j7\Omega_0}z^{-4} + \sqrt{2}h(11)e^{j11\Omega_0}z^{-8} + \cdots \\
&= -j\sqrt{2}\left[h(3) + h(7)z^{-4} + h(11)z^{-8} + \cdots\right] \\
&= -j\sqrt{2}H_3(z^4).
\end{aligned}$$

The coefficients of filters $G_0(z^4)$ and $G_2(z^4)$ are purely real whereas the coefficients of filters $G_1(z^4)$ and $G_3(z^4)$ are purely imaginary. Because filter input sequence is real, the outputs of filters $G_0(z^4)$ and $G_2(z^4)$ are combined to form the real part of the filter output and the outputs of filters $G_1(z^4)$ and $G_3(z^4)$ are combined to form the imaginary part of the filter output. Applying the Noble identities, the downsampled-by-4 operation is moved to the input side of the filter to create the system illustrated in Figure 10.1.11 (b).

Observe that no multiplications by the complex exponential are required because the required frequency translations occur for free. The initial downsample operation is performed at 1/4 the IF sample rate. Any of the applicable techniques of Section 9.3 may be used for the resampling filter.

Case 3: The IF Sample Rate Aliases the Intermediate Frequency to a Simple Rational Multiple of Sample Rate and is Not an Integer Multiple of the Symbol Rate
When it is not possible to sample the IF signal and alias the samples to the quarter sample rate, multirate processing combined with complex-valued signal processing can still be exploited to reduce the computation burden required to perform the demodulator tasks. A complex-valued band-pass sequence — with spectral content only on the positive Ω axis — can be created by filtering the real-valued IF samples using a complex-valued filter. The frequency response of this filter is "one-sided" and centered at the discrete-time frequency corresponding to the intermediate frequency, that is, at $\Omega_0 = (\omega_0 T) \bmod (2\pi)$ rad/sample. This filtering reduces the bandwidth of the signal so that it can be downsampled. Because the signal is complex-valued, the downsampling can be performed without concern for the positive and negative frequency components overlapping as a result of aliasing.

The starting point is the system shown in Figure 10.1.9 (c) and the spectra of the corresponding signals illustrated in Figure 10.1.10 (c). In this case,

$$\Omega_0 = \text{integer} \times \frac{2\pi}{D}. \quad (10.38)$$

As a consequence
$$D\Omega_0 = 0 \mod (2\pi) \tag{10.39}$$
and the spectra in the second and third plots of Figure 10.1.10 (c) are the same and centered at baseband. Thus, the residual heterodyne required to produce the spectrum in the third plot from the spectrum in the second plot is not required.

Examination of this process in the time domain reveals another advantage. Let $h(n)$ be the prototype FIR low-pass filter designed to operate at the IF sample rate. The desired band-pass filter, centered at Ω_0 rad/sample is created from $h(n)$ using

$$g(n) = h(n)\sqrt{2}e^{j\Omega_0 n}. \tag{10.40}$$

Downsampling the output of the filter $G(z)$ by a factor D can be accomplished efficiently using the polyphase partition of $G(z)$ (see Section 3.2):

$$G(z) = G_0\left(z^D\right) + z^{-1}G_1\left(z^D\right) + \cdots + z^{-m}G_m\left(z^D\right) + \cdots + z^{-(D-1)}G_{D-1}\left(z^D\right) \tag{10.41}$$

where the m-polyphase subfilter may be expressed as

$$\begin{aligned}G_m(z^D) &= g(m) + g(m+D)z^{-D} + g(m+2D)z^{-2D} + \cdots \\ &= \sqrt{2}h(m)e^{j\Omega_0 m} + \sqrt{2}h(m+D)e^{j\Omega_0(m+D)}z^{-D} + \sqrt{2}h(m+2D)e^{j\Omega_0(m+2D)}z^{-2D} + \cdots \\ &= \sqrt{2}e^{j\Omega_0 m}\left[h(m) + h(m+D)e^{j\Omega_0 D}z^{-D} + h(m+2D)e^{j2D\Omega_0}z^{-2D} + \cdots\right].\end{aligned} \tag{10.42}$$

Applying this polyphase partition to the system shown in Figure 10.1.9 (c) produces the system shown in Figure 10.1.12 (a). In general, the output of the polyphase filterbank is a band-pass signal centered at $D\Omega_0 \mod (2\pi)$ rad/sample as shown by the second plot in Figure 10.1.10 (c). The complex heterodyne is used to translate this signal to baseband where it is resampled by a real-valued, low-pass resampling filter in preparation for processing by the matched filter operating at N samples/symbol.

In this case, Ω_0 is an integer multiple of $2\pi/D$ and the downsample-by-D operation aliases the IF signal directly to baseband as described above. (There is no interference resulting from spectral overlap with the negative frequency component because it was eliminated by the band-pass filter.) The other advantage is a reduction in the complexity required to implement the polyphase filter bank. The complex exponentials multiplying the filter coefficients inside the square brackets of (10.42) are unity. Hence, the powers of z^{-D} inside the square brackets are purely real so that

$$G_m(z^D) = \sqrt{2}e^{j\Omega_0 m}H_m\left(z^D\right) \tag{10.43}$$

where $H_m(z^D)$ is the z-transform of the m-th subfilter in the polyphase decomposition of the real-valued, low-pass filter $H(z)$.

The resulting system is illustrated in Figure 10.1.12 (b). There are two noteworthy features of this system. First, the polyphase filterbank is composed of real-valued subfilters. The

Section 10.1 Advanced Discrete-Time Architectures

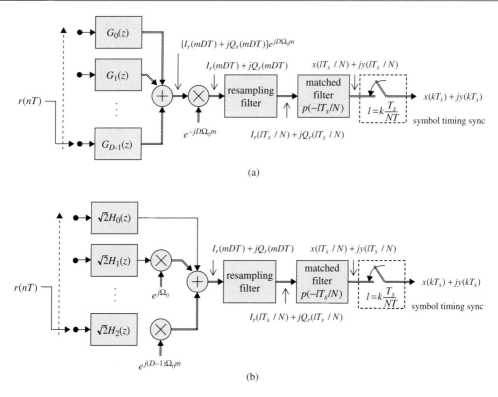

Figure 10.1.12 A QAM demodulator based on a polyphase filter implementation of the system illustrated in Figure 10.1.9 (c): (a) A direct application of the polyphase filter to the filter $G(z)$; (b) the simplification that follows when Ω_0 is an integer multiple of $2\pi/D$.

filterbank input is real. As such, only real-valued arithmetic is required for the filtering. The signal does not become complex valued until multiplication by the complex exponentials at the *output* of each subfilter. This represents a substantial reduction in required resources over the direct application suggested by the system in Figure 10.1.12 (a). The second feature is that heterodyne operation using multiplication by $e^{jD\Omega_0 m}$ is not required in Figure 10.1.12 (b) because $D\Omega_0$ is a multiple of 2π. (In other words, the downsample-by-D operation aliases the sampled IF signal directly to baseband.) Note that the systems of Figure 10.1.11 are special cases of Figure 10.1.12 (b) that exploit the special properties of $\Omega_0 = \pi/2$.

Case 4: The IF Sample Rate is Arbitrary and is Not an Integer Multiple of the Symbol Rate
When the IF sample rate is such that IF does not alias to a simple rational ratio of the sample rate, complex-valued signal processing and multirate processing may still be used to advantage. In this case, the system of Figure 10.1.12 (a) may be used. The downsample factor D is chosen to reduce the sample rate as close as possible to the minimum sample rate determined by the sampling theorem. The resampling filter is then used to change the

sample rate to an integer multiple of the symbol rate. The frequency domain illustrations of Figure 10.1.10 (c) apply to this case.

Using the polyphase filterbank, the complex-valued band-pass filter $G(z)$ can operate at the reduced sample rate. A detailed diagram of the demodulator is illustrated in Figure 10.1.13. Observe that the downsample-by-D operation creates a complex-valued band-pass sequence centered at $D\Omega_0 \mod (2\pi)$. The signal is translated to complex baseband by the complex-valued heterodyne shown using four multipliers and two adders. Even though this residual complex-valued heterodyne is still required, it operates at the reduced sample rate. The resulting baseband signals are resampled to N times the symbol rate and matched filtered to produce the desired outputs.

10.1.3 Putting It All Together

The four cases outlined above demonstrate that when the ADC is placed at IF, more options for efficient signal-processing are available to the system designer. These efficiencies are realized by exploiting multirate signal processing to perform frequency translation, sample rate conversion, and matched filtering. An additional advantage is that carrier phase and symbol timing synchronization can be incorporated into the multirate processing. This is possible only when the ADC clock rate is high enough to sample the IF signal at the desired rate. This section outlines the basic architectures for QAM demodulators as a function of ADC placement and the use of multirate processing.

First Generation Architectures. First generation sampled-data systems were based on the I/Q baseband sampling architecture described previously. The basic architecture is illustrated by the block diagram in Figure 10.1.14. The system includes a section to perform analog preconditioning, an interface segment to convert the preconditioned data to a sampled data representation, a digital post conditioning segment to minimize the contributions of channel distortion and noise to the output signal, and a detector—implemented using DSP—wto estimate the parameters of the modulated signal. The detector also supplies feedback information to a carrier recovery loop, a timing recovery loop, and an optional equalizer controller. The carrier recovery loop aligns the frequency and phase of the voltage-controlled oscillator (in the final downconverter) to the carrier frequency and phase of the received IF signal. Similarly, the timing recovery loop aligns the frequency and phase of the sampling clock so that the position of the data samples at the output of the discrete-time matched filter coincide with the time location of the maximum eye opening.

This architecture presents some challenges to the system designer. The first set of challenges is associated with I/Q baseband sampling outlined earlier in this section. The second set of challenges are due to the inclusion of the continuous-time components in the carrier phase and symbol timing PLLs. Continuous-time signals are required to control the VCOs of the carrier phase PLL (and possibly the symbol timing PLL depending on implementation details). The error signals are generated by phase detectors and loop filters that reside in the DSP. Thus, digital-to-analog conversion must take place in the feedback portion of these loops. The support hardware required to operate the ADCs include analog smoothing filters, registers for word framing and latching, and data and control signal lines.

Section 10.1 Advanced Discrete-Time Architectures

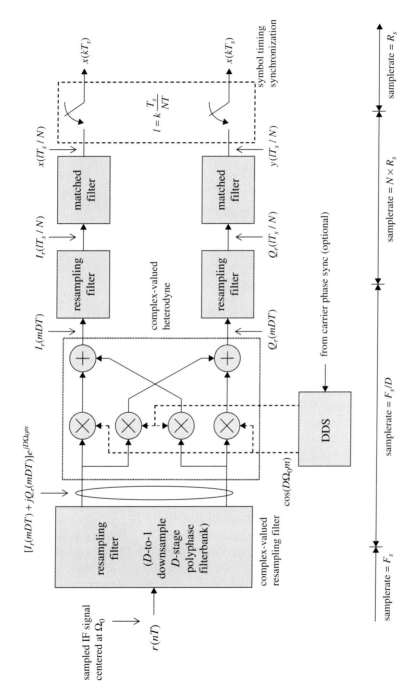

Figure 10.1.13 A detailed block diagram of the QAM demodulator based on the system illustrated in Figure 10.1.12 (a).

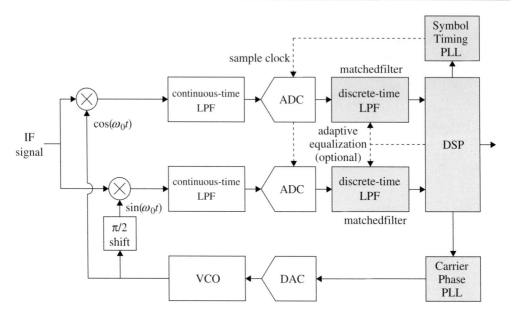

Figure 10.1.14 A block diagram of a first generation discrete-time detector. The shaded areas reside in the continuous-time domain and the unshaded areas reside in the discrete-time domain.

This additional hardware complicates system design by blurring the boundary between the continuous-time circuits and discrete-time circuits.

This architecture emerged primarily due to the limited sample rates of ADCs available at the time. As ADC rates increased, it became feasible to move the ADC to the IF signal and sample the IF signal directly. This lead to the second generation discrete-time detectors.

Second Generation Architectures. Second generation architectures, such as the one illustrated in Figure 10.1.15, sampled the IF signal directly, and were little more than discrete-time equivalents of the established continuous-time designs. Because the sampling and quantization occur prior to the quadrature downconversion, the downconversion process can take place in discrete-time where there are no gain and phase balance issues (it is just multiplication by numbers). After the digital downconversion, the sampled signal is filtered to the signal bandwidth (plus any frequency offset due to Doppler or local oscillator drift) and resampled to reduce the clock rate required to process the samples. Another advantage is that the carrier phase PLL functions reside solely in the discrete-time domain. As such, the complexity of the DAC in the feedback path of the first generation detector is removed.

These advantages come at a price: the ADC and the discrete-time processing that follows must operate at a much higher clock rate. In addition, the symbol timing synchronizing PLL is closed at the IF ADC. This not only complicates the system design (by requiring a signal

Section 10.1 Advanced Discrete-Time Architectures 635

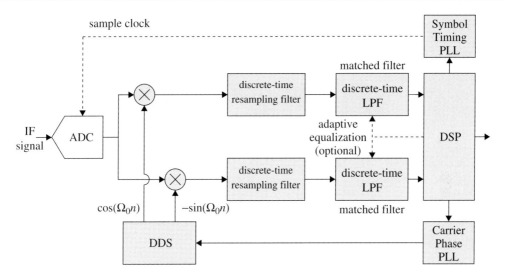

Figure 10.1.15 A block diagram of a second generation discrete-time detector. The shaded areas reside in the continuous-time domain and the unshaded areas reside in the discrete-time domain. (Compare with the first generation detector illustrated in Figure 10.1.14.)

path from the DSP to the ADC) but also substantially increases clock jitter on the ADC. The effects of clock jitter on ADC performance is discussed in Section 9.1.1.

The clock rates required for IF sampling were discussed in Section 10.1.2. Band-pass subsampling can be used to reduce the clock rate and symbol timing synchronization can be performed exclusively in discrete-time. These features are characteristic of the third-generation receivers described next.

Third Generation Architectures. The common theme in the second-generation receiver is that the sampled signal is translated, filtered, and downsampled after being collected and sampled as a real-valued signal from the IF stage. This architecture is motivated by the structure of the continuous-time systems. Third generation receivers exploit DSP-unique solutions based on multirate processing. The basic structure is illustrated in Figure 10.1.16. The most common application of multirate processing combines band-pass subsampling with resampling to perform the required frequency translations. When possible, the sample rate is selected to produce a real-valued, band-pass sequence centered at the quarter sample rate. With this sample rate, multirate processing can be used to perform the required frequency translations for free. The sample rate is a function of the intermediate frequency as described previously. Because the intermediate frequency is only indirectly related to the symbol rate, this sample rate is rarely an integer multiple of the symbol rate. As a consequence, some resampling must be performed prior to the matched filtering operation.

A unique attribute of the downsampling operation is that phase of the sample locations in the downsampled time series can be changed relative to the epochs in the series. For example,

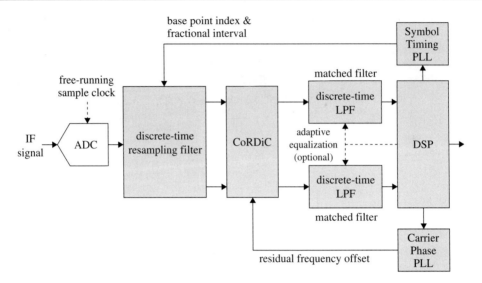

Figure 10.1.16 A block diagram of a third generation discrete-time detector. The shaded areas reside in the continuous-time domain and the unshaded areas reside in the discrete-time domain. (Compare with the first and second generation detectors illustrated in Figures 10.1.14 and 10.1.15, respectively.)

if D-to-1 downsampling is being performed, the polyphase filter bank provides access to D different sample phase offsets. This access can be exploited in the symbol timing synchronizer. Rather than have the timing recovery loop modify the locations of sample points during the ADC process, as done in first and second generation receivers, the phase of the resampling process can be controlled in the resampling filter. Because the phase of the resampling filter is defined by the selection of the phase weights in the subfilter, the timing recovery process defaults to controlling an index pointer in the filter's coefficient space. Note that when the polyphase filter is used as a part of the timing loop, the number of stages in the partition must be increased to satisfy the timing granularity requirements of the timing loop rather than the downsampling requirement.

The resampling in the polyphase filter is used to effect both a spectral translation and a timing recovery process. The result is a complex-valued signal at baseband at the reduced sample rate with the correct timing, but not with the correct carrier frequency and phase correction. Using a complex heterodyne to perform the carrier phase and frequency correction requires four multiplications and two additions (it is a complex-by-complex multiplication) as opposed to the two multiplications when performed on the real-valued band-pass IF signal samples. This complexity increase is offset by the fact that the complex-by-complex multiplications occur at a lower clock rate than the pair of real-by-real multiplications at IF. The complexity of the complex baseband heterodyne can be reduced by using the CoRDiC algorithm described in Chapter 9. In this way, the de-rotation required by the final heterodyne can be performed without the need of a complex-by-complex multiply. This option offers marked advantages when the system is implemented in a Field programmable gate array (FPGA).

The final result is a structure that "looks" very different from the continuous-time designs. The discrete-time processing may reside in digital application-specific integrated circuits (ASICs) which may have some functionality parameterizable, FPGAs, and programmable processors (usually DSPs). At the time of this writing, much of the high-rate signal processing resides in FPGAs while many of the lower rate functions (speech encoding/decoding, for example) reside in specialized programmable processors such as DSPs. The cost of custom ASIC design is tremendous, thus only high volume applications (such as mobile telephony) are economically viable at the present. The look will continue to change as the cost and processing capabilities evolve.

10.2 CHANNELIZATION

Most receivers are required to select the energy in a specified bandwidth centered around a desired frequency. This requirement implies that the spectral content outside the desired frequency band must be rejected. In many applications, there are several bands of potential interest each centered at a different carrier frequency. These bands are usually called *channels* and the function of channel selection is often called *tuning*.

The concept of using different carrier frequencies to multiplex different signals in the RF spectrum is called *frequency division multiplexing* (FDM), and is perhaps the multiplexing technique most familiar to students new to communications. Examples of FDM include commercial broadcast AM and FM radio, television (broadcast and cable, analog and digital), first and second generation cellular telephony, aviation communication systems, public safety radio (police, fire, emergency response), and cordless telephones. Most satellite transponders are channelized as specified by the transponder "frequency plan." Examples are illustrated in Table 10.2.1.

With analog communications, FDM is the most practical multiplexing technique. For digital communications such as QAM, time division multiplexing is also a practical method. Code division multiplexing is the method used in spread spectrum systems. Recent advances in multiple-antenna systems have made spatial-multiplexing a viable method as well. The reality is almost all systems use a combination of multiplexing techniques to meet system requirements and almost all include some form of FDM. Although multiplexing and multiple access techniques are not a focus of this text, FDM plays a fundament role in the overall architecture of the receiver.

The process by which the receiver isolates the desired bandwidth centered around the desired carrier frequency is called *channelization* (i.e., the receiver channelizes the RF spectrum for its own purposes). How this is accomplished is strongly dependent on whether continuous-time processing or discrete-time processing is used.

10.2.1 Continuous-Time Techniques: The Superheterodyne Receiver

The most intuitive approach to channel selection using continuous-time processing is the use of an adjustable (or tunable) band-pass filter as illustrated in Figure 10.2.1 (a). Such

Table 10.2.1 Some channelization plans for some commonly familiar wireless communication networks.

System	Frequency Range	Channel Bandwidth	Number of Channels
Commercial broadcast AM	535–1605 kHz	10 kHz	107
Commercial broadcast FM	88–108 MHz	200 kHz	100
Commercial broadcast TV			
Channels 2–4	54–72 MHz	6 MHz	3
Channels 5–6	76–82 MHz	6 MHz	2
Channels 7–13	174–216 MHz	6 MHz	7
Channels 14–69	470–806 MHz	6 MHz	56
AMPS (Advanced Mobile Phone Service): First generation cellular			
Mobile to base-station	824–849 MHz	30 kHz	832
Base-station to mobile	869–894 MHz	30 kHz	832
IS-54: First generation digital cellular			
Mobile to base-station	824–849 MHz	30 kHz	832
Base-station to mobile	869–894 MHz	30 kHz	832
GSM (Groupe Spécial Mobile): European digital cellular			
Mobile to base-station	890–915 MHz	200 kHz	125
Base-station to mobile	935–960 MHz	200 kHz	125
Aviation communication links			
Navigation	108–118 MHz	50 kHz	200
Air-traffic control	118–137 MHz	25 kHz	760
Cordless telephone standards			
CT2 (Europe)	846.15–868.05 MHz	100 kHz	40
DECT (Europe)	1880–1900 MHz	1.782 MHz	10
PACS (USA)	1850–1910 MHz	300 kHz	200
	1930–1990 MHz	300 kHz	200
Aeronautical telemetry standard (IRIG 106)			
L-band channels	1435–1525 MHz	1 MHz	100
S-band channels	2200–2290 MHz	1 MHz	90
	2360–2390 MHz	1 MHz	30

systems are usually referred to as "tuned radio frequency," or TRF systems. The basic concept underlying the operation is straightforward as illustrated in Figure 10.2.2. Suppose the spectra of the desired signals are arranged as shown in Figure 10.2.2 (a) and that the signal centered at carrier frequency f_1 is the desired signal. The center frequency of the adjustable band-pass filter is adjusted (or tuned) as shown in Figure 10.2.2 (b) to isolate the desired spectrum as illustrated in Figure 10.2.2 (c). This signal is then processed by the demodulator/detector. Likewise if the signal centered at frequency f_2 is desired, then the band-pass filter is "tuned" to f_2 as illustrated in Figure 10.2.2 (d). The signal illustrated in Figure 10.2.2 (e) is then presented to the demodulator/detector.

Section 10.2 Channelization

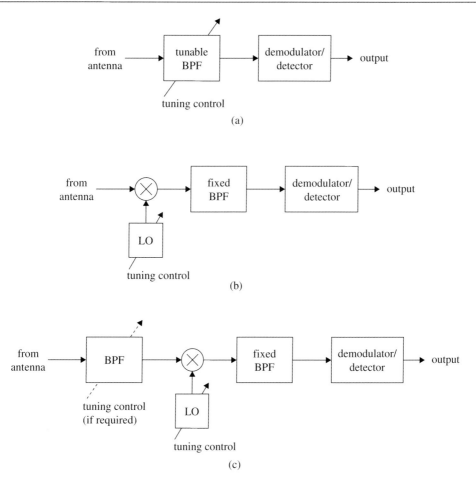

Figure 10.2.1 Channelizers based on continuous-time processing: (a) a simple tuned radio frequency (TRF) channelizer based on a tunable band-pass filter; (b) a simple superheterodyne channelizer; (c) the superheterodyne channelizer including the image rejection filter.

There are two challenges with this approach. The first involves the difficulties in designing tunable band-pass filters with constant bandwidth and sufficient frequency selectivity over the entire tunable range. This difficulty is illustrated by the simple example in Exercise 10.11. The second difficulty is the design of a demodulator/detector capable of operating over a range of center frequencies. This may or may not be a problem, depending on the modulation type and the demodulation method employed.

The earliest AM radio receivers, such as the one illustrated in Figure 10.2.3, were TRF receivers. The desired radio station was selected by adjusting variable capacitors and/or inductors in each stage of a multistage band-pass filter. As a consequence, many early radios,

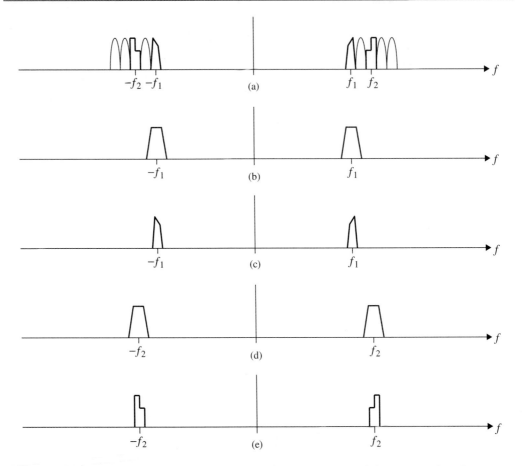

Figure 10.2.2 A frequency domain illustration of the operation of the TRF receiver in an FDM system.

like the one in Figure 10.2.3, was equipped with multiple tuning knobs: one knob dedicated to each tuning stage.

The challenges of the TRF receiver were overcome by examining the problem from a different point of view. Instead of moving the filter to signal as the TRF receiver does, the signal could be moved to fixed filter by using a tunable oscillator as illustrated in Figure 10.2.1 (b). This receiver, invented by Howard Armstrong during World War I, is called the *superheterodyne*[1] receiver. The frequency of the tunable oscillator, called a *local oscillator*

---

[1] The term *heterodyne* is derived from the Greek *heteros* (ετερος) meaning "other" and *dynamis* (δυναμις) meaning "force." It was first coined by Reginald Fessenden in a 1901 patent application where the well-known property of a "beat frequency"—the difference frequency resulting from the simultaneous presence of two audio tones—was applied to radio frequency signals. The original concept envisioned the transmission of two RF signals each received with its own antenna and combined at the detector. Later,

Figure 10.2.3 The "American Beauty" built by American Auto & Radio Mfg. Co. of Kansas City, Missouri from 1925 to 1926. The tuner consisted of two stages of tunable circuits based on vacuum tubes. (Image courtesy of Western Historic Radio Museum, Virginia City Nevada, http://www.radioblvd.com)

or LO, is adjusted to translate the center frequency of the desired signal to a fixed frequency called the *intermediate frequency* or IF. A fixed band-pass filter isolates the desired signal for processing by the demodulator/detector. Because the band-pass filter does not have to be tuned, it can be designed to have excellent band-pass characteristics. As an additional benefit, the demodulator/detector only has to operate at a single center frequency and can thus be optimized for operation at IF.

The operation of the superheterodyne receiver is best described in the frequency domain. Suppose the desired signal is centered at f_c. As such, it may be expressed as

$$s(t) = u(t) \cos(2\pi f_c t) \qquad (10.44)$$

$$S(f) = \frac{1}{2} U(f + f_c) + U(f - f_c). \qquad (10.45)$$

Using the modulation theorem from Chapter 2, the Fourier transform of the product $x(t) = s(t) \cos(2\pi f_{LO} t)$ is

$$X(f) = \frac{1}{2} U(f + f_c + f_{LO}) + \frac{1}{2} U(f + f_c - f_{LO}) + \frac{1}{2} U(f - f_c + f_{LO}) + \frac{1}{2} U(f - f_c - f_{LO}). \qquad (10.46)$$

---

one of the two RF transmitter-antenna systems was replaced by an RF source generated at the detector (a local oscillator) to produce the version of the heterodyne receiver described in most introductory texts on communication systems today. Fessenden (who was introduced in Chapter 1) was awarded U.S. Patent 706,740 issued on 12 August 1902.

If the third term on the right-hand-side of (10.46) is to be centered at the IF f_0, then the relationship $f_0 = f_c + f_{LO}$ must hold and it follows that the LO frequency must be

$$f_{LO} = f_c - f_0. \tag{10.47}$$

In other words, the LO frequency must be set to the difference between the carrier frequency and the IF to translate the desired spectrum from f_c to the IF f_0. Use of (10.47) for the LO frequency is called *low-side mixing*. The operation of the superheterodyne receiver using low-side mixing is illustrated in Figure 10.2.4. The desired signal spectrum, centered at f_c as shown in Figure 10.2.4 (a), is translated to f_0 by the mixer whose frequency is set to $f_{LO} = f_c - f_0$, as shown in Figure 10.2.4 (b). A fixed band-pass filter, with frequency response illustrated in Figure 10.2.4 (c), isolates the desired signal centered at IF as illustrated in Figure 10.2.4 (d).

As an alternative, the second term of (10.46) could be used to define the IF. If this spectral copy is to be centered at IF, then the relationship $-f_0 = f_c - f_{LO}$ must hold from which the relationship

$$f_{LO} = f_c + f_0 \tag{10.48}$$

follows. When the LO frequency is defined using (10.48), *high-side mixing* results. Operation of the superheterodyne receiver using high-side mixing is illustrated in Figure 10.2.5. The spectrum of the desired spectrum, centered at f_c as shown in Figure 10.2.5 (a), is translated to

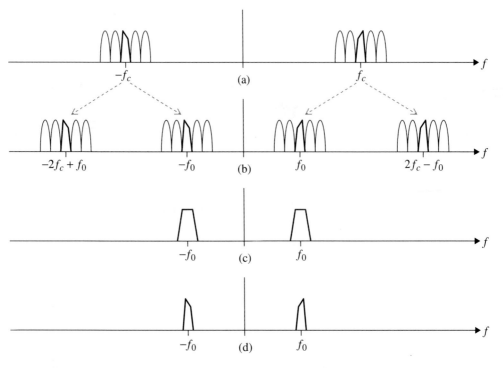

Figure 10.2.4 Spectra associated with the superheterodyne receiver using low-side mixing.

Section 10.2 Channelization

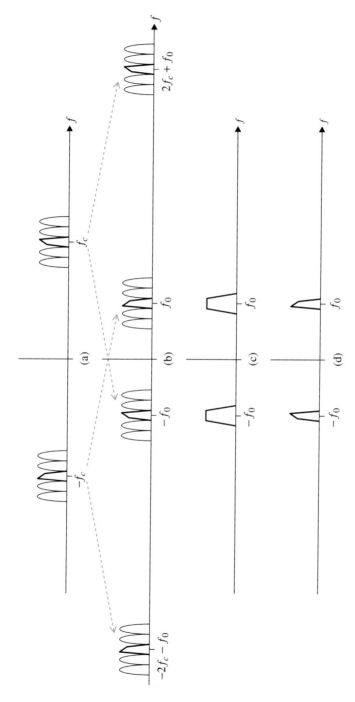

Figure 10.2.5 Spectra associated with the superheterodyne receiver using high-side mixing.

the IF by the mixer using an LO set to $f_{LO} = f_c + f_0$. The resulting translations are shown in Figure 10.2.5 (b). The fixed band-pass filter illustrated in Figure 10.2.5 (c) isolates the desired signal at IF to produce the spectrum illustrated in Figure 10.2.5 (d). Observe that when high-side mixing is used, the negative and positive frequency components are swapped at IF. As a consequence, the conjugate of the signal is presented to the demodulator/detector. Contrast this with the case of low-side mixing where the positive frequency component is translated to IF as illustrated in Figure 10.2.4 (b).

The fact that negative or positive frequency components can be translated to the IF by mixing poses a problem for the superheterodyne receiver. This problem is illustrated in Figure 10.2.6, which is a modified version of Figure 10.2.4 for low-side mixing. Using low-side mixing the spectrum centered at f_c is translated to f_0. Another band of frequencies centered at $-f_i$, however, is also translated to f_0 when f_i satisfies the condition $-f_i + f_{LO} = f_0$. Substituting (10.47) shows that this frequency is

$$f_i = f_c - 2f_0. \tag{10.49}$$

This undesired frequency is called an *image frequency*. The spectral content centered at the image frequency must be removed prior to the mixer. This is accomplished using an image

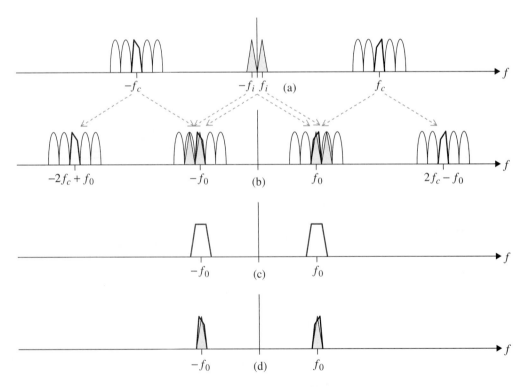

Figure 10.2.6 Spectra associated with the superheterodyne receiver using low-side mixing. In this figure, the image frequency is shown.

rejection filter as illustrated in Figure 10.2.1 (c). If the image frequencies fall outside the band of desired signals, the image rejection filter can be a fixed filter. If, on the other hand, the image frequencies fall inside the band of desired signals, the image rejection filter must be tunable. While at first glance this may seem to defeat the purpose of using the superheterodyne receiver in place of the TRF receiver, the image rejection filter can be a coarsely tunable band-pass filter. Its only requirements are to pass the desired signal with as little distortion as possible and reject the spectral content centered at the image frequency. These requirements are much less stringent than channel selection requirements for the TRF receiver. The exact bandwidth requirements for this filter are explored in Exercise 10.13. The image frequency given by (10.49) applies to the superheterodyne receiver using low-side-mixing. The problem for high-side mixing is explored in Exercise 10.12. Some common intermediate frequencies used in the United States are summarized in Table 10.2.2.

The characteristic that makes a particular frequency a good intermediate frequency is a trade-off between competing demands.

- First, and most importantly, if the IF is nonzero, it should be larger than half the RF bandwidth. This ensures that the spectral copies centered at the IF frequency f_0 illustrated in Figure 10.2.4 (d) or Figure 10.2.5 (d) do not overlap each other at baseband.
- High quality IF band-pass filters—the fixed band-pass filter of Figure 10.2.1 (c)—with good out-of-band rejection are easier to produce at lower frequencies. This suggests the intermediate frequency be chosen as *low* as possible.
- The image rejection band-pass filter—the variable band-pass filter of Figure 10.2.1 (c)—must reject spectral energy centered at the image frequency that is separated from the RF carrier by twice the IF—see (10.49) and Exercise 10.13. With most continuous-time band-pass filters, the out-of-band attenuation increases as the frequency moves further away from the center of the pass band. This suggests the intermediate frequency be chosen as *high* as possible.

The RF image rejection band-pass filter and the IF channel selection band-pass filter impose competing constraints on the intermediate frequency. Sometimes, these competing

Table 10.2.2 Common intermediate frequencies used in the United States

IF	Application
262.5 kHz	Commercial broadcast AM receivers in automobiles
455 kHz	Commercial broadcast AM receivers
10.7 MHz	Commercial broadcast FM receivers
21.4 MHz	Two-way FM radios
30 MHz	RADAR receivers
43.75 MHz	Video IF in NTSC television receivers
45 MHz	Surveillance and monitoring receivers
60 MHz	RADAR receivers
70 MHz	Satellite receivers
140 MHz	Satellite receivers

constraints make the structure of Figure 10.2.1 (c) unworkable. One way to address this issue is to perform the translation from RF to the desired IF in more than one step. Receivers that use two steps are called *dual-conversion* receivers. The characteristics of the dual-conversion receivers are examined in Exercises 10.26 and 10.27.

Another method that can be used to address the competing demands is to remove the image frequency issue. The image frequency is a result of mixing a band-pass signal with a real-valued sinusoid. Multiplication by a real-valued sinusoid produces frequency translations in both the positive and negative frequency directions. The image frequency issue can be eliminated (along with its competing constraint on the intermediate frequency) by using complex-valued signal processing. Mixing with a complex exponential produces frequency translations in only one direction (see Appendix B). This eliminates the possibility for interference by an image frequency.

The block diagram of a system that performs a "one-sided" frequency translation is illustrated in Figure 10.2.7. In Figure 10.2.7 (a), the received RF signals are translated from the RF carrier frequency $\omega_c = 2\pi f_c$ rad/s to an intermediate frequency $\omega_0 = 2\pi f_0$ rad/s that is greater than zero. A frequency-domain representation of this system is illustrated in Figure 10.2.8. Multiplication by the complex exponential translates all the spectral content to the left by $\omega_c - \omega_0$ rad/s as shown in Figures 10.2.8 (a) and (b). The resulting spectrum no longer possesses conjugate symmetry. Hence, the corresponding time-domain signal is complex-valued. Observe that because $\omega_0 > 0$, a "one-sided" band-pass filter is required as illustrated in Figure 10.2.8 (c). This filter isolates the positive-frequency component of the desired signal as shown in Figure 10.2.8 (d). Implementing a complex-valued heterodyne requires two mixers and an adder as illustrated in Figure 10.2.7 (a). The mixers require quadrature versions of the LO: $\cos(\omega_{LO}t)$ and $\sin(\omega_{LO}t)$ as shown.

The main challenges with this approach are the I/Q-imbalance issue associated with producing quadrature sinusoids and the difficulties associated with constructing a complex-valued band-pass filter in continuous-time. The second issue can be overcome by reducing the intermediate frequency to baseband and replacing the complex-valued band-pass filter with a real-valued low-pass filter. Such a receiver is called a "zero-IF" receiver. The block diagram of a zero IF receiver is illustrated in Figure 10.2.7 (b). A frequency domain interpretation of the zero-IF receiver is shown in Figure 10.2.9. In this case, multiplication by the complex exponential translates the positive frequency component of the desired signal, centered at $\omega_c = 2\pi f_c$ rad/s to baseband. These translations are illustrated in Figures 10.2.9 (a) and (b). The frequency translations are in only one direction. Consequently, there is no interference resulting from sign negative frequency component of the desired signal. A low-pass filter is now required to isolate the desired signal and reject the unwanted spectral energy as shown in Figures 10.2.9 (c) and (d). Because the low-pass filter can possess conjugate symmetry, its impulse response is purely real, and a real-valued filter can be used. Processing a complex-valued signal with a real-valued low-pass filter is accomplished using a pair of identical real-valued low-pass filters such as the system illustrated in Figure 10.2.7 (b). In this system, the requirement of a complex-valued continuous-time band-pass filter has been replaced by the requirement of a pair of identical continuous-time low-pass filters. The filters must be exactly

Section 10.2 Channelization

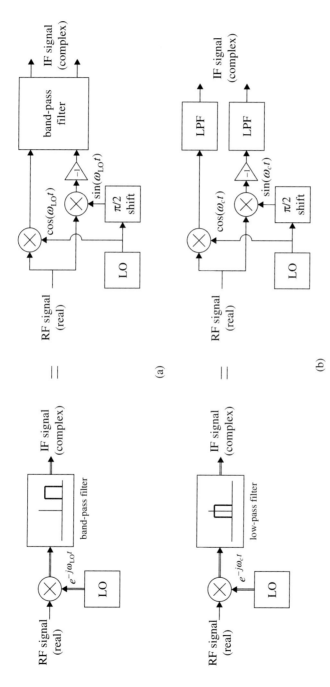

Figure 10.2.7 The superheterodyne receiver based on complex-valued signal processing: (a) The receiver architecture based on a nonzero intermediate frequency; (b) the receiver architecture based on a zero intermediate frequency.

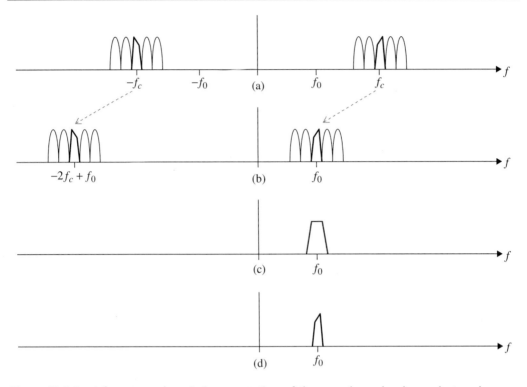

Figure 10.2.8 A frequency domain interpretation of the complex-valued superheterodyne receiver of Figure 10.2.7 (a): (a) The spectrum of the received signal; (b) the spectrum of the signal at the output of the complex-valued mixer; (c) the transfer function of the complex-valued band-pass IF filter; (d) the output of the band-pass IF filter.

the same to preserve the phase and amplitude relationship of the real and imaginary parts of the complex-valued signal.

The demodulator architecture based on I/Q baseband sampling, such as the one illustrated in Figure 10.1.5 (a) or the first generation architecture of Figure 10.1.14 are examples of the zero-IF superheterodyne channelizer. Manufacturing tolerances in the continuous-time components impose gain and phase imbalances that can distort the received signal and degrade performance. These shortcomings do not carry over into discrete-time processing. Consequently, channelization in discrete-time processing is almost always based on the zero-IF architecture.

The QAM detectors discussed in Section 5.3 were assumed to operate directly on the IF output of a superheterodyne receiver. In principle, these receivers could operate directly at RF with channel selection being performed in discrete-time using the zero-IF architecture. IF sampling is the most common approach and is due largely to the limitations of the ADC.

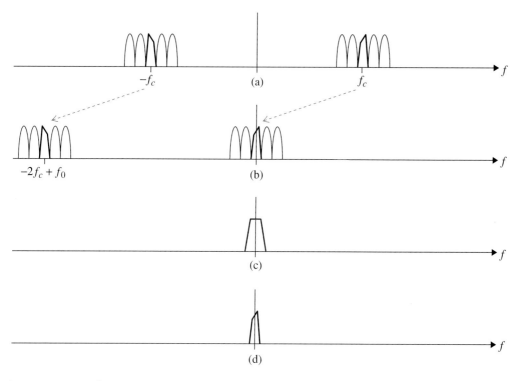

Figure 10.2.9 A frequency domain interpretation of the zero-IF receiver of Figure 10.2.7 (b): (a) The spectrum of the received signal; (b) the spectrum of the signal at the output of the complex-valued mixer; (c) the transfer function of the low-pass IF filter; (d) the output of the low-pass IF filter.

1. Operation at IF allows the ADC to operate at a lower frequency. In theory, the ADC could sample the RF signal directly and take advantage of the principles of bandpass sampling to reduce the sampling rate at which the ADC must operate. This approach works best when there is only one signal present in the RF spectrum. In other words, this approach requires that some form of channelization has already been performed. In applications where the sampling is applied directly to the RF spectrum, channelization is performed using discrete-time techniques described in Section 10.2.2. The maximum sample rate limits the bandwidth (and consequently the RF center frequencies) over which this approach is feasible.
2. Quantization effects (described in Section 9.1.1) limit the dynamic range of the output of the ADC. This issue can pose serious limitations if the ADC is applied directly to an RF signal composed of several signals in an FDM arrangement. If the amplitude of the desired signal is small and the amplitudes of the undesired signals centered at adjacent frequencies are large, most of the ADC dynamic range is consumed quantifying the larger signals (which will be eliminated by subsequent filtering in the discrete-time

processing that follows). The desired signal winds up buried in the quantization noise and is all but unrecoverable. Dynamic range is also an issue with continuous-time processing, but not as severe an issue as it is with ADCs and discrete-time processing. Thus, when the RF spectrum will present signals with a wide variety of amplitudes in the different channel slots, continuous-time processing in the form of a superheterodyne channelizer is often used to isolate the desired signal and adjust its amplitude in preparation for sampling.

10.2.2 Discrete-Time Techniques Using Multirate Processing

The discrete-time QAM demodulators introduced in Chapter 5 are examples of a discrete-time superheterodyne concept where the intermediate frequency is 0 (i.e., the IF signal is the complex baseband equivalent). Using the property that a band-pass FIR filter can be generated from a low-pass FIR filter by heterodyning the filter coefficients (see Section 2.7), both the superheterodyne and RTF channelizers may be applied in sampled data receivers in a straightforward way. The concept is illustrated in Figure 10.2.10 where the complex-valued equivalent (see Appendix B) is used for notational convenience. Assuming that the desired signal is centered at Ω_0 rad/sample, the superheterodyne receiver of Figure 10.2.10 (a) uses multiplication by the complex exponential $e^{-j\Omega_0 n}$ to center the spectrum of the desired signal at baseband and a real-valued low-pass filter $h(n)$ to eliminate the unwanted spectral content. The RTF receiver shown in Figure 10.2.10 (b) uses a complex-valued band-pass filter with coefficients

$$g(n) = h(n)e^{j\Omega_0 n} \tag{10.50}$$

to eliminate the unwanted spectral content first, followed by multiplication by $e^{-j\Omega_0 n}$ to translate the output to baseband. These two systems produce exactly the same output as discussed in Section 2.7.

Discrete-time filters are just a set of numbers (i.e., the filter coefficients). Thus, the problems associated with adjustable filters in continuous-time processing do not carry over to discrete-time processing. (In other words, the center frequency of a filter is changed by using a different set of numbers for the filter coefficients.) Note that in either case, the bandwidth of the signal $\tilde{x}(nT)$ is much less than the bandwidth of the input signal $r(nT)$. For this reason, a downsample operation is almost always included in discrete-time channelizers as illustrated in Figure 10.2.10 for both systems.

Consider the spectrum of a real-valued continuous-time signal $r(t)$ shown in Figure 10.2.11 (a). (Only the positive frequency portion of the spectrum is shown.) The signal consists of uniformly spaced carriers separated by Δf Hz. This scenario is a common feature of many real applications as suggested by the summary of Table 10.2.1. Let the desired signal be the signal whose spectrum is centered at $k\Delta f$ Hz for some integer k. The desired signal may be represented as

$$\text{desired signal} = \text{Re}\left\{\tilde{u}_k(t)e^{j2\pi k\Delta f t}\right\}. \tag{10.51}$$

The goal is to extract samples of $u_k(t)$ from samples of $r(t)$. To achieve this, the input signal $r(t)$ is filtered and sampled at T-spaced intervals to produce the corresponding discrete-time

Section 10.2 Channelization

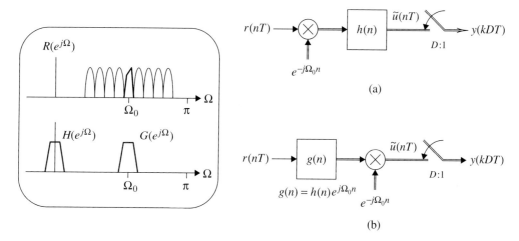

Figure 10.2.10 Two equivalent methods for channelization using discrete-time processing: (a) Shift the data to the filter (this is the discrete-time equivalent of the superheterodyne receiver); (b) shift the filter to the data (this is the discrete-time equivalent of the TRF receiver).

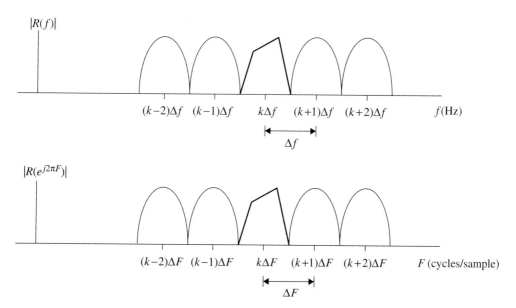

Figure 10.2.11 Spectral representation of regularly spaced channels: (a) Fourier transform of the continuous-time signal; (b) DTFT of the sampled signal.

spectrum shown in Figure 10.2.11 (b). The carrier spacing in discrete-time is $\Delta F = \Delta f T$ cycles/sample. The desired signal has a spectrum centered at $k\Delta F$ cycles/sample.

The starting point for developing the multirate channelizer is the adjustable band-pass filter approach introduced in Figure 10.2.10 (b). The application of this system to the problem at hand is illustrated in Figure 10.2.12 (a). The complex-valued band-pass FIR filter $g(n)$ is obtained from a real-valued low-pass FIR prototype filter $h(n)$ using (10.50). By adjusting the pass band of the FIR filter $g(n)$—that is, by adjusting Ω_0 in (10.50)—the desired channel is isolated and then translated to baseband via multiplication by the complex exponential. Because the baseband signal has a narrow bandwidth and has been bandlimited by the filter $g(n)$, it may be downsampled by a factor D as shown. The first step is to slide the downsample operation to the input side of the complex exponential multiplier. The resulting system is illustrated in Figure 10.2.12 (b). Observe that the complex exponential operates at a lower sample rate and advances the phase by $D\Delta F$ cycles/sample instead of ΔF cycles/sample.

The band-pass filter may be replaced by an M-stage polyphase filter

$$G(z) = G_0(z^M) + z^{-1}G_1(z^M) + \cdots + z^{-m}G_m(z^M) + \cdots + z^{-(M-1)}G_{M-1}(z^M) \quad (10.52)$$

as shown in Figure 10.2.12 (c). Note that M and D are *different*. The z-transform of the m-th subfilter may be expressed as

$$\begin{aligned}
G_m(z^M) &= g(m) + g(m+M)z^{-M} + g(m+2M)z^{-2M} + \cdots \\
&= h(m)e^{j\Omega_0 m} + h(m+M)e^{j\Omega_0(m+M)}z^{-M} + h(m+2M)e^{j\Omega_0(m+2M)}z^{-2M} + \cdots \\
&= e^{j\Omega_0 m}\left[h(m) + h(m+M)e^{j\Omega_0 M}z^{-M} + h(m+2M)e^{j2\Omega_0 M}z^{-2M} + \cdots\right] \\
&= e^{j\Omega_0 m} H_m\left(e^{-j\Omega_0 M}z^M\right)
\end{aligned}$$
(10.53)

where $\Omega_0 = k \times 2\pi \Delta F/M$. The block diagram using this notation is illustrated in Figure 10.2.13 (a). The term $e^{j2\pi k \Delta F M}$ is a result of the complex heterodyne needed to convert the low-pass filter prototype to a band-pass filter centered at $2\pi k\Delta F$ rad/sample. Observe that when

$$M = \frac{1}{\Delta F}, \quad (10.54)$$

$e^{j2\pi k \Delta F M} = 1$ and the heterodyne operation required to create the band-pass filter from the low-pass filter is obtained for free. The complexity of the filter bank is reduced because the coefficients of each filter in the polyphase filterbank are samples from the real-valued low-pass filter $h(n)$. As such, the filters in the partition require real multiplications instead of the complex multiplications required by a direct implementation. A block diagram of the system of Figure 10.2.13 (a) when M satisfies (10.54) is shown in Figure 10.2.13 (b).

Using the Noble identities, the downsample operation can be moved from the output side of the subfilters to the input side as illustrated in Figure 10.2.14 (a). To apply the Noble identities, the downsample factor D must divide the number of partitions M. That is,

$$\frac{M}{D} = C \quad C \text{ is an integer.} \quad (10.55)$$

Section 10.2 Channelization

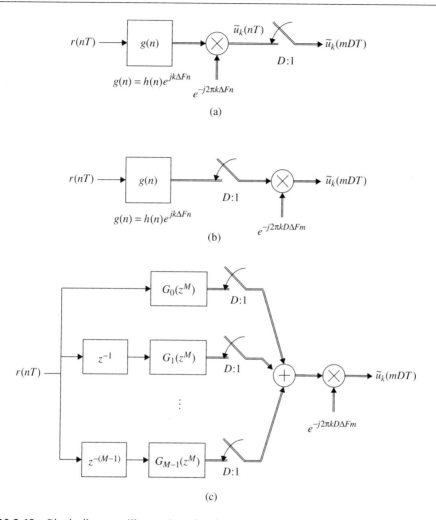

Figure 10.2.12 Block diagram illustrating the development of the multirate discrete-time channelizer: (a) A channelizer based on an adjustable band-pass filter; (b) same as (a) except that the downsample operation has been moved to the other side of the heterodyne operation; (c) same as (b) except that the band-pass filter has been replaced by its corresponding polyphase filterbank representation.

In this system, the subfilters operate in parallel at a sample rate D times smaller than the input sample rate. The delays on the input side of the downsample operation function at the high rate and define how the data are shifted into each stage of the filterbank. Normally the delay/downsample operation is performed by a commutator that shuffles the input data stream into the subfilters in the proper order. This function can be thought of as a generalized serial-to-parallel conversion as illustrated in Figure 10.2.14 (b).

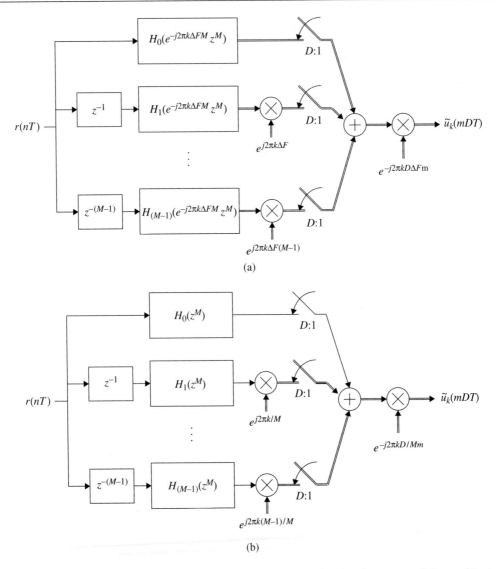

Figure 10.2.13 Continuation of Figure 10.2.12 illustrating the development of the multirate discrete-time channelizer: (a) Same as Figure 10.2.12 (c) except that the subfilters are expressed in terms of the polyphase representation of the real-valued low-pass filter prototype; (b) the simplification of (a) when $M = 1/\Delta F$.

The downsample operation aliases all of the channels to one of C center frequencies relative to the output sample rate. The low-pass subfilters in the polyphase filterbank impose a different phase profile on each of these copies. These different phase profiles are exploited by the bank of phase shifters (the complex exponential multipliers at the outputs of the

Section 10.2 Channelization

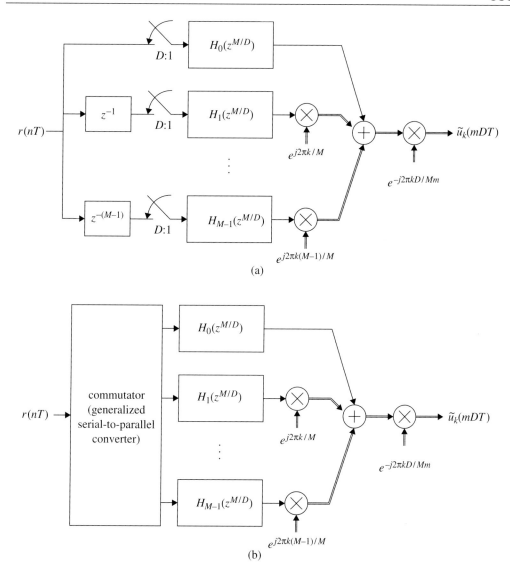

Figure 10.2.14 Continuation of Figure 10.2.13 illustrating the development of the multirate discrete-time channelizer: (a) Same as Figure 10.2.13 (b) except that the downsample operation has been moved to the input side of the subfilters in the filterbank; (b) an abstract representation of (a).

subfilters) to cophase the M copies of the desired signal at the outputs of the filters and combine the M copies of all the other channels so that they cancel out. The last multiplication by the complex exponential is a residual heterodyne required to translate the desired signal from its aliased center frequency to baseband.

Using the relationships (10.54) and (10.55), the complex exponential used for the residual heterodyne may be expressed as

$$e^{-j2\pi k \Delta F m} = \left(e^{-j\frac{2\pi k}{C}}\right)^m. \quad (10.56)$$

There are three special values of C for which the residual heterodyne can be accomplished without multiplications:

$$C = 1: \quad \left(e^{-j\frac{2\pi k}{C}}\right)^m = 1 \quad (10.57)$$

$$C = 2: \quad \left(e^{-j\frac{2\pi k}{C}}\right)^m = \begin{cases} 1 & k \text{ even} \\ (-1)^m & k \text{ odd} \end{cases} \quad (10.58)$$

$$C = 4: \quad \left(e^{-j\frac{2\pi k}{C}}\right)^m = \begin{cases} 1 & k = 0 \mod 4 \\ (-j)^m & k = 1 \mod 4 \\ (-1)^m & k = 2 \mod 4 \\ j^m & k = 3 \mod 4 \end{cases}. \quad (10.59)$$

The aliasing resulting from the downsample operations is illustrated conceptually by the plots in Figures 10.2.15–10.2.17 for these three values of C.

This remarkable architecture is made possible by the frequency translations that come for free when downsampling a band-pass signal. Because there is no continuous-time counterpart for sample-rate changes, the channelizers of Figures 10.2.15–10.2.17 have no counterpart in continuous-time processing. As a consequence, the discrete-time channelizer looks very different than its continuous-time counterpart.

An even more remarkable feature of the discrete-time channelizer becomes evident when more than one of the channels is desired simultaneously. Close inspection of the system in Figure 10.2.14 (b) shows that channel selection is controlled by the phase shifters operating on the subfilter outputs and the residual heterodyne. This fact is also evident by considering that all of the channels are present in the downsampled subfilter inputs as illustrated in Figures 10.2.15–10.2.17. Thus, the processing path for each channel can share the polyphase filterbank but requires its own set of phase shifters and residual heterodyne. As an example, consider the receiver illustrated in Figure 10.2.18 designed to isolate two channels, centered at $k\Delta F$ and $r\Delta F$ cycles/sample simultaneously.

Additional insight into the operation of the multirate multichannel receiver is available by examining the equations defining its operation. Let $y_l(mDT)$ be the output of subfilter $H_l\left(z^{M/D}\right)$ in the polyphase filterbank. The desired signal may be expressed as

$$\tilde{u}_k(mDT) = \left[\sum_{l=0}^{M-1} y_l(mDT)e^{j\frac{2\pi k}{M}l}\right] e^{-j\frac{2\pi k}{C}m}. \quad (10.60)$$

The summation inside the square brackets is the inverse FFT of the filterbank outputs. This observation suggests the architecture illustrated in Figure 10.2.19. The channelizer is able to

Section 10.2 Channelization

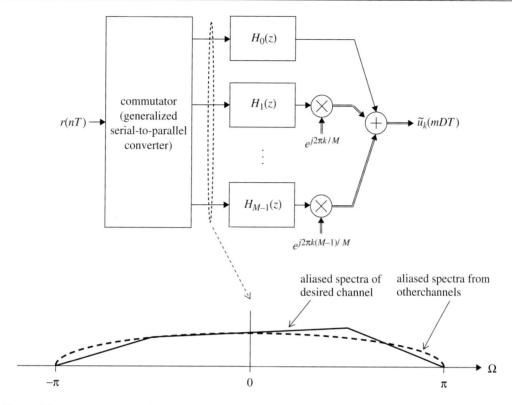

Figure 10.2.15 Conceptual representation of the aliasing that occurs for the case $M/D = 1$. For this case, all the channels are critically sampled (hence each channel occupies the entire bandwidth). All of the channels alias to baseband.

produce all M FDM channels as parallel time-series using a polyphase filterbank, an M-point inverse FFT, and some residual heterodynes. If the ratio M/D is chosen to be 1, 2, or 4, then the residual heterodyne operations do not require any multiplications as discussed above.

In general, if the simultaneous output of $1 < K < M$ channels is desired, two choices are available to the system designer. The first choice is a straightforward generalization of the system illustrated in Figure 10.2.18. One filterbank provides data to K phase-shift-heterodyne subsystems to produce the desired output. The other option is to produce all M channels using the FFT-based system of Figure 10.2.19 and discard the channels that are not required. Which option is best is usually determined by the complexity which, in turn, is dominated by the number of multiplications. The approach requiring the fewest number of multiplications depends on M, D, and K. These ideas are explored in the exercises.

Practical constraints imposed by ADCs do not allow the technique described in this section to be applied directly at RF for many applications. In these cases, a hybrid approach is commonly used. The band of frequencies is translated to an intermediate frequency using a continuous-time superheterodyne receiver described in Section 10.2.1. The IF filter is

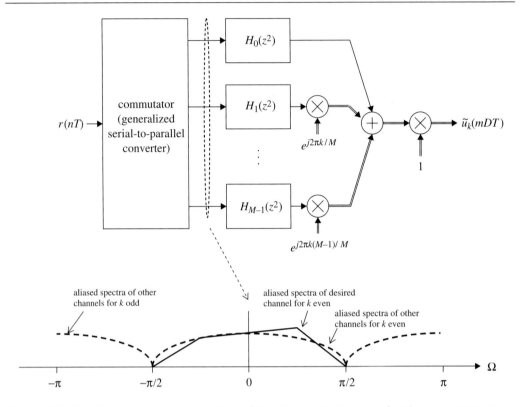

Figure 10.2.16 Conceptual representation of the aliasing that occurs for the case $M/D = 2$. For this case, all channels are oversampled by a factor 2 (hence each channel occupies one-half of the entire bandwidth). All channels centered at an even multiple of ΔF alias to baseband whereas those channels centered at an odd multiple of ΔF alias to $\Omega = \pi$. In this example, the spectrum of the desired signal is centered at an even multiple of ΔF.

designed to isolate the band of possible desired frequencies from the other signals in the radio spectrum as well as to bandlimit the signal prior to sampling. After sampling by an ADC operating on the IF signal, channel selection proceeds as outlined above.

10.3 NOTES AND REFERENCES

10.3.1 Topics Covered

The application of discrete-time processing to demodulation and detection has received considerable attention since the 1990s and is the basis for the software defined radio concept [229]. Examples of the options outlined in this chapter have been described in Refs. [230]–[258].

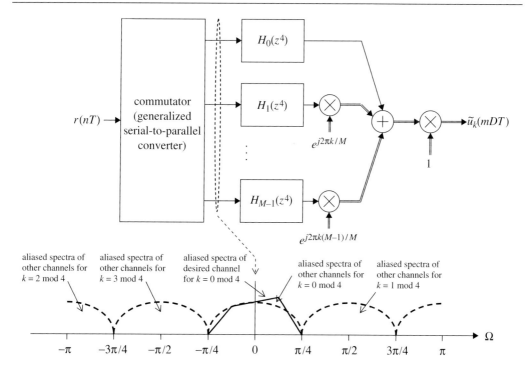

Figure 10.2.17 Conceptual representation of the aliasing that occurs for the case $M/D = 4$. For this case, all channels are oversampled by a factor 4 (hence each channel occupies one-quarter of the entire bandwidth). All channels alias to $\Omega = 0$, $\Omega = \pi/2$, $\Omega = \pi$, or $\Omega = 3\pi/2$ as determined by $k \mod 4$ as shown. In this example, the spectrum of the desired signal is centered at $k \Delta F$ where k is a multiple of 4.

The continuous-time channelizer in the form of the superheterodyne receiver was developed by Howard Armstrong in 1918 while serving in the Army Signal Corps in France during World War I. He presented a paper on the superheterodyne receiver at an IRE meeting in December of 1919. See Brittain [259] and Douglas [260]. Other articles on the history of its development appear in the November 1990 issue of *Proceedings of the Radio Club of America*: a special issue entitled "The Legacies of Edwin Howard Armstrong." The superheterodyne is a standard topic in senior-level textbooks on communications. Examples include Carlson [261], Couch [262], Proakis and Salehi [11], Roden [14], and Stremler [263]. In many instances the superheterodyne receiver is described in the context of commercial broadcast AM radio. But it is much more widely applicable than this context inadvertently suggests.

Discrete-time channelizers have been described by Zangi and Koilpillai [264] and Harris, Dick and Rice [244]. The channelizer presented in this chapter required the ratio M/D to be an integer. As it turns out, this is not a strict requirement. This more general case is described by Harris, Dick, and Rice [244]. Discrete-time channelizers are special cases of

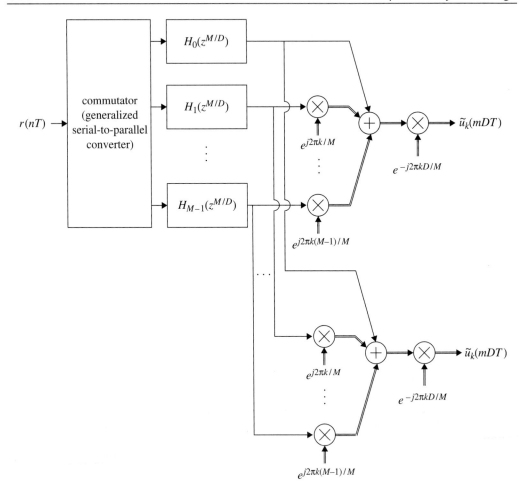

Figure 10.2.18 Discrete-time channelizer that selects two channels simultaneously. The channelizer is based on a single polyphase filterbank but uses a different set of phase shifters to extract the channel centered at $k\Delta F$ cycles/sample and the channel centered at $r\Delta F$ cycles/sample.

a more general function that has come to be known as "transmultiplexing." See the survey article by Scheuermann and Göckler [265] and its references for more information.

10.3.2 Topics Not Covered

The issues that limit performance of zero-IF receivers were briefly described in Section 10.2.1. The zero-IF receiver was introduced as a method of eliminating the image frequency problem by using continuous-time complex-valued signal processing. An overview article on this issue was written by Mirabbasi and Martin [266]. In the literature that deals with the practical

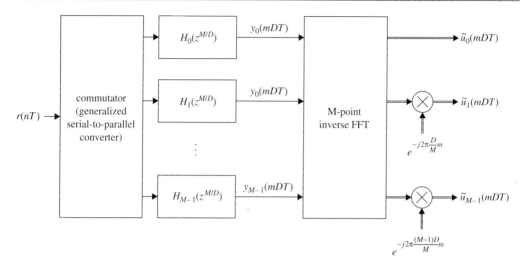

Figure 10.2.19 Discrete-time channelizer that selects *M* channels simultaneously. The channelizer is based on a single polyphase filterbanks together with an *M*-point inverse FFT that computes the phase shifts required to isolate a particular channel.

limitations of this approach and solutions using continuous-time processing, the problem is frequently cast as an image rejection problem. See Razavi [267], Namgoong and Meng [268], and Chen and Huang [269]. In discrete-time processing, most of the effort has been devoted to the "I/Q-imbalance" problem introduced in this text in the context of architectures that use a pair of ADCs to perform I/Q baseband sampling. A variety of discrete-time techniques have been analyzed in the open literature. Examples include Harris [270], Aschwanden [271], Razavi [267], Glas [272], Yu and Snelgrove [273], Valkama, et al. [274–278], Cetin, Kale, and Morling [279], and Liu, Golden, and Askar [280].

For transmitters, the linearity of the RF power amplifier is a very important design consideration. Nonlinear effects distort the transmitted waveform and increase its bandwidth as illustrated by the simple example illustrated in Figures 5.4.1 and 5.4.2. An easy-to-read tutorial of nonlinear power amplifiers and some compensation techniques was written by Katz [281]. A more in-depth treatment is the book by Kenington [282]. Designing and constructing a more linear RF power amplifier circuit is a very challenging task, and often does not result in a particularly power efficient (ratio of DC power in to RF power out) solution. Most of the effort has been devoted to "linearizing" power-efficient nonlinear amplifier designs. The three most popular categories of power amplifier linearization are feedback, feedforward, and predistortion. Feedback linearization is the simplest, but can suffer from stability problems. See the book by Kenington [282] and the work by Akaiwa and Nagata [283]. Feedforward systems divide the amplifier input into two paths. One path passes through the amplifier and the other, appropriately scaled and delayed, serves as a reference. An error signal, obtained by computing the difference between the two, is used to cancel the distortion at the amplifier output. The advantages and disadvantages of this approach are discussed by Kenington [282] and the papers by Seidel [284], Bennett and Clements [285], Kumar and Wells [286], Narahashi and Nojima [287], and Cavers [288]. Predistortion techniques distort the amplifier input signal in an attempt to create a composite transfer characteristic (the cascade of the predistorter and the nonlinear amplifier) that is linear [281, 282]. This technique has received considerable attention [289–318]. Another, less common, technique known as LINC (Linear Amplification with Nonlinear Components) is described by Cox [319] and Kenington [282] and was applied in the system described by Bateman, Haines, and Wilkinson [320].

10.4 EXERCISES

10.1 This exercise explores the effects of I/Q imbalance on a QAM modulator. In Chapter 5, it was shown that QAM signals may be expressed as

$$s(t) = I(t)\sqrt{2}\cos(\omega_0 t) - Q(t)\sqrt{2}\sin(\omega_0 t)$$

where $I(t)$ and $Q(t)$ are appropriately weighted pulse trains. The sinusoids $\sqrt{2}\cos(\omega_0 t)$ and $\sqrt{2}\sin(\omega_0 t)$ are perfectly balanced in both phase and amplitude. The most common method for modeling phase and amplitude imbalance is to assume $\sqrt{2}\cos(\omega_0 t)$ and $\sqrt{2}(1+\epsilon)\sin(\omega_0 t + \phi)$ are available for mixing with the pulse trains. Note that the phase error ϕ and the amplitude error ϵ are applied to the quadrature term. In this case, the QAM signal may be expressed as

$$s(t) = I(t)\sqrt{2}\cos(\omega_0 t) - Q(t)\sqrt{2}(1+\epsilon)\sin(\omega_0 t + \phi).$$

The phase trajectory may be used to visualize the effect of the phase and amplitude imbalance in the time domain. In the absence of phase and amplitude imbalance, the phase trajectory is a plot of $Q(t)$ versus $I(t)$ such as those shown in Figure 5.3.14 for QPSK and 16-QAM. In the presence of phase and amplitude imbalance, the phase trajectory is obtained by expressing the modulated QAM signal as

$$s(t) = \tilde{I}(t)\sqrt{2}\cos(\omega_0 t) - \tilde{Q}(t)\sqrt{2}\sin(\omega_0 t)$$

and producing a plot of $\tilde{Q}(t)$ versus $\tilde{I}(t)$.

(a) Show that the relationship between $I(t), Q(t)$ and $\tilde{Q}(t), \tilde{I}(t)$ is

$$\begin{bmatrix} \tilde{I}(t) \\ \tilde{Q}(t) \end{bmatrix} = \begin{bmatrix} 1 & (1+\epsilon)\sin(\phi) \\ 0 & (1+\epsilon)\cos(\phi) \end{bmatrix} \begin{bmatrix} I(t) \\ Q(t) \end{bmatrix}.$$

(b) Produce the inphase and quadrature pulse trains $I(nT)$ and $Q(nT)$ corresponding to 8PSK for 1000 random symbols using the SRRC pulse shape with 100% excess bandwidth. Produce the pulse trains using a sample rate equivalent to 8 samples/symbol. Plot the phase trajectory.

(c) Compute the distorted inphase and quadrature pulse trains $\tilde{I}(nT)$ and $\tilde{Q}(nT)$ from the pulse trains in part (b) and using $\epsilon = 0.05$ and $\phi = 20°$. Plot the phase trajectory.

(d) Compare the two phase trajectories and comment on the effect of phase and amplitude imbalance.

(e) Experiment by producing phase trajectory plots using different values of ϵ and ϕ. How does ϵ change the phase trajectory? How does ϕ change the phase trajectory?

10.2 Repeat Exercise 10.1 (b)–(e) except using 16-QAM.

10.3 This exercise explores the effects of I/Q imbalance on the performance of a demodulator/detector. Assume a QAM demodulator of the form used in Figure 10.1.5 (b). Neglecting carrier phase offset, timing offset, and noise, the received signal may be expressed as

$$s(t) = I(t)\sqrt{2}\cos(\omega_0 t) - Q(t)\sqrt{2}\sin(\omega_0 t)$$

as explained in Chapter 5. When operating in perfect synchronism, the matched filter outputs may be expressed as

$$x(t) = I(t) * p(-t) = \sum_m a_0(m) r_p(t - mT_s)$$

$$y(t) = Q(t) * p(-t) = \sum_m a_1(m) r_p(t - mT_s)$$

where $a_0(m)$ and $a_1(m)$ are the symbols in the signal constellation and $r_p(t)$ is the autocorrelation function of the pulse shape. In ideal noiseless operation, the sampled outputs of the matched filters are (assuming the pulse shape satisfies the Nyquist no-ISI condition) $x(kT_s) = a_0(k)$ and $y(kT_s) = a_1(k)$ and coincide exactly with the points in the signal constellation. Now suppose the quadrature sinusoids have a phase and amplitude imbalance. Instead of using $\sqrt{2}\cos(\omega_0 t)$ and $\sqrt{2}\sin(\omega_0 t)$ with the quadrature mixers, $\sqrt{2}\cos(\omega_0 t)$ and $\sqrt{2}(1 + \epsilon)\sin(\omega_0 t + \phi)$ are used. In this model, the amplitude error ϵ and the phase error ϕ are applied to the quadrature component. Let the outputs of the matched filters in the presence of phase and amplitude imbalance be $\tilde{x}(t)$ and $\tilde{y}(t)$.

(a) Show that the relationship between $x(t), y(t)$ and $\tilde{x}(t), \tilde{y}(t)$ may be expressed as

$$\begin{bmatrix} \tilde{x}(t) \\ \tilde{y}(t) \end{bmatrix} = \begin{bmatrix} 1 & 0 \\ (1+\epsilon)\sin(\phi) & (1+\epsilon)\cos(\phi) \end{bmatrix} \begin{bmatrix} x(t) \\ y(t) \end{bmatrix}.$$

(b) What is the relationship between $\tilde{x}(kT_s), \tilde{y}(kT_s)$ and $a_0(k), a_1(k)$?

(c) Produce sampled versions of the matched filter outputs $x(nT)$ and $y(nT)$ corresponding to 1000 random 8-PSK symbols using the SRRC pulse shape with 100% excess bandwidth. Generate the pulse trains at a sample rate equivalent to 8 samples/symbol. Obtain the signal space projections $x(kT_s)$ and $y(kT_s)$. Plot the signal space projections.

(d) Use the signals $x(nT)$ and $y(nT)$ from part (c) to generate $\tilde{x}(nT)$ and $\tilde{y}(nT)$ for $\epsilon = 0.05$ and $\phi = 20°$. Obtain the signal space projections $\tilde{x}(kT_s)$, $\tilde{y}(kT_s)$ and plot the signal space projections.

(e) Compare the two signal space projections and comment on the effect of phase and amplitude imbalance.

(f) Experiment by producing signal space projection plots using different values of ϵ and ϕ. How does ϵ change the signal space projection? How does ϕ change the signal space projection?

10.4 Repeat Exercise 10.3 (c)–(f) except using 16-QAM.

10.5 Show that the QAM signal given by (10.12) is the real part of (10.11).

10.6 Show that a QAM modulator can be constructed from the system illustrated in Figure 10.1.3 (d) by using polyphase partitions of $p_c(nT)$ and $p_s(nT)$ to produce the system shown below.

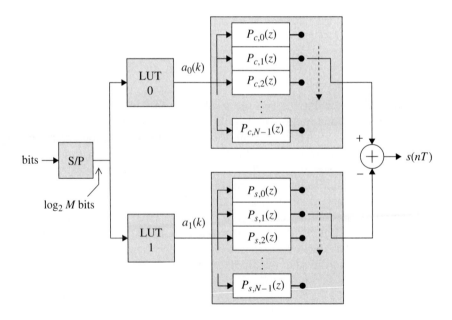

10.7 Starting with the QAM modulator illustrated in Figure 10.1.4, show that by computing only the real part, the system shown below produces the desired signal $s(nT)$.

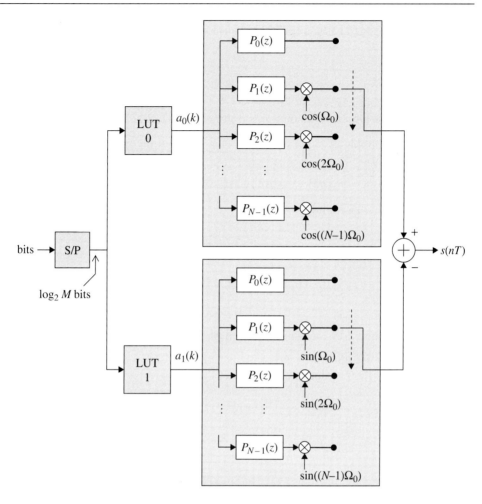

10.8 The QAM modulator of Exercise 10.6 was derived from the QAM modulator of Figure 10.1.3 (d) by using polyphase partitions of $p_c(nT)$ and $p_s(nT)$. The QAM modulator of Exercise 10.7 was derived from the QAM modulator of Figure 10.1.3 (c) by using a polyphase partition of $g(nT)$. Show that the filterbanks in these two modulators are exactly identical.

10.9 Let the sampled real-valued, band-pass signal at the input to a discrete-time QAM demodulator be

$$r(nT) = I_r(nT)\sqrt{2}\cos(\Omega_0 n) - Q_r(nT)\sqrt{2}\sin(\Omega_0 n)$$

and let the baseband matched filter outputs be

$$x(nT) = p(-nT) * I_r(nT)$$
$$y(nT) = p(-nT) * Q_r(nT)$$

where $p(-nT)$ is the impulse response of the matched filter. Let the filter $g(nT)$ be defined as
$$g(nT) = p(-nT)\sqrt{2}e^{j\Omega_0 n}.$$
Show that
$$r(nT) * g(nT) = [x(nT) + jy(nT)]e^{j\Omega_0 n}.$$
(Hint: use either of the methods outlined in Exercise B.18.)

10.10 Let the sampled real-valued, band-pass signal at the input to a discrete-time QAM demodulator be
$$r(nT) = I_r(nT)\sqrt{2}\cos(\Omega_0 n) - Q_r(nT)\sqrt{2}\sin(\Omega_0 n)$$
and let the baseband matched filter outputs be
$$x(nT) = p(-nT) * I_r(nT)$$
$$y(nT) = p(-nT) * Q_r(nT)$$
where $p(-nT)$ is the impulse response of the matched filter. Let the filter $g(nT)$ be defined as
$$g(nT) = p(-nT)\sqrt{2}e^{-j\Omega_0 n}.$$
Derive an expression, similar to that of Exercise 10.9, for $r(nT) * g(nT)$.

10.11 This exercise illustrates the challenges of using a tuned circuit for channel selection using the RLC circuit shown below.

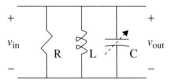

(a) Show that the resonant frequency (the frequency that renders the transfer function purely real) is
$$\omega_0 = \frac{1}{\sqrt{LC}}.$$

(b) For the purposes of illustration, the 3-dB bandwidth will be used as the measure of bandwidth. The 3-dB bandwidth is based on the frequencies ω_1 and ω_2 defined as shown below:

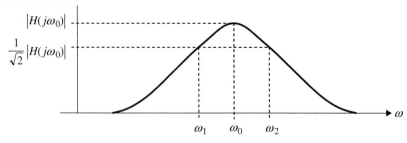

Show that these frequencies are

$$\omega_1 = -\frac{1}{2RC} + \sqrt{\left(\frac{1}{2RC}\right)^2 + \frac{1}{LC}}$$

$$\omega_2 = \frac{1}{2RC} + \sqrt{\left(\frac{1}{2RC}\right)^2 + \frac{1}{LC}}.$$

(c) Derive an expression for the 3-dB bandwidth $W_{3\text{-dB}}$.

(d) Tunable circuits are usually built using a variable capacitor as shown. If the circuit is to be tuned over the carrier frequencies corresponding to the commercial broadcast AM band (540–1600 kHz), determine the minimum and maximum values for the variable capacitor using $R = 157$ kΩ and $L = 0.25$ mH.

(e) Plot the 3-dB bandwidth (in kHz) as a function of center frequency (in kHz). Describe the relationship between the tuned frequency and the bandwidth. To select a channel in the commercial broadcast AM band, the bandwidth of the filter should be 10 kHz. Does your plot indicate that this circuit is suitable for this application?

10.12 Derive the image frequency for a superheterodyne receiver using high-side mixing.

10.13 Consider the application of the superheterodyne receiver to a band-pass signal centered at f_c Hz with an RF bandwidth of B Hz. A block diagram of the superheterodyne receiver and the spectrum of the band-pass input signal are shown below.

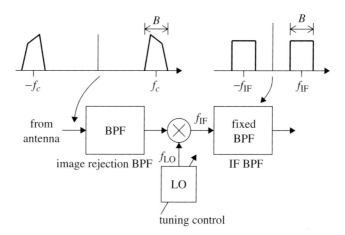

The intermediate frequency (IF) filter is an ideal BPF, centered at f_{IF} with a bandwidth of B.

(a) For a system that uses low-side mixing, show that the bandwidth of an ideal image rejection band-pass filter is $4f_{\text{IF}} - B$.

(b) Determine the bandwidth of an ideal image rejection band-pass filter for a system that uses high-side mixing.

(c) Compare your answers from parts (a) and (b). What conclusions may be drawn?

10.14 The commercial broadcast AM band extends from 535 to 1605 kHz and consists of 107 channels, spaced 10 kHz apart. The carrier frequencies are 540 kHz, 550 kHz, and so on. The IF is 455 kHz.
(a) Determine if the image rejection filter must be tunable for a superheterodyne receiver using low-side mixing.
(b) Determine if the image rejection filter must be tunable for a superheterodyne receiver using high-side mixing.

10.15 The commercial broadcast FM band extends from 88 to 108 MHz and consists of 100 channels spaced 200 kHz apart. The carrier frequencies are 88.1 MHz, 88.3 MHz, and so on. The IF is 10.7 MHz.
(a) Determine if the image rejection filter must be tunable for a superheterodyne receiver using low-side mixing.
(b) Determine if the image rejection filter must be tunable for a superheterodyne receiver using high-side mixing.

10.16 The AMPS mobile to base-station link occupies the band from 824 to 849 MHz and consists of 832 channels spaced 30 kHz apart. The carrier frequencies are 824.04 MHz, 824.07 MHz, and so on. The IF is 10.7 MHz.
(a) Determine if the image rejection filter must be tunable for a superheterodyne receiver using low-side mixing.
(b) Determine if the image rejection filter must be tunable for a superheterodyne receiver using high-side mixing.

10.17 Repeat Exercise 10.16 except using an IF of 70 MHz.

10.18 The AMPS base-station to mobile link occupies the band from 869 to 894 MHz and consists of 832 channels spaced 30 kHz apart. The carrier frequencies are 869.04 MHz, 869.07 MHz, and so on. The IF is 10.7 MHz.
(a) Determine if the image rejection filter must be tunable for a superheterodyne receiver using low-side mixing.
(b) Determine if the image rejection filter must be tunable for a superheterodyne receiver using high-side mixing.

10.19 Repeat Exercise 10.18 except using an IF of 70 MHz.

10.20 The GSM base-station to mobile link occupies the band from 890 to 915 MHz and consists of 125 channels spaced 200 kHz apart. The carrier frequencies are 890.2 MHz, 890.4 MHz and so on. The IF is 10.7 MHz.
(a) Determine if the image rejection filter must be tunable for a superheterodyne receiver using low-side mixing.

(b) Determine if the image rejection filter must be tunable for a superheterodyne receiver using high-side mixing.

10.21 Repeat Exercise 10.20 if the IF is 70 MHz.

10.22 The GSM mobile to base-station link occupies the band from 935 to 960 MHz and consists of 125 channels spaced 200 kHz apart. The carrier frequencies are 935.2 MHz, 935.4 MHz and so on. The IF is 10.7 MHz.
(a) Determine if the image rejection filter must be tunable for a superheterodyne receiver using low-side mixing.
(b) Determine if the image rejection filter must be tunable for a superheterodyne receiver using high-side mixing.

10.23 Repeat Exercise 10.22 if the IF is 70 MHz.

10.24 Communications between aircraft and air-traffic control operate in the band from 118 to 137 MHz and consists of 760 channels spaced 25 kHz apart. The carrier frequencies are 118 MHz, 118.025 MHz, and so on. The IF is 10.7 MHz.
(a) Determine if the image rejection filter must be tunable for a superheterodyne receiver using low-side mixing.
(b) Determine if the image rejection filter must be tunable for a superheterodyne receiver using high-side mixing.

10.25 Repeat Exercise 10.24 if the IF is 21.4 MHz.

10.26 Consider the dual-conversion superheterodyne receiver shown below. The received band-pass signal, centered at f_c Hz with an RF bandwidth B Hz, is to be converted to the intermediate frequency f_{IF} Hz using two oscillators operating at f_1 Hz and f_2 Hz and three band-pass filters. Assume $f_2 > f_1 > f_{LO}$.

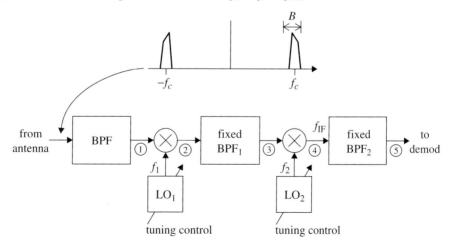

(a) Sketch the magnitude of the Fourier transform at points 1, 2, 3, 4, and 5.
(b) Derive an expression for f_{LO} in terms of f_c, f_1, f_2.

(c) What is the relationship between the bandwidth B and the frequencies f_c, f_1, f_2?

(d) Sketch the requirements of three band-pass filters.

10.27 Consider a 5 Msymbol/s QAM signal, centered at 12 GHZ, with an RF bandwidth of 6 MHz. A superheterodyne receiver is to be constructed that translates the desired carrier from 12 GHz to 70 MHz. Suppose that practical constraints limit the center-frequency-to-bandwidth ratio of all band-pass filters to 10 or less. For example, the image rejection filter of the superheterodyne receiver of Figure 10.2.1 (c) requires a band-pass filter centered at 12 GHz with a bandwidth of $4 \times 70 + 6 = 286$ MHz. This ratio is

$$\frac{\text{center frequency}}{\text{bandwidth}} = \frac{12000}{286} \approx 42$$

which is much larger than the maximum allowable value of 10. In this case, the dual-conversion super-heterodyne receiver, such as the one illustrated by the block diagram of Exercise 10.26 should be used. Specify a set of local oscillator frequencies f_1 and f_2, along with the three band-pass filter specifications that produce a final IF of $f_{\text{IF}} = 70$ MHz and meet the filter constraints. For simplicity, assume $f_c > f_1 > f_2$.

10.28 This exercise explores the why the real-valued superheterodyne receiver of Figure 10.2.1 (c) cannot be used with zero IF. Let the received signal be

$$r(t) = I_r(t)\sqrt{2}\cos(\omega_c t) - Q_r(t)\sqrt{2}\sin(\omega_c t)$$

and let the LO output be

$$\text{LO} = \sqrt{2}\cos(\omega_c t).$$

(a) Express the product $r(t)\cos(\omega_c t)$ in terms of baseband signals and double frequency signals (centered at $2\omega_c$ rad/s).

(b) The low-pass filter eliminates the double frequency terms. What is the output of the low-pass filter? Is it possible to recover $I_r(t)$ and $Q_r(t)$ from this output?

10.29 Repeat Exercise 10.28 using

$$r(t) = I_r(t)\sqrt{2}\cos(\omega_c t + \theta) - Q_r(t)\sqrt{2}\sin(\omega_c t + \theta)$$

and

$$\text{LO} = \sqrt{2}\cos(\omega_c t).$$

10.30 This exercise explores the conditions that determine when the discrete-time channelizer of Figure 10.2.18 should be used or when the discrete-time channelizer of Figure 10.2.19 should be used.

(a) Determine the number of complex-valued multiplications required by the phase-shift network for channel k for the discrete-time channelizer of Figure 10.2.18.

(b) Using your answer in part (a), derive an expression for the number of complex-valued multiplications required to produce K channels in parallel.

Section 10.4 Exercises

(c) The discrete-time channelizer of Figure 10.2.19 requires an M-point inverse FFT. Let $M = p_1 p_2 \cdots p_\nu$ where the p's are prime factors that are not necessarily distinct. The number of complex-valued multiplications is approximately [321]

$$N(p_1 + p_2 + \cdots + p_\nu - \nu).$$

Using your answer from part (b), determine the condition for K that must be satisfied for the discrete-time channelizer of Figure 10.2.19 to require fewer multiplications than the channelizer of Figure 10.2.18.

10.31 Consider a 4 Mbit/s QPSK system using the SRRC pulse shape with 50% excess bandwidth. The receiver presents to the ADC this signal centered at an intermediate frequency of 70 MHz. Two constraints are placed on sample rate selection: First the sample rate must be such that the IF aliases to the quarter-sample-rate frequency. The second constraint is that the sample rate must be an integer multiple of the symbol rate (to provide an integer number of samples/symbol at the output of the ADC).

(a) What is the lowest sample rate that satisfies the two requirements?

(b) As it turns out, there are four sample rates that satisfy the two requirements. What are the other three?

10.32 A commonly used IF used in surveillance and monitoring applications is 45 MHz. This exercise explores sample rates for this IF that produces band-pass discrete-time signals centered at the frequency corresponding to the quarter sample rate as discussed in option 2 in Section 10.1.3.

(a) Construct a table, similar to Table 10.1.1, for this IF.

(b) A commonly transmitted signal in these applications is video, which has a bandwidth at IF of approximately 8 MHz. What is the lowest sample rate that satisfies the sampling theorem and the quarter-sample-rate property?

10.33 The IF used in demodulators for commercial broadcast FM is 10.7 MHz. This exercise explores sample rates for this IF that produce band-pass discrete-time signals centered at the frequency corresponding to the quarter sample rate as discussed in option 2 in Section 10.1.3.

(a) Construct a table, similar to Table 10.1.1, for this IF.

(b) The bandwidth of a commercial broadcast FM signal is 200 kHz. What is the lowest sample rate that satisfies the sampling theorem and the quarter-sample-rate property?

10.34 The IF used in demodulators for commercial broadcast AM is 455 kHz. This exercise explores sample rates for this IF that produce band-pass discrete-time signals centered at the frequency corresponding to the quarter sample rate as discussed in option 2 in Section 10.1.3.

(a) Construct a table, similar to Table 10.1.1, for this IF.

(b) The bandwidth of a commercial broadcast AM signal is 10 kHz. What is the lowest sample rate that satisfies the sampling theorem and the quarter-sample-rate property?

10.35 A commonly used IF used in radio astronomy and amateur radio is 21.4 MHz. Construct a table, similar to Table 10.1.1 of sample rates that produce band-pass discrete-time signals centered at the frequency corresponding to the quarter sample rate for this IF.

Appendix A Pulse Shapes

A.1 FULL-RESPONSE PULSE SHAPES

Full-Response pulse shapes are pulse shapes whose time support is equal to the symbol time T_s. Usually, $T_1 = 0$ and $T_2 = T_s$. The four most commonly used full-response pulse shapes are the non-return-to-zero (NRZ), the return-to-zero (RZ), the Manchester (MAN), and the half-sine (HS). These pulse shapes are illustrated in Figure A.1.1. The time-domain characteristics for these pulse shapes are listed in Table A.1.1. The frequency-domain characteristics of these pulses are listed in Table A.1.2.

The magnitudes of their Fourier transforms are plotted in Figure A.1.2. The Fourier transforms of the NRZ, RZ, and HS pulse shapes have some common features. Each has a "main lobe" that exhibits a maximum at $f = 0$ and decreases to zero at f equals some multiple of $1/T_s$. The width of the main lobe is $1/T_s$ for the NRZ pulse shape, $2/T_s$ for the RZ pulse shape, and $1.5/T_s$ for the HS pulse shape. As frequency increases, a second lobe, called a "side lobe" appears. For the NRZ and HS pulse shapes, the width of each side lobe is $1/T_s$, which means the Fourier transform of these pulse shapes has $1/T_s$-spaced nulls. The side lobe width for the RZ pulse shape is $2/T_s$ and the Fourier transform has $2/T_s$-spaced nulls. Observe that the Fourier transforms of the NRZ, RZ, and HS pulse shapes are maximum at DC. Thus, baseband signaling based on these pulse shapes has a significant DC component.

The structure is somewhat different for the MAN pulse shape. Because the Fourier transform for the MAN pulse shape is zero at $f = 0$, there is no main lobe in the same sense there was for the other three pulse shapes. There, however, there are alternating lobes and nulls where the nulls are spaced $2/T_s$ in frequency. In many applications, it is convenient to think of the main lobe as extending from $f = 0$ to $f = 2/T_s$. Note that the Fourier transform of the MAN pulse shape is zero at DC. Thus, baseband signaling based on the MAN pulse shape has no DC component.

Each of these pulse shapes present time-domain and frequency-domain trade-offs. For example, the timing error detectors described in Chapter 8 only produce nonzero output in the presence of a waveform transition. The only way to produce a waveform transition using the NRZ and HS pulse shapes is to have a data transition. The RZ and MAN pulse shapes, on the other hand, possess a waveform transition midway through the symbol period that is independent of the data. Thus, for timing synchronization, the RZ and MAN pulse shapes are desirable. Figure A.1.2, however, shows that there is a bandwidth penalty associated with this choice. To quantify this trade-off, a more precise definition of "bandwidth" is required.

Appendix A Pulse Shapes

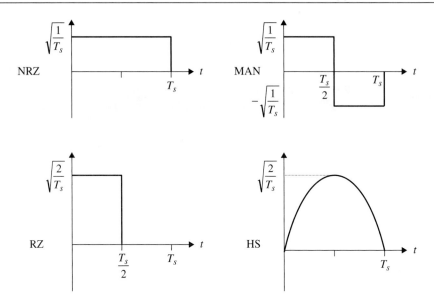

Figure A.1.1 The non-return-to-zero (NRZ), the return-to-zero (RZ), Manchester (MAN), and half-sine (HS) pulse shapes as shown.

Table A.1.1 Time-domain properties of the NRZ, RZ, MAN, and HS pulse shapes

Name	$p(t)$	$r_p(\tau)$
NRZ	$p(t) = \begin{cases} \sqrt{\frac{1}{T_s}} & 0 \le t \le T_s \\ 0 & \text{otherwise} \end{cases}$	$r_p(\tau) = \begin{cases} 1 - \frac{|\tau|}{T_s} & |\tau| \le T_s \\ 0 & |\tau| > T_s \end{cases}$
RZ	$p(t) = \begin{cases} \sqrt{\frac{2}{T_s}} & 0 \le t \le \frac{T_s}{2} \\ 0 & \text{otherwise} \end{cases}$	$r_p(\tau) = \begin{cases} 1 - 2\frac{|\tau|}{T_s} & |\tau| \le \frac{T_s}{2} \\ 0 & |\tau| > \frac{T_s}{2} \end{cases}$
MAN	$p(t) = \begin{cases} \sqrt{\frac{1}{T_s}} & 0 \le t \le \frac{T_s}{2} \\ -\sqrt{\frac{1}{T_s}} & \frac{T_s}{2} \le t \le T_s \\ 0 & \text{otherwise} \end{cases}$	$r_p(\tau) = \begin{cases} 1 - 3\frac{|\tau|}{T_s} & |\tau| \le \frac{T_s}{2} \\ \frac{|\tau|}{T_s} - 1 & \frac{T_s}{2} < |\tau| \le T_s \\ 0 & |\tau| > T_s \end{cases}$
HS	$p(t) = \begin{cases} \sqrt{\frac{2}{T_s}} \sin\left(\frac{\pi t}{T_s}\right) & 0 \le t \le T_s \\ 0 & \text{otherwise} \end{cases}$	$r_p(\tau) = \begin{cases} \left(1 - \frac{|\tau|}{T_s}\right)\cos\left(\frac{\pi|\tau|}{T_s}\right) \\ + \frac{1}{\pi}\sin\left(\frac{\pi|\tau|}{T_s}\right) & |\tau| \le T_s \\ 0 & |\tau| > T_s \end{cases}$

There are several measures of bandwidth in use. All of the definitions are based on $|P(f)|^2$, the magnitude-squared of the Fourier transform of the pulse shape $p(t)$. Some of the more common definitions are

Absolute Bandwidth: The absolute bandwidth is the value B_{abs} such that $|P(f)|^2 = 0$ for $f \ge B_{\text{abs}}$. The absolute bandwidths for the NRZ, RZ, MAN, and HS pulse shapes are infinite.

Appendix A Pulse Shapes

Table A.1.2 Frequency-domain properties of the NRZ, RZ, MAN, and HS pulse shapes

Name	$|P(f)|^2$	B_{abs}	$B_{90\%}$	$B_{99\%}$	$B_{-60\,dB}$
NRZ	$|P(f)|^2 = T_s \dfrac{\sin^2(\pi f T_s)}{(\pi f T_s)^2}$	∞	$0.85/T_s$	$10.29/T_s$	$318.50/T_s$
RZ	$|P(f)|^2 = \dfrac{T_s}{2}\dfrac{\sin^2\left(\frac{\pi f T_s}{2}\right)}{\left(\frac{\pi f T_s}{2}\right)^2}$	∞	$1.70/T_s$	$20.57/T_s$	$637/T_s$
MAN	$|P(f)|^2 = T_s \dfrac{\sin^4\left(\frac{\pi f T_s}{2}\right)}{\left(\frac{\pi f T_s}{2}\right)^2}$	∞	$3.05/T_s$	$30.75/T_s$	$635/T_s$
HS	$|P(f)|^2 = \dfrac{T_s}{2\pi^2}\left(\dfrac{\cos(\pi f T_s)}{(f T_s)^2 - \frac{1}{4}}\right)^2$	∞	$0.78/T_s$	$1.18/T_s$	$15/T_s$

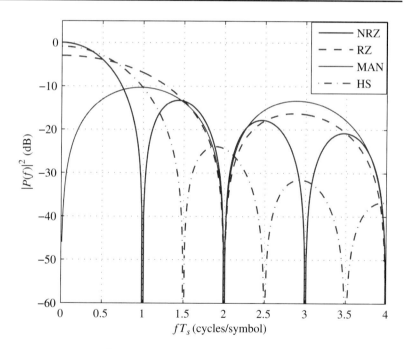

Figure A.1.2 A plot of $|P(f)|^2$ versus fT_s for the NRZ, RZ, MAN, and HS pulse shapes. The mathematical expressions are listed in Table A.1.2.

90% Bandwidth: The 90% bandwidth is the value $B_{90\%}$ that accounts for 90% of the total energy in $p(t)$. Expressed mathematically $B_{90\%}$ is

$$\int_0^{B_{90\%}} |P(f)|^2 df = 0.90 \int_0^{\infty} |P(f)|^2 df. \tag{A.1}$$

99% Bandwidth: The 99% bandwidth is the value $B_{99\%}$ that accounts for 99% of the total energy in $p(t)$:

$$\int_0^{B_{99\%}} |P(f)|^2 df = 0.99 \int_0^{\infty} |P(f)|^2 df. \qquad (A.2)$$

−60 dB Bandwidth: The −60 dB bandwidth is the value $B_{-60\text{ dB}}$ such that $|P(f)|^2 < 10^{-6}$ for $f \geq B_{-60\text{ dB}}$.

A.2 PARTIAL-RESPONSE PULSE SHAPES

Each of the pulse shapes in the previous section had infinite absolute bandwidth. This is because each pulse shape had finite support in the time domain (infinite support in one domain has a transform that has finite support in the other domain). Finite support in the time domain is desirable because the matched-filter outputs corresponding to adjacent symbols do not overlap or interfere with each other. A natural question to ask is whether or not it is possible to relax the constraint on finite support in the time domain (to reduce the bandwidth) and simultaneously impose a constraint that the matched filter outputs corresponding to different symbols do not interfere with each other. As it turns out, this is possible. The conditions under which this is achievable are summarized by the Nyquist no-ISI theorem presented below.

In Chapter 5, it was shown that if $\sum_i a(i)p(t - iT_s)$ is the input to a matched filter with impulse response $h(t) = p(-t)$, then the output is

$$x(t) = \sum_i a(i) r_p(t - iT_s) \qquad (A.3)$$

where $r_p(\tau)$ is the pulse shape autocorrelation function

$$r_p(\tau) = \int_{-\infty}^{\infty} p(t) p(t - \tau) dt. \qquad (A.4)$$

Symbol decisions are based on T_s-spaced samples of the matched filter output. The matched filter output at $t = kT_s$ is

$$x(kT_s) = \sum_i a(i) r_p((k-i)T_s) = a(k) + \sum_{i \neq k} a(i) r_p((k-i)T_s). \qquad (A.5)$$

The first term is the desired symbol and the second term (the summation) represents the "intersymbol interference" or ISI. The ISI is zero when the pulse shape autocorrelation function satisfies

$$r_p(kT_s) = \begin{cases} 1 & k = 0 \\ 0 & k \neq 0 \end{cases}. \qquad (A.6)$$

Appendix A Pulse Shapes

In other words, $r_p(t)$ has zero crossings at nonzero integer multiples of the symbol time. The Nyquist no-ISI theorem defines the conditions on the pulse shape to make this so.

> **Theorem A.1 (Nyquist No-ISI)** Let $r_p(\tau)$ and $R_p(f)$ be Fourier transform pairs. Then
> $$r_p(kT_s) = \begin{cases} 1 & k = 0 \\ 0 & k \neq 0 \end{cases} \quad \text{if and only if} \quad \sum_{m=-\infty}^{\infty} R_p\left(f + \frac{m}{T_s}\right) = T_s. \tag{A.7}$$

The Nyquist no-ISI theorem can be proved using the Poisson sum formula from Fourier analysis. The Poisson sum formula is

$$\sum_{m=-\infty}^{\infty} R_p\left(f + \frac{m}{T_s}\right) = T_s \sum_{k=-\infty}^{\infty} r_p(kT_s) e^{-j2\pi kT_s f}. \tag{A.8}$$

Assuming $r_p(t)$ satisfies the no-ISI condition (A.6), the right-hand side of (A.8) reduces to the $k = 0$ term, which proves that the right-hand side of (A.6) is a sufficient condition. Assuming the right-hand side of (A.6) is true, the Poisson sum formula is satisfied when $r_p(t)$ satisfies the no-ISI condition (A.6). This proves that the right-hand side of (A.6) is necessary and the proof is complete. This proof relies on the Poisson sum formula. The derivation of the Poisson sum formula is explored in Exercise A.4.

To apply this result to the design of pulse shapes suitable for use with linear modulation, it is assumed that a baseband channel with absolute bandwidth B is available. Three cases, illustrated in Figure A.2.1, need to be considered.

Case 1 ($1/T_s > 2B$): In this case, the symbol rate is greater than twice the bandwidth. The sum

$$\sum_{m=-\infty}^{\infty} R_p\left(f + \frac{m}{T_s}\right)$$

consists of nonoverlapping spectra spaced by $1/T_s$ as illustrated in Figure A.2.1 (a). As such, there is no choice for $R_p(f)$ that satisfies (A.7).

Case 2 ($1/T_s = 2B$): In this case, the symbol rate is equal to twice the bandwidth and the spectral copies of the sum

$$\sum_{m=-\infty}^{\infty} R_p\left(f + \frac{m}{T_s}\right)$$

touch, as illustrated in Figure A.2.1 (b). The only way (A.7) can be satisfied is if

$$R_p(f) = \begin{cases} T_s & |f| < B \\ 0 & |f| > 0 \end{cases}. \tag{A.9}$$

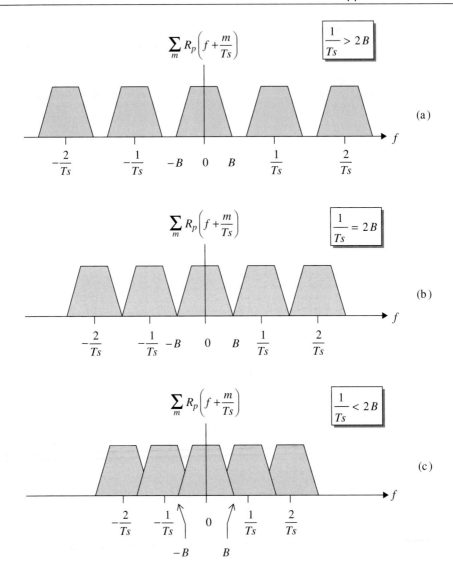

Figure A.2.1 Graphical interpretation of the Nyquist no-ISI theorem: (a) $1/T_s > 2B$; (b) $1/T_s = 2B$; (c) $1/T_s < 2B$.

This corresponds to

$$r_p(\tau) = \frac{\sin\left(\dfrac{\pi t}{T_s}\right)}{\dfrac{\pi t}{T_s}}. \tag{A.10}$$

Appendix A Pulse Shapes

Case 3 ($1/T_s < 2B$): In this case, the symbol rate is less than twice the bandwidth and the spectral copies of the sum $\sum_{m=-\infty}^{\infty} R_p\left(f + \frac{m}{T_s}\right)$ overlap as illustrated in Figure A.2.1 (c). There are many pulse shapes that satisfy the Nyquist no-ISI condition. The right-hand side of (A.7) requires $P(f)$ to have an odd-symmetric taper about $f = 1/(2T_s)$. Examples of $R_p(f)$ that satisfy this condition are illustrated in Figure A.2.2. The support of $R_p(f)$ extends to $f = 1/T_s$ for the examples illustrated in Figure A.2.2 (a) and (d). In the other cases, the support of $R_p(f)$ extends to a fraction of $1/T_s$. This fraction is given by $(1 + \alpha)/2$. In these pulse shapes, the parameter α controls the bandwidth of the pulse. The minimum bandwidth of a pulse is $1/(2T_s)$—see Case 2 above—and this corresponds to $\alpha = 0$. The symmetry about $f = 1/(2T_s)$ limits $\alpha \leq 1$. The convention is to call the parameter $0 \leq \alpha \leq 1$ the *excess bandwidth*. Its value is often given as a percent: 0% excess bandwidth means the pulse has the minimum bandwidth allowed by the Nyquist no-ISI theorem: $1/(2T_s)$. A 100% excess bandwidth means the pulse has the maximum allowable bandwidth specified by the Nyquist no-ISI theorem: $1/T_s$.

The corresponding time-domain autocorrelation function $r_p(t)$ also possesses structure. For spectra $R_p(f)$ that are continuous in f, the time-domain function may be expressed in the form

$$r_p(t) = \frac{\sin\left(\frac{\pi T}{T_s}\right)}{\frac{\pi T}{T_s}} \times \phi(t) \qquad (A.11)$$

where $\phi(t)$ is a real even function. This can be shown by examining some properties of $R_p(f)$. Close examination of Figures A.2.2 (a), (b), (d), and (e) shows that $R_p(f)$ may be expressed as

$$R_p(f) = \begin{cases} \dfrac{T_s}{2}\left[1 + G(f)\right] & 0 \leq f \leq \dfrac{1+\alpha}{2T_s} \\ \dfrac{T_s}{2}\left[1 + G(-f)\right] & -\dfrac{1+\alpha}{2T_s} \leq f \leq 0 \end{cases} \qquad (A.12)$$

where $G(f)$ is an auxiliary frequency-domain function given by

$$G(f) = \begin{cases} 1 & 0 \leq |f| \leq \dfrac{1-\alpha}{2T_s} \\ \text{constrained} & \dfrac{1-\alpha}{2T_s} \leq |f| \leq \dfrac{1+\alpha}{2T_s} \\ -1 & |f| > \dfrac{1+\alpha}{2T_s} \end{cases} \qquad (A.13)$$

and the constraint is any shape that produces an $R_p(f)$ that satisfies (A.7). A necessary condition from (A.7) is that

$$R_p(f) + R_p\left(f - \frac{1}{T_s}\right) = T_s \quad \text{for} \quad \frac{1-\alpha}{2T_s} \leq f \leq \frac{1+\alpha}{2T_s} \qquad (A.14)$$

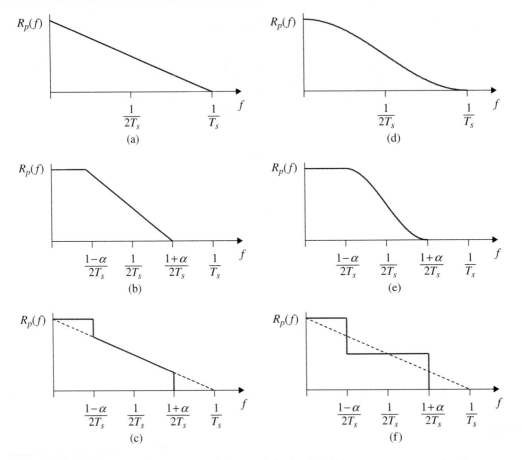

Figure A.2.2 Examples of spectra $R_p(f)$ that satisfy the Nyquist no-ISI condition for the case $1/T_s < 2B$. The spectra in (d) and (e) are examples of the spectral raised-cosine taper described below for 100% excess bandwidth (d) and less than 100% excess bandwidth (e). See Section A.3 for additional explanations regarding the other pulse shapes.

from which

$$G(f) + G\left(\frac{1}{T_s} - f\right) = 0 \qquad (A.15)$$

follows. Note that

$$G\left(f + \frac{1}{2T_s}\right) = -G\left(-f + \frac{1}{2T_s}\right) \quad \text{for} \quad \frac{1-\alpha}{2T_s} \leq |f| \leq \frac{1+\alpha}{2T_s}. \qquad (A.16)$$

Appendix A Pulse Shapes

The time-domain function $r_p(t)$ is the inverse Fourier transform of $R_p(f)$:

$$r_p(t) = \frac{T_s}{2} \int_{-\frac{1+\alpha}{2T_s}}^{0} [1 + G(-f)] e^{j2\pi ft} df + \frac{T_s}{2} \int_{0}^{\frac{1+\alpha}{2T_s}} [1 + G(f)] e^{j2\pi ft} df \quad \text{(A.17)}$$

$$= \frac{T_s}{2} \int_{-\frac{1+\alpha}{2T_s}}^{\frac{1+\alpha}{2T_s}} e^{j2\pi ft} df + \frac{T_s}{2} \int_{-\frac{1+\alpha}{2T_s}}^{0} G(-f) e^{j2\pi ft} df + \frac{T_s}{2} \int_{0}^{\frac{1+\alpha}{2T_s}} G(f) e^{j2\pi ft} df. \quad \text{(A.18)}$$

Using the definition (A.13), the expression takes on the form

$$r_p(t) = \frac{T_s}{2\pi t} \sin\left(\pi \frac{1+\alpha}{T_s} t\right) + \frac{T_s}{2\pi t} \sin\left(\pi \frac{1-\alpha}{T_s} t\right)$$

$$+ \frac{T_s}{2} \int_{-\frac{1+\alpha}{2T_s}}^{-\frac{1-\alpha}{2T_s}} G(-f) e^{j2\pi ft} df + \frac{T_s}{2} \int_{\frac{1-\alpha}{2T_s}}^{\frac{1+\alpha}{2T_s}} G(f) e^{j2\pi ft} df \quad \text{(A.19)}$$

$$= \frac{T_s}{\pi t} \sin\left(\frac{\pi t}{T_s}\right) \cos\left(\frac{\alpha \pi t}{T_s}\right) + T_s \int_{\frac{1-\alpha}{2T_s}}^{\frac{1+\alpha}{2T_s}} G(f) \cos(2\pi ft) df. \quad \text{(A.20)}$$

Applying the method of integration by parts, the integral may be expressed as

$$\int_{\frac{1-\alpha}{2T_s}}^{\frac{1+\alpha}{2T_s}} G(f) \cos(2\pi ft) df = -\frac{1}{\pi t} \sin\left(\frac{\pi t}{T_s}\right) \cos\left(\frac{\alpha \pi t}{T_s}\right) - \frac{1}{2\pi t} \int_{\frac{1-\alpha}{2T_s}}^{\frac{1+\alpha}{2T_s}} G'(f) \sin(2\pi ft) df \quad \text{(A.21)}$$

where $G'(f) = dG(f)/df$. Substituting the expression (A.21) for the integral in (A.20) produces

$$r_p(t) = -\frac{T_s}{2\pi t} \int_{\frac{1-\alpha}{2T_s}}^{\frac{1+\alpha}{2T_s}} G'(f) \sin(2\pi ft) df. \quad \text{(A.22)}$$

Using the change of variables $x = f - 1/(2T_s)$, the integral expression (A.22) may be expressed as

$$r_p(t) = \frac{\sin\left(\frac{\pi t}{T_s}\right)}{\frac{\pi t}{T_s}} \times \left(-\frac{1}{2}\right) \int_{-\frac{\alpha}{2T_s}}^{\frac{\alpha}{2T_s}} G'\left(x + \frac{1}{2T_s}\right) \cos(2\pi xt) dx. \quad \text{(A.23)}$$

Defining $\phi(t)$ as

$$\phi(t) = -\frac{1}{2} \int_{-\frac{\alpha}{2T_s}}^{\frac{\alpha}{2T_s}} G'\left(x + \frac{1}{2T_s}\right) \cos(2\pi xt) dx \qquad (A.24)$$

produces the result (A.11). Because $G\left(x + \frac{1}{2T_s}\right)$ is an odd function of x over the interval of integration, $G'\left(x + \frac{1}{2T_s}\right)$ is an even function of x over the interval of integration. As $\cos(2\pi xt)$ is an even function of x for all x, the integrand is an even function of x. Consequently, the resulting time-domain function $\phi(t)$ is an even function of t.

This rather remarkable result is illustrated graphically in Figure A.2.3. The desired response $r_p(t)$ is the product of a sinc function and a bandlimited even function of time as shown. The sinc function guarantees the desired zero crossings and the function $\phi(t)$ controls the temporal side-lobe decay. The faster the side lobe decay, the more bandwidth is required. In the frequency domain, $R_p(f)$ is the convolution of a rectangle and a real, even, and bandlimited function of f as shown. The shape of $\Phi(f)$ defines the nature of the transition band whereas the width of $\Phi(f)$ defines the excess bandwidth. Any real, even function $\phi(t)$ can be used to produce the desired response that posseses a Fourier transform that is continuous in f. Note that spectra that are not continuous in f, such as those illustrated in Figure A.2.2 (c) and (f), may not have a time-domain representation of the form (A.11).

Once the desired spectrum $R_p(f)$ is identified using the Nyquist no-ISI theorem, the pulse shape, $p(t)$, that produces the desired $R_p(f)$ is obtained using the identity (see Exercise A.5)

$$R_p(f) = |P(f)|^2. \qquad (A.25)$$

The pulse shape is thus the inverse Fourier transform of $\sqrt{R_p(f)}$. The most commonly used pulse shape that satisfies the Nyquist no-ISI condition is the square-root raised-cosine pulse shape described in the next section.

A.2.1 Spectral Raised-Cosine Pulse Shape

The most popular form for $R_p(f)$ that has an odd symmetric taper about $f = 1/2T_s$ is the *spectral raised-cosine*. This shape is defined by

$$R_p(f) = \begin{cases} T_s & 0 \leq |f| \leq \frac{1-\alpha}{2T_s} \\ \frac{T_s}{2}\left[1 + \cos\left(\frac{\pi |f| T_s}{\alpha} - \frac{\pi(1-\alpha)}{2\alpha}\right)\right] & \frac{1-\alpha}{2T_s} \leq |f| \leq \frac{1+\alpha}{2T_s} \\ 0 & |f| > \frac{1+\alpha}{2T_s} \end{cases} \qquad (A.26)$$

where $0 \leq \alpha \leq 1$ is the excess bandwidth.[1] The transition band follows half a period of a cosine, hence the name. The absolute bandwidth of $R_p(f)$ is $B_{\text{abs}} = \frac{2+\alpha}{2T_s}$. Note that for

---

[1]The parameter α is also called *roll-off factor* because it controls the transition band taper as illustrated in Figure A.2.4.

Appendix A Pulse Shapes

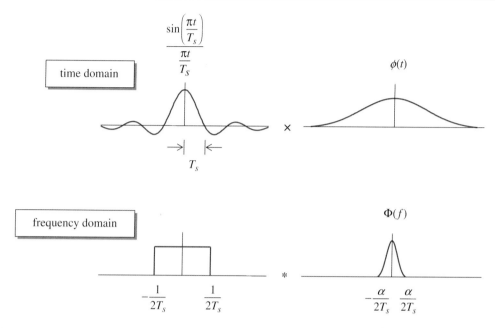

Figure A.2.3 A graphical interpretation of the property (A.11). A Nyquist pulse can be generated in the time domain by multiplying a sinc function with a real and even function of time. In the frequency domain, this is equivalent to convolving the Fourier transforms of the two.

$\alpha = 0$, $R_p(f)$ is equal to a rectangle extending from $-1/2T_s$ to $1/2T_s$. Thus, the spectral raised-cosine pulse shape reduces the spectral rectangle as required by case 2 of the Nyquist no-ISI theorem. For $\alpha > 0$ the bandwidth of $R_p(f)$ is wider than the rectangle by an amount $\alpha/2T_s$.

In the time-domain, the spectral raised-cosine is

$$r_p(t) = \frac{\sin\left(\frac{\pi t}{T_s}\right)}{\frac{\pi t}{T_s}} \times \frac{\cos\left(\frac{\pi \alpha t}{T_s}\right)}{1 - \left(\frac{2\alpha t}{T_s}\right)^2}. \tag{A.27}$$

The spectral raised-cosine is a special case of (A.11) where $\phi(t)$ is

$$\phi(t) = \frac{\cos\left(\frac{\pi \alpha t}{T_s}\right)}{1 - \left(\frac{2\alpha t}{T_s}\right)^2}. \tag{A.28}$$

The expression (A.27) is plotted in the lower portion of Figure A.2.4. Note the T_s-spaced zero crossings. This plot shows that $\alpha = 0$ produces a rectangle in the frequency domain which corresponds to the sinc function in the time domain.

The pulse shape that produces the raised-cosine spectrum is the inverse Fourier transform of $\sqrt{R_p(f)}$. For this reason, the pulse shape is called the *square-root raised-cosine* (SRRC) pulse shape. The SRRC has a Fourier transform $P(f)$ and time-domain expression $p(t)$ as follows:

$$P(f) = \begin{cases} \sqrt{T_s} & 0 \leq |f| \leq \dfrac{1-\alpha}{2T_s} \\ \sqrt{T_s} \cos\left(\dfrac{\pi |f| T_s}{2\alpha} - \dfrac{\pi(1-\alpha)}{4\alpha}\right) & \dfrac{1-\alpha}{2T_s} \leq |f| \leq \dfrac{1+\alpha}{2T_s} \\ 0 & |f| > \dfrac{1+\alpha}{2T_s} \end{cases} \quad (A.29)$$

$$p(t) = \frac{1}{\sqrt{T_s}} \frac{\sin\left(\pi(1-\alpha)\dfrac{t}{T_s}\right) + \dfrac{4\alpha t}{T_s}\cos\left(\pi(1+\alpha)\dfrac{t}{T_s}\right)}{\dfrac{\pi t}{T_s}\left[1 - \left(\dfrac{4\alpha t}{T_s}\right)^2\right]}. \quad (A.30)$$

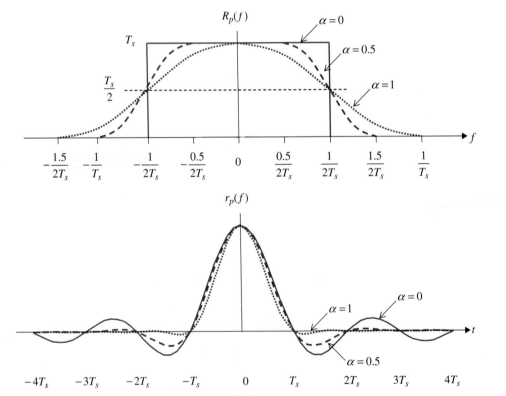

Figure A.2.4 Raised-cosine spectrum (top) and corresponding time-domain waveform (bottom) for excess bandwidths of 0%, 50%, and 100%.

Appendix A Pulse Shapes

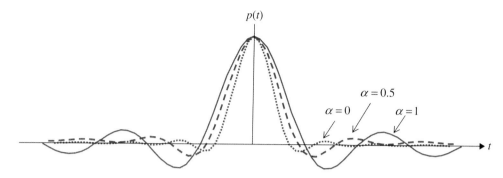

Figure A.2.5 Square-root raised-cosine (SRRC) pulse shape for excess bandwidths of 0%, 50%, and 100%.

Note that there are two values of t for which $p(t)$ must be evaluated in the limit $t = 0$ and $t = \pm \frac{T_s}{4\alpha}$ (see Exercise A.6). The SRRC pulse shape $p(t)$ is plotted in Figure A.2.5 for excess bandwidths of 0%, 50%, and 100%. Observe that the zero crossings of $p(t)$ do not occur at nonzero integer multiples of the symbol time.

The discrete-time version of the SRRC pulse shape can be obtained using the impulse invariance technique outlined in Chapter 3. Because the pulse shape (A.30) is bandlimited, the impulse invariance technique produces an exact discrete-time version. The sample rate $1/T$ must exceed twice the highest frequency; that is

$$\frac{1}{T} > \frac{1+\alpha}{T_s}. \tag{A.31}$$

It is standard practice to select the sample rate to be N times the symbol rate; that is

$$\frac{T_s}{T} = N. \tag{A.32}$$

The sampled version of the square-root raised-cosine filter is produced by substituting nT for t in (A.30) and scaling the amplitude by T. Using (A.32) produces

$$p(nT) = \frac{1}{\sqrt{N}} \frac{\sin\left(\pi(1-\alpha)\frac{n}{N}\right) + \frac{4\alpha n}{N}\cos\left(\pi(1+\alpha)\frac{n}{N}\right)}{\frac{\pi n}{N}\left[1 - \left(\frac{4\alpha n}{N}\right)^2\right]}. \tag{A.33}$$

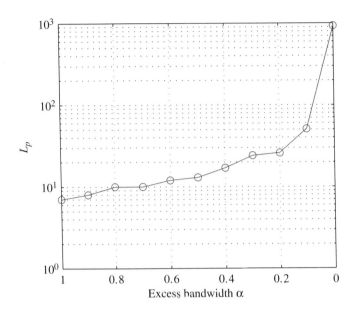

Figure A.2.6 Required truncation length ($\pm L_p$ symbols) to achieve -40 dB stop-band attenuation as a function of the excess bandwidth α.

The SRRC pulse shape has infinite support in time. Practical implementations truncate the pulse shape to the interval $-L_pT_s \leq t \leq L_pT_s$ (or $-L_pN \leq n \leq L_pN$) to span $2L_p + 1$ symbols. As a consequence of this truncation, the $p(t)$ now has finite support which means $|P(f)|^2$ has infinite support. This infinite support in the frequency domain takes the form of side lobes. The larger L_p is, the lower the side lobes. In practice, L_p is chosen to force the side lobe levels to an acceptably low level. A simple truncation of the SRRC pulse shape (A.30) is adequate as long as the desired side lobe level is not lower than -40 dB. The value of L_p required to achieve a stop-band attenuation of 40 dB is illustrated in Figure A.2.6. For large excess bandwidths, L_p is less than 10. As α decreases, L_p increases, but only moderately until α drops below 20%. For small excess bandwidth, the required filter length is quite large.

A side lobe level of about -40 dB is all that is achievable by truncating the SRRC pulse shape. This is due to the fact that the transition band of $P(f)$ is of a quarter cycle of a cosine. As a consequence, there is a discontinuity in $P(f)$ at the stop-band that produces the relatively high side lobes when $p(t)$ is truncated. Another type of transition-band-taper must be used to achieve lower side lobe levels. An example of an alternate taper is described in the next section.

The second consequence of truncation is that the truncated pulse shape only approximately satisfies the Nyquist no-ISI condition. The effect of this is best quantified using $r_p(kT_s)$. For truncated pulse shapes, $r_p(kT_s)$ is small, but not zero for $k \neq 0$. This is illustrated in Figure A.2.7 for an SRRC pulse shape with $\alpha = 0.5$ and $L_p = 3, 6,$ and 12.

Appendix A Pulse Shapes

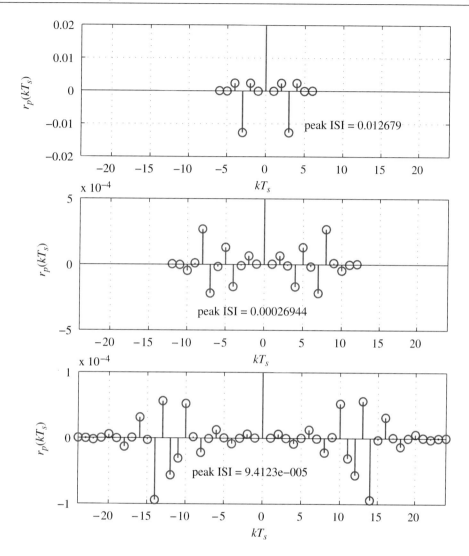

Figure A.2.7 An illustration of peak ISI as a function of L_p for the SRRC pulse shape with $\alpha = 0.5$: (top) $L_p = 3$, (middle) $L_p = 6$, (bottom) $L_p = 12$.

The nonzero values for $k \neq 0$ contribute ISI and perturb the location of the projection on to the signal space. A first-order approximation to the effect of ISI on performance is to quantify the maximum magnitude of $r_p(kT_s)$ for $k \neq 0$. This value is called the *peak ISI*. Figure A.2.7 illustrates that the peak ISI decreases as L_p increases. In fact, the peak ISI $\rightarrow 0$ as $L_p \rightarrow \infty$.

A.2.2 Other Nyquist Pulse Shapes

A variety of other pulse shapes have been proposed. Among these are the "flipped exponential" pulse shape whose corresponding $R_p(f)$ is

$$R_p(f) = \begin{cases} T_s & |f| \leq \frac{1-\alpha}{2T_s} \\ T_s \exp\left\{\beta\left(\frac{1-\alpha}{2T_s} - |f|\right)\right\} & \frac{1-\alpha}{2T_s} < |f| \leq \frac{1}{2T_s} \\ T_s\left(1 - \exp\left\{\beta\left(|f| - \frac{1+\alpha}{2T_s}\right)\right\}\right) & \frac{1}{2T_s} < |f| \leq \frac{1+\alpha}{2T_s} \\ 0 & \frac{1+\alpha}{2T_s} < |f| \end{cases} \quad (A.34)$$

where

$$\beta = \frac{\ln(2)}{\frac{\alpha}{2T_s}} \quad (A.35)$$

and whose time-domain function is

$$r_p(t) = \frac{\sin\left(\frac{\pi t}{T_s}\right)}{\frac{\pi t}{T_s}} \times \frac{4\beta\pi t \sin\left(\frac{\pi \alpha t}{T_s}\right) + 2\beta^2 \cos\left(\frac{\pi \alpha t}{T_s}\right) - \beta^2}{(2\pi t)^2 + \beta^2}. \quad (A.36)$$

Another is the "flipped-hyperbolic secant" pulse shape whose corresponding $R_p(f)$ is

$$R_p(f) = \begin{cases} T_s & |f| \leq \frac{1-\alpha}{2T_s} \\ T_s \text{sech}\left(\gamma\left(|f| - \frac{1-\alpha}{2T_s}\right)\right) & \frac{1-\alpha}{2T_s} < |f| \leq \frac{1}{2T_s} \\ T_s\left[1 - \text{sech}\left(\gamma\left(\frac{1+\alpha}{2T_s} - |f|\right)\right)\right] & \frac{1}{2T_s} < |f| \leq \frac{1+\alpha}{2T_s} \\ 0 & \frac{1+\alpha}{2T_s} < |f| \end{cases} \quad (A.37)$$

where

$$\gamma = \frac{\ln\left(\sqrt{3}+2\right)}{\frac{\alpha}{2T_s}} \quad (A.38)$$

and whose time-domain function is

$$r_p(t) = \frac{\sin\left(\frac{\pi t}{T_s}\right)}{\frac{\pi t}{T_s}} \times \left[8\pi t \sin\left(\frac{\pi \alpha t}{T_s}\right) F_1(t) + 2\cos\left(\frac{\pi \alpha t}{T_s}\right)\left[1 - 2F_2(t)\right] + 4F_3(t) - 1\right] \quad (A.39)$$

where

$$F_1(t) = \sum_{k=0}^{\infty} (-1)^k \frac{(2k+1)\gamma}{[(2k+1)\gamma]^2 + (2\pi t)^2} \tag{A.40}$$

$$F_2(t) = \sum_{k=0}^{\infty} (-1)^k \frac{(2\pi t)^2}{[(2k+1)\gamma]^2 + (2\pi t)^2} \tag{A.41}$$

$$F_3(t) = \sum_{k=0}^{\infty} (-1)^k \frac{(2\pi t)^2}{[(2k+1)\gamma]^2 + (2\pi t)^2} e^{(2k+1)\frac{\alpha\gamma}{2T_s}}. \tag{A.42}$$

A third pulse is the "flipped-inverse hyperbolic secant" pulse shape whose corresponding $R_p(f)$ is

$$R_p(f) = \begin{cases} T_s & |f| \leq \frac{1-\alpha}{2T_s} \\ T_s \left[1 - \frac{T_s}{\alpha\gamma} \text{arcsech}\left(\frac{T_s}{\alpha}\left(\frac{1+\alpha}{2T_s} - |f|\right)\right)\right] & \frac{1-\alpha}{2T_s} < |f| \leq \frac{1}{2T_s} \\ \frac{T_s}{\alpha\gamma} \text{arcsech}\left(\frac{T_s}{\alpha}\left(|f| - \frac{1+\alpha}{2T_s}\right)\right) & \frac{1}{2T_s} < |f| \leq \frac{1+\alpha}{2T_s} \\ 0 & \frac{1+\alpha}{2T_s} < |f| \end{cases} \tag{A.43}$$

where γ is defined by (A.38). The corresponding time-domain response $r_p(t)$ must be evaluated using numerical techniques.

These pulse shapes produce better symbol error probabilities in the presence of symbol timing sampling errors than the spectral raised-cosine pulse shape. As before, the actual pulse shape used by the transmitter is the inverse Fourier transform of $\sqrt{R_p(f)}$. Unfortunately, for these three spectra closed-form expressions do not exist.

A.2.3 Nyquist Pulse Shapes with Improved Stop-Band Attenuation

Figure A.2.6 shows that the side lobe levels fall very slowly with increased filter length and increases with reduced transition bandwidth. These levels of attenuation will not meet realistic spectral mask requirements for out-of-band attenuation that are typically on the order of 60–80 dB. Some mechanism must be invoked to control the out-of-band side lobe levels of the truncated pulse shape. Whatever process is invoked should preserve the ISI levels obtained by convolving the fixed length square-root Nyquist filter with itself.

Truncating the ideal pulse shape $p(t)$ given by (A.30) is equivalent to multiplying $p(t)$ by a rectangular window $w(t)$ given by

$$w(t) = \begin{cases} 1 & -L_p T_s \leq t \leq L_p T_s \\ 0 & \text{otherwise} \end{cases}. \tag{A.44}$$

In the frequency domain, this has the effect of convolving $P(f)$ with $W(f)$ which is given by

$$W(f) = 2L_p T_s \frac{\sin(2\pi f L_p T_s)}{2\pi f L_p T_s}. \quad (A.45)$$

(This is the cause of the side lobes observed in the spectrum of the truncated pulse shape.) Attempting to control the spectral side lobes by only applying other windows to $p(t)$ results in significant increase in the ISI levels at the receiver output. This is illustrated in Figure A.2.8, which demonstrates the effect on spectral side lobes and ISI levels as a result of applying windows to the prototype impulse response.

The Nyquist no-ISI condition requires symmetry about $fT_s = \pm 1/2$. This forces the 3-dB points of the pulse shape to occur exactly at $fT_s = \pm 1/2$. The increase in ISI observed in Figure A.2.8 is traced to the shift of the 3-dB point away from $fT_s = \pm 1/2$ as illustrated in Figure A.2.9. A design technique must control side lobe levels while maintaining the 3-dB frequency at $fT_s = 1/2$ is required.

A design technique can be based on commonly available FIR low-pass filter design techniques. Most FIR low-pass filter designs require the specification of the filter length, pass-band, and stop-band frequencies as discussed in Chapter 3. In addition, the pass-band ripple and stop-band ripple can also be defined. In general, for a fixed filter length, the stop-band attenuation increases as the pass band and/or transition band become wider. The square-root raised-cosine pulse shape is an example of a low-pass filter where the pass band extends to $fT_s = (1-\alpha)/2$ and a stop band beginning at $fT_s = (1+\alpha)/2$.

A very simple iterative algorithm based on the Parks–McClellan algorithm (described in Chapter 3) transforms an initial low-pass filter to a square-root-Nyquist spectrum with the specified roll-off while preserving the ability to independently control pass-band ripple and stop-band ripple. Figure A.2.10 illustrates the form of the algorithm by starting the Parks–McClellan algorithm with pass-band and stop-band edges matched to the roll-off boundaries of the Nyquist spectrum. The resulting filter crosses the band edge ($fT_s = 1/2$) with more attenuation than the desired -3 dB level. The attenuation level can be raised toward the desired -3 dB level by increasing the frequency of the pass-band edge. The algorithm performs the successive shifts to the right of the pass-band edge (frequency F_1) until the error

$$e = \left|P\left(\frac{1}{2T_s}\right)\right| - \frac{1}{\sqrt{2}} \quad (A.46)$$

is reduced to zero. The value of the shifts are controlled using a gradient descent method based on the equations

$$e(n) = \frac{1}{\sqrt{2}} - \left|P\left(\frac{1}{2T_s}\right)\right|$$
$$F_1(n+1) = F_1(n) \times (1 + \mu e(n)) \quad (A.47)$$

where μ is a positive constant that defines the step size and n indexes the steps in the iteration. The step size controls the convergence rate and the approximation error.

An example of this example is illustrated in Figures A.2.11–A.2.13. for the design of a square-root Nyquist pulse shape with $\alpha = 0.5$ and $L_p = 6$. A square-root Nyquist pulse

Appendix A Pulse Shapes

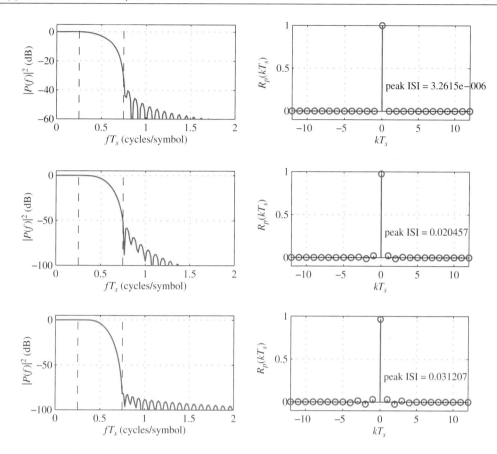

Figure A.2.8 The effect of windows on the square-root raised-cosine pulse shape. The vertical dashed lines in each plot indicate the frequencies corresponding to $fT_s = (1-\alpha)/2$ and $fT_s = (1+\alpha)/2$. (Top) $|P(f)|^2$ and $r_p(kT_s)$ for a square-root raised-cosine pulse shape with $\alpha = 0.5$ and $L_p = 6$. $|P(f)|^2$ shows that the stop-band attenuation is about 40 dB. The plot of $r_p(kT_s)$ shows the peak ISI is very modest. (The peak ISI $\to 0$ as $L_p \to \infty$.) (Middle) $|P(f)|^2$ and $r_p(kT_s)$ for a square-root raised-cosine pulse shape with $\alpha = 0.3$, $L_p = 6$ multiplied by a Hann window. (The excess bandwidth of the prototype square-root raised-cosine filter had to be reduced to account for the effect of the window.) $|P(f)|^2$ shows that the stop-band attenuation is about 60 dB and the plot of $r_p(kT_s)$ shows the peak ISI is about 0.02. Note that the increased stop-band attenuation is achieved at the expense of peak ISI. (Bottom) $|P(f)|^2$ and $r_p(kT_s)$ for a square-root raised-cosine pulse shape with $\alpha = 0.2$, $L_p = 6$ multiplied by a Kaiser window with $\beta = 6$. Again α had to be reduced to account for the effect of the windowing. $|P(f)|^2$ shows that the stop-band attenuation is about 80 dB and the plot of $r_p(kT_s)$ shows the peak ISI is about 0.03. Again observe that increased stop-band attenuation is achieved at the expense of peak ISI.

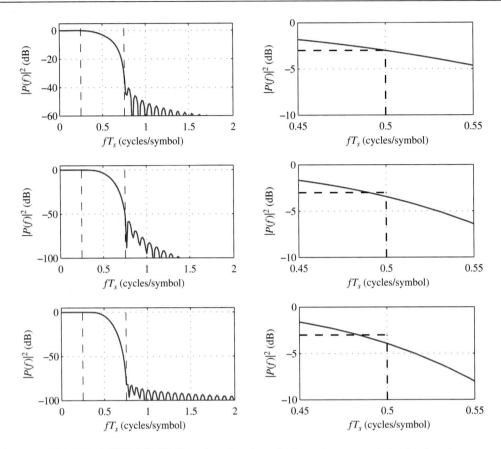

Figure A.2.9 Close-ups of the 3-dB points for the windowed square-root raised-cosine designs illustrated in Figure A.2.8. Observe that the further away the 3-dB point is from the desired $fT_s = 1/2$, the more severe the peak ISI.

shape at $N = 8$ samples/symbols was designed using the iterative algorithm described above. For the purposes of comparison, a square-root raised-cosine pulse shape was designed with the same parameters ($\alpha = 0.5$, $L_p = 6$, and $N = 8$). Both filters have the same length: $2 \times L_p \times N + 1 = 97$. Figure A.2.11 shows the spectra of the two pulse-shape designs. Note that the magnitude of both pulse shapes is $1/\sqrt{2}$ at $fT_s = 1/2$ as required. As a consequence, the Nyquist spectrum and raised-cosine spectrum are $1/2$ at $fT_s = 1/2$. It appears at first glance that the spectrum for the square-root Nyquist pulse shape has a steeper slope and, consequently, a reduced roll-off factor. In fact, the two filters have the same roll-off of 0.5 but the square-root Nyquist pulse shape has a smoother transition. This smoother transition is required to achieve reduced side lobe levels and increased out-of-band attenuation.

Appendix A Pulse Shapes

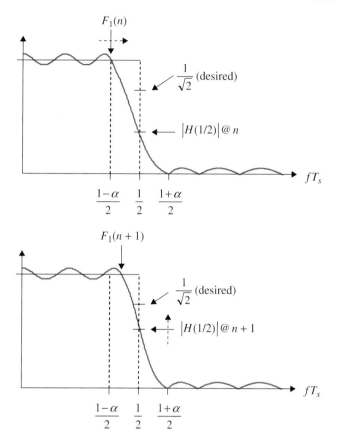

Figure A.2.10 Spectrum of the Parks–McClellan algorithm output at iteration n. The algorithm is initialized by designing a low-pass filter with the upper pass-band frequency set to $fT_s = (1-\alpha)/2$ and the lower stop-band frequency set to $fT_s = (1+\alpha)/2$. The upper edge of the pass band (frequency F_1) is increased to increase $|H(1/2)|$ to the desired level.

The increased out-of-band attenuation of the square-root Nyquist pulse shape is illustrated in Figure A.2.12. Here the spectra of the two pulses are plotted using a decibel scale for the magnitude to illustrate the out-of-band signal levels or, stop-band attenuation. The stop-band attenuation for the square-root raised-cosine pulse shape is barely 40 dB whereas that of the square-root Nyquist pulse shape is more than 80 dB.

Figure A.2.13 shows a detailed plot of the pulse autocorrelation function $r_p(kT_s)$ for both the square-root Nyquist pulse shape (top) and the square-root raised-cosine pulse shape (bottom). Observe that the peak ISI is about the same in both cases. This illustrates that the iterative algorithm is able to produce pulse shapes that, relative to the square-root raised-cosine pulse shape, have vastly improved spectral properties without an increase in the ISI.

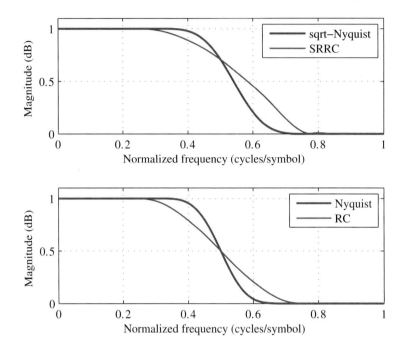

Figure A.2.11 A plot of the spectra for (top) the square-root Nyquist pulse shape (solid line) and square-root raised-cosine pulse shape (dashed line); (bottom) the Nyquist pulse shape (solid line) and raised-cosine pulse shape (dashed line). Both filters were designed using $\alpha = 0.5$, $L_p = 6$, and $N = 8$.

A.3 NOTES AND REFERENCES

A.3.1 Topics Covered

The NRZ pulse shape has been used since the earliest days of digital communications. This is because the amplitudes created by "keying" a source on or off produced a pulse train with the NRZ pulse shape in a very natural way. As data rates increased and timing issues became important, the RZ and MAN pulse shapes also became popular, due in large measure to the ease with which they could be generated from the NRZ pulse shape.

Nyquist pulse shapes follow from the Nyquist no-ISI theorem that was first published by Nyquist in 1928 [322]. The derivation of the form (A.11) is due to Scanlan [323]. The examples shown in Figure A.2.2 (c) and (f) are taken from Franks [324]. The spectrum of Figure A.2.2 (c) was shown by Franks to correspond to the pulse that minimizes the RMS distortion in the presence of small timing errors. The spectrum of Figure A.2.2 (f) was shown by Franks to correspond to the pulse that minimizes the RMS distortion due to uncompensated carrier phase offset in a QAM system. The spectra of Figures A.2.2 (a) and (b) were shown by Sousa and Pasupathy [95] to minimize the peak ISI in the presence of a small timing error. The

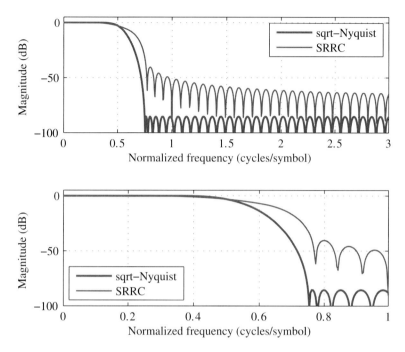

Figure A.2.12 A plot of the spectra for the square-root Nyquist pulse shape (solid line) and square-root raised-cosine pulse shape (dashed line). The lower plot is a close-up of the upper plot. Both filters were designed using $\alpha = 0.5$, $L_p = 6$, and $N = 8$.

raised-cosine taper has been in use for a very long time and is described in most textbooks on digital communications. See, for example, Proakis [10], Rappaport [12], and Stüber [16]. The "flipped-exponential" pulse was first suggested by Beaulieu, Tan, and Damen [325], the "flipped-hyperbolic secant" and "flipped-inverse hyperbolic secant" pulses were introduced by Assalini and Tonello [326]. The iterative method for generating FIR square-root Nyquist pulses with improved stop-band attenuation is due to Harris [53, 327]. Another iterative method has been described by Jayasimha and Singh [328].

A.3.2 Topics Not Covered

The partial-response pulse shapes outlined in this appendix were designed to eliminate intersymbol interference. When the pulse train is accompanied by noise, minimizing the effect of the noise has also been of interest. Usually, the optimization criterion is mean squared error rather than probability of symbol error. This approach produces a more mathematically tractable solution. For full-response pulse shapes, the receiving filter that minimizes the distortion due to noise is the matched filter [82]. Pulse shapes and receiving filters that eliminate ISI and minimize the noise distortion or jointly minimize ISI and noise

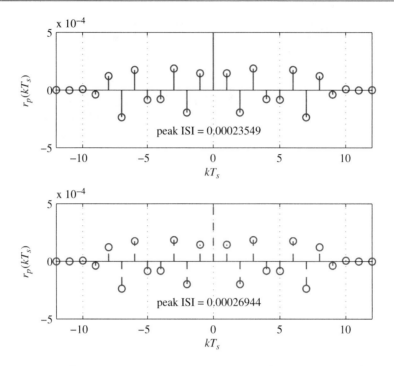

Figure A.2.13 A plot of the symbol-spaced autocorrelation function $r_p(kT_s)$ for the square-root Nyquist pulse shape (top plot) and the square-root raised-cosine pulse shape (lower plot).

distortion for partial-response pulse shapes have been investigated by Tufts [329], Berger and Tufts [330], and Hill [331], among others.

There are some other pulse shapes that are popular in the literature but are not covered in this text. They are a partial-response version of the NRZ pulse shape and the *temporal raised-cosine* (both full response and partial response). For the most part, these pulse shapes are used with continuous phase modulation (CPM), which is not covered in this text. The text by Anderson, Aulin, and Sundberg [31] is the standard reference for CPM. Chapters 4 and 5 of Proakis [10] also provide an excellent summary. Another popular pulse shape used with CPM is the "Gaussian" pulse shape. The Gaussian pulse shape is used with MSK to form GMSK, the modulation used for GSM and for many satellite communication systems. The term "Gaussian" is used because the weighted pulse train used for modulation is an NRZ pulse train that is filtered by a low-pass filter whose impulse response is the Gaussian function. Thus, the pulse shape is really a low-pass filtered NRZ and is a partial-response pulse shape. This modulation was first proposed by Murota and Hirade [109]. Partial-response pulse shapes that do not satisfy the Nyquist no-ISI condition have also been investigated. In this case, the ISI is controlled and known. This information is exploited by the detector to make the symbol decisions. This method is called "duo-binary" signaling and was invented by Linder [332]

Appendix A Pulse Shapes

generalized by Kretzmer [333]. An excellent survey article on this approach was published by Kabal and Pasupathy [334]. Chapter 9 of Proakis [10] is also an excellent resource.

Another important focus of research has been on pulse shapes that satisfy the no-ISI condition both with and without matched filtering. (The square-root raised-cosine pulse shape, by itself, possesses ISI. This is evident from the fact that the zero crossings of the pulse shape do not occur at multiples of T_s as illustrated in Figure A.2.5. It is not until it is processed by a filter matched to the square-root raised-cosine pulse that no ISI is present.) Pulse shapes that satisfy this additional no-ISI constraint are called "generalized Nyquist pulses." Examples of these pulse shapes are described by Xia [335], Demeechai [336], Kisel [337,338], Alagha and Kabal [339], and Tan and Beaulieu [340,341].

Following the theme of this text, discrete-time realizations of Nyquist pulse shapes have been examined. Considerable effort has been devoted to continuous-time realization of the Nyquist pulse shapes. One method, described by Kesler and Taylor [342], is based on a phase equalized fourth-order Butterworth filter. Other approaches are described in the relatively recent work by Baher and Beneat [343], Hassan and Ragheb [344], Mneina and Martens [345], and the references cited therein.

A.4 EXERCISES

A.1 Plot the autocorrelation functions $r_p(\tau)$ for the four full-response pulse shapes summarized in Table A.1.1. Use the plots to explain why the use of these pulse shapes in a binary PAM system using a matched filter detector does not produce any ISI.

A.2 Consider the "triangular" pulse shape defined by

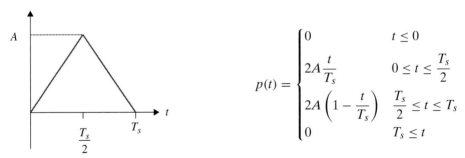

$$p(t) = \begin{cases} 0 & t \leq 0 \\ 2A\dfrac{t}{T_s} & 0 \leq t \leq \dfrac{T_s}{2} \\ 2A\left(1 - \dfrac{t}{T_s}\right) & \dfrac{T_s}{2} \leq t \leq T_s \\ 0 & T_s \leq t \end{cases}$$

(a) Compute the value of A required for $p(t)$ to be a unit-energy pulse shape.

(b) Compute

$$r_p(\tau) = \int_0^{T_s} p(t)p(t-\tau)dt.$$

(c) Compute $|P(f)|^2$.

(d) Compute B_{abs}, B_{90}, B_{99}, and $B_{-60\,\text{dB}}$. (Note: you will have to compute most of these numerically.)

A.3 Consider the "temporal raised-cosine" pulse shape defined by

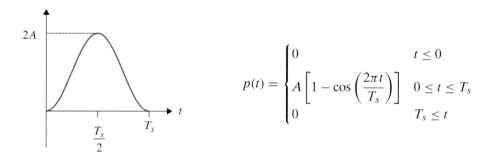

$$p(t) = \begin{cases} 0 & t \leq 0 \\ A\left[1 - \cos\left(\dfrac{2\pi t}{T_s}\right)\right] & 0 \leq t \leq T_s \\ 0 & T_s \leq t \end{cases}$$

(a) Compute the value of A required for $p(t)$ to be a unit-energy pulse shape.

(b) Compute

$$r_p(\tau) = \int_0^{T_s} p(t)p(t-\tau)dt.$$

(c) Compute $|P(f)|^2$.

(d) Compute B_{abs}, B_{90}, B_{99}, and $B_{-60 \text{ dB}}$. (Note: you will have to compute most of these numerically.)

A.4 This exercise explores the proof of the Poisson sum formula (A.8). The proof is centered on the two ways to represent the Fourier transform of the product

$$g(t) = r_p(t) \sum_{k=-\infty}^{\infty} \delta(t - kT_s).$$

(a) To produce the left-hand side of the Poisson sum formula, compute the Fourier transform of $g(t)$ by using the Fourier transforms of $r_p(t)$ and the impulse train together with the property that multiplication in the time-domain is convolution in the frequency domain.

(b) To produce the right-hand side of the Poisson sum formula, apply the Fourier transform integral directly to $g(t)$ and use the "sifting property" of the impulse function.

A.5 Let $p(t)$ be a real-valued pulse shape and let

$$P(f) = \int_{-\infty}^{\infty} p(t)e^{-j2\pi ft}dt$$

Appendix A Pulse Shapes

be the Fourier transform of $p(t)$. The autocorrelation function of $p(t)$ is

$$r_p(\tau) = \int_{-\infty}^{\infty} p(t)p(t-\tau)dt$$

and its Fourier transform is

$$R_p(f) = \int_{-\infty}^{\infty} r_p(\tau) e^{-j2\pi f \tau} d\tau.$$

Prove that

$$R_p(f) = |P(f)|^2.$$

A.6 The square-root raised-cosine pulse shape given by (A.30) has two values of t which produce an indeterminant form. This problem explores the evaluation of $p(t)$ at these values.

(a) Show that $p(t)$ at $t = 0$ is an indeterminant of the form 0/0. Using L'Hôpital's rule, show that

$$\lim_{t \to 0} p(t) = \frac{1}{\sqrt{T_s}} \left[1 - \alpha + \frac{4\alpha}{\pi} \right].$$

(b) Show that $p(t)$ at $t = \pm \dfrac{T_s}{4\alpha}$ is an indeterminant of the form 0/0. Using L'Hôpital's rule, show that

$$\lim_{t \to \frac{T_s}{4\alpha}} p(t) = \frac{\alpha}{\sqrt{2T_s}} \left[\left(1 + \frac{2}{\pi}\right) \sin\left(\frac{\pi}{4\alpha}\right) + \left(1 - \frac{2}{\pi}\right) \cos\left(\frac{\pi}{4\alpha}\right) \right].$$

A.7 Derive an expression for the 3-dB bandwidth of the raised-cosine spectrum. The expression should be in terms of the excess bandwidth α and the symbol time T_s.

A.8 Derive an expression for the 3-dB bandwidth of the square-root raised-cosine pulse shape. The expression should be in terms of the excess bandwidth α and the symbol time T_s.

Appendix B
The Complex-Valued Representation for QAM

B.1 INTRODUCTION

Complex numbers provide compact representation of real-valued band-pass signals, such as QAM signals. Many of the benefits result from the fact that the bookkeeping required to track the trigonometric identities in the mathematical representation of the mixing process is much easier with the equivalent complex-valued representation. For example, many of the rotations involved in carrier phase synchronization are more efficiently described as rotations of a complex number by a phase. In addition, the signal processing is sometimes simplified using complex-valued signal processing on the complex-valued equivalent of a signal. Signals with a one-sided spectrum (i.e., a signal whose Fourier transform is nonzero only for positive (or negative) frequencies) can be frequency translated without need to track potential overlap of positive and negative frequency components. Signals with a one-sided spectrum correspond to complex-valued time-domain signals.

B.1.1 Key Identities

Most of the concepts summarized in this appendix follow from just a few mathematical identities, all of which are familiar to a senior electrical engineering student. The first set of identities are the Euler identities

$$\cos(\theta) = \frac{e^{j\theta} + e^{-j\theta}}{2}$$
$$\sin(\theta) = \frac{e^{j\theta} - e^{-j\theta}}{j2} \tag{B.1}$$
$$e^{j\theta} = \cos(\theta) + j\sin(\theta).$$

Appendix B The Complex-Valued Representation for QAM

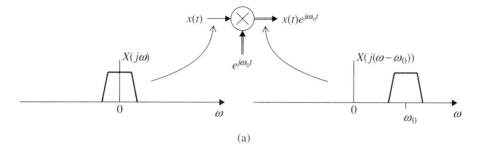

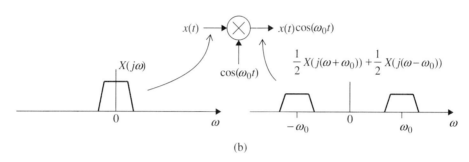

Figure B.1.1 An illustration of frequency translations resulting from multiplication by sinusoids: (a) The "one-sided" frequency translation resulting from multiplication by a complex exponential. Because the product is a complex-valued signal, the Fourier transform of the product does not possess conjugate symmetry. (b) The "two-sided" frequency translation resulting from multiplication by a real-valued sinusoid. The product is a real-valued signal. Hence, the Fourier transform of the product does possess conjugate symmetry.

The second set of identities involve the Fourier transform of the complex exponential $e^{j\omega_0 t}$:

$$e^{j\omega_0 t} \leftrightarrow 2\pi \delta(\omega - \omega_0) \tag{B.2}$$

$$x(t)e^{j\omega_0 t} \leftrightarrow X\left(j(\omega - \omega_0)\right) \tag{B.3}$$

where $X(j\omega)$ is the Fourier transform of $x(t)$ and the double arrow indicates a Fourier transform pair. This shows that multiplication by a complex exponential can be interpreted as performing a frequency translation in one direction, as illustrated in Figure B.1.1 (a). This is in contrast with the multiplication by $\cos(\omega_0 t)$, which performs a frequency translation in two directions as illustrated in Figure B.1.1 (b).

The final concepts used in this appendix are related to some important properties of the Fourier transform. Let $x(t)$ be a time-domain signal whose Fourier transform is $X(j\omega)$.

- If $x(t)$ is real, then $X(j\omega) = X^*(-j\omega)$. That is, the Fourier transform of a real-valued signal is conjugate symmetric. The conjugate symmetry of $X(j\omega)$ implies that the real part of $X(j\omega)$ is even and the imaginary part of $X(j\omega)$ is odd:

$$\text{Re}\left\{X(j\omega)\right\} = \text{Re}\left\{X(-j\omega)\right\}$$
$$\text{Im}\left\{X(j\omega)\right\} = -\text{Im}\left\{X(-j\omega)\right\}.$$
(B.4)

Another way of expressing the same property is that the magnitude of $X(j\omega)$ is even and the phase of $X(j\omega)$ is odd:

$$|X(j\omega)| = |X(-j\omega)|$$
$$\angle X(j\omega) = -\angle X(-j\omega).$$
(B.5)

- It follows immediately that if $X(j\omega)$ is not conjugate symmetric, then $x(t)$ is not a real-valued signal. Thus, if $X(j\omega)$ is nonzero only for $\omega > 0$, then $x(t)$ is a complex-valued signal.
- If, in addition to being real, $x(t)$ is even, then $X(j\omega)$ is real and even.
- If $x(t)$ is real and odd, then $X(j\omega)$ is purely imaginary and odd.

B.1.2 A Complex-Valued Signal: Does It Really Exist?

The terms "real," "imaginary," and "complex" are unfortunate because they imply that a "real" signal exists in nature and an "imaginary" signal does not and that a "complex" signal is somehow complicated.[1] A complex-valued signal should be thought of as a two-part signal consisting of a pair of real-valued signals called the "real" signal and the "imaginary" signal. That is, the complex-valued waveform $z(t)$ may be expressed as

$$z(t) = x(t) + jy(t)$$
(B.6)

where $j = \sqrt{-1}$ is the "imaginary operator" and serves as a place holder to identify the imaginary component of the two-part (complex-valued) signal. The real part of $z(t)$ is the real-valued waveform $x(t)$ and the imaginary part of $z(t)$ is the real-valued waveform $y(t)$. In the sections that follow, the following notation will be used to express these concepts:

$$\text{Re}\{z(t)\} = x(t)$$
(B.7)
$$\text{Im}\{z(t)\} = y(t).$$
(B.8)

It should be pointed out that both $x(t)$ and $y(t)$ are real-valued waveforms even though $y(t)$ is the "imaginary" component of $z(t)$.

---

[1] In 1831, Carl Friedrich Gauss wrote, "If this subject [complex numbers] has hitherto been considered from the wrong viewpoint and thus enveloped in mystery and surrounded by darkness, it is largely an unsuitable terminology which should be blamed. Had $+1$, -1, and $\sqrt{-1}$, instead of being called positive, negative and imaginary (or worse still impossible) unity, been given the names, say, of direct, inverse and lateral unity, there would hardly have been any scope for such obscurity." See Section B.4 for a brief summary of how and why the unfortunate terms "imaginary" and "complex" came to be.

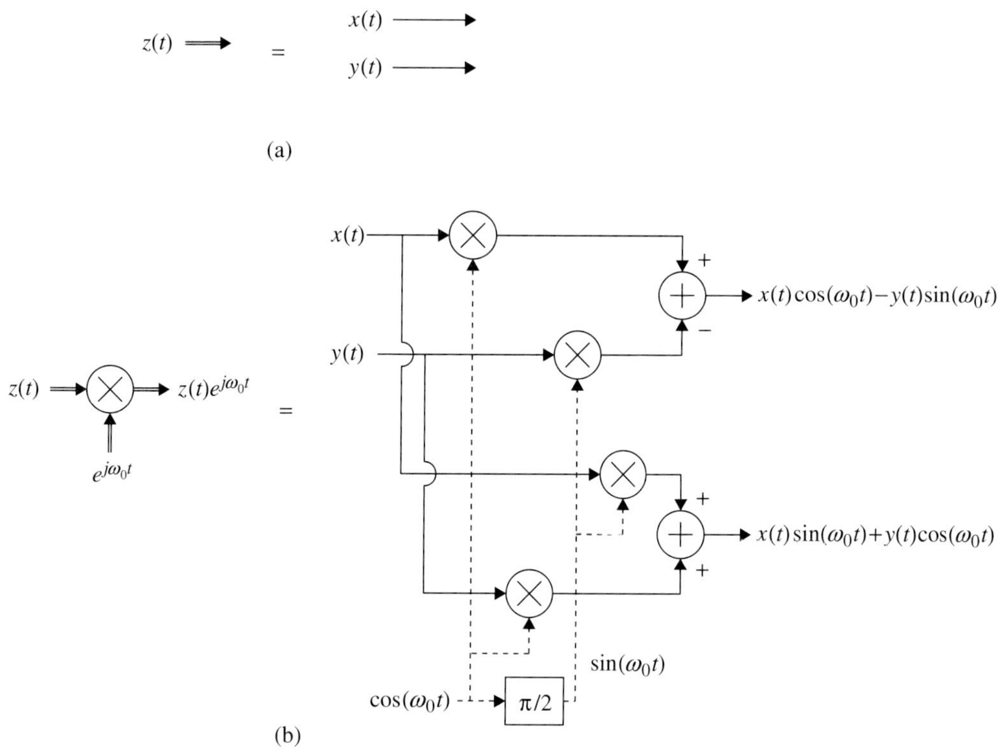

Appendix B The Complex-Valued Representation for QAM

Figure B.1.2 (a) Representation of complex-valued signals in terms of real-valued signals. (b) Representation of a complex-valued mixing in terms of real-valued mixing.

In continuous-time processing, a real-valued signal requires one wire or trace (and a corresponding ground) to convey the signal from one part of a circuit to another. A complex-valued signal requires two wires (and a ground) to convey the signal from one part of a circuit to another. This idea is conceptualized in Figure B.1.2 (a) where the convention is that single lines are used to represent real-valued signals and double lines are used to represent complex-valued signals. In discrete-time processing, the real-valued signal requires one location in memory to store the signal for use in processing. A complex-valued signal requires two locations in memory to store the signal for use in processing.

One of the most common functions performed in communications is mixing. Mixing is used to perform frequency translations and is represented mathematically by multiplication by a sinusoid. Multiplication of a complex signal, such as $z(t)$ given by (B.6), by a complex exponential $e^{j\omega_0 t}$ can be represented in terms of real-valued operations by following using the Euler identity (B.1) and the rules for complex multiplication:

$$z(t)e^{j\omega_0 t} = [x(t) + jy(t)][\cos(\omega_0 t) + j\sin(\omega_0 t)] \qquad (B.9)$$
$$= x(t)\cos(\omega_0 t) - y(t)\sin(\omega_0 t) + j[x(t)\sin(\omega_0 t) + y(t)\cos(\omega_0 t)]. \qquad (B.10)$$

A block diagram of the system that performs this multiplication is illustrated in Figure B.1.2 (b). Note that this system requires four real-valued mixers and two real-valued adders.

B.1.3 Negative Frequency: Does It Really Exist?

Many students new to communication theory do not believe that negative frequency exists. This erroneous notion is reinforced by laboratory test instruments that appear to ignore the concept of negative frequency. For example, frequency counters report their estimates as positive values and spectrum analyzers only display the positive frequency axis. But negative frequency does exist and plays an important role in complex-valued signal processing.

The first clue that this might be true is the property that the Fourier transform of a real-valued signal is conjugate symmetric. If the negative frequency component did not exist then the only way for $X(\omega) = X^*(-\omega)$ to be true is for $X(\omega) = 0$ for all ω.

A simple example illustrates the role of negative frequency. Consider the Fourier transform of the cosine

$$\cos(\omega_0 t) \rightarrow \underbrace{\pi\delta(\omega + \omega_0)}_{\text{negative frequency component}} + \underbrace{\pi\delta(\omega - \omega_0)}_{\text{positive frequency component}} . \quad (B.11)$$

The negative frequency component is needed to create a real-valued signal. This is seen by examining the cosine using the Euler identity:

$$\cos(\omega_0 t) = \underbrace{\frac{1}{2}e^{-j\omega_0 t}}_{\text{negative frequency component}} + \underbrace{\frac{1}{2}e^{j\omega_0 t}}_{\text{positive frequency component}} . \quad (B.12)$$

Interpreting $e^{j\omega_0 t}$ as a phasor rotating in the counterclockwise (CCW) direction at ω_0 rad/s and $e^{-j\omega_0 t}$ as a phasor rotating in the clockwise (CW) direction at ω_0 rad/s, the cosine can be interpreted as the phasor sum illustrated in Figure B.1.3. The phasor sum is always on the real

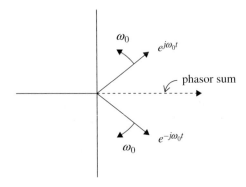

Figure B.1.3 A phasor representation of the two complex exponentials required to form the cosine using the Euler identity (B.1).

axis, thus producing a real-valued signal. This example shows that the concept "positive" and "negative" frequency provides a sense of direction (CCW or CW) to the rotation. Frequency counters report the magnitude of the rotation (which is always nonnegative) but not the direction. This value is related to frequency in the same way speed is related to velocity.

For a real-valued signal $x(t)$, the conjugate symmetry of the Fourier transform $X(j\omega)$ means that all the information about $X(j\omega)$ for $\omega < 0$ can be obtained from $X(j\omega)$ for $\omega > 0$. For this reason, spectrum analyzers only display the positive frequency portion of the frequency axis.

B.2 MODULATORS AND THE COMPLEX-VALUED REPRESENTATION

A QAM signal may be represented as (see Section 5.3)

$$s(t) = I(t)\sqrt{2}\cos(\omega_0 t) - Q(t)\sqrt{2}\sin(\omega_0 t) \tag{B.13}$$

where

$$I(t) = \sum_k a_0(k) p(t - kT_s)$$
$$Q(t) = \sum_k a_1(k) p(t - kT_s) \tag{B.14}$$

where $p(t)$ is a unit-energy pulse shape and the symbol pair $(a_0(k), a_1(k))$ defines a point in the signal constellation. This signal may also be expressed as

$$s(t) = \sqrt{2}\mathrm{Re}\left\{[I(t) + jQ(t)]e^{j\omega_0 t}\right\}. \tag{B.15}$$

To see that this is so, the substitution (B.1) is applied to the complex exponential and the multiplication is carried out to produce four terms. Collecting terms produces

$$s(t) = I(t)\sqrt{2}\cos(\omega_0 t) - Q(t)\sqrt{2}\sin(\omega_0 t) + j\left[I(t)\sqrt{2}\sin(\omega_0 t) + Q(t)\sqrt{2}\cos(\omega_0 t)\right]. \tag{B.16}$$

The first term on the right-hand side, the real part of the expression, is the desired QAM signal. An example of a block diagram that generates this signal is illustrated in Figure B.2.1. The block diagram uses the notational convention that complex-valued signals are represented by a double line whereas real-valued signals are represented by a single line. Observe that the complex-valued signals are composed of a pair of real-valued signals as explained in the previous section.

It is instructive to see the relationship between (B.13) and $I(t) + jQ(t)$ in the frequency domain. These relationships are illustrated in Figure B.2.2. The spectrum of $I(t) + jQ(t)$ is illustrated in Figure B.2.2 (a). Because the time-domain signal is complex-valued, the spectrum does not possess conjugate symmetry. Using the relationship (B.3), the spectrum of $[I(t) + jQ(t)]e^{j\omega_0 t}$—the term inside the real operator of (B.15)—is a translated version of

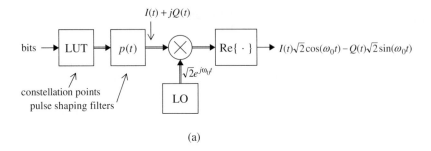

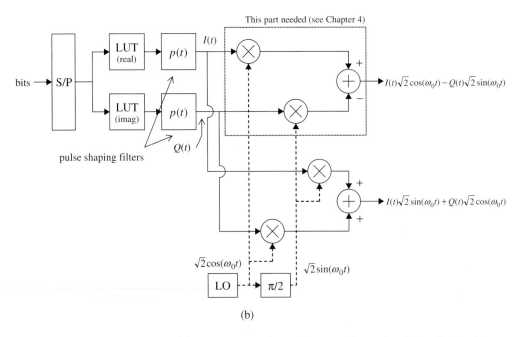

Figure B.2.1 QAM modulator: (a) A representation using complex-valued signals. Double lines denote complex-valued signals and single lines denote real-valued signals. (b) The interpretation of the complex-valued modulator. The complex-valued signal is a pair of real-valued signals with the proper phase relationship. Note that only the real part is retained to create a real-valued, band-pass signal.

the spectrum of $I(t) + jQ(t)$. The spectrum is illustrated in Figure B.2.2 (b). Note that as a consequence of $[I(t) + jQ(t)]e^{j\omega_0 t}$ being complex-valued, the spectrum does not possess conjugate symmetry. In fact, the spectrum is "one-sided" as shown. The spectrum of the real-valued signal (B.15), illustrated in Figure B.2.2 (c), possesses conjugate symmetry[2] as

---

[2]Often, students expect the spectrum to be symmetric about $\omega = \omega_0$ as a consequence of the conjugate symmetry condition. Conjugate symmetry, however, requires symmetry about $\omega = 0$ as shown. If, in addition, the signal is to possess symmetry about $\omega = \omega_0$, then $I(t) + jQ(t)$ must be real.

Appendix B The Complex-Valued Representation for QAM

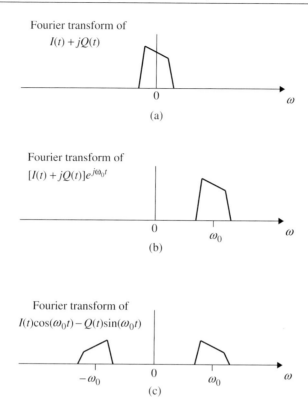

Figure B.2.2 The relationship between the spectra of the signals involved in (B.15).

explained in the previous section. A method for deriving the spectrum Figure B.2.2 (c) from that of Figure B.2.2 (b) is outlined in Exercise B.17.

The aforementioned discussion shows that a QAM modulator can be thought of as a system that generates a weighted pulse train, where the weights are complex-valued, performs a "one-sided" frequency translation of the pulse train using multiplication by a complex exponential, and retains the real part of the resulting signal for transmission. The complex-valued frequency translation requires four real-valued mixers and two real-valued two-input adders. All of these components must have precisely the same gains and delays in order to produce the desired result. Limitations in manufacturing tolerances make it difficult to produce four mixers and two adders with equal gains and delays in continuous-time hardware. For this reason, complex-valued signal processing is rarely used in continuous-time processing. In discrete-time processing, the mixers and adders are replaced by arithmetic operations and the issues of equal gains and delays do not exist. As a consequence, complex-valued signal processing is often used in discrete-time signal processing.

B.3 DEMODULATORS/DETECTORS AND THE COMPLEX-VALUED REPRESENTATION

B.3.1 Demodulators

QAM demodulators can also be interpreted as performing complex-valued signal processing. The main feature of the QAM demodulator is the pair of mixers operating on the received signal $r(t)$ using the sinusoids $\cos(\omega_0 t)$ and $-\sin(\omega_0 t)$. Multiplying the received signal by these sinusoids and keeping track of the inphase and quadrature components is equivalent to multiplying the received signal by the complex exponential $e^{-j\omega_0 t}$ and using the imaginary number operator j to keep track of the imaginary components in the subsequent processing. The real part of the product is the desired inphase term while the imaginary part of the subsequent processing is the quadrature component. These ideas are illustrated by the block diagram in Figure B.3.1.

To see that this is so, let the received signal be

$$r(t) = I_r(t)\sqrt{2}\cos(\omega_0 t) - Q_r(t)\sqrt{2}\sin(\omega_0 t) \tag{B.17}$$

$$= \text{Re}\left\{[I_r(t) + jQ_r(t)]e^{j\omega_0 t}\right\}. \tag{B.18}$$

Multiplying $r(t)$ by the complex exponential $\sqrt{2}e^{-j\omega_0 t}$ produces (see Exercise B.18)

$$r(t)e^{-j\omega_0 t} = [I_r(t) + jQ_r(t)] + [I_r(t) - jQ_r(t)]e^{-j2\omega_0 t}. \tag{B.19}$$

The second term is the "double frequency" term first encountered in Section 5.3. The double frequency term may be eliminated by a low-pass filter as shown in Figure B.3.1 (a). The output of the low-pass filter is the complex-valued baseband signal $I_r(t) + jQ_r(t)$. The real part is the inphase component whereas the imaginary part is the quadrature component.

The next processing step is the matched filter. The output of the matched filter is the complex-valued baseband signal $x(t) + jy(t)$. The matched filter output is sampled to produce the desired signal space projections. The equivalent processing in terms of real-valued signals is illustrated in Figure B.3.1 (b).

The matched filter is also a low-pass filter. As such, it can also be used to eliminate the "double frequency" term of (B.19). Consequently, the first low-pass filter may not be necessary. This is why it is indicated as an optional filter in Figure B.3.1.

The spectra of the important signals in Figure B.3.1 are illustrated in Figure B.3.2. The received signal is a real-valued, band-pass signal whose spectrum is shown in Figure B.3.2 (a). For this reason, the spectrum is conjugate symmetric. The product (B.19) is a one-sided frequency translation of the spectrum of $r(t)$ by $-\omega_0$ rad/s. This is shown in Figure B.3.2 (b) where the spectral content of $r(t)$ centered at ω_0 has been translated to baseband, whereas the spectral content of $r(t)$ centered at $-\omega_0$ has been translated to $-2\omega_0$. The low-pass filter passes the spectral content at baseband and eliminates the spectral content at $-2\omega_0$. The output is the complex-valued baseband signal $I_r(t) + jQ_r(t)$ whose spectrum is illustrated in Figure B.3.2.

Appendix B The Complex-Valued Representation for QAM

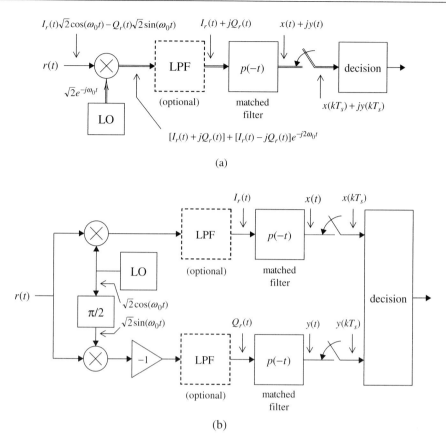

Figure B.3.1 QAM demodulator: (a) A representation using complex-valued signals. Double lines denote complex-valued signals and single lines denote real-valued signals. (b) The interpretation of the complex-valued modulator. The complex-valued signal is a pair of real-valued signals with the proper phase relationship.

The interpretation suggested by Figure B.3.1 (a) uses complex-valued signal processing operating on a real-valued signal. The complex-valued mixing operation requires two real-valued mixers. Both mixers must have the same gain and delay to produce the desired result. Manufacturing tolerances limit the ability to produce two mixers with exactly the same gain and delay. The consequences of this limitation in the context of QAM demodulators is summarized in Section 10.1.2. As a consequence, this approach has found limited success in demodulators based on continuous-time processing.

The mathematical foundations of the one-sided frequency translation using complex-valued signal processing are the same for discrete-time processing. In this case, the limitations due to manufacturing tolerances do not exist because mixers and adders are replaced by arithmetic computations. (Finite precision arithmetic and clock speed are the limitations in the discrete-time world.) As such, the interpretation offered by Figure B.3.1 (a) is widely used

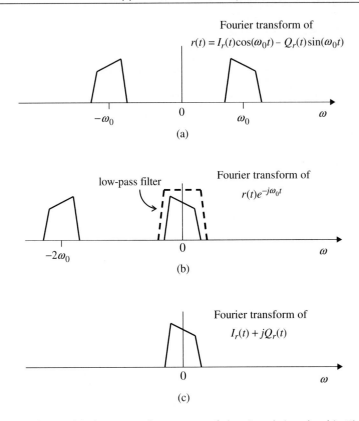

Figure B.3.2 The relationship between the spectra of the signals involved in Figure B.3.1.

in discrete-time processing. Often, this interpretation is taken one step further in discrete-time processing. Samples of the real-valued band-pass signal are converted to a complex-valued signal of the form

$$\left[I_r(nT) + jQ_r(nT)\right] e^{j\Omega_0 n}.$$

The subsequent frequency translation operates on a complex-valued signal instead of a real-valued signal. The advantages of this approach are the frequency translations that accompany multirate processing as discussed in Section 10.1.2. This type of processing is almost impossible to perform in continuous-time hardware because of the gain and delay mismatches resulting from limited manufacturing tolerances as discussed above. Again, these limitations do not exist in discrete-time realizations and have proved immensely popular.

B.3.2 Synchronization

The error signals used for symbol timing synchronization and carrier phase synchronization may also be expressed in terms of complex-valued signals. Referring to Figure B.3.1, let

$$\tilde{x}(t) = x(t) + jy(t) \qquad (B.20)$$

be the matched filter outputs and let

$$\tilde{a}(k) = a_0(k) + ja_1(k) \tag{B.21}$$

be the k-th symbol. The data-aided timing error detectors of Section 8.5 (except for the GTED, which only has a non-data-aided form) may be expressed as

$$\text{MLTED:} \quad e(k) = \text{Re}\left\{\tilde{a}^*(k)\dot{\tilde{x}}\left(kT_s + \hat{\tau}\right)\right\} \tag{B.22}$$

$$\text{ELTED:} \quad e(k) = \text{Re}\left\{\tilde{a}^*(k)\left[\tilde{x}\left((k+1/2)T_s + \hat{\tau}\right) - \tilde{x}\left((k-1/2)T_s + \hat{\tau}\right)\right]\right\} \tag{B.23}$$

$$\text{ZCTED:} \quad e(k) = \text{Re}\left\{\tilde{x}\left((k-1/2)T_s + \hat{\tau}\right)\left[\tilde{a}^*(k-1) - \tilde{a}^*(k)\right]\right\} \tag{B.24}$$

$$\text{GTED:} \quad e(k) = \text{Re}\left\{\tilde{x}\left((k-1/2)T_s + \hat{\tau}\right)\left[\tilde{x}^*\left((k-1)T_s + \hat{\tau}\right) - \tilde{x}^*\left(kT_s + \hat{\tau}\right)\right]\right\} \tag{B.25}$$

$$\text{MMTED:} \quad e(k) = \text{Re}\left\{\tilde{a}^*(k-1)\tilde{x}\left(kT_s + \hat{\tau}\right) - \tilde{a}^*(k)\tilde{x}\left((k-1)T_s + \hat{\tau}\right)\right\}. \tag{B.26}$$

The data-aided maximum likelihood carrier phase error detector of Section 7.4 may be expressed as

$$e(k) = \text{Im}\left\{\tilde{a}^*(k)\tilde{x}'(kT_s)\right\} \tag{B.27}$$

where $\tilde{x}'(kT_s) = x'(kT_s) + jy'(kT_s)$ is the rotated version of the signal space projection as explained in Section 7.4. In Chapter 7 the relationship between the matched filter outputs $x(kT_s), y(kT_s)$ and the rotated matched filter outputs $x'(kT_s), y'(kT_s)$ was expressed in matrix-vector form using the rotation matrix:

$$\begin{bmatrix} x'(kT_s) \\ y'(kT_s) \end{bmatrix} = \begin{bmatrix} \cos(\theta) & \sin(\theta) \\ -\sin(\theta) & \cos(\theta) \end{bmatrix} \begin{bmatrix} x(kT_s) \\ y(kT_s) \end{bmatrix}. \tag{B.28}$$

In words, the point $(x'(kT_s), y'(kT_s))$ is the rotation of the point $(x(kT_s), y(kT_s))$ in the clockwise direction by θ (or, in the counterclockwise direction by $-\theta$). When the points $(x'(kT_s), y'(kT_s))$ and $(x(kT_s), y(kT_s))$ are treated as the complex numbers $\tilde{x}'(kT_s) = x'(kT_s) + jy'(kT_s)$ and $\tilde{x}(kT_s) = x(kT_s) + jy(kT_s)$, the rotation is accomplished via multiplication by $e^{-j\theta}$. As a consequence, the relationship (B.28) may be expressed as

$$\tilde{x}'(kT_s) = \tilde{x}(kT_s)e^{-j\theta}. \tag{B.29}$$

B.4 NOTES AND REFERENCES

The reservations many have about complex-valued signals has its origins in lingering doubts about complex numbers. The square root of a negative number puzzled many great mathematicians before the 1800s for whom numbers were abstractions of geometric lengths and mathematical manipulations were abstractions of geometric procedures (e.g., bisecting an angle using a compass and straight edge). This view of mathematics was inherited from

Greek thought and was the source of their skepticism regarding 0 (no length—nothing to measure) and negative numbers (a negative length?). The square root of a positive number had a perfectly rational geometric interpretation in that there is a geometric procedure to produce from a line of length L another line whose length is $\sqrt{L}$. But the square root of a negative number (a negative length) just didn't make any sense. Descartes, in his *La Geometrie* published in 1637, used the word "imaginary" to describe the root of a negative number because this same geometric construction problem was "impossible." The views of the time were summarized by Leonhard Euler, who wrote in his 1770 *Algebra*, "All such expressions as $\sqrt{-1}, \sqrt{-2}$, etc., are consequently impossible or imaginary numbers, since they represent roots of negative quantities; and of such numbers we may truly assert that they are neither nothing, nor greater than nothing, nor less than nothing, which necessarily constitutes them imaginary or impossible." The square root of a negative number shows up in intermediate calculations when applying Cardan's formula for finding the roots of a cubic polynomial. Although not understood, many contemporary mathematicians plowed along using the well-established rules of arithmetic to produce an inexplicable "real" solution for the root. Such results prompted the Italian mathematician Bombelli to declare, "The whole matter seemed to rest on sophistry rather than on truth."

Negative numbers in general and the square root of negative numbers in particular began to be accepted once their geometrical interpretation was understood. Things began to clear when the English mathematician John Wallis revealed the proper geometric interpretation of negative numbers in 1685. The Norwegian-Danish mathematician Caspar Wessel who spent his career as a surveyor provided the required geometric interpretation of $\sqrt{-1}$ in *Om Directionens Analytiske Betegning* in 1799. Just as negative numbers indicated a place to the left of the origin on a line, Wessel's interpretation was that $\sqrt{-1}$ indicated a place *above* the number line. Carl Fredrich Gauss independently developed the same interpretation but did not publish it until 34 years later. It was during this era that the notation used today was introduced. Leonhard Euler, for a long time confused why $\sqrt{ab} = \sqrt{a}\sqrt{b}$ implies the incorrect conclusion $-1 = \sqrt{(-1)(-1)} = \sqrt{-1}\sqrt{-1} = 1$, used $i = \sqrt{-1}$ to avoid repeating the mistake in 1777. Carl Friedrich Gauss also used $i = \sqrt{-1}$ and introduced the word "complex" for numbers of the form $a + bi$ in 1831. He used the term "complex plane" to describe the geometrical interpretation of a complex number as a point in a Cartesian coordinate system.

For an in-depth treatment of the history of complex numbers, see the delightful book by Nahin [346].

B.4.1 Topics Covered

Most of the material in this section is a straightforward application of complex variables and the Fourier transform. The properties of the Fourier transform are derived in many textbooks devoted to linear systems and transform theory. Oppenheim and Willsky [40] and Bracewell [39] are representative examples.

The complex-valued signal processing described applies to either continuous-time processing or the discrete-time processing (even though all of the development was in the continuous-time domain). For wireless communications, much of the complex-valued signal processing is devoted to frequency translations and thus involves mixers and quadrature

Appendix B The Complex-Valued Representation for QAM

sinusoids. The challenge of performing complex-valued signal processing of this nature lies in the manufacturing tolerances that limit the ability to produce mixers with exactly the same gain and a pair of sinusoids with exactly the same amplitudes and precisely 90° out of phase. This is called the "I/Q imbalance problem" in the technical literature and is discussed in more detail in Chapter 10. The paper by Mirabbasi and Martin [266] provides a nice overview of complex-valued continuous-time signal processing. The challenges are not present in discrete-time processing. As a consequence, complex-valued processing is very popular in the discrete-time domain as explained in Chapter 10.

B.4.2 Topics Not Covered

The complex-valued equivalents of real-valued signals are closely related to the Hilbert transform. The Hilbert transform of $x(t)$ is

$$\hat{x}(t) = \frac{1}{\pi} \int_{-\infty}^{\infty} \frac{x(u)}{u-t} du.$$

From the point of view of a filter, the Hilbert transform leaves the amplitude of $X(j\omega)$ unchanged but alters the phase of $X(j\omega)$ by $\pm\pi/2$ depending on the sign of ω. A nice introduction to the Hilbert transform, in the context of its relationship to the continuous-time Fourier transform, is given by Bracewell [39]. The discrete-time Hilbert transform and its relationship to causality is provided by Oppenheim and Schafer [41]. In the context of wireless communications, the Hilbert transform is used to represent single side band signals in the time domain. See Carlson [261], Couch [262], Proakis and Salehi [11], Roden [14], and Stremler [263]. These same concepts could have been applied to the complex-valued baseband or band-pass signals that are related to the real-valued band-pass signal of interest. This application is a very simple application of the Hilbert transform.

B.5 EXERCISES

B.1 Express $\cos(-\omega_0 t)$ in terms of complex exponentials, similar to the form of the Euler identities (B.1).

B.2 Express $-\cos(\omega_0 t)$ in terms of complex exponentials, similar to the form of the Euler identities (B.1).

B.3 Express $\sin(-\omega_0 t)$ in terms of complex exponentials, similar to the form of the Euler identities (B.1).

B.4 Express $-\sin(\omega_0 t)$ in terms of complex exponentials, similar to the form of the Euler identities (B.1).

B.5 Express $e^{-j\omega_0 t}$ in terms of cosine and sine in a form similar to that of (B.1).

B.6 Express $\cos(\omega_0 t + \theta)$ in terms of complex exponentials in a form similar to the Euler identities (B.1).

B.7 Express $\sin(\omega_0 t + \theta)$ in terms of complex exponentials in a form similar to the Euler identities (B.1).

B.8 Express $e^{-j(\omega_0 t + \theta)}$ in terms of cosine and sine in a form similar to that of (B.1).

B.9 Compute the Fourier transform of $e^{-j\omega_0 t}$.

B.10 Derive the Fourier transform for $\cos(\omega_0 t)$ by using the Euler identity (B.1) to express the cosine in terms of complex exponentials then using the Fourier transform of the complex exponentials.

B.11 Derive the Fourier transform for $\sin(\omega_0 t)$ by using the Euler identity (B.1) to express the sine in terms of complex exponentials then using the Fourier transform of the complex exponentials.

B.12 Compute the Fourier transform of $\cos(\omega_0 t + \theta)$.

B.13 Compute the Fourier transform of $\sin(\omega_0 t + \theta)$.

B.14 Compute the Fourier transform of $e^{-j(\omega_0 t + \theta)}$.

B.15 Compute the Fourier transform of $x(t) e^{-j\omega_0 t}$.

B.16 In an earlier course, the notion of a *phasor* was introduced. The phasor associated with $A\cos(\omega_0 t + \theta)$ is $Ae^{j\theta}$. The phasor concept resulted from writing

$$A\cos(\omega_0 t + \theta) = \mathrm{Re}\left\{Ae^{j(\omega_0 t + \theta)}\right\} = \mathrm{Re}\left\{Ae^{j\theta} e^{j\omega_0 t}\right\}.$$

The term $e^{j\omega_0 t}$ is the "rotational operator" and the term $e^{j\theta}$ is the "phase operator." When the phase operator is coupled with the magnitude A, the "phasor" $Ae^{j\theta}$ results. In a linear system, no new frequencies are produced so the rotational operator is dropped and all analysis is performed using manipulations of the phasor. The same concept can be applied to a QAM signal. The QAM signal can be expressed in the form

$$I(t)\cos(\omega_0 t) - Q(t)\sin(\omega_0 t) = A(t)\cos(\omega_0 t + \theta(t))$$

and the corresponding phasor is $A(t) e^{j\theta(t)}$. Determine expressions for $A(t)$ and $\theta(t)$ in terms of $I(t)$ and $Q(t)$.

B.17 This exercise investigates a way to derive the spectrum in Figure B.2.2 (c) from the spectrum of Figure B.2.2 (b). For notational convenience, let $\tilde{u}(t) = I(t) + jQ(t)$ and let $\tilde{U}(j\omega)$ be the Fourier transform of $\tilde{u}(t)$.
(a) Show that the Fourier transform of $\tilde{u}(t) e^{j\omega_0 t}$ is $\tilde{U}(j(\omega - \omega_0))$.
(b) Show that the Fourier transform of $\tilde{u}^*(t) e^{-j\omega_0 t}$ is $\tilde{U}^*(-j(\omega + \omega_0))$.
(c) Using the relationship

$$\mathrm{Re}\{Z\} = \frac{Z + Z^*}{2}$$

Appendix B The Complex-Valued Representation for QAM

for a complex variable Z, show that the Fourier transform of

$$\text{Re}\left\{\tilde{u}(t)e^{j\omega_0 t}\right\}$$

is

$$\frac{1}{2}\tilde{U}\left(j(\omega-\omega_0)\right) + \frac{1}{2}\tilde{U}^*\left(-j(\omega+\omega_0)\right).$$

B.18 Let the real-valued, band-pass signal at the input to a QAM demodulator be given by

$$r(t) = I_r(t)\sqrt{2}\cos(\omega_0 t) - Q_r(t)\sin(\omega_0 t).$$

This problem investigates two ways to show that

$$r(t)\sqrt{2}e^{-j\omega_0 t} = I_r(t) + jQ_r(t) + [I_r(t) - jQ_r(t)]e^{-j2\omega_0 t}.$$

(a) Use the Euler identities

$$\cos(\theta) = \frac{e^{j\theta} + e^{-j\theta}}{2}$$

$$\sin(\theta) = \frac{e^{j\theta} - e^{-j\theta}}{j2}$$

to express the cosine and sine in terms of complex exponentials. Multiply through by $e^{-j\omega_0 t}$ and organize the terms in the resulting product.

(b) Express $r(t)$ as

$$r(t) = \text{Re}\left\{[I_r(t) + jQ_r(t)]e^{j\omega_0 t}\right\}$$

and apply the relationship

$$\text{Re}\{Z\} = \frac{Z + Z^*}{2}$$

to this expression for $r(t)$. Multiply through by $e^{-j\omega_0 t}$ and organize the terms in the resulting product.

B.19 Show that the timing-error-detector error signals (B.22)–(B.26) are equivalent to the QAM timing error detectors (8.98)–(8.102), respectively, in Chapter 8.

B.20 Show that the carrier phase-error-detector error signal (B.27) is equivalent to the error signal (7.54) in Chapter 7.

B.21 Show that the matrix form (B.28) is equivalent to (B.29). Hint: express the right-hand side of (B.29) in rectangular form and equate the real and imaginary parts.

B.22 Let the received QAM signal be

$$r(t) = I_r(t)\sqrt{2}\cos(\omega_0 t) - Q_r(t)\sqrt{2}\sin(\omega_0 t).$$

In Section B.3.1 it was shown that the I/Q system below is equivalent to a system that performs a one-sided frequency translation using complex-valued signal processing (such as the one shown below).

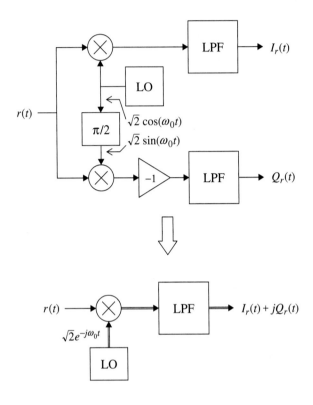

The sign of the quadrature sinusoid $\sqrt{2}\sin(\omega_0 t)$ needs to be negative in order to perform the proper signal space projections as described in Chapter 5. Because continuous-time I/Q demodulators use $\cos(\omega_0 t)$ and $\sin(\omega_0 t)$, a negative sign must be inserted as shown. The resulting frequency translations are illustrated in Figure B.3.2.

Now suppose the sign change block is not included. The resulting system is illustrated by the following block diagram.

Appendix B The Complex-Valued Representation for QAM

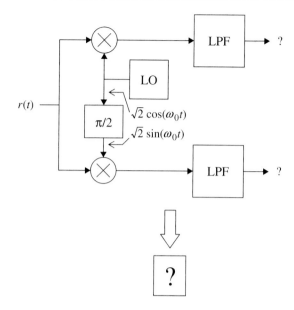

(a) Determine a mathematical expression for the low-pass filter outputs.
(b) Sketch a block diagram of the equivalent complex-valued system. What is the relationship between the output of this system and the output of the system that includes the negative sign on $\sin(\omega_0 t)$.
(c) Sketch the frequency translations that correspond to this system.

Appendix C Phase Locked Loops

The phase locked loop (PLL) is the core component in many synchronization subsystems. This appendix summarizes the functionality of the PLL and the methods commonly used to analyze a PLL. The material presented here is designed to complement the material contained in Chapters 8 and 7. The continuous-time PLL is analyzed first in Section C.1. These results are then applied to the discrete-time PLL analyzed in Section C.2.

C.1 THE CONTINUOUS-TIME PLL

PLLs are used in digital communication systems to synchronize the local oscillators in the receiver to the oscillators used by the transmitter (carrier phase synchronization) or to synchronize the data clock in the receiver to the data clock used at the data source (symbol timing synchronization). In either case, the PLL can be thought of as a device that tracks the phase and frequency of a sinusoid. A simple PLL that is designed to track the phase of an input sinusoid is illustrated in Figure C.1.1. The PLL consists of three basic components: the phase detector, the loop filter, and the voltage-controlled oscillator (VCO).

The input to the loop is the sinusoid

$$x(t) = A\cos(\omega_0 t + \theta(t)) \tag{C.1}$$

and the output of the VCO is

$$y(t) = \cos(\omega_0 t + \hat{\theta}(t)). \tag{C.2}$$

The phase detector is a device whose output is a function $g(\cdot)$ of the phase difference between the two inputs. Because the loop input and VCO output form the inputs to the phase detector, the phase detector output is $g(\theta(t) - \hat{\theta}(t))$. The difference $\theta(t) - \hat{\theta}(t)$ is called the *phase error* and is denoted $\theta_e(t)$. The phase error is filtered by the loop filter to produce a control voltage $v(t)$ that is used to set the phase of the VCO. The VCO output $y(t) = \cos(\omega_0 t + \hat{\theta}(t))$ is related to the input $v(t)$ via the phase relationship

$$\hat{\theta}(t) = k_0 \int_{-\infty}^{t} v(x)dx \tag{C.3}$$

where k_0 is a constant of proportionality, called the VCO gain, that has units radians/volt.

Appendix C Phase Locked Loops

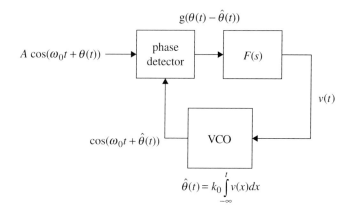

Figure C.1.1 Basic structure of a phase locked loop showing the three basic components: (1) the phase detector, (2) the loop filter, and (3) the voltage-controlled oscillator or VCO.

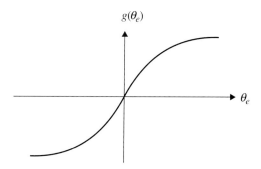

Figure C.1.2 Typical phase detector input/output relationship. This curve is called an "S curve" because the shape of the curve looks like the letter "S."

When functioning properly, the loop adjusts the control voltage $v(t)$ to produce a phase estimate $\hat{\theta}(t)$ that drives the phase error to zero. To see why this is so, consider the plot of $g(\theta_e)$ versus θ_e for a typical phase detector shown in Figure C.1.2. When the phase error is positive, $\theta(t) > \hat{\theta}(t)$ which means the phase of the VCO output lags the loop input and must be increased. The positive phase error produces a control voltage $v(t)$ that is also positive at the loop filter output. The positive control voltage increases the phase of the VCO output $\hat{\theta}(t)$, thus producing the desired result. Likewise, when the phase error is negative, $\theta(t) < \hat{\theta}(t)$. The negative phase error produces a negative control voltage which decreases the phase of the VCO output.

The PLL of Figure C.1.1 has a *phase equivalent* representation as shown in Figure C.1.3. The phase equivalent PLL is derived from the actual PLL by writing down the phases of all the sinusoids and tracking the operations on those phases through the loop. Because the phase

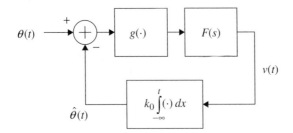

Figure C.1.3 Phase equivalent PLL corresponding to the PLL illustrated in Figure C.1.1.

detector is, in general, a nonlinear function of the phases of its inputs, the phase equivalent loop is a nonlinear feedback control system that can be analyzed using advanced techniques.

A common method of analyzing a nonlinear feedback control system is to linearize the system at a desired operating point. When this is done, the loop can be analyzed using standard linear system techniques. The desired operating point is at $\theta_e = 0$. In most cases, $g(\theta_e) \approx k_p \theta_e$ for small θ_e, so the loop is a linear feedback system when θ_e is small. In this case, the phase equivalent loop assumes the form shown in Figure C.1.4 (a). The phase equivalent loop is a linear system. Thus, it can be analyzed using Laplace transform techniques. Computing the Laplace transform of all signals and systems in the loop of Figure C.1.4 (a) produces the frequency domain version of the loop shown in Figure C.1.4 (b).

PLL operation is characterized using both the phase error $\theta_e(t)$ and the VCO output $\hat{\theta}(t)$. After some Laplace domain algebra, the transfer functions for the phase error and VCO output phase are

$$G_a(s) = \frac{\Theta_e(s)}{\Theta(s)} = \frac{s}{s + k_0 k_p F(s)} \tag{C.4}$$

$$H_a(s) = \frac{\hat{\Theta}(s)}{\Theta(s)} = \frac{k_0 k_p F(s)}{s + k_0 k_p F(s)}. \tag{C.5}$$

Ideally, the PLL should produce a phase estimate that has zero phase error. This characteristic, together with the phase error transfer function, will be used to determine the desirable properties of the loop filter. The phase estimate transfer function, or loop transfer function, is used to characterize the performance of the PLL.

C.1.1 PLL Inputs

Two PLL inputs are of special interest in characterizing PLL performance. The first represents the case where the loop input differs from the VCO output by a simple phase difference $\Delta\theta$. This is modeled by setting

$$\theta(t) = \Delta\theta u(t) \tag{C.6}$$

where $u(t)$ is the unit step function. This case is illustrated in Figure C.1.5 (a) where the argument of the input sinusoid, or instantaneous phase, is plotted as a function of time. The slope of the line is the frequency of the input sinusoid. Note that it does not change. This

Appendix C Phase Locked Loops

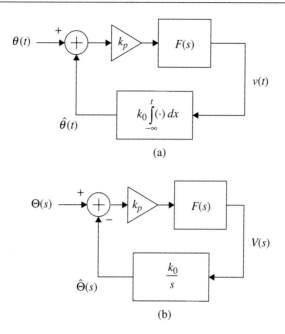

Figure C.1.4 Linearized phase equivalent PLL showing (a) time-domain signals and (b) frequency-domain signals.

means the frequency does not change. At time $t = 0$, a phase step occurs that changes the intercept point of the line. The loop outputs corresponding to the phase step input are called the *step responses* of the loop. The Laplace transform of the input is

$$\Theta(s) = \frac{\Delta\theta}{s}. \tag{C.7}$$

The second PLL input of interest is the case where the input sinusoid differs from the VCO output by a frequency shift of $\Delta\omega$ rad/s. The frequency offset is modeled by observing that

$$\cos((\omega_0 + \Delta\omega)t) = \cos(\omega_0 t + \underbrace{\Delta\omega t}_{\theta(t)}). \tag{C.8}$$

Thus,

$$\theta(t) = (\Delta\omega)t u(t) \tag{C.9}$$

which is called a "ramp" input. This case is illustrated in Figure C.1.5 (b) where the argument of the input sinusoid is plotted as a function of time. The slope of the instantaneous phase is instantaneous frequency. Prior to $t = 0$, the slope of the line is ω_0, and after $t = 0$ the slope of the line is $\omega_0 + \Delta\omega$. This change in slope is the frequency change at $t = 0$. The loop outputs corresponding to the ramp input are called the *ramp responses* of the loop. The Laplace

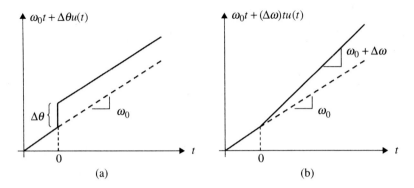

Figure C.1.5 Two phase inputs to be analyzed: (a) a phase step of $\Delta\theta$; (b) a frequency step of $\Delta\omega$ which is equivalent to a phase ramp with slope $\omega + \Delta\omega$.

transform of the ramp input is

$$\Theta(s) = \frac{\Delta\omega}{s^2}. \tag{C.10}$$

C.1.2 Phase Error and the Loop Filter

The properties of the phase error are used to determine the desirable characteristics of the loop filter. In particular, it will be shown that to achieve a zero steady-state phase error for a constant phase offset, the loop filter must have a nonzero DC gain and to achieve a zero steady-state phase error for a constant frequency offset, the loop filter must have an infinite DC gain.

Because a constant phase offset is modeled by a step function for $\theta(t)$, the starting point is the derivation of an expression for the phase error step response. Substituting (C.7) in (C.4), and solving for $\Theta_e(s)$, the Laplace transform of the phase error step response is

$$\Theta_{e,\text{step}}(s) = \frac{\Delta\theta}{s + k_0 k_p F(s)}. \tag{C.11}$$

The steady-state phase error may be obtained by computing the inverse Laplace transform of (C.11) and taking the limit as $t \to \infty$. Unfortunately, this approach requires knowledge of the loop filter transfer function $F(s)$. An alternate approach is to use the final value theorem for Laplace transforms. Applying the final value theorem to (C.11), the steady-state phase error is

$$\theta_{e,\text{step}}(\infty) = \lim_{s \to 0} \{s\Theta_e(s)\} \tag{C.12}$$

$$= \lim_{s \to 0} \left\{ \frac{s\Delta\theta}{s + k_0 k_p F(s)} \right\} \tag{C.13}$$

$$= 0 \quad \text{if } F(0) \neq 0. \tag{C.14}$$

Appendix C Phase Locked Loops

This means that as long as the loop filter has a nonzero DC gain, the steady-state phase error for a step input is zero.

The expression for the steady-state phase error for a constant frequency offset is obtaining following the same procedure except that a phase ramp is used instead of a phase step for the input. Substituting (C.10) into (C.4) and solving for $\Theta_e(s)$, the Laplace transform of the phase error ramp response is

$$\Theta_{e,\text{ramp}}(s) = \frac{\Delta\omega}{s^2 + sk_0k_pF(s)}. \tag{C.15}$$

Using the final value theorem, the steady-state phase error ramp response is

$$\theta_{e,\text{ramp}}(\infty) = \lim_{s\to 0} \{s\Theta_e(s)\} \tag{C.16}$$

$$= \lim_{s\to 0} \left\{ \frac{\Delta\omega}{s + k_0k_pF(s)} \right\} \tag{C.17}$$

$$= 0 \quad \text{if } F(0) = \infty. \tag{C.18}$$

This means that as long as the loop filter has an infinite DC gain, the steady-state phase error for a ramp input is zero.

The two conditions (C.14) and (C.18) indicate that the loop filter must have a nonzero DC gain to drive the steady-state phase error to zero in the presence of a phase offset and an infinite DC gain to drive the steady-state phase error to zero in the presence of a frequency offset. A loop filter that satisfies these conditions is the *proportional-plus-integrator* loop filter[1] with transfer function

$$F(s) = k_1 + \frac{k_2}{s}. \tag{C.19}$$

The first term in (C.19) is a simple gain of k_1 and contributes, to the filter output, a signal that is proportional to the filter input. Hence the word "proportional" in the filter name. The second term in (C.19) is an ideal integrator with infinite memory and a gain k_2. This portion of the loop filter contributes, to the filter output, a signal that is proportional to the integral of the input signal. Hence the word "integrator" in the filter name.

Not all PLLs use a proportional-plus-integrator loop filter. With continuous-time processing, it is challenging to produce an integrator with an infinite DC gain. The VCO is characterized by a single-pole element in the Laplace domain. Thus, the use of a loop filter with p poles produces a PLL whose linear phase equivalent system is a $p+1$ order feedback control system. A loop filter with no poles leads to a first-order feedback system and a loop filter with one pole leads to a second-order feedback control system. Because second-order systems are easy to understood and have well-developed design procedures, loop filters with one pole are preferred in most applications. As a consequence, loop filters of the form

$$F(s) = \frac{a_1s + a_0}{b_1s + b_0} \tag{C.20}$$

---

[1] A commonly used filter in feedback control is the "proportional-integrator-differentiator" or PID filter. Feedback control systems that use a PID filter are called PID controllers. PID controllers are the most common form of linear feedback control.

are examined. Although all first- and second-order PLLs are special cases of the PLL that uses (C.20) as a loop filter, four special cases are of interest:

$F(s) = k$ $\quad (a_1 = b_1 = 0)$: This filter is a simple gain and produces a first-order PLL.

$F(s) = \dfrac{k}{s+k}$ $\quad (a_1 = 0)$: This filter is a first-order low-pass filter where $1/k$ is the time constant. It is often called a "leaky integrator." It produces a second-order PLL.

$F(s) = k_1 + \dfrac{k_2}{s}$ $\quad (b_0 = 0)$: This is the proportional-plus-integrator filter introduced earlier. It produces a second-order PLL.

$F(s) = \dfrac{k_1 + s}{k_2 + s}$ $\quad (k_1 = a_1/a_0, k_2 = b_1/b_0)$: This is the general first-order filter expressed in a more convenient form. It produces a second-order PLL.

The properties of these filters and the resulting steady-state phase errors for step and ramp inputs are summarized in Table C.1.1. All of the filters produce a PLL with zero steady-state phase error for a step input. Only the proportional-plus-integrator loop filter, however, produces a PLL with zero steady-state phase error for a ramp input.

There are design techniques for higher order systems, but are more difficult than the design techniques for second-order systems. One application where higher order systems are preferred is GPS receivers. The signal received from a satellite in low-earth orbit experiences a Doppler shift (or carrier frequency offset) that is not constant in time, but linear in time (i.e., $\theta(t) \sim t^2$). To track the frequency offset with zero phase error, a loop filter with two poles is required. The resulting PLL is modeled by a third-order feedback control system.

Table C.1.1 Summary of loop filter characteristics and steady-state phase errors

$F(s)$	$F(0)$	$\theta_{e,\text{step}}(\infty)$	$\theta_{e,\text{ramp}}(\infty)$
k	k	0	$\dfrac{\Delta\omega}{k_0 k_p k}$
$\dfrac{k}{s+k}$	1	0	$\dfrac{\Delta\omega}{k_0 k_p}$
$k_1 + \dfrac{k_2}{s}$	∞	0	0
$\dfrac{k_1 + s}{k_2 + s}$	$\dfrac{k_1}{k_2}$	0	$\dfrac{\Delta\omega k_2}{k_0 k_p k_1}$

C.1.3 Transient Analysis

The loop performance is characterized by the properties of the loop output $\hat{\theta}(t)$. The Laplace transform of the loop output is given by

$$\hat{\Theta}(s) = \frac{k_0 k_p F(s)}{s + k_0 k_p F(s)} \Theta(s) \tag{C.21}$$

where

$$\Theta(s) = \begin{cases} \dfrac{\Delta\theta}{s} & \text{for a phase offset} \\ \dfrac{\Delta\omega}{s^2} & \text{for a frequency offset} \end{cases} \tag{C.22}$$

and $F(s)$ is one of the loop filters in Table C.1.1. When $F(s) = k$, the loop transfer function is

$$H_a(s) = \frac{k_0 k_p k}{s + k_0 k_p k}. \tag{C.23}$$

The step response is

$$\hat{\Theta}(s) = \frac{k_0 k_p k \Delta\theta}{s(s + k_0 k_p k)} \tag{C.24}$$

$$\hat{\theta}(t) = \Delta\theta \left(1 - e^{-k_0 k_p k t}\right) u(t) \tag{C.25}$$

and the ramp response is

$$\hat{\Theta}(s) = \frac{k_0 k_p k \Delta\omega}{s^2 (s + k_0 k_p k)} \tag{C.26}$$

$$\hat{\theta}(t) = \left[\Delta\omega t - \frac{\Delta\omega}{k_0 k_p k}\left(1 - e^{-k_0 k_p k t}\right)\right] u(t). \tag{C.27}$$

The loop transfer functions for the other three filters are of the form

$$H_a(s) = \frac{b_2 s^2 + b_1 s + b_0}{s^2 + 2\zeta\omega_n s + \omega_n^2} \tag{C.28}$$

and are thus second-order systems. The poles of the system are

$$p_1, p_2 = -\zeta\omega_n \pm \omega_n\sqrt{\zeta^2 - 1}. \tag{C.29}$$

The characteristics of the loop output are governed by the value of ζ. When $\zeta < 1$, the poles are complex conjugate pairs and the loop response exhibits damped oscillations. For this reason, ζ is called the *damping factor* and ω_n is called the *natural frequency*. A second-order system with damped sinusoidal transients is termed *underdamped*. When $\zeta > 1$, the poles are real and distinct and loop response is the sum of decaying exponentials. Such a system is *overdamped*. When $\zeta = 1$, the poles are real and repeated and the loop responses are on the boundary between damped oscillations and decaying exponentials. The PLL is *critically damped* in this case.

As an example, the loop transfer function for the "proportional-plus-integrator" loop filter is

$$H_a(s) = \frac{k_0 k_p k_1 s + k_0 k_p k_2}{s^2 + k_0 k_p k_1 s + k_0 k_p k_2} \quad \text{(C.30)}$$

which may be expressed in the form

$$H_a(s) = \frac{2\zeta \omega_n s + \omega_n^2}{s^2 + 2\zeta \omega_n s + \omega_n^2} \quad \text{(C.31)}$$

where

$$\zeta = \frac{k_1}{2}\sqrt{\frac{k_0 k_p}{k_2}} \quad \text{(C.32)}$$

$$\omega_n = \sqrt{k_0 k_p k_2}. \quad \text{(C.33)}$$

The step response as a function of ζ is summarized in Table C.1.2. The frequency response is plotted in Figure C.1.6 for four different values of the damping factor ($\zeta = 0.5, 0.7071, 1, 2$). The frequency response shows that the PLL acts as a low-pass filter when operating as a linear system. The damping factor controls the magnitude of the frequency response at $\omega = \omega_n$. For underdamped systems ($\zeta < 1$), the frequency response displays a peak at $\omega = \omega_n$. The amplitude of this peak grows as ζ decreases. In fact, as $\zeta \to 0$, the peak grows to an impulse and the loop response is purely sinusoidal.[2] For overdamped systems, the peak at $\omega = \omega_n$ disappears and the loop response looks more like a traditional low-pass filter.

The effect of the spectral peak in $H_a(j\omega)$ is best observed by examining the step response in the time domain as illustrated in Figure C.1.7. The step response for the same four values of ζ are shown by the heavy lines and the step input is the thin dotted line. The ideal loop response is one that follows the input exactly. For small values of ζ, the step response overshoots and then oscillates about the ideal response. The amplitudes of the oscillations decrease with time so that in the steady state the loop output matches the loop input. Thus, the presence of a spectral peak in $H_a(j\omega)$ produces a damped sinusoidal component in the time-domain loop response. For overdamped systems ($\zeta > 1$), the step response does not oscillate about the ideal response but gradually approaches the ideal response as time increases. This demonstrates the classic trade-off for second-order systems: underdamped systems have a fast settling time, but exhibit overshoot and oscillations; overdamped systems have a slow settling time, but no oscillations.

---

[2]This is how an oscillator is designed using a second-order feedback system. The goal, however, is to build synchronizers, not oscillators. Underdamped systems with $\zeta = 0$ will not be examined further.

Table C.1.2 Summary of pole locations and corresponding step responses for a second-order loop using a proportional-plus-integrator loop filter.

Damping Factor	Poles	Step Response
$\zeta < 1$ (underdamped)	complex conjugates at $-\zeta\omega_n \pm j\omega_n\sqrt{1-\zeta^2}$	$\Delta\theta\left(1 - \left[\cos\left(\sqrt{1-\zeta^2}\omega_n t\right) - \dfrac{\zeta}{\sqrt{1-\zeta^2}}\sin\left(\sqrt{1-\zeta^2}\omega_n t\right)\right]e^{-\zeta\omega_n t}\right)u(t)$
$\zeta = 1$ (critically damped)	real and repeated at $-\omega_n$ (2)	$\Delta\theta\left[1 - e^{-\omega_n t} + \omega_n t e^{-\omega_n t}\right]u(t)$
$\zeta > 1$ (overdamped)	real and distinct at $-\zeta\omega_n$	$\Delta\theta\left(1 - \left[\cosh\left(\sqrt{1-\zeta^2}\omega_n t\right) - \dfrac{\zeta}{\sqrt{1-\zeta^2}}\sinh\left(\sqrt{1-\zeta^2}\omega_n t\right)\right]e^{-\zeta\omega_n t}\right)u(t)$

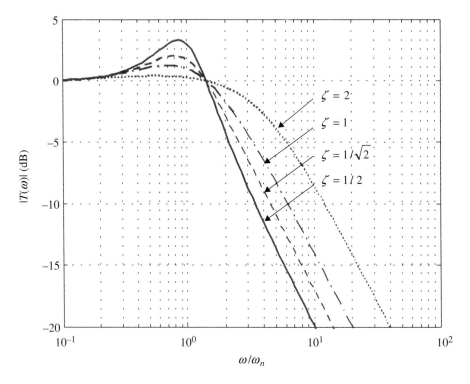

Figure C.1.6 The frequency response of the second-order PLL with transfer function (C.31) for $\zeta = 0.5, 1/\sqrt{2}, 1, 2$.

C.1.4 PLL Bandwidth

Another way to characterize the PLL is to compute the bandwidth of the frequency response $H_a(j\omega)$. The loop bandwidth is a function of both ζ and ω_n. The 3-dB bandwidth $\omega_{3\text{dB}}$ is obtained by setting $|H_a(j\omega)|^2 = 1/2$ and solving for ω. For example, the 3-dB bandwidth for a second-order loop using a proportional-plus-integrator loop filter is

$$\omega_{3\text{dB}} = \omega_n \sqrt{1 + 2\zeta^2 + \sqrt{\left(1 + 2\zeta^2\right)^2 + 1}}. \tag{C.34}$$

While the 3-dB bandwidth is a familiar concept, it is not a very useful measure of bandwidth for a PLL. A more useful measure of bandwidth is the equivalent noise bandwidth B_n. The equivalent noise bandwidth of a linear system with transfer function $H_a(j2\pi f)$ is the bandwidth (measured in hertz) of a fictitious rectangular low-pass filter with the same area as $|H_a(j2\pi f)|^2$. Calculation of the noise bandwidth is illustrated in Figure C.1.8. The magnitude-squared of the transfer function $|H_a(j2\pi f)|^2$ is represented by the solid line and the area under

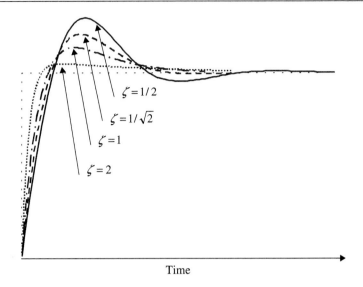

Figure C.1.7 Step response for the second-order PLL for $\zeta = 0.5, 1/\sqrt{2}, 1, 2$.

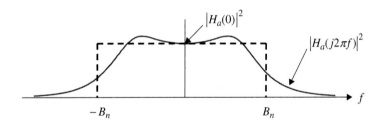

Figure C.1.8 Conceptual diagram for the calculation of noise bandwidth.

$|H_a(j2\pi f)|^2$ is obtained by integrating $|H_a(j2\pi f)|^2$ over the frequency variable f:

$$\text{area} = \int_{-\infty}^{\infty} |H_a(j2\pi f)|^2 df. \tag{C.35}$$

The fictitious rectangular low-pass filter is represented by the dashed line in Figure C.1.8. Fixing the height of this filter at $|H_a(0)|^2$ and the width at $2 \times B_n$, the area under this filter is $2B_n|H_a(0)|^2$. Equating the two areas and solving for B_n produces the desired expression for noise bandwidth:

$$B_n = \frac{1}{2|H_a(0)|^2} \int_{-\infty}^{\infty} |H_a(j2\pi f)|^2 df. \tag{C.36}$$

Table C.1.3 Transfer function, loop parameters, and equivalent noise bandwidth for a phase locked loop using the four loop filters of interest

Loop Filter $F(s)$	ζ	ω_n	$H_a(s)$	B_n (Hz)
k	—	—	$\dfrac{k_0 k_p k}{s + k_p k_0 k}$	$\dfrac{k_0 k_p k}{4}$
$\dfrac{k}{k+s}$	$\dfrac{1}{2}\sqrt{\dfrac{k}{k_0 k_p}}$	$\sqrt{k_0 k_p k}$	$\dfrac{\omega_n^2}{s^2 + 2\zeta\omega_n s + \omega_n^2}$	$\dfrac{\omega_n}{8\zeta}$
$k_1 + \dfrac{k_2}{s}$	$\dfrac{k_1}{2}\sqrt{\dfrac{k_0 k_p}{k_2}}$	$\sqrt{k_0 k_p k_2}$	$\dfrac{2\zeta\omega_n s + \omega_n^2}{s^2 + 2\zeta\omega_n s + \omega_n^2}$	$\dfrac{\omega_n}{2}\left(\dfrac{1}{4\zeta} + \zeta\right)$
$\dfrac{k_1 + s}{k_2 + s}$	$\dfrac{1}{2}\dfrac{k_0 k_p + k_2}{\sqrt{k_0 k_p k_1}}$	$\sqrt{k_0 k_p k_1}$	$\dfrac{\omega_n^2 + \omega_n\left(2\zeta - \dfrac{k_2}{\omega_n}\right)}{s^2 + 2\zeta\omega_n s + \omega_n^2}$	$\dfrac{\omega_n}{8\zeta}\left[1 + \left(2\zeta - \dfrac{k_2}{\omega_n}\right)^2\right]$

For the "proportional-plus-integrator" loop filter, the noise bandwidth is

$$B_n = \frac{\omega_n}{2}\left(\zeta + \frac{1}{4\zeta}\right). \tag{C.37}$$

The noise bandwidth B_n is a nonlinear function of ζ. For fixed ω_n, B_n assumes a minimum of $\omega_n/2$ at $\zeta = 1/2$ and is larger for all other values of ζ. For $\zeta \geq 1/2$, B_n grows monotonically with ζ.

The transfer functions and equivalent noise bandwidth of PLLs using the four loop filters in Table C.1.1 are summarized in Table C.1.3

C.1.5 Acquisition

Any PLL requires a nonzero period of time to reduce a phase error to zero. During the initial stages of acquisition, the input voltage to the VCO is adjusted to produce an output whose frequency matches that of the input. This initial phase of the acquisition process is called *frequency lock*. Once frequency lock is achieved, an additional period of time is required to reduce the loop phase error to an acceptably low level. This phase of the acquisition process is called *phase lock*. Consequently, the acquisition time T_{LOCK} is well approximated by the sum of the time to achieve frequency lock T_{FL} and the time to achieve phase lock T_{PL}:

$$T_{\text{LOCK}} \approx T_{\text{FL}} + T_{\text{PL}} \tag{C.38}$$

where

$$T_{\text{FL}} \approx 4\frac{(\Delta f)^2}{B_n^3} \tag{C.39}$$

$$T_{\text{PL}} \approx \frac{1.3}{B_n}. \tag{C.40}$$

Appendix C Phase Locked Loops

The Δf in (C.39) is the frequency offset (in hertz) and B_n in (C.39) and (C.40) is the noise bandwidth. It is possible for the frequency offset Δf to be so large that the loop can never acquire lock. The range of frequency offsets for which the loop can acquire lock is called the *pull-in* range $(\Delta f)_{\text{pull-in}}$, and is well approximated by

$$(\Delta f)_{\text{pull-in}} \approx \left(2\pi\sqrt{2}\zeta\right) B_n. \tag{C.41}$$

C.1.6 Tracking

Tracking performance is quantified by the variance of the phase error. Conceptually, the phase error variance, $\sigma_{\theta_e}^2$, is

$$\sigma_{\theta_e}^2 = \mathrm{E}\left\{\left|\theta - \hat{\theta}\right|^2\right\}. \tag{C.42}$$

For a linear PLL that has a sinusoidal input with power P_{in} W together with additive white Gaussian noise with power spectral density $N_0/2$ W/Hz, the phase error variance is

$$\sigma_{\theta_e}^2 = \frac{N_0 B_n}{P_{\text{in}}}. \tag{C.43}$$

Because the noise power at the PLL input (within the frequency band of interest to the PLL) is $N_0 B_n$, the ratio $P_{\text{in}}/N_0 B_n$ is often called the loop signal to noise ratio. Thus, for a linear PLL with additive white Gaussian noise, the phase error variance is inversely proportional to the loop signal-to-noise ratio.

Equations (C.39) and (C.40) indicate that acquisition time is inversely proportional to a power of B_n. This suggests that the larger the equivalent loop bandwidth, the faster the acquisition. Equation (C.43) shows that the tracking error is proportional to B_n. This suggests that the smaller the equivalent loop bandwidth, the smaller the tracking error. Thus, fast acquisition and good tracking place competing demands on PLL design. Acquisition time can be decreased at the expense of increased tracking error. Tracking error can be decreased at the expense of increased acquisition time. A good design balances the two performance criteria. Where that balance is depends on the application, the signal to noise level, and system-level performance specifications.

C.1.7 Loop Constant Selection

Appropriate choice of loop constants is a critical step in PLL design. Most often, PLL design specifications define the nature of the response (i.e., underdamped, critically damped, or overdamped) and the noise bandwidth B_n. The procedure is demonstrated using the proportional-plus-integrator loop filter. The resulting second-order loop, with transfer function given by (C.30), or the equivalent canonical form (C.31), offers the designer two degrees of freedom: the damping constant and the natural frequency. But there are four loop constants (k_0, k_p, k_1, k_2) that must be selected. The loop constants, however, are not all independent. Solving (C.32) and (C.33) for the loop constants gives

$$\begin{aligned} k_0 k_p k_1 &= 2\zeta \omega_n \\ k_0 k_p k_2 &= \omega_n^2. \end{aligned} \tag{C.44}$$

Observe that the VCO sensitivity k_0 and the phase detector gain k_p always occur as a pair. Because ζ is a parameter chosen by the designer, all that remains is to find an expression for ω_n. Solving (C.37) for ω_n gives

$$\omega_n = \frac{2B_n}{\zeta + \dfrac{1}{4\zeta}}. \tag{C.45}$$

Substituting (C.45) for ω_n in (C.44) produces the desired expression for the loop constants in terms of the damping constant and the equivalent noise bandwidth:

$$k_p k_0 k_1 = \frac{4\zeta B_n}{\zeta + \dfrac{1}{4\zeta}}$$

$$k_p k_0 k_2 = \frac{4B_n^2}{\left(\zeta + \dfrac{1}{4\zeta}\right)^2}. \tag{C.46}$$

In many cases, the phase detector is the most complicated component in the loop and possesses a gain, k_p, which cannot be adjusted. So, for a given phase detector, k_p can be treated as fixed. The other three loop constants should be adjusted to satisfy the relationships (C.46). One approach is to fix the VCO sensitivity at a value that ensures good VCO performance and adjust the filter constants k_1 and k_2 to obtain the desired loop response.

As an example, consider the design of a critically damped PLL with a noise bandwidth of 25 Hz that uses a proportional-plus-integrator loop filter. Using the relations (C.46) with $\zeta = 1$ and $B_n = 25$ produces the desired conditions for the loop constants:

$$\begin{aligned} k_0 k_p k_1 &= 80 \\ k_0 k_p k_2 &= 1600. \end{aligned} \tag{C.47}$$

C.2 DISCRETE-TIME PHASE LOCKED LOOPS

Discrete-time PLLs are used in sampled data systems. The structure of a discrete-time PLL is essentially the same as that of a continuous-time PLL. An example of the basic system architecture is illustrated in Figure C.2.1. A discrete-time phase detector, discrete-time loop filter, and direct digital synthesizer (DDS) are arranged in a feedback loop in the same way the phase detector, loop filter, and VCO were for the continuous-time PLL. Samples of a sinusoid with frequency Ω_0 rad/sample and phase samples $\theta(nT)$ form the input to the PLL. The discrete-time phase detector computes a function of the phase difference between the input and the DDS output. This phase difference is the discrete-time phase error. The phase error is filtered by the loop filter and input to the DDS. The DDS is a discrete-time version of the VCO and is described in detail in Section 9.2.2. The input/output relationship for the DDS is

$$\text{output} = \cos\left(\Omega_0 n + \hat{\theta}(nT)\right) \tag{C.48}$$

Appendix C Phase Locked Loops

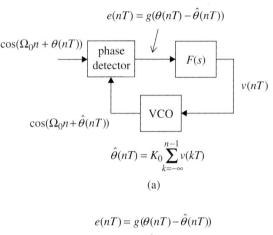

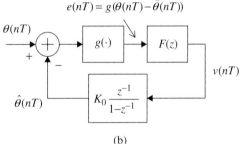

Figure C.2.1 Basic structure of a discrete-time PLL (a) and the corresponding phase equivalent PLL (b).

where

$$\hat{\theta}(nT) = K_0 \sum_{k=-\infty}^{n-1} v(kT). \tag{C.49}$$

Typically, the design of a discrete-time PLL begins with a continuous-time PLL design and uses a continuous-time to discrete-time transformation to produce the discrete-time PLL. The design process begins with the continuous-time PLL because there are well-developed design techniques for continuous-time systems. The continuous-time to discrete-time transformations are intended to produce a discrete-time system with the same behavior as the continuous-time system. Some of the more popular continuous-time to discrete-time transformations, such as impulse invariance, step invariance, and integral approximations, were summarized in Section 2.6.

The problem with applying these approaches directly to the linearized phase equivalent PLL is that the linear phase equivalent PLL is only a model of the actual PLL. For example, in a real system the actual phases $\theta(nT)$ and $\hat{\theta}(nT)$ are not available. What is available to the

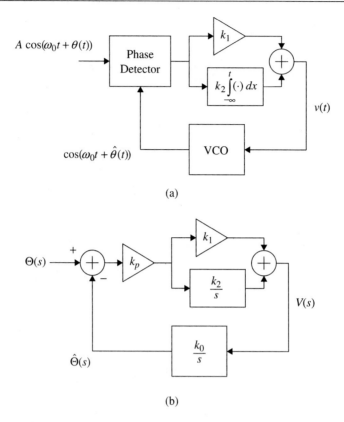

Figure C.2.2 Second-order continuous-time phase locked loop with proportional-plus-integrator loop filter (a) and the linearized phase equivalent PLL (b).

system are samples of the input sinusoid and the discrete-time DDS output. A hybrid approach is often used to perform the continuous-time to discrete-time transformation. This hybrid approach is illustrated here for the proportional-plus-integrator loop filter. (The application of this approach to PLLs using other loop filters is straightforward.)

The starting point is the continuous-time PLL and its phase-equivalent loop. These loops are illustrated in Figure C.2.2. The discrete-time PLL that mimics the continuous-time PLL of Figure C.2.2 is illustrated in Figure C.2.3. In Figure C.2.3 (a), the integral portion of the loop filter uses a simple filter with a pole at $z = 1$. The discrete-time PLL uses a direct digital synthesizer (DDS) in place of the VCO. A single-pole filter is used to integrate the DDS input to calculate the instantaneous phase. Note the use of uppercase letters for the phase detector gain, loop filter constants, and DDS gain to distinguish them from their counterparts in the continuous-time PLL. The phase equivalent discrete-time loop is illustrated in Figure C.2.3 (b).

Appendix C Phase Locked Loops

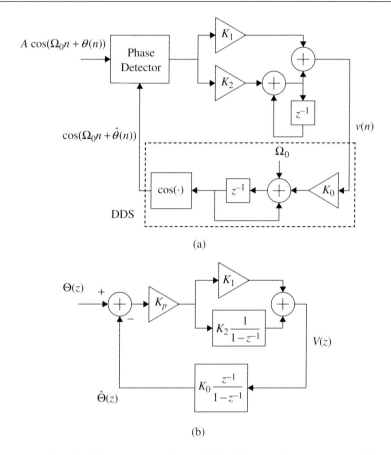

Figure C.2.3 Second-order discrete-time phase locked loop with proportional-plus-integrator loop filter that mimics the continuous-time PLL illustrated in Figure C.2.2: (a) The discrete-time PLL and DDS; (b) the linearized phase equivalent discrete-time PLL.

The loop transfer functions for the continuous-time loop of Figure C.2.2 and the discrete-time loop of Figure C.2.3 are

$$H_a(s) = \frac{2\zeta\omega_n s + \omega_n^2}{s^2 + 2\zeta\omega_n s + \omega_n^2} \tag{C.50}$$

$$H_d(z) = \frac{K_p K_0 (K_1 + K_2) z^{-1} - K_p K_0 K_1 z^{-2}}{1 - 2\left(1 - \frac{1}{2} K_p K_0 (K_1 + K_2)\right) z^{-1} + \left(1 - K_p K_0 K_1\right) z^{-2}} \tag{C.51}$$

where ζ and ω_n in (C.50) are given by (C.32) and (C.33), respectively. Applying Tustin's equation

$$\frac{1}{s} \rightarrow \frac{T}{2} \frac{1 + z^{-1}}{1 - z^{-1}}. \tag{C.52}$$

to $H_a(s)$ produces a discrete-time version of the continuous-time loop denoted $H_a\left(\frac{2}{T}\frac{1-z^{-1}}{1+z^{-1}}\right)$. After some algebra, $H_a\left(\frac{2}{T}\frac{1-z^{-1}}{1+z^{-1}}\right)$ may be expressed as

$$H_a\left(\frac{2}{T}\frac{1-z^{-1}}{1+z^{-1}}\right) = \frac{\frac{2\zeta\theta_n + \theta_n^2}{1+2\zeta\theta_n+\theta_n^2} + 2\frac{\theta_n^2-\zeta\theta_n}{1+2\zeta\theta_n+\theta_n^2}z^{-1} + \frac{\theta_n^2}{1+2\zeta\theta_n+\theta_n^2}z^{-2}}{1 - 2\frac{\theta_n^2-1}{1+2\zeta\theta_n+\theta_n^2}z^{-1} + \frac{1-2\zeta\theta_n+\theta_n^2}{1+2\zeta\theta_n+\theta_n^2}z^{-2}} \quad \text{(C.53)}$$

where

$$\theta_n = \frac{\omega_n T}{2}. \quad \text{(C.54)}$$

Equating the denominator polynomials in $H_d(z)$ and $H_a\left(\frac{2}{T}\frac{1-z^{-1}}{1+z^{-1}}\right)$ produces the following relationships:

$$1 - \frac{1}{2}K_p K_0(K_1 + K_2) = \frac{\theta_n^2 - 1}{1 + 2\zeta\theta_n + \theta_n^2}$$

$$1 - K_p K_0 K_1 = \frac{1 - 2\zeta\theta_n + \theta_n^2}{1 + 2\zeta\theta_n + \theta_n^2}. \quad \text{(C.55)}$$

Finally, solving for the loop constants gives

$$K_p K_0 K_1 = \frac{4\zeta\theta_n}{1 + 2\zeta\theta_n + \theta_n^2}$$

$$K_p K_0 K_2 = \frac{4\theta_n^2}{1 + 2\zeta\theta_n + \theta_n^2}. \quad \text{(C.56)}$$

The expressions for the loop constants are a function of the damping ratio ζ, the natural frequency ω_n, and the sampling period T. (The functional dependence on ω_n and T is through θ_n.) The loop constants can be expressed in terms of the ζ, T, and equivalent loop bandwidth B_n by solving the expression for B_n for ω_n, expressing θ_n in terms of B_n and substituting into (C.56). For the proportional-plus-integrator loop filter, B_n is related to ω_n by (C.37). Solving for ω_n and substituting produces

$$\theta_n = \frac{B_n T}{\zeta + \frac{1}{4\zeta}}. \quad \text{(C.57)}$$

Substituting into the relations (C.56) produces

$$K_0 K_p K_1 = \frac{4\zeta \left(\dfrac{B_n T}{\zeta + \dfrac{1}{4\zeta}} \right)}{1 + 2\zeta \left(\dfrac{B_n T}{\zeta + \dfrac{1}{4\zeta}} \right) + \left(\dfrac{B_n T}{\zeta + \dfrac{1}{4\zeta}} \right)^2}$$

$$K_0 K_p K_2 = \frac{4 \left(\dfrac{B_n T}{\zeta + \dfrac{1}{4\zeta}} \right)^2}{1 + 2\zeta \left(\dfrac{B_n T}{\zeta + \dfrac{1}{4\zeta}} \right) + \left(\dfrac{B_n T}{\zeta + \dfrac{1}{4\zeta}} \right)^2}.$$

(C.58)

Note that when the equivalent loop bandwidth is small relative to the sample rate, $B_n T \ll 1$ so that the relations (C.58) are well approximated by

$$K_0 K_p K_1 \approx \frac{4\zeta}{\zeta + \dfrac{1}{4\zeta}} (B_n T)$$

$$K_0 K_p K_2 \approx \frac{4}{\left(\zeta + \dfrac{1}{4\zeta}\right)^2} (B_n T)^2.$$

(C.59)

Comparing (C.59) with (C.46) shows that for the case where the sample rate is large relative to the loop equivalent bandwidth, the expressions for the loop filter constants for the discrete-time loop are the same as those for the continuous-time loop except that the loop bandwidth is normalized by the sample rate.

This design procedure requires the noise bandwidth to be specified relative to the sample rate $1/T$. In digital communications, it is common practice to specify the noise bandwidth relative to the symbol rate $1/T_s$ (i.e., $B_n T_s$ is specified instead of $B_n T$ or B_n). The expressions (C.58) can be adjusted to account for this by using $N = T_s/T$ so that

$$\theta_n = \frac{B_n T_s}{N \left(\zeta + \dfrac{1}{4\zeta}\right)}.$$

(C.60)

Substituting (C.60) into (C.56) produces

$$K_0 K_p K_1 = \frac{\dfrac{4\zeta}{N}\left(\dfrac{B_n T_s}{\zeta + \dfrac{1}{4\zeta}}\right)}{1 + \dfrac{2\zeta}{N}\left(\dfrac{B_n T_s}{\zeta + \dfrac{1}{4\zeta}}\right) + \left(\dfrac{B_n T_s}{N\left(\zeta + \dfrac{1}{4\zeta}\right)}\right)^2}$$

$$K_0 K_p K_2 = \frac{\dfrac{4}{N^2}\left(\dfrac{B_n T_s}{\zeta + \dfrac{1}{4\zeta}}\right)^2}{1 + \dfrac{2\zeta}{N}\left(\dfrac{B_n T_s}{\zeta + \dfrac{1}{4\zeta}}\right) + \left(\dfrac{B_n T_s}{N\left(\zeta + \dfrac{1}{4\zeta}\right)}\right)^2}. \tag{C.61}$$

C.2.1 Examples

Two examples are presented to illustrate ways a phase detector could be realized in a discrete-time system. The first example is the PLL illustrated in Figure C.2.4. This PLL is designed to track the phase of a complex exponential. The DDS output is also a complex exponential whose phase is the loop estimate of the input phase. The DDS produces a complex exponential by using a cosine and sine look-up table to output the real and imaginary parts, respectively. The phase detector computes the phase error by first computing the product of the input and the complex conjugate of the DDS output. This product is given by $\exp\{j(\theta(n) - \hat{\theta}(n))\}$. The arg$\{\cdot\}$ function computes the phase of the complex exponential using a four-quadrant arctangent. The output of the phase detector is thus $g(\theta_e(n)) = \theta_e(n) = \theta(n) - \hat{\theta}(n)$. The corresponding S-curve is shown in the lower portion of Figure C.2.4. Note that this phase detector is linear over the interval $-\pi \leq \theta_e \leq \pi$ and has a slope of one. For this reason, $K_p = 1$. The loop filter is the discrete-time version of the familiar proportional-plus-integrator loop.

The loop responses are illustrated in Figure C.2.5 for an input with

$$\Omega_0 = \frac{2\pi}{10} \tag{C.62}$$

$$\theta(n) = \pi u(n). \tag{C.63}$$

The real part of the loop input is illustrated by the dashed line in the top plot of Figure C.2.5. The loop filter constants were chosen to produce a critically damped loop with $B_n T = 0.05$. Using (C.58) produces $K_1 = 0.1479$ and $K_2 = 0.0059$. The real part of the loop output (which is the DDS output) is illustrated by the solid line in the top plot of Figure C.2.5. Note that in

Appendix C Phase Locked Loops

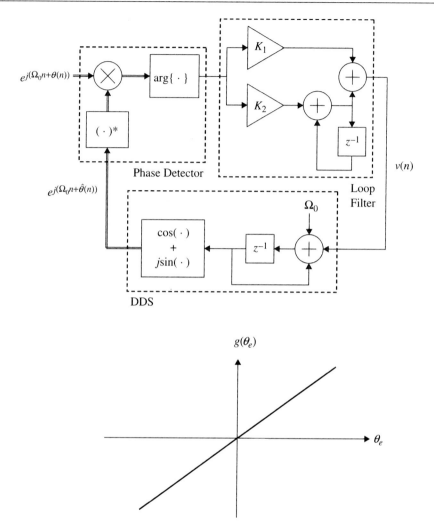

Figure C.2.4 A second-order discrete-time PLL with complex input and output. The phase detector is realized by computing the argument (or phase) of the conjugate product of the loop input and the DDS output. A four-quadrant arctangent function can be used to compute the argument. The S-curve corresponding to this phase detector is shown below the block diagram.

the beginning, the two sinusoids are π radians out of phase. As time progresses, DDS output becomes phase-aligned with the PLL input. The phase error $\theta_e(n)$ is plotted in the middle plot of Figure C.2.5. The phase error starts at π radians, decreases to -0.5 radians then slowly approaches 0. The bottom plot of Figure C.2.5 illustrates the arguments of the two complex exponentials. The dashed line is the argument of the input sinusoid, $\Omega_0 n + \pi u(n)$, which is

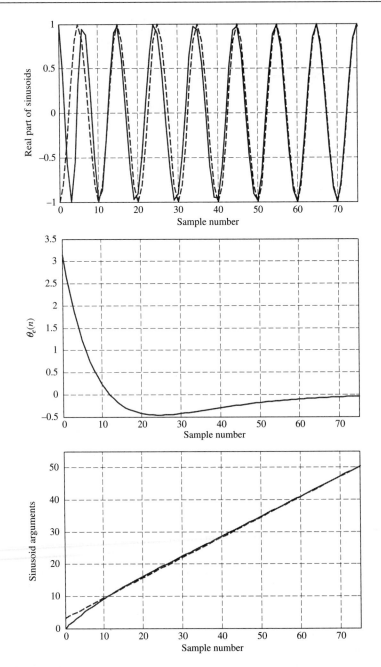

Figure C.2.5 A plot of loop inputs and outputs for the second-order discrete-time PLL illustrated in Figure C.2.4: (top) The real part of the loop input (dashed line) and the real part of the DDS output (solid line); (middle) the phase error $\theta_e(n)$; (bottom) the argument of the input complex exponential (dashed line) and the argument of the DDS output complex exponential (solid line).

the equation of a straight line with slope Ω_n and intercept π. The solid line is the argument of the complex exponential at the output of the DDS. It starts at zero and eventually aligns itself with the PLL input. The difference between these two curves is the phase error, which is illustrated in the middle plot of this figure. Note that the solid line is below the dashed line for samples 0 through 12. This means the phase error is positive as illustrated in the phase error plot. For samples 13 though 75, the solid line is above the dashed line, which means the phase error is negative. This is also confirmed by the phase error plot. The fact that the DDS output matches the PLL input is confirmed in all three plots in Figure C.2.5.

A variation on this loop architecture is illustrated in Figure C.2.6. In this loop, the arg{·} function has been replaced by the Im{·} which is the imaginary part of the complex value. This is desirable because it avoids the complexity of computing a four-quadrant arctangent. The S-curve, however, is different. The S-curve for this phase detector is obtained by letting the input be

$$x(n) = A \exp\left\{j\left(\Omega_0 n + \theta(n)\right)\right\}. \tag{C.64}$$

If the complex-valued DDS output is

$$y(n) = \exp\left\{j\left(\Omega_0 n + \hat{\theta}(n)\right)\right\}, \tag{C.65}$$

then the conjugate product is

$$x(n)y^*(n) = A \exp\left\{j\left(\theta(n) - \hat{\theta}(n)\right)\right\}. \tag{C.66}$$

The imaginary part of this product is

$$\theta_e(n) = A \sin\left(\theta(n) - \hat{\theta}(n)\right). \tag{C.67}$$

The S-curve is a sine function with period 1 and amplitude A. Thus $K_p = A$, the amplitude of the input complex sinusoid.

The loop responses for the scenario (C.62) and (C.63) are plotted in Figure C.2.6. For $A = 1$, $K_p = 1$ and the loop filter coefficients are the same as before. Observe that the loop response is somewhat different for this case. This difference is most noticeable when comparing the phase errors of Figures C.2.5 and C.2.7. The initial phase error is π, which corresponds to a zero crossing in the S-curve of the phase detector in Figure C.2.6. Because the slope of this zero crossing is negative, this zero crossing is not a stable lock point and the phase error migrates from π toward 0 as time increases. The phase error reaches $\pi/2$ at approximately sample 17 as evidenced by the peak of θ_e at one in Figure C.2.7. The phase error overshoots and eventually settles to zero. Because of the nonlinear shape of the S-curve, the phase error starts small, grows large, and then settles to zero. As the phase error decreases, the PLL behaves more like the linear system of Figure C.2.4. This behavior is also evident by inspection of the sinusoidal plots in Figure C.2.7 (a) and the phase trajectory plots in Figure C.2.7 (c).

In the second example, a real-valued discrete-time PLL is explored. This PLL is illustrated in Figure C.2.8. The input to the loop is the sinusoid

$$x(n) = A \cos\left(\Omega_0 n + \theta(n)\right). \tag{C.68}$$

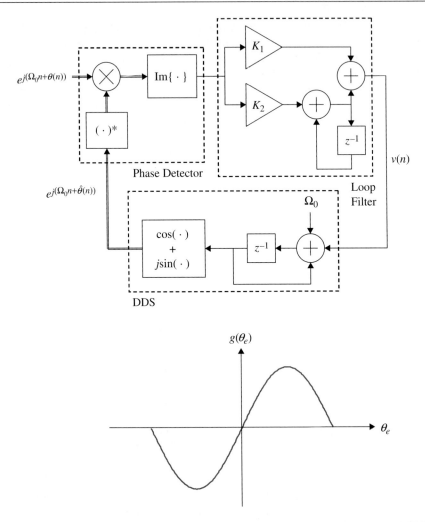

Figure C.2.6 A second-order discrete-time PLL with complex input and output. The phase detector is realized by computing the imaginary part of the conjugate product of the loop input and the DDS output. The S-curve corresponding to this phase detector is shown below the block diagram.

The DDS is designed to output two sinusoids in phase quadrature:

$$y_c(n) = \cos\left(\Omega_0 n + \hat{\theta}(n)\right) \quad \text{(C.69)}$$

$$y_s(n) = -\sin\left(\Omega_0 n + \hat{\theta}(n)\right). \quad \text{(C.70)}$$

Appendix C   Phase Locked Loops

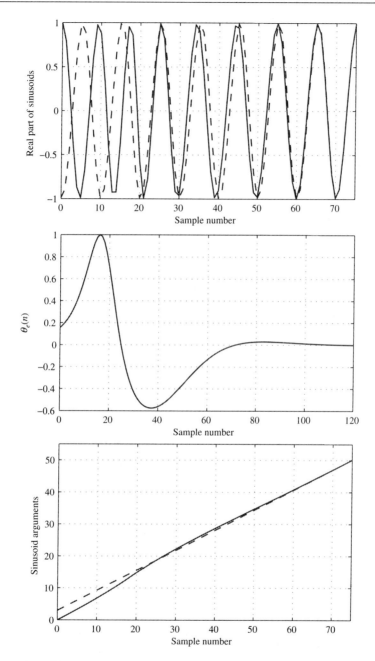

Figure C.2.7 A plot of loop inputs and outputs for the second-order discrete-time PLL illustrated in Figure C.2.6: (top) The real part of the loop input (dashed line) and the real part of the DDS output (solid line); (middle) the phase error $\theta_e(n)$; (bottom) the argument of the input complex exponential (dashed line) and the argument of the DDS output complex exponential (solid line).

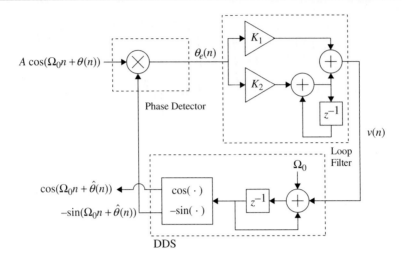

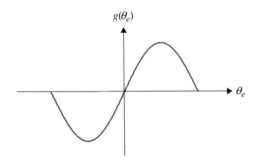

Figure C.2.8 A second-order discrete-time PLL with complex input and output. The phase detector is realized by forming the product between the input sinusoid and the quadrature DDS output. The S-curve corresponding to this phase detector is shown below the block diagram.

The phase detector forms the product

$$x(n)y_s(n) = -A\cos(\Omega_0 n + \theta(n))\sin\left(\Omega_0 n + \hat{\theta}(n)\right) \quad (C.71)$$

$$= \frac{A}{2}\sin\left(\theta(n) - \hat{\theta}(n)\right) - \frac{A}{2}\sin\left(2\Omega_0 n + \theta(n) + \hat{\theta}(n)\right). \quad (C.72)$$

The loop filter is a low-pass filter. Hence, only the low-pass component of the product $x(n)y_s(n)$ needs to be tracked. Thus, the S-curve is

$$g(\theta_e) = g\left(\theta(n) - \hat{\theta}(n)\right) \quad (C.73)$$

$$= \frac{A}{2}\sin\left(\theta(n) - \hat{\theta}(n)\right) \quad (C.74)$$

and is illustrated in the lower part of Figure C.2.8. First note that the S-curve is given by the sine of the phase error. As a consequence of the approximation $\sin(X) \approx X$ for small X, the phase detector is approximately linear for small values of the phase error:

$$g(\theta_e) \approx \frac{A}{2}\theta_e \quad \text{for small } \theta_e. \tag{C.75}$$

From this we see that the phase detector gain K_p is $\frac{A}{2}$, which depends on the amplitude of the sinusoid at the PLL input. This illustrates an important caveat in PLL design: Care must be taken to insure that the input signal levels are controlled. Otherwise, the PLL may not perform as designed.

The loop responses are illustrated in Figure C.2.9 for an input with

$$\Omega_0 = \frac{2\pi}{10} \tag{C.76}$$

$$\theta(n) = \pi u(n). \tag{C.77}$$

The loop input for $A = 1$ is illustrated by the dashed line in the top plot of Figure C.2.9. As in the previous example, the loop filter constants were chosen to produce a critically damped loop with $B_n T = 0.05$. Using (C.58) produces $K_1 = 0.2958$ and $K_2 = 0.0118$. The loop output (which is the inphase component of the DDS output) is illustrated by the solid line in the top plot of Figure C.2.9. Note that in the beginning, the two sinusoids are π radians out of phase. As time progresses, DDS output becomes phase-aligned with the PLL input. The phase error $\theta_e(n)$ is plotted in the middle plot of Figure C.2.9. In this case, the phase error can be thought of as a slowly varying DC component plus a constant amplitude sinusoid. The sinusoid, whose frequency is $2 \times \Omega_0$, is a result of the double frequency component that results from the multiplication performed in the phase detector as predicted by (C.72). This double frequency term is eliminated by the low-pass filter action of the loop filter so that the loop responds only to the average value of the phase error.[3] The average phase error starts small and increases from samples 0 to 20. From samples 20 to 40, the average phase error decreases, passing through zeros at sample 27. It increases again from samples 40 to 60 thereafter settling to an average value of zero. The shape of the average phase error is different from that in the previous example (see the middle plot of Figure C.2.5) even though the loop input is the same. This difference is due to the nonlinear nature of the phase detector. Because the S-curve is nonlinear, the loop response to large phase errors is also nonlinear. The nonlinear characteristics of the response are not predicted by the linear PLL analysis.

The bottom plot of Figure C.2.9 illustrates the arguments of the input sinusoid (dashed line) and the DDS output (solid line). As before, the dashed line is a plot of $\Omega_0 n + \pi u(n)$,

---

[3] A low-pass filter *could* be inserted after the mixer to eliminate the double frequency term. This filter would then be considered part of the phase detector, but the phase detector would not be memoryless (i.e., it has its own poles and zeros). The poles and zeros of this low-pass filter increase the order of the system. Although higher order systems are not necessarily bad, the design procedure is much more involved and loop stability harder to achieve. When the inclusion of such a filter is unavoidable, designers usually force the loop bandwidth to be much smaller than the bandwidth of the low-pass filter. This forces the transients due to the loop filter to dominate the transients of the low-pass filter so that the poles and zeros of the low-pass filter can be ignored.

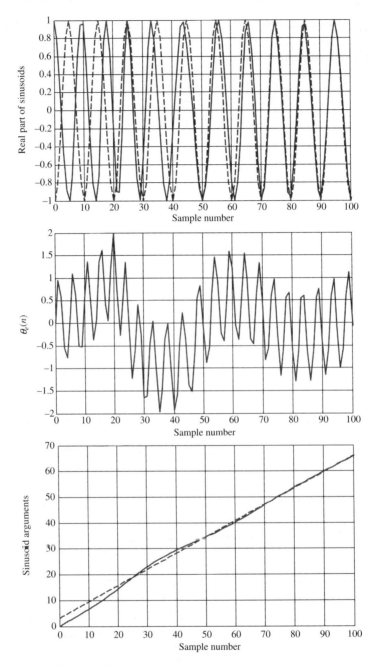

Figure C.2.9 A plot of loop inputs and outputs for the second-order discrete-time PLL illustrated in Figure C.2.8: (top) The loop input (dashed line) and the DDS output (solid line); (middle) the phase error $\theta_e(n)$; (bottom) the argument of the input sinusoid (dashed line) and the argument of the quadrature DDS output (solid line).

which is the equation of a straight line with slope Ω_n and intercept π. The solid line is the argument of the quadrature DDS output which *is* the DDS input. The solid line tracks the behavior of the average phase error. It starts at zero with a slope slightly less than Ω_0. The slope then becomes greater than Ω_0 as the PLL output tries to catch up with the PLL input. Note that the two lines cross at sample 27, precisely where the average phase error crossed zero. The solid line and dashed line eventually merge as the DDS output aligns itself with the PLL input. The fact that the DDS output matches the PLL input is confirmed in all three plots in Figure C.2.9.

C.3 NOTES AND REFERENCES

C.3.1 Topics Covered

The classic text on phase locked loops is Garnder's [127]. The basics of continuous-time phase locked loops are also covered in reasonably well-written texts by Best [347] and Stephens [348]. Waggener [174] provides a nice applications-oriented explanation of PLLs in digital communication systems. Discrete-time PLLs and their relationship to continuous-time PLLs are discussed in the seminal paper by Lindsey and Chie [349]. The conversion from continuous-time PLLs to discrete-time PLLs is described in [348].

C.3.2 Topics Not Covered

The effect of additive noise at the PLL input requires thorough familiarity with continuous-time random processes and as such is beyond the scope of this text. Viterbi [125] and Gardner [127] provide a thorough treatment of the subject.

Two important characteristics of PLL performance, when used for carrier phase synchronization, were not covered. The first characteristic is *hang-up*. Hang-up occurs when the PLL locks temporarily in one of the unstable zero-crossings in the S-curve. Hang-up, in the context of digital communications, has been examined by Gardner [350,351], Meyr and Popken [352], and Fitz [353]. The other PLL characteristic that was not covered is the *cycle slip*. Cycle slips occur when the phase error becomes large enough to move the lock point by 2π radians. When this happens, an entire cycle of the input sinusoid has been skipped. Cycle slips are particularly destructive in applications where every cycle counts, such as timing synchronization. Fortunately, cycle slips are low probability events at normal operating signal-to-noise ratios. Cycle slips have been examined by Sannemann and Rowbotham [354], Tausworthe [355], Ascheid and Meyr [356], Moeneclaey [357], and Chie [358].

C.4 EXERCISES

C.1 Consider a linear continuous-time PLL with loop filter $F(s) = k$.
(a) Determine the time-domain expression for the output $\hat{\theta}(t)$ for a step input, i.e., $\theta(t) = \Delta\theta u(t)$.

(b) Determine the time-domain expression for the output $\hat{\theta}(t)$ for a ramp input, i.e., $\theta(t) = \Delta\omega t u(t)$.

C.2 Consider a linear continuous-time PLL with loop filter $F(s) = \dfrac{k}{s+k}$.

(a) Determine the time-domain expression for the output $\hat{\theta}(t)$ for a step input, i.e., $\theta(t) = \Delta\theta u(t)$. Note that there are three cases to consider: $\zeta < 1, \zeta = 1,$ and $\zeta > 1$.

(b) Determine the time-domain expression for the output $\hat{\theta}(t)$ for a ramp input, i.e., $\theta(t) = \Delta\omega t u(t)$. Note that there are three cases to consider: $\zeta < 1, \zeta = 1,$ and $\zeta > 1$.

C.3 Consider a linear continuous-time PLL with loop filter $F(s) = k_1 + \dfrac{k_2}{s}$.

(a) Determine the time-domain expression for the output $\hat{\theta}(t)$ for a step input, i.e., $\theta(t) = \Delta\theta u(t)$. Note that there are three cases to consider: $\zeta < 1, \zeta = 1,$ and $\zeta > 1$.

(b) Determine the time-domain expression for the output $\hat{\theta}(t)$ for a ramp input, i.e., $\theta(t) = \Delta\omega t u(t)$. Note that there are three cases to consider: $\zeta < 1, \zeta = 1,$ and $\zeta > 1$.

C.4 Consider a linear continuous-time PLL with loop filter $F(s) = \dfrac{k_1 + s}{k_2 + s}$.

(a) Determine the time-domain expression for the output $\hat{\theta}(t)$ for a step input, i.e., $\theta(t) = \Delta\theta u(t)$. Note that there are three cases to consider: $\zeta < 1, \zeta = 1,$ and $\zeta > 1$.

(b) Determine the time-domain expression for the output $\hat{\theta}(t)$ for a ramp input, i.e., $\theta(t) = \Delta\omega t u(t)$. Note that there are three cases to consider: $\zeta < 1, \zeta = 1,$ and $\zeta > 1$.

C.5 Determine the 3-dB bandwidth of a continuous-time PLL with the following loop filters

(a) $F(s) = k$.

(b) $F(s) = \dfrac{k}{s+k}$.

(c) $F(s) = k_1 + \dfrac{k_2}{s}$.

(d) $F(s) = \dfrac{k_1 + s}{k_2 + s}$.

In each case, compare the 3-dB bandwidth with the noise equivalent bandwidth.

C.6 Consider a linear continuous-time PLL with a proportional-plus-integrator loop filter $F(s) = k_1 + \dfrac{k_2}{s}$. Why is the proportional component necessary? (Hint: Determine the transfer function for this loop in terms of both k_1 and k_2 and set $k_1 = 0$.)

C.7 Consider a continuous-time to discrete-time transformation of a first-order continuous-time linear PLL with loop filter $F(s) = k$.

(a) Derive the transfer function of the continuous-time loop $H_a(s)$.

(b) Use Tustin's equation to convert the continuous-time transfer function $H_a(s)$ to a discrete-time transfer function $H_a\left(\frac{2}{T}\frac{1-z^{-1}}{1+z^{-1}}\right)$.

(c) Derive the transfer function $H_d(z)$ for the linearized discrete-time PLL shown below.

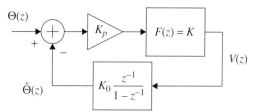

i. Equate the denominators of $H_a\left(\frac{2}{T}\frac{1-z^{-1}}{1+z^{-1}}\right)$ and $H_d(z)$ and solve for the loop filter constant K.

ii. Express the equation for K in terms of the equivalent noise bandwidth of the linearized continuous-time PLL.

C.8 Consider a continuous-time to discrete-time transformation of a second-order continuous-time linear PLL with loop filter $F(s) = \dfrac{k}{s+k}$.

(a) Derive the transfer function of the continuous-time loop $H_a(s)$. Express $H_a(s)$ in the standard form using $s^2 + 2\zeta\omega_n s + \omega_n^2$ for the denominator.

(b) Use Tustin's equation to convert the continuous-time transfer function $H_a(s)$ to a discrete-time transfer function $H_a\left(\frac{2}{T}\frac{1-z^{-1}}{1+z^{-1}}\right)$.

(c) Derive the transfer function $H_d(z)$ for the linearized discrete-time PLL shown below.

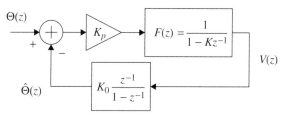

(d) Equate the denominators of $H_a\left(\frac{2}{T}\frac{1-z^{-1}}{1+z^{-1}}\right)$ and $H_d(z)$ and solve for the loop filter constant K.

(e) Express the equation for K in terms of the equivalent noise bandwidth and damping factor ζ of the linearized continuous-time PLL.

C.9 Consider a continuous-time to discrete-time transformation of a second-order continuous-time linear PLL with loop filter $F(s) = \dfrac{s+k_1}{s+k_2}$.

(a) Derive the transfer function of the continuous-time loop $H_a(s)$. Express $H_a(s)$ in the standard form using $s^2 + 2\zeta\omega_n s + \omega_n^2$ for the denominator.

(b) Use Tustin's equation to convert the continuous-time transfer function $H_a(s)$ to a discrete-time transfer function $H_a\left(\frac{2}{T}\frac{1-z^{-1}}{1+z^{-1}}\right)$.

(c) Derive the transfer function $H_d(z)$ for the linearized discrete-time PLL shown below.

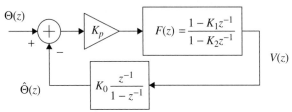

(d) Equate the denominators of $H_a\left(\frac{2}{T}\frac{1-z^{-1}}{1+z^{-1}}\right)$ and $H_d(z)$ and solve for the loop filter constants K_1 and K_2.

(e) Express the equation for K_1 and K_2 in terms of the equivalent noise bandwidth and damping factor ζ of the linearized continuous-time PLL.

Bibliography

[1] T. Sarker, R. Mailloux, A. Oliner, M. Salzar-Palma, and D. Sengupta, *History of Wireless*, John Wiley & Sons, Hoboken, NJ, 2006.
[2] IEEE Communications Society, *A Brief History of Communications: IEEE Communications Society—A Fifty Year Foundation for the Future*, IEEE Communications Society, New York, 2002.
[3] J. Anderson, *Digital Transmission Engineering*, IEEE Press, Piscataway, NJ, 2005.
[4] L. Couch, *Digital and Analog Communication Systems*, Pearson Prentice Hall, Upper Saddle River, NJ, 2006.
[5] M. Frerking, *Digital Signal Processing in Communication Systems*, Kluwer Academic Press, Norwell, MA, 1994.
[6] H. Harada and R. Prasad, *Simulation and Software Radio for Mobile Communications*, Artech House, Boston, MA, 2002.
[7] S. Haykin, *An Introduction to Analog and Digital Communications*, John Wiley & Sons, New York, 1989.
[8] S. Haykin and M. Moher, *Modern Wireless Communications*, Pearson Prentice Hall, Upper Saddle River, NJ, 2005.
[9] T. Lewis, *Empire of the Air: The Men Who Made Radio*, HarperCollins, New York, 1991.
[10] J. Proakis, *Digital Communications*, McGraw Hill, New York, 4th edition, 2000.
[11] J. Proakis and M. Salehi, *Communication Systems Engineering*, Prentice Hall, Upper Saddle River, NJ, 2002.
[12] T. Rappaport, *Wireless Communications: Principles and Practice*, Prentice Hall, Upper Saddle River, NJ, 2002.
[13] J. Reed, *Software Radio: A Modern Approach to Radio Engineering*, Prentice Hall, Upper Saddle River, NJ, 2002.
[14] M. Roden, *Analog and Digital Communication Systems*, Prentice Hall, Upper Saddle River, NJ, 1996.
[15] B. Sklar, *Digital Communicaitons: Fundamentals and Applications*, Prentice Hall, Upper Saddle River, NJ, 2001.
[16] G. Stüber, *Principles of Mobile Communication*, Kluwer Academic Press, Norwell, MA, 2001.
[17] J. Tsui, *Digital Techniques for Wideband Receivers*, Scitech Publishing, Inc., Raleigh, NC, 2004.
[18] W. Tuttlebee, *Software Defined Radio: Enabling Technologies*, John Wiley & Sons, New York, 2002.
[19] S. Wilson, *Digital Modulation and Coding*, Prentice Hall, Upper Saddle River, NJ, 1996.
[20] R. Ziemer and W. Tranter, *Principles of Communications: Systems, Modulation, and Noise*, Houghton Mifflin, Boston, MA, 1995.

[21] L. Ippolito, R. Kaul, and R. Wallace, *Propagation Effects Handbook for Satellite Systems Design*, NASA, 1983 (NASA Reference Publication 1082(03)).

[22] R. Blahut, *Principles and Practice of Information Theory*, Addison Wesley, Boston, MA, 1987.

[23] T. Comer and J. Thomas, *Elements of Information Theory*, John Wiley & Sons, New York, 1991.

[24] R. Gallager, *Information Theory and Reliable Communication*, John Wiley & Sons, New York, 1968.

[25] W. Jakes, *Microwave Mobile Communications*, IEEE Press, Piscataway, NJ, 1993.

[26] J. Parsons, *The Mobile Radio Propagation Channel*, John Wiley & Sons, New York, 1992.

[27] S. Lin and D. Costello, *Error Control Coding*, Pearson Prentice Hall, Upper Saddle River, NJ, 2004.

[28] T. Moon, *Error Correction Coding: Mathematical Methods and Algorithms*, John Wiley & Sons, Hoboken, NJ, 2005.

[29] C. Schlegel, *Trellis Coding*, IEEE Press, Piscataway, NJ, 1997.

[30] S. Wicker, *Error Control Systems for Digital Communication and Storage*, Prentice Hall, Upper Saddle River, NJ, 1995.

[31] J. Anderson, T. Aulin, and C. Sundberg, *Digital Phase Modulation*, Plenum Press, New York, 1986.

[32] A. Goldsmith, *Wireless Communications*, Cambridge University Press, New York, 2005.

[33] E. Larsson and P. Stoica, *Space-Time Block Coding for Wireless Communications*, Cambridge University Press, Cambridge, UK, 2003.

[34] M. Jankiraman, *Space-Time Codes and MIMO Systems*, Artech House, Norwood, MA, 2004.

[35] A. Paulraj, R. Nabar, and D. Gore, *Introduction to Space-Time Wireless Communications*, Cambridge Univeristy Press, New York, 2003.

[36] G. Giannakis, Z. Liu, X. Ma, and S. Zhou, *Space Time Coding for Broadband Wireless Communications*, John Wiley & Sons, New York, 2003.

[37] G. Giannakis, Y. Hua, P. Stoica, and L. Tong, *Signal Processing Advances in Wireless and Mobile Communications, Volume 2: Trends in Single- and Multi-User Systems*, Prentice Hall, Upper Saddle River, NJ, 2000.

[38] B. Vucetic and J. Yuan, *Space-Time Coding*, John Wiley & Sons, Hoboken, NJ, 2003.

[39] R. Bracewell, *The Fourier Transform and Its Applications*, McGraw-Hill, New York, 1986.

[40] A. Oppenheim and A. Willsky, *Signals & Systems*, Prentice-Hall, Upper Saddle River, NJ, 1997.

[41] A. Oppenheim and R. Schafer, *Discrete-Time Signal Processing*, Prentice Hall, Upper Saddle River, NJ, second edition, 1999.

[42] J. McClellan, R. Schafer, and M. Yoder, *DSP First: A Multimedia Approach*, Prentice Hall, Upper Saddle River, NJ, 1998.

[43] R. Churchill and J. Brown, *Complex Variables and Applications*, McGraw-Hill, 6th edition, 1995.

[44] W. Cochran et al., "What is the fast Fourier transform?," *Proceedings of the IEEE*, vol. 55, no. 10, pp. 1664–1674, October 1967.

[45] J. Cooley and J. Tukey, "An algorithm for the machine calculation of complex Fourier series," *Mathematics of Computation*, vol. 19, pp. 297–301, April 1965.

[46] W. Gentleman and G. Sande, "Fast Fourier transforms—For fun and profit," *Proceedings of the American Federation of Information Processing Societies (AFIPS) Conference*, 1966, pp. 563–578.

[47] I. Good, "The interaction algorithm and practical Fourier analysis," *Journal of the Royal Statistical Society. Series B.*, vol. 20, no. 2, pp. 361–272, 1958.

[48] L. Thomas, "Using a computer to solve problems in physics," *Applications of Digital Computers*, Ginn, Boston, MA, 1963.

[49] N. Brenner and C. Rader, "A new principle for fast Fourier transformation," *IEEE Transactions on Acoustics, Speech, and Signal Processing*, vol. 24, no. 3, pp. 264–266, June 1976.

[50] S. Winograd, "On computing the discrete Fourier transform," *Mathematics of Computation*, vol. 32, pp. 175–199, January 1978.

[51] C. Rader, "Discrete Fourier transforms when the number of data samples is prime," *Proceedings of the IEEE*, vol. 56, no. 6, pp. 1107–1108, June 1968.

[52] M. Heideman, D. Johnson, and C. S. Burrus, "Gauss and the history of the fast Fourier transform," *IEEE Acoustics, Speech, and Signal Processing (ASSP) Magazine*, vol. 1, no. 4, pp. 14–21, January 1984.

[53] f. harris, *Multirate Signal Processing for Communication Systems*, Prentice Hall, Upper Saddle River, NJ, 2004.

[54] P. Vaidyanathan, *Multirate Systems and Filter Banks*, PTR Prentice Hall, Upper Saddle River, NJ, 1993.

[55] J. Proakis, C. Rader, F. Ling, and C. Nikias, *Advanced Digital Signal Processing*, Macmillan, New York, 1992.

[56] D. Humpherys, *The Analysis, Design, and Synthesis of Electrical Filters*, Prentice Hall, Englewood Cliffs, NJ, 1970.

[57] f. harris, "On the use of windows for harmonic analysis with the discrete Fourier transform," *Proceedings of the IEEE*, vol. 66, pp. 51–84, January 1978.

[58] T. Parks and J. McClellan, "Chebyshev approximation for nonrecursive digital filters with linear phase," *IEEE Transactions on Circuit Theory*, vol. 19, no. 2, pp. 189–194, March 1972.

[59] J. McClellan and T. Parks, "A unified approach to the design of optimum FIR linear-phase digital filters," *IEEE Transaction on Circuit Theory*, vol. 20, no. 6, pp. 697–701, November 1973.

[60] T. Parks and J. McClellan, "A program for the design of linear phase finite impulse response digital filters," *IEEE Transactions on Audio and Electroacoustics*, vol. 20, no. 3, pp. 195–199, August 1972.

[61] J. McClellan, T. Parks, and L. Rabiner, "A computer program for designing optimum FIR linear phase digital filters," *IEEE Transactions on Audio and Electroacoustics*, vol. 21, no. 6, pp. 506–526, December 1973.

[62] O. Herrmann, L. Rabiner, and D. Chan, "Practical design rules for optimum finite impulse response lowpass digital filters," *Bell System Technical Journal*, vol. 52, no. 6, pp. 769–799, July–August 1973.

[63] J. Kaiser, "Nonrecursive digital filter design using the i_0-sinh window function," *Proceedings of the IEEE International Symposium on Circuits and Systems*, San Francisco, CA, April 1974, pp. 20–23.

[64] J. Kaiser, "Digital filters," in *System Analysis by Digital Computer*, F. Kuo and J. Kaiser, editors, John Wiley & Sons, New York, 1966.

[65] R. Hogg and E. Tanis, *Probability and Statistical Inference*, Macmillan, New York, 1983.

[66] S. Ross, *A First Course in Probability*, Prentice Hall, Upper Saddle River, NJ, 2002.

[67] P. Beckmann, *Probability in Communication Engineering*, Harcourt, Brace and World, Inc., New York, 1967.

[68] W. Davenport and W. Root, *An Introduction to the Theory of Random Signals and Noise*, IEEE Press, New York, 1987.

[69] R. Gray and L. Davisson, *An Introduction to Statistical Signal Processing*, Cambridge University Press, Cambridge, United Kingdom, 2004.

[70] C. Helstrom, *Probability and Stochastic Processes for Engineers*, MacMillan, New York, 1991.

[71] A. Leon-Garcia, *Probability and Random Processes for Electrical Engineering*, Addison-Wesley, Reading, MA, 1994.

[72] A. Papoulis, *Probability, Random Variables, and Stochastic Processes*, McGraw-Hill, New York, 1991.

[73] H. Stark and J. Woods, *Probability and Random Processes with Applications to Signal Processing*, Prentice Hall, Upper Saddle River, NJ, 2002.

[74] R. Ziemer, *Elements of Engineering Probability and Statistics*, Prentice Hall, Upper Saddle River, NJ, 1997.

[75] H. Tucker, *A Graduate Course in Probability*, Academic Press, New York, 1967.

[76] A. Karr, *Probability*, Springer-Verlag, New York, 1993.

[77] E. Wong, *Stochastic Processes in Information and Dynamical Systems*, McGraw-Hill, New York, 1971.

[78] R. Gray and L. Davisson, *Random Processes: A Mathematical Approach for Engineers*, Prentice Hall, Englewood Cliffs, NJ, 1986.

[79] ETSI Secretariat, "Digital video broadcasting (DVB); second generation framing structure, channel coding and modulation systems for broadcasting, interactive services, news gathering and other broadband satellite applications," Technical Report. ETSI EN 302 307 V1.1.1, European Telecommunications Standards Institute, France, 2004.

[80] IEEE, "International standard ISO/IEC 8802-11:1999/Amd 1:2000(E), IEEE Std 802.11a-1999," 2000.

[81] J. Wozencraft and I. Jacobs, *Principles of Communication Engineering*, Waveland Press, Prospect Heights, IL, Reprint Edition (June 1990), 1965.

[82] D. North, "An analysis of the factors which determine signal/noise discrimination in pulse-carrier systems," Tech. Rep., No. 6 PTR-6C, RCA, 1943.

[83] W. Davenport and W. Root, *An Introduction to the Theory of Random Signals and Noise*, Wiley-IEEE Press, Piscataway, NJ, Reprint (October 1987), 1958.

[84] C. Helstrom, *Statistical Theory of Signal Detection*, Pergamon, New York, 1960.

[85] H. Van Trees, *Detection, Estimation, and Modulation Theory, Part I*, John Wiley & Sons, New York, 1968.

[86] H. Meyr, M. Oerder, and A. Polydoros, "On samping rate, analog prefiltering, and sufficient statistics for digital receivers," *IEEE Transactions on Communications*, vol. 42, no. 12, pp. 3208–3214, December 1994.

[87] C. Campopiano and B. Glazer, "A coherent digital amplitude and phase modulation scheme," *IRE Transactions on Communication Systems*, vol. CS-10, pp. 90–95, March 1962.

[88] R. Lucky and J. Hancock, "On the optimum performance of N-ary systems having two degrees of freedom," *IRE Transactions on Communication Systems*, vol. 10, pp. 185–192, June 1962.

[89] C. Weber, "New solutions to the signal design problem for coherent channels," *IEEE Transactions on Information Theory*, vol. 12, pp. 161–167, April 1966.

[90] J. Salz et al., "Data transmission by combined AM and PM," *Bell System Technical Journal*, vol. 50, pp. 2399–2419, September 1971.

[91] C. Thomas, M. Weidner, and S. Durrani, "Digital amplitude-phase keying with M-ary alphabets," *IEEE Transactions on Communications*, vol. 22, no. 2, pp. 168–180, February 1974.

[92] K. Kawai et al., "Optimum combinations of amplitude and phase modulation scheme and its application to data transmission modem," *Proceedings of the IEEE International Conference on Communications*, June 1972, pp. 29.6–29.11.

[93] G. Foschini, R. Gitlin, and S. Weinstein, "Optimization of two-dimensional signal constellations in the presence of Gaussian noise," *IEEE Transactions on Communications*, vol. 22, no. 1, pp. 28–38, January 1974.

[94] N. Blachman, "A comparison of the informational capacities of amplitude- and phase-modulation communication systems," *Proceedings of the IRE*, vol. 41, pp. 748–759, June 1953.

[95] E. Sousa and S. Pasupathy, "Pulse shape design for teletext data transmission," *IEEE Transactions on Communications*, vol. 31, no. 7, pp. 871–878, July 1983.
[96] B. Saltzberg, "Performance of an efficient parallel data transmission system," *IEEE Transactions on Communication Technology*, vol. 15, no. 6, pp. 805–811, December 1967.
[97] R. Chang and R. Gibbey, "A theoretical study of performance of an orthogonal multiplexing data transmission scheme," *IEEE Transactions on Communication Technology*, vol. 16, no. 4, pp. 529–540, August 1968.
[98] S. Weinstein and P. Ebert, "Data transmission by frequency-division multiplexing using the discrete Fourier transform," *IEEE Transactions on Communication Technology*, vol. 19, no. 5, pp. 628–634, October 1971.
[99] B. Hirosaki, "An orthogonally multiplexed QAM system using the discrete Fourier transform," *IEEE Transactions on Communication Technology*, vol. 29, no. 7, pp. 982–989, July 1981.
[100] B. Hirosaki, S. Hasegawa, and A. Sabato, "Advanced group-band modem using orthogonally multiplexed QAM techniques," *IEEE Transactions on Communications*, vol. 34, pp. 587–592, June 1986.
[101] L. Cimini, "Analysis and simulation of a digital mobile channel using orthogonal frequency division multiplexing," *IEEE Transactions on Communications*, vol. 33, no. 7, pp. 665–675, July 1985.
[102] H. Sari, G. Karam, and J. Jeanclaude, "Transmission techniques for digital terrestrial TV broadcasting," *IEEE Communications Magazine*, vol. 33, no. 2, pp. 100–109, February 1995.
[103] J. Chow, J. Tu, and J. Cioffi, "A discrete multitone transceiver system for HDSL applications," *IEEE Journal on Selected Areas in Communications*, vol. 9, pp. 895–908, August 1991.
[104] I. Kalet, "The multitone channel," *IEEE Transactions on Communications*, vol. 37, no. 2, pp. 119–124, February 1989.
[105] J. Bingham, "Multicarrier modulation for data transmission: An idea whose time has come," *IEEE Communications Magazine*, vol. 28, no. 5, pp. 5–14, May 1990.
[106] R. Prasad, *OFDM for Wireless Communication Systems*, Artech House, Boston, MA, 2004.
[107] A. Bahai, B. Saltzberg, and M. Egen, *Multi-Carrier Digital Communications: Theory and Applications of OFDM*, Springer, New York, 2004.
[108] L. Hanzo and T. Keller, *OFDM and MC-CDMA: A Primer*, John Wiley & Sons, Hoboken, NJ, 2006.
[109] K. Murota and K. Hirade, "GMSK modulation for digital mobile radio telephony," *IEEE Transactions on Communications*, vol. 29, no. 7, pp. 1044–1050, July 1981.
[110] V. Johannes, "Improving on bit error rate," *IEEE Communications Magazine*, vol. 22, no. 12, pp. 18–20, December 1984.
[111] C. Waff, "The road to the deep space network," *IEEE Spectrum Magazine*, vol. 30, no. 4, pp. 50–57, April 1993.
[112] M. Hata, "Empirical formula for propagation loss in mobile radio services," *IEEE Transactions on Vehicular Technology*, vol. 29, no. 1, pp. 317–325, January 1980.
[113] B. Sklar, "A structured overview of digital communications—A tutorial review—part I," *IEEE Communications Magazine*, vol. 21, no. 4, pp. 4–17, August 1983.
[114] B. Sklar, "A structured overview of digital communications—a tutorial review—part II," *IEEE Communications Magazine*, vol. 21, no. 7, pp. 4–17, October 1983.
[115] D. Pozar, *Microwave Engineering*, John Wiley & Sons, New York, 2004.
[116] G. Gonzalez, *Microwave Transistor Amplifiers: Analysis and Design*, Prentice Hall, Upper Saddle River, NJ, 1997.

[117] U. Rohde and D. Newkirk, *RF/Microwave Circuit Design for Wireless Applications*, John Wiley & Sons, New York, 2000.

[118] T. Lee, *The Design of CMOS Radio-Frequency Integrated Circuits*, Cambridge University Press, Cambridge, UK, 1998.

[119] R. Gagliardi, *Satellite Communications*, Van Nostrand Reinhold, New York, 1991.

[120] T. Pratt and C. Bostian, *Satellite Communications*, John Wiley & Sons, New York, 1986.

[121] W. Pritchard, H. Suyderhoud, and R. Nelson, *Satellite Communication Systems Engineering*, Prentice Hall, Englewood Cliffs, NJ, 1993.

[122] M. Richharia, *Satellite Communications Systems*, McGraw-Hill, New York, 1995.

[123] U. Mengali and A. D'Andrea, *Synchronization Techniques for Digital Receivers*, Plenum Press, New York, 1997.

[124] J. Proakis, P. Drouilhet, and R. Price, "Performance of coherent detection systems using decision-directed channel measurement," *IEEE Transactions on Communication Systems*, vol. 12, pp. 54–63, March 1964.

[125] A. Viterbi, *Principles of Coherent Communications*, McGraw-Hill, New York, 1966.

[126] F. Natali and W. Walbesser, "Phase-locked loop detection of binary PSK signals utilizing decision feedback," *IEEE Transactions on Aerospace and Electronic Systems*, vol. 5, pp. 83–90, January 1969.

[127] F. Gardner, *Phaselock Techniques*, John Wiley & Sons, New York, 1979.

[128] W. Lindsey and M. Simon, *Telemcommunication Systems Engineering*, Prentice Hall, Englewood Cliffs, NJ, 1973.

[129] H. Kobayashi, "Simultaneous adaptive estimation and decision algorithm for carrier modulated data transmission systems," *IEEE Transactions on Communications*, vol. 19, pp. 268–280, June 1971.

[130] W. Lindsey, *Sychronization Systems in Communication and Control*, Prentice Hall, Englewood Cliffs, NJ, 1972.

[131] M. Simon and J. Smith, "Carrier synchronization and detection of QASK signal sets," *IEEE Transactions on Communications*, vol. 22, pp. 98–106, February 1974.

[132] D. Falconer, "Jointly adaptive equalization and carrier recovery in two-dimensional digital communication systems," *Bell Systems Technical Journal*, vol. 55, pp. 317–334, March 1976.

[133] U. Mengali, "Joint phase and timing acquisition in data transmission," *IEEE Transactions on Communications*, vol. 25, pp. 1174–1185, October 1977.

[134] D. Falconer and J. Salz, "Optimal reception of digital data over the Gaussian channel with unknown delay and phase jitter," *IEEE Transactions on Information Theory*, vol. 23, pp. 117–126, January 1977.

[135] M. Simon, "Optimum receiver structures for phase-multiplexed modulations," *IEEE Transactions on Communications*, vol. 26, pp. 865–872, June 1978.

[136] M. Meyers and L. Franks, "Joint carrier and symbol timing recovery for PAM systems," *IEEE Transactions on Communications*, vol. 28, pp. 1121–1129, August 1980.

[137] A. Leclerc and P. Vandamme, "Universal carrier recovery loop for QASK and PSK signal sets," *IEEE Transactions on Communications*, vol. 31, pp. 130–136, January 1983.

[138] S. Moridi and H. Sari, "Analysis of four decision feedback carrier recovery loops in the presence of intersymbol interference," *IEEE Transactions on Communications*, vol. 33, pp. 543–550, June 1985.

[139] L. Franks, "Carrier and bit synchronization in data communications—A tutorial review," *IEEE Transactions on Communications*, vol. 28, pp. 1107–1120, August 1980.

[140] W. Weber, "Differential encoding for multiple amplitude and phase shift keying systems," *IEEE Transactions on Communications*, vol. 26, no. 3, pp. 385–391, March 1978.

[141] Range Commanders Council Telemetry Group, Secretariat, Range Commanders Council, U.S. Army White Sands Missile Range, NM, *IRIG Standard 106-04*, 2004.

[142] K. Feher, *Digital Communications: Satellite/Earth Station Engineering*, Prentice Hall, Upper Saddle River, NJ, 1983.

[143] M. Rice, "Differential encoding revealed: An explanation of the Tier-1 differential encoding in IRIG-106," *Proceedings of the International Telemetering Conference*, Las Vegas, NV, October 2007.

[144] A. J. Viterbi and A. M. Viterbi, "Nonlinear estimation of PSK-modulated carrier phase with application to burst digital transmission," *IEEE Transactions on Information Theory*, vol. 29, pp. 543–551, July 1983.

[145] M. Moeneclaey and G. de Jonghe, "ML-oriented NDA carrier synchronization for generally rotationally symmetric signal constellations," *IEEE Transactions on Communications*, vol. 42, pp. 2531–2533, August 1994.

[146] J. Frazier and J. Page, "Phase-lock loop frequency, acquisition study," *IRE Transactions on Space Electronics and Telemetry*, vol. 8, pp. 210–227, September 1962.

[147] D. Messerschmitt, "Frequency detectors for PLL acquisition in timing and carrier recovery," *IEEE Transactions on Communications*, vol. 27, no. 9, pp. 1288–1295, September 1979.

[148] U. Mengali, "Acquisition behaviour of tracking loops operating in frequency search modes," *IEEE Transactions on Aerospace and Electronic Systems*, vol. 10, pp. 583–587, September 1974.

[149] C. Cahn, "Improving frequency acquisition of a Costas loop," *IEEE Transactions on Communications*, vol. 25, pp. 1453–1459, December 1977.

[150] F. Natali, "AFC tracking algorithms," *IEEE Transactions on Communications*, vol. 32, no. 8, pp. 935–947, August 1984.

[151] F. Natali, "Noise performance of a cross-product AFC with decision feedback for DPSK signals," *IEEE Transactions on Communications*, vol. 34, no. 3, pp. 303–307, March 1986.

[152] F. Gardner, "Properties of frequency difference detectors," *IEEE Transactions on Communications*, vol. 33, no. 2, pp. 131–138, February 1985.

[153] H. Sari and S. Moridi, "New phase and frequency detectors for carrier recovery in PSK and QAM systems," *IEEE Transactions on Communications*, vol. 36, pp. 1035–1043, September 1988.

[154] H. Meyr and G. Ascheid, *Synchronization in Digital Communications*, vol. 1, John Wiley & Sons, New York, 1990.

[155] A. D'Andrea and U. Mengali, "Noise performance of two frequency-error detectors derived from maximum likelihood estimation methods," *IEEE Transactions on Communications*, vol. 42, pp. 793–802, February/March/April 1994.

[156] G. Karam, F. Daffara, and H. Sari, "Simplified versions of the maximum likelihood frequency detector," *Proceedings of the IEEE Global Communications Conference*, Orlando, FL, December 6–9, 1992.

[157] S. Kay, "A fast and accurate single frequency estimator," *IEEE Transactions on Acoustics, Speech, and Signal Processing*, vol. 37, pp. 1987–1990, December 1989.

[158] M. Fitz, "Planar filtered techniques for burst mode carrier synchronization," *Proceedings of the IEEE Global Communications Conference*, Phoenix, AZ, December 2–5, 1991.

[159] M. Fitz, "Further results in the fast estimation of a single frequency," *IEEE Transactions on Communications*, vol. 42, pp. 862–864, March 1994.

[160] M. Luise and R. Reggiannini, "Carrier frequency recovery in all-digital modems for burst-mode transmissions," *IEEE Transactions on Communications*, vol. 1995, pp. 1169–1178, February/March/April 1995.

[161] J. Chuang and N. Sollenberger, "Burst coherent demodulation with combined symbol timing, frequency offset estimation, and diversity selection," *IEEE Transactions on Communications*, vol. 39, pp. 1157–1164, July 1991.

[162] F. Classen, H. Meyr, and P. Sehier, "Maximum likelihood open loop carrier synchronizer," *Proceedings of the IEEE International Conference on Communications*, Geneva, Switzerland, May 23–26, 1993.

[163] T. Vaidis and A. Polydoros, "Effects of large frequency offset in digital receivers and related algorithms," *Proceedings of the IEEE Global Communications Conference*, San Antonio, TX, November 25–29, 2001, pp. 1349–1355.

[164] L. Erup, F. Gardner, and R. Harris, "Interpolation in digital modems—Part II: Implementation and performance," *IEEE Transactions on Communications*, vol. 41, no. 6, pp. 998–1008, June 1993.

[165] F. Gardner, "Demodulator reference recovery techniques suited for digital implementation," Final report, European Space Agency, August 1988, ESTEC Contract No. 6847/86/NL/DG.

[166] J. Bergmans and H. Wong-Lam, "A class of data-aided timing recovery schemes," *IEEE Transactions on Communications*, vol. 43, pp. 1819–1827, February/March/April 1995.

[167] F. Gardner, "A BPSK/QPSK timing error detector for sampled receivers," *IEEE Transactions on Communication*, vol. 34, pp. 423–429, May 1986.

[168] K. Mueller and M. Müller, "Timing recovery in digital synchronous data receivers," *IEEE Transactions on Communications*, vol. 24, pp. 516–531, May 1976.

[169] R. Shafer and L. Rabiner, "A digital signal processing approach to interpolation," *Proceedings of the IEEE*, vol. 61, pp. 692–702, June 1973.

[170] F. Gardner, "Interpolation in digital modems—Part I: Fundamentals," *IEEE Transactions on Communications*, vol. 41, no. 3, pp. 501–507, March 1993.

[171] f. harris and M. Rice, "Multirate digital filters for symbol timing synchronization in software radios," *IEEE Journal on Selected Areas in Communications*, vol. 19, no. 12, pp. 2346–2357, December 2001.

[172] J. Bingham, *The Theory and Practice of Modem Design*, John Wiley & Sons, New York, 1983.

[173] X. Qin, H. Wang, L. Zeng, and F. Xiong, "An all digital clock smoothing technique—Counting-prognostication," *IEEE Transactions on Communications*, vol. 51, no. 2, pp. 166–169, February 2003.

[174] W. Waggener, *Pulse Code Modulation Systems Design*, Artech House, Boston, MA, 1999.

[175] J. Peek, "Communications aspects of the compact disc digital audio system," *IEEE Communications Magazine*, vol. 23, no. 2, pp. 7–15, February 1985.

[176] L. Franks and J. Bubrouski, "Statistical properties of timing jitter in PAM recovery schemes," *IEEE Transactions on Communications*, vol. 22, pp. 913–920, July 1974.

[177] M. Moeneclaey, "Comparisons of two types of symbol synchronizers for which self noise is absent," *IEEE Transactions on Communications*, vol. 31, pp. 329–334, March 1983.

[178] A. D'Andrea, U. Mengali, and M. Moro, "Nearly optimum prefiltering in clock recovery," *IEEE Transactions on Communications*, vol. 34, pp. 1081–1088, November 1986.

[179] A. D'Andrea and M. Luise, "Design and analysis of jitter-free clock recovery scheme for QAM systems," *IEEE Transactions on Communications*, vol. 41, pp. 1296–1299, September 1993.

[180] A. D'Andrea and M. Luise, "Optimization of symbol timing recovery for QAM data demodulators," *IEEE Transactions on Communications*, vol. 44, pp. 299–406, March 1996.

[181] W. Cowley, "The performance of two symbol timing recovery algorithms for PSK demodulators," *IEEE Transactions on Communications*, vol. 42, pp. 2345–2355, June 1994.

[182] C. Dick and F. Harris, "On the structure, performance, and applications of recursive all-pass filters with adjustable and linear group delay," *Proceedings of the IEEE International Conference on Acoustics, Speech, and Signal Processing*, Orlando, FL, May 14–17, 2002, pp. 1517–1520.

[183] f. harris, "On the structure, performance, and application to modem timing recovery of recursive all-pass filters with adjustable equal-ripple group-delay," *Proceedings of the IEEE Asilomar*

Conference on Signals, Systems, and Computers, Pacific Grove, CA, November 4–7 2001, pp. 592–596.

[184] M. Oerder and H. Meyr, "Digital filter and square timing recovery," *IEEE Transactions on Communications*, vol. 36, pp. 605–612, May 1988.

[185] f. harris, "Implementing waveform shaping filters to pre-equalize gain and phase distortion of the analog signal processing path in DSP based modems," *Proceedings of the IEEE Military Communications Conference*, October 2–5, 1994, pp. 633–638.

[186] H. Nicholas, H. Samueli, and B. Kim, "The optimization of direct digital frequency synthesizer performance in the presence of finite word length effects," *Proceedings of the 42nd Annual Frequency Control Symposium*, 1988, pp. 357–363.

[187] J. Bermudez, R. Seara, and S. Filho, "Correction of the $(\sin x)/x$ distortion in discrete-time/continous-time signal conversions," *Electronics Letters*, vol. 24, pp. 1559–1560, December 1988.

[188] S. Filho, R. Seara, and J. Bermudez, "A new method for the compensation of the $(\sin x)/x$ distortion in discrete-time to continous-time signal conversions," *Proceedings of the IEEE International Symposium on Circuits and Systems*, May 1989, pp. 1668–1671.

[189] J. Bermudez, S. Filho, R. Seara, and J. Mayer, "A new improved iterative method for the compensation of the $(\sin x)/x$ frequency response distortion," *Proceedings of the IEEE International Symposium on Circuits and Systems*, May 1990, pp. 2793–2796.

[190] R. Seara, S. Filho, J. Bermudez, and J. Mayer, "On the compensation of the $(\sin x)/x$ distortion in discrete-time to continuous-time signal conversions," *IEEE Transactions on Circuits and Systems—I: Fundamental Theory and Applications*, vol. 42, no. 6, pp. 343–351, June 1995.

[191] H. Samueli, "The design of multiplierless FIR filters for compensating D/A converter frequency response distortion," *IEEE Transactions on Circuits and Systems*, vol. 35, no. 8, pp. 1064–1066, August 1988.

[192] T. Lin and H. Samueli, "A CMOS bit-level pipelined implementation of an FIR $x/\sin(x)$ predistortion digital filter," *Proceedings of the IEEE International Symposium on Circuits and Systems*, May 8–11, 1989, pp. 351–354.

[193] T. Lin and H. Samueli, "A 200-MHz CMOS $x/\sin(x)$ digital filter for compensating D/A converter frequency response distortion in high-speed communication systems," *Proceedings of the IEEE Global Communications Conference*, December 2–5, 1990, pp. 1722–1726.

[194] H. Samueli and B. Wong, "A VLSI architecture for a high-speed all-digital quadrature modulator and demodulator for digital radio applications," *IEEE Journal on Selected Areas in Communications*, vol. 8, no. 8, pp. 1512–1519, October 1990.

[195] B. Henriques and J. Franca, "High-speed D/A conversion with linear phase $\sin x/x$ compensation," *Proceedings of the IEEE International Symposium on Circuits and Systems*, May 3–6, 1993, pp. 1204–1207.

[196] J. Tierney, C. Rader, and B. Gold, "A digital frequency synthesizer," *IEEE Transactions on Audio Electroacoustics*, vol. 19, no. 1, pp. 48–57, March 1971.

[197] G. Sunderland, R. Strauch, S. Wharfield, H. Peterson, and C. Cole, "CMOS/SOS frequency synthesizer LSI circuit for spread spectrum communications," *IEEE Journal of Solid-State Circuits*, August 1984, pp. 497–506.

[198] S. Mehrgardt, "Noise spectra of digital sine-generators using the table-lookup method," *IEEE Transactions on Acoustics, Speech, and Signal Processing*, vol. 31, no. 4, pp. 1037–1039, August 1983.

[199] H. Nicholas and H. Samueli, "An analysis of the output spectrum of direct digital frequency synthesizers in the presence of phase-accumulator truncation," *Proceedings of the 41st Annual Frequency Control Symposium*, 1987, pp. 495–502.

[200] V. Kroupa, "Spectral properties of DDFS: Computer simulations and experimental verifications," in *Direct Digital Frequency Synthesizers*, V. Kroupa, editor, pp. 152–164. IEEE Press, Piscataway, NJ, 1999.

[201] J. Garvey and D. Babitch, "An exact spectral analysis of a number controlled oscillator based synthesizer," *Proceedings of the 44th Annual Frequency Control Symposium*, 1990, pp. 511–521.

[202] S. Jasper, "Frequency resolution in a digital oscillator," U. S. Patent 4 652 832, March 1987.

[203] M. Flanagan and G. Zimmerman, "Spur-reduced digital sinusoidal synthesis," *IEEE Transactions on Communications*, vol. 43, no. 7, pp. 2254–2262, July 1995.

[204] f. harris and B. McKnight, "Error feedback loop linearizes direct digital synthesizers," *Proceedings of the 29th Asilomar Conference on Signals, Systems, and Computers*, October 30–November 2, 1995, vol. 1, pp. 98–102.

[205] E. Hogenauer, "An economical class of digital filters for decimation and interpolation," *IEEE Transactions on Acoustics, Speech, and Signal Processing*, vol. 29, no. 2, pp. 155–162, April 1981.

[206] S. Chu and C. Burrus, "Multirate filter designs using comb filters," *IEEE Transactions on Circuits and Systems*, vol. 31, no. 11, pp. 913–924, November 1984.

[207] P. Vaidyanathan and T. Ngyuen, "A trick for the design of FIR half-band filters," *IEEE Transactions on Circuits and Systems–II*, vol. 34, pp. 297–300, March 1987.

[208] J. Proakis and D. Manolakis, *Digital Signal Processing: Principles, Algorithms, and Applications*, 4th edition, Prentice Hall, Upper Saddle River, NJ, 2006.

[209] T. Ramstad, "Digital methods for conversion between arbitrary sampling frequencies," *IEEE Transactions on Acoustics, Speech, and Signal Processing*, vol. 32, no. 3, pp. 577–591, June 1984.

[210] J. Volder, "The CORDIC trigonometric computing technique," *IRE Transactions on Electronic Computers*, vol. 8, no. 3, pp. 330–334, September 1959.

[211] J. Walther, "A unified algorithm for elementary functions," *Proceedings of the AFIS Spring Joint Computer Conference*. vol. 38, pp. 279–385, American Federation of Information Processing Societies, Inc., 1971.

[212] H. Ahmed, J.-M. Delosme, and M. Morf, "Highly concurrent computing structures for matrix arithmetic and signal processing," *Computer*, vol. 15, no. 1, pp. 65–82, January 1982.

[213] J.-M. Muller, "Discrete-basis and computation of elementary functions," *IEEE Transactions on Computers*, vol. 34, no. 9, pp. 857–862, September 1985.

[214] C. Schelin, "Calculator function approximation," *American Mathematical Monthly*, vol. 90, no. 5, May 1983.

[215] J. Bajard, S. Kla, and J. Muller, "BKM: A new hardware algorithm for complex elementary functions," *IEEE Transactionson Computers*, vol. 43, no. 8, pp. 955–963, August 1994.

[216] B. Oliver, "Automatic volume control as a feedback problem," *Proceedings of the IRE*, vol. 36, no. 4, pp. 466–473, April 1948.

[217] W. Victor and M. Brockman, "The application of linear servo theory to the design of AGC loops," *Proceedings of the IRE*, vol. 48, no. 2, pp. 234–238, February 1960.

[218] E. Banta, "Analysis of automatic gain control," *IEEE Transactions on Automatic Control*, vol. 9, no. 2, pp. 181–182, April 1964.

[219] R. Simpson and W. Tranter, "Baseband AGC in an AM-FM telemetry system," *IEEE Transactions on Communication Technology*, vol. 18, no. 1, pp. 59–63, February 1970.

[220] J. Ohlson, "Exact dynamics of automatic gain control," *IEEE Transactions on Communications*, vol. 22, no. 1, pp. 72–75, Jaunary 1974.

[221] C. Mercy, "A review of automatic gain control theory," *The Radio and Electronic Engineer*, vol. 51, no. 11/12, pp. 579–590, November/December 1981.

[222] H. Schachter and L. Bergstein, "Noise analysis of an automatic gain control system," *IEEE Transactions on Automatic Control*, vol. 9, no. 3, pp. 249–255, July 1964.

[223] H. Meyr, *Synchronization in Digital Systems*, vol. 1: Phase-,Frequency-Locked Loops, and Amplitude Control, Wiley Interscience, New York, 1990.
[224] D. Morgan, "On discrete-time AGC amplifiers," *IEEE Transactions on Circuits and Systems*, vol. 22, no. 2, pp. 135–146, February 1975.
[225] W. Weber III, "Decision-directed automatic gain control for MAPSK systems," *IEEE Transactions on Communications*, vol. 22, no. 2, pp. 135–146, February 1975.
[226] C. Farrow, "A continuously variable digital delay element," *Proceedings of the IEEE International Symposium on Circuits and Systems*, Espoo, Finland, June 8–9, 1988, pp. 2641–2645.
[227] f. harris, M. d'Oreye de Lantremange, and A. Constantinides, "Design and implementation of efficient resampling filters using polyphase recursive all-pass filters," *Proceedings of the Asilomar Conference on Signals, Systems, and Computers*, Pacific Grove, CA, November 4–6, 1991, pp. 1031–1036.
[228] f. harris, I. Gurantz, and S. Tzukerman, "Digital T/2 Nyquist filtering using recursive all-pass two-stage resampling filters for a wide range of selectable signalling rates," *Proceedings of the Asilomar Conference on Signals, Systems, and Computers*, Pacific Grove, CA, October 26–28 1992, pp. 676–680.
[229] J. Mitola, "The software radio architecture," *IEEE Communications Magazine*, vol. 33, pp. 26–38, May 1995.
[230] F. Takahata, M. Yasunaga, Y. Jirata, T. Ohsawa, and J. Namiki, "A PSK group modem for satellite communications," *IEEE Journal on Selected Areas in Communications*, vol. 5, pp. 648–661, May 1987.
[231] J. Chamberlin, C. Hester, J. Meyers, T. Mock, F. Moody, R. Simons, E. Bahm, and J. Ritchie, "Design and field test of a 256-QAM DIV modem," *IEEE Journal on Selected Areas in Communications*, vol. 5, no. 3, pp. 349–356, April 1987.
[232] Y. Nakamura, Y. Saito, and S. Aikawa, "256 QAM modem for multicarrier 400 Mbit/s digital radio," *IEEE Journal on Selected Areas in Communications*, vol. 5, no. 3, pp. 329–335, April 1987.
[233] G. Acheid, M. Oerder, J. Stahl, and H. Meyr, "An all digital receiver for bandwidth efficient transmission at high data rates," *IEEE Transactions on Communications*, vol. 37, pp. 804–813, August 1989.
[234] G. Benelli, A Fioravanti, A. Garzelli, and P. Matteini, "Some digital receivers for the GSM pan-European cellular communication system," *IEE Proceedings—Communications*, vol. 141, no. 3, pp. 168–176, June 1994.
[235] H. Meyr and R. Subramanian, "Advanced digital receiver principles and technologies for PCS," *IEEE Communications Magazine*, vol. 23, no. 1, pp. 68–78, January 1995.
[236] W. Song, "A new 3-GSPS 65-GOPS UHF digital radar receiver and its performance characteristics," *Proceedings of the IEEE Asilomar Conference on Signals, Systems & Computers*, Pacific Grove, CA, November 2–5, 1997, pp. 1542–1546.
[237] H. Meyr, M. Moeneclaey, and S. Fechtel, *Digital Communication Receivers*, John Wiley & Sons, New York, 1998.
[238] R. Makowitz, A. Turner, J. Gledhill, and M. Mayr, "A single-chip DVB-T receiver," *IEEE Transactions on Consumer Electronics*, vol. 3, no. 44, pp. 990–993, August 1998.
[239] S. Wenjun and E. Sanchez-Sinencio, "Next generation wideband multi-standard digital receiver design," *Proceedings of the IEEE Midwest Symposium on Circuits and Systems*, Lancing, MI, August 8–11, 2000, pp. 424–427.
[240] C. Dodley and R. Erving, "In-building software radio architecture, design and analysis," *Proceedings of the IEEE International Symposium on Personal, Indoor and Mobile Radio Communication*, London, UK, September 18–21, 2000, pp. 479–483.

[241] R. Makowitz, M. Mayr, C. Patzelt, and D. Hoheisel, "First IF sampling techniques for DVB-T receivers," in *Proceedings of the IEEE International Caracas Conference on Devices, Circuits and Systems*, Cancún, México, March 15–17, 2000, pp. T89/1–T89/4.

[242] J. Gagne, J. Gauthier, K. Wu, and R. Bosisio, "High-speed low-cost direct conversion digital receiver," *IEEE International Microwave Symposium Digest*, Phoenix, AZ, May 20–25, 2001, pp. 1093–1096.

[243] J. Thor and D. Akos, "A direct RF sampling multifrequency GPS receiver," *Proceedings of the IEEE Position Location and Navigation Symposium*, Palm Springs, CA, April 15–18, 2002, pp. 44–51.

[244] f. harris, C. Dick, and M. Rice, "Digital receivers and transmitters using polyphase filter banks for wireless communications," *IEEE Transactions on Microwave Theory and Techniques*, vol. 51, no. 4, pp. 1395–1412, April 2003.

[245] D. Bruckmann, "Optimized digital receiver for flexible wireless terminals," *Proceedings of the IEEE Workshop on Signal Processing Advances in Wireless Communications*, Rome, Italy, June 15–18, 2003, pp. 649–652.

[246] C. Chen, K. George, W. McCormick, and J. Tsui, "Design and measurement of 2.5 Gsps digital receiver," *Proceedings of the IEEE Instrumentation and Measurement Technology Conference*, Vail, CO, May 20–22, 2003, pp. 258–263.

[247] S. Levantino, C. Samori, M. Banu, J. Glas, and V. Boccuzzi, "A CMOS GSM IF-sampling circuit with reduced in-channel aliasing," *IEEE Journal of Solid-State Circuits*, vol. 98, no. 6, pp. 150–141, June 2003.

[248] L. Zhou, S. Shetty, R. Spring, H. Ariak, W. Zheng, J. Hyun, M. Tofighi, and A. Daryoush, "IC based broadband digital receiver for 4G wireless communications," *Proceedings of the IEEE Radio and Wireless Conference*, Atlanta, GA, September 19–22, 2004, pp. 339–342.

[249] C. Haskins and W. Millard, "X-band digital receiver for the New Horizons spacecraft," *Proceedings of the IEEE Aerospace Conference*, Big Sky, MT, March 6–13, 2004, pp. 1479–1488.

[250] N. Scolari and C. Enz, "Digital receiver architectures for the IEEE 802.15.4 standard," *Proceedings of the IEEE International Symposium on Circuits and Systems*, Vancouver, British Columbia, May 23–26, 2004, IV-345–IV-348.

[251] A. Kiyono, K. Minseok, K. Ichige, and H. Arai, "Jitter effect on digital downconversion receiver with undersampling scheme," *Proceedings of the IEEE Midwest Symposium on Circuits and Systems*, Hiroshima, Japan, July 25–28, 2004, pp. 677–680.

[252] C. Devries and R. Mason, "A 0.18 μm CMOS 900 MHz receiver front-end using RF Q-enhanced filters," *Proceedings of the IEEE International Symposium on Circuits and Systems*, Vancouver, British Columbia, May 23–26, 2004, pp. 325–328.

[253] H. Pekau, J. Nakaska, J. Kulyk, G. McGibney, and J. Haslett, "SOC design of an IF subsampling terminal for a gigabit wireless LAN with asymmetric equalization," *Proceedings of the IEEE International Workshop on System-on-Chip for Real-Time Applications*, Banff, Alberta, Canada, July 19–21, 2004, pp. 307–313.

[254] D. de Souza, I. Krikidis, L. Naviner, J. Danger, M. de Barros, and B. Neto, "Implementation of a digital receiver for DS-CDMA communication system using HW/SW codesign," *Proceedings of the IEEE Midwest Symposium on Circuits and Systems*, Cincinnati, OH, August 7–10, 2005, pp. 587–590.

[255] H. Pekau and J. Haslett, "A comparison of analog front end architectures for digital receivers," *Proceedings of the IEEE Canadian Conference on Electrical and Computer Engineering*, Saskatoon, Saskatchewan, Canada, May 1–4 2005, pp. 1073–1077.

[256] J. Tang, Z. Xu, and B. Sadler, "Digital receiver for TR-UWB systems with inter-pulse interference," *Proceedings for the IEEE Workshop on Signal Processing Advances in Wireless Communications*, New York, June 5–8, 2005, pp. 420–424.

[257] C. Tseng and H. Chien, "Digital quadrature demodulation of multiple RF signals," *Proceedings for the IEEE Workshop on Signal Processing Advances in Wireless Communications*, New York, June 5–8, 2005, pp. 37–41.

[258] H. Wang, Y. Lu, Y. Wan, W. Tang, and C. Wang, "Design of a wideband digital receiver," *Proceedings of the IEEE International Conference on Communications, Circuits and Systems*, Hong Kong, China, May 27–30, 2005, pp. 677–681.

[259] J. Brittain, "Electrical Engineering Hall of Fame—Edwin H. Armstrong," *Proceedings of the IEEE*, vol. 92, no. 3, pp. 575–578, March 2004.

[260] A. Douglas, "Who invented the superheterodyne?," *Proceedings of the Radio Club of America*, vol. 64, no. 3, pp. 123–142, November 1990.

[261] B. Carlson, *Communication Systems*, McGraw-Hill, New York, 1986.

[262] L. Couch, *Digital and Analog Communication Systems*, Prentice-Hall, Upper Saddle River, NJ, 2001.

[263] F. Stremler, *Introduction to Communication Systems*, Addison-Wesley, Reading, MA, 1990.

[264] K. Zangi and R. Koilpillai, "Software radio issues in cellular base stations," *IEEE Journal on Selected Areas in Communications*, vol. 17, no. 4, pp. 561–573, April 1999.

[265] H. Scheuermann and H. Göckler, "A comprehensive survey of digital transmultiplexing methods," *Proceedings of the IEEE*, vol. 69, pp. 1419–1450, November 1981.

[266] S. Mirabbasi and K. Martin, "Classical and modern receiver architectures," *IEEE Communications Magazine*, vol. 38, no. 11, pp. 132–139, November 2000.

[267] B. Razavi, "Design considerations for direct-conversion receivers," *IEEE Transactions on Circuits and Systems—II: Analog and Digital Signal Processing*, vol. 44, no. 6, pp. 428–435, June 1997.

[268] W. Namgoong and T. Meng, "Direct-conversion RF receiver design," *IEEE Transactions on Communications*, vol. 49, no. 3, pp. 518–529, March 2001.

[269] C. Chen and C. Huang, "On the architecture and performance of a hybrid image rejection receiver," *IEEE Journal on Selected Areas in Communications*, vol. 19, no. 6, pp. 1029–1040, June 2001.

[270] f. harris, "Digital filter equalization of analog gain and phase mismatch in I-Q receivers," *Proceedings of the IEEE International Conference on Universal Personal Communications*, Cambridge, MA, September 29–October 2, 1996, pp. 793–796.

[271] F. Aschwanden, "Direct conversion—How to make it work in TV tuners," *IEEE Transactions on Consumer Electronics*, vol. 42, no. 3, pp. 729–738, August 1996.

[272] J. Glas, "Digital I/Q imbalance compensation in a low-IF receiver," *Proceedings of the IEEE Global Communications Conference*, Sydney, Australia, November 8–12, 1998, pp. 1461–1466.

[273] L. Yu and M. Snelgrove, "A novel adaptive mismatch cancellation system for quadrature IF radio receiver," *IEEE Transactions on Circuits and Systems—II: Analog and Digital Signal Processing*, vol. 46, pp. 789–801, June 1999.

[274] M. Valkama, M. Renfors, and V. Koivunen, "Blind signal estimation in conjugate signal models with application to I/Q imbalance compensation," *IEEE Signal Processing Letters*, vol. 12, no. 11, pp. 733–736, November 2005.

[275] M. Valkama, M. Renfors, and V. Koivunen, "Advanced methods for I/Q imbalance compensation in communication receivers," *IEEE Transactions on Signal Processing*, vol. 49, no. 10, pp. 2335–2344, October 2001.

[276] M. Valkama, K. Salminen, and M. Renfors, "Digital I/Q-imbalance compensation in low-IF receivers: Principles and practice," *Proceedings of the IEEE International Conference on Digital Signal Processing*, Santorini, Greece, July 1–3, 2002, pp. 1179–1182.

[277] M. Valkama, M. Renfors, and V. Koivunen, "Compensation of frequency-selective I/Q imbalances in wideband receivers: Models and algorithms," *Proceedings of the IEEE Workshop on Signal Processing Advances in Wireless Communications*, Taoyuan, Taiwan, March 20–23, 2001, pp. 42–45.

[278] M. Valkama and M. Renfors, "Advanced DSP for I/Q-imbalance compensation in a low-IF receiver," *Proceedings of the IEEE International Conference on Communications*, New Orleans, LA, June 18–22, 2000, pp. 768–772.

[279] E. Cetin, I. Kale, and R. Morling, "Adaptive digital receivers for analog front-end mismatch correction," *Proceedings of the IEEE Vehicular Technology Conference*, Atlantic City, NJ, October 7–11, 2001, pp. 2519–2522.

[280] T. Liu, S. Golden, and N. Askar, "A spectral correction algorithm for I-Q channel imbalance problem," *Proceedings of the IEEE Global Communications Conference*, San Antonio, TX, November 25–29, 2001, pp. 334–338.

[281] A. Katz, "Linearization: Reducing distortion in power amplifiers," *IEEE Microwave Magazine*, vol. 2, no. 4, pp. 37–49, December 2001.

[282] P. Kenington, *High Linearity RF Amplifier Design*, Artech House, Norwood, MA, 2000.

[283] Y. Akaiwa and Y. Nagata, "Highly efficient digital mobile communications with a linear modulation method," *IEEE Journal on Selected Areas in Communication*, vol. 5, pp. 890–895, June 1987.

[284] H. Seidel, "A microwave feed-forward experiment," *Bell Systems Technical Journal*, vol. 50, no. 9, pp. 2879–2918, November 1971.

[285] T. Bennett and R. Clements, "Feedforward—An alternative approach to amplifier linearization," *Radiation and Electronic Engineering*, vol. 44, no. 5, pp. 257–262, May 1974.

[286] S. Kumar and G. Wells, "Memory controlled feedforward lineariser suitable for MMIC implementation," *IEE Proceedings—Part H, Microwaves, Antennas and Propagation*, vol. 138, no. 1, pp. 9–12, February 1991.

[287] S. Narahashi and T. Nojima, "Extremely low-distortion multi-carrier amplifier—self adjusting feed-forwarded (SAFF) amplifier," *Proceedings of the IEEE International Conference on Communications*, Denver, CO, June 23–26, 1991, pp. 1485–1490.

[288] J. Cavers, "Adaptation behavior of a feedforward amplifier linearizer," *IEEE Transactions on Vehicular Technology*, vol. 44, no. 1, pp. 31–40, February 1995.

[289] A. Saleh and J. Salz, "Adaptive linearization of power amplifiers in digital radio systems," *Bell Systems Technical Journal*, vol. 62, no. 4, pp. 1019–1033, April 1983.

[290] J. Graboski and R. Davis, "An experimental MQAM MODEM using amplifier linearization and baseband equalization techniques," *Proceedings of the National Communications Conference*, 1982, pp. E3.2.1–E3.2.6.

[291] H. Girard and K. Feher, "A new baseband linearizer for more efficient utilization of earth station amplifiers used for QPSK transmission," *IEEE Journal on Selected Areas in Communications*, vol. 1, no. 1, pp. 46–56, January 1983.

[292] J. Namiki, "An automatically controlled predistorter for multilevel quadrature amplitude modulation," *IEEE Transactions on Communications*, vol. 31, no. 5, pp. 707–712, May 1983.

[293] M. Nannicini, P. Magni, and F. Oggionni, "Temperature controlled predistortion circuits for 64 QAM microwave power amplifiers," *IEEE MTT-S International Microwave Symposium Digest*, St. Louis, MO, vol. 85, pp. 99–102, June 1985.

[294] S. Cheung and A. Aghvami, "Performance of a 16-ary DEQAM modem employing a baseband or RF predistorter over a regenerative satellite link," *IEE Proceedings—Radar and Signal Processing*, vol. 135, no. 6, pp. 547–557, December 1998.

[295] N. Benvenuto, F. Piazza, A. Uncini, and M. Visintin, "Generalised backpropagation algorithm for training a data predistorter with memory in radio systems," *Electronics Letters*, vol. 32, no. 20, pp. 1925–1926, September 1996.

[296] M. Ghaderi, S. Kumar, and D. Dodds, "Fast adaptive polynomial I and Q predistorter with global optimisation," *IEE Proceedings—Communications*, vol. 143, no. 2, p. 78, April 1996.

[297] A. D'Andrea, V. Lottici, and R. Reggiannini, "RF power amplifier linearization through amplitude and phase predistortion," *IEEE Transactions on Communications*, vol. 44, no. 11, pp. 1477–1484, November 1996.

[298] J. Han, T. Chung, and S. Nam, "Adaptive predistorter for power amplifier based on real-time estimation of envelope transfer characteristics," *Electronics Letters*, vol. 35, no. 25, pp. 2167–2168, December 1999.

[299] T. Liu, S. Boumaiza, and F. Ghannouchi, "Augmented Hammerstein predistorter for linearization of broad-band wireless transmitters," *IEEE Transactions on Microwave Theory and Techniques*, vol. 54, no. 4, pp. 1340–1349, April 2006.

[300] K. Lee and P. Gardner, "Adaptive neuro-fuzzy inference systems (ANFIS) digital predistorter for RF power amplifier linearization," *IEEE Transactions on Vehicular Technology*, vol. 55, no. 1, pp. 43–51, January 2006.

[301] H. Lai, "An adaptive procedure on envelope statistics for predistorter designs based on statistical modeling methods," *IEEE Transactions on Circuits and Systems—II: Analog and Digital Signal Processing*, vol. 52, no. 11, pp. 756–760, November 2005.

[302] J. Cavers, "Amplifer linearization using a digital predistorter with fast adaptation and low memory requirements," *IEEE Transactions on Vehicular Technology*, vol. 39, no. 4, pp. 374–382, November 1990.

[303] J. Cavers, "New methods for adaptation of quadrature modulators and demodulators in amplifier linearization circuits," *IEEE Transactions on Vehicular Technology*, vol. 46, no. 3, pp. 706–717, August 1997.

[304] J. Cavers, "The effect of quadrature modulator and demodulator errors on adaptive digital predistorters for amplifier linearization," *IEEE Transactions on Vehicular Technology*, vol. 46, no. 2, pp. 456–466, May 1997.

[305] S. Stapleton, G. Kandola, and J. Cavers, "Simulation and analysis of an adaptive predistorter utilizing a complex spectral convolution," *IEEE Transactions on Vehicular Technology*, vol. 41, no. 4, pp. 387–394, November 1992.

[306] S. Stapleton and F. Costescu, "An adaptive predistorter for a power amplifier based on adjacent channel emissions," *IEEE Transactions on Vehicular Technology*, vol. 41, no. 1, pp. 49–56, February 1992.

[307] C. Eun and E. Powers, "A new Volterra predistorter based on the indirect learning architecture," *IEEE Transactions on Signal Processing*, vol. 45, no. 1, pp. 223–227, January 1997.

[308] J. de Mingo and A. Valdovinos, "Performance of a new digital baseband predistorter using calibration memory," *IEEE Transactions on Vehicular Technology*, vol. 50, no. 4, pp. 1169–1176, July 2001.

[309] J. Kim, M. Jeon, J. Lee, and Y. Kwon, "A new active predistorter with high gain and programmable gain and phase characteristics using cascode-FET structures," *IEEE Transactions on Microwave Theory and Techniques*, vol. 50, no. 11, pp. 2459–2466, November 2002.

[310] K. Lee and P. Gardner, "A novel digital predistorter technique using an adaptive neuro-fuzzy inference system," *IEEE Communications Letters*, vol. 7, no. 2, pp. 55–57, February 2003.

[311] W. Jung, W. Kim, K. Kim, and K. Lee, "Digital predistorter using multiple lookup tables," *Electronics Letters*, vol. 39, no. 19, pp. 1386–1388, September 2003.

[312] R. Raich, H. Qian, and G. Zhou, "Orthogonal polynomials for power amplifier modeling and predistortion design," *IEEE Transactions on Vehicular Technology*, vol. 53, no. 5, pp. 1468–1479, September 2004.

[313] L. Ding, G. Zhou, D. Morgan, Z. Ma, J. Kenney, J. Kim, and G. Giardina, "A robust digital baseband predistorter constructed using memory polynomials," *IEEE Transactions on Communications*, vol. 52, no. 1, pp. 159–165, January 2004.

[314] S. Mahil and A. Sesay, "Rational function based predistorter for traveling wave tube amplifiers," *IEEE Transactions on Broadcasting*, vol. 51, no. 1, pp. 77–83, March 2005.

[315] R. Iommi, G. Macchiarella, A. Meazza, and M. Pagani, "Study of an active predistorter suitable for MMIC implementation," *IEEE Transactions on Microwave Theory and Techniques*, vol. 53, no. 3, pp. 874–880, March 2005.

[316] O. Hammi, S. Boumaiza, M. Jaidane-Saidane, and F. Ghannouchi, "Digital subband filtering predistorter architecture for wireless transmitters," *IEEE Transactions on Microwave Theory and Techniques*, vol. 53, no. 5, pp. 1643–1652, May 2005.

[317] G. Acciari, F. Giannini, E. Limiti, and M. Rossi, "Baseband predistorter using direct spline computation," *IEE Proceedings—Circuits, Devices, and Systems*, vol. 152, no. 3, pp. 259–265, June 2005.

[318] N. Naskas and Y. Papananos, "Non-iterative adaptive baseband predistorter for RF power amplifier linearisation," *IEE Proceedings—Microwaves, Antennas and Propagation*, vol. 152, no. 3, pp. 103–110, April 2005.

[319] D. Cox, "Linear amplification with non-linear components," *IEEE Transactions on Communications*, vol. 22, no. 12, pp. 1942–1945, December 1974.

[320] A. Bateman, D. Haines, and R. Wilkinson, "Linear transceiver architectures," *Proceedings of the IEEE Vehicular Technology Conference*, Philadelphia, PA, June 15–17, 1988, pp. 478–484.

[321] A. Oppenheim and R. Schafer, *Digital Signal Processing*, Prentice Hall, Upper Saddle River, NJ, first edition, 1975.

[322] H. Nyquist, "Thermal agitation of electric charge in conductors," *Physical Review*, vol. 32, pp. 110–113, July 1928.

[323] J. Scanlan, "Pulses satisfying the Nyquist criterion," *Electronic Letters*, vol. 28, no. 1, pp. 50–52, January 1992.

[324] L. Franks, "Futher results on Nyquist's problem in pulse transmission," *IEEE Transactions on Communication Technology*, vol. 16, no. 2, pp. 337–340, April 1968.

[325] N. Beaulieu, C. Tan, and M. Damen, "A 'Better Than' Nyquist pulse," *IEEE Communications Letters*, vol. 5, no. 9, pp. 367–368, September 2001.

[326] A. Assalini and A. Tonello, "Improved Nyquist pulses," *IEEE Communications Letters*, vol. 8, no. 2, pp. 87–89, February 2004.

[327] f. harris, C. Dick, S. Seshagiri, and K. Moerder, "An improved square-root Nyquist shaping filter," *Proceedings of the Software Defined Radio Conference*, Orange County, CA, November 2005.

[328] S. Jayasimha and P. Singh, "Design of Nyquist and near-Nyquist pulses with spectral constraints," *Proceedings of the IEEE International Conference on Personal Wireless Communications*, Hyderbad, India, December 17–20, 2000, pp. 38–42.

[329] D. Tufts, "Nyquist's problem—The joint optimization of transmitter and receiver in pulse amplitude modulation," *Proceedings of the IEEE*, vol. 53, no. 3, pp. 248–259, March 1965.

[330] T. Berger and D. Tufts, "Optimum pulse amplitude modulation part I: Transmitter-receiver design and bounds from information theory," *IEEE Transactions on Information Theory*, vol. 13, no. 2, pp. 196–208, April 1967.

[331] F. Hill, "A unified approach to pulse design in data transmission," *IEEE Transactions on Communications*, vol. 25, no. 3, pp. 346–354, March 1977.

[332] A. Linder, "The duo-binary technique for high-speed data transmission," *IEEE Transactions on Communication Electronics*, vol. 82, pp. 214–218, May 1963.

[333] E. Kretzmer, "An efficient binary data transmission system," *IEEE Transactions on Communication Systems*, vol. 12, no. 2, pp. 250–251, June 1964.

[334] P. Kabal and S. Pasupathy, "Partial-response signaling," *IEEE Transactions on Communications*, vol. 23, no. 9, pp. 921–934, September 1975.

[335] X. Xia, "A family of pulse-shaping filters with ISI-free matched and unmatched filter properties," *IEEE Transactions on Communications*, vol. 45, no. 10, pp. 1157–1158, October 1997.

[336] T. Demeechai, "Pulse-shaping filters with ISI-free matched and unmatched filter properties," *IEEE Transactions on Communications*, vol. 46, no. 8, p. 992, August 1998.

[337] A. Kisel, "An extension of pulse shaping filter theory," *IEEE Transactions on Communications*, vol. 47, no. 5, pp. 645–647, May 1999.

[338] A. Kisel, "Nyquist 1 universal filters," *IEEE Transactions on Communications*, vol. 48, no. 7, pp. 1095–1099, July 2000.

[339] N. Alagha and P. Kabal, "Generalized raised-cosine filters," *IEEE Transactions on Communications*, vol. 47, no. 7, pp. 989–997, July 1999.

[340] C. Tan and N. Beaulieu, "An investigation of transmission properties of Xia pulses," *Proceedings of the IEEE International Conference on Communications*, Vancouver, British Columbia, June 6–10, 1999, pp. 1197–1201.

[341] C. Tan and N. Beaulieu, "Transmission properties of conjugate-root pulses," *IEEE Transactions on Communications*, vol. 52, no. 4, pp. 553–558, April 2004.

[342] S. Kesler and D. Taylor, "Research and evaluation of the performance of digital modulations in satellite communications systems," Technical Report. CRL Report No. 92, McMaster University, Hamilton, Ontario, Canada, 1981.

[343] H. Baher and J. Beneat, "Design of analog and digital data transmission filters," *IEEE Transactions on Circuits and Systems—I: Fundamental Theory and Applications*, vol. 40, no. 7, pp. 449–460, July 1993.

[344] E. Hassan and H. Ragheb, "Design of linear phase Nyquist filters," *IEE Proceedings—Circuits, Devices and Systems*, vol. 143, no. 3, pp. 139–142, June 1996.

[345] S. Mneina and G. Martens, "Maximally flat delay Nyquist pulse design," *IEEE Transactions on Circuits and Systems—II: Express Briefs*, vol. 51, no. 6, pp. 294–298, June 2004.

[346] P. Nahin, *An Imaginary Tale: The Story of $\sqrt{-1}$*, Princeton University Press, Princeton, NJ, 1998.

[347] R. Best, *Phase-Locked Loops: Design, Simulation, and Applications*, McGraw-Hill Professional, New York, 1999.

[348] D. Stephens, *Phase-Locked Loops for Wireless Communications*, Kluwer Academic Publishers, Norwell, MA, 2002.

[349] W. Lindsey and C. Chie, "A survey of digital phase-locked loops," *Proceedings of the IEEE*, vol. 69, no. 4, pp. 410–430, April 1981.

[350] F. Gardner, "Hangup in phase-lock loops," *IEEE Transactions on Communications*, vol. 25, pp. 1210–1214, October 1977.

[351] F. Gardner, "Equivocation as a cause of PLL hangup," *IEEE Transactions on Communications*, vol. 30, pp. 2242–2243, October 1982.

[352] H. Meyr and L. Popken, "Phase acquisition statistics for phase-locked loops," *IEEE Transactions on Communications*, vol. 28, pp. 1365–1372, August 1980.

[353] M. Fitz, "Equivocation in nonlinear digital carrier synchronizers," *IEEE Transactions on Communications*, vol. 39, pp. 1672–1682, November 1991.

[354] R. Sannemann and J. Rowbotham, "Unlock characteristics of the optimum type II phase-locked loop," *IEEE Transactions on Aerospace and Navigational Electronics*, vol. 11, pp. 14–24, March 1964.

[355] R. Tausworthe, "Cycle slipping in phase-locked loops," *IEEE Transactions on Communication Technology*, vol. 15, pp. 417–421, June 1967.

[356] G. Ascheid and H. Meyr, "Cycle slips in phase-locked loops: A tutorial survey," *IEEE Transactions on Communications*, vol. 30, pp. 2228–2241, October 1982.

[357] M. Moeneclaey, "The influence of phase-dependent loop noise on the cycle slipping of symbol synchronizers," *IEEE Transactions on Communications*, vol. 33, pp. 1234–1239, December 1985.

[358] C. Chie, "New results on mean time-to-first-slip for a first-order loop," *IEEE Transactions on Communications*, vol. 33, no. 9, pp. 897–903, September 1985.

Index

Note: Page numbers followed by *t* refer to tables.

A

Additive white Gaussian noise (AWGN), 202–208, 233
    continuous-time random processes, 202–204
    in sampled-data system, 206–208
    thermal noise and, 204–206
Aeronautical telemetry, 8
"AM/AM" curves, 260
American Institute of Electrical Engineers (AIEE), 9, 10
Amplitude phase shift keying (APSK), 244–245
Analog modulation, 12, 13, 14
Analog-to-digital converter (ADC), 519, 555
    clock-jitter effects, 527–529
    discrete-time demodulators and, 611–614
    quantization effects, 520–527
Analysis equation, 218, 219
    detection and, 223–226
Arbitrary resampling, 565–577
Armstrong, Howard, 3, 604, 640, 659
Automatic gain control (AGC), 372, 451, 588–593, 614

B

Band-pass signal, 38, 39, 49, 375, 386, 588, 589, 609
    conversion of, 534–537
    discrete-time processing of, 67–70
    sampling, 60, 62–65
Baseband pulse amplitude modulation, 215
Baseband signal, 38, 49, 68, 673, 708
    conversion of, 529–534
    sampling, 60
Bell, Alexander Graham, 3
Bilinear transform, 134, 154
Binary PAM, 438–443, 478–494
Binary phase shift keying, *see* BPSK
Bivariate Gaussian probability density function, 196
Blackman window, 142, 143, 153
Boolean algebra, 180
Boot, Henry, 4
BPSK
    carrier phase synchronization, 375–381
        continuous-time techniques, 391–394
        Costas loop, 394
        discrete-time techniques, 375–381
        phase ambiguity resolution, 396–397, 400–402
    description, 242
    performance
        bandwidth, 313–314
        probability of bit error, 319–321, 325–331
    symbol timing synchronization, 494–500

C

Carrier phase synchronization, 359
    basic problem formulation, 360–365

Carrier phase synchronization (*continued*)
   for BPSK, 375–381
     Costas loop, 394
     in phase ambiguity resolution, 396–397, 400–402
     using continuous-time techniques, 391–394
     maximum likelihood phase estimation, 409–421
   for MQAM, 381–382
   for offset QPSK, 382–391
   phase ambiguity resolution, 394–409
   for QPSK
     Costas loop, 394
     examples, 374–375
     heuristic phase error detector, 365–370
     maximum likelihood phase error detector, 370–373
     in phase ambiguity resolution, 397–398, 402–405
     using continuous-time techniques, 391–394
     using phase adjusted quadrature sinusoids, 360–362
     using post-matched filter de-rotation operation, 362–365
Carson, John, 3, 8–9
Cascade-integrator-comb (CIC) filter, 557–562
CCITT V.29 constellations, 245
Cellular telephony link budget, 343–344
Channelization
   using continuous-time techniques, 637–650
   using discrete-time techniques, 650–658
Circular normal distribution, 197
Clock-jitter, 527–529
Comb filter, *see* Cascade-integrator-comb (CIC) filter
Communications Satellite Corporation (Comsat), 5
Complementary error function, 190, 191
Complex-valued representations, 700, 702–704
   demodulators and, 708–710
   modulators and, 705–707
   synchronization and, 710–711
Continuous-time discrete-time interface, 519–537
   analog-to-digital converter (ADC), 520–529
   digital-to-analog converter (DAC), 529–537
Continuous-time Fourier transform, 37–40
   inverse transform, 37, 38, 39
   properties, 39t
   transform pairs, 40t
Continuous-time phase locked loops, 718–732
   acquisition, 730–731
   loop constant selection, 731–732
   loop filter, 722–724
   phase error, 722–724
   PLL bandwidth, 728–730
   PLL inputs, 720–722
     ramp responses, 721
     step responses, 721
   tracking, 731
   transient analysis, 725–728
Continuous-time processing, 16, 17, 67, 703
   MQAM modulator and, 246–251
   PAM modulator and, 229–231
Continuous-time random processes, 202–204
Continuous-time signals, 24–26
   aperiodic signal, 25
   discrete-time processing of, 65–67
   energy signal, 24–25
   impulse function, 26
   periodic signal, 25
   power signal, 24, 25
   unistep function, 26
Continuous-time systems, 28–29
Continuous-time techniques
   carrier phase synchronization and, 391–394
   channelization and, 637–650
   for M-ary PAM, 438–443
   for MQAM, 443–445
   superheterodyne receiver, 637–650

Index   **771**

    symbol timing synchronization and, 438–443
Continuous-wave (CW), 10
Contour of constant probability density, 197
Coordinate Rotation Digital Computer, *see* CoRDiC algorithm
CoRDiC algorithm, 596
    moving along other shapes, 585–587
    rotations, 578–584
Costas loop
    for BPSK carrier phase synchronization, 394
    for QPSK carrier phase synchronization, 391
Cross MQAM, 243–244
Cubic interpolator, 467
Cumulative distribution function, 181, 182

D

De Forest, Lee, 2
Delay axis ambiguity, 405–406
Descartes, 712
Differential encoding, 398–409
    for BPSK, 400–402
    for offset QPSK, 405–409
    for QPSK, 402–405
Digital modulation, 9, 12, 15
    advantages of, 13–14
    disadvantages of, 14
Digital video broadcast (DVB) link budget, 341
Digital-to-analog converter (DAC), 519, 555
    band-pass signals, conversion of, 534–537
    baseband signals, conversion of, 529–534
    discrete-time modulators and, 605–606
Direct digital synthesizer (DDS), 159, 538, 542–555
    finite precision arithmetic, 545–549
    fundamental relationships, 542–545
    spurious spectral lines and, 549–555
Discrete Fourier transform (DFT), 50–55

    fast Fourier transform (FFT) and, 52–55
    inverse, 50
Discrete-multitone modulation (DMT), 271
Discrete-time architectures, 604
    for QAM demodulators
        ADC placement, 611–614
        multirate processing, 616–632
        sampling rates, 614–616
    for QAM modulators
        DAC placement, 605–606
        multirate processing, 606–611
Discrete-time differentiator, 149–152
Discrete-time filter design methods, 127–159
    discrete-time differentiator, 149–152
    discrete-time integrator, 152–159
    FIR filter designs, 134–149
    IIR filter designs, 129–134
    parameters, 128
Discrete-time Fourier transform (DTFT), 31–32, 46–50
    attributes, 47–49
    inverse transform, 46
    pairs, 47t
    properties, 46t
Discrete-time integrator, 152–159
    bilinear transform, 154
    recursion and, 154–155, 156
Discrete-time low-pass filter, 66–67
Discrete-time oscillators, 537
    direct digital synthesizer and, 542–555
    LTI system and, 538–542
Discrete-time phase locked loops, 732–747
Discrete-time processing, 15, 606, 637, 707, 709–710
    of band-pass signals, 67–70
    of continuous time signals, 65–67
    PAM detector and, 235
    PAM modulator and, 233
    popularity of, 14–19
    trends of, 16–17
Discrete-time signals, 26–28
    aperiodic signal, 27

Discrete-time signals (*continued*)
   energy signal, 27
   impulse function, 28
   periodic signal, 27
   power signal, 27
   unistep function, 28
Discrete-time systems, 29
Discrete-time techniques
   channelization and, 650, 658
   for M-ary PAM, 445
   for MQAM, 494–497
   for offset QPSK, 497–500
Dithering technique, 551–552
Downsampling, 118–120, 560, 630, 635–636
Dual-conversion receivers, 646

E

Early-late gate detector, 442
Early-late timing error detector (ELTED), 453–455
Echo, 5, 338
Eglund, Carl, 8
Electronic communications, 1
   digital trend, 7–8
   mathematical trend, 8–10
   wireless trend, 6–7
Energy signal
   in continuous-time signals, 24–25
   in discrete-time signals, 27
Equiripple property, 146
Equivalent noise temperature, 334–335
Error feedback method, 552–555
Error feedforward method, 551
Error function, 188–189
Euler, Leonhard, 712
Event, 178
Event space, 178
Explorer I, 5

F

Farnsworth, Philo T., 4
Farrow interpolator structure, 470
Fast Fourier transform (FFT), 52–55
Federal Communications Commission (FCC), 4

Fressenden, Reginald, 2, 3, 640n1
Finite impulse response (FIR) filter designs
   basic structure, 134–138
   discrete-time differentiator, 152
   as linear phase filter, 136
   using approximations, 143–149
   windowing and, 138–143
      Blackman window, 142, 143
      Kaiser-Bessel window, 142
Fleming, Ambrose, 2
Flipped exponential pulse shape, 688
Flipped-hyperbolic secant pulse shape, 688–689
Flipped-inverse hyperbolic secant pulse shape, 689
Fourier transform
   continuous-time, 37–40
   discrete, 50–55
   discrete-time, 31–32, 46–50
   fast, 52–55
Frequency division multiple access, 11
Frequency division multiplexing (FDM), 637
Frequency domain characterization
   for continuous-time LTI systems
      in frequency domain using Fourier transform, 30, 37–40
      in s-domain using Laplace transform, 30, 32–36
      in time domain, 30
   for discrete-time LTI systems
      in frequency domain using Fourier transform, 31–32, 46–50
      in time domain, 30–31
      in z-domain using z-transform, 31, 40–45
Frequency lock, 730
Friis equation, 331–334
Full-response pulse shapes, 673–676

G

Gain method, 339
Gardner timing error detector (GTED), 458–459
Gauss, Carl Friedrich, 702n1, 712
Gaussian random sequence, 199, 200

Gaussian random variables
  density and distribution functions, 188–192
  product moments, 192–193
  random variables, functions of, 193–194
  rectangular to polar conversion, 194

H

Half-band filters, 562–565
Half-sine (HS) pulse shape, 306, 313, 673
  frequency-domain properties, 675t
  time-domain properties, 674t
Hartley, R. V. L., 9
Hertz, Heinrich, 2
Heterodyne, 2, 618, 636
  complex, 619, 624, 627, 630
  residual, 656, 657
Hogenauer filters, 557, 560–562

I

Impulse function, 26, 27–280
Impulse invariance filter design, 132–133
Impulse-train sampling, 115–118
Infinite impulse response (IIR) filter designs
  basic structure, 129–132
  bilinear transform, 134
  impulse invariance, 132–133
Institute of Electrical and Electronics Engineers, 10
Institute of Radio Engineers (IRE), 9
INTELSAT, 5
Interpolation, 122, 234, 257, 448, 462–475
  control
    modulo-1 counter, 475–477
    recursive control, 477–478
  piecewise polynomial interpolation, 465–470
  polyphase filterbank interpolation, 470–475
Inverse image, 180
IRIG-106 standard, 422, 503

J

Jansky, Karl, 4

K

Kaiser–Bessel window, 142
Kraus, John, 4

L

Laplace transform, 30, 32–36, 37, 38, 721
  inverse transform, 32, 34, 722
  loop output, 725
  partial fraction expansion and, 34–36
  phase error step, 722
  properties, 33t
  region of convergence and, 36
Linear interpolator approximation, 568, 570, 572, 573
Linear modulation, 214, 305
  comparison of, 325–331
  link budgets, 331–344
    equation, 339–344
    equivalent noise temperature, 334–335, 337–339
    Friis equation, 331–334
    noise figure, 336, 339
  M-ary baseband pulse amplitude modulation, 227–238
  M-ary quadrature amplitude modulation (MQAM), 238
  maximum likelihood detection, 273–279
  multicarrier modulation (MCM), 265–273
  offset QPSK, 260–265
  PAM performance, 306–313
    bandwidth, 306
    probability of error, 307–313
  QAM performance, 313–325
    bandwidth, 313–314
    probability of error, 314–325
  signal spaces, 215–227
  white Gaussian noise projection, 345–347
Linear time-invariant system, *see* LTI system
Link budget
  equation, 339–344
    cellular telephony link budget, 343–344

Link budget (*continued*)
    satellite link budget, 341–343
    space exploration link budget, 341
    equivalent noise temperature, 334–335, 337–339
    Friis equation, 331–334
    noise figure, 336, 339
Link margin, 341
Look-up table (LUT), 545, 548
LTI system, 28, 29, 30–32, 537
    continuous-time systems, 30
    discrete-time oscillators and, 538–542
    discrete-time systems, 30–32
    random sequences and discrete-time systems, 200–202

M

Manchester (MAN) pulse shape, 306, 313, 501, 673
    frequency-domain properties, 675t
    time-domain properties, 674t
Marconi, Guglielmo, 2, 3, 4
Mariner 2, 6
Martin, W. H., 5
M-ary baseband pulse amplitude modulation, *see* M-ary PAM
M-ary PAM, 227–238
    continuous-time realization, 229–233
    discrete-time realization, 233–238
    symbol timing synchronization
        using continuous-time techniques, 438–443
        using discrete-time techniques, 445–494
M-ary phase shift keying, *see* MPSK
M-ary Quadrature Amplitude Modulation, *see* MQAM
MATLAB, 191
Maximum likelihood (ML)
    decision rule, 273–279
        carrier phase estimation, 409–421
        phase error detector, 370–373
        symbol timing estimation, 503–514
        timing error detector, 449–453, 478–484
Maxwell, James, 2

Minimum probability of error constellations, 245–246
Modulo-1 counter interpolation control, 475–477
Moments, 186
Morse, Samuel, 1
MPSK, 242, 319–321, 330
MQAM, 238
    carrier phase synchronization, 381–382
    constellations and, 242–246
    continuous-time realization, 246–256
    description, 238–247
    discrete-time realization, 256–260
    orthogonal basis functions, 238–239
    performance, 313–331
        bandwidth, 313–314
        probability of error, 314–331
    polar form, 242
    rectangular form, 242
    symbol timing synchronization
        using continuous-time techniques, 443–445
        using discrete-time techniques, 494–497
Mueller and Müller timing error detector (MMTED), 459–462
Multicarrier modulation (MCM), 265–273
Multipath fade, 344
Multipath propagation, 344
Multiply-accumulate, 136
Multirate signal processing, 635
    discrete-time techniques and, 606–611, 616–632, 650–658
    downsampling, 118–120
    impulse-train sampling, 115–118
    Noble identities, 122
    polyphase filterbanks, 122–127
    upsampling, 120–122
Multivariate Gaussian random variable, 195–198
    bivariate Gaussian distribution, 196–197
    linear operations and, 197–198

N

Nearest neighbor approximation, 567, 573

Negative frequency, 704–705
Noble identities, 122, 621, 623, 627, 629, 652
Noise figure, 336–337
Non-return-to-zero (NRZ) pulse shape, 280, 306, 313, 673
    frequency-domain properties, 675t
    time-domain properties, 674t
Nyquist, Harry, 9, 205
Nyquist no-ISI theorem, 677–682
Nyquist pulse shapes, 688–695

O

Offset QPSK, 260–265
    carrier phase synchronization, 382–391
    discrete-time technique for, 497–500
Open interval, 181
Orthogonal frequency division multiplexing (OFDM), 272

P

Pairwise error probability, 322
PAM performance
    bandwidth, 306
    probability of error, 307–313
Parks–McClellan algorithm, 146–148, 534, 595, 690
Parseval's theorem, 38, 570, 571
Partial-response pulse shapes, 676
    Nyquist pulse shapes, 688–695
    spectral raised-cosine pulse shape, 682–687
Parts obsolescence, 16
Peak-to-average ratio, 524
Penzias, Arno, 4, 338, 339
Phase accumulation register, 545
Phase accumulator truncation, 548, 551
Phase ambiguity resolution, 394
    differential encoding, 398–409
    unique word method
        with BPSK, 396–397
        with QPSK, 397–398
Phase lock, 730
Phase locked loops (PLL)
    carrier phase PLL, 362, 364, 369, 375, 381, 382, 386, 394, 396, 397
    continuous-time, 718–732
        acquisition, 730–731
        loop constant selection, 731–732
        loop filter, 722–724
        phase error, 722–724
        PLL bandwidth, 728–730
        PLL inputs, 720–722
        tracking, 731
        transient analysis, 725–728
    discrete-time, 732–747
Phase trajectory, 254, 261, 262
Piecewise polynomial interpolation, 465–470
Pole-zero plot, 34, 42
Polyphase filterbanks, 563–564, 611, 632
    arbitrary resampling and, 565–577, 630, 632
    interpolation, 470–475
    upsample-filter and, 125–127
    z-transform and, 122–125
Power signal
    in continuous-time signals, 24, 25
    in discrete-time signals, 27
Power spectral density, 199–200
    of band-pass QAM signal, 313
    of PAM pulse train, 306
    of random process, 204
    of thermal noise, 205
Priori probability density function, 274
Probability density function, 182, 185, 187
Probability space, 179, 180
Professional Group on Communication Systems (PGCS), 9–10
Pseudorandom scrambling, 501–503
Pull-in range, 731
Pulse shape, 227, 229, 231, 233–238, 241, 246–260, 261–264, 277, 280, 291–293, 295–302, 305–306, 310, 313–316, 325, 330, 352, 355, 360, 362, 375, 386, 399, 410, 415, 417, 420, 435–437, 445, 447, 450–452, 454–456, 458–462, 468, 471, 473, 478, 484, 494, 497, 499, 501, 503, 511, 524–525, 555, 598, 606, 608–609, 613–614, 662–663, 671, 673–699

Q

QAM demodulators, 12, 13, 260
  complex-valued representations for, 700, 708–710
  discrete-time architectures for
    ADC placement, 611–614
    multirate processing, 616–632
    sampling rates, 614–616
  resampling filters and, 555
QAM modulators
  complex-valued representations for, 700, 705–707
  discrete-time architectures for
    DAC placement, 605–606
    multirate processing, 606–611
  resampling filters and, 555
QAM performance
  bandwidth, 313–314
  probability of error, 314–325
    MPSK, 319–321
    square MQAM, 317–319
    union bound, 321–325
QPSK
  carrier phase synchronization
    Costas loop, 391
    examples, 374–375
    heuristic phase error detector, 365–370
    maximum likelihood phase error detector, 370–373
    phase ambiguity resolution, 397–398, 402–405
    using continuous-time techniques, 391–394
  description, 238–242
  performance
    bandwidth, 313–314
    probability of error, 321, 325–326, 329–330
  symbol timing synchronization, 494–497, 503–513
Quadrature amplitude modulation (QAM), 215, see also QAM demodulators; QAM modulators; QAM performance
Quaternary phase shift keying, see QPSK

R

RADAR (Radio Detection and Ranging), 4–5
Radio astronomy, 4
Radix-2 FFT, 52
Ramp input, 721
Randall, John, 4
Random sequences, 198–202
  discrete-time LTI systems and, 200–202
  power spectral density, 199–200
Random variable, 180
  continuous, 187
  discrete, 186
  functions, 193–194
  Gaussian, 188–194
  multivariate Gaussian, 195–198
  variance of, 186
Reber, Grote, 4
Recursive interpolation control, 477–478
Region of convergence (ROC), 32, 36, 44, 45
Resampling filters, 555
  arbitrary resampling, 565–577
  CIC filters, 557–562
  half-band filters, 562–565
  Hogenauer filters, 557–562
Return-to-zero (RZ) pulse shape, 306, 313, 673
  frequency-domain properties, 675t
  time-domain properties, 674t
Rice, Stephen O., 9

S

Sample space, 178
Sampling theorem, 56–65, 615
  band-pass signal, 60, 62–65
  baseband signal, 60
Satellite communication, 5–6
Satellite link budget, 341–343
Saturation, 260
Shannon, Claude, 9
Signal amplitude to peak distortion level ratio, 573–577
Signal set, 215
Signal space, 214

analysis equation and detection, 223–226
energy, 215
linear independence, 215
matched filter, 226–227
orthogonality, 216
orthonormal set, 216–218
span, 215
synthesis equation and linear modulation, 222–223
Signal-to-distortion power ratio, 570–573
Signal-to-noise ratio, 305
Signal-to-quantization-noise ratio, 521–522, 523, 524, 527
Software defined radios, 17, 448
Software radio, 17
Space exploration link budget, 341
Spectral raised-cosine pulse shape, 682–687
Spectral regrowth, 261
Spurious-free dynamic range (SFDR), 527
Spurious spectral lines, 549–555
   dithering technique and, 551–552
   error feedback method and, 552–555
   error feedforward method and, 551
Sputnik I, 5
Square MQAM, 242–243, 317–319
Square-root raised-cosine (SRRC) pulse shape, 306, 314, 684–687, 690
Superheterodyne receiver, 637–650, 657, 659
Symbol timing synchronization, 434
   basic problem formulation, 436–438
   for M-ary PAM
      using continuous-time techniques, 438–443
      using discrete-time techniques, 445
      maximum likelihood estimation, 503–514
   for MQAM
      using continuous-time techniques, 443–445
      using discrete-time techniques, 494–497
   for offset QPSK
      using discrete-time techniques, 497–500
      transition density, 501–503
SYNCOM, 5
Synthesis equation, 218, 219, 222–223
System components
   automatic gain control, 588–593
   continuous-time discrete-time interface, 519–537
   CoRDiC, 578–587
   discrete-time oscillators, 537–555
   resampling filters, 555–577
System design, 604
   channelization
      using continuous-time techniques, 637–650
      using discrete-time techniques, 650–658
   discrete-time QAM demodulators, 611–632
   discrete-time QAM modulators, 605–611
   first generation architectures, 632–634
   second generation architectures, 634–635
   third generation architectures, 635–637

T

Tapped delay-line filter, 136
Telegraph, 1–3, 7
Television, 4, 6, 7
Telstar, 5, 338
TELSTAR402R, 341
Thermal noise, 204–206
Thompson, William, 2, 8
Timing error detector (TED), 437, 438, 449
   early–late timing error detector (ELTED), 453–455
   Gardner timing error detector (GTED), 458–459
   maximum likelihood timing error detector (MLTED), 449–453
   Mueller and Müller timing error detector (MMTED), 459–462

Timing error detector (TED) (*continued*)
   zero-crossing timing error detector
      (ZCTED), 455–458
Transition density, 501–503
Transversal filter, 136
Tuned radio frequency (TRF)
   systems, 638–640
Tustin's equation, 156, 735
Two's complement method, 546–547
Two-way mobile radio, 5

U

Unique word method
   for phase ambiguity resolution
      for BPSK, 396–397
      for QPSK, 397–398
Unit-step function, 26, 27–28
Universal mobile telecommunications
   system (UMTS) link budget, 343
Upper tail probability normal
   (UTPN), 191–192
Upsampling, 120–122, 562

V

Van-Allen, James, 4
Voltage controlled oscillator (VCO), 542,
   543, 718

W

Wallis, John, 712
Warner, Sidney, 5

Watson-Watt, Robert, 4
Wessel, Caspar, 712
White Gaussian noise, 204–206
   additive, 202–208
      continuous-time random process,
         202–204
      in sampled data system, 206–208
      projection onto orthonormal basis set,
         345–347
Williams, E. M., 5
Wilson, Robert, 4, 338, 339
Windowing method
   Blackman window, 142, 143
   Kaiser–Bessel window, 142
Wireless communication, 6–7
   basics of, 10–12
   benefits of, 2–3

Y

Y method, 337–339

Z

Z-transform, 31, 40–45, 46, 557, 652
   inverse transform, 40, 43–44
   pairs, $42t$
   partial fraction expansion and, 43–44
   properties, $41t$
   of recursion, 155
   region of convergence and, 44
Zero-crossing timing error detector
   (ZCTED), 455–458, 484–494
"Zero-IF" receiver, 646